Next Generation of International Chemical Additives

Next Generation of International Chemical Additives

A Critical Review of Current US Patents

Thomas F. DeRosa

ELSEVIER

AMSTERDAM • BOSTON • HEIDELBERG • LONDON
NEW YORK • OXFORD • PARIS • SAN DIEGO
SAN FRANCISCO • SINGAPORE • SYDNEY • TOKYO

Elsevier
The Boulevard, Langford Lane, Kidlington, Oxford OX5 1GB, UK
Radarweg 29, PO Box 211, 1000 AE Amsterdam, The Netherlands

British Library Cataloguing in Publication Data
A catalogue record for this book is available from the British Library

Library of Congress Cataloging-in-Publication Data
A catalog record for this book is available from the Library of Congress

ISBN: 978-0-444-53788-1

For information on all Elsevier publications
visit our web site at store.elsevier.com

Printed and bound in Great Britain
12 13 14 10 9 8 7 6 5 4 3 2 1

This Book is Dedicated to My Mother and Mother-In-Law

Gloria DeRosa
and
Marguerite Schuster

Contents

Preface

Within a typical week, approximately 3,000 US Patents are issued and 7,000 US Patent Applications filed. Within the 9,000 issued patents, 1,800 are directly involved with chemical sciences associated with new synthetic methods, preparation of biologically active agents, polymers, chemical additives, plastics, and improved or enhanced chemical or polymer processes. It was the objective of this treatise to focus our efforts on new and novel or improved chemical additives.

A chemical additive is broadly defined as an agent that can impede, enhance, or stabilize a chemical process through a unique and anticipated chemical reaction or physical interaction. Moreover, the amount of additive needed to achieve a specified goal is completely immaterial.

By contrast additive mixtures or additive packages are defined as the mixture of two or more chemical agents that achieve the identical objectives as a single addition but do so by (a) forming an insitu intermediate; (b) become synergistic in the presence of each other; or (c) act as individual chemical additives. The description of chemical additives in this text has been separated into 31 categories including: 1 Adhesives; 2 Antioxidants; 3 Automotive Additives; 4 Automotive Tire Additives; 5 Battery And Charge Extenders; 6 Carpets; 7 Colorants; 8 Composites; 9 Cosmetics; 10 Dental Adhesives; 11 Disinfectants; 12 Dispersants; 13 Drilling Additives; 14 Drug Delivery; 15 Electronic Additives; 16 Extreme Pressure Additives; 17 Films; 18 Fire Retardants; 19 Flame Suppressants; 20 Food Additives; 21 Ink; 22 Laundry; 23 Medical; 24 Ophthalmology; 25 Paint Additives; 26 Photography; 27 Polymers; 28 Powder Coating Additives; 29 Surfactants; 30 Textile Additives; 31 Water Treatment.

I urge the reader not to limit their review of chemical additives to sections they are currently actively involved. In many instances chemical additives described in an unrelated industrial may be potentially useful agents in your own field. Casual brosing throughout this text is also strongly encouraged since each entry within every category is accompanied by at least four similar (or dissimilar) chemical additives along with easy to access references. It is my belief that by reviewing this book in areas unrelated to your own to identify next generation chemical additives will pay off handsomely.

Thomas F. DeRosa

Introduction

In order to make a useful polymer product, a variety of additives must be added to the polymer itself. The commercial success and usefulness is determined by the additives required for processing, enhancement of specific properties, lifetimes control, and particular use requirements and forms. Thus the great importance of additives in polymer systems. This book has great interests for scientists and engineers doing research and development on additives and to individuals involved in modifying polymers for specific end use requirements. This book alerts all to the latest developments and possibilities for improvements in all of these areas. Those involved in these areas should have this book on their desk! The author has done a most useful service to all.

Eli M. Pearce, PhD
University Research Professor
Herman Mark Polymer Research Institute
Polytechnic Institute of NYU
2005 American Chemical Society President

Chapter 1

Adhesives

A. CROSSLINKABLE ADDITIVE PACKAGE

Trimethoxysiloxane Copolymers

Title:	Curable composition
US Patent:	7,772,332 (August 10, 2010)
Author:	Katsuyu Wakabayashi
Assignee:	Kaneka Corporation (Osaka, JP)
Significance:	Curable resins having high mechanical strength, good adhesive strength and rubber elasticity as well as good workability without the use of organotin reagents were prepared. The method entails reacting polypropylene oxide containing propenyl and trimethoxysilyl termini with a crosslinkable siloxy and 3-(diethylamino)propyl amine.

Additive Components The additive composition consists of three components

Adhesive Additive	Polypropylene oxide containing propenyl and trimethoxysilyl terminus (I)
CAS Number	Unreported
Safety Concerns	Unreported
Additive Component	DMSV-42 (II)
CAS Number	68083-19-2
Safety Concerns	Mild-to-moderate skin irritant
Additive Component	3-Aminopropyltriethoxysilane (III)
CAS Number	68083-19-2
Safety Concerns	Causes eye and skin burns. May cause severe respiratory tract irritation with possible burns. May cause severe digestive tract irritation with possible burns. May be harmful if swallowed.

Additive Structure The crosslinked adhesive resin is the reaction product comprising

Next Generation of International Chemical Additives. DOI: http://dx.doi.org/10.1016/B978-0-444-53788-1.00001-2

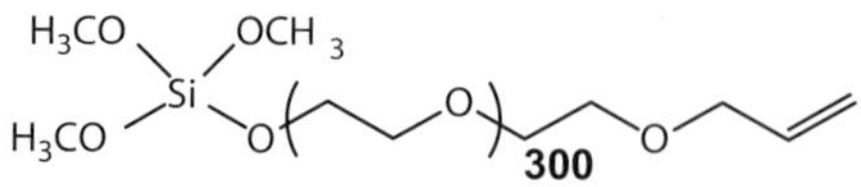

(I)
Adhesive Additive

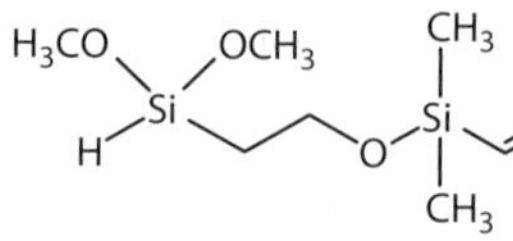

Adhesive Component
(II)

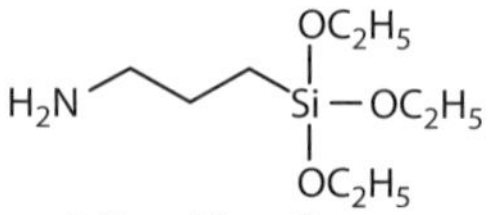

Adhesive Component
(III)

Safety Concerns Since no purification was used in the final preparation of the curable additive, residual amounts of the following chemical reagents may be present in varying amounts in the product.

Name Polypropylene oxide diol
CAS Number 25322-69-4
Safety Concerns May be harmful by inhalation, ingestion, and through skin absorption. May act as an irritant.

Name Zinc hexacyanocobaltate glyme complex
CAS Number 116912-63-1
Safety Concerns Residual zinc hexacyanocobaltate glyme is a moderate skin irritant.

Name Trimethoxy silane
CAS Number 2487-90-3
Safety Concerns Very toxic. May be fatal if inhaled. May cause blindness. Corrosive. Harmful if swallowed and in contact with the skin. Very destructive of mucous membranes.

Name Allyl chloride
CAS Number 107-05-1
Safety Concerns While there is evidence that exposure to allyl chloride may result in reversible enzymatic evidence of liver damage, one report described impaired kidney function in workers exposed to allyl chloride in unknown concentrations.

Name Platinum(0)-1,3-divinyl-1,1,3,3-tetramethyldisiloxane
CAS Number 68478-92-2
Safety Concerns Contact with the eyes and the skin may cause irritation. In susceptible persons, skin reactions may occur. This is a short-term effect and should go away after washing with soap and water.

Additive Preparation Commercial supplies of reagent grade additive components can be obtained from the following suppliers:

Polypropylene Oxide Diol.
Lone Star Chemical
214 E. College Street
Grapevine, TX 76051
888-575-9195

Trimethoxy Silane
Voster Chemicals
415 Huguenot Street
New Rochelle, NY 10801
800-282-3982

Zinc Hexacyanocobaltate Glyme
Bayer Material Sciences Limited
Berthold Alfes
Hong Kong, China
852-2894-7357

Allyl Chloride
Nanjing Fubang Chemical Co., Ltd.
5F Tianzheng International Square
No. 399 Zhongyang
Nanjing of Jiangsu, China
86-25-83179277

Platinum (0)-1,3-divinyl-1,1,3,3-tetra-methyldisiloxane
United Technical Technologies
2731 Bartram Road
Bristol, PA 19007
800-541-0559

DMSV-42
Gelest, Inc.
11 East Steel Road
Morrisville, PA 19067
215-547-1015

Additive Preparation

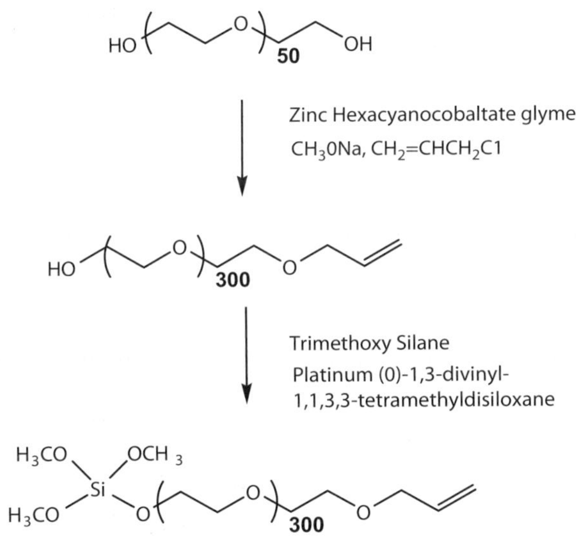

Additive Synthesis

1. Preparation of high molecular polypropylene oxide containing an allyl and hydroxyl terminus

A mixture containing polyoxypropylene diol, Mn of approximately 2000 Da, propylene oxide, and zinc hexacyanocobaltate glyme was reacted at ambient temperature to produce polypropylene oxide having hydroxyl groups at the termini with an Mn of approximately 25,500 Da. The reaction product was then treated with 1.2 equivalents of sodium methoxide dissolved in methyl alcohol and then heated to a sufficient temperature to distill off methyl alcohol. The solution was then treated with 1.5 equivalents of allyl chloride, and unreacted allyl chloride was distilled from the reaction vessel. The allyl intermediate was then isolated by dissolving 300 parts by weight of *n*-hexane and water, and water was subsequently removed by centrifuging. Hexane and any remaining water were then removed by distillation at a reduced pressure and the product isolated had an Mn of approximately 26,200 Da.

2. *Preparation of high molecular polypropylene oxide containing a trimethoxysilyl terminus and allyl terminus*

A reactor containing 100 parts by weight of the Step 1 product was treated with 1.1 parts by weight of trimethoxy silane in a solution of 150 ppm platinum(0)-1,3-divinyl-1,1,3,3-tetramethyldisiloxane dissolved in isopropyl alcohol, and the mixture was heated to 90 °C for 2 h. The product that was then isolated had trimethoxysilyl groups at one end of the termini.

Optimum Treatment Level

Optimum silane coupling entailed using five parts by weight of DMSV-42, 100 parts by weight of the Step 2 experimental additive and 2–3 parts by weight based of 3-amino-1-triethoxysilyl-propane. Using this weight ratio, the blend can be expected to cure in approximately 15 min.

Advantages over Prior Art

The curable additive composition of the current invention contains a reactive silicon hydrolyzable group bonded to a silicon atom which is curable without the use of dibutyltin bis(acetylacetonate). The absence of organotin as the curing catalyst provides a less toxic adhesive and a resin with improved strength and enhanced adhesiveness.

Notes

1. In an earlier investigation by Wakabayashi,[1] the reaction product of trifluoropropyl dimethoxy silane and bis(pentafluorophenyl)dimethoxysilane was used in preparing organotin-free adhesive resins for use in water-resistant applications.
2. Nishiwaki[2] developed a high yielding and general method for silylating nonsilanol hydroxyl groups by dehydrative condensation using t-butyldimethylsilanol with ytterbium(III) trifluoro-methanesulfonate and obtained conversions exceeding 95%.
3. The silane coupling chemical additive (IV) was used by Tsumura[3] to prepare adhesive resins and light-shielding pastes when blended with titanium oxide.

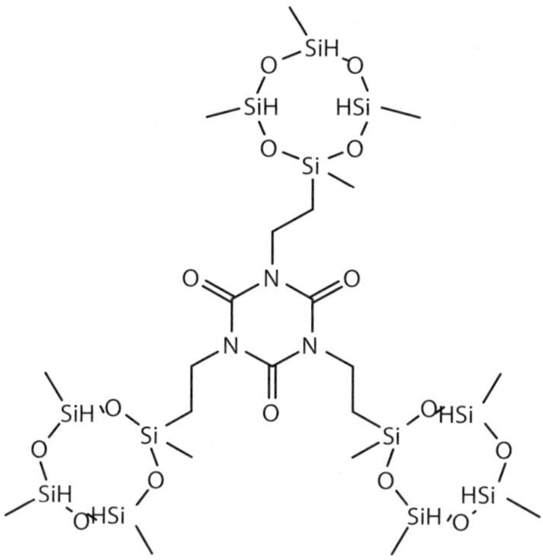

(IV)

4. Saito[4] prepared adhesive compositions containing γ-methacryloxypropyltrimethoxysilane (V), which were very workable, had excellent adhesion, rubber-like properties, good storage stability, and were quick curing.

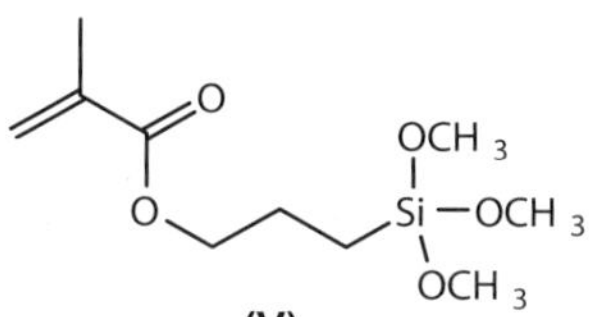

(V)

REFERENCES

1. Katsuyu Wakabayashi, US Patent Application 20070265409 (November 15,2007) **Kaneka Corporation (Osaka, JP)**
2. Hiromi Nishiwaki et al, US Patent 6,875,880 (April 5, 2005) **Shin-Etsu Chemical, Co., Ltd. (Tokyo, JP)**
3. Manabu Tsumura et al, US Patent 7,785,715 (August 31, 2010) **Kaneka Corporation (Kita-ku, Osaka-shi, JP)** **Nichia Corporation (Anan-shi, Tokushima, JP)**
4. Atsushi Saito et al, US Patent 7,781,559 (August 24, 2010) **Cemedine Co., Ltd. (Shinagawa-ku, Tokyo, JP)**

B. GLASS ADHESIVES

Poly(Acrylic Acid-co-4-Vinylpyridine *N*-oxide)

Title:	Glass Adhesion Promoter
US Patent:	7,829,627 (November 9, 2010) 627
Author:	Klin A. Rodrigues et al.
Assignee:	Akzo Nobel N.V. (Arnhem, NL)

Significance: Copolymers of acrylic acid containing up to 1.6 wt% silanol or 4-vinylpyridine *N*-oxide are effective as glass adhesive additives. Copolymers were prepared in quantitative yields by reacting acrylic acid with either vinyltriethoxy or tri-I-propoxy and hydrolyzing the product or copolymerizing with 4-vinyl-pyridine *N*-oxide.

Additive Name Poly(acrylic acid-co-*N*,*N*-dimethylacrylamide) (I)
Poly(acrylic acid-co-vinyl trihydroxylsilane) (II)
Poly(acrylic acid-co-4-vinylpyridine *N*-oxide) (III)

Safety Concerns Polyacrylic acid is extremely toxic and may cause cancer. In addition, it may potentially cause genetic damage

Additive Structures

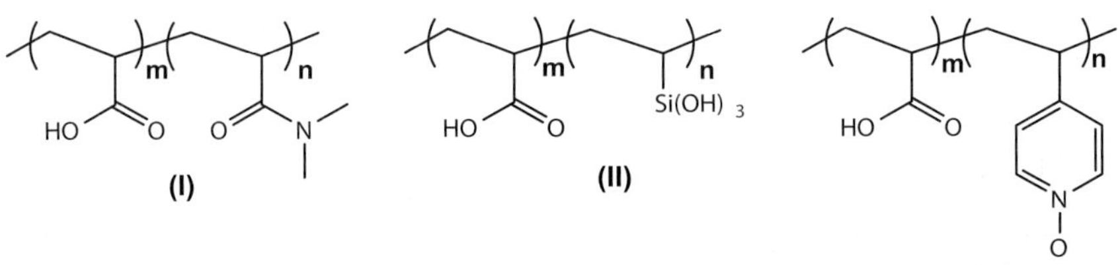

Additive Preparation

Case I

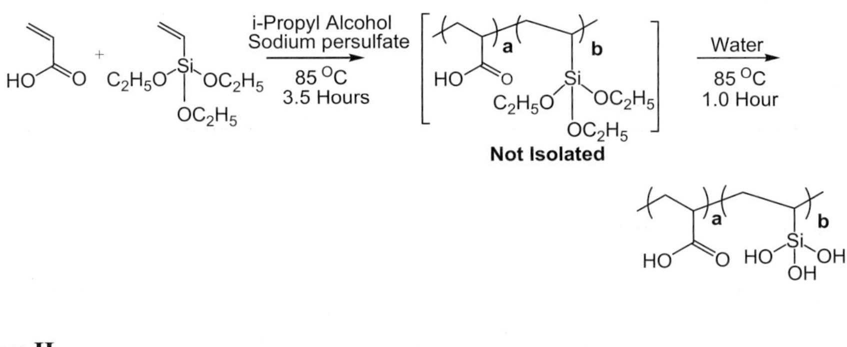

Case II

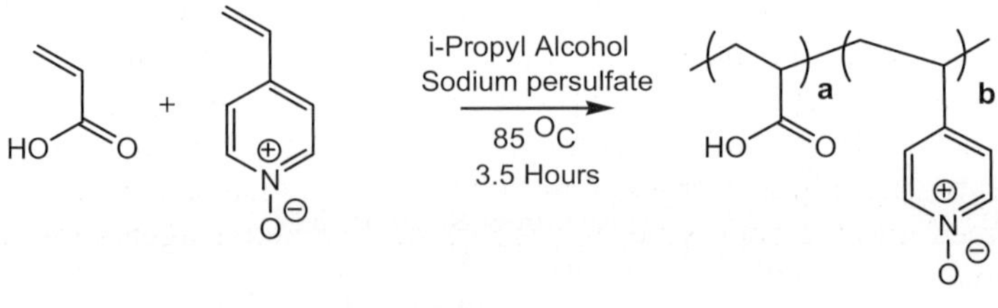

Additive Synthesis

1. Preparation of poly(acrylic acid-co-vinyl trihydroxylsilane)

A reactor was charged with 200 g of water and 244 g of isopropanol; it was then warmed to 85 °C and treated with 295 g of acrylic acid and 5 g of vinyl triethoxy silane over a period of 3 h. An initiator solution consisting of 15 g of sodium persulfate dissolved in 100 g water was then added to the reaction pot incrementally over a period of 3.5 h. After the addition was completed, the reaction mixture was heated to 85 °C for one additional hour followed by isopropanol distillation using a Dean–Stark trap, and the product was isolated.

Additive Testing

2. Glass Adhesion Evaluation

A 10 wt% aqueous solution of a selected acrylic acid copolymer was sprayed onto a glass microscope slide to form a film and was then weighed. The glass slide was then soaked in a 1.0% sodium hydroxide solution for 10 min, dried, and the adhesion effectiveness was measured by the amount of polymer still remaining on the glass surface. Test results are provided in Table 1.

ADDITIVE TEST RESULTS

TABLE 1 Glass adhesive properties of experimental polyacrylate copolymers

Sample	Polymer	Adhesion to Glass
Control	Polyacrylic acid (Alcosperse® 602A)	Good
(I)	Poly(acrylic acid-co-*N*,*N*-dimethylacrylamide)	Good
(II)	Poly(acrylic acid-co-vinyl trihydroxylsilane)	Excellent
(III)	Poly(acrylic acid-co-4-vinylpyridine *N*-oxide)	Excellent

Advantages over Prior Art

While polyacrylic acid has been derivatized with numerous surface active agents and used in glass adhesion formulations, three aspects of the current invention are unique and distinguish it over prior art. First, the current application requires only up to 1.6 wt% comonomer to generate a high-performance glass adhesion additive. Second, the synthetic route used to prepare these surface active agents generates polymers with reproducible molecular and physical properties. This characteristic streamlines commercialization of the additive. Finally, other components, including corrosion inhibitors, antioxidants, etc., can be readily incorporated into the copolymer to prepare surface-active agents for customized applications.

Notes

1. Fehn[1] prepared a two-component self-adhesive crosslinking silicone composition for glass consisting of vinyl-dimethylsiloxy-terminated polydimethylsiloxane having a viscosity of 20,000 Da and dimethylvinylsiloxy-endcapped polydimethylsiloxane having a vinyl content of 2.5 mmol/g and then compounded these agents with platinum divinyltetramethyldisiloxane.
2. Chaussade[2] prepared an unusually active glass-adhering coating by catalyzing the reaction of vinyl-di-methoxy-polydimethylsiloxane oil (IVa) having a viscosity of approximately 135,000 mPas. In addition, α,ω-trimethylsilylated polydimethylsiloxane oil (VIb) having a viscosity of 100 mPas and vinyltrimethoxysilane (V) were compounded with dibutyltin dilaurate (VI) to avoid the use of aminosilanes.

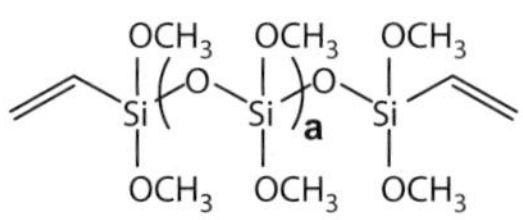

(IVa) Viscosity ~ 135,000 mPas

(IVb) Viscosity ~ 100 mPas

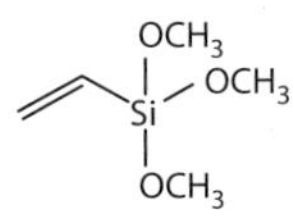

(V)

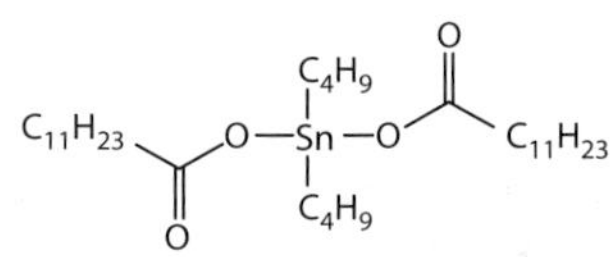

(VI)

3. Schoeley[3] prepared glass adhesive additives consisting of the reaction product of α,ω-dihydroxypolydimethylsiloxane, trimethylsilyl-polydimethylsiloxane, α-hydroxy-ω-trimethylsilyl-polydimethylsiloxane, ethyltriacetoxysilane, and methyltriacetoxysilane catalyzed without the use of aminosilanes.
4. Polypropylene-g-maleic anhydride (VII) was prepared by Tse[4] and was effective as both a glass and metal adhesive.

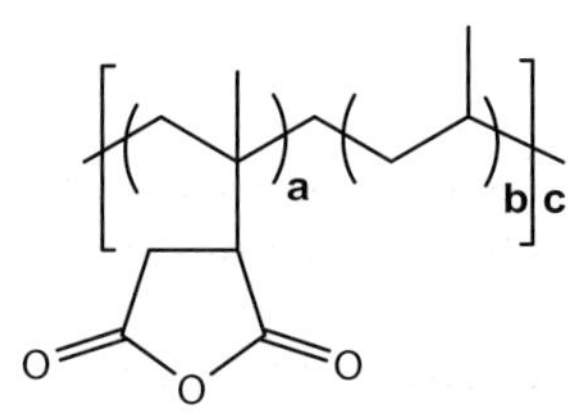

(VII)

REFERENCES

1. Armin Fehn, US Patent 7,842,771 (November 30, 2009)
   Wacker Chemie AG (Munich, DE)
2. Marc Chaussade et al, US Patent 7,838,613 (November 23, 2010)
   Rhodia Chimie (Boulogne-Billancourt Cedex, FR)
3. Peter Schoeley et al, US Patent 7,820,770 (October 26, 2010)
   Wacker Chemie AG (Munich, DE)
4. Mun-Fu Tse et al, US Patent 7,812,085 (October 12, 2009)
   ExxonMobil Chemical Patents Inc. (Houston, TX)

C. HOT MELT ADHESIVES

Kraton Terpolymer–Paraffin Blends

Title:	Low application temperature hot melt adhesive
US Patent:	7,795,336 (September 14, 2010)
Author:	Charles W. Paul et al.
Assignee:	Henkel AG & Co. KGaA (Duesseldorf, DE)

Significance: The addition of up to 3 wt% of wax in rubber-based hot melt adhesive compositions containing poly(styrene-b-butadiene-b-styrene), poly(styrene-b-isoprene-b-styrene), or poly(styrene-b-ethylene-b-propylene-b-styrene) has been found to be effective in lowering the rubber-based hot melt processing temperature.

Additive Name Paraffin wax[1] ($20 \leq n \leq 40$ range alkane mixture)
C AS Number 8002-74-2 and 64742-43-4
Safety Concerns Very slight skin irritant

Additive Name Microcrystalline wax (clay treated)
C AS Number 64742-42-3
Safety Concerns Prolonged or repeated exposure may cause irritation

Additive Structure

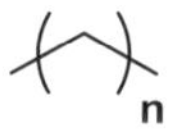

n ~ 20 - 40

Analogs None

Additive Preparation Commercial supplies of both reagent grade additives can be obtained from the following suppliers:

Paraflint C-80 Wax
Sasol Sasol Wax GmbH
Worthdamm 13-27 20457
Hamburg, Germany
49 40 78115-0

Microsere 5999 Microcrystalline Wax
Honeywell International Ltd.
Specialty Wax & Additives
101 Columbia Road, P.O. Box 1053
Morristown, NJ 07962-1053
814-827-3609

Formulation

Half of the tackifying resin Wingtack® from Goodyear or Escorez® from Exxon was placed into a jacketed mixing kettle equipped with rotors, and the temperature was increased from 100 °C to 200 °C. The precise temperature, however, depends on the softening point of the block copolymer resin. After the tackifying resin melted, stirring was initiated, and the block copolymer, wax, oil, and an antioxidant were blended together and further heated until a smooth homogeneous mass was obtained.

OPTIMUM ADDITIVE TREATMENT

TABLE 2 The effect of selected waxes on the physical properties of the adhesive. All formulations contained a tackifying resin, block copolymer, oil, and an antioxidant. The top two formulations are particularly useful in preparing laminates for disposable diapers

Wax	Crystallization Temperature (°C)	Cube Flow @ 130 °F (% Flow)	Viscosity @ 120 °C	Storage Modulus (dyn/cm²)
1.5% Paraflint C-80 Fisher–Tropsch wax	60.3	0	3925	3.0×10^6
2 5% of Microsere 5999	66.2	0	3015	7.0×10^6
1% Paraflint H-4	88.9	>500	8500	0.5×10^6

Advantages over Prior Art

The use of nominal amounts of wax diluents as crystalline reinforcing agents requires longer polymer processing times to bond to laminates, and processing temperatures between 130 °C and 150 °C are usually required. The advantage of the current invention is that processing temperatures of ≤120 °C were required and that the formulation improved the thermal aging in processing equipment while minimizing concerns associated with heat sensitive materials or substrates Table 2.

Notes

1. Wax diluent additives suitable for use in adhesive formulations are those that provide the adhesive with a viscosity <10,000 centipoises at 120 °C, a cube flow of <300% at 130 °F, a DSC crystallization temperature of <75 °C, and a storage modulus of $<1.0 \times 10^7$ dynes/cm² at 10 rads/sec at 25 °C.
2. A low application temperature thermoplastic hot melt adhesive additive package useful in preparing laminates for disposable diapers was prepared by He[1] and consisted of the following:

Ingredient	Composition (%)
Kraton D 1124K (SIS block copolymer with 30% styrene)	20
Escorez 5400 (petroleum hydrocarbon resin)	55.8
Kristallex F115 (α-methyl-styrene-based hydrocarbon resin)	7
Nyflex (mineral oil)	16.5
Irganox 1010FF/565/PS800 (antioxidants)	0.4/0.1/0.2

3. Disposable absorbent articles were prepared by Kong[2] by ultraviolet curing of 2-ethylhexyl acrylate bonded to a pendant benzophenone on an acrylic polymer and containing Foral® 85 as tackifier. Polybutadiene-grafted benzophenone photoinitiators prepared by Herr[3] were also effective as initiators in preparing disposable absorbent articles.
4. A low application temperature hot melt adhesive consisting of poly(ethylene-co-methacrylic acid) [Surlyn®] sodium or calcium salt ionomer resin, the thermoplastic elastomer poly(styrene-b-isoprene-b-styrene), and a tackifying resin was prepared by He[4] and used in disposable absorbent garments.

REFERENCES

1. Qiwei He et al, US Patent 7,799,863 (September 21, 2003)
   Henkel AG & Co. KGaA (Duesseldorf, DE)
2. Wei Kong US 7,754,327 (July 13, 2010)
   Henkel AG & Co. KGaA (Duesseldorf, DE)
3. Donald E. Herr et al, US Patent 7,799,884 (September 21, 2010)
   Henkel AG & Co. KGaA (Duesseldorf, DE)
4. Qiwei He et al, US Patent 7,795,341 (September 14, 2010)
   Henkel AG & Co. KGaA (Duesseldorf, DE)

D. UV CURABLE ADHESIVES

Knoevenagel Replacement Reaction

Application:	Adhesive Additive
Title:	Imines and methods of preparing electron-deficient olefins using such novel imines
US Patent:	8,053,589 (November 8, 2011)
Author:	Ciaran B. McArdle
Assignee:	Henkel Ireland Ltd. (Whitestown, IE)

Significance: A high yielding low temperature single-pot method for both preparing and polymerizing cyanoacrylate monomer independent of the Knoevenagel condensation is described. This method entails condensing electron-deficient ethyl cyanoacetate with an aliphatic or aromatic imine followed by stirring the two-component mixture for about 20 min at ambient temperature. In this process, in situ generated cyanoacrylate is released as a gas and immediately coats the interior walls of the glass apparatus.

Additive Names

1-Ethyl-3-(2,2-dimethyl-2-iminepropyl) imidazolium iodide (I)
3,3,-Dimethyl-3-imine pyridinium iodide (II)
Poly(4-t-butylimine-α-methylstyrene) (III)
1,4-Dimethyl-1,4-diiminecyclohexane (IV)

Safety

Ethylcyanoacetate is considered a strong eye irritant. It is a considered a moderate lachrymator which causes moderate irritation of the digestive tract.

Ethyl cyanoacrylate is considered unstable since it polymerizes rapidly especially in the presence of moisture. The material is combustible and incompatible with strong oxidizing agents or moisture. It is also considered a severe eye and respiratory hazard. Safety glasses, gloves, and adequate ventilation must be provided when blended in formulations. It is essential that this material does not come into contact with the skin or the eyes.

Additive Structures

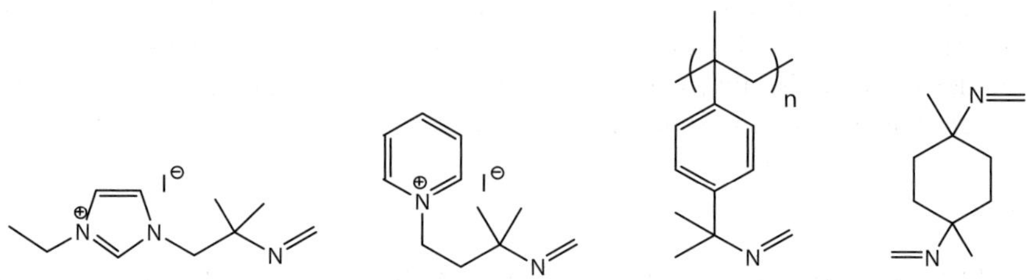

(I) (II) (III) (IV)

In situ Preparation of Ethylcyanoacrylate and Monomer Polymerization

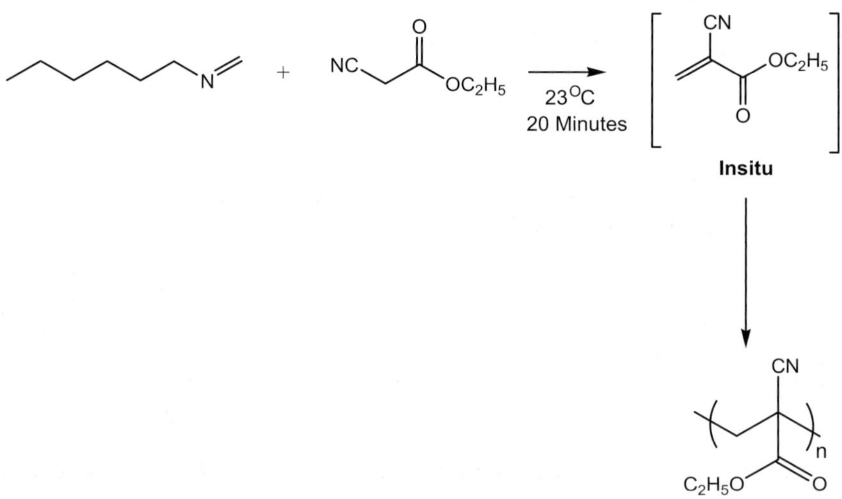

Additive Synthesis

1. Preparation of poly(ethyl cyanoacrylate)

A two-necked flask was charged with a 1.0:1.1 molar ratio of *n*-octyl imine and ethyl cyanoacetate, and then the contents were stirred at ambient temperature. This resulted in a strong exotherm. Thereafter, it was observed that ethyl cyanoacrylate monomer had polymerized inside the condenser of the distillation apparatus to give a thick cream-colored coating brittle polymer.

IR [film] (KBr) cyano, 2250 cm^{-1} (weak), carbonyl, 1750 cm^{-1} (strong)

Additive Testing

Other than demonstrating the formation and characterization of poly(ethyl cyanoacrylate) at ambient temperature, no testing data were supplied by the authors.

Advantages over Prior Art

There are several advantages of the current invention over the existing methods for preparing. First, it provides an alternative to the Knoevenagel condensation, which uses dinitromethane or related chemical analogs. Second, adhesive compositions for consumer or industrial applications will not be required to use the cyanoacrylate monomer directly. Finally, when the adhesive additive is applied using two tubes, each component has an infinite shelf life with the absence of clogging either nozzle.

Notes

1. 3-(2-Cyanacryloyloxy)-butyric acid ethyl ester (V) and octyl-cyanoacrylate (VI) adhesive monomers prepared by Quintero[1] were effective in retarding blood flow from wounds and aiding repair and regrowth of living tissue.

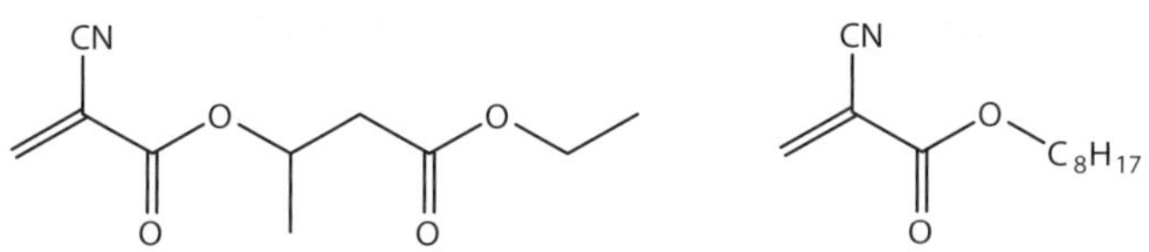

(V) (VI)

2. Biocompatible tissue adhesives were prepared by Hadba[2] by condensing polyethylene oxide diisocyanate (VII) with sorbitol (VIII).

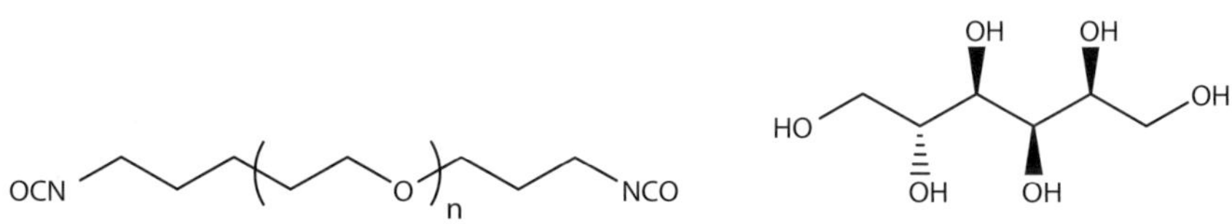

(VI) (VII)

3. Fritz[3] prepared a diisocyanate-terminated macromer (VIII) for use as an internal biological adhesive.

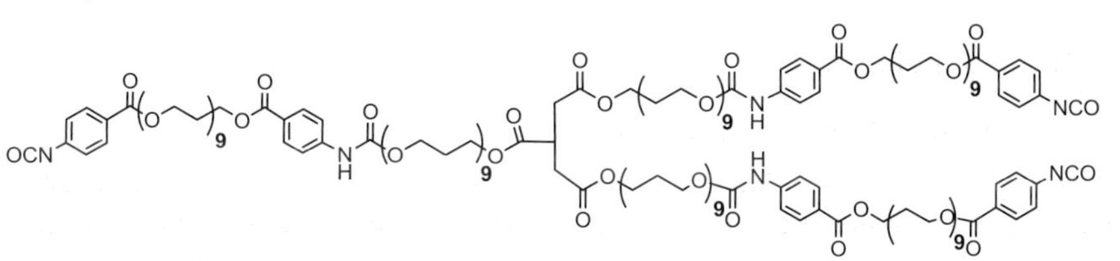

(VIII)

4. Additional cyanoacrylate generators (IX) and (X) were previously prepared by McArdle[4] and used in adhesive applications.

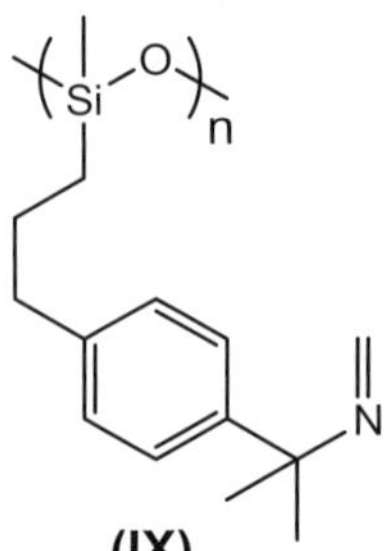

(IX)

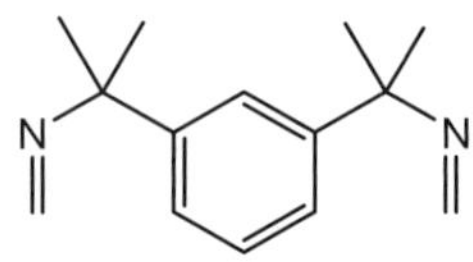

(X)

REFERENCES

1. Julian A. Quintero et al, US Patent Application 20080241249 (October 2, 2008)
Closure Medical Corporation (Raleigh, NC)
2. Ahmad R. Hadba et al, US Patent 7,998,466 (August 16, 2011)
Tyco Healthcare Group LP (North Haven, CT)
3. Benjamin D. Fitz et al, US Patent 7,968,668 (June 28, 2011)
Ethicon Inc. (Somerville, NJ)
4. Ciaran B. McArdle et al, US Patent 7,973,119 (July 5, 2011)
Loctite (R&D) Limited (Dublin, IE)

E. UV CURABLE ADHESIVES

Oxetane Epoxy Alcohols

Title: Ultraviolet-curable resin composition containing a polyfunctional oxetanyl group
US Patent: 7,977,401 (July 12, 2011)
Author: Kouichi Yokota et al.
Assignee: DIC Corporation (Tokyo, JP)

Significance: Ionically curable polymerization of urethane resins containing oxetanyl termini were at least 90% curable in compositions containing up to 45% of the curable oxetane termini. Polymerized resins were flexible when bent to ≥90° and showed no evidence of cracking after curing.

Additives 1,3-Bis-(3-ethyl-oxetane-3-ylmethoxy)-propane-2-ol (I)
3-Ethyl-3-hydroxy methyl oxetane (II)

Safety Oxetane and its derivatives are eye, respiratory system, and skin irritants. They are especially harmful if ingested.

Additive Structures

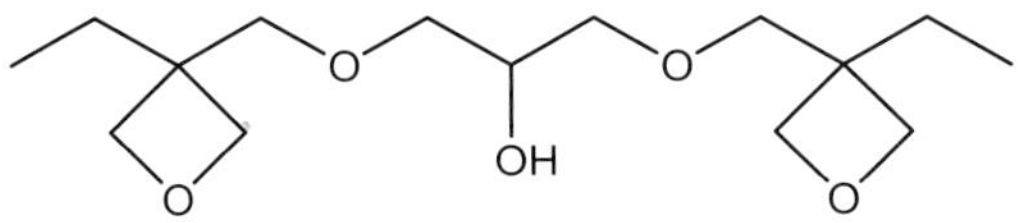

(I)

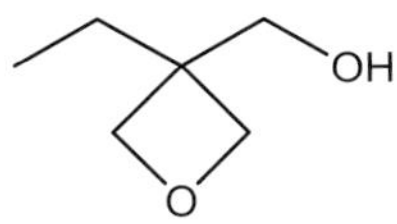

(II)

Additive Preparation

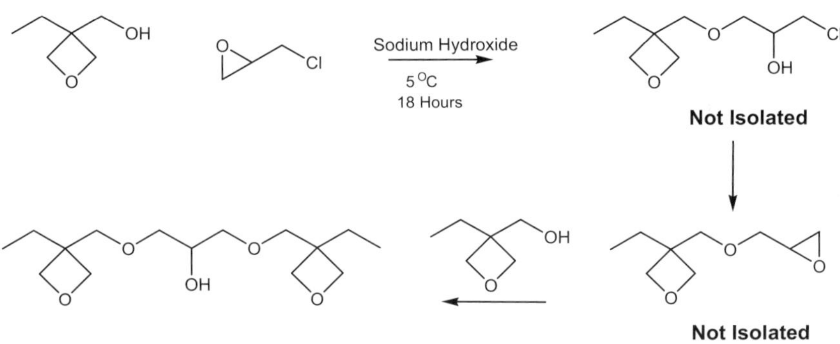

Additive Synthesis

1. Preparation of 1,3-bis(3-ethyl-oxetane-3-yl methoxy)-propane-2-ol

A round-bottom beaker was charged with 818 g of 3-ethyl-3-hydroxy methyl oxetane and 90 g of sodium hydroxide and then cooled with stirring to 5 °C and treated with the dropwise addition of 182 g of 1-chloro-2,3-epoxy propane over 2 h while stirring for 18 h. Thereafter, the product was isolated by distillation at 140 °C and 20 mbar, and the structure was confirmed by ^{1}H- and ^{13}C-NMR analyses.

Testing

A. Photocurable Composition

A polypropylene urethane prepolymer was prepared by blending 496 g of polypropylene glycol having a hydroxyl group equivalent of 1000 with 186 g of 4,4′-diphenylmethane diisocyanate for 3 h at 90 °C and then postreacting it with 318 g of the Step 1 product for 3 h at 90 °C.

A mixture consisting of 100 g of this composition and 40 g of acetone was then blended in a sealed type planetary mixer until a uniform consistency was obtained.

The mixture was then treated with 5 g of CPI-100P® [diphenyl-4-(phenylthio) phenyl sulfonium hexafluorophosphate] (Note 1) in propylene carbonate. Testing results are provided in Table 3.

B. Curability

A polypropylene plate was coated with a 100-mm coating of the photocurable composition at 23 °C and then air dried for 5 min to volatilize acetone. Thereafter, the plate was irradiated with an ultraviolet dose of between 700 and 800 mJ/cm2 using a high-pressure mercury lamp having an intensity of 120 W/cm and then left undisturbed for 7 days. Test results are provided in Table 3.

TEST RESULTS

TABLE 3 **Photocurability of urthane prepolymers terminated with oxetane derivatives from Additives 1 and 2 on polypropylene plates**

Formulation	Example 1	Example 2	Example 3	Example 4	Example 5
Photocurable composition	Additive 1	Additive 2	Additive 1	Additive 1	Additive 1
Photocurable composition (g)	100	100	50	50	50
Cationic photopolymerization initiator	CPI-100P	CPI-100P	CPI-100P	CPI-100P	CPI-100P
Cationic photopolymerization initiator (g)	5	5	5	5	5
Cationic polymerizable	X	X	EX-214L	UVR-6110	OXT-221
Cationic polymerizable (g)	X	X	50	50	50
Solvent	Acetone	Acetone	X	X	X
Solvent (g)	40	40	X	X	X
Curability (%)	95	93	91	90	96

Advantages over Prior Art

Polymerization of ultraviolet-curable resins may be less than desirable to use when continuous exposure to ultraviolet rays is difficult, since the degree of curability has consistently been problematic. Oxetanyl-terminated resins described in the current invention have addressed this concern by producing 90–95% cured resins with very high flexibility with no evidence of material cracking in a short UV exposure time.

Notes

1. CPI-100P® (III) is a cationic photoinitiator or photoacid generators of triarylsulfonium salt-type with PF_6^-.

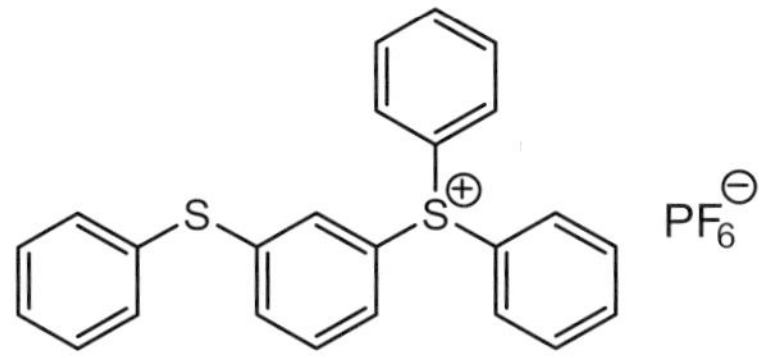

(III)

2. Adamantane-containing photocurable analogs (IV) were previously prepared by Takenaka[1] and used to encapsulate light-emitting diodes.

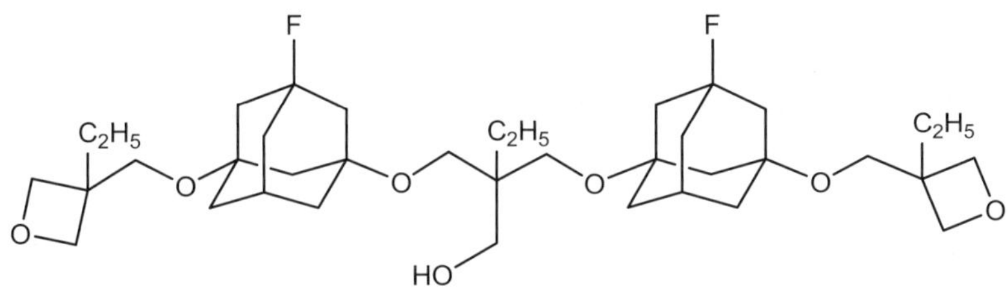

(IV)

3. Kennedy[2] prepared photocurable epoxide prepolymers (V), which were used in contact and intraocular lens formulations.

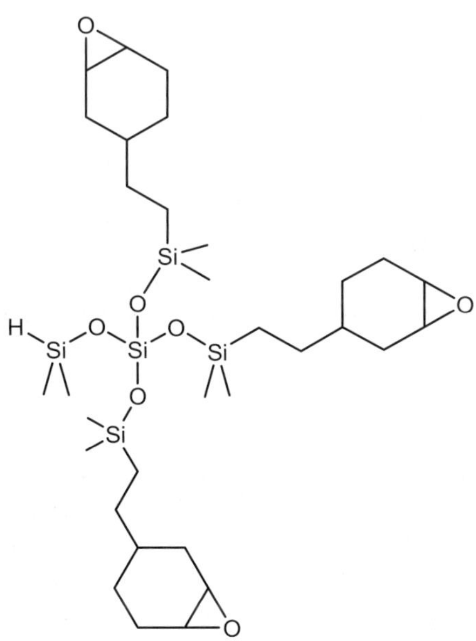

(V)

4. Eckert[3] prepared photocurable halogenated epoxy functionalized derivatives (VI), which were used in restorative dental compositions.

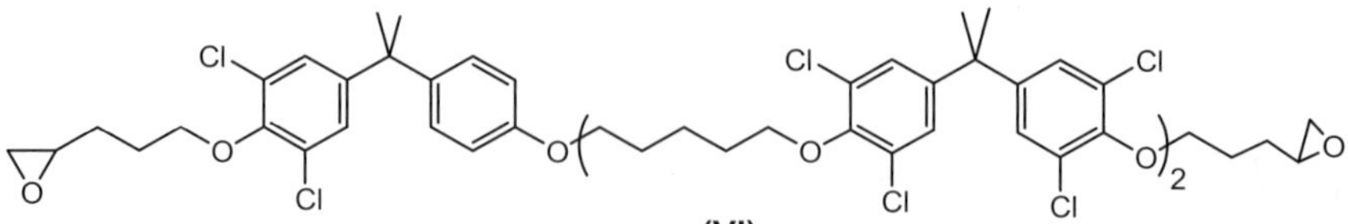

(VI)

5. Photopolymerizable methyl methacrylate compositions containing a 2-pyrimidine terminus (VII) were prepared by Suzuki[4] and used in coloring formulations for use in LCD display devices.

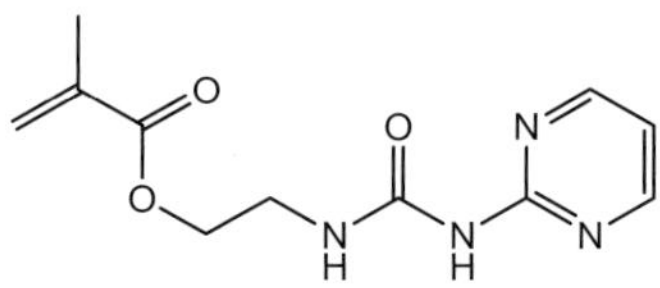

(VII)

REFERENCES

1. Junji Takenaka et al, US Patent 7,754,903 (July 13, 2010)
Tokuyama Corporation (Mikage-Cho, Shunan-Shi, Yamaguchi, JP)

2. Joseph P. Kennedy et al, US Patent 7,964,693 (June 21, 2011)
The University of Akron (Akron, OH)

3. Adrian S. Eckert et al, US Patent 7,939,580 (May 10, 2011)
3M Innovative Properties Company (St. Paul, MN)

4. Shigekazu Suzuki, US Patent 7,927,772 (April 19, 2011)
Fujifilm Corporation (Minato-Ku, Tokyo, JP)

Chapter 2

Antioxidants

A. ANTIOXIDANTS TO MINIMIZE RESIN DISCOLORATION

Phenolic/Phosphite Hybrids

Title:	Macromolecular antioxidants based on sterically hindered phenols and phosphites
US Patent:	7,956,153 (June 7, 2011)
Author:	Ashok L. Cholli et al.
Assignee:	Polnox Corporation (Lowell, MA)
Significance:	A phenolic phosphite antioxidant was inexpensively prepared, which has an oxidative induction time of 25 min longer than those of the Irganox® antioxidant product line. The additive is designed to be used in polymer blends to prevent material discoloration during high-temperature extrusion.
Additive	Tris[*N*-(4-hydroxyphenyl)-3-(2,6-di-t-butyl-4-hydroxyphenyl) propionamide]-phosphite (I)
Safety	This additive is analogous to Irganox®-related antioxidants. Irganox-based antioxidants may cause mechanical irritation to the eyes, skin, and respiratory system. Liver and thyroid effects have been reported after repeated exposure to these agents. The use of a respirator is strongly encouraged to avoid breathing dust.

Next Generation of International Chemical Additives. DOI: http://dx.doi.org/10.1016/B978-0-444-53788-1.00002-4

Additive Structure (Note 1)

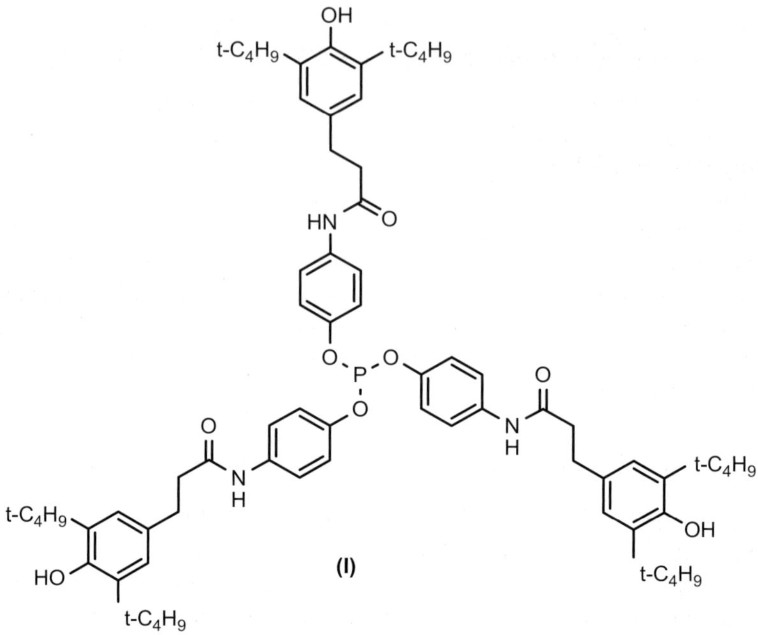

(I)

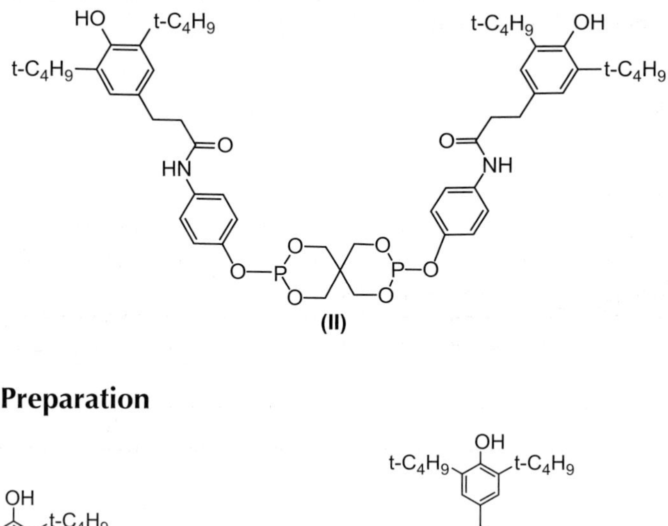

(II)

Additive Preparation

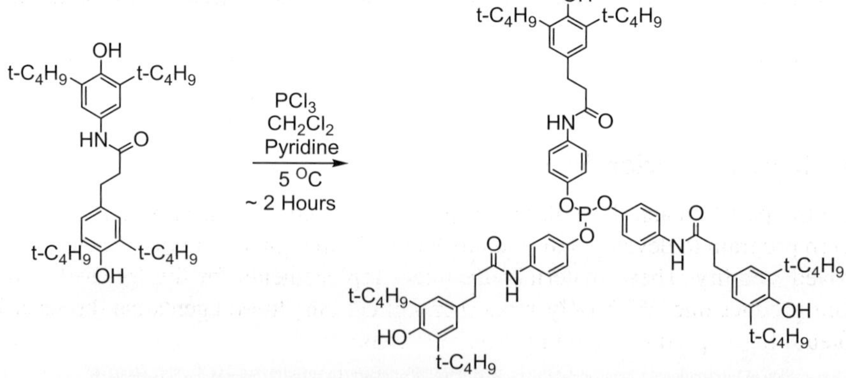

Additive Synthesis

1. *Preparation of tris[N-(4-hydroxyphenyl)-3-(2,6-di-t-butyl-4-hydroxyphenyl)-propionamide]phosphite*

A 1-liter two-neck round bottom flask equipped with a thermometer, dropping funnel, and a nitrogen inlet was charged with 100 g of *N*-(4-hydroxyphenyl)-3-(2,6-di-t-butyl-4-hydroxy phenyl) propionamide and 22 ml of pyridine dissolved in 200 ml of methylene chloride and then treated with 7.86 ml of phosphorous trichloride at 0–5 °C over a period of 30–40 min. Thereafter, the reaction was monitored by TLC/HPLC, and the contents were cooled to ambient temperature after the reaction was completed. Methylene chloride was then removed by distillation, and the residue was treated with 100 ml of methanol. The white crystalline precipitate which was formed was isolated and then washed several times with cold methanol and the product isolated after drying had an MP = 224–245 °C.

Testing and Treatment Levels

Unadditized polypropylene powder was additized with 1000 ppm of the Step 1 product and then extruded through a single-screw extruder, and the product was isolated in the form of pellets. Polypropylene pellets were then subjected to an accelerated oxidative stability test using differential scanning calorimetry according to the testing protocol of ASTM D3895. A reference sample additized with 1000 ppm Irganox® 1010 was also prepared. Test results are provided in Table 1.

Test Results

TABLE 1 Comparison of oxidative onset and termination of polypropylene additized with 1000 ppm of the Step 1 product or Irganox® 1010 and then extruded through a single extruded at 230 °C

Antioxidant	Oxidation Onset (min)	Oxidation Completion (min)
Step 1 Product	27	59
Irganox® 1010*	12	32

**Tetrakis(methylene-(3,5-di-t-butyl-1-hydrocinnamate))methane.*

Advantages over Prior Art

Within the past 5 years, the Polnox Corporation has been engaged in an aggressive research program to develop antioxidants having both higher antioxidant activity and improved stability. These materials are likely replacements for the Irganox® antioxidant product line. With only modest modifications, these agents can be used in petroleum-based products and in processed foods.

Notes

1. Phenolic-based antioxidants (III) having an Mn of approximately 40,000 Da were prepared by Cholli[1] by polymerization of phenolic esters using horse radish peroxidase.

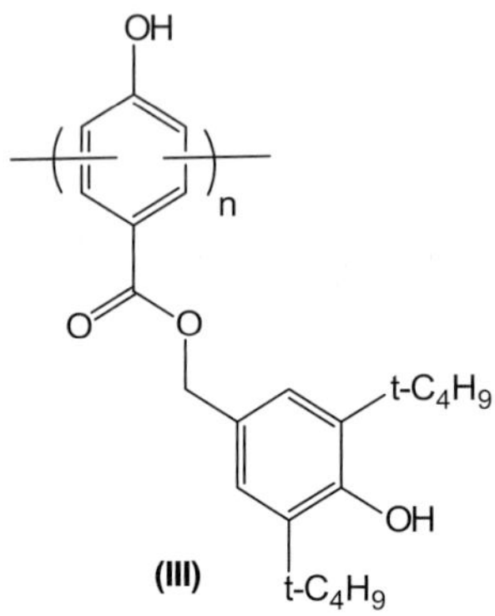

(III)

2. Cholli[2] prepared pentaerythritol-based tetraester antioxidants (IV), which were used in polymer-blending formulations to minimize resin discoloration during extrusion processing.

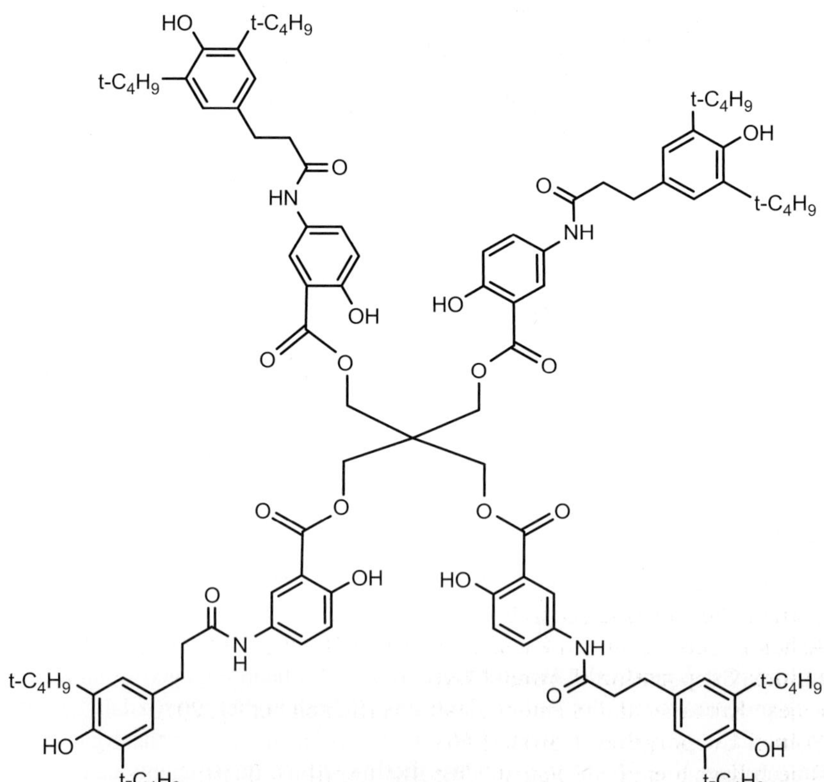

(IV)

3. Kumar[3] prepared nitrogen and hindered phenol-containing antioxidants (V), which had high oxidative induction times and were used in propylene-blending formulations to minimize resin discoloration.

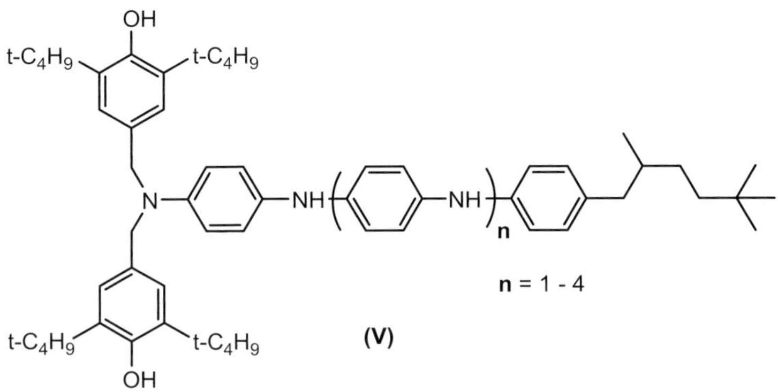

(V)

4. Rajesh[4] prepared gemini antioxidants (VI), which were used in polypropylene, elastomers, and lubricant-blending formulations.

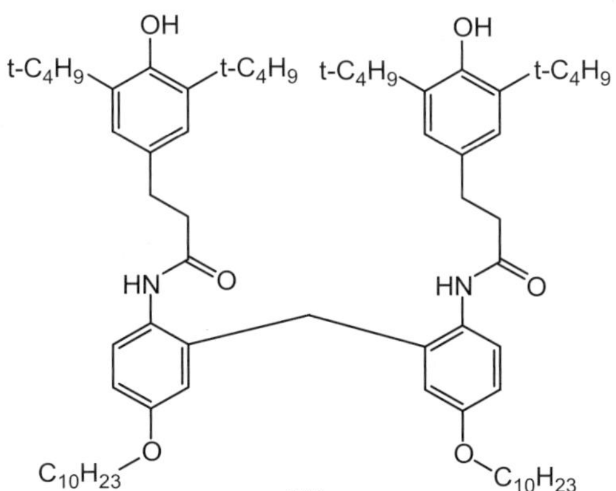

(VI)

REFERENCES

1. Ashok L Cholli, US Patent 7,923,587 (April 12, 2011)
   Polnox Corporation (Lowell, MA)
2. Ashok L Cholli et al, US Patent Application 20090184294 (July 23, 2009)
   Polnox Corporation (Lowell, MA)
3. Rajesh Kumar et al, US Patent 7,799,948 (September 21, 2010)
   Polnox Corporation (Lowell, MA)
4. Rajesh Kumar et al, US Patent 7,705,185 (April 27, 2010)
   Polnox Corporation (Lowell, MA)

B. ANTIOXIDANTS TO MINIMIZE RESIN DISCOLORATION IN THERMOSETS

Bisphenol 2-Carboxyethyl Acrylates

Title:	Bisphenol monoester compound
US Patent:	8,076,402 (December 13, 2011)
Author:	Ryoji Soma et al.
Assignee:	Sumitomo Chemical Company, Limited (Tokyo, JP)

Significance: A polymerizable analog of bisphenol A was prepared by condensing 2,4-di-t-pentylphenol with acetaldehyde and then converting it into the monoester by reacting with 2-carboxyethyl acrylate. When compounded with poly(ethylene-b-propylene) and then heated to 180 °C in air for 2 h, the resulting resin had an improved degree of discoloration over an acrylate analog.

Additive Name 2,4-Di-t-pentyl-6-[1-(3,5-di-t-pentyl-2-hydroxyphenyl) ethyl]-2-carboxyethyl acrylate (I)

Safety Although little safety information is available on the current material, it is virtually identical to the antioxidant Sumilizer® GS (II) manufactured by Sumitomo Chemical Company, Ltd. Sumilizer® GS has very low skin and oral toxicity and is a very mild eye irritant. The agent is non–genotoxic in vitro gene cell mutation and chromosome aberration testing.

Additive Structure

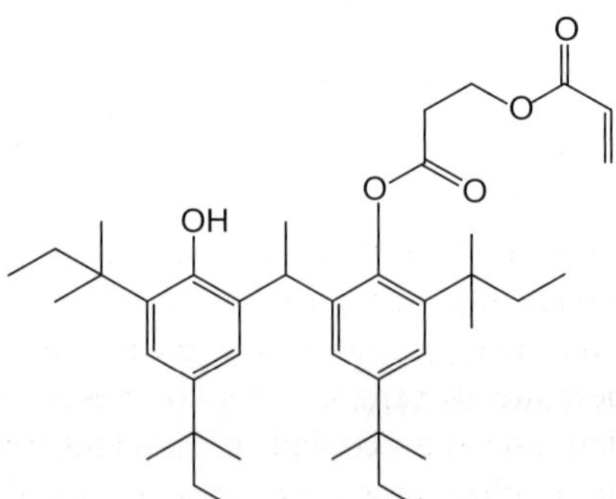

(I)

Additive Preparation

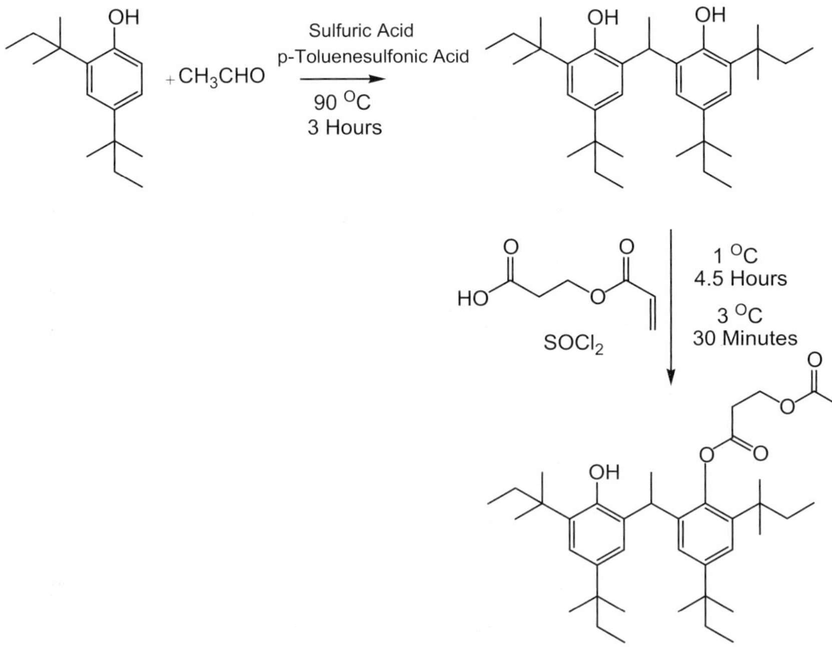

Additive Synthesis

1. *Preparation of 2,4-di-t-pentyl-6-[1-(3,5-di-t-pentyl-2-hydroxyphenyl)ethyl]*

A 250-ml round-bottom flask was initially charged with 117.5 g of 2,4-di-t-pentylphenol, 0.9 g of *p*-toluene sulfonic acid monohydrate, and 6.9 g of a 78% aqueous solution of sulfuric acid and then treated with the dropwise addition of 12.1 g of acetaldehyde and 28 g of xylene over 2 h at 35 °C. The mixture was then heated for 3 h at 90 °C and then cooled and treated with 157 g of xylene, 59 g of water, and 11.7 g of 25% aqueous sodium hydroxide. After removing the aqueous phase, the organic phase was washed with water and then azeotropically distilled, and 292 g of a yellow oil was isolated.

2. *Preparation of 2,4-di-t-pentyl-6-[1-(3,5-di-t-pentyl-2-hydroxyphenyl)ethyl]-2-carboxyethyl acrylate*

A large reaction vessel was charged with the entire Step I product and 28 g of 50% aqueous potassium hydroxide and then heated to 98 °C for 3 h. After cooling to 50 °C, hexane was added, and the mixture was azeotropically distilled at 106 °C and then cooled to ambient temperature. A separate reaction vessel was charged with 35.3 g of 2-carboxyethyl acrylate and then cooled to 1 °C and treated with the dropwise addition of 26.7 g of thionyl chloride over 4.5 h. After the thionyl chloride addition was completed, the contents of the second vessel were added to the contents of the first vessel and then stirred for 30 min at 3 °C. Thereafter, the vessel was treated with

150 ml of water and then stirred for 10 min at 30 °C, and the organic layer was isolated. The product was isolated using the procedure described in Step 1, and 303 g of the brown oil was isolated having a purity of 93%.

TESTING

Antioxidancy

A blend consisting of 11.6 mg of the Step 2 product dissolved in 5 ml of chloroform was added to 2 g of poly(ethylene-b-propylene) having an MI of 9–10 g/10 min @ 230 °C under 2.16 kg, and the mixture was stirred until the chloroform vaporized off. The residue was then heated in a gear oven for 2 h at 180 °C in air. Test results are provided in Table 2.

Test Results

TABLE 2 Oxidative stability of poly(ethylene-b-propylene) additized with 5.8 wt% of the Step 2 product or the comparative additive and then heated for 2 h at 180 °C in air

Additive	Discoloration Degree (YI Value)	Discoloration Degree (Improvement Index) (%)
Step 2 Product	4.03	0
Comparative*	12.22	7.1

*2,4-Di-t-pentyl-6-[1-(3,5-di-t-pentyl-2-hydroxyphenyl)ethyl]phenyl acrylate (II).

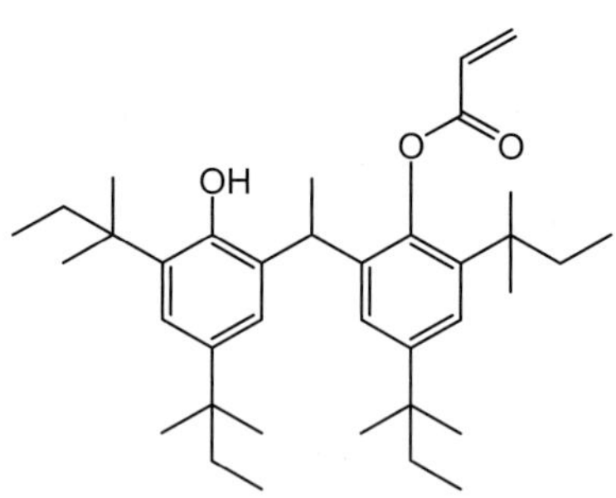

(II)

Advantages over Prior Art

The bisphenol precursors of the current invention and Sumilizer GS® are manufactured by Sumitomo Chemical Company, Ltd. are essentially identical. The improved antioxidancy of the carboxyethyl acrylate-functionalized compound most likely reflects higher thermal grafting as a result of an extended chain length into the poly(ethylene-b-propylene) matrix when using the acrylated-functionalized analog.

This suggests that the thermal grafting efficacy of related materials may also be increased by judiciously extending chain lengths.

Notes

1. Additional hindered bisphenol antioxidant monoester analogs (III) and (IV) were previously prepared by Kimura[1] and used as antioxidants in engineered materials to minimize resin discoloration when processed at elevated temperatures.

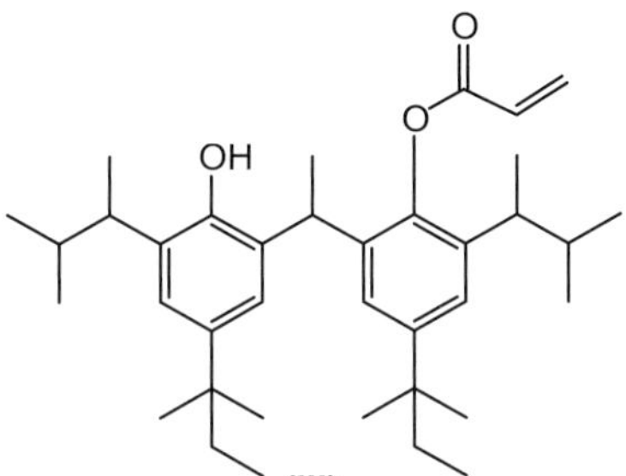

(III)

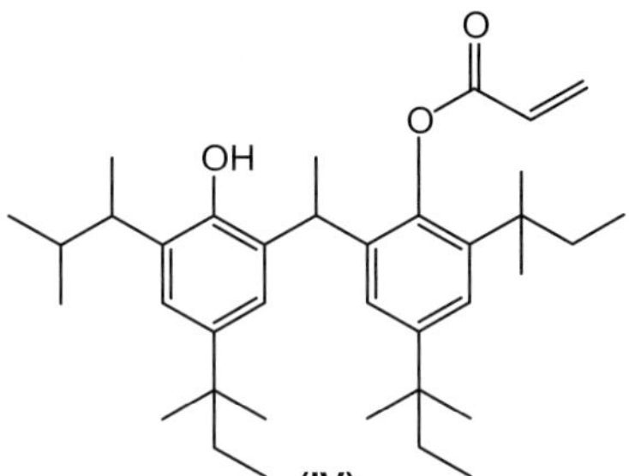

(IV)

2. Macromolecular antioxidants (V) prepared by Cholli[2] were used in thermoplastics, automotive additives, and cosmetics.

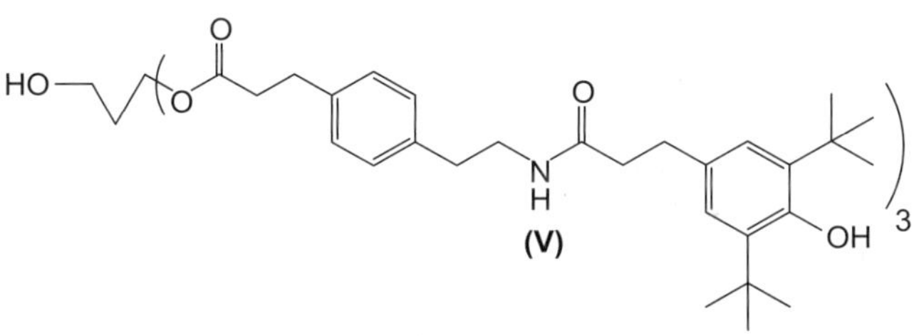

(V)

3. The polymeric phenolic antioxidant (VI) prepared by Cholli[3] was more effective at lower treatment rates than Irgonox® 1010 as a thermal stabilizer when processing polypropylene at elevated temperatures.

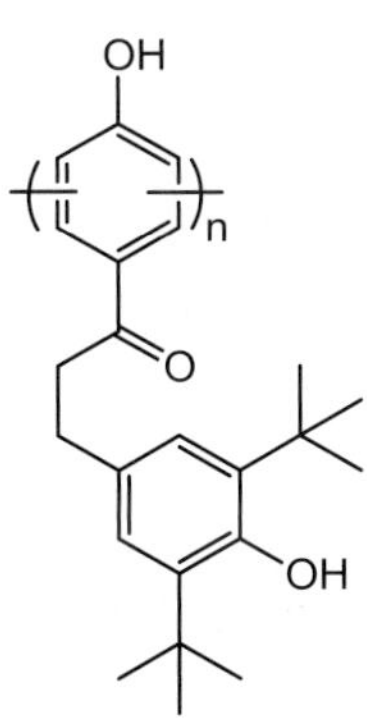

(VI)

4. Cholli[4] used Coprinus cinereus peroxidase and Arthromyces ramosus peroxidase to prepare high molecular weight aromatic antioxidants (VII) and (VIII) from the corresponding monomers.

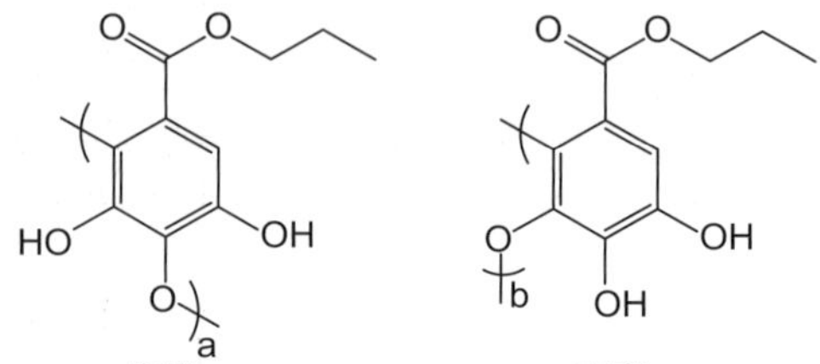

(VII) (VIII)

REFERENCES

1. Kenji Kimura et al, US Patent 7,977,418 (July 12, 2011)
Sumitomo Chemical Company, Limited (Tokyo, JP)
2. Ashok L. Cholli et al, US. Patent 8,039,673 (October 18, 2011)
Polnox Corporation (Lowell, MA)
3. Ashok L. Cholli et al, US. Patent 7,923,587 (April 12, 2011)
Polnox Corporation (Lowell, MA)
4. Ashok L. Cholli et al, US. Patent 7,754,267 (July 13, 2011)
The United States of America as represented by the Secretary of the Army (Washington, DC) University of Massachusetts Lowell (Lowell, MA)

C. PHOTOCHEMICAL ANTIOXIDANTS TO MINIMIZE RESIN DISCOLORATION

Benzotriazole–Benzophenone Hybrids

Title:	Ultraviolet light absorbing ketones of 2-(2-hydroxyphenyl) benzotriazole
US Patent:	7,847,103 (December 7, 2010)
Author:	Chia-Hu Chang
Assignee:	Chia-Hu Chang (Tappan, NY)
Significance	Hybrid antioxidants containing both benzotriazole and benzophenone chromophores were prepared in high yields and in a single step. The hybrid ultraviolet stabilizers have molar and mass extinction coefficients of 1.13 and 1.52, respectively, which are significantly higher than those of existing commercially available materials.

Additive Name 2-(2-Hydroxy-3-benzoyl-5-methylphenyl)benzotriazole (I), 2-(2-hydroxy-3-benzoyl-5-t-butylphenyl)benzotriazole) (II), 2-(2-hydroxy-3-(3-methylbenzoyl)-5-methylphenyl) benzotriazole (III)

Safety Concerns Inhalation of material may be harmful and may cause mild to severe respiratory distress. Contact with unprotected skin may cause burns. Containers may explode when heated.

Additive Structures

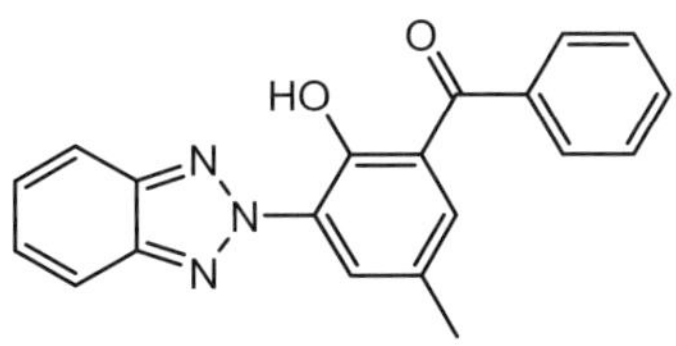

(I)

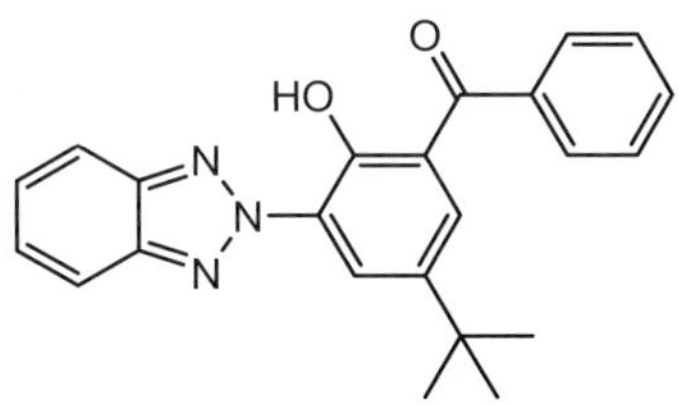

(II)

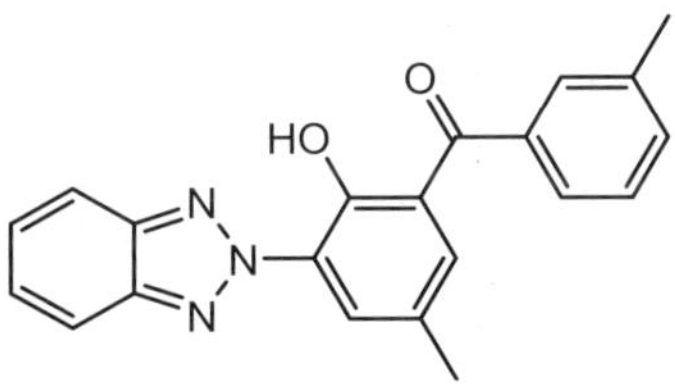

(III)

Additive Preparation

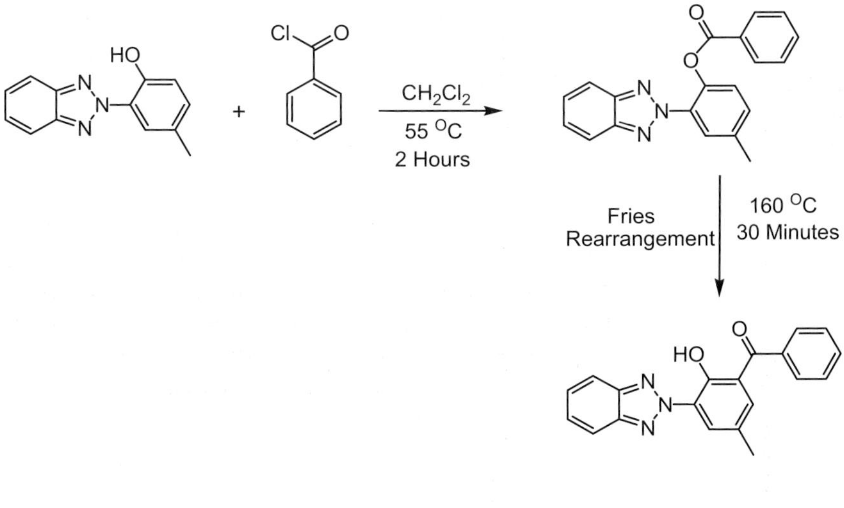

Additive Synthesis

1. Preparation of 2-(2-hydroxy-3-benzoyl-5-methylphenyl) benzotriazole

A solution containing 112.5 g of 2-(2-hydroxy-5-methylphenyl)benzotriazole, 84 g of benzoyl chloride, and 133 g of aluminum chloride was dissolved in 1000 ml of methylene chloride and then refluxed for 2 h. The solution was then concentrated by distillation, and an ester intermediate was isolated. This material was then subjected to Fries rearrangement conditions by heating to 160 °C for 30 min with mechanical mixing and was then cooled to 90 °C and treated with water until hydrogen chloride evolution stopped. Thereafter, an additional 500 ml of water containing 50 g of concentrated hydrochloric acid was added, and the mixture was refluxed for 1 h before cooling to ambient temperature. The organic layer was isolated and treated with 500 ml of water containing 50 g of 45% sodium hydroxide and then refluxed for an additional hour and cooled to ambient temperature. The organic phase was reisolated and retreated with an additional 500 ml of water containing 50 g of concentrated hydrochloric acid and then refluxed for an additional hour, and the organic phase was reisolated. The organic phase was then washed with water several times, dried, and the solid isolated had an mp of 166–167.8 °C in a 65% yield. The UV spectrum of the product showed two separate absorbances, indicative of benzotriazole and benzophenone chromophores.

Additive Testing

Extinction coefficients of an experimental and commercial ultraviolet stabilizer are provided in Table 3.

TABLE 3 **Molar and mass extinction coefficients for TINUVIN® 938 and the Step 1 product performed in chloroform. The mass extinction coefficient of the benzotriazole–benzophenone hybrid was approximately 35% higher than that of the unmodified benzotriazole precursor, 2-(2-hydroxy-5-methylphenyl)-benzotriazole**

Sample	Extinction Coefficient (molar)	Extinction Coefficient (mass)
TINUVIN® 928 [2-(2-Hydroxy-3-α-cumyl-5-t-octylphenyl)-2H-benzotriazole]	1.0	1.0
Step 1 Product	1.13	1.52

Advantages over Prior Art

The current art provides a method for producing ultraviolet light absorbers having a high extinction coefficient by the coupling of benzotriazole and benzophenone derivatives. The ultraviolet absorbance spectra of these hybrids were very strong and pronounced, while not displaying extensive individual ultraviolet absorbances for either hybrid component.

Notes

1. Deo[1] prepared a series of benzotriazole methylmethacrylate monomers (IV), which were subsequently incorporated into polymer compositions to absorb ultraviolet light in eyeglasses and contact lenses.

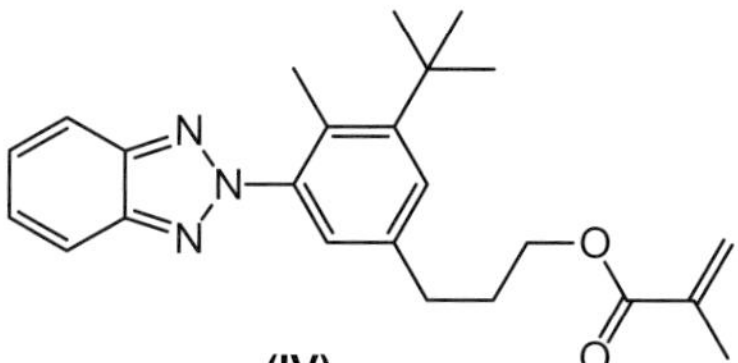

(IV)

2. Gugumus [2] used a three-component additive package containing 2,2,6,6-tetramethyl-piperidine *N*-oxide derivatives (V), (VI), and (VII) as a method for providing ultraviolet stabilization for polypropylene films at elevated temperatures.

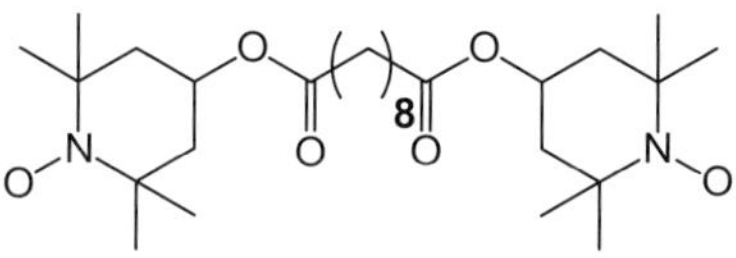

(V)

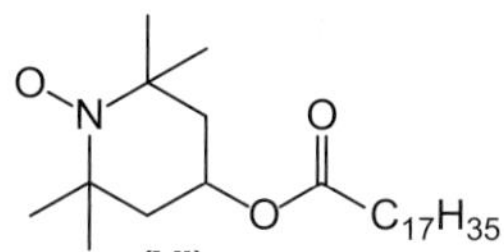

(VI)

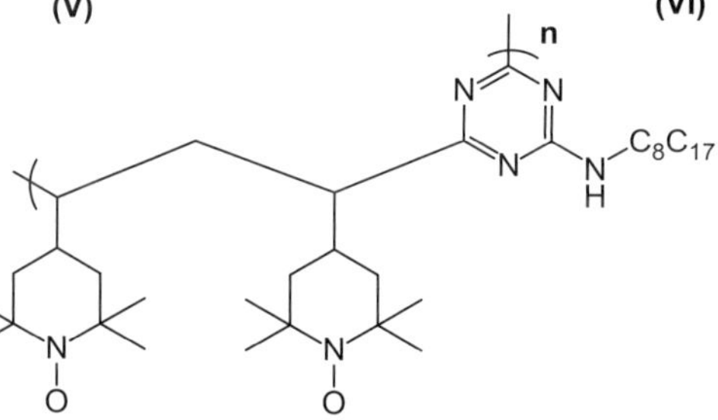

(VII)

3. Triaryl-s-triazines (VIII) prepared by Gupta[3] imparted high weathering and UV resistance to polycarbonates without compromising their physical strength properties. Rudiger[4] prepared bis-biphenyl-s-triazines (IX), which were coextruded with polycarbonate for making articles that required improved weatherability and UV resistance.

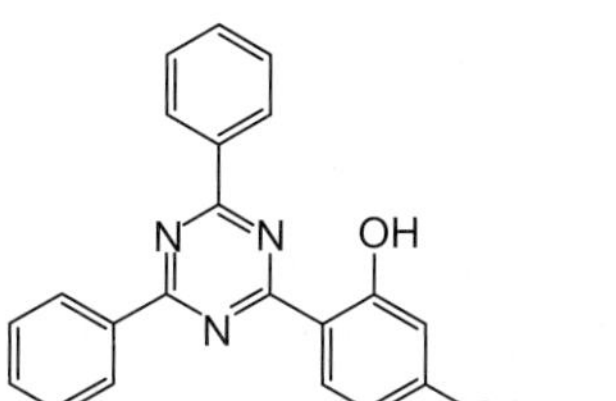

(VIII)

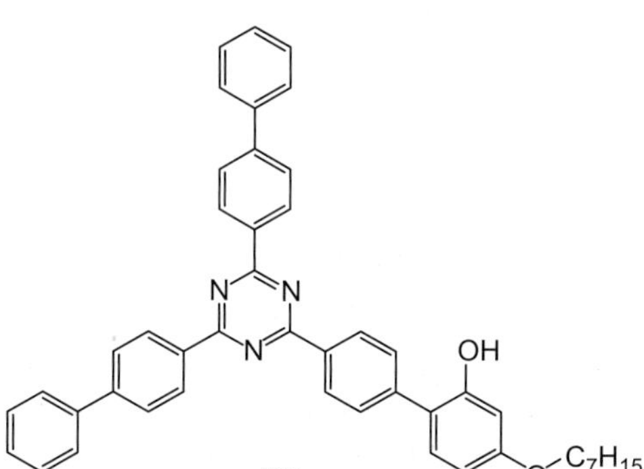

(IX)

4. Davis[5] incorporated α,ω-(3-(4-hydroxy-3-methoxyphenyl)-1-propyl)decadimethylsiloxane (XII) into polycarbonate films to improve the weatherability and obtain improved flow properties and UV resistance for thermosets.

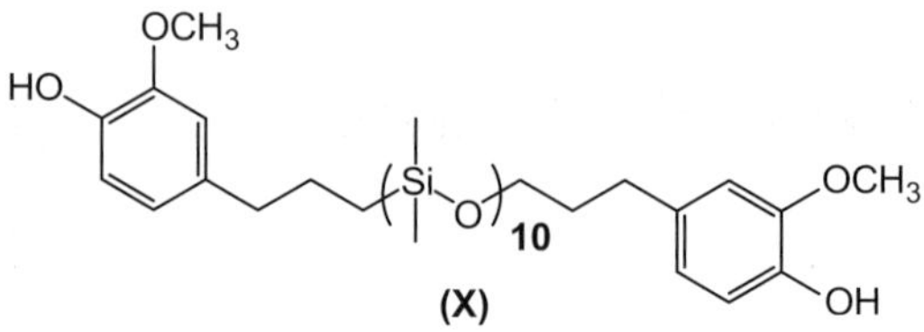

(X)

REFERENCES

1. Keshav Deo et al, US Patent 7,825,257 (November 2, 2010)
Bausch & Lomb Incorporated (Rochester, NY)
2. Francois Gugumus, US Patent 7,595,008 (September 29, 2009)
Ciba Specialty Chemicals Corporation (Tarrytown, NY)
3. Ram B. Gupta et al, US Patent 6,867,250 (March 15, 2005)
Cytec Technology Corp. (Wilmington, DE)
4. Claus Rudiger et al, US Patent 7,652,082 (January 26, 2010)
BASF SE (Ludwigshafen, DE)
5. Gary C Davis et al, US Patent 7,838,602 (November 23, 2010)
Sabic Innovative Plastics IP B.V. (NL)

Chapter 3

Automotive Additives

A. DETERGENTS TO CLEAN INTAKE VALVES TO IMPROVE VEHICLE FUEL ECONOMY

Polyisobutylenephenolic–Styrene Oxide Ammomium Acetate

Title:	Quaternary ammonium salt of a Mannich compound
US:	7,906,470 (March 15, 2011)
Author:	Paul R. Stevenson et al.
Assignee:	The Lubrizol Corporation (Wickliffe, OH)

Significance: An automotive detergent additive was prepared by quarternizing the Mannich base, 4-polyisobutylene-2-aminomethylphenol with styrene oxide. At fuel treatment levels of 12.5 ptb's, BMW™ intake valve deposits were reduced by approximately 115%.

Additive Name 4-Polyisobutylene-2-[(*N*,*N*-dimethyl-*N*-oligomeric styrene oxide)aminomethyl]-phenol acetate (I)

Safety Concerns No explicit MSDS exists for this experimental product. Since it is produced using industrial chemicals, batch-to-batch reagent stoichiometry will be inconsistent. The presence of 4-polyisobutylenephenol and formaldehyde are of special concern.

4-Polyisobutylenephenol is a moderate skin, lung, and eye irritant. It may affect liver and kidney functions as well as male and female reproductive organs.

Formaldehyde is a cancer hazard. It is a very strong eye, mucus membrane, and skin irritant. Repeated or prolonged breathing may cause shortness of breath.

Next Generation of International Chemical Additives. DOI: http://dx.doi.org/10.1016/B978-0-444-53788-1.00003-6

Additive Structure

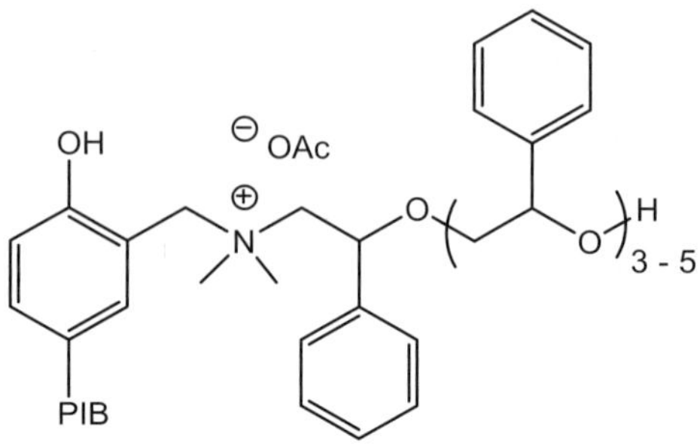

Polyisobutylene (PIB) Mn ~ 1,000 daltons **(I)**

Additive Preparation

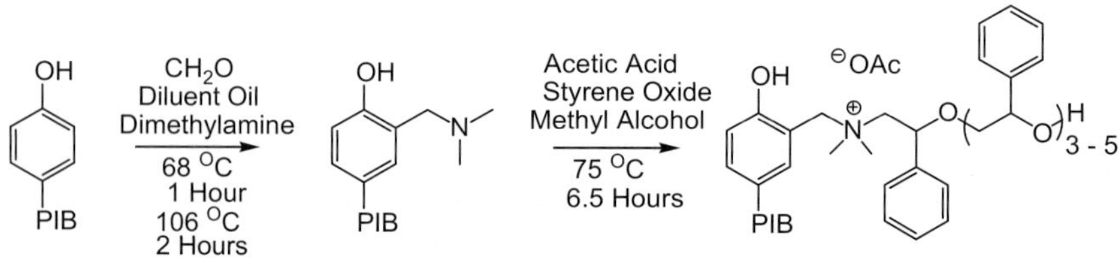

Additive Synthesis

1. Preparation of 4-polyisobutylene-2-N,N-dimethyl-methylamine

A reactor was charged with 800 g of 4-polyisobutylenephenol having an Mn of approximately 1000 Da and 240 g of diluent oil SO®-44 and then stirred until the mixture was thoroughly blended. The solution was then treated with the dropwise addition of 55.9 g of formaldehyde over 50 min and then treated with the incremental addition of 73.3 g of dimethylamine over 50 min. The mixture was then heated to 68 °C for 1 h and then further heated to 106 °C for 2 h. Thereafter, the mixture was heated to 130 °C for 30 min and then cooled overnight. The reaction mixture was purified by vacuum distillation at 130 °C @ –0.9 bar to remove any remaining water and unreacted reagents.

2. Preparation of 4-polyisobutylene-2-[(N,N-dimethyl-N-oligomeric styrene oxide)-aminomethyl]-phenol acetate

A reactor was charged with 1700 g of the Step 1 product, 263 g of styrene oxide, 66 g of acetic acid, and 4564 g of methyl alcohol and then refluxed at 75 °C for 6.5 h under a nitrogen atmosphere. The reaction product was purified by distillation at 130 °C @ –0.9 bar, and the product was isolated.

Testing

BMW™ Intake Valve Deposit Test

The ASTM D550 Driving Cycle Test Dose Rate Detergent Active Testing Fuel protocol was used to evaluate the effectiveness of the experimental detergent additive.

In this test, fuel was additized with the Step 2 product dissolved in an inert diluent oil in a ratio of experimental agent-to-diluent oil of about 75:25 by weight. Additized fuel was also prepared containing three commercially available fuel packages at the identical treatment level as that of references.

BMW™ Intake Valve Deposit test results are provided in Table 1.

Test Results

TABLE 1 BMW™ Intake Valve Deposit Test results demonstrating the high detergency of the experimental additive over three commercial gasoline detergent packages

Detergent	Dosage (PTB)	Intake Valve Deposits (mg/valve)
Commercial Gasoline Package-1	12.5	55.0
Commercial Gasoline Package-2	12.5	84.0
Commercial Gasoline Package-3	25.0	18.1
Step 2 Additive	12.5	5.0

Advantages over Prior Art

Quaternary amine detergent additives are not novel. In two earlier investigations, polyisobutylene-containing quaternary amine succinimides were prepared but were then abandoned because of the high treatment level required to optimize their detergency. Neither case, however, incorporated oligomeric styrene oxide. Although the dynamics for the high detergency of the experimental additive remain unclear, the presence of oligomeric styrene oxide appears to have a synergistic effect.

Notes

1. Wollenberg[1] prepared ionic overbased carbonate Mannich condensation products (II) consisting of C_{10}–C_{15} alkylphenol with ethylene carbonate, which were effective as lubricant detergents, antiwear agents, and antioxidants when used to additize fuels and lubricating oils.

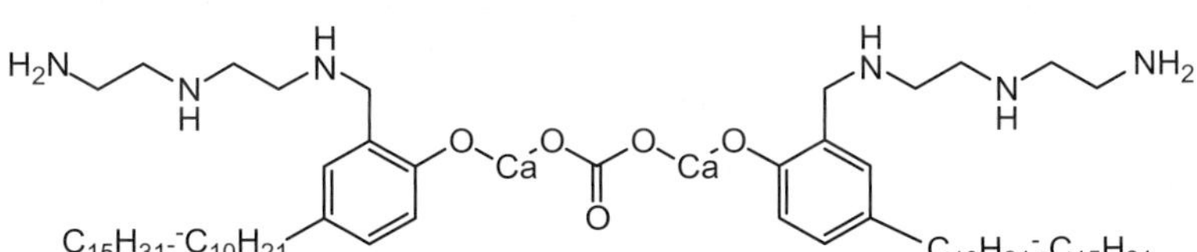

(II)

2. Ionic overbased calix[4]arenes (III) were prepared by Svarcass[2] and used as a lubricant additive in direct fuel injection two-cycle engines.

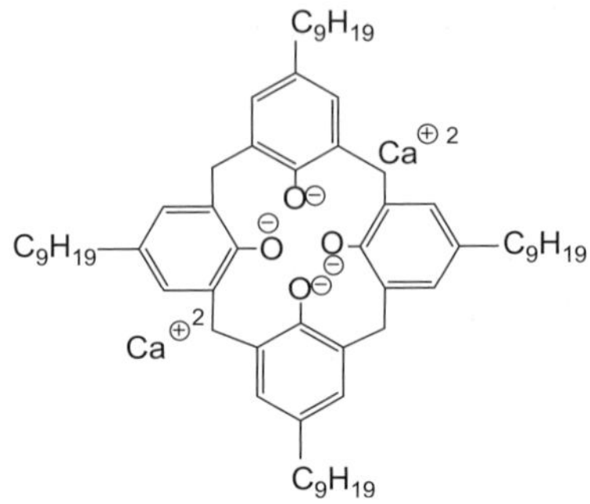

(III)

3. Hammond[3] prepared the first quaternary detergent, benzyl 3-(polyisobutenyl-succinimido)pyridinium chloride (IV), which was moderately effective as a lubricant detergent additive.

PIB Mn ~ 900 daltons

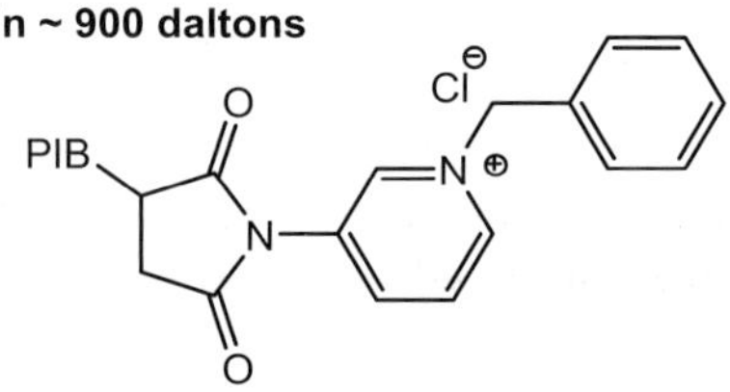

(IV)

4. Kurek[4] prepared a quaternary succinimide fuel detergent composition (V), which minimized deposits on intake valves and on metallic surfaces.

PIB Mn ~ 1,500 daltons

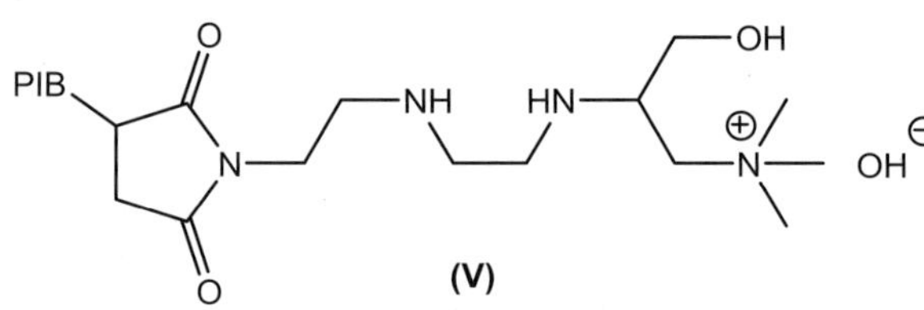

(V)

REFERENCES

1. Robert H. Wollenberg et al, US Patent Application 20050107266 (May 19, 2005)
   Chevron Texaco Corporation (San Ramon, CA.)
2. Laimute R Svarcas et al, US Patent Application 20050130856 (June 16, 2005)
   Lubrizol Corporation (Wickliffe, OH)
3. Kenneth G. Hammond, US Patent 4,338,206 (July 6, 1982)
   Texaco Inc. (White Plains, NY)
4. Paul R. Kurek, US Patent 5,254,138 (October 19, 1991))
   UOP (Des Plaines, IL)

B. GAS GENERATORS FOR AIR BAGS

5-Amino-tetrazoles

Title:	Gas generant compositions
US Patent:	7,914,632 (March 29, 2011)
Author:	Sudhakar R. Ganta et al.
Assignee:	TK Holdings, Inc. (Armada, MI)

Significance: The gas generant (1H-tetrazol-5-yl)-carbamic acid methyl ester was prepared by reacting 5-amino tetrazole with methyl chloroformate. When used as an inflation additive in automotive air bags containing phase stabilized ammonium nitrate formulations, higher overall propellant decomposition rates, greater gas production, and substantially reduced ash formation were observed.

Additive (1H-Tetrazol-5-yl)-carbamic acid methyl ester (I)

Safety The precursor, 5-amino-tetrazole, as well as (1H-tetrazol-5-yl)-carbamic acid methyl ester are nitrogen gas generators and explode when heated.

Additive Structure

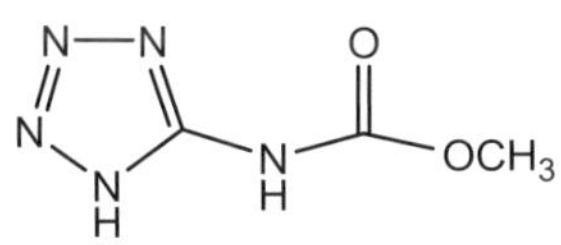

(I)

Additive Preparation

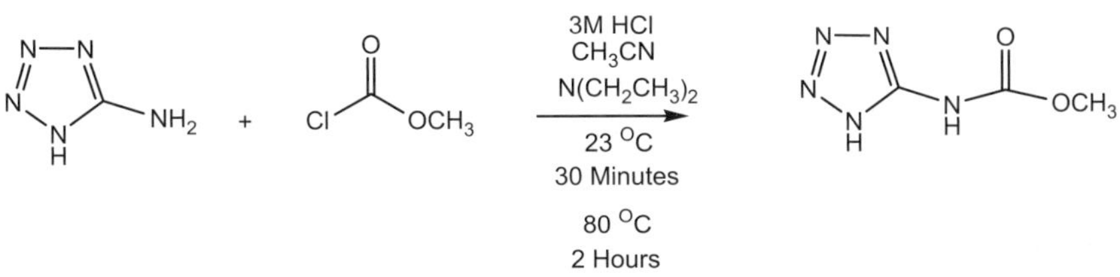

Additive Synthesis

1. Preparation of (1H-tetrazol-5-yl)-carbamic acid methyl ester

A reaction flask containing 10.00 g of 5-amino tetrazole and 16.48 ml of triethyl amine dissolved in acetonitrile was treated with 9.14 ml of methyl chloroformate at 0 °C and then gradually brought to ambient temperature and stirred for about 30 min. The mixture was then heated to reflux for 2 h and then recooled to ambient temperature. The reaction mixture was then concentrated under reduced pressure, and a solid was isolated. The residue was washed with dilute hydrochloric acid, dried at 105 °C for 6 h, and the product was isolated as a white-colored solid in a 92% yield. Differential scanning calorimetry testing indicated a sharp exotherm at 236 °C.

Additive Testing Results

A. Burn Rate

A mixture of (1H-tetrazol-5-yl)-carbamic acid methyl ester and phase stabilized ammonium nitrate, 24:76, respectively, had a burn rate of 0.4 inches per second.

B. Gas Generation

A propellant mixture of (1H-tetrazol-5-yl)-carbamic acid methyl ester and phase stabilized ammonium nitrate, 24:76, respectively, had an oxygen balance of −0.76 resulting in a 96.5% gas yield and produced 4.04 moles of gas per 100 g of propellant.

Advantages over Prior Art

While accelerated thermal decomposition of phase stabilized ammonium nitrate with a metal-free gas generator has been reported, the current method provides a streamlined and high yielding method for generating the reagent. In addition, the current agent has a very high exotherm, which results in high gas generation with minimum ash formation.

Notes

1. Although effective in accelerating the thermal decomposition of phase-stabilized ammonium nitrate, a propellant composition was prepared by Kubo[1] consisting of the following:
   i. Ammonium nitrate
   ii. Copper (II) nitrate
   iii. 5-Amino tetrazole

This generated very large quantities of ash byproducts.

2. A gas generant composition prepared by Ganta[2] consisting of the nonmetallic additive, tris(5-amino tetrazolo) triazine (II), mixed with phase stabilized ammonium nitrate exhibited a good decomposition rate, high gas generation while generating moderately low ash formation.

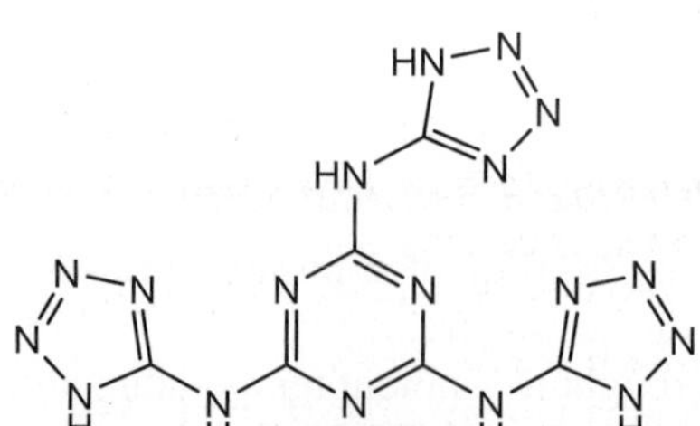

(II)

3. Gas generant compositions prepared by Ganta[3] containing one of the nonmetallic additives bis(tetrazole) urea (III) or 1-triazolyl-3-tetrazolyl urea (IV) were blended with phase stabilized ammonium nitrate and were extremely effective as decomposition rate enhancers while generating little ash byproducts.

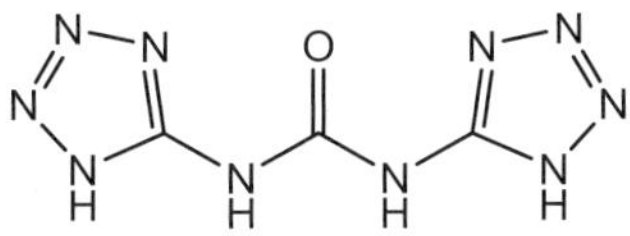

(III)

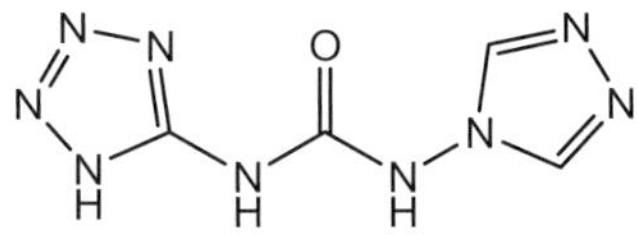

(IV)

4. Blends of the metal-free additive gas generator consisting of 3,4-bis(tetrazolo)-nitrobenzene (V) with phase stabilized ammonium nitrate was prepared by Ganta[4] and this generated minimum ash.

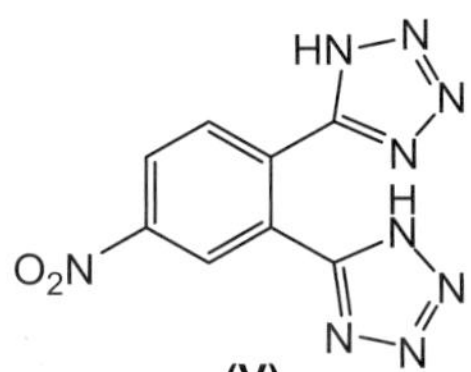

(V)

REFERENCES

1. Dairi Kubu et al, US Patent 7,918,949 (April 5, 2011)
Nippon Kayaku Kabushiki Kaisha (Tokyo, JP)
2. Sudhakar R Ganta et al, US Patent Application 20080169051 (July 17, 2008)
TK Holdings, Inc. (Armada, MI)
3. Sudhakar R Ganta et al, US Patent Application 7,847,102 (December 7, 2010)
TK Holdings, Inc. (Armada, MI)
4. Sudhakar R Ganta et al, US Patent Application 20080110536 (May 15, 2008)
TK Holdings, Inc. (Armada, MI)

C. LUBRICATING OIL ANTIOXIDANTS

Diphenylaminophenolics

Title:	Nitrogen and hindered phenol containing dual functional macromolecular antioxidants: synthesis, performances, and applications
US Patent:	7,799,948 (September 21, 2010)
Author:	Rajesh Kumar et al.
Assignee:	Polnox Corporation (Lowell, MA)

Significance:	A single-step method for preparing dual functional antioxidants containing ≥one sterically hindered phenols and diphenylamines is described. The presence of mixed antioxidant units within a single molecule suggests their use in a wide range of materials including plastics, elastomers, lubricants, petroleum based products, cosmetics, and processed food products.
Additive Name	*N*-(2,6-Di-t-butyl-1-hydroxybenzene-4-methylene)-*N*-tri(phenyleneamine)phenyl] (I)
CAS Number	Unregistered
Safety Concerns	The presence of *N*-phenyl-phenylenediamine as a component in the antioxidant suggests that it may be harmful if inhaled or absorbed through the skin, eye contact, or is inhaled through the respiratory route.
Additive Name	*N*,*N*-Di(2,6-di-t-butyl-1-hydroxybenzene-4-methylene)-*N*-phenylenediamine (II)
CAS Number	Unregistered
Safety Concerns	The presence of *N*-phenyl-phenylenediamine as a component in the antioxidant suggests that it may be harmful if inhaled or absorbed through the skin, eye contact, or is inhaled through the respiratory route.
Additive Name	*N*-(2,4-Di-t-butyl-1-hydroxybenzene-4-methylene)-*N*-1,3-dimethylbutyl-*N*-phenylenediamine (III)
CAS Number	Unregistered
Safety Concerns	The presence of *N*-phenyl-phenylenediamine as a component in the antioxidant suggests that it may be harmful if inhaled or absorbed through the skin, if it comes in eye contact, or is inhaled through the respiratory route.
Additive Name	*N*-(2-t-Butyl-1-hydroxybenzene-6-methyl-6-methylene)-*N*-1,3-dimethylbutyl-*N*-phenylenediamine (IV)
CAS Number	Unregistered
Safety Concerns	The presence of *N*-phenyl-phenylenediamine as a component in the antioxidant suggests that it may be harmful if inhaled or absorbed through the skin, if it comes in eye contact, or is inhaled through the respiratory route.
Additive Name	*N*-(2,4-Di-t-butyl-1-hydroxybenzene-6-methylene)-*N*-[tri(phenyleneamine)-phenyl] (V)
CAS Number	Unregistered
Safety Concerns	The presence of *N*-phenyl-phenylenediamine as a component in the antioxidant suggests that it may be harmful if inhaled or absorbed through the skin, if it comes in eye contact, or is inhaled through the respiratory route.

Additive Structures

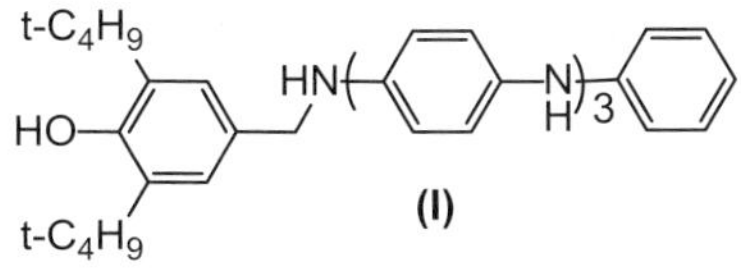
(I)

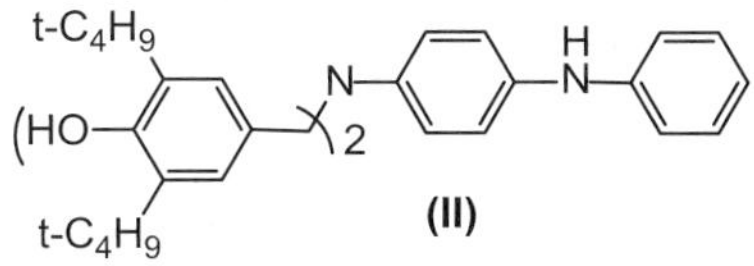
(II)

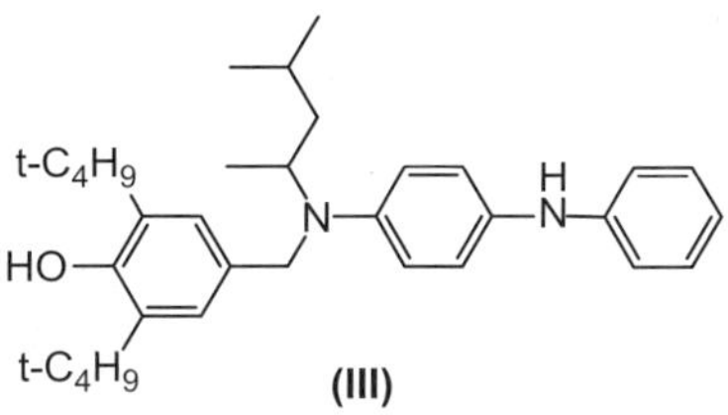
(III)

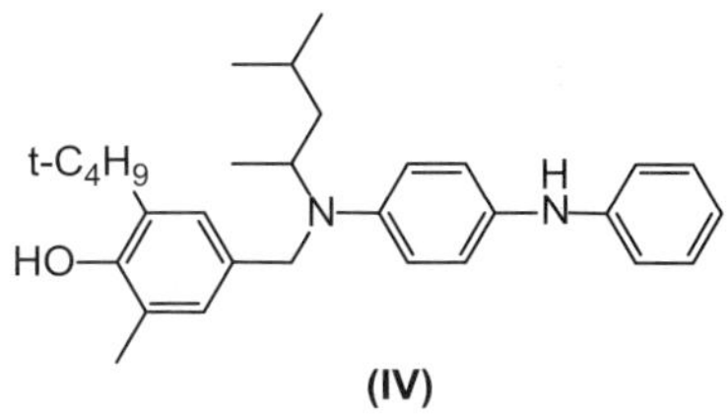
(IV)

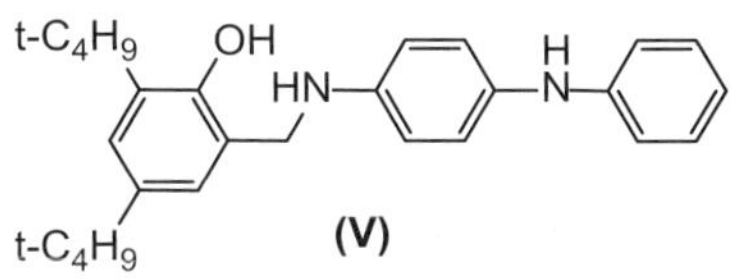
(V)

Additive Preparation

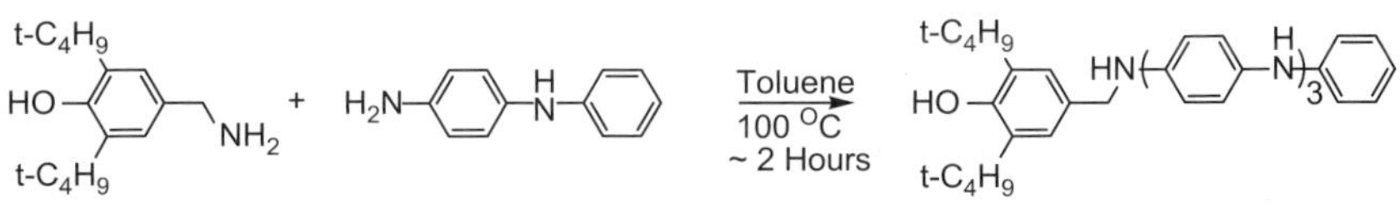

Additive Synthesis

Step 1 Preparation of N-(2,6-Di-t-butyl-1-hydroxybenzene-4-methylene)-N-[tri(phenylene amine)-phenyl]

A three-neck round bottom flask containing a Dean–Stark trap equipped with a condenser was charged with 3 g of 2,6-di-t-butyl-4-aminomethylphenol and 2 g of *N*-phenyl-1,4-phenylenediamine dissolved in 100 ml of toluene. The solution was then heated to reflux at 100 °C while the reaction extent was monitored by thin layer chromatography. When the reaction was completed, the mixture was concentrated, purified by column chromatography, and the product was characterized by using spectroscopic analytical techniques.

Additive Formulation

Experimental additives were evaluated for their antioxidancy by dissolving 200 ppm of a selected experimental antioxidant in lubricant oil and then evaluating by their oxidation induction times at 200 °C. Antioxidants were mixed in oil and then stirred for 5–15 min at 60 °C and then evaluated immediately.

Test Results

A performance summary of Additives I, II, III, and V with commercial antioxidants Irganox® L 57 and 6PPD in a synthetic polyol ester–based oil, and reference antioxidants are provided in Table 2.

TABLE 2 Oxidation induction times for lubricant oil additized with 200 ppm of selected experimental antioxidants compared with commercial antioxidants Irganox® L57 and Irganox® 6PPD

Antioxidant	Oxidation Induction Times (min)
Irganox® L57 (Dioctylated diphenylamine)	3
Irganox® 6PPD (*N*-Hexyl phenyl-1,4-phenylenediamine)	22
Experimental Agent I	115
Experimental Agent II	28
Experimental Agent III	32
Experimental Agent V	25

Advantages over Prior Art

While individual aromatic amines and sterically hindered phenols antioxidants exist, difunctional antioxidants and oligomers are unreported. The preparation of these duel functional antioxidants is routine and versatile, and antioxidant properties of these products are consistently superior to those of commercially available antioxidants.

Notes

1. In an earlier investigation by the Kumar[1], di- and tri- sterically hindered phenolic additives (VI) were prepared and used as antioxidants in resins, lubricating oils, and related petroleum applications.

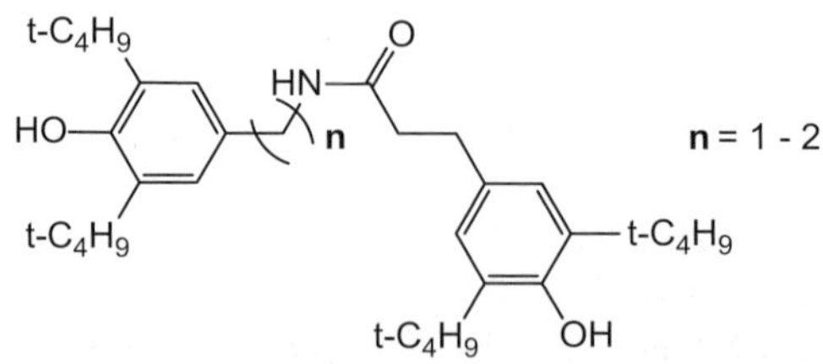

(VI)

2. Sterically hindered phenolic antioxidants (VII) and polymers (VIII) were previously prepared by Ashok[2] and used in meat products, dairy products, cereals, beverages, crackers, and bakery products as food additives.

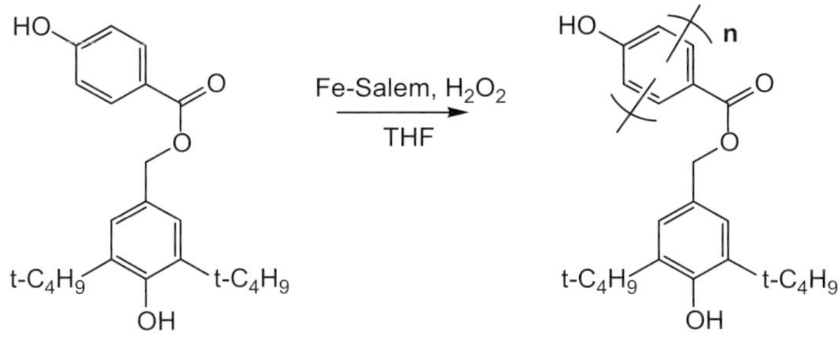

(VII) (VIII)

3. Loper[3] prepared the dispersant/antioxidant lubricant oil additive polyisobutentyl-g-*N*-phenylphenylene succinimide (IX) by condensing polyisobutenyl-g-succinic anhydride (X) with *N*-phenyl-4-phenylenediamine (XI) as illustrated in Eq. 1.

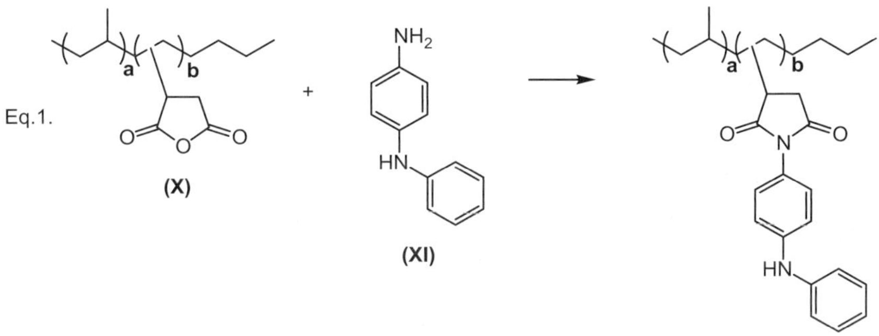

4. Cholliet[4] prepared sterically hindered polymeric phenolic antioxidants (XI) for use as a food additive in unsaturated oils to minimize oxidation during processing and storage.

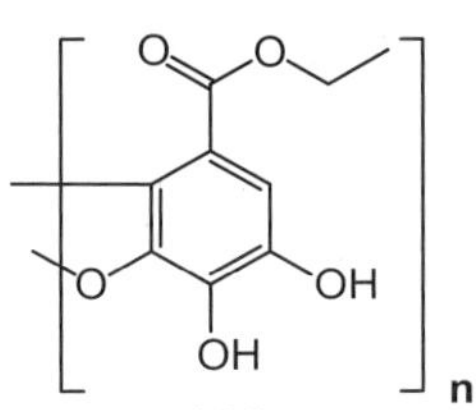

(XII)

REFERENCES

1. Rajesh Kumar et al, US Patent Application 20090184294 (July 23, 2009)
Polnox Corporation (Lowell, MA)
2. Ashok L. Cholli, US Patent 20080311065 (December 18, 2008)
Polnox Corporation (Lowell, MA)
3. John T. Loper et al, US Patent 7,361,629 (April 22, 2008)
Afton Chemical Corporation (Richmond, VA)
4. Rokok L. Cholliet al, US Patent 7,223,432 (May 29, 2007)
University of Massachusetts Lowell (Lowell, MA)
United States of America, as represented by the Secretary of the Army (Washington, DC)

D. LUBRICANT ANTIOXIDANTS THAT ARE COLORLESS AND AMINE-REAGENT-FREE

N-Phenylamine and α-Naphthylamine Alkylated with α-Methylstyrene

Title:	Alkylated PANA and DPA compositions
US Patent:	8,030,259 (October 4, 2011)
Author:	Michael Aebli et al.
Assignee:	Ciba Speciality Chemicals Corp. (Tarrytown, NY)

Significance: *N,N*-Diphenylamine and *N*-phenyl-α-naphthyl amine were alkylated with α-methylstyrene and/or tripropylene using a FULCAT 22B® clay catalyst. Antioxidants prepared in this process were colorless, catalyst-free after filtration, and contained <0.25 wt% of residual aromatic amines.

Additive Names

N-4-(2-Phenylisopropyl) diphenylamine (I)
4,4′-(1,1,3-Trimethylhexyl) diphenylamine (II)
N-4-(2-Phenylisopropyl)phenyl-4-*N*-(2-phenylisopropyl)-α-naphthyl-amine (III)
N-4-(2-Phenylisopropyl)phenyl-*N*-α-naphthylamine (IV)
N-4-(2-Phenylisopropyl)phenyl-*N*-4′-(1,1,3-trimethylhexyl) phenylamine (V)

Safety

N-Phenylnaphthylenes and diphenylamines are skin, eye, and mucous membrane irritants. Exposure by inhalation causes respiratory irritation characterized by the presence of a sore nose, throat, shortness of breath, headache, bluish skin, and methemoglobinemia. Although there is no evidence of carcinogenicity in these agents, they are suspected of being human carcinogens because of the presence of residual aromatic amines.

Additive Structures

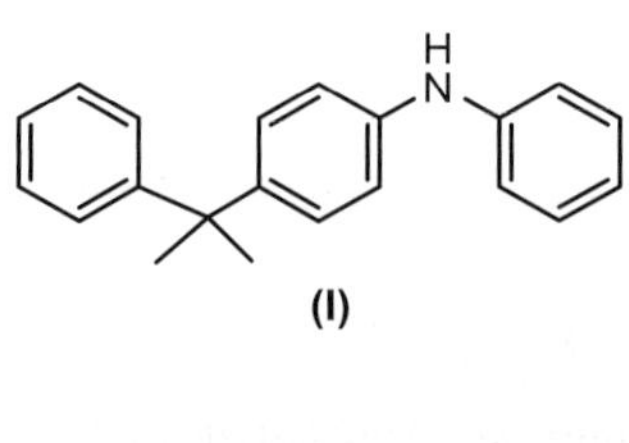

(I)

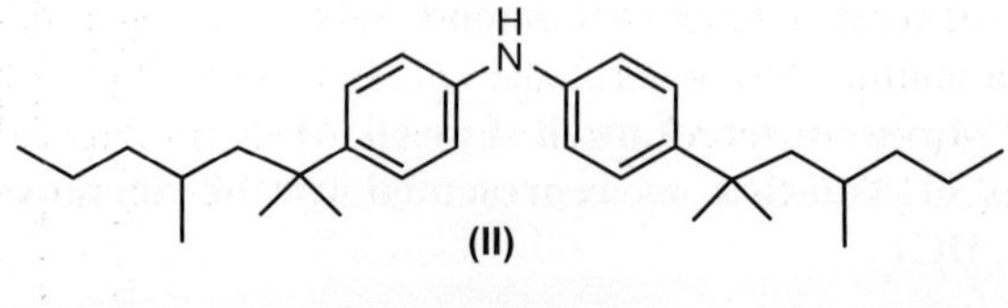

(II)

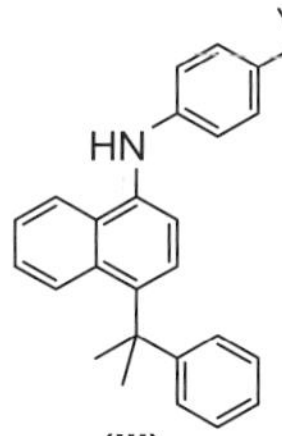

(III)

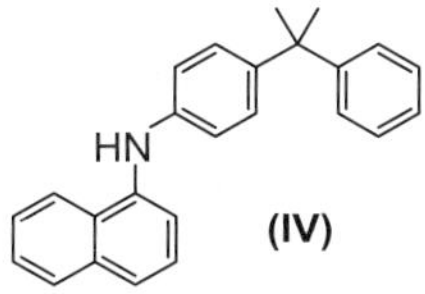

(IV)

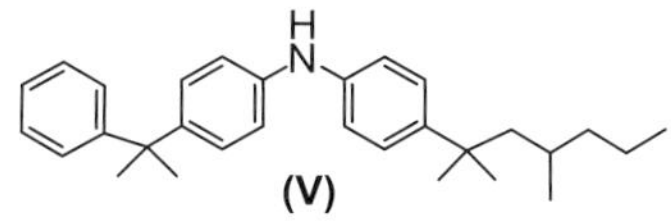

(V)

Additive Preparation

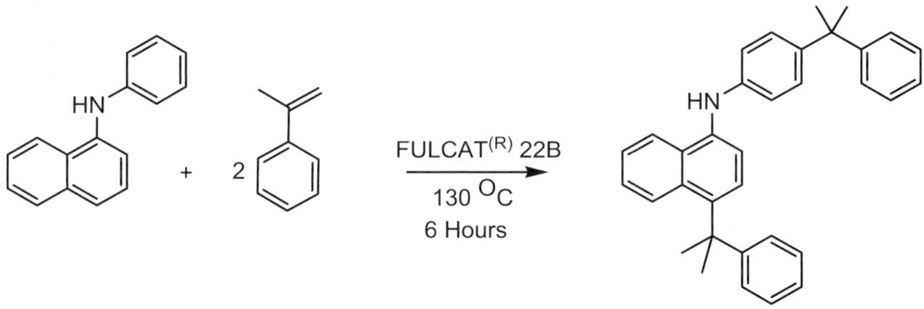

Additive Synthesis

1. Preparation of N-4-(2-phenylisopropyl)phenyl-4-(2-phenylisopropyl)-α-naphthylamine

A 1000-ml glass reaction vessel equipped with a high-temperature oil heating device, reflux cooler with water separator, feed unit for α-methylstyrene, propeller stirrer, and sampling device was charged with 403 g of *N*-α-naphthyl-*N*-phenylamine and 40.3 g of FULCAT® 22B clay catalyst (Rockwell Additives) and then evacuated to 20 mbar and flushed with nitrogen. The surrounding temperature was then set to 130 °C, and then this was treated with 477.8 g of α-methylstyrene over 3 h. After the addition was completed, the inner temperature of the reactor was kept at 130–135 °C for an additional 3 h and then cooled to 110 °C to remove the catalyst by filtration. Unreacted α-methylstyrene was removed by distilling the mixture at 225 °C at 20 mbar, and 660 g of the product was isolated.

Testing

Lubricant antioxidancy testing data are not provided by the author.

Advantages over Prior Art

Previous methods for preparing alkylated *N*,*N*-diphenyl- and *N*-phenyl-α-naphthylamines generated black antioxidant mixtures containing at least 2% of residual aromatic amine. Moreover, these methods also alkylated aromatic amines in the 2- and 2′-positions, thereby generating viscous product mixtures. The judicious use of FULCAT 22B® as a reaction catalyst has eliminated these reaction limitations.

Notes

1. Ryan[1] additized lubricants used in steam and gas turbines with an antioxidant mixture consisting of *N,N*-bis(2-ethylhexyl)-ar-methyl-1H-benzotriazole-1-methanamine (VII) and 0.15–0.5wt% of alkylated phenyl-α-naphthyl amine to enhance the oil oxidative stability and to reduce sludge formation.

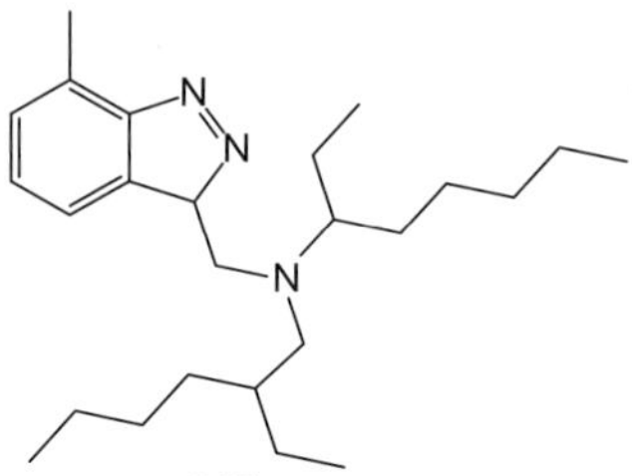

(VII)

2. N,N′-Bis(2,6-diisopropylphenyl)-*p*-phenylenediamine (VIII) prepared by Patil[2] was effective as an antioxidant when used in lubricating compositions and hydraulic fluids.

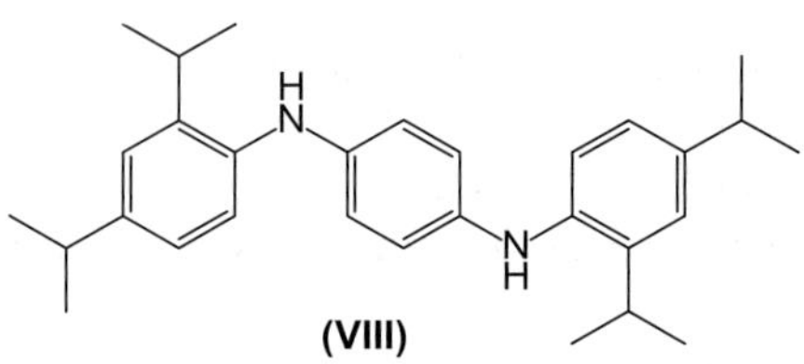

(VIII)

3. Habeeb[3] improved the antioxidancy of base oil by additizing with chromium acetylacetonate (IX), which behaved as a peroxide scavenger.

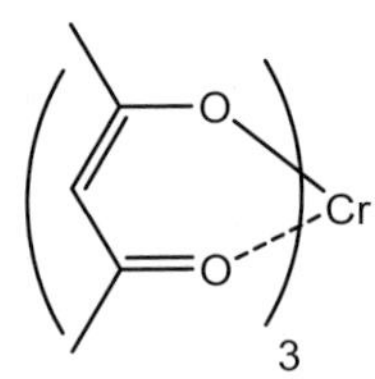

(IX)

4. Tetraaromatics (X) prepared by Ma[4] were effective antioxidants and used in automotive lubrication formulations.

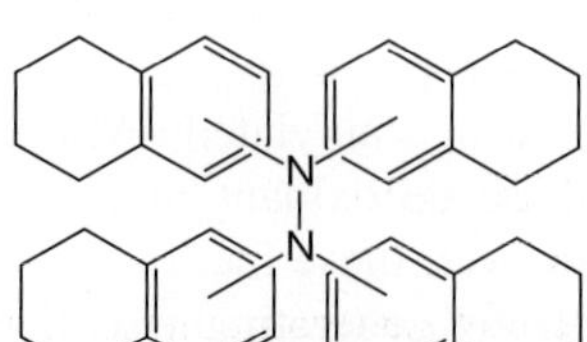

(X)

REFERENCES

1. Helen Ryan et al, US. Patent Application 20100099589 (April 22, 2010) **Afton Chemical Corporation (Richmond, VA)**
2. Abhimanyu O. Patil et al, US. Patent 8,017,806 (September 13, 2011) **Exxonmobil Research and Engineering Company (Annandale, NJ)**
3. Jacob Joseph Habeeb et al, US. Patent 8,030,257 (October 11, 2011) **Exxonmobil Research and Engineering Company (Annandale, NJ)**
4. Qinggao Ma et al, US. Patent 7,608,568 (October 27, 2009) **Chemtura Corporation (Middlebury, CT)**

E. LUBRICATING OIL BOUNDARY FRICTION MODIFIERS WITH MINIMUM OIL

Discoloration

Molybdenum Alkyl Phenolics

Title:	Oil-soluble molybdenum derivatives derived from hydroxyethyl-substituted Mannich bases
US Patent:	7,884,059 (February 8, 2011)
Author:	Vincent J. Gatto et al.
Assignee:	Afton Chemical Corporation (Richmond, VA)
Significance:	An oil-soluble molybdenum derivative was prepared, which was effective as a friction modifier. When used to additize processed oil or fully formulated SAE 5W-45 oil at 800 ppm, minimum oil discoloration was observed.
Additive Name	4-Dodecyl-2-[aminomethyl-2-(2-aminoethylamine)ethoxy molybdenum (VI) oxide]-phenol (I)
Safety Concerns	4-Dodecenylphenol, their Mannich products, and molybdenum derivatives cause eye, skin, gastrointestinal, and respiratory tract chemical burns where their effect may be delayed. Animal data from chronic inhalation studies have shown inflammatory changes in the lungs of rats with cellular aggregates resembling nodules consistent with earlier clinical reports of lung problems among exposed workers. In addition, inhalation of unreacted molybdenum oxide is associated with an increased risk of pulmonary tumors.
Name	Paraformaldehyde
CAS Number	30525-89-4
Safety Concerns	Experiments with animals suggest that this material may act as a carcinogen. Ingestion may be fatal. It may cause heritable genetic damage and irreversible damage to the eyesight. It is corrosive.

Additive Structure

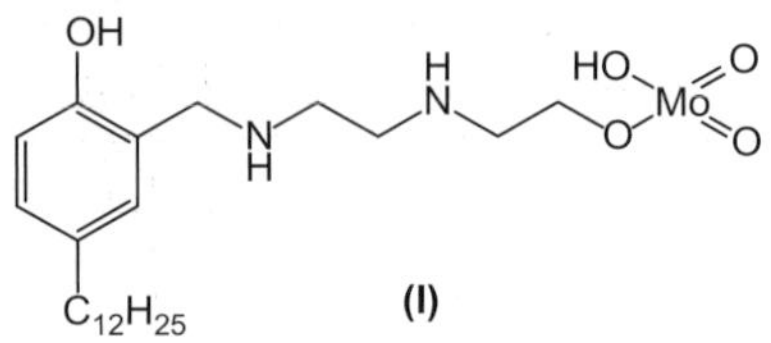

(I)

Additive Preparation

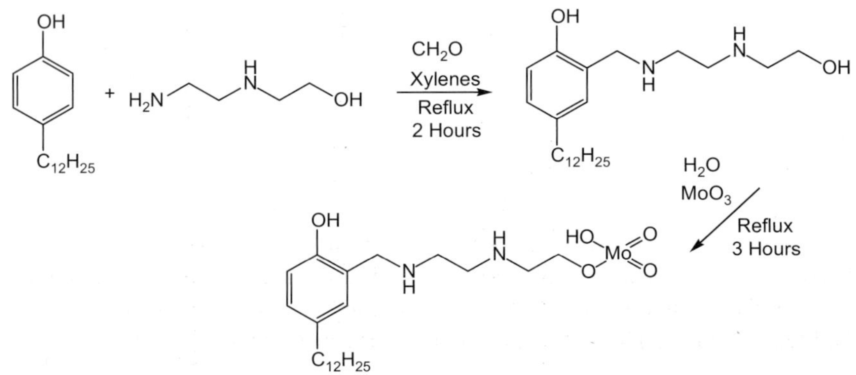

Additive Synthesis

1. Preparation of the Mannich reaction product

A reactor containing a Dean–Stark trap and heating mantle was charged with 52.5 g of 4-dodecylphenol, 20.8 g of 2-(2-aminoethylamino)ethanol, 7.9 g of paraformaldehyde, and 43 g of xylenes. The mixture was refluxed for approximately 2 h, and then the water was azeotropically removed.

2. Preparation of molybdenum complex

The entire Step 1 product was heated to 60 °C and then treated with 14.4 g of molybdenum (VI) oxide and 15 ml of deionized water. The reaction mixture was heated to reflux for 3 h with the azeotropic removal of water and then diluted with 100 g of Process Oil #5 and filtered. The filtrate was concentrated in vacuo, and 188.9 g of product was isolated.

Testing

1. Boundary Friction Coefficients

The boundary coefficients for finished oils containing the experimental molybdenum additive were determined by measuring the boundary friction coefficient in a high-frequency reciprocating rig (HFRR). Testing results are provided in Table 3.

TABLE 3 Boundary friction coefficient test results determined using the high-frequency reciprocating rig test. Testing was conducted using fully formulate SAE 5W-30 GF-3 containing 800 ppm of molybdenum

Material	Friction Coefficient (@ 300 ppm Mo)	Friction Coefficient (@ 600 ppm Mo)	Friction Coefficient (@ 800 ppm Mo)
Step 2 Product	0.091	0.065	0.064
Comparative-1	0.089	0.084	0.089
Comparative-2	0097	0.078	0.079
Comparative-3	0.089	0.092	0.074

2. SAE 5W-30 Color

Molybdenum additives in base oil can darken the oil color of the finished oil product. Although dark finished oils are purely an esthetic issue and do not impact performance, darkened oil products are negatively perceived by consumers. The ASTM D 1500 color scale was used to determine the effects of molybdenum-containing additives in a fully formulated SAE 5W-30 containing 800 ppm of molybdenum (1 wt%) of the Step 2 product. The test results are provided in Table 4.

TABLE 4 Oil color for oils containing 1wt% of the Step 2 additive determined using the ASTM D 1500 color test. Although darkened oil color does not impact the oil performance, the product appearance or esthetics may be negatively perceived by consumers

Material	Processed Oil Color	SAE 5W-30 Color
Molybdenum Free	0.5	3.0
Step 2 Product	4.0	4.1
Comparative-1	4.0	5.0
Comparative-2	5.2	6.5
Comparative-3	6.5	7.5

Test Results

1. Boundary Friction Coefficients

2. SAE 5W-30 Color

Advantages over Prior Art

Molybdenum-containing lubricant additives prepared using alkyl phenols, which condensed with aldehydes or polyamines, are effective as friction modifiers. Upon dissolution of these materials in either base or fully formulated oils, however, the lubricant darkens or blackens. The molybdenum derivative of the current invention addresses esthetic concerns associated with darkened or black of base or fully formulated oils while remaining an effective friction modifier additive.

Notes

1. Tynik[1] prepared a polyisobutylene monosuccinimide ammonium tungstate (II), which was effective as an automotive lubricant dispersant, antioxidant, and friction modifier.

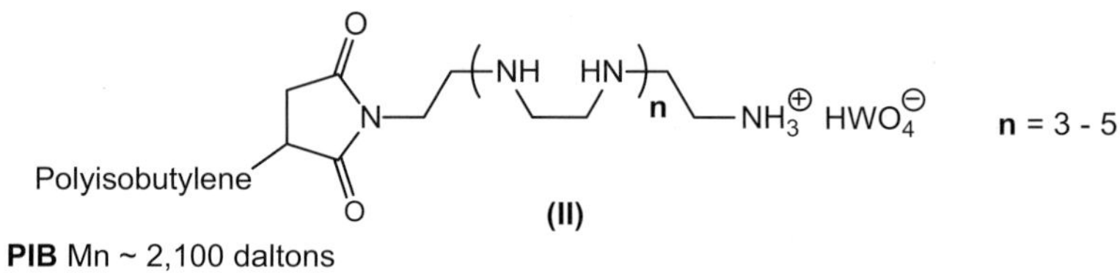

(II)

2. Ramanathan[2] prepared a series of amine organotungstate salts (III)–(V), which were effective when used in lubricant oil compositions as friction modifiers.

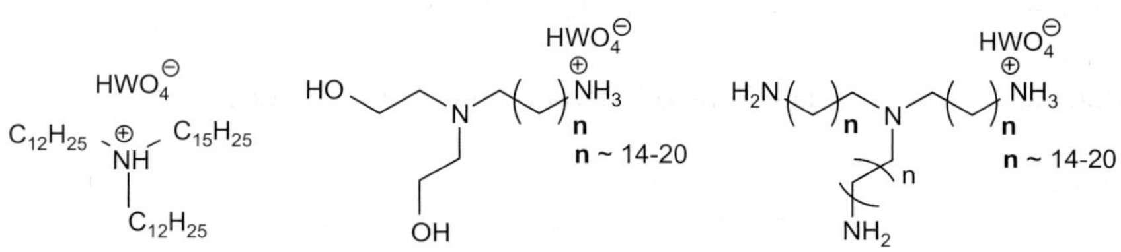

(III) (IV) (V)

3. Esch[3] prepared titanium tetraoleate (VI), which was effective as an antiwear and friction modifiers in lubricating oil compositions.

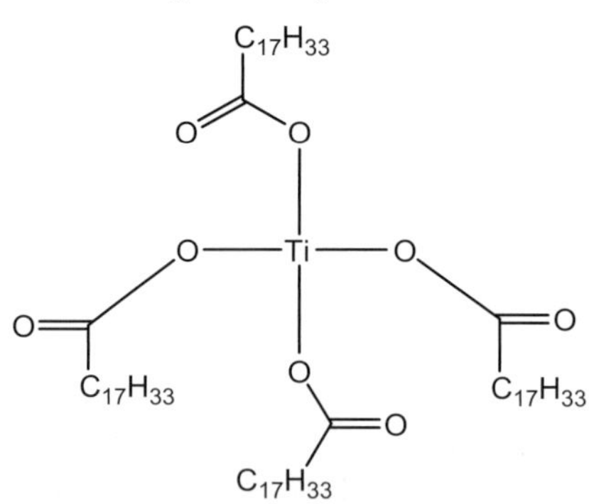

(VI)

4. Jao[4] prepared oil-soluble nanoparticles having a core diameter ranging from about 10 to 100 nm by photocrosslinking a poly(2-cinnamoyloxyalkyl acrylate) core and a diblock acrylate. The two-component nanoparticles were effective in lowering the coefficient of friction when used in lubricating oil formulations.

REFERENCES

1. Robert John Tynik et al, US Patent 7,858,565 (December 28, 2010)
R.T. Vanderbilt Company, Inc. (Norwalk, CT)
2. Ramanathan Ravichandran et al, US Patent 7,820,602 (October 26, 2010)
King Industries, Inc. (Norwalk, CT)
3. Karl Esch, Jr. et al, US Patent 7,767,632 (August 3, 2010)
Afton Chemical Corporation (Richmond, VA)
4. Tze-Chi Jao et al, US Patent 7,632,788 (December 15, 2009)
Afton Chemical Corporation (Richmond, VA)

F. LUBRICATING OIL BOUNDARY FRICTION MODIFIERS TO IMPROVE FUEL ECONOMY

Tartaric Acids

Title:	Tartaric acid derivatives as fuel economy improvers and antiwear agents in crankcase oils and preparation thereof
US Patent:	7,807,611 (October 5, 2010)
Author:	Jody Kocsis et al.
Assignee:	The Lubrizol Corporation (Wickliffe, OH)
Significance:	Automotive lubricant packages containing a mixture of oleyl tartimide and mono- and di-tartiamides were effective as friction modifiers when used in automotive lubricant oil packages. An improvement in vehicle fuel economy and reduction in the engine wear scar diameter was immediately observed using these agents in lubricating oil formulations.

Safety Concerns

Additive Name	Tartaric acid oleylimide (I) containing varying amounts of mono- and dioleylamides.
Safety Concerns	Initial studies to determine whether a dose-dependent increase in postimplantation loss, embryo lethality, and developmental abnormalities or variations is currently under investigation. As to whether this chemical poses a significant risk to human health is undetermined.
Additive Name	Tartaric acid oleylamide (II) containing varying amounts of monooleylamide.

Safety Concerns

Initial studies to determine whether a dose-dependent increase in postimplantation loss, embryo lethality, and developmental abnormalities or variations are currently under investigation. As to whether this chemical poses a significant risk to human health is undetermined.

Additive Name

Tartaric acid dioleylamide (III) containing varying amounts of dioleylamide.

Safety Concerns

Initial studies to determine whether a dose-dependent increase in postimplantation loss, embryo lethality, and developmental abnormalities or variations are currently under investigation. As to whether this chemical poses a significant risk to human health is undetermined.

Additive Structures

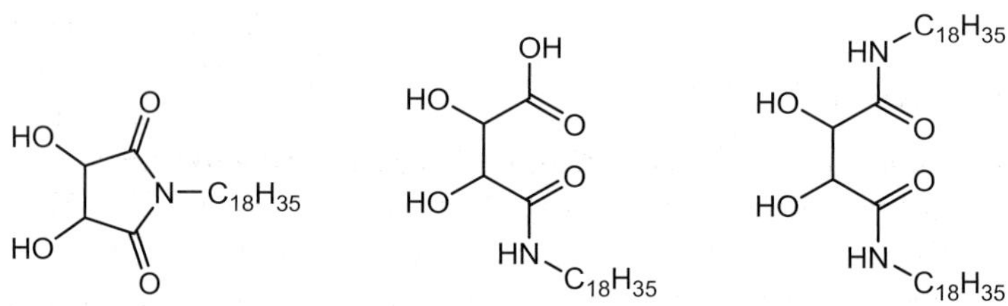

(I) (II) (III)

Additive Preparation

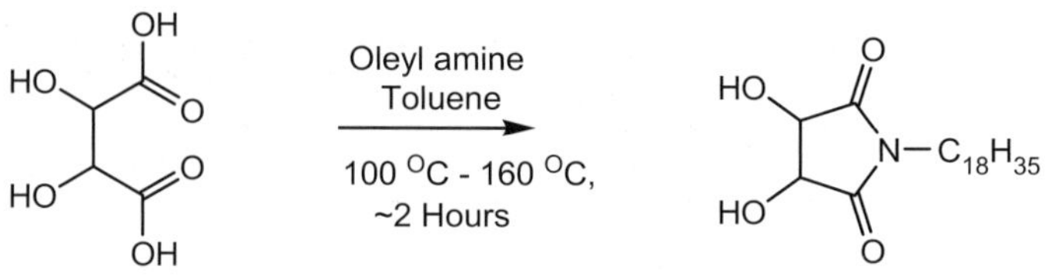

Additive Sources

Reagent grade supplies of these additives are available from the following supplier.

Tartaric acid, oleylimide

Tartaric acid oleylamide

Tartaric acid dioleylamide

Sargeant-Welch
P.O. Box 4130
Buffalo, NY 14217
800-727-4368

1. *Preparation of tartaric acid oleylimide mixture using the method of Barrer*[1]

A reactor was charged with 150 g of tartaric acid and 173 g of toluene and then heated to 100 °C and treated portionwise with 281 g of oleylamine. Once the addition of oleylamine was completed, the reactor temperature was increased 130 °C, and water azeotroping was removed. The reactor temperature was then increased to 160 °C for 1 h to distill off toluene. The liquid residue was filtered through diatomaceous earth, and the isolated product had a nitrogen content of 3.42% and hydroxyl content of 8.33%.

Testing

Scar Reduction and Fuel Economy

The average scar diameters were reduced by approximately 16%, while wear scar diameters were reduced by approximately 9% using lubricant additive packages containing between 0.05% and 0.1 wt% of the Step 1 product. In addition, lubricant additive packages additized with approximately 0.5 wt% of the experimental friction modifier had an improved fuel economy of between 1.4% and 1.9%.

Advantages over Prior Art

Tartaric acid–imide/amide-based additive mixtures potentially represent the next generation of ashless lubricant friction modifiers as replacements for those derived from diethanolamine (IV), glycerol (V), and fatty acid esters. While friction modifier lubricant additives derived from tartaric acid, diethanolamine, or glycerol are all readily prepared, imide- and amide-based derivatized friction modifiers are resistant to hydrolysis and have enhanced thermal stability.

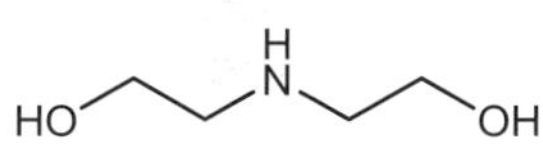

(IV)

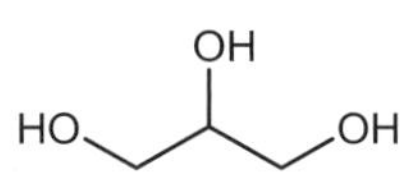

(V)

Notes

1. Fujitsu[2] demonstrated that lubricant packages additized with a mixture of 1 wt% glycerol monooleate and 0.05 wt% trimethylol propane monooleate as ashless friction modifiers outperformed additive packages consisting of 1 wt% glycerol monooleate and 0.05 wt% oleyl amides under a variety of loads.
2. Roby[3] prepared an ashless friction modifier lubricant package containing 0.4 wt% of the transesterification product of canola oil with trimethylol propane triheptanoate (Cargill® AP560). Additive packages were also prepared containing the reaction product of trimethylol propane and tricaprylate/caprate (Lexolube® 3 N-310).

3. Kasai[4] prepared an ashless friction modifier consisting of polyglycerin (VI).

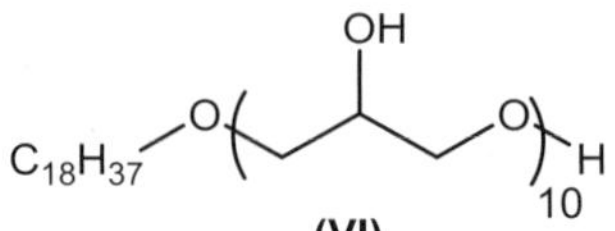

(VI)

REFERENCES

1. Daniel E. Barrer, US Patent 4,237,022 (December 2, 1980)
The Lubrizol Corporation (Wickliffe, OH)
2. Takashi Fujitsu et al, US 7,741,258 (June 22, 2010)
Shell Oil Company (Houston, TX)
3. Stephen H. Roby et al, US Patent 7,678,747 (March 16, 2010)
Cherron Oronite Company LLC (San Ramon, CA)
4. Moritsugu Kasai US Patent 7,803,745 (September 28, 2010)
Idemitsu Kosan Co., Ltd. (Tokyo, JP)

G. MARINE LUBRICANT DISPERSANT DESIGNED TO SHED WATER

Polyisobutylene-g-Succinic Acid Containing Oligomeric Naphthyl Esters

Title:	Lubricating oil composition
US Patent:	7,781,385 (August 24, 2010)
Author:	James C. Dodd et al.
Assignee:	Infineum International Limited (GB)

Significance: A nonphenolic overbase marine engine lubricant additive which does not emulsify oil contaminated with water and which is effective as an asphaltene dispersant was prepared by condensing polyisobutylene-g-succinic anhydride with oligomeric 2-naphthol.

Additive Name Poly(2-(2-naphthyloxy) ethanol)-co-formaldehyde) polyisobutylene monosuccinate

Safety Concerns The stoichiometry used to prepare this marine lubricating oil dispersant is seldom accurate since reagent purity varies from batch-to-batch and chemical supplier. In addition, because of the high product volumes produced in a single reaction, product purification becomes impossible. Thus, it can be expected that the final product contains all the chemical reagents used to prepare the additive.

Summary of Safety Concerns of Chemical Agents Used to Prepare Polymeric Additives

Name 2-Naphthol
CAS Number 135-19-3
Safety Concerns Harmful if swallowed or inhaled. Strong eye and skin irritant.

Name Ethylene carbonate
CAS Number 96-49-1
Safety Concerns Hazardous in the case of skin contact and an eye irritant. Hazardous in the case of ingestion and a moderately strong lung irritant.

Name Xylene
CAS Number 106-42-3
Safety Concerns Very hazardous in the case of skin contact (irritant) or eye contact (irritant). Inflammation of the eye is characterized by redness, watering, and itching. Skin inflammation is characterized by itching, scaling, reddening, or occasionally blistering. Toxic if ingested or inhaled.

Name Paraformaldehyde
CAS Number 30525-89-4
Safety Concerns Experiments with animals suggest that this material may act as a carcinogen. Ingestion may be fatal. May cause heritable genetic damage. May cause irreversible damage to the eyesight. Corrosive.

Name Polyisobutenyl-g-succinic anhydride
CAS Number 64741-88-4
Safety Concerns Mildly irritating to the skin and the eyes.

Additive Structure

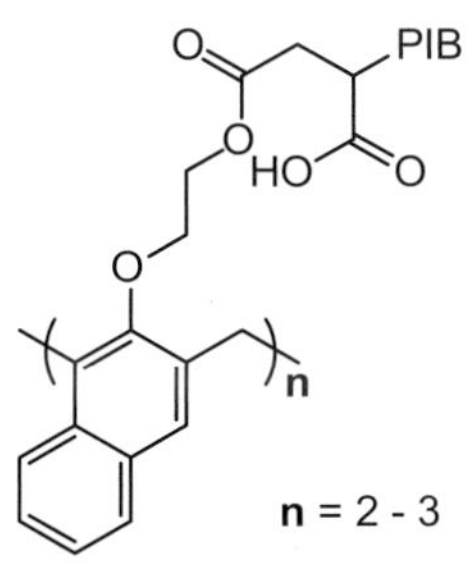

Additive

Additive Analogs None

Additive Preparation

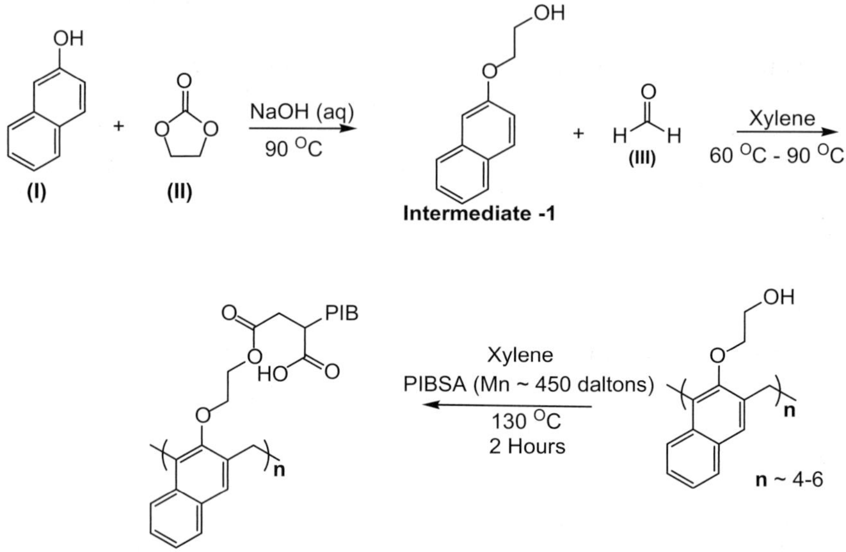

Additive Synthesis

Step 1 Preparation of 2-(2-naphthyloxy)ethanol

A 2-liter resin kettle equipped with a mechanical stirrer, condenser, Dean–Stark trap, and inlets for nitrogen was charged with 600 g of 2-naphthol, 372 g of ethylene carbonate, and 200 g of xylene and then heated to 90 °C. The mixture was then treated with 3.0 g of 50% aqueous sodium hydroxide, and water was azeotropically removed at 165 °C for 2 h. Carbon dioxide also evolved as the reaction progressed. When carbon dioxide evolution stopped, the reaction pot was cooled to ambient temperature, and a solid product was isolated. The structure of the intermediate was confirmed using FT-IR, HPLC, ^{1}H- and ^{13}C-NMR.

Step 2 Oligomerization of 2-(2-naphthyloxy) ethanol

A 2-liter resin kettle equipped with a mechanical stirrer, condenser, Dean–Stark trap, and inlets for nitrogen was charged with the entire Step 1 product, 200 g of toluene, 60.0 g of SA® 117 (an industrial hyperbranched polyether solvent), and the mixture was heated to 70 °C. The mixture was then treated with paraformaldehyde over a 15 mi period at 70–80 °C and then heated to 90 °C for 1 h. The temperature was then gradually increased to 110 °C for 3 h, and 83 ml of water was collected by azeotropic distillation. The product was isolated where it solidified while cooling to an ambient temperature and had an Mn of approximately 1000 Da. The material was characterized by ^{1}H- and ^{13}C-NMR, FDMS, MALDI-TOF.

Step 3 Acylation of poly(2-(2-naphthyloxy) ethanol)-co-formaldehyde) with PIBSA

A 5-liter resin kettle equipped with a mechanical stirrer, condenser, Dean–Stark trap, inlets for nitrogen, and an additional funnel was charged with the entire Step 2 product, 200 g of toluene, and the mixture was heated to 120 °C. The mixture was then treated portionwise with 2500 g of polyisobutenyl succinic anhydride having an Mn of approximately 450 Da and heated to 120 °C for 2 h. It was then further heated to 140 °C for an additional 2 h to remove the reaction solvent and unreacted reagents. The semisolid was then dissolved in 1100 g of base oil (AMEXOM® 100 N), and the product was isolated and characterized using GPC and FT-IR.

Testing

Additive Treatment Level

Lubricating oil containing between 0.5 and 5.0 mass% of the experimental agent was effective in shedding water contained in the lubricating oil without any evidence of emulsification.

Advantages over Prior Art

A nonmetallic nonphenolic overbase marine trunk piston engine oil lubricating oil dispersant with water shedding properties capable of keeping finely divided solids in suspension and effective as asphaltene dispersant is not reported.

Notes

1. Oil-soluble lubricating oil overbase dispersant additives have consistently utilized alkyl phenols (V), as described by Song[1]. Disulfide derivatives (VI) prepared by Emert[2] were also effective as overbased dispersants in lubricant oils. Neither of these dispersants was effective in dispersing asphaltenes or in preventing oil/water emulsions when used in marine lubricating oils, however.

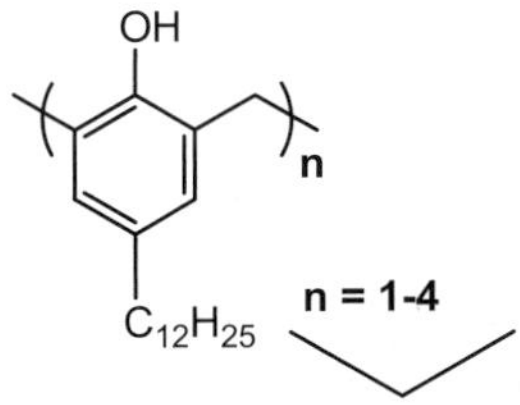

(V)

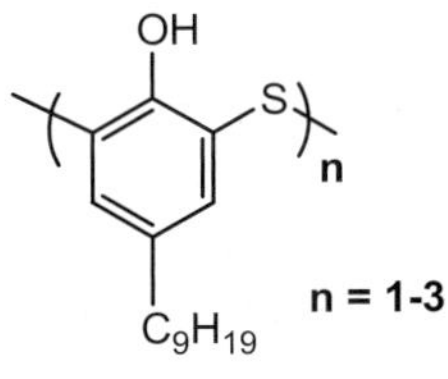

(VI)

2. Cook[3] developed a Gemini-PIB-based (VII) emulsifying agent having a pH > 10 for use as a marine lubricant dispersant contaminated with water designed to neutralize inorganic acids generated by sulfur present in the fuel. This agent was not, however, effective as an asphaltene dispersant.

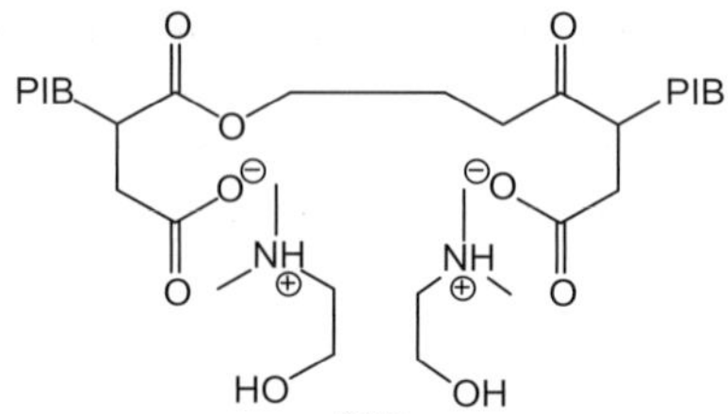

(VII)

3. Nelson[4] prepared a diester marine crankcase lubricant dispersant by reacting PIBSA with 1,3-dioxolan-2-one, ethylene glycol, di(ethylene glycol), tri(ethylene glycol), di(propylene glycol), tri(butylene glycol), or penta(ethylene glycol), which was effective as foam and corrosion inhibitors in diesel marine applications. Water-shedding properties were not observed, however.

REFERENCES

1. Won R. Song; et al, US Patent 5,277,833 (January 11, 1994)
   Exxon Chemical Patents Inc. (Linden, NJ)
2. Jacob Emert et al, US Patent 6,030,930 (February 29, 2000)
   Exxon Chemical Patents Inc. (Linden, NJ)
3. Stephen J. Cook et al, US Patent 7,651,984 (January 26, 2010)
   The Lubrizol Corporation (Wickliffe, OH)
4. Kenneth D Nelson, US Patent 7,256,242 (August 14, 2007)
   Chevron Oronite Company, LLC (San Ramon, CA)

H. TRANSMISSION OIL ADDITIVE TO IMPROVE TRACTION AND HEAT RESISTANCE

Dicyclohexyl Norbornene Diester

Title:	Lubricating oil
US:	7,902,406 (March 8, 2011)
Author:	Suguru Ohara et al.
Assignee:	Kyowa Hakko Chemical Co., Ltd. (Tokyo, JP)
Significance:	Automotive transmission fluids additized with 2,5-dicychlhexyl-carboxylatehydromethyl-3-methy-5-norbornene showed improvement in both traction and heat-resistant properties in the transmission oil. The additive was prepared by the oxo hydroformulation synthesis using inexpensive industrial chemicals and chemical intermediates.
Additive Name	2,5-Dicychexyl-carboxylatehydromethyl-3-methy-5-norbornene (I)

Safety Concerns While no reported safety concerns have been registered with the diester derivative, 2-formyl-3-methyl-5-norbornene, 2,5-diformyl-3-methy-5-norbornene, and 2,5-dihydoxy methyl-3-methy-5-norbornene are considered to be skin and eye irritants.

Additive Structure

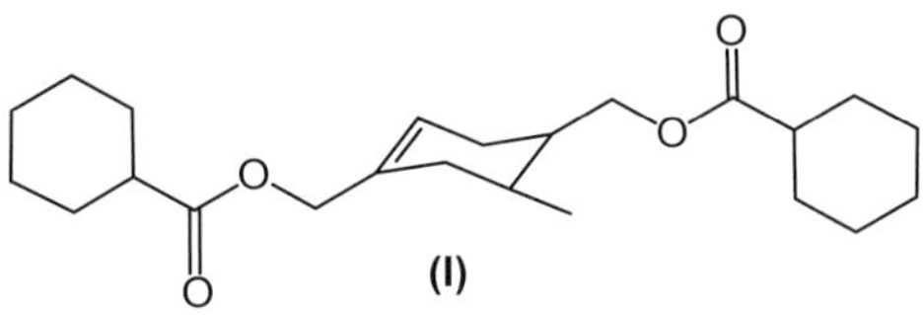

(I)

Additive Preparation

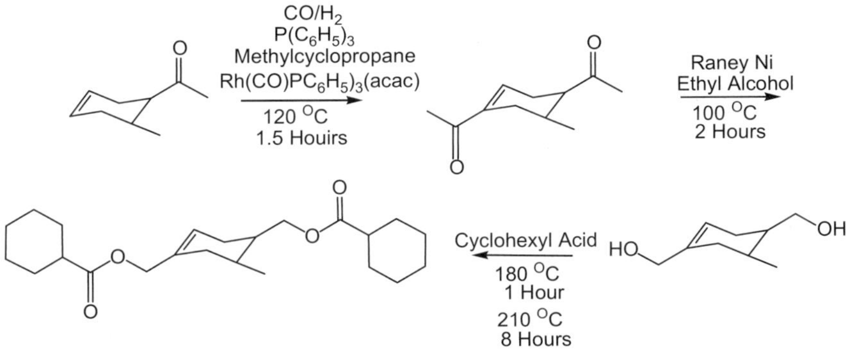

Additive Synthesis

1. Preparation of 2,5-diformyl-3-methy-5-norbornene

A stainless electromagnetic stirring-type autoclave having an internal volume of 1 liter was charged with 136 g of 2-formyl-3-methyl-5-norbornene, 200 ml of methylcyclohexane, 12 mg of Rh(CO)(PPh_3)(acac), and 655 mg of triphenylphosphine and then purged with mixed gases consisting of carbon monoxide and hydrogen in a 1:1 molar ratio. The mixture was then heated under stirring, and the temperature inside the system was raised to 120 °C in about 30 min while maintaining the pressure inside the system at 4 MPa using the gas mixture. The mixture was further stirred for four additional hours while still maintaining the temperature at 120 °C while the carbon monoxide/hydrogen gas mixture remained in a 1/1 molar ratio at 5 MPa. After the reaction was completed, the mixture was taken out from the autoclave, and the product was concentrated. The mixture was then distilled under reduced pressure and 167 g of product isolate. The distillate was analyzed by FT-IR and ^{1}H-NMR to confirm that it was the desired intermediate.

2. *Preparation of 2,5-dihydroxy methyl-3-methy-5-norbornene*

A stainless electromagnetic stirring-type autoclave having an internal volume of 1 liter was charged with 167 g of the Step 1 product, 334 g of ethanol, and 25 g of a Raney nickel catalyst and then purged with hydrogen gas. The mixture was then heated under stirring, and the reaction temperature was raised to 100 °C in about 30 min while maintaining the hydrogen pressure at 5 MPa. The mixture was further stirred for an additional 2 h at 100 °C in an autoclave pressure maintained at 5 MPa of hydrogen gas. The reaction mixture was then filtered and concentrated, and the residue was distilled under reduced pressure so that 155 g of product was isolated. The distillate was analyzed by FT-IR and ^{1}H-NMR to confirm that it was the desired intermediate.

3. *Preparation of 2,5-dicychlhexylcarboxylatehydromethyl-3-methy-5-norbornene*

A reaction flask was charged with 85 g of the Step 2 Product and 192 g of cyclohexanecarboxylic acid, and the mixture was then heated to 180 °C for 1h and then to 210 °C for 8 h. Thereafter, the mixture was cooled to ambient temperature, and unreacted cyclohexanecarboxylic acid was removed by distillation at 150 °C under reduced pressure. The reaction mixture was neutralized with a saturated aqueous solution of sodium bicarbonate, washed with water, the water was removed under reduced pressure at 130 °C, and 173 g of product was isolated.

Testing

1. *Traction Coefficient Testing*

Traction coefficient testing was performed using the Step 3 product, two comparisons and SANTOTRAC®-50 (Nippon Oil Corporation) where the traction coefficient was determined using a four-cylinder type traction tester. The results are provided in Table 5.

2. *Heat Resistance Testing*

Heat resistance of the Step 3 product and SANTOTRAC® 50 was determined by differential thermal analysis and the residual sample remaining after a decrease of 5%, 10% and 50% was recorded. Test results are provided in Table 6.

Test Results

1. *Traction Coefficient Testing*

2. *Heat Resistance Testing*

Advantages over Prior Art

Resins containing dimethanolnorbornane or perhydronaphthalene have improved heat resistance derived from the rigid skeleton of norbornane or perhydronaphthalene. This is the first instance of using these chemical agents to improve both the traction and heat resistance properties of a fluid. Moreover, the presence of monounsaturation in norbornane further 'hardened' the skeleton.

TABLE 5 Traction coefficient testing suggesting that the traction coefficient of the Step 3 product outperformed Comparisons 1 and 2 as well as Santotrac® 50

Variable	Step 3 Product	Comparison 1*	Comparison 2**	SANTO-TRAC®-50***
Oil Temperature (°C)	40	38	40	41
Viscosity Index	39	40	7	123
Kinetic Viscosity @ 40°C(cST)	242	236	20	30
Traction coefficient	0.1025	0.0984	0.0982	0.1025

**Comparison–1:*

***2Comparison–2:*

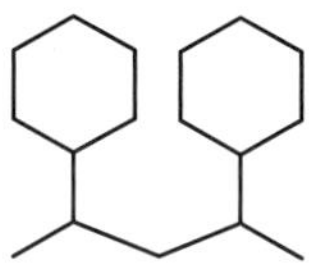

****3SANTOTRAC®-50 (Low viscosity synthetic hydrocarbon-base fluid).*

TABLE 6 Heat resistance testing results indicating the reduction in sample weight of the Step 3 product was far superior than that of commercial SANTOTRAC®-50 in all thermal testing phases

Reduction in Weight (%)	Step 3 Product (°C)	SANTOTRAC®-50 (°C)
5	239	146
10	255	161
15	297	197

Notes

1. Shimura[1] additized lubricating oil compositions with perfluoropolyether oil, $CF_3CF_2CF_2CF_2CF_2O(CF(CF_3)CF_2O)_{200})OCF_2CF_2CF_2CF_2CF_3$, which dramatically improved the heat properties of the liquid.
2. Bartley[2] prepared cyclic oligomers from the copolymerization of isobutylene and isoprene that were hydrogenated forming 1-(2,2,4-trimethylpentyl)-2-isopro-

pyl-3,3,5,5-tetramethyl-cyclohexane (II) and related compounds, which are useful as traction fluid lubricants but had limited thermal properties.

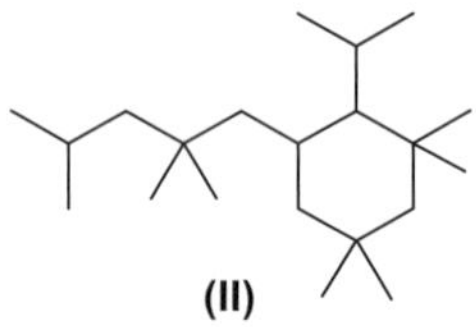

(II)

3. Unusually effective traction drive fluid additives consisting of rigid perhydronaphthalene (III), perhydroindane (IV), and dibicyclo[3.2.1]heptane (V), were prepared by Ishida,[3] which consistently outperformed cyclohexane analogs but which had only marginal thermal improved properties.

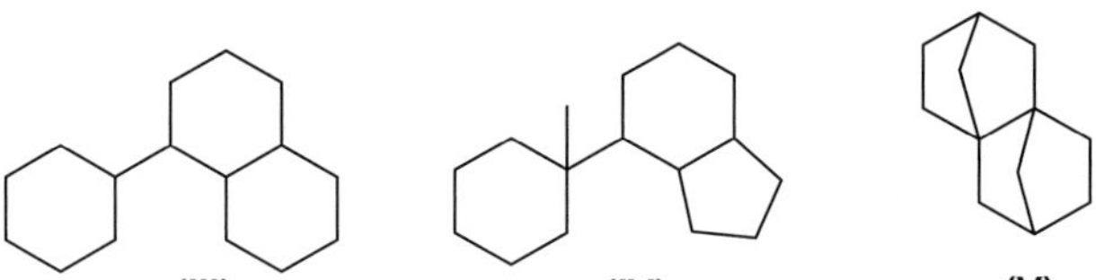

(III) (IV) (V)

4. Ishida[4] prepared fully formulated lubricating additive packages containing dihexane- or di-perhydronaphthalene dioxolates to enhance traction properties.

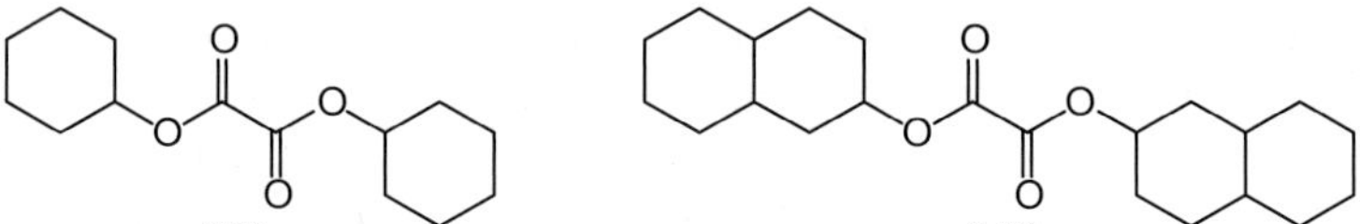

(VI) (VII)

REFERENCES

1. Akihiko Shimura et al, US Patent 7,709,424 (May 2, 2010)
   NOK Kluber Co., Ltd. (Tokyo, JP)
2. Stuart L. Bartley et al, US Patent 7,045,488 (May 16, 2006)
   Lubrizol Corporation (Wickliffe, OH)
3. Noboru Ishida et al, US Patent 6,638,417 (October 8, 2003)
   Nippon Mitsubishi Oil Corporation (Tokyo, JP)
4. Noboru Ishida et al, US Patent 6,242,393 (July 5, 2001)
   Nippon Mitsubishi Oil Corporation (Tokyo, JP)

Chapter 4

Automotive Tire Additives

A. IMPROVEMENT OF DYNAMIC MODULUS STORAGE WITHOUT HYSTERESIS PENALTY

Polystyrene–butadiene Nanoparticles

Title:	Rubber composition containing a polymer nanoparticle
US Patent:	7,649,049 (July 19, 2010)
Author:	James P. Kleckner et al.
Assignee:	Bridgestone Corporation (JP)
Significance:	Automotive tires compounded with up to 14% of polymeric nanoparticles having an average diameter of ≤50 nm have been found to be effective in improving the dynamic storage modulus of the tire without significantly impacting the tire hysteresis. The nanoparticle tire additive consists of a poly(styrene-b-butadiene) core with a poly(styrene-co-butadiene-co-1,4-divinyl benzene) surface.

Name	Styrene
CAS Number	97-88-1
Safety Concerns	Sensitizer in the case of skin contact. Eye and inhalation irritant. Slightly hazardous in the case of skin contact or ingestion. Possible teratogenic in humans.

Additive Name	1,3-Butadiene
CAS Number	106-99-0
Safety Concerns	Irritant to the eyes and the skin. Inhalation of moderate concentrations may cause dizziness, drowsiness, blurring of vision, and nausea. At higher concentrations, product becomes anasphyxiant. Extremely flammable.

Next Generation of International Chemical Additives. DOI: http://dx.doi.org/10.1016/B978-0-444-53788-1.00004-8

Additive Structure

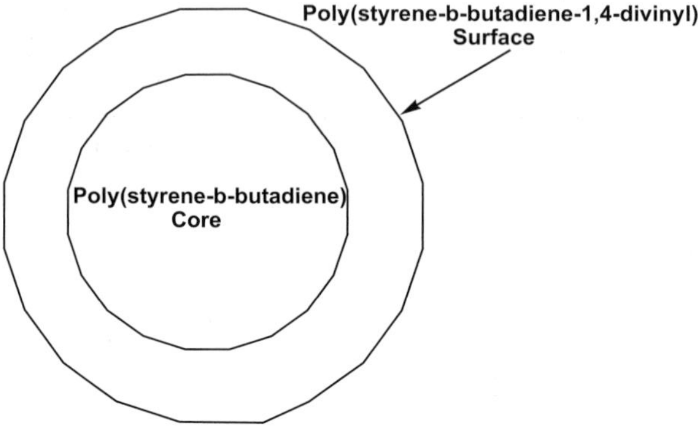

(I)

Additive Preparation

Additive Synthesis Using the Method of Wang[1]

1. Preparation of the poly(styrene/butadiene) living polymer nanoparticle core

A reactor was charged with 1.03 kg of 22% butadiene dissolved in hexane, 0.55 kg of hexane, and 0.70 kg of 33 wt% styrene dissolved in hexane and then heated to 50 °C. The mixture was initiated using 5.0 ml of 1.6 M *n*-butyl lithium diluted in 30 ml hexane while maintaining the reaction temperature at 57 °C for the duration of the polymerization. The living polymer was used directly in Step 2 with purification.

2. Preparation of polymer nanoparticle core coated with poly(styrene/butadiene/1,3-divinyl benzene)

A reactor containing the Step 1 living polymer was mixed with 1.36 kg of 33 wt% of styrene dissolved in hexane and then further diluted with 0.90 kg hexane, 1.36 kg 33 wt% of butadiene, and 50 ml of 1,4-divinyl benzene containing 1.0 ml of bis(2-oxolamyl) propane (Note 1) and then heated to 50 °C. The reaction mixture was then poured into bottles under a protective atmosphere for 1h. The polymerizations were then terminated by treating each vessel with 1 ml of isopropanol and 3 ml of 2,6-di-*t*-butylphenol. Transmission electron microscopy indicated that the average particle diameter was about ≤50 nm with polydispersity of about 1.1. Physical properties of nanoparticles are provided in Table 1.

Testing

Dynamic Storage Modulus Testing

The formulation used to assess the dynamic storage modulus of two separate blends containing the Step 2 product and an unadditized control is provided in Table 2. Dynamic storage modulus testing results are provided in Table 3.

Additive Blending Package

TABLE 1 Physical properties of nanoparticles containing a poly(styrene-b-butadiene) core with a poly(styrene-co-butadiene-co-4-divinyl benzene) surface. The average particle size was approximately ≤50 nm

Property	Measurement
Surface Layer Tg	−37
Core of Total Molecular Weight (%)	31.3
Core Composition	86% Styrene 14% Divinyl Benzene
Surface Layer Composition	Styrene–Butadiene
Surface Later Molecular Weight (Da)	1.9×10^5
Styrene in Surface Layer (%)	25

TABLE 2 Automotive tire formulations additized with the Step 2 nanoparticle product. An unadditized control was included to determine their nanoparticle effect on the dynamic storage modulus on automotive tires

Component (phr)	Unadditized Blend (Control)	Step 1 Product Blend 1	Step 1 Product Blend 2
Solution SBR	67	67	67
Emulsion SBR	33	33	33
Experimental Nanoparticles	**None**	7	14
Carbon Black	49.5	49.5	49.5
Silica	30	30	30
Bis-3-trimethoxypropyl-disulfide	2.4	2.4	2.4
Hydrocarbon Resin	8	8	8
Oil	25.13	25.13	25.13
Wax	2	2	2
Antioxidant	1.8	1.8	1.8
Stearic Acid	1	1	1
Processing Aid	0.8	0.8	0.8
Zinc Oxide	2.5	2.5	2.5
Sulfur	1.85	1.85	1.85
N-Cyclohexyl-2-benzothiazole-sulfenamide	1.8	1.8	1.8
Diphenylguanidine	0.8	0.8	0.8
2,2-Dibenzothiazole disulfide	0.4	0.4	0.4

Testing Results

TABLE 3 Effect of dynamic storage modulus on automotive tire compositions containing the Step 2 product nanoparticles and an unadditized control. Higher values are preferred, which suggests improved tire cornering and steering response

Property	Unadditized Control	Step 1 Product Blend 1	Step 1 Product Blend 2
Dynamic storage modulus @ −20 °C (Mpa)	997	1182	1277
Dynamic storage modulus @ 0 °C (Mpa)	48.6	51.9	58.3
Dynamic storage modulus @ 30 °C (Mpa)	19.5	21.6	24.4
Tan δ @ 0 °C	0.749	0.733	0.725
Tan δ @ 30 °C	0.436	0.442	0.454
Tan δ @ 60 °C	0.314	0.320	0.345

Advantages over Prior Art

The present invention addresses the synthetic and blending requirements needed for increasing the stiffness of a tread rubber to improve vehicle steering response, high level of grip as well as low fuel consumption without incurring a hysteresis penalty. The synthetic methods are direct, versatile, and prepared in high yields. By contrast, other methods designed to increase tire stiffness using high filler loading, fillers with a high surface area, and styrene–butadiene polymers with a high styrene content have each had separate performance tradeoffs.

NOTES

1. Bis(2-oxolamyl) propane (II) was initially prepared by Hall[2] by condensing 1.5 moles of furan and 1.0 mole of acetone and are used as polymerization modifiers.

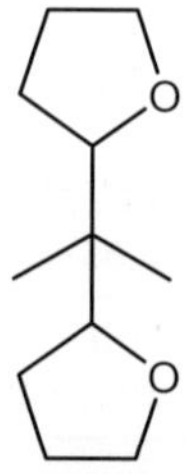

(II)

2. Zheng[3] prepared nanoparticles consisting of a polystyrene core containing polybutadiene 'hair' on the surface. Nanoparticles were then used to prepare automotive tires having an increased dynamic storage modulus. The preparation of these nanoparticles from this method is given below.

   A quart jar was charged with 50 g of dry polystyrene and 500 ml of hexanes and then sealed with a rubber septum containing a Teflon liner. The suspension was sonicated for 30 min using a 600-Watt sonicator with a 10-inch wave extender whereupon the aggregates of polystyrene particles were broken. The solution was then purged with nitrogen for 30 min, resealed, and injected with 10 ml of 1.6 M *n*-butyl lithium solution in hexane containing 1 ml of *N,N,N′,N′*-tetramethylethylenediamine. The sealed bottle was then placed in a 65 °C water bath and rotated at 20 rpm for 30 min. Lithiated polystyrene particles then became orange in color. The solution was then treated with 1,3-butadiene dissolved in hexanes while maintaining the temperature at 65 °C. When the full conversion of butadiene was observed, the reaction mixture was injected with isopropanol and then stabilized with 1% butylated hydroxytoluene. The molecular weight of the 'hairs' attached to the polystyrene core was determined to be 7100 Da.
3. Hollow nanoparticles consisting of poly(styrene-b-1,3-butadiene-b-styrene) were prepared by Wang[4] and used as an additive component in automotive tires since they improved the compatibility with rubbers and enhanced performance.
4. Polymeric nanoparticles having flower-like structures were prepared using a dianionic initiator, and poly(1,3-butadiene-b-styrene) was prepared by Wang[5] and used as a high performance tire additive.

REFERENCES

1. Xiaorong Wang et al, US Patent Application 20090048390 (February 19, 2009)
   Bridgestone Corporation (JP)
2. James E Hall, US Patent 4,647,635 (March 3, 1987)
   The Firestone Tire & Rubber Company (Akron, OH)
3. Lei Zheng et al, US Patent 7,408,005 (August 5, 2080)
   The Goodyear Tire & Rubber Company (Akron, OH)
4. Xiaorong Wang et al, US Patent Application 20080286374 (November 20, 2008)
   Bridgestone Corporation (JP)
5. Xiaorong Wang et al, US Patent 7,786,236 (August 31, 2010)
   Bridgestone Corporation (JP)

B. REDUCTION OF ROLLING RESISTANCE IN AUTOMOTIVE TIRES

Chloromethylaromatic Dioxasilinanes

Title:	Halofunctional silane, process for its preparation, rubber composition containing the same, and articles manufactured therefrom
US Patent:	7,816,435 (October 19, 2010)

Author:	Ping Jiang et al.
Assignee:	Momentive Performance Materials Inc. (Albany, NY)
Significance:	Rubber compositions containing up to 4 parts per hundred of the additive, 2-(3-chloropropenyl)-2-ethoxy-5-methyl-[1,3,2]dioxasilinane were determined to be effective in reducing the rolling resistance in automotive tires by ≥10%. The additive was prepared in a high yield by reacting 3-chloropropenyltriethoxysilane with 2-methyl-1,3-propanediol.

Additive Name	2-(3-Chloro-propenyl)-2-ethoxy-5-methyl-[1,3,2] dioxasilinane (I)
Safety Concerns	Severe skin and respiratory irritant. Flammable.
Additive Name	2-[2-(4-Chloromethyl-phenyl)-ethyl]-2-ethoxy-5-methyl-[1,3,2]dioxasilinane (II)
Safety Concerns	Severe skin and respiratory irritant. Flammable.

Additive Structures

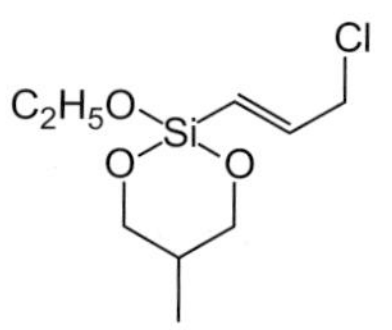

(I)

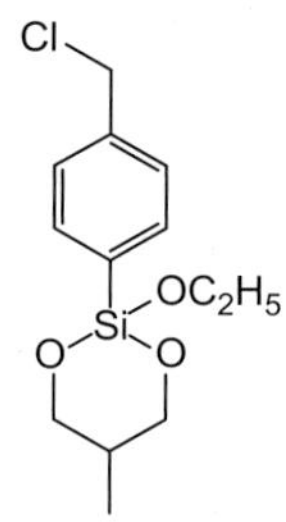

(II)

Additive Preparation

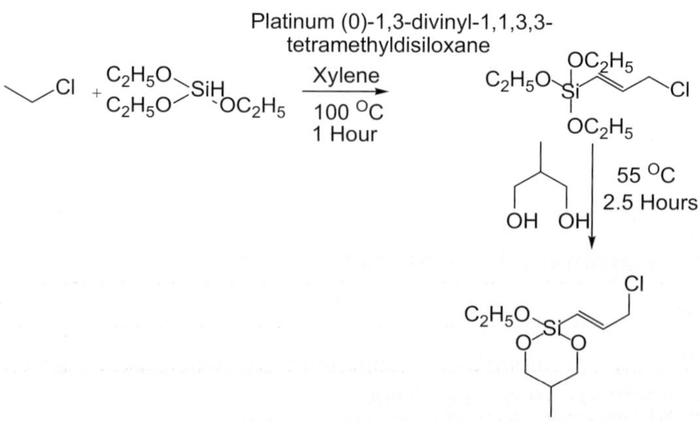

Additive Synthesis

Preparation of 3-chloropropenyltriethoxysilane

At ambient temperature, a 250-ml 3-neck round-bottom flask equipped with a reflux condenser, addition funnel, and a stir bar was charged with 50 g of propargyl chloride and 0.15 g of 3 wt% platinum(0)-1,3-divinyl-1,1,3, 3-tetramethyldisiloxane in xylenes. The mixture was then treated with the dropwise addition of 115.6 g of triethoxysilane and then heated to 100 °C for 1h. After distillation at 108 °C @ 49 mmHg, 103 g of the product was isolated as a 1.00:0.13 isomeric mixture.

Preparation of 2-(3-chloro-propenyl)-2-ethoxy-5-methyl-[1,3,2] dioxasilinane

A 3-neck 250-ml round-bottom flask fitted with a magnetic stirrer, distillation head, two thermocouples, receiver, dry ice trap, and vacuum pump was charged with 46.7 g of the Step 1 product and 17.8 g of 2-methyl-1,3-propanediol and then heated to 55 °C for 3 h. After cooling to ambient temperature, the product was isolated as a clear liquid.

Additive Formulation

A summary of additive formulations is provided in Table 4, where all additive components are in parts per hundred.

TABLE 4 Summary of rubber formulations used to evaluate Step 1 and Step 2 products

Component	Experimental Blend 1	Experimental Blend 2	Experimental Blend 3	Experimental Blend 4
Natural Rubber	100	100	100	100
Silica	58	58	58	58
Process Oil	3	3	3	3
Zinc Oxide	4.0	5.0	4.0	4.0
Stearic Acid	2.0	2.0	2.0	2.0
Naugurd Q (antioxidant)	2.0	2.0	2.0	2.0
Comparative Additive 1*	5.0	X	X	X
Comparative Additive 2**	X	4.4	X	X
Step 1 Product	X	X	7.0	X
Step 2 Product	X	X	X	5.0

*3-Chloropropenyltriethoxysilane.
**(p-Chloromethylphenylethyl)triethoxysilane.

Additive Testing

Formulations were cured before testing using Curing Procedure/Measurement Testing Standard Mooney viscosity and scorch ASTM D1646. A summary of physical property testing of rubber blends after curing is provided in Table 5.

TABLE 5 Summary of physical testing for four formulations. Modulus value under high deformation (M300) and (M300/M100) ratio is the highest for the Step 2 additive suggesting better tire performance

Physical Property	Formulation Using Comparative Additive 1	Formulation Using Comparative Additive 2	Formulation Using the Step 1 Product	Formulation Using the Step 2 Product
Hardness (Shore A)	64	61	64	70
Tensile (MPa)	29.4	30.3	28.1	28.8
Elongation (%)	605	528	519	632
25% Modulus (MPa)	1.03	0.74	0.83	1.64
100% Modulus (MPa)	2.74	2.12	2.26	2.53
300% Modulus (MPa)	12.77	12.8	13.93	10.40
RI (300%/25%)	12.46	10.08	16.84	6.36
RI (300%/100%)	4.66	6.98	6.16	4.11

Advantages over Prior Art

Although polysulfurized alkoxysilanes and mercaptofunctionalized are considered effective in improving rolling resistance in automotive tires, the reinforcing and abrasion resistance of these fully formulated blends is poor. The current additive is designed to address this limitation.

Notes

1. Rubber compositions containing 3-mercaptopropyl[1.3.2]-dioxasilane (III) were previously prepared by Chaves[1] and used to enhance the rolling resistance of automotive tires.

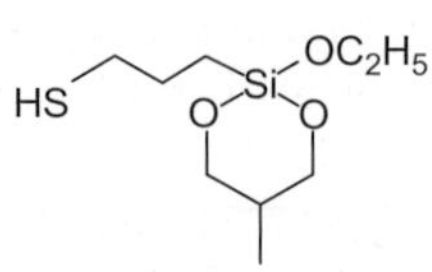

(III)

2. Cruse[2] prepared automotive tires with improved rolling resistance by replacing sulfurized silane (Silquest®-1891) (IV) with 2-triethoxysilylether-bis-(3-thia-4-oxohexyl)cyclohexane (V).

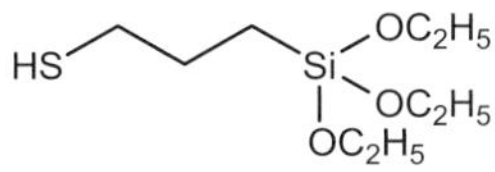

(IV)

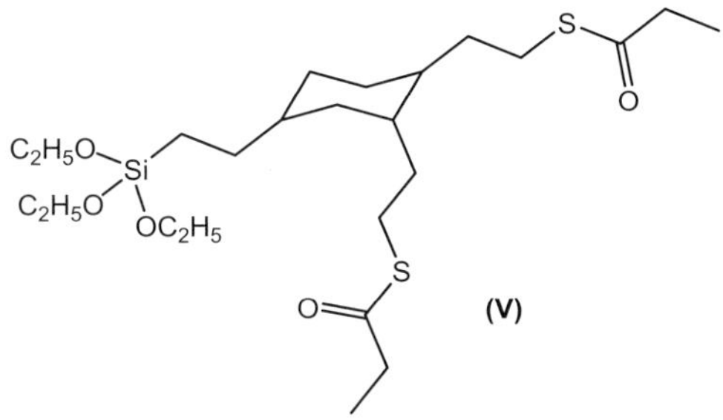

(V)

3. Coupling agents containing polythiosulfenamide (VI) were prepared by DaSilva[3] and used to improve the rolling resistance and enhance road grip wear resistance.

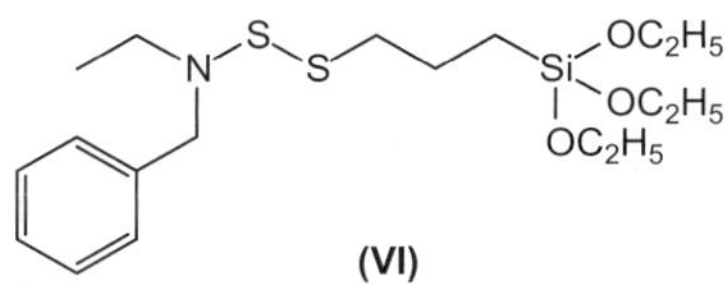

(VI)

4. Fuso[4] prepared thiosubstituted silane coupling agents (VII), which were more effectively bound to the prevulcanized automotive filler and elastomer, which subsequently improved rolling resistance.

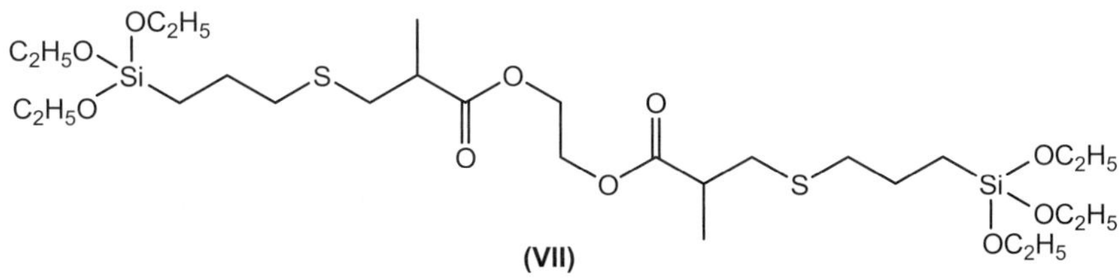

(VII)

REFERENCES

1. Antonio Chaves et al, US Patent Application 20080039645 (February 4, 2008) **Momentive Performance Materials Inc. (Albany, NY)**
2. Richard Cruse et al, US Patent 7,781,606 (August 24, 2010) **Momentive Performance Materials Inc. (Albany, NY)**
3. Araujo Da Silva, et al, US Patent 7,576,153 (August 18, 2009) **Michelin Recherche et Technique S.A. (Granges-Paccot, CH)**
4. Francesco Fuso et al, US Patent 7,723,409 (May 25, 2010) **Ciba Specialty Chem. Corp. (Tarrytown, NY)**

C. VULCANIZATION ACCELERATORS

Dialiphatic Dithiophosphates

Title:	Dithiophosphate composition and utility in rubber
US:	7,902,403 (March 8, 2011)
Author:	Thomas J. Karol et al.
Assignee:	R.T. Vanderbilt Company, Inc. (Norwalk, CT)
Significance:	Three novel vulcanization accelerators were prepared by modifying existing synthetic methods used to prepare dithiophosphates. In the first method, phosphorous pentasulfide was condensed with a mixture of aliphatic mono- and diols producing a mixture of monomeric and dimeric dithiophosphates. This mixture was then polymerized using hydrogen peroxide. In the second method, oligomers generated from the first method were reacted with zinc oxide or zinc oxide generating zinc dithiophosphate. In either case, poly(dithiophosphate) and zinc dithiophosphate performed as well as commercial vulcanization accelerators, Vanax® 196 and Vocal®.
Additive Name	2-Butoxy-2′-ethylhexoxy-1,6-dihexoxy)dithiophosphate and butoxy-ethylhexoxy–dithiophosphate (mixture) (I) Poly(2,2′-diethylhexoxy-tri-propyleneoxy-zinc dithiophosphate) (II) 2,2′-Ethylhexoxy-1,6-dihexoxy)zinc dithiophosphate and di(ethylhexoxy–zinc dithiophosphate (mixture) (III) Poly(2,2′-diethylhexoxy-tri-propyleneoxy dithiophosphate) (IV)
Safety Concerns	Zinc dialkyldithiophosphates and dialkyldithiophosphates are toxic agents. Signs of toxicity include moderate to severe skin, eye, and mucosal Irritation. Repeated exposure can result in irreversible eye damage, respiratory distress, dermatitis, and body weight loss. Studies have demonstrated that these agents have a threshold to be toxic to genetic material in cells but do not present a significant risk for mutagenicity or carcinogenicity in humans.

Additive Structures

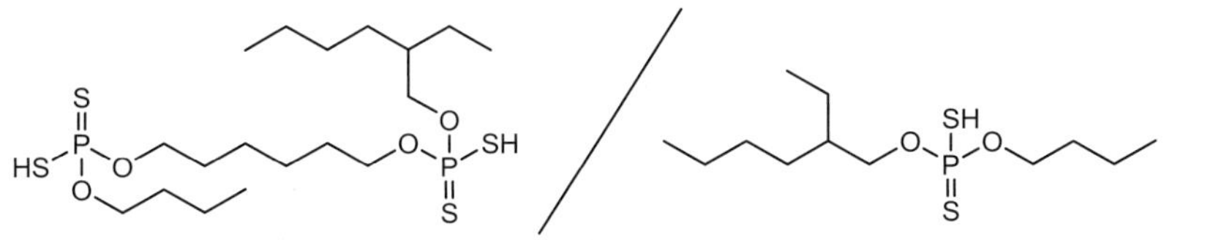

(I)

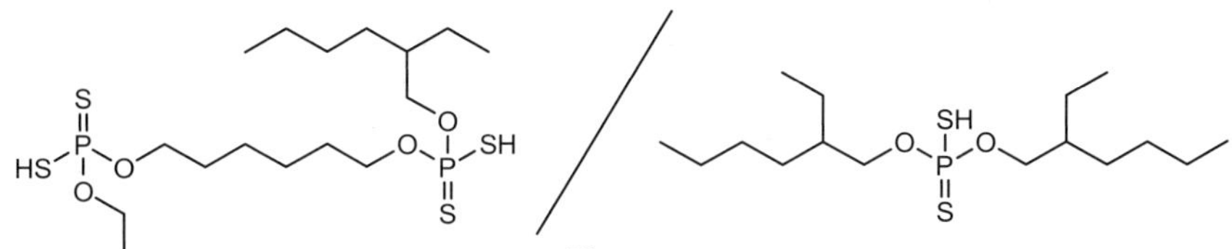

(II)

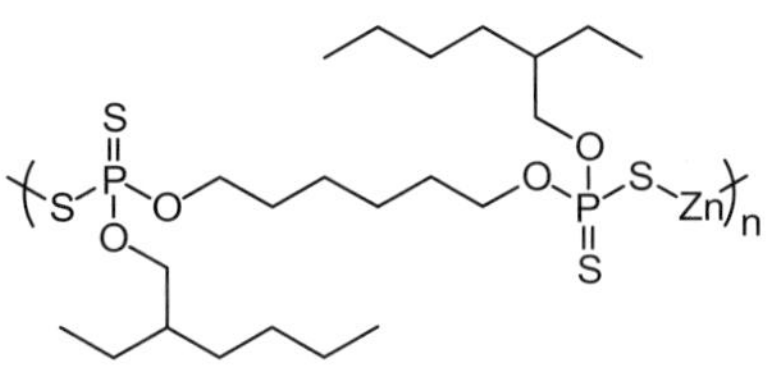

(III)

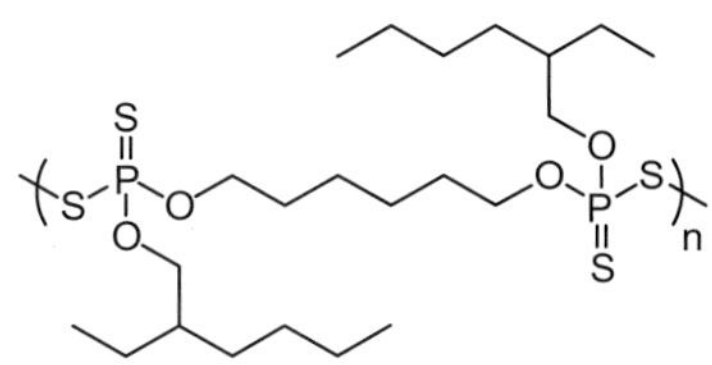

(IV)

Part I. Additive Preparation Using Hydrogen Peroxide

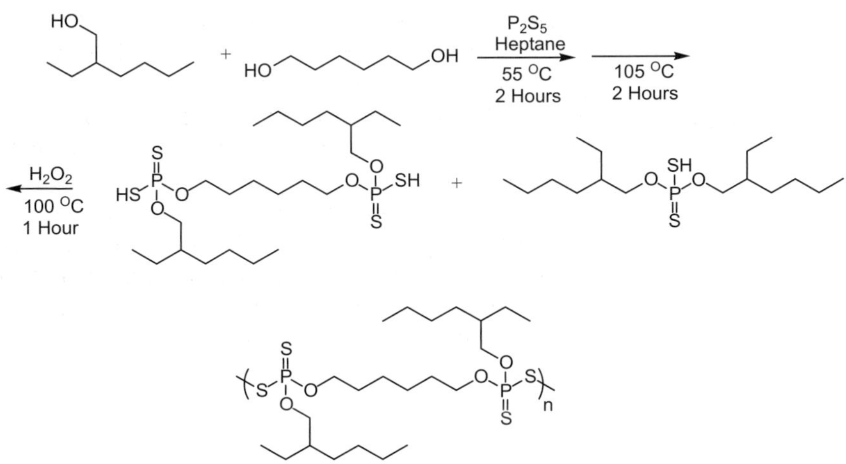

Part II. Zinc Inclusion Route

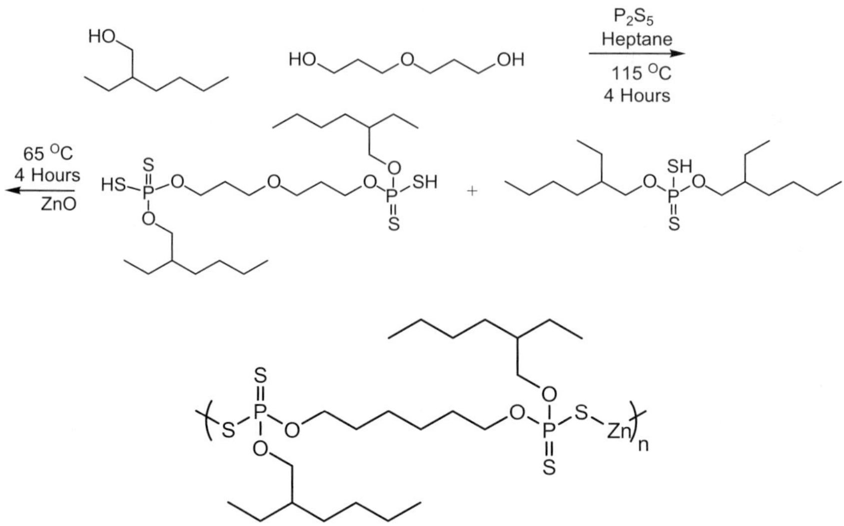

ADDITIVE SYNTHESIS

Hydrogen Peroxidation Route

1. Preparation of dithiophosphoric acid

A three-neck flask charged with 102.1 g of 2-ethylhexanol and 31.0 g of 1,6-hexanediol was heated until all 1,6-hexanediol had melted and then treated with 72.0 g of phosphorous pentasulfide over a period of 10 min. The flask was then rinsed with 30 ml of heptane to wash all remaining phosphorous pentasulfide into the reaction flask. Thereafter, the mixture was heated to 55 °C for 2 h and then further heated to 105 °C for an additional 2 h. The temperature was finally increased to 120 °C for an additional 3.5 h and then the mixture was filtered through diatomaceous earth and the product was isolated.

2. Preparation of polymeric dithiophosphate by oxidation

A reactor was charged with 145.9 g of the Step 1 product and then treated with 27.8 g of 30% hydrogen peroxide. The reaction temperature was then increased rapidly to 100 °C and then cooled in an ice water bath. Thereafter, an additional 1.9 g of 30% hydrogen peroxide was added, and the product was blended with 58.8 g of Hi-Sil® ABS, dried in an oven at 55 °C, and a uniform powder was isolated.

Zinc Inclusion Route

1. Preparation of dithiophosphoric acid

A flask was charged with 74.1 g of phosphorous pentasulfide, and 60.2 g of heptanes was then placed in an ice water bath. A separatory funnel was then charged with 37.2 g of dipropylene glycol and 104.4 g of 2-ethylhexanol and then added to the phosphorous pentasulfide hexane solution over a period of 15 min. Thereafter, the rcaction mixture was slowly heated to 115 °C, and heptane was removed by distillation. The remaining mixture was heated to 115 °C for 5 h and then filtered through diatomaceous earth, and the product was isolated.

2. Preparation of polymeric zinc dithiophosphate by zinc insertion

A three-neck flask was charged with 204.3 g of the Step 1 product and 27.1 g of zinc oxide and then heated to 65 °C for 4 h. The temperature was then further increased to 110 °C for 1h to distill off water, and after filtering through diatomaceous earth, the product was isolated.

Testing

1. Vulcanization acceleration testing

The Moving Die Rheometer test was conducted at 160 °C with a 0.5° Arc according to the ASTM D5289 testing protocol to evaluate vulcanization acceleration using EPDM rubber. Testing formulations are provided in Table 6.

TABLE 6 Formulations used to evaluate the vulcanization acceleration of selected additives at 160 °C according to the ASTM D5289 test method. All components are in parts by weight. The commercial vulcanization accelerator VANAX® 196 was provided as a reference

Component	VANAX® 196*1 Blend	Step 1 Product Blend	Step 2 Product Blend	Step 2′ Product Blend
SMR-L	100	100	100	100
VANPLAST® R	2	2	2	2
VANFRE® AP-2	2	2	2	2
Zinc Oxide	5	5	5	5
Stearic Acid	2	2	2	2
N990 Carbon	35	35	35	35
Black	X	X	X	X
N330 Carbon	30	30	30	30
Black Sulfur	2.5	2.5	2.5	2.5
ALTAX® (MBTS)	1.5	1.5	1.5	1.5
VANAX® 196*	1.5	X	X	X
Additive 2	X	1.5	X	X
Additive 3	X	X	1.5	X
Additive 6	X	X	X	1.5

*2,2-Di-ethylhexoxydithiophosphate.

Testing Results

The results of the Moving Die Rheometer testing using VANAX® 196* and selected additives are provided in Table 7.

TABLE 7 Mechanical testing results for EPDM rubber containing selected experimental vulcanization accelerators. VANAX® 196 and experimental agents were essentially equivalent in acceleration properties

Component	VANAX® 196* Blend	Step 1 Product Blend	Step 2 Product Blend	Step 2′ Product Blend
Minimum Torque ML (dN-m)	0.40	0.41	0.50	0.42
Maximum Torque MH (dN-m)	17.81	17.91	18.50	17.52
Ts1(min)	1.73	1.51	1.70	1.41
T90 (min)	8.30	7.26	8.32	6.54
Cure Rate Index (min^{-1})	15.2	17.4	15.1	19.5
Cure Rate (dN-m/min)	2.65	3.04	2.72	3.33
Tan Delta at ML	1.63	1.56	1.45	1.53
Tan Delta at MH	0.04	0.04	0.04	0.04

* = V

Advantages over Prior Art

The use of polymeric dithiophosphates and zinc dithiophosphates as vulcanization accelerates is unreported. These agents offer high activity, excellent vulcanization properties, and ease of handling and preparation. Finally, these agents appear to be targeted as replacements for nitrosoamine-generating accelerants because of the severe toxicity associated with these agents.

Notes

1. Heiliger[1] previously prepared vulcanization accelerators consisting of dithiophosphoric acid polysulfide mixtures (V) by oxidizing O,O-di-2-ethylhexyldithiophosphoric acid octonate with hydrogen peroxide.

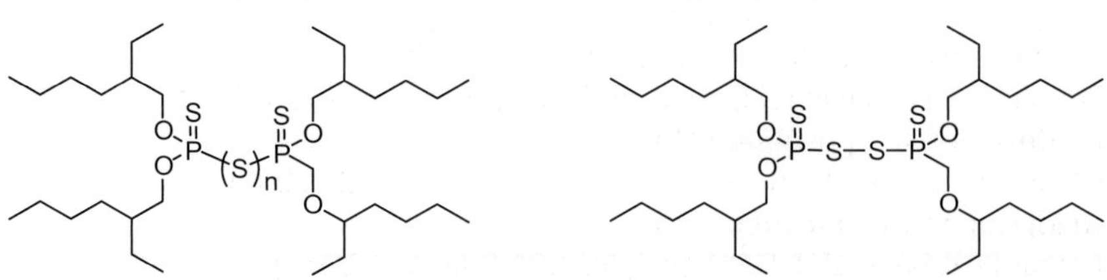

(V)

2. Hogan[2] prepared functionalized oligo(styrene-co-1,3-butadiene) (VI), which was then used in automotive tire formulations as vulcanates.

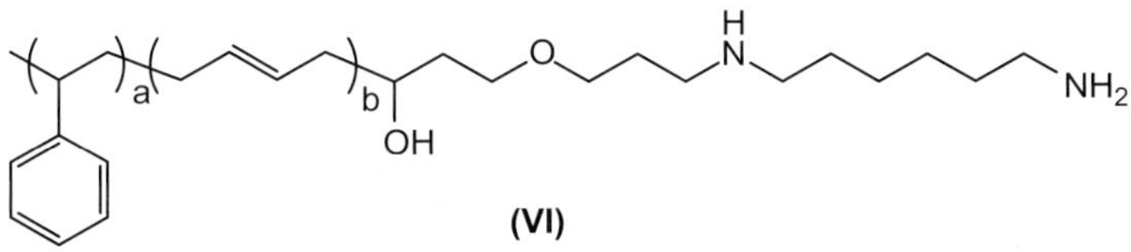

(VI)

3. To reduce hysteresis, Lin[3] formulated automotive tires by initially using elastomeric compounds and then compounded them with reinforcing agents such as bis[3-(triethoxysilyl) propyl]tetrasulfide and then vulcanizing the mixture with *N*-cyclohexyl-2-benzothiazolesulfenamine (VII).

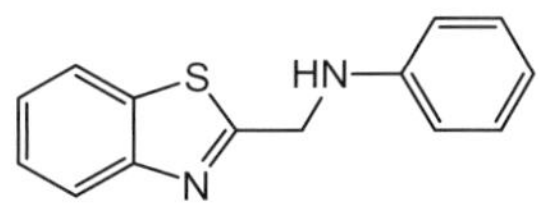

(VII)

4. Cruise[4] prepared a novel vulcanization accelerator by coupling bis(3-triethoxysilyl-1-propyl)-tetrasulfide (VIII) with ethylene glycol to form the vulcanization dimer (IX).

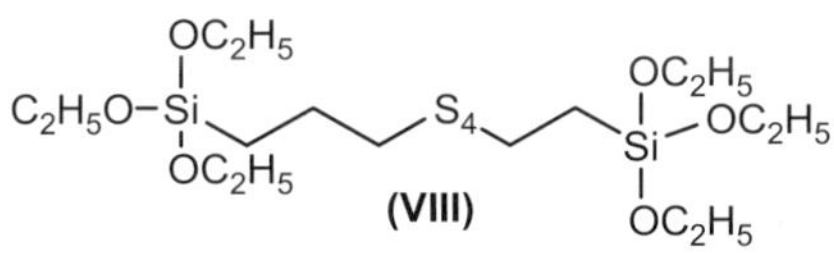

(VIII)

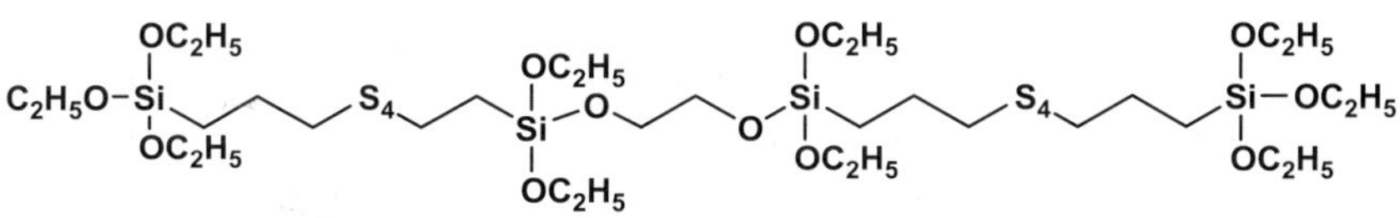

(IX)

REFERENCES

1. Ludger Heiliger et al, US Patent 6,761,832 (July 13, 2004)
Rhein Chemie Rheinau GmbH (Mannheim, DE)
2. Terrence E. Hogan et al, US Patent 7,902,278 (March 8, 2011)
Bridgestone Corporation (JP)
3. Chen-Chy Lin, et al US Patent 6,194,509 (February 22, 2001)
Bridgestone Corporation (JP)
4. Richard Cruise et al, US Patent 7,902,391 (March 8, 2011)
Momentive Performance Materials Inc. (Albany, NY)

Chapter 5

Battery and Charge Extenders

A. CHARGE CONTROL EXTENDERS

Oxidized Thia-calix[4]- and [8]arenes

Title:	Oxidized mixed cyclic phenol sulfides and charge control agents using the same toners
US Patent:	7,820,832 (October 26, 2010)
Author:	Masateru Yasumura et al.
Assignee:	Hodogaya Chemical Co., Ltd. (Tokyo, JP)

Significance: The oxidation products of thia-*p*-*t*-butylcalix[4]- and -[8]arene were prepared in two steps and were effective as toner charge control agents. When used in compositions containing either a noncoated or silicon-coated ferrite carrier, both saturated charge and environmental stability exceeded those of metallic salicylates.

Additive Name Oxidized thia-*p*-*t*-butylcalix[4]arene (I)
Safety Concerns Eye shield, gloves, and dust mask type N95 recommended when handling

Additive Name Oxidized thia-*p*-*t*-butylcalix[8]arene (II)
Safety Concerns Eye shield, gloves, and dust mask type N95 recommended when handling

Additive Structures

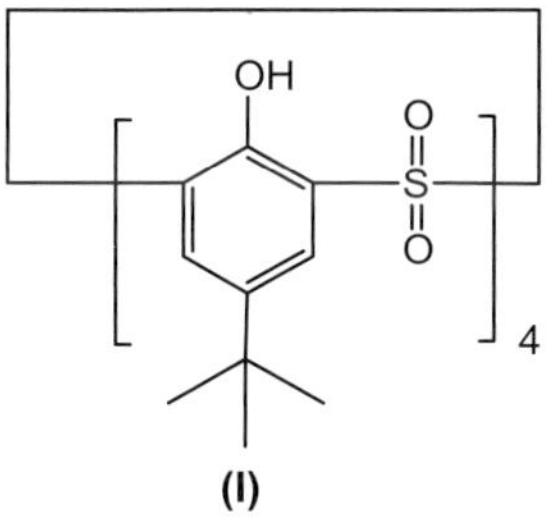

(I)

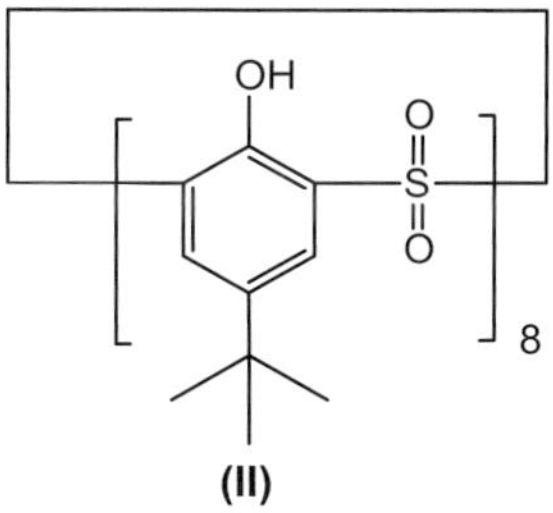

(II)

Next Generation of International Chemical Additives. DOI: http://dx.doi.org/10.1016/B978-0-444-53788-1.00005-X

Additive Preparation

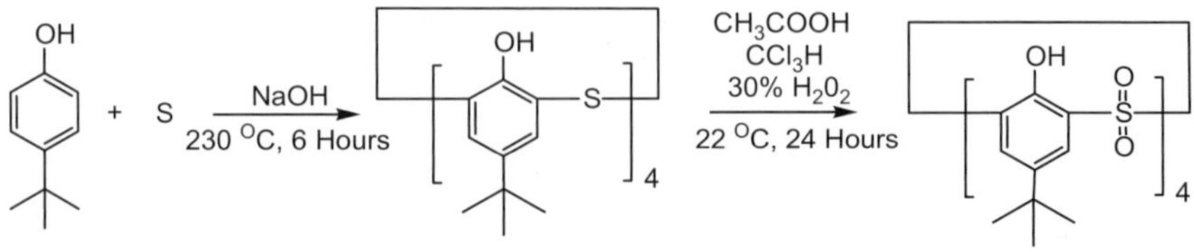

Additive Synthesis

1. *Preparation of thia-p-t-butylcalix[4]arene according to the method of Hori*[1]

A reaction vessel was charged with 45.2 g of 4-t-butylphenol, 14.4 g of elemental sulfur, and 3.0 g of sodium hydroxide and was then gradually heated to 230 °C while stirring for 6 h. The mixture was then cooled to ambient temperature and treated with 500 ml of diethyl ether and 1 M of sulfuric acid. The diethyl ether layer was then isolated, concentrated, and then purified using silica gel column chromatography with hexane/chloroform. The crude product was further purified by recrystallization using chloroform/acetone, and the product was isolated as a colorless solid having an mp = 320–322 °C.

2. *Preparation of oxidized thia-p-t-butylcalix[4]arene*

A reactor was charged with 9.0 g of the Step 1 product dissolved in 150 ml of chloroform and then treated with the dropwise addition of 114 g of 30% hydrogen peroxide and 500 ml of acetic acid at an ambient temperature. After 30 min, the addition was completed, and the mixture was stirred for 48 h at an ambient temperature. Thereafter, the reaction mixture was treated with 300 ml of water and extracted three times using 100 ml of chloroform. The combined chloroform layers were washed with 100 ml of water, dried with anhydrous magnesium sulfate, and then concentrated. The product was isolated as a white powder in a 41% yield.

Additive Formulation

A mixture consisting of 94 parts by weight of polypropylene wax (FC-316® Mitsubishi Rayon Co., Ltd.), one weight part of a selected experimental additive, and 5 weight parts of a coloring agent was blended by melting at 110 °C. The cooled mixture was then crushed in a hammer mill and then finely crushed with a jet mill so that an average particle diameter of 10 ± 0.5 μm was obtained. Thereafter, four weight parts of the mixture was blended with 100 weight parts of a carrier and evaluated after shaking at 25 °C at 50% humidity. Test results are provided in Table 1.

Advantages over Prior Art

Most charge transfer agents consist of zinc of chromium complexes or their salts so that the handling and disposal of these materials are problematic. Color concerns,

TABLE 1 Test results indicating that both environmental and saturated charges of the Step 2 product were superior to commercial charge control agents

		F-150*		F96-150**	
Charge Control Agent	Coloring Agent	Saturated Charge (−μC/g)	Environmental Stability	Saturated Charge (−μC/g)	Environmental Stability
Step 2 Product	Carbon Black	40.4	Good	21.7	Excellent
Aluminum Salicylate	Carbon Black	22.2	Poor	16.2	Unsatisfactory
Zinc Salicylate	Carbon Black	21.1	Unsatisfactory	15.1	Poor
Step 2 Product	C.I. Pigment Red 57:1	33.0	Excellent	20.0	Good
Aluminum Salicylate	C.I. Pigment Red 57:1	13.1	Unsatisfactory	11.0	Unsatisfactory
Zinc Salicylate	C.I. Pigment Red 57:1	20.8	Unsatisfactory	14.1	Unsatisfactory

**Cu-Zn ferrite particles having a mean particle diameter of 80μm with a magnetization saturation of 65emu/g (Powdertech Co., Ltd.).*
***Silicon-coat type ferrite particles (Powder Tech K.K.).*

late charging rise times, and the destabilizing effect in hot humid environments can dramatically impair the performance of these charge control agents. Oxidized thiocalix[4]- and -[8]arenes were both high performing components in toner packages and address all concerns associated with metal-containing materials.

Notes

1. Nakamura[2] prepared 4-phenyl-4-(4-chloro)-calix[8]arene, (III) which was effective as a component in color imaging toner packages.

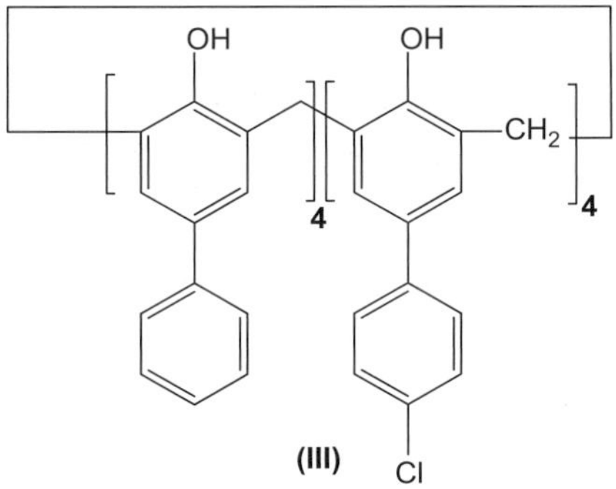

(III)

2. Bender[3] prepared a series of naphthalene tetracarboxylic diimide dimers (IV), which were capable of functioning as both binders and electron-transporting materials to enhance toner images.

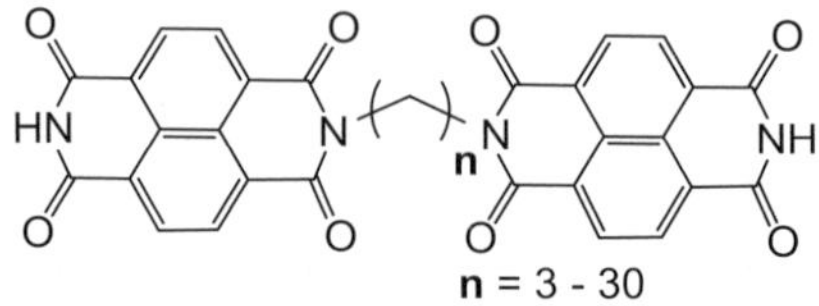

(IV)

3. Moriyama[4] prepared a toner composition containing a 4:6 mixture of crystalline poly(fumaric acid-co-1,6-hexanediol) and bisphenol acetone-polyethylene oxide adduct, respectively, which was able to continuously reproduce excellent toner images without image quality deterioration even when used in high-speed printers.
4. *p*-Terphenyl derivatives (V) were prepared by Katsumi[5] with improved organic solvent solubility and were used as charge transporting agents in electrophotographic toner compositions.

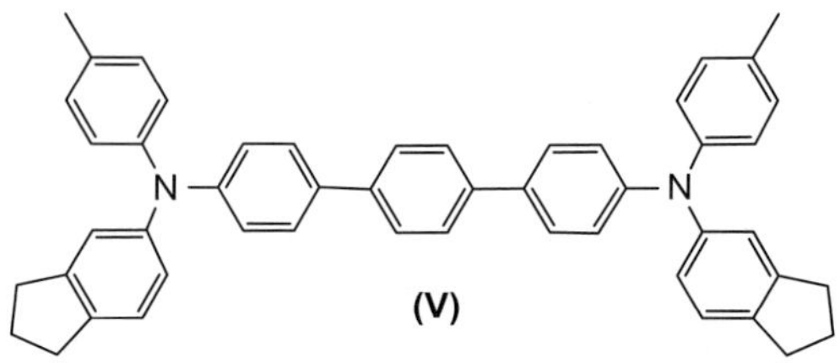

(V)

REFERENCES

1. Takashi Hori et al, US Patent 5,824,808 (October 20, 1998)
Cosmo Research Institute (Tokyo, JP)
2. Yasushige Nakamura et al, US Patent Application 20030017407 (January 23, 2003)
Fujitsu Limited (Kawasaki, JP)
3. Timothy P. Bender et al, US Patent 7,820,780 (October 26, 2010)
Xerox Corporation (Norwalk, CT)
4. Shinji Moriyama et al, US Patent 7,803,509 (September 28, 2010)
Kao Corporation (Tokyo, JP)
5. Abe Katsumi et al, US Patent 7,799,492 (September 21, 2010)
Hodogaya Chemical Co., Ltd. (Kawasaki-shi, JP)

B. EXTENDED BATTERY LIFETIME ADDITIVES

Pentacyclic or Tetrazapentalene Anionic Salts

Title:	Pentacyclic anion salts or tetrazapentalene derivatives and their uses as ionic conducting materials
US Patent:	7,906,235 (March 15, 2011)

Author: Christophe Michot et al.
Assignee: Hydro-Quebec (Montreal, CA)

Significance: Ionic conductors consisting of lithium or potassium salts of cyclopentadienium, imidazolium or triazonium derivatives containing di-cyano, trifluoromethyl, or perflouroalkyl were prepared and extended the battery lifetime. When used as a battery additive component, the electrostatic discharge exceeded 600 cycles at approximately 62% use.

Additive Name
2-Amino-5-trifluoromethyl-1,3,4-triazolium potassium (I)
Perfluorobutanesulfonyl-4,5-dicyanoimidazolium potassium (II)
2,5-Trifluoromethyl-3,4-dicyanocyclopentadienium potassium (III)
4,4′-Azobis(4-cyano-4-methylbutyl)-2,5-trifluoromethyl-3,4-dicyanocyclopentadienium ketone (IV)
2-Di-(2-ethylhexyl)aminosulfonyl-4,5-dicyanoimidazolium lithium (V)
Poly(styrene-4-(sulfonyl-4,5-dicyano-2,5-di-trifluoromethylcyclopentadenium potassium-co-acrylonitrile) (VI)

Safety The materials of the current invention are ionic salts intended as replacement additives for highly conducting ionic salts. They are considerably stabler than ionic salts containing coordination anions BF_4^-, PF_6^-, or AsF_6^-. They are considered mildly irritating to eyes, respiratory system, and skin. The effects of chronic exposure to these agents are unknown.

Additive Structures

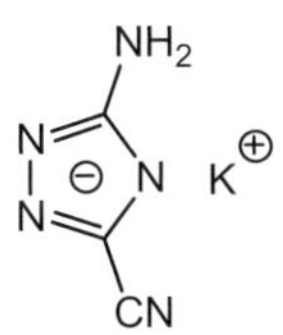

(I)

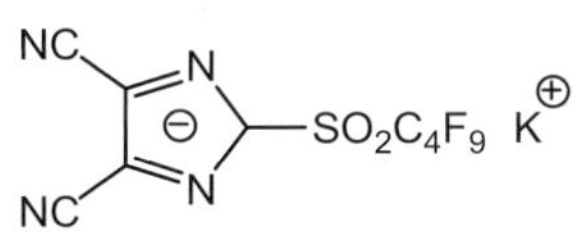

(II)

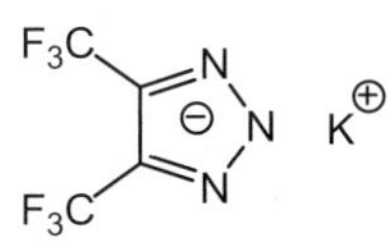

(III)

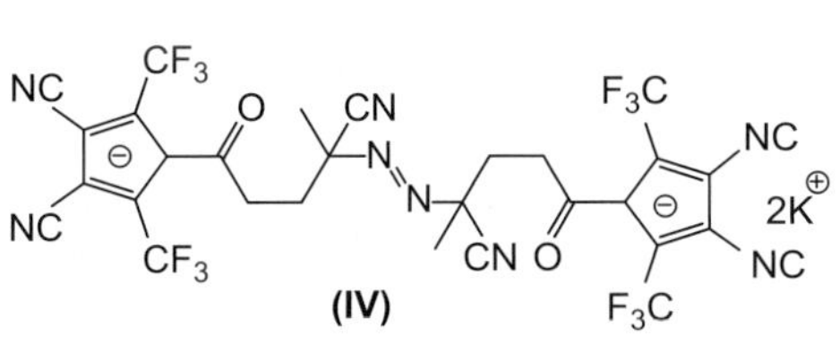

(IV)

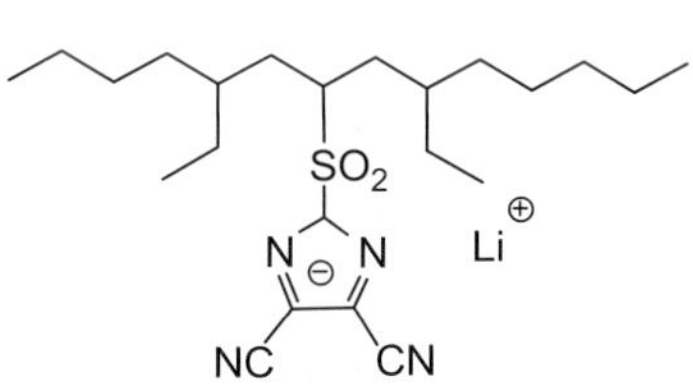

(V)

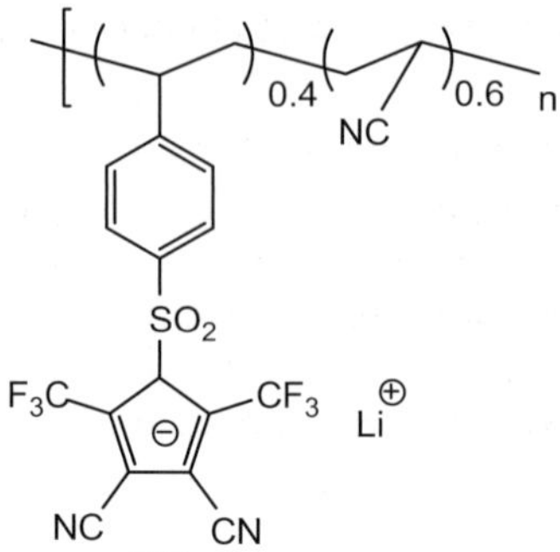

(VI)

Additive Preparation

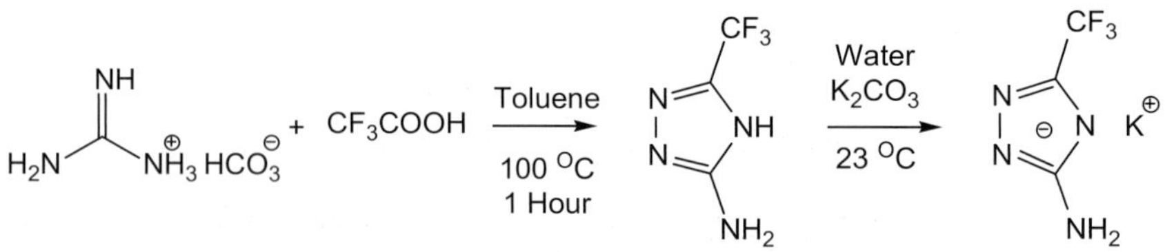

Additive Synthesis

1. Preparation of 2-amino-5-trifluoromethyl-1,3,4-triazole

A 1000-ml round bottom flask was charged with 136.11 g of aminoguanidine bicarbonate in 500 ml of toluene and then stirred and treated with 119.73 g of trifluoroacetic acid. After the evolution of carbon dioxide ceased, the mixture was heated to reflux for 24 h to azeotropically remove 54.5 ml of water. After cooling to an ambient temperature, the mixture was filtered, and the product was isolated in a 92% yield.

2. Preparation of 2-amino-5-trifluoromethyl-1,3,4-triazolium potassium

The Step 1 product was treated with 20 mol% excess of potassium carbonate dissolved in water at an ambient temperature. After concentrating and drying the reaction mixture, a solid was extracted with acetonitrile, filtered to remove excess potassium carbonate, and the product was isolated after vacuum distillation.

TESTING

Battery Testing

A battery was produced by superposing a current collector consisting of stainless steel having a thickness of 2 mm, a cathode consisting of a film pastil composite material having a thickness of 72 μm and comprising 45 vol% vanadium dioxide, 5 vol% Shawinigan black, and 50 vol% polyethylene oxide having an Mn of approximately 3×10^5 Da. An electrolyte consisting of a pastil of a poly(ethylene oxide) film having an Mn of approximately 5×10^6 Da containing a selected lithium salt derivative with an O/Li = 15/1.0 as well as an anode made of a sheet of metallic lithium having a thickness of 50 μm was used as the current collector. The pastils constituting the electrode and electrolyte were cut out in a glove box and piled in the order indicated above. The collectors were then placed on either side of the pile obtained. The assembly was sealed in a housing to form a button-shaped battery, which was connected to a generator from the atmosphere to limit mechanical stress on the films. The assembly was then placed in an enclosure under argon at 60°C and cycled between 1.8 and 3.3 V at a rate of charge/discharge, where the nominal capacity charged or discharged was in 10 h.

Test Results

A) The following potassium salts were tested for 600 cycles and percent utilization and were considered acceptable at an O/K = 14/1.

Potassium salts of

2-Amino-5-trifluoromethyl-1,3,4-triazole
2-Cyano-5-trifluoromethyl-1,3,4-triazole

B) The following lithium salts were tested for 600 cycles and percent utilization and were considered acceptable at an O/Li = 15/1.

Lithium salts of

4,5-Trifluoromethyl-1,2,3-triazole
4-Trifluoromethyl-5-cyano-1H-1,2,3-triazole
2-Cyano-5-trifluoromethyl-1,3,4-triazole
2-Trifluoromethanesulfonyl-5-trifluoromethyl-1,3,4-triazole
2-Dimethylaminosulfonyl-4,5-dicyanoimidazole
1-Fluoro-2,5-trifluoromethyl-3,4-dicyanocyclopentadiene
Poly(2-(3,4-epoxy-1,1-difluorobutane)-5-cyano-1,3,4-triazole)

C) The following lithium salts were tested for 600 cycles and percent utilization and were considered acceptable at an O/Li = 15/1

Lithium salts of

4-Trifluoromethyl-5-cyano-1H-1,2,3-triazole
2-Cyano-5-trifluoromethyl-1,3,4-triazole
2-Trifluoromethanesulfonyl-5-trifluoromethyl-1,3,4-triazole
2-Dimethylaminosulfonyl-4,5-dicyanoimidazole
1-Fluoro-2,5-trifluoromethyl-3,4-dicyanocyclopentadiene
Poly(2-(3,4-epoxy-1,1-difluorobutane)-5-cyano-1,3,4-triazole)

Advantages over Prior Art

The current invention has several distinctions over existing battery additive technology. The first improvement is that salts obtained from this method dissociate in water or in an organic medium without side reactions associated with dissociated salts containing coordination anions BF_4^-, PF_6^-, or AsF_6^-. The second improvement is that ionic properties can be enhanced in a cyclopentadienium anion by either (a) introducing a more electronegative atom into the ring or (b) by introducing electronegative substituents into cyclopendiene or its precursor before converting it into an anionic salt. Both of these improvements result in the formation of stable anionic salts that extended the batter discharge.

Notes

1. In a subsequent investigation by Michot[1], 1,2,4-triazolium sulfonyl ammonium salts (VII) and (VIII) were prepared and found to be effective as battery additives.

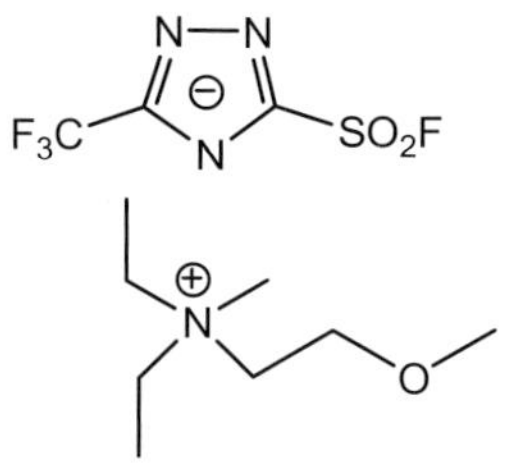

(VII)

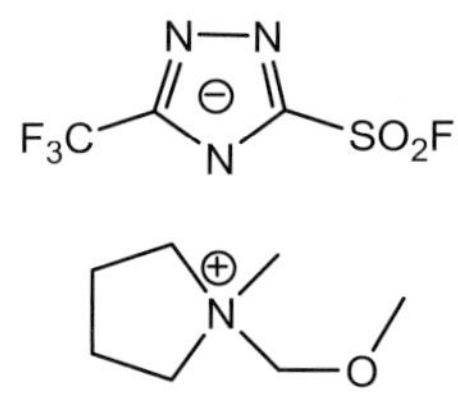

(VIII)

2. The electrolytic battery additive, lithium vanadyl (IV) orthophosphate, $LiVOPO_4$, prepared by Barker[2] was effective for use in rechargeable batteries.
3. Zaghib[3] prepared polymer electrolytes for use in electrochemical generators consisting of either poly(vinylidenefluoride), poly(vinyldiene fluoro-co-hexafluoropropene), poly(tetrafluoro-ethylene), or poly(ethylene-co-propylene-co-5-methylene-2-norbornene) blended with SiO_2, Al_2O_3, or TiO_2 nanoparticles.
4. Solvent-free noncorrosive proton-transfer ionic liquids consisting of ethylammonium nitrate or ethylammonium formate doped imidazole were prepared by Angell[4] and had excellent stability at elevated temperatures and were utilized in solid electrolyte compositions.

REFERENCES

1. Christophe Michot et al, US Patent 7,919,629 (April 5, 2011)
   Phostech Lithium Inc. (Montreal, CA)
   Universite de Montreal (Montreal, CA)
2. Jeremy Barker et al, US Patent 7,901,810 (March 8, 2011)
   Valence Technology, Inc. (Las Vegas, NV)
3. Karim Zaghib et al, US Patent 7,897,674 (March 1, 2011)
   Hydro-Quebec (Montreal, Quebec, CA)
4. C. Austen Angell, et al, US Patent 7,867,658 (January 11, 2011)
   Arizona Board of Regents for and on behalf of Arizona State University (Scottsdale, AZ)

C. IMPROVING PEAK, REPAIR COMPLETE AND WITHSTAND VOLTAGES USING IONIC LIQUIDS

Perfluoroalkylimidazolium Hexafluorophosphates, Sulfonamides, or Sulfates

Title:	Ionic liquid and method for producing the same, method for forming oxide film on a metal surface, electrolyte capacitor, and electrolyte
US:	Patent 8,034,956 (October 11, 2011)
Author:	Yasuhiro Tsukada et al.
Assignee:	Kaneka Corporation (Osaka, JP)

Significance: Perfluoroalkylimidazolium salts containing hexafluorophosphate, bis(trifluoromethane)sulfonamide or 2,2,3,3,4,4,5,5-octafluoro-pentane sulfate counterions were prepared by the addition of perfluoroalkyl iodide to *N*-alkylimidazole followed by anion exchange. When these ionic liquids were used as electrolytes in capacitor peak, repair, and withstand voltages were dramatically improved.

Additive Names

1-Ethyl-3-perfluorobutylmethylimidazolium hexafluorophosphate (I)
1-Methyl-3-perfluorobutylmethylimidazolium hexafluorophosphate (II)
1-Methyl-3-perfluorobutylmethylimidazolium bis(trifluoromethane)-sulfonamide (III)
1-Methyl-3-perfluorobutylethylimidazolium-2,2,3,3,4,4-octafluoro-pentane sulfate, (IV)

Safety

Ionic solvents, particularly those containing the hexafluorophosphate counterion, are mild skin irritants but moderate to severe eye irritants. Although the vapor pressure for these agents is considered very low, users are cautioned to avoid using them in confined or enclosed areas.

Additive Structures

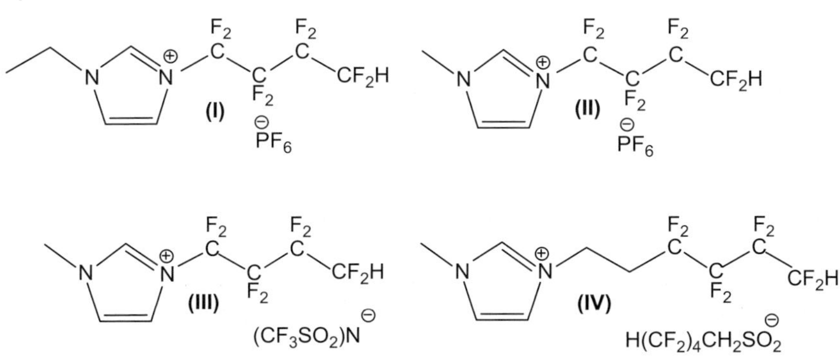

Additive Preparation

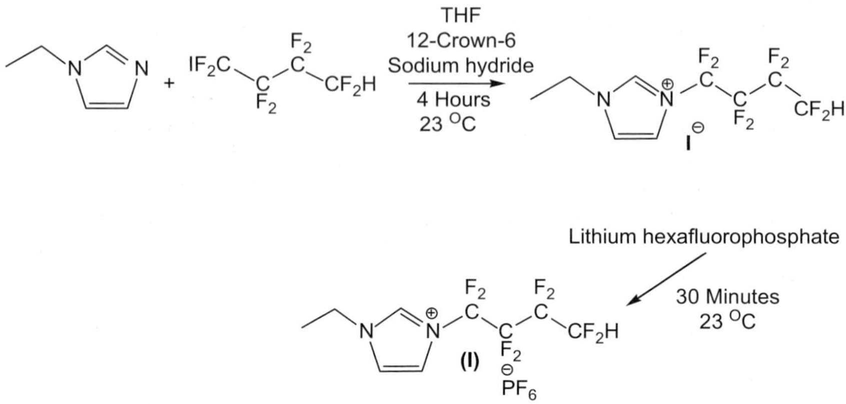

Additive Synthesis

1. *Preparation of 1-ethyl-3-perfluorobutylmethylimidazolium bromide*

A 1-liter reactor was charged with 9.9 g of sodium hydride, 200 ml of tetrahydrofuran, and 3.85 g of 10 wt% 12-crown-6 and then treated with 9.9 g of *N*-ethylimidazole and stirred in an ice bath. After 2 h, the mixture was treated with 50 g of 2-perfluorobutylmethyl iodide and then stirred at an ambient temperature for 4 h. When the reaction solution was transparent, it was treated with 15.9 g of ethyl bromide and then heated to 120 °C for 7 h and concentrated. After drying, 42.21 g of product was isolated as a brown liquid.

2. *Preparation of 1-ethyl-3-perfluorobutylmethylimidazolium hexafluorophosphate*

A reactor was charged with 7.82 g of the Step 1 product, 3.0 g of lithium hexafluorophosphate, and 300 ml of water at an ambient temperature and then stirred for 30 min and treated with anhydrous magnesium sulfate. After vacuum drying at 60 °C, 4.72 g of the purified product was isolated as a brownish-red liquid having a viscosity of 6.3 P at 25 °C

^{1}H-NMR (d_6-DMSO, δ)	1.33 (Triplet, J = 5Hz, 3H), 4.21 (quartet, J = 5Hz, 2H), 4.52 (singlet, 2H), 7.07 (doublet, J = 5Hz, 1H), 7.44 (doublet, J = 5Hz, 1H), 8.23 (singlet, 1H), 9.04 (singlet, 1H)

TESTING

Evaluation of Oxide Film–Repairing Ability

A 50 mm × 5 mm × 0.5 mm aluminum plate having a purity of 99.99% was cleaned by immersing in a mixture of 15 parts of 85% nitric acid and 85% phosphoric acid at

85 °C and then washed with purified water and immersed in 1 M sodium hydroxide for 10 min. The plate was immersed in boiling water containing 1 g/L of aqueous diammonium adipate solution and a cylindrical copper electrode used as a cathode. The anode was then placed 7 mm from the center of the copper electrode, and a constant current of 10 mA/cm^2 was applied between the anode and the cathode. An oxide film was then formed by applying a voltage of 100 V between the anode and the cathode in boiling water for 3 min

To assess the effectiveness of the experimental agents, the aluminum plate having the oxide film was immersed to a depth of 0.5 cm in a selected ionic liquid at 25 °C, and changes in the current were measured when the applied voltage was gradually raised to 200 V at a rate of 1 V/sec. Test results are provided in Table 2.

Test Results

TABLE 2 Oxide film–repairing properties of selected ionic solvents and their effects on peak, repair, and withstanding voltages

Additive	Viscosity (25 °C) (P)	Peak Voltage (V)	Repair Complete Voltage (V)	Withstand Voltage (V)
I	6.3	15	40–50	180
II	5.9	15	40–50	170
III	5.8	20	40–50	150
IV	6.0	20	40–50	190
Reference-1 (1-ethyl-3-methyl-*p*-toluenesulfonic acid)	X	X	X	90–100
Reference-2 (1-ethyl-3-methyl-*p*-toluenesulfonic bromide	X	45	120	~20

Advantages over Prior Art

Electrolytic systems for capacitors typically consist of mixtures of ethylene glycol or γ-butyrolactone and a salt of an organic or inorganic acid. These compositions are, however, both inadequate for realizing a low impedance capacitor and can evaporate at relatively moderate temperatures. Imidazolium-based ionic liquids completely dissociate, have very, very low vapor pressures, and can be used in low impedance capacitor applications.

Notes

1. Miles[1] developed high cell voltages and improvements in energy and power density in thermal batteries using a mixture of molten $LiNO_3/KNO_3$ as electrolytes.

2. Kweon[2] determined that a lithium battery could be recharged by applying a surface-treatment layer consisting of aluminum oxide, aluminum hydroxide, boron oxide ammonium vanadate, or vanadium oxide on the battery.
3. Takamuku[3] determined the optimum negative active voltage needed for a rechargeable lithium battery using a vanadium oxide/lithium carbonate mixture having a Li:V ratio of 1.1:0.9 mixed with 3 wt% of anhydrous zirconium oxide.
4. Ohzuku[4] used a mixture of artificial graphite, acetylene black, and nickel powder bound in either polytetrafluoroethylene or polyvinylidene fluoride and then used this mixture to prepare a nonaqueous electrolytic high capacity secondary cell. The secondary cell consisted of a negative electrode that absorbs and releases lithium ions or metallic lithium and a nickel, manganese, or cobalt positive electrode.

REFERENCES

1. Melvin H. Miles et al, US. Patent 8,039,138 (October 11, 2011)
Millennium Engineering and Integration Company (Arlington, VA)
2. Ho-Jin Kweon et al, US. Patent 8,034,486 (October 11, 2011)
Samsung SDI Co., Ltd. (Yongin-si, KR)
3. Akira Takamuku et al, US. Patent 7,981 (July 19, 2011)
Samsung SDI Co., Ltd. (Gongse-dong, Giheung-gu, Yongin-si, Gyeonggi-do, KR)
4. Tsutomu Ohzuku et al, US. Patent 7,935,443 (May 3, 2011)
Panasonic Corporation (Osaka, JP)

Chapter 6

Carpets

A. STAIN RESISTANT CARPET SPRAY

Hyperbranched Maleic Anhydride Octadiene Copolymers

Title:	Hyperbranched maleic anhydride–diene polymers
US:	Patent 7,833,585 (November 16, 2010)
Author:	Stephen Ernest Jacobson et al.
Assignee:	E.I. du Pont de Nemours and Company (Wilmington, DE)

Significance: Hyperbranched copolymers prepared by reacting either 1,7-octadiene or 1,9-decadiene with maleic anhydride were found to be effective as textile stain repellants against acid dyes found in coffee stains and soft drink preparations containing FD&C Red Dye No. 40. The effectiveness of these antistaining agents is due to the presence of a high density of succinic anhydride groups present at the surface of the fabric substrate.

Additive Name Poly[(1,7-octadiene-co-maleic anhydride)-g-(1,7-octadiene-co-succinic anhydride)] (I)
Poly[(1,9-decadiene-co-maleic anhydride)-g-(1,9-decadien-co-succinic anhydride)] (II)

Safety Concerns Both copolymers are mild skin and eye irritants because of the presence of maleic anhydride and its hydrolysis product succinic acid. The presence of unreacted maleic anhydride in either copolymer can potentially cause skin burns after prolonged exposure.

Next Generation of International Chemical Additives. DOI: http://dx.doi.org/10.1016/B978-0-444-53788-1.00006-1

Additive Structures

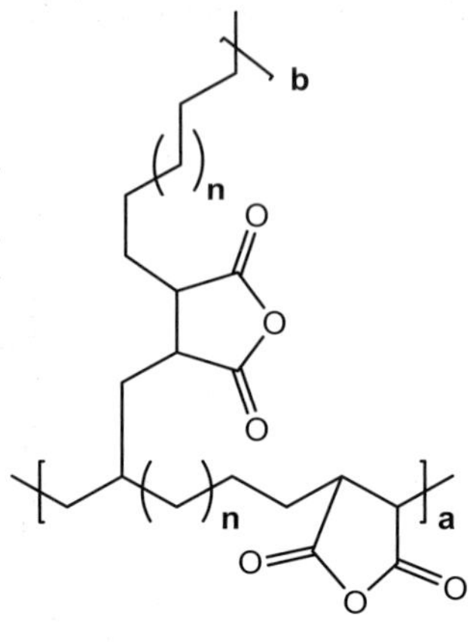

(I) (n = 4)

(II) (n = 6)

Additive Preparation

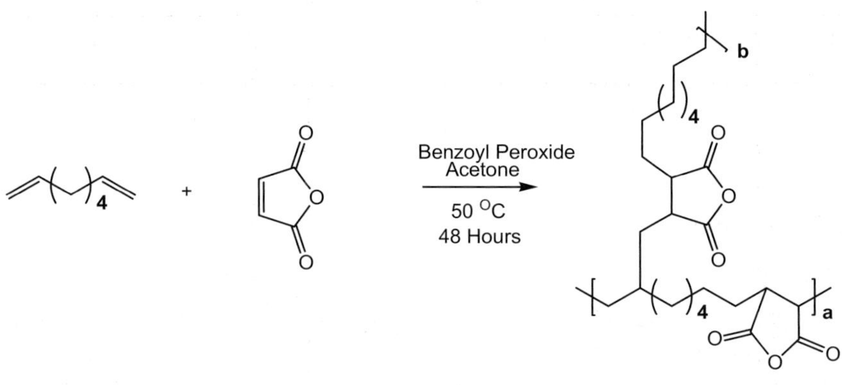

Additive Synthesis

1 Preparation of poly[(1,7-octadiene-co-maleic anhydride)-g-(1,7-octadiene-co-maleic anhydride)]

A 2-liter round bottom flask was charged with 88.0 g of 1,7-octadiene, 78.4 g of maleic anhydride, and 800 parts of acetone and then stirred at 23 °C until all the solids dissolved. The solution was then heated to 50 °C and purged with nitrogen for 1 h and then incrementally treated with 29.2 g of benzoyl peroxide over 30 min. The mixture was then heated to 50 °C for 48 h while monitoring with gas chromatography, and a clear brown solution was isolated. Approximately 650 g of acetone was then distilled from the solution, and the product was isolated by precipitating the concentrate into 2 liters of a 1:1 mixture of toluene/hexane. The gray precipitate was washed three times a piece with 400 ml toluene and then hexane, and 132 g of the product was isolated. A summary of the reaction stoichiometry and physical properties of copolymers is provided in Table 1.

TABLE 1 **Physical properties of poly[(1,7-octadiene- and 1,9-decadiene co-succinic anhydride)-g-(1,7-octadiene-co-maleic anhydride)]**

Sample	Diene	Initiator (g)	Reaction Solvent (g)	Mw	Mw/Mn	[η] (ml/g)
1	1,7-Octadiene	Benzoyl Peroxide (29.2)	Acetone (800)	370,000	12	9.2
7	1,7-Octadiene	Benzoyl Peroxide (29.2)	THF (800)	130,000	6.4	7.8
8	1,7-Octadiene	VASO-64*	Acetone (200)	X	X	X
9	1,7-Octadiene	Benzoyl Peroxide (7.3)	MIBE (200)	140,000	15	10
12	1,9-Decadiene	Benzoyl Peroxide (7.3)	Acetone (200)	155,000	2.5	6.6

**2,2′-Azobis(isobutyronitrile).*

Additive Testing

1. Coffee Test

Clean nylon carpet samples containing a spray coating of a selected experimental additive was stained with 50 ml of coffee at 62 °C and was then made to remain undisturbed for 4 h. The fabric was then thoroughly rinsed in cold water for at least 10 min. The effectiveness of the stain repellency was determined by measuring the carpet color using a Minolta Chroma Meter CR-410. Test results are provided in Table 2.

2. Kool Aid™ Stain Test

Clean nylon carpet samples containing a spray coating of a selected experimental additive were stained with 20 ml of an 8.3 wt% aqueous solution of cherry flavored presweetened Kool Aid™. The carpets were then rinsed with tap water after 24 h and visually rated from 10 (complete stain removal) to 1 (unchanged stain) using the AATCC Red 40 Stain ScaleTest Method 175. Test results are provided in Table 2.

Additive Test Results

TABLE 2 Effectiveness of poly[(1,7-octadiene-co-maleic anhydride)-g-(1,7-octadiene-co-maleic anhydride)] in repelling coffee and Kool Aid™ Stains

Surface Coating	Coffee Blocked (%)	Kool Aid™ Test (24h)
Control*	36	6
Step 1 Product	45	9
Step 1 Product	44	9
Step 1 Product	38	9
Step 1 Product	51	10

Poly(methylvinyl ether-co-succinic anhydride).

Advantages over Prior Art

Hyperbranched poly(1,7-octadiene or 1,9-decadiene-co-succinic anhydride) have a dentrite-like structure but with a less controlled architecture. Their effectiveness as textile stain repulsing agents is due to the presence of a high density of surface-active groups from the polymer on the fabric surface and in internal cavities. This high surface/cavity density covering is absent in linear 1-alkene/maleic copolymers.

Notes

1. Hopkins[1] prepared a carpet stain protective additive (III) consisting of poly(styrene-co-succinic anhydride) amidated with 4-aminosalicylic acid and hydrolyzed with water, which was coffee and Kool Aid™ stain resistant.

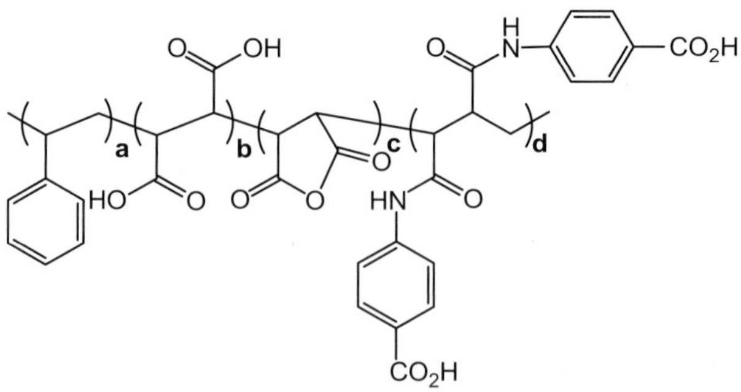

(III)

2. Poly(1-octene-co-succinic anhydride) diamidated with a 1.2:1 molar ratio of 3-amino-4-hydroxy-benzenesulfonic acid hydrate and triethylamine (IV) was prepared by Hopkins[2] and was effective as a carpet stain–resistant additive when applied as a surface coating.

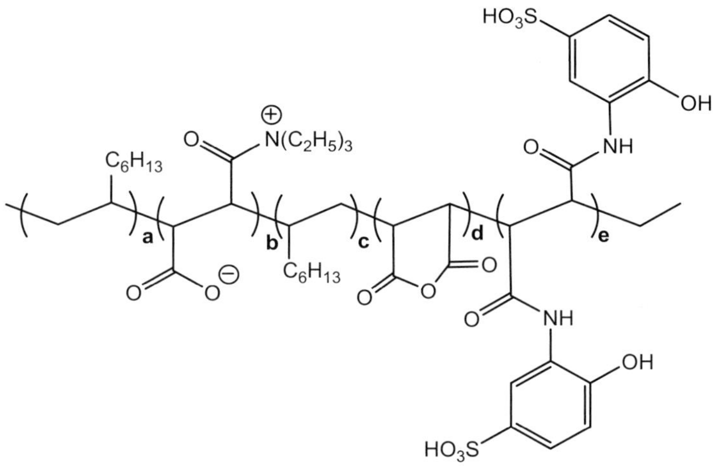

(IV)

3. Yamamoto[3] prepared a terpolymer (V) containing a low molecular weight perfluorinated acrylate, which was effective as a water- and oil-repellent protective carpet additive.

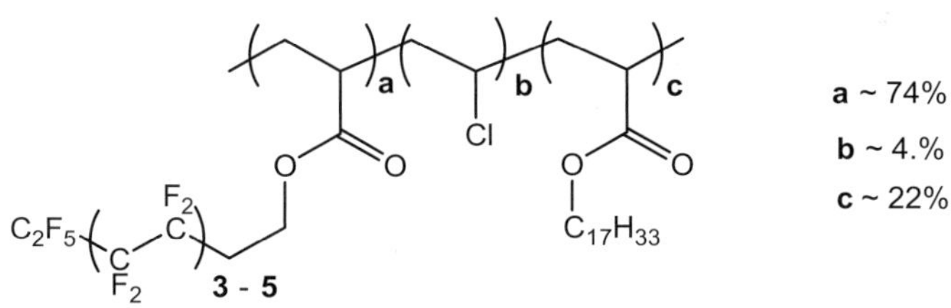

(V)

4. Carpeting dip coated with the diurethane oligomer consisting of the reaction product of 1,6-hexamethylene diisocyanate, methoxy polyethylene glycol having an Mn of approximately 750 Da and 3,3,4,4,5,5,6,6,7,7,8,8,8-tridecafluorooctan-1-ol was prepared by Wang,[4] and this improved the stain release properties of the fabric.

REFERENCES

1. Timothy Hopkins et al, US Patent 7,816,005 (October 19, 2010)
   E.I. du Pont de Nemours and Company (Wilmington, DE)
2. Timothy Hopkins et al, US Patent 7,740,913 (June 22, 2010)
   E.I. du Pont de Nemours and Company (Wilmington, DE)
3. Ikuo Yamamoto et al, US Patent 7,678,155 (March 16, 2010)
   Daikin Industries, Ltd. (Osaka, JP)
4. Ying Wang et al, US Patent 7,494,511 (February 24, 2009)
   E.I. du Pont de Nemours and Company (Wilmington, DE)

Chapter 7

Colorants

A. HEAT-STABLE PIGMENTS WITH IMPROVED FASTNESS AND GOOD WEATHERABILITY

Diketopyrrolopyrrole Pigment Mixtures

Title:	Heat-stable diketopyrrolopyrrole pigment mixtures
US Patent:	7,786,313 (August 31, 2010)
Author:	Olof Wallquist et al.
Assignee:	Ciba Specialty Chemicals Corp. (Tarrytown, NY)

Significance: A method for preparing heat-stable and tinctorially strong diketopyrrolopyrrole pigment mixtures for use in the coloring of organic material, particularly resins, is described. These chemical additives address concerns with existing diketopyrrolopyrrole pigments including fastness to migration, light, weathering, and heat resistance. Blends containing these additives are characterized as a having a high degree of transparency and transparency or contrast.

Additive Name

3-(4-Methyl-phenyl)-6-(4-decylthiophenyl)-2,5-dihydropyrrolo[3,4-c]pyrrole-1,4-dione (I)
3-(4-Methyl-phenyl)-6-(4-octadecylthiophenyl)-2,5-dihydropyrrolo[3,4-c]pyrrole-1,4-dione (II)
3-(4-Methyl-phenyl)-6-(4-(1-thio-(3-sulfonate sodium) propyl)2,5-dihydropyrrolo[3,4-c]pyrrole-1,4-dione (III)
3-(4-Methyl–phenyl)-6-(4-(1-oxy-(3-methoxy) propyl)2,5-dihydropyrrolo[3,4-c]pyrrole-1,4-dione (IV)

CAS number Experimental additives

Next Generation of International Chemical Additives. DOI: http://dx.doi.org/10.1016/B978-0-444-53788-1.00007-3

Safety Concerns

Diketopyrrolopyrrole pigments are not considered to be hazardous substances as defined by the OSHA hazardous protocol. They may, however, form explosive dust clouds in air.

Additive Structures

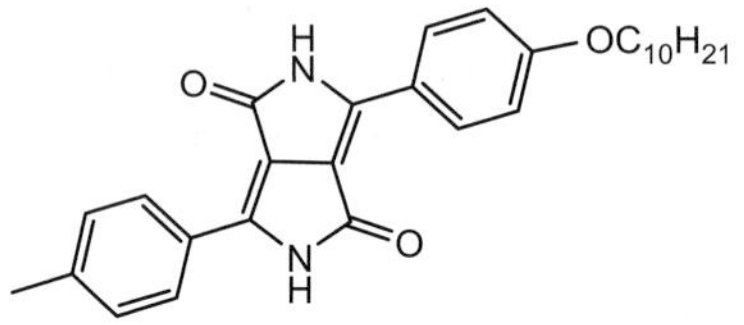

(I)

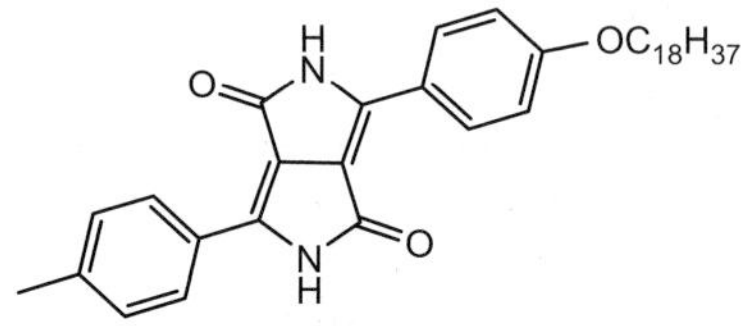

(II)

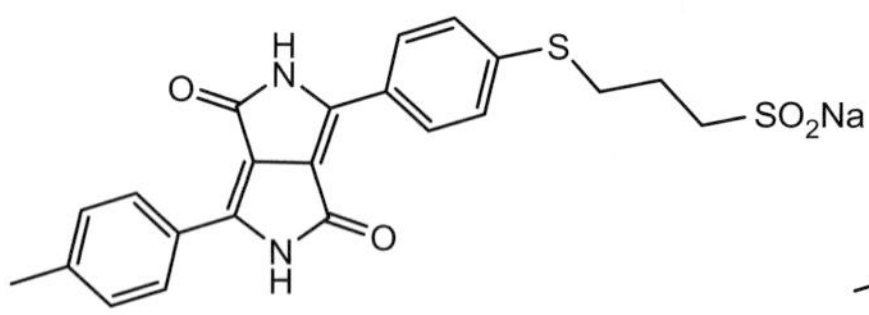

(III)

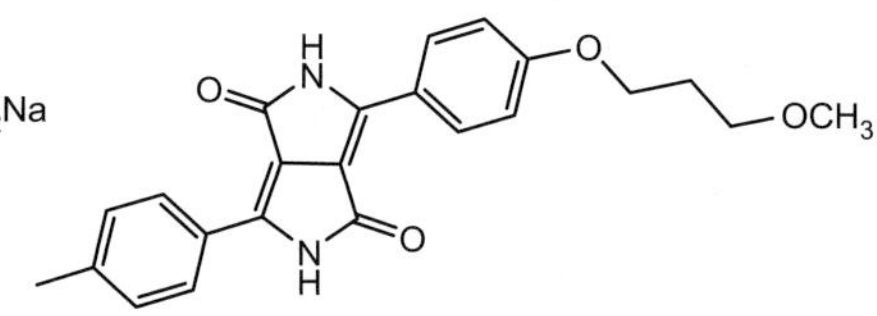

(IV)

Additive Preparation and Formulation

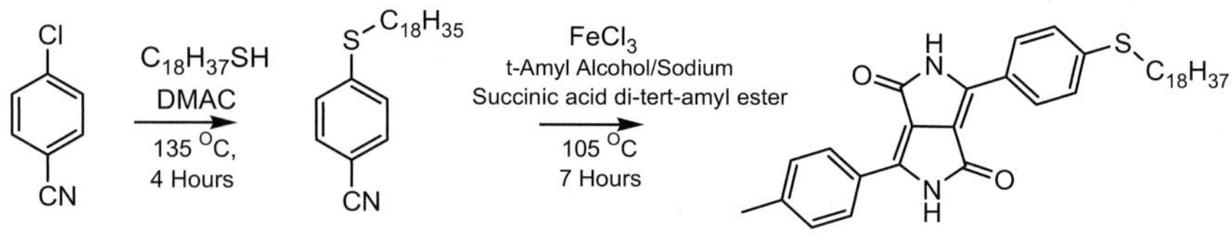

Target Additive

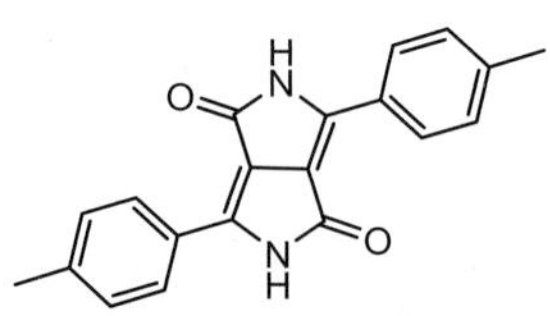

Reaction Side Product

Additive Synthesis

Step 1. Preparation of p-octadecylthiobenzonitrile

A five-necked 750-ml flask containing a white suspension of 48.18 g of *p*-chlorobenzonitrile and 67.72 g of finely ground potassium carbonate in 200 ml of dimethylacetamide was treated with 100.00 g of 1-octadecanethiol dissolved in 200 ml of dimethylacetamide at 30 °C over 25 min. The mixture was then heated to 135 °C and stirred for 4 h. The temperature was then reduced to 80 °C, and the contents were stirred for an additional 16 h and then precipitated by pouring into 1600 g of water at an ambient temperature. The mixture was then filtered, and the product was isolated in a 90.4% yield as a white crystalline solid.

Step 2. Preparation of diketopyrrolopyrrole colorant mixture

A three-neck 750-ml sulfonating flask containing a thermometer, a mechanical stirrer, a dropping funnel, and a reflux condenser was charged with 400 ml of t-amyl alcohol and a catalytic amount of $FeCl_3$ at 112 °C and was then treated with 24.8 g of sodium over 40 min. This solution was then treated with 62.0 g of *p*-toluene nitrile and 4.2 g of the Step 1 product and heated for 5 h at 105 °C. Thereafter, the reaction mixture was treated with 92.2 g of succinic acid di-*t*-amyl ester and stirred for an additional 2 h at 105 °C. The mixture was then cooled to an ambient temperature and stirred overnight and then precipitated in a solution of 450 ml of water and 450 ml of methanol. The mixture was then heated to 32 °C for 5h and then filtered and washed with 500 ml of methanol. After drying at 80 °C, the product was isolated in an 87% yield as a red solid.

Advantages over Prior Art

The current method for preparing diketopyrrolopyrrole colorants produces heat-fast and tinctorially strong transparent pigments in the form of pigment mixtures that have excellent fastness and weatherability. These physical properties make these experimental agents particularly useful for crystalline resin blending or for use on weather-exposed surfaces.

Notes

1. Diketopyrrolopyrrole colorants (V) and (VI) were blended into highly crystallized polylactic acid biodegradable resin compositions by Sato.[1] Rigid molded plastics had excellent transparency and good weatherability properties.

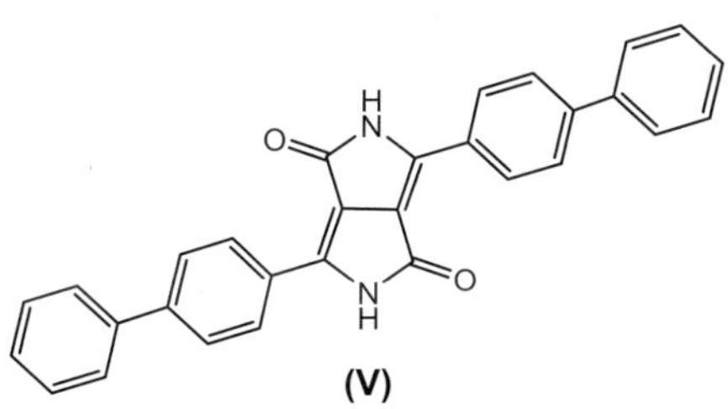

(V)

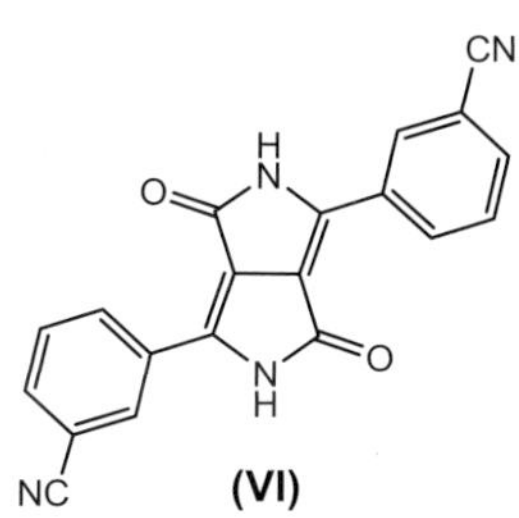

(VI)

2. Adams[2] observed that by blending diketopyrrolopyrrole colorants with pentaerythritol tetrakis(3-(3,5-di-t-butyl-4-hydroxyphenol)propionate (IRGANOX®-1010) (VII), low dust colorant levels were generated. The efficacy of the diketopyrrolopyrrole pigment was unchanged.

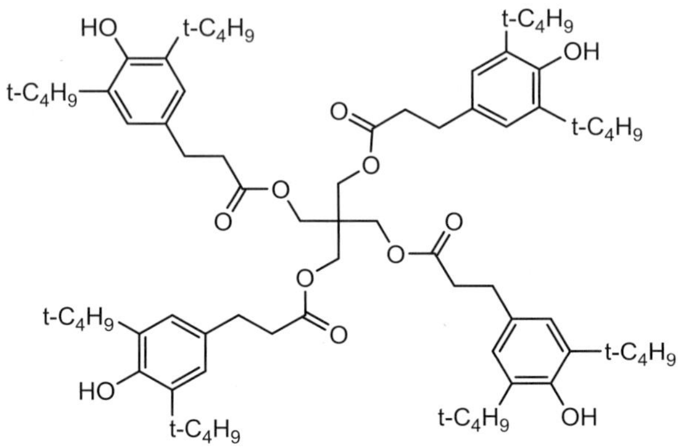

(III)

3. A solvent-free method of preparing diketopyrrolopyrrole derivatives was devised by Kaul[3] and is as described below

   Solvent-free Preparation of 3,6-bis(4-chlorophenyl)-2,5-dihydropyrrolo-[3,4-c]pyrrole-1,4-dione

   At ambient temperature, a 10-liter reactor was charged with 2475 g of *p*-chlorobenzonitrile, 2181.6 g of di-isopropyl succinate, and 2869.2 g of sodium t-butyrate and then heated to 100 °C for 1 h under nitrogen while thoroughly mixing. The temperature was then lowered to 80 °C, and the reaction mixture was noticeably thicker and finally became a paste. With continued heating at 80–85 °C, alcohol began to distill. Thereafter, the temperature was maintained at 100 °C for 3 h to distill any remaining isopropyl and t-butyl alcohols. The crude reaction product then became crumbly, and ultimately, a semipowdery material was formed. The mixture was further heated to 120 °C for 60 min, cooled to 50 °C, emptied in a polyethylene sack, and the product isolated in a 95.54% yield as an intense red solid.
4. Ganschow[4] developed a process for preparing diketopyrrolopyrrole pigments with enhanced light and weather fastness, improved thermal stability, and heightened fastness by limiting the pigment crystal size to be <400 nm. The process entailed hydrolyzing the diketopyrrolopyrrole salt pigment at 80 °C in warm water followed by pressure finishing at 120 °C for 5 h.

REFERENCES

1. Noritaka Sato et al, US Patent 7,790,782 (September 7, 2010)
   Sony Corporation (JP)
2. John Adams et al, US 7,789,957 (September 7, 2010)
   Ciba Corporation (Tarrytown, NY)

3. Bansi Lal Kaul, US Patent 7,728,139 (June 1, 2010)
   MCA Technologies GmbH (Biel-Benken, CH)
4. Matthias Ganschow et al, US Patent 7,695,558 (April 13, 2010)
   Clariant Finance (BVI) Limited (Tortola, VG)

Chapter 8

Composites

A. ORGANIC/INORGANIC SILICON COMPOSITES CONTAINING MOLECULAR AGGREGATES

Poly(Silsesquioxane-co-Methyl Methacrylate)

Title:	Polymer obtained by the addition–polymerization initiated by a silicon compound
US Patent:	7,964,692 (June 21, 2011)
Author:	Mikio Yamahiro et al.
Assignee:	JNC Corporation (Tokyo, JP)

Significance: Ring closure of sodium alkyl (aryl) silsesquioxane with acetoxyalkyltrichloro-silane was used to prepare a well-characterized monofunctionalized organic-inorganic composite material intermediates. Two composite materials were prepared by transfer radical polymerization of 2-bromo-2-methylpropionyl-containing silsesquioxane with methyl methacrylate and styrene. Both polymeric silsesquioxanes had predictable number average molecular weights and very low polydispersities.

Additives

Acetoxyethyl-heptaphenyloctasilsesquioxane) (I)
Acetoxyethyl-heptatrifluoropropyloctasilsesquioxane (II)
(2-Bromo-2-methylpropionyloxyethyl)-hexacyclohexyloctsilsesquioxane (III)
Hydroxyethyl-heptaisooctyloctasilsesquioxane (IV)
(2-Methylpropionyloxyethyl)-heptaphenyloctasilsesquioxane–polymethyl methacrylate bromide (V)
(2-Methylpropionyloxyethyl)-heptaphenyloctasilsesquioxane–polystyrene bromide (VI)

Safety

Poly(phenyl silsesquioxane) is a mild to moderate eye irritant which causes redness and pain. This material is both a skin and respiratory irritant.
Accidental ingestion causes nausea, vomiting, and diarrhea.

Next Generation of International Chemical Additives. DOI: http://dx.doi.org/10.1016/B978-0-444-53788-1.00008-5

Additive Structures

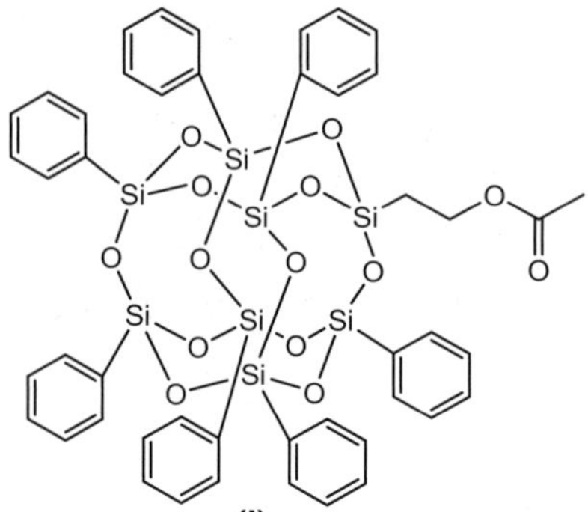

(I)

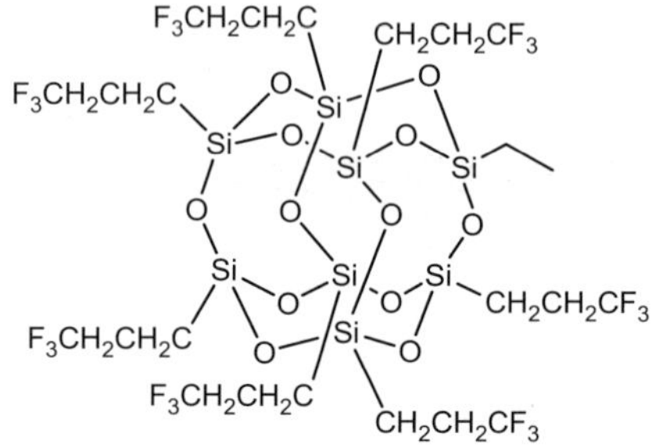

(II)

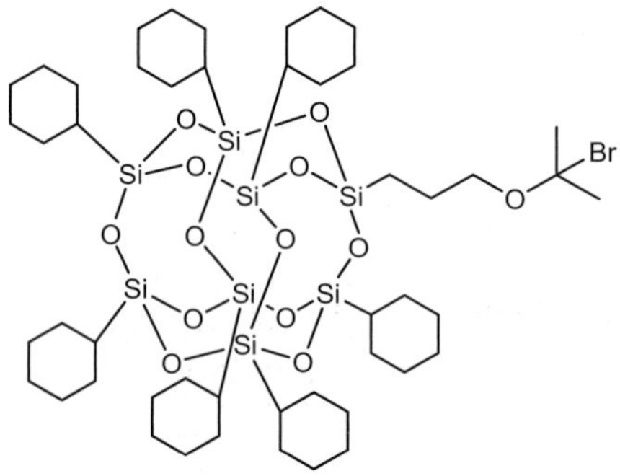

(III)

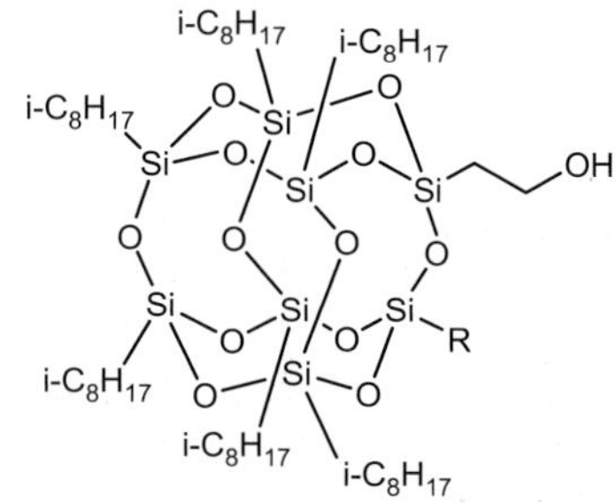

(IV)

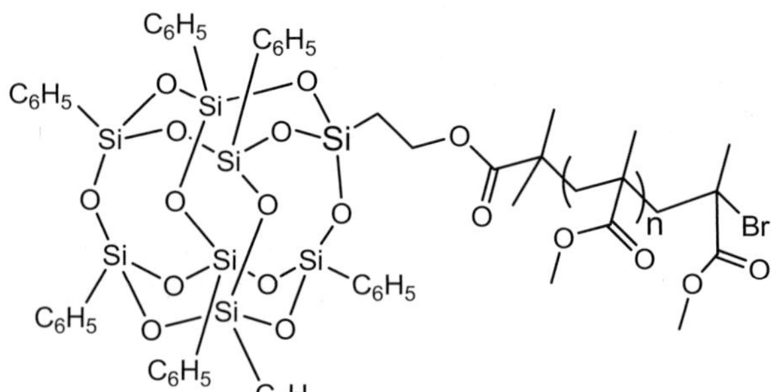

(V)

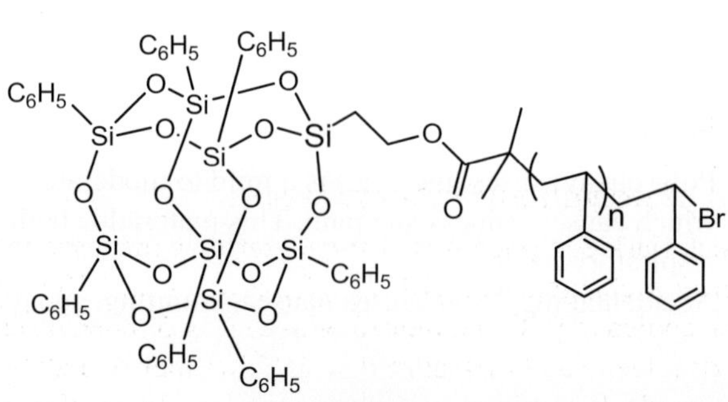

(VI)

Additive Preparation

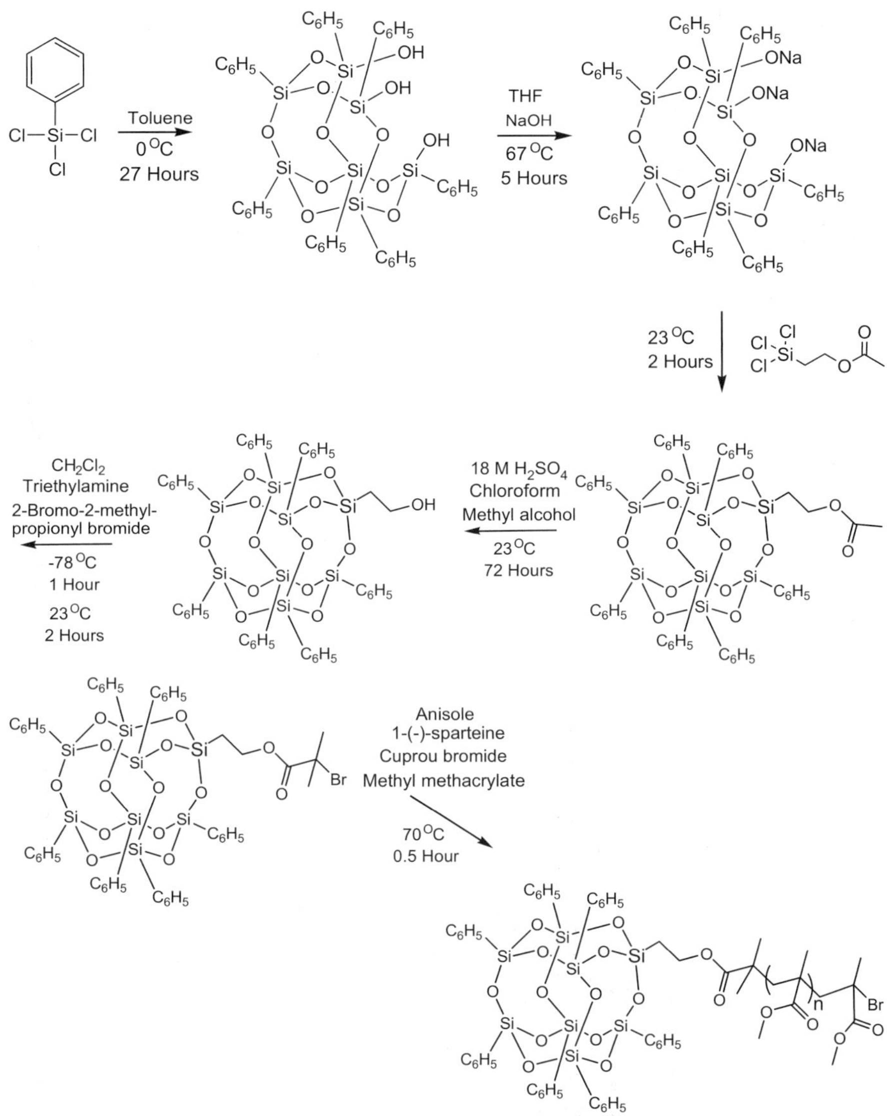

Additive Synthesis

1. Preparation of polyphenyl silsesquioxane

A 4-liter four-neck flask equipped with a stirrer, reflux condenser, thermometer, and dropping funnel was charged with 640.7 g of ice/water and 200 g of toluene. The mixture was then cooled to 0 °C and treated with 211.5g of phenyltrichlorosilane dissolved in 130 g of toluene and then stirred at 0 °C for approximately 26 h. The solution was then warmed to an ambient temperature and washed with purified water, and 120.7 g of product was isolated having a molecular weight of approximately 3000 Da.

2. Preparation of sodium phenylsilsesquioxane salt

A four-neck 500-ml flask equipped with a reflux condenser and a thermometer was charged with 12.9 g of the Step 1 product, 250 ml of anhydrous tetrahydrofuran, and 4.0 g of sodium hydroxide and then heated to 67 °C for 5 h. A precipitate that was formed was isolated by filtration and then washed twice with tetrahydrofuran. The precipitate was then dried under vacuum, and 10.1 g of the product was isolated.

3. Preparation of acetoxyethyl-heptaphenyloctasilsesquioxane

A reactor was charged with 10 g of the Step 2 product and 200 ml tetrahydrofuran and then stirred and treated with 3.3 g of acetoxyethyltrichlorosilane for 2 h at an ambient temperature, and the mixture was precipitated by pouring into 1200 ml of hexane. A solid was isolated by filtration and then redissolved in 90 g of toluene, and the organic layer was washed three times with 330 ml of water. The organic layer was then dried using magnesium sulfate, filtered, and reprecipitated by pouring into 90 g of ethanol. After drying, 6.88 g of the product was isolated.

IR (KBr, cm^{-1})	1740 (C=O); 1430 (Si–C_6H_5); 1240 (C–O); 1135–1090 (Si–C_6H_5); 1090–1000 (Si–O–Si)
^{1}H-NMR (400 MHz, TMS standard)	δ 7.82–7.72, 7.46–7.31 (m, 35H, Ph–Si); 4.32–4.28 (t,2H, –O–CH_2); 1.84 (s, 3H, CH_3–(C=O); 1.37–1.33 (t, 2H, –CH_2–Si)
^{13}C-NMR (100 MHz, TMS standard)	δ 171.15 (C=O); 134.4–134.3, 131.1–131.0, 130.2, 128.12 (Ph–C_6H_5); 60.6 (–O–CH_2–), 20.8 (CH_3–(C=O); 13.2 (–CH_2–Si)

4. Preparation of hydroxyethyl-heptaphenyloctasilsesquioxane

A 500-ml Kjeldahl flask was charged with 2.58 g of the Step 3 product containing 175 ml of methanol, 175 ml chloroform, and 0.7 ml of 18 M sulfuric acid and then stirred for 72 h at an ambient temperature. The solution was then concentrated, and the concentrate was dissolved in 500 ml of ethyl acetate. The organic solution was then washed with 500 ml of water and dried using anhydrous magnesium sulfate. After filtering and concentrating, 2.37 g of crude product was isolated as a colorless solid. The residue was then recrystallized from toluene, and 0.48 g of product was isolated as a colorless solid matter.

IR (KBr, cm^{-1})	3600–3200 (OH); 1420 (Si–C_6H_5), 1135–1090 (Si–C_6H_5), 1090–1000 (Si–O–Si)
^{1}H-NMR (400MHz, TMS standard)	δ 7.82–7.72, 7.46–7.31 (m, 35H, C_6H_5–Si); 3.85–3.87 (t,2H, –O–CH_2); 1.42–1.62 (broad, 1H, –OH), 1.26–1.31 (t,2H, Si–O–CH_2–)

^{13}C-NMR (100MHz, TMS standard)	δ 134.5–134.1, 131.1–131.0, 130.3, 128.11–127.9 (C_6H_5–Si); 58.6; (–O–CH_2H), 17.5 (Si–CH_2–)
^{29}Si-NMR (79MHz, TMS standard)	δ −67.31 (–CH_2–$SiO_{1.5}$); −78.42, −78.79 (C_6H_5–$SiO_{1.5}$)

5. *Preparation of (2-bromo-2-methylpropionyloxyethyl)-heptaphenyloctasilsesquioxane*

A 25-ml Kjeldahl flask was charged with 1.21g of the Step 4 product, 0.12 g of triethylamine, and 6.41 of methylene chloride and then cooled on a dry ice–methanol bath to maintain a solution temperature at −78 °C. The mixture was then quickly treated with 0.3 g of 2-bromo-2-methylpropionyl bromide and then stirred at −78 °C for 1 h and for 2 h at an ambient temperature. The solution was then filtered to remove triethylamine–hydrobromic acid salt, and an additional 100 ml of methylene chloride was added to the solution. The organic solution was then washed once with 300 ml of water, twice with 300 ml of 1% sodium bicarbonate solution, and finally twice with 300 ml of purified water. After drying the solution with magnesium sulfate, the mixture was concentrated to about 20 ml. The solution was then treated with 400 ml of methanol and transferred to a freezing chamber at −35 °C where the product precipitated from solution and was isolated in an 81.4% yield.

IR (KBr, cm^{-1})	1740 (C=O), 1430 (Si–C_6H_5), 1270 (C–O), 1135–1090 (Si–C_6H_5), 1090–1000 (Si–O–Si)
^{1}H-NMR (400MHz, TMS standard)	δ 7.82–7.72, 7.46–7.31 (m, 35H, C_6H_5–Si), 4.41–4.37 (t, 2H, –O–CH_2–), 1.79 (s, 6H, –C(Br) ((CH_3)$_2$), 1.43–1.39 (t, 2H, –CH_2–Si)
^{13}C-NMR (100MHz, TMS standard)	δ 171.7 (C=O), 134.3, 131.1–131.1, 131.2–130.1, 128.1–128.0 (C_6H_5–Si), 62.5 (–CH_2–O–), 55.8 (–C(Br)), 30.6 ((–CH_3)$_2$), 12.9 (Si–CH_2–).
^{29}Si-NMR (79MHz, TMS standard)	δ −68.27 (–CH_2–$SiO_{1.5}$); −78.4, −78.7 (C_6H_5–$SiO_{1.5}$)

6. *Preparation of (2-methylpropionyloxyethyl)-heptaphenyloctasilsesquioxane-polymethyl methacrylate bromide*

A heat-resistant glass ampul was charged with cuprous bromide, the Step 5 product, methyl methacrylate, l-(−)-sparteine, and anisole solution and then quickly cooled using liquid nitrogen. After freeze vacuum deaeration was carried out three times, the ampul was sealed. The sealed heat resistant ampul was set in a constant temperature shaking bath and a transfer radical polymerization carried for 30 min at 70 °C. A brown viscous solution as obtained, which was isolated by precipitation in hexane and then redissolved in tetrahydrofuran solution and passed through a column filled with activated carbon to remove residual complexed copper. The solid was then reprecipitated in hexane, and the polymer was isolated. Physical properties of (2-methylpropionyloxyethyl)-heptaphenyloctasilsesquioxane-polymethylmethacrylate bromide are provided in Table 1.

TABLE 1 Physical properties of silsesquioxane-polymethylmethacrylate prepared by phase transfer polymerization

Experiment	Reaction Time (h)	Conversion (mol%)	Polymer Mn (Da)	Polydispersity
1	1.0	6.1	3700	1.14
2	2.0	12.3	7600	1.11
3	3.2	17.8	11,400	1.09
4	3.7	22.5	13,200	1.11
5	4.0	29.1	17,000	1.11
6	5.0	28.9	17,000	1.14
7	6.5	34.9	20,700	1.15
8	9.0	45.9	24,900	1.20
9	13.0	64.1	36,500	1.23
10	18.0	75.7	45,000	1.30

Testing

Test data are not provided.

Advantages over Prior Art

The synthetic methods used to prepare silsesquioxane intermediates and polymers have limitless synthetic and application possibilities. Silsesquioxanes can be tailor-made to address a wide variety of industrial/commercial concerns while sodium alkyl-silsesquioxane can be monofunctionalized with virtually any trichlorosilane derivative. Finally, virtually any well-defined mono-, di-, and tri-block polymers anchored to a silsesquioxane core can be readily prepared using transfer radical polymerization.

Notes

1. Chen[1] prepared and polymerized octraacrylic acid silsesquioxane (VII), which was used to prepare nanostructure monolayers in memory devices

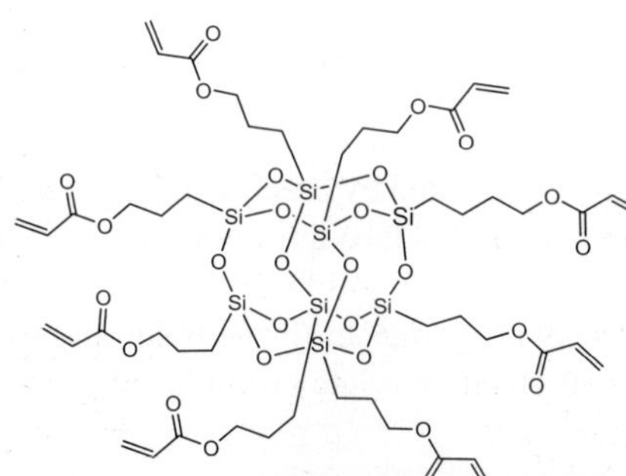

(VII)

2. Polymerized tetrafunctional silsesquioxane epoxides (VIII) prepared by Ootake[2] were used in heat resistance and in electrical insulating applications.

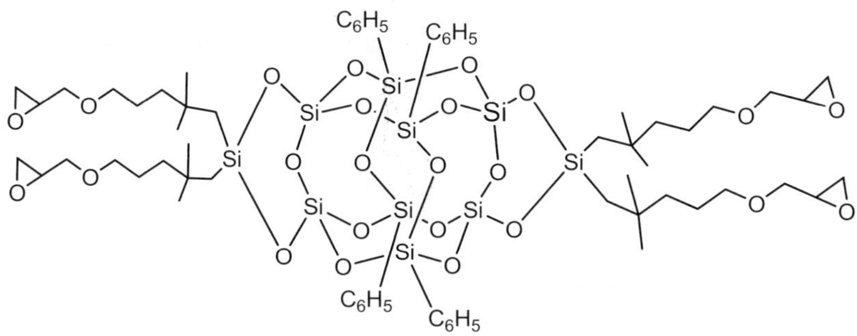

(VIII)

3. Photoconductors prepared by Wu[3] were overcoated and then crosslinked with a mixture of silsesquioxane epoxide (IX) and silsesquioxane diol (X) to minimize low background and charge deficient spots.

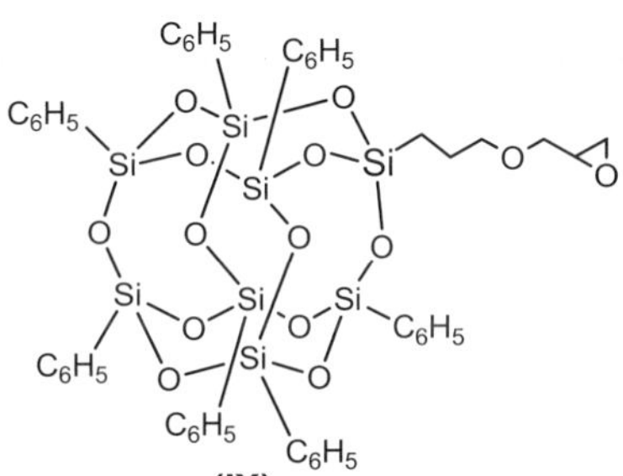

(IX)

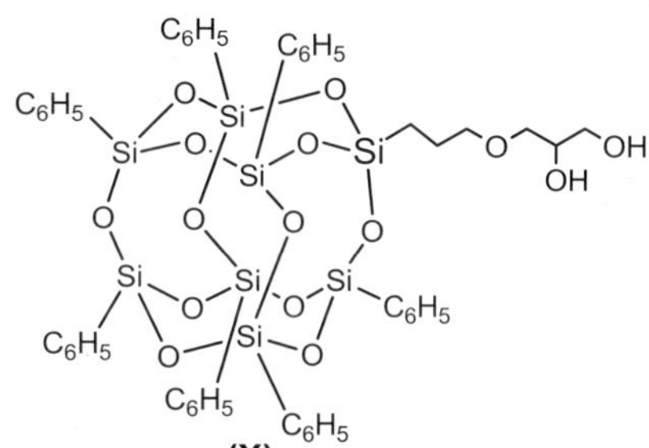

(X)

4. Ohrlein[4] prepared dye-functionalized silsesquioxane clusters (XI) for use as colorants in materials, goods, formulations, hair, nails, and skin.

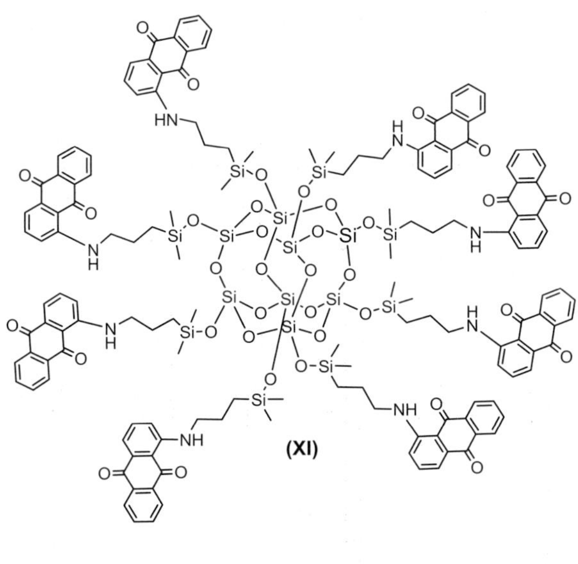

(XI)

REFERENCES

1. Jian Chen et al, US Patent 7,968,273 (June 28, 2011)
   Nanosys, Inc. (Palo Alto, CA)
2. Nobumasa Ootake et al, US Patent 7,939,617 (May 10, 2011)
   Chisso Corporation (Osaka, JP)
3. Jin Wu, et al, US Patent 7,897,314 (March 1, 2011)
   Xerox Corporation (Norwalk, CT)
4. Reinhold Ohrlein et al, US Patent 7,947,848 (May 24, 2011)
   BASF SE (Ludwigshafen, DE)

Chapter 9

Cosmetics

A. BODY PERSPIRATION TREATMENT

Cyclohexylethyl Carboxylic Acid Esters

Title:	Cyclohexylethyl carboxylic acid ester compositions and method for using the same for reducing malodors
US Patent:	(August 9, 2011)
Author:	Richard K. Payne et al.
Assignee:	International Flavors & Fragrances Inc. (New York, NY)

Significance: (1R)- and (1S)-1-Cyclohexyl acetate and (1R)- and (1S)-1-cyclohexylethanol were prepared and found to be effective in eliminating malodors by binding to sensory receptors. In this way, receptors became inaccessible to malodor-causing chemical agents, particularly those associated with body perspiration, mildew, and mold.

Additive

(1R)-1-Cyclohexylethyl butyrate (I)
(1S)-1-Cyclohexylethyl butyrate (II)
(1S)-1-Cyclohexylethyl acetate (III)
(1R)-1-Cyclohexylethyl acetate (IV)
(1R)-1-Cyclohexanol (V)
(1S)-1-Cyclohexanol (VI)

Safety

Cyclohexanol as well as acetate and butyrate esters are moderate to severe skin and eye irritants. Typical TLV is 200 mg/m^3 (50 ppm). In addition, they may cause irritation if ingested and respiratory tract irritation if chronically inhaled.

Next Generation of International Chemical Additives. DOI: http://dx.doi.org/10.1016/B978-0-444-53788-1.00009-7

Additive Structures

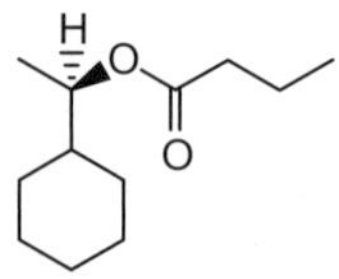

(I)

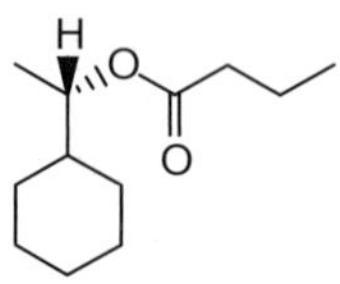

(II)

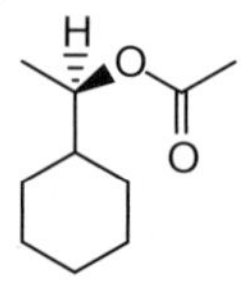

(III)

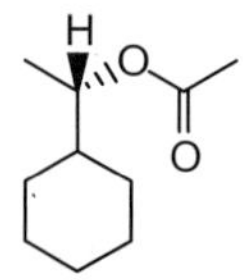

(IV)

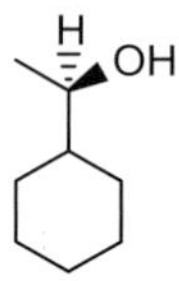

(V)

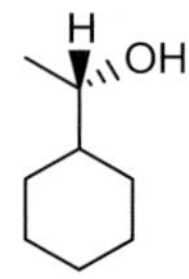

(VI)

Additive Preparation

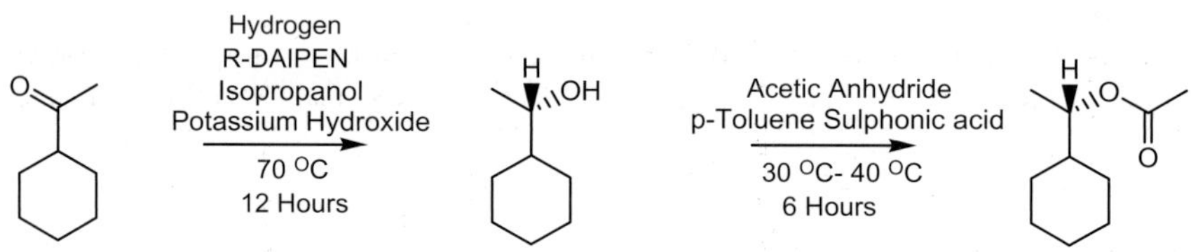

Additive Synthesis

1. Preparation of (R)-α-methylbenzyl alcohol

An autoclave was charged with 1.03 kg of acetophenone, 300 g of isopropanol, and 100 g of KOH and then treated with 150 mg R-DAIPEN and hydrogenated at 30 bar at 70 °C for 12 h. The mixture was then transferred into a separatory funnel, and the aqueous layer was discarded. The organic layer was washed with 200 ml of water and then distilled first at 100 mmHg to remove isopropanol and then at 5–10 mmHg to yield 980 g of (R)-α-methylbenzyl alcohol.

2. Preparation of (R)-1-cyclohexylethyl acetate

A 2-liter reaction flask was charged with 980 g of the Step 1 product, 1 kilogram of acetic anhydride, and 0.4 g of *p*-toluene sulfonic acid and then heated from between 30 °C and 40 °C for 6 h. The mixture was then transferred into a separatory funnel and washed twice with 200 ml of water and then once with 200 ml of 10% sodium hydroxide. The organic mixture was distilled at 5–10 mmHg and 1.3 kg of the product was isolated having an optical rotation $[\alpha]_D = +2.59^{O}$ with an optical purity of 80.5% ee.

Testing

Malodorous odor testing

In the testing procedure, 1 g of a selected experimental agent was placed into a 1-gallon jar containing isovaleric acid and then sealed by securely capping for 24 h. Internal panelists consisting of trained females having a mean age of 40 years were instructed to (a) uncap the jar; (b) place their noses immediately above the opening; (c) take short sniffs for 2–3 s; (d) recap the jar; and (e) repeat steps (a) through (d) steps after 30 s. Each panelist then rated the intensity of the odor using the intensity scales 1–5, where 1 represents the "Least Intense Malodor" and 5 represents the "Most Intense Maloror." Test results are provided in Table 1.

Test Results

TABLE 1 Effectiveness of (1R,1S)-,(1R)-, and (1S)-1-cyclohexylethyl butyrate, Additive 1, countering the malodor from isovaleric acid. Lower numbers are preferred

Malodor Concentration	Additive 1 R,S (5%	Additive 1 S (5%)	Additive 1 R (2%	Additive 1 R,S (2%)	Additive 1 R,S (1%)	Additive 1 S (1%)	Additive 1 R (1%)
1%	2.50	2.08	1.42	2.15	–	–	–
2%	–	–	–	–	2.46	1.92	1.50

Advantages over Prior Art

Most methods designed to counter offensive odors use perfumes and fragrances that blend with the malodor to generate a more desirable odor. This method, however, requires large quantities of fragrance material to overwhelm the malodorous compounds. The current method is unique and distinguished from other methods since the chemical additives are designed to bind directly with sensory receptors. This association renders receptors inaccessible to malodorous compounds thereby eliminating the perceived malodor. Finally cyclohexyl ester treatment levels to eliminate malodors appear to be extremely low.

Notes

1. Fujii[1] prepared a four-component mixture consisting of 2-amino-2-hydroxymethyl-1,3-propanediol (I), polyoxyethylene containing eight oxyethylene groups (II), 2-ethylhexyl monoglyceryl ether (III), and lauryldimethylamine oxide (V), which was effective against malodors from perspiration.

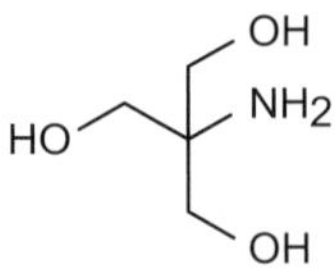

(I)

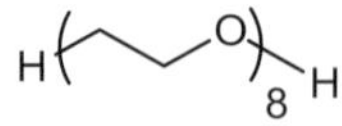

(II)

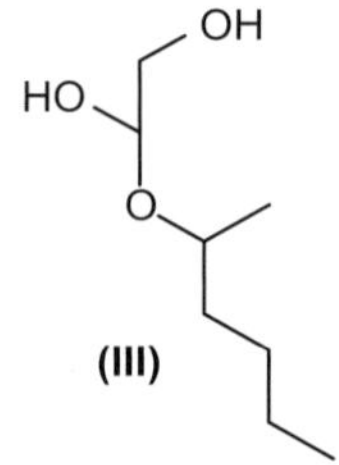

(III)

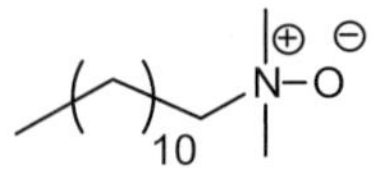

(IV)

2. Rayms-Keller[2] empirically determined that filters containing a mixture of 3-hydroxy-3-methylhexanoic acid (V) and 2-hydroxy-3-methylhexanoic acid (VII) could be used to detect the presence of human beings by monitoring fatty acid metabolism caused by bacteria and volatile organic compound malodors.

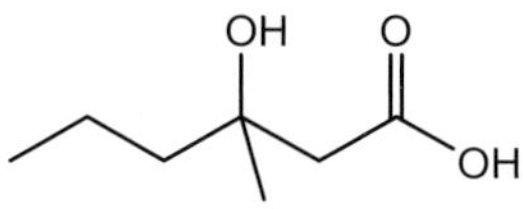

(V)

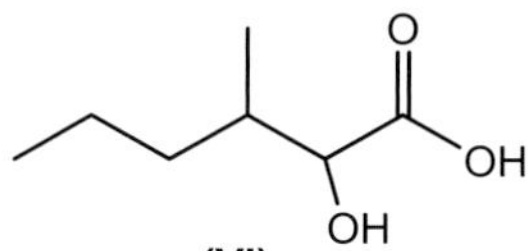

(VI)

3. Fiber material having improved malodor scavenger properties was prepared by Wood[3] by blending a mixture of zinc nanoparticles and triacetyl α-cyclodextrin into nonwoven mats.
4. Man[4] eliminated malodors from dirty or contaminated surfaces by washing with an alkaline mixture of octanoic acid (VII) and octanoic peracid (VIII) dissolved in poly(ethylene oxide).

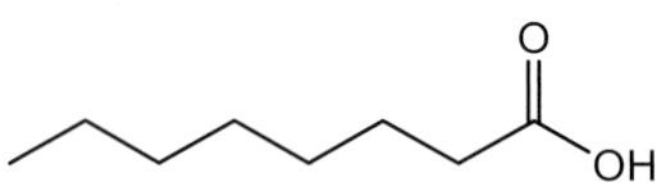

(VII)

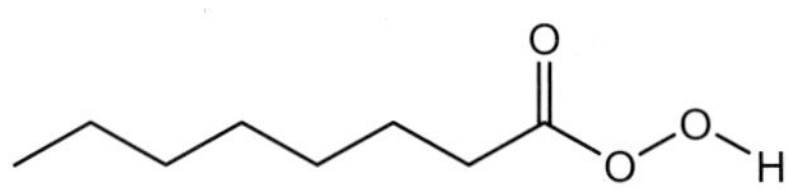

(VIII)

REFERENCES

1. Motoko Fujii et al, US. Patent 8,007,545 (August 30, 2011)
   Kao Corporation (Tokyo, JP)
2. Alfredo Rayms-Keller, US. Patent H2,256 (June 7, 2011)
   United States of America as represented by the Secretary of the Navy (Washington, DC)
3. Will Wood et al, US Patent 7,931,9578 (April 26, 2011)
   Cellresin Technologies, LLC (Minneapolis, MN)
4. Victor Fuk-Pong Man et al, US Patent 7,887,641 (February 15, 2011)
   Ecolab USA Inc. (St. Paul, MN)

B. BODY PERSPIRATION MALODOR TREATMENT

4,4′-Bis(dimethylamino)benzhydrol

Title:	Odor controlling article including a visual indicating device for monitoring odor absorption
US Patent:	7,837,663 (November 23, 2010)
Author:	John G. MacDonald et al.
Assignee:	Kimberly-Clark Worldwide, Inc. (Neenah, WI)
Significance:	Silica or alumina nanoparticles containing zinc chloride covered with 4,4′-bis(dimethylamino)benzhydrol have been found to be effective in both eliminating malodorous vapors and providing the user with a visual indicator of their presence. Malodorous odors include diallylthiosulfonate from garlic, hydrogen, and ethyl sulfide and acetaldehyde.
Additive Name	4,4′-Bis(dimethylamino)benzhydrol, [Michler's hydrol] (I)
Safety Concerns	Blenders are encouraged to wear gloves and face mask since the material is irritating to the eyes, skin, and respiratory system.

Additive Structure

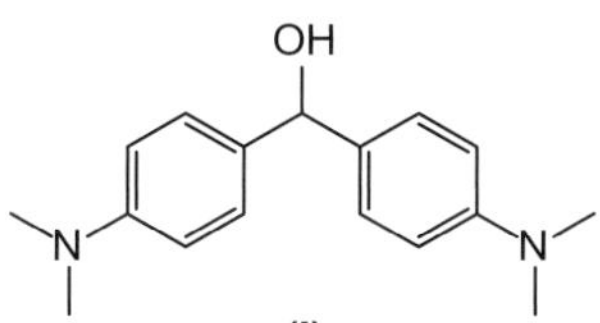

(I)

Additive Sources

4,4′-Bis(dimethylamino)benzhydrol can be purchased from suppliers provided below.

Sigma-Aldrich Chemical Company
PO Box 14508 10777
St. Louis, MO 63178
800-325-3010

Nissan Chemical America Corporation
Westheimer, Suite 830
Houston, TX 77042
713-532-4745

Additive Testing

Odor Reduction Testing

A 1–10mg/ml solution of 4,4′-bis(dimethylamino)benzhydrol was prepared by dissolving the additive in water, ethyl alcohol, or acetonitrile and then applying to cellulose, nonwoven cotton, or glass fiber by dipping or spraying or by inkjet printing. Selected samples of 4,4′-bis(dimethylamino)benzhydrol were also blended with silica nanoparticles covered with $ZnCl_2$ to determine the potential synergistic effects. In all cases, however, excess solution was removed from the substrate, and the material was air or oven dried.

Odor-reducing properties of dried materials were determined using headspace gas by chromatographic testing with an Agilent 5890 Series II gas chromatograph with an Agilent 7694 headspace sampler. In this procedure, a 0.005–0.006 g of sample of the coated material was placed into in a 20-ml vial and treated with ethyl mercaptan. Thereafter, a 1-ml sample vial headspace aliquot was removed and placed in a gas chromatography oven at 37 °C for 10 min for evaluation. Test results are provided in Table 2.

Test Results

TABLE 2 Effectiveness of 4,4′-bis(dimethylamino)benzhydrol as an odor-absorbing agent for ethyl mercaptan as indicated by gas chromatograph using a 1-ml headspace. Silica nanoparticles covered with $ZnCl_2$ and then additized with 4,4′-bis(dimethylamino)-benzhydrol demonstrated moderate synergistic effects

Concentration of 4,4′-Bis(dimethylamino) benzhydrol (mg/ml)	Area of Ethyl Mercaptan Peak	Odor Reduction (%)
Control	1 061 270	0
$ZnCl_2$ (Silica nanoparticles)	870 978	17.9
0.5	712 649	32.9
2.0	667 661	37.1
0.5 + $ZnCl_2$ (Silica nanoparticles)	638 016	39.9

Advantages over Prior Art

Two approaches have been used to address the presence of malodorous vapors. The first method uses a variety of perfumes to mask the offensive odor. While effective, some fragrances can be offensive or irritating to some, however, The current method uses materials that behave as both odor-absorbing agents and odor indicators at very low treatment levels.

Notes

1. 4,4′-Bis(dimethylamino)benzhydrol indicates the presence of urine (amines) or sulfur-containing compounds by initially changing to a blue color, and then by absorbing the chemical agent and becoming colorless (III), as indicated in Eq. 1.

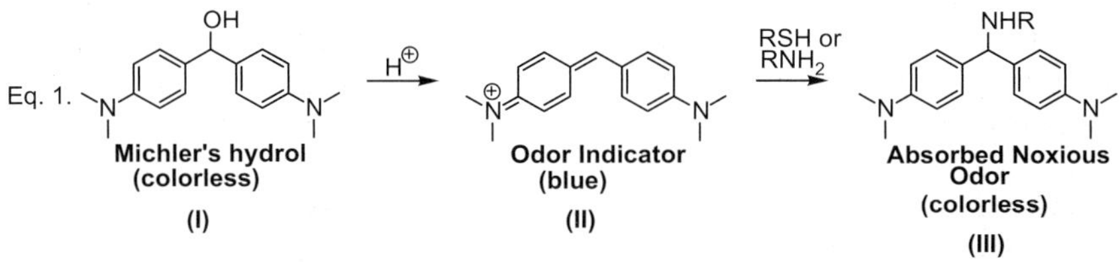

2. In an earlier investigation by the author,[1] 4,4′-bis(dimethylamino)benzhydrol was used as a visual detector to detect bad breath caused by diallyl thiosulfinate vapor (allicin) contained in garlic.
3. McGrath[2] prepared a series of $CuCl_2$ coordination complexes containing carminic acid, Acid Blue 45, Alizarin Red S, Acid Blue 40 (iV), or Acid Green 41 blended with 5–20 μm of silica coated with alumina having a surface area of about 1000 m^2/g, which became colored after it absorbed the malodor of diallyl thiosulfinate vapor.

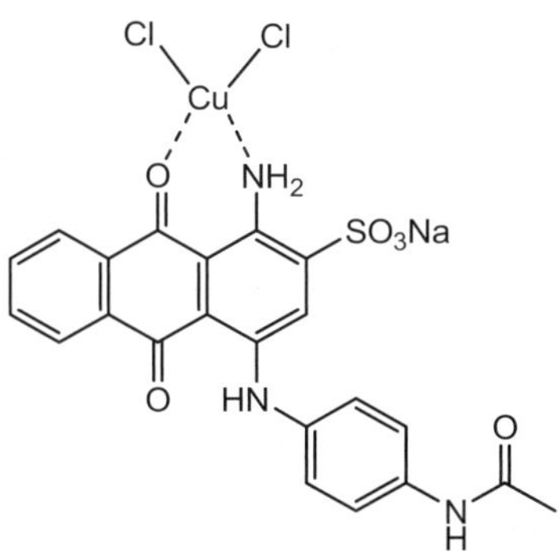

Acid Blue 40-$CuCl_2$ Coordination Complex

(IV)

4. Boga[3] prepared a visual detector and an odor eliminator associated with vaginitis consisting of α-naphtholbenzein (VI) or pararosanilin carbinol (VI), blended with cellulose materials.

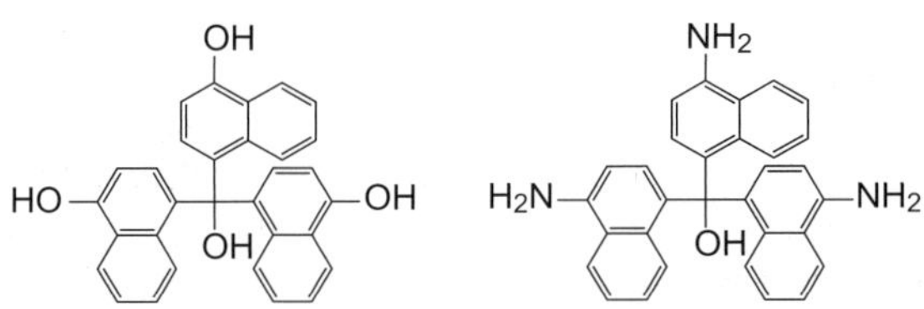

α-Naphtholbenzein (V) Pararosanilin Carbinol (VI)

5. Urlaub[4] prepared nanoparticles having a negative Zeta Potential consisting of silica or alumina with a diameter of <500 nm, which had a surface area of 500 m^2/g with at least 50% of the surface modified with $CuCl_2$ and where the modified silica was effective in adsorbing ammonia from urine, acetaldehyde, and *n*-butyric acid.

REFERENCES

1. John G. MacDonald et al, US Patent 7,413,550 (April 19, 2008) **Kimberly-Clark Worldwide, Inc. (Neenah, WI)**
2. Kevin McGrath et al, US Patent 7,678,367 (March 16, 2010) **Kimberly-Clark Worldwide, Inc. (Neenah, WI)**
3. RameshBabu Boga et al, US Patent 7,592,020 (September 22, 2009) **Kimberly-Clark Worldwide, Inc. (Neenah, WI)**
4. John Jerald Urlaub et al, US Patent 7,488,520 (February 10, 2009) **Kimberly-Clark Worldwide, Inc. (Neenah, WI)**

C. DETERGENT POWDER TO ENHANCER FABRIC FRAGRANCE AND SHAMPOO

cis/trans 4-Isoamylcyclohexanol

Title:	4-Isoamylcyclohexanol as an odiferant
US Patent:	7,884,065 (February 8, 2011)
Author:	Aurelia Reckziegel et al.
Assignee:	Symrise AG (Holzminden, DE)

Significance: A commercially viable method for preparing a 1:3 mixture of cis- and trans 4-isoamylcyclohexanol, respectively, is described. When used to additize clothing detergent powder, fabric softener, or shampoo, it imparts a lasting pleasant roselike with geranium notes fragrance to fabrics and hair.

Additive Names

cis-4-Isoamylcyclohexanol (I)
trans-4-Isoamylcyclohexanol (II)

Safety Concerns

Although data for either of these materials are unavailable, no negative toxicity of C_2–C_6-alkylcyclohexanols or 4-*t*-amylcyclohexanol in personal hygiene consumer products containing these odoriferous substances has been reported.

Additive Structure

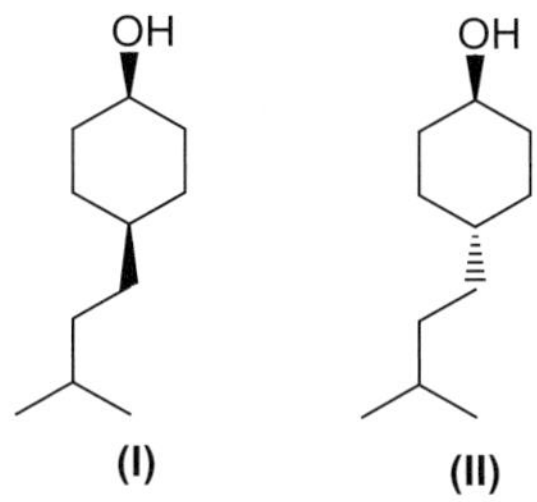

Additive Preparation

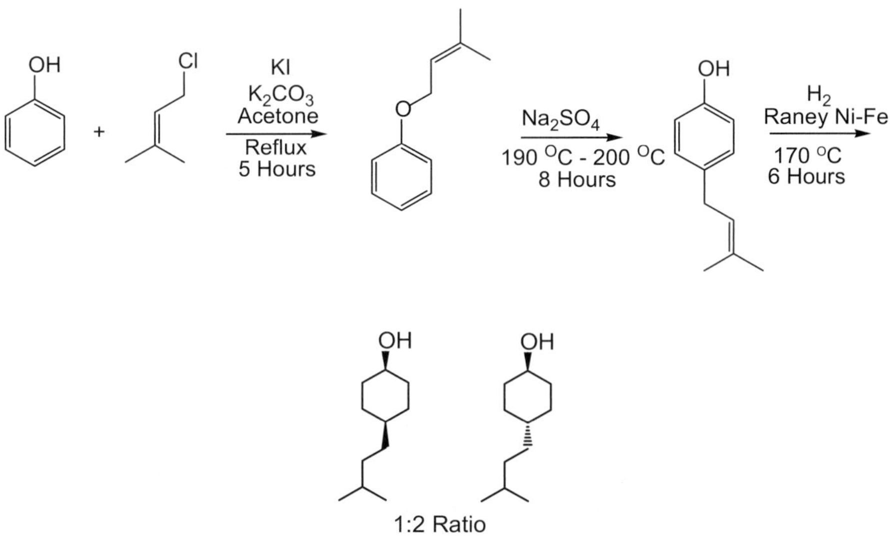

Additive Synthesis

1. Preparation of phenyl prenyl ether

A reactor was charged with 391 g of phenol, 532 g of potassium carbonate, and 45 g of potassium iodide dissolved in 1000 ml of acetone and then treated with 478 g of prenyl chloride at an ambient temperature. The mixture was then refluxed for 5 h and then cooled and treated with water. The aqueous phase was extracted thrice with diethyl ether, and the combined organic phases were washed with sodium chloride solution and then dried over Na_2SO_4. After the mixture was filtered and concentrated, 444 g of the product was isolated by distillation.

2. Preparation of 4-isoamylphenol

A reactor charged with 837 g of the Step 1 product and 453 g of anhydrous sodium acetate containing an overhead stirrer under a nitrogen atmosphere was heated to 190–200 °C 8 h. The mixture was then cooled to an ambient temperature and heated with water. The phases were then separated, and the aqueous phase was extracted thrice with 500 ml of diethyl ether. The combined organic phases were dried over Na_2SO_4, filtered, and concentrated. After distillation, 413 g of the product was isolated.

3. Preparation of cis/trans-4-isoamylcyclohexanol

A hydrogenation autoclave was charged with 165 g of the Step 2 product and 2 wt% of Raney nickel consisting of 45% nickel, 3% aluminum, 8% iron with a 44% water content, and the hydrogenation conducted at a hydrogen pressure of 20 bar for 6 h at 170 °C. The reaction mixture was then cooled, filtered, and 158 g of 4-isoamylcyclohexanol was isolated consisting of 33.1% cis-4-isoamylcyclohexanol and 66.3% of trans-4-isoamylcyclohexanol.

Testing

1. Powder Detergent Additive

An 80 g detergent sample was treated with 1 g of a 50% strength mixture of isopropyl myristate consisting of either cis/trans 4-isoamylcyclohexanol mixture or citronellol. Two sets of cotton hand towels were then blended with the fabrics and then washed separately with 80 g portions of washing powder at 40 °C in Miele washing machines. The materials were then dried and evaluated by testers. Test results are summarized below.

2. Fabric Softener Additive

A 40-g sample of a fabric softener was mixed thoroughly with 0.5 g of a 50% strength mixture of isopropyl myristate with either a cis/trans 4-isoamylcyclohexanol mixture or citronellol. In each case, three hand towels of cotton and three hand towels of blended fabric were washed in water at 40 °C in Miele washing machines first with 80 g of a nonperfumed standard washing powder and then separately with a fabric softener. Test results are summarized below.

3. Shampoo

A 30 g of shampoo was mixed thoroughly with 0.3 g of a 50% strength solution of isopropyl myristate with either cis/trans mixture 4-isoamylcyclohexanol or citronellol. In each case, 20 g of hair was washed separately by hand for 1 min with 1 g of shampoo that foamed in 2 g of water. Each shampooed hair was then separately rinsed under water at 30 °C for 30 s. Test results are summarized below.

Test Results

1. Powder Detergent Additive

Hand towels evaluated by a group of 16 experts gave a significantly higher fragrance for the 4-isoamylcyclohexanol cis/trans mixture in a blindfold evaluation with a significant difference ($p<0.05$) than when using citronellol.

2. *Fabric Softener Additive*

The olfactory evaluation of hand towels blow dried with the fabric softener additized with the cis/trans mixture of 4-isoamylcyclo-hexanol had a significantly higher fragrance ($p<0.05$) than towels dried with fabric softener containing citronellol.

Shampoo Additive

The olfactory evaluation of hair washed with shampoo additized with the cis/trans mixture of 4-isoamylcyclohexanol was perceived to be more intense on damp and dry hair and preferred by experts for its olfactory properties ($p<0.01$) over citronellol.

Advantages over Prior Art

Perfumes having a pleasant roselike with geranium notes fragrance are novel. In addition, the cis/trans mixture appears sufficiently robust to be used in a wide variety of consumer-oriented products. Finally, the synthetic method used to prepare the agent is readily amenable to commercialization.

Notes

1. Dirk[1] prepared the odoriferous agent, 3-methylbenzylisobutyrate (III), which when mixed with methyl dihydrojasmonate, diethyl phthalate, indole, geraniol, and l-citronellol had a pleasant Lily of the Valley flowery freshness.

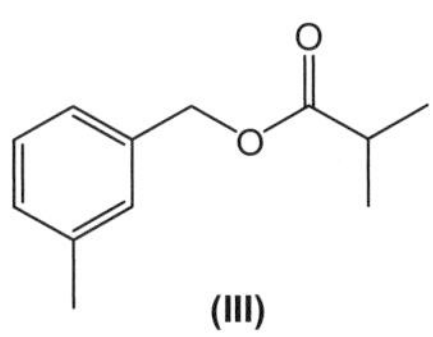

(III)

2. Hoelscher[2] prepared furanoid (IV) and pyranoid (V), C_{14}—C_{18}-oxabicycloalkanones as odoriferous and/or aroma substances suitable as food additives to enhance taste.

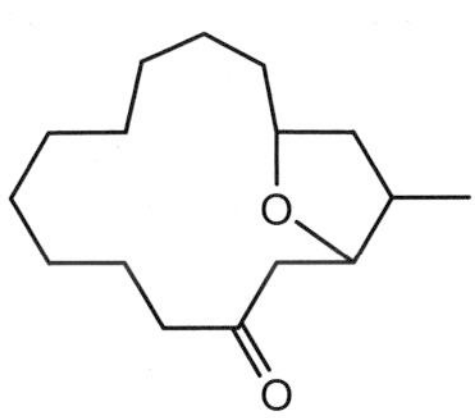

(IV)

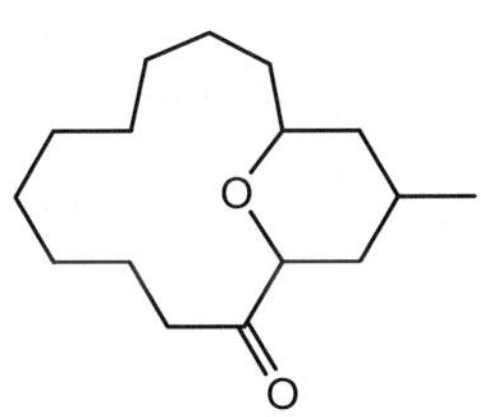

(V)

3. Granier[3] prepared a series of cyclooctene derivatives (VI) having an armoise, wormwood, thujone, natural fragrance that were used in personal hygiene formulations.

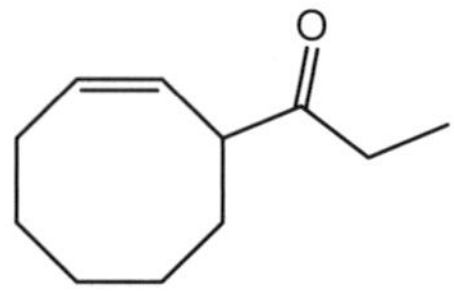

(VI)

4. By condensing campholenaldehyde with two equivalents of 3-methyl-3-buten-1-ol in toluene containing a catalytic amount of *p*-toluene sulfonic acid, Plessis[4] prepared a three-component pyran mixture (VII) through (IX) having a pleasant orange, woody, a bit dusty/musky fragrance.

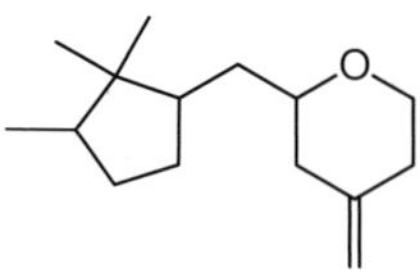

(VII)

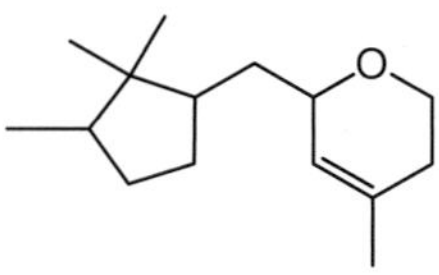

(VIII)

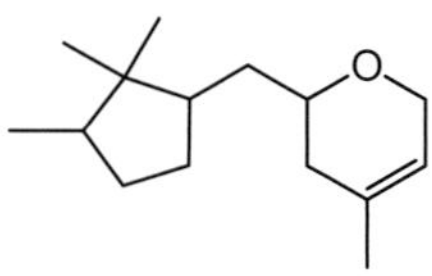

(IX)

REFERENCES

1. Erich Dilk et al, US Patent 7,638,479 (December 29, 2009)
   Symrise GmbH & Co. KG (Holzminden, DE)
2. Bernd Hoelscher, US Patent 7,615,526 (November 10, 2009)
   Symrise GmbH & Co. KG (Holzminden, DE)
3. Thierry Granier et al, US Patent 7,888,309 (February 15, 2011)
   Givaudan SA (Vernier CH, CH)
4. Caroline Plessis et al, US Patent 7,875,737 (January 25, 2011)
   V. Mane Fils (Bar-sur-Loup, FR)

C. FRAGRANCE DELIVERY DEVICE

1-(4-Hydroxyl-3,5-di-*t*-butylphenyl)-2-iodoethanone

Title:	Photolatent systems
US Patent:	7,820,865 (October 26, 2010)
Author:	Katharina Fritzsche; et al
Assignee:	Ciba Corporation (Tarrytown, NY)

Significance: A fragrance delivery system has been prepared by attaching perfume to 1-(4-hydroxyl-3,5-dialkyl)-2-iodoethanone. When this composition was exposed to sunlight, the composition cleaves apart from releasing the perfume and generating the nontoxic photodecomposition product, 4-hydroxyl-3,5-dialkylacetic acid.

Additive Name 1-(4-Hydroxyl-3,5-di-*t*-butylphenyl)-2-iodoethanone (I)
1-(4-Hydroxyl-3,5-di-cyclohexylphenyl)-2-iodoethanone (II)
1-(4-Hydroxyl-3,5-di-*t*-isipropylphenyl)-2-iodoethanone (III)

Safety Concerns 1-(4-Hydroxy-3,5- and 2,4-dialkyl)-haloethanone derivatives are primary irritants

Additive Name 1-(4-Hydroxyl-3,5-di-*t*-butylphenyl)-2-iodopethanone

Safety Concerns Eye and skin irritant. Dust/vapors may cause respiratory tract and mucous membrane irritation.

Additive Structures

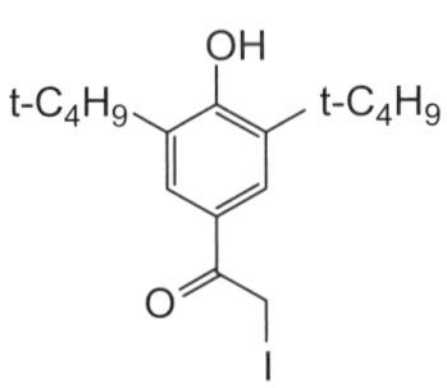

(I)

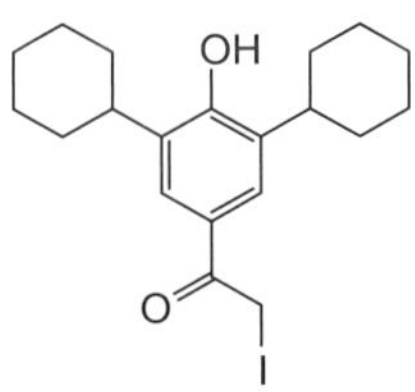

(II)

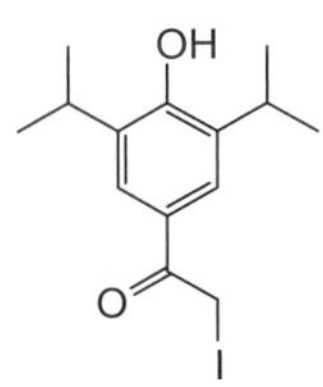

(III)

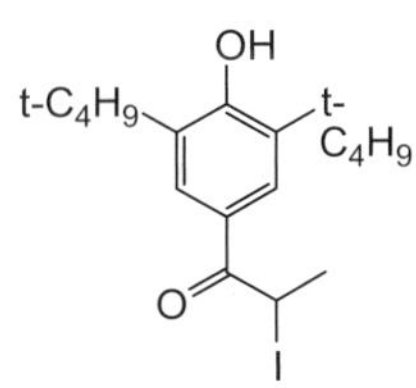

(IV)

Additive Preparations

Case I. Preparation of fragrance delivery agent

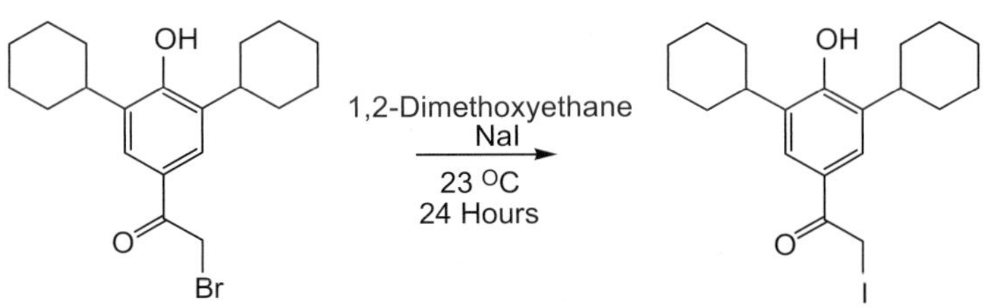

Case II. Preparation of fragrance delivery system using fragrance of vanilla and 1-(4-hydroxyl-3,5-di-*t*-butylphenyl)-2-iodoethanone as the delivery agent

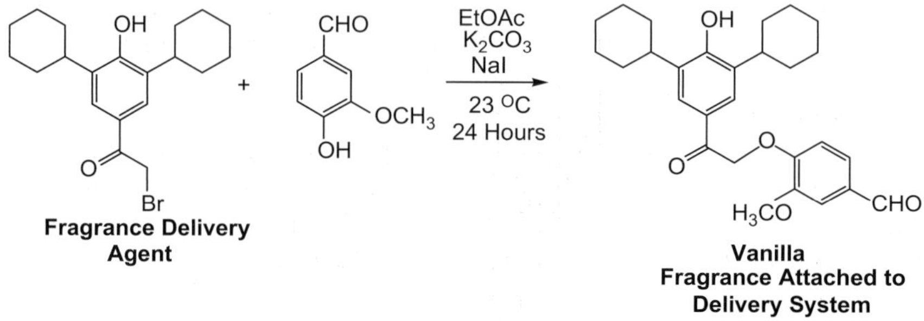

ADDITIVE SYNTHESIS

Case 1

1. Preparation of 1-(4-hydroxy-3,5-di-cyclohexyphenyl)-2-iodoethanone

A reactor was charged with 19.47 g of 1-(4-hydroxy-3,5-di-cyclohexylphenyl)-2-bromoethanone dissolved in 75 ml of 1,2-dimethoxyethane and then treated with the dropwise addition of 7.87 g of sodium iodide dissolved in 75 ml of 1,2-dimethoxyethane at an ambient temperature. The mixture was then stirred for 16 h at an ambient temperature and then treated with 25 ml of hexane and stirred for an additional 15 min. Inorganic precipitates were removed by filtration, and the solution was concentrated. The residue was purified by recrystallization from hexane, and the product was isolated in a 57% yield and had an mp = 183-185 °C.

Case 2

2. Preparation of 4-[2-(3,5-di-t-butyl-4-hydroxyphenyl)-2-oxo-ethoxy]-3-methoxybenzaldehyde by the in situ generation of 1-(4-hydroxy-3,5-t-butylphenyl)-2-iodoethanone

A mixture of 19.4 g of 4-hydroxy-3-methoxybenzaldehyde, 1.88 g of sodium iodide, and 34.5 g of potassium carbonate dissolved in 1000 ml of ethyl acetate was stirred for 30 min and then treated with 43.6 g of 1-(4-hydroxy-3,5-di-*t*-butylphenyl)-2-bromoethanone. The reaction mixture continued to be stirred at an ambient temperature for an additional 16 h and was then acidified with 2 M HCl and extracted with ethyl acetate. The extracts were combined, washed with water, dried, and evaporated. Crystallization of the crude product from dichloromethane/hexane provided 36.85 g of the target compound having an mp = 99–100 °C.

Additive Formulation

1. Photochemical Release of Vanilla in Polypropylene

A mixture consisting of 0.50 g of Irganox® B 215, two parts of Irgafos®, and 30 g of the Step 2 product was blended in a turbo mixer with 969.5 g of polypropylene powder and then extruded through a twin screw extruder at 160–170 °C producing tapes with a width of 20 mm and a thickness of 1 mm. The tapes were then cut into pieces with a length of about 50 mm and irradiated on a PPG UV-curing unit using a 2 Hg-arc bulbs at 80 W/cm with a line speed of 20 m/min where the irradiation process was repeated many times. The release of vanilla from the samples was then determined by smelling the surface, and the strength of the vanilla smell is summarized in Table 3.

2. Irradiation with UV-Fluorescence

The Step 2 product was placed under an artificial daylight lamp according to the DIN 6173 testing protocol and irradiated 5 days/week for 4 weeks with an interruption of 2 days between weeks. At various intervals, the release of vanilla on the strips was detected by smelling the surface. Test results are provided below.

Additive Test Results

1. Photochemical Release of Vanilla in Polypropylene

TABLE 3 Detection of vanillin from the photochemical release of the Step 2 product, 4-[2-(3,5-di-*t*-butyl-4-hydroxyphenyl)-2-oxo-ethoxy]-3-methoxybenzaldehyde

Number of Irradiation Passes Under the Mercury Lamp	Vanilla Smell
0	Very Slight
2	Very Slight
5	Weak
10	Strong

2. Irradiation with UV-Fluorescence

The Step 2 product was placed under an artificial daylight lamp according to the DIN 6173 protocol and then irradiated for 5 days/week for 4 weeks with an interruption of 2 days between weeks. In intervals, the release of vanilla on the strips was detected by smelling the surface.

A mild fragrance of vanillin was detected after the Step 2 product, 4-[2-(3,5-di-*t*-butyl-4-hydroxyphenyl)-2-oxo-ethoxy]-3-methoxybenzaldehyde, had been irradiated with either UV-fluorescence or natural daylight.

Advantages over Prior Art

A photolatent fragrance delivery system was previously prepared using 2-nitrobenzyl derivatives. Inherent in this process was the photogeneration of potentially toxic and yellow staining 2-nitrosobenzaldehyde derivatives. The additive of the current invention does not suffer from these limitations. In addition, since the photodegradation products of the current invention absorb at considerably shorter wavelengths, higher overall fragrance delivery from the photochemical intermediate occurs.

Notes

1. Corzani[1] prepared a fragrance delivery system consisting of a mixture of 50% poly(tetramethylene glycol) having an Mn of approximately 2000 Da with 25 wt% of a rosin ester plasticizer (Foralyn® 5020F) and 25 wt% benzyl acetate with a selected perfume.
2. Yang[2] prepared a fragrance delivery system consisting of 20% limonene and 3% γ-terpinene compounded with an ionic stabilizer, dihydropropyltrimonium chloride (Glycerol Quat™ (V), fatty esters, and a silicon fluid.

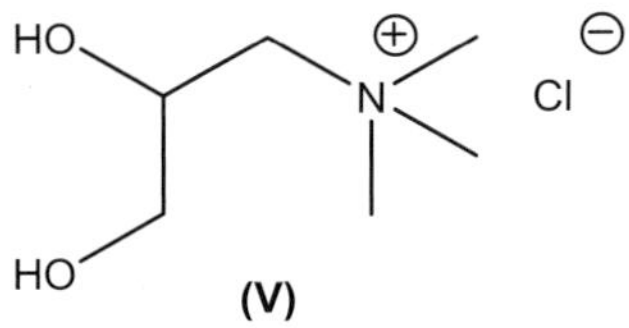

(V)

3. Polymeric nanoparticles consisting of polystyrene crosslinked with divinylbenzene having a Tg > 50 °C and containing an olfactive component were prepared by Quellet[3] and used in cosmetic applications.
4. A fragrance delivery system consisting of a nanostructured microcapsule shell consisting of urea–formaldehyde or melamine–formaldehyde and containing an olfactive component was prepared by Lei[4] and had at least a 6-month shelf life when stored at 37 °C.

REFERENCES

1. Italo Corzani et al, US Patent 7,795,476 (September 14, 2010)
The Procter & Gamble Company (Cincinnati, OH)
2. Lin Yang et al, US Patent 7,794,741 (September 14, 2010)
Conopco, Inc. (Englewood Cliffs, NJ)
3. Christian Quellet et al, US Patent 7,776,939 (August 17, 2010)
Givauden SA (Vemler, CH)
4. Yabin Lei et al, US Patent 7,833,960 (November 16, 2010)
International Flavors & Fragrances Inc. (New York, NY)

D. HAIR BLEACH ACTIVATORS FOR BOTH DYEING AND LIGHTENING HAIR

Cationic Acyl Ionic Liquids

Title:	Cationic acylpyridinium derivatives for use as bleach activators
US Patent:	8,021,438 (September 20, 2011)
Author:	Wibke Gross et al.
Assignee:	Henkel AG & Co. KGaA (Duesseldorf, DE)

Significance: The ionic salt of 4-acetyl-1-methylpyridinium *p*-toluene sulfonate was prepared in a high yield by condensing 4-acetylpyridine and *p*-toluene sulfonic acid methyl ester. When used in hair dyeing formulations, hair lightening and dyeing were simultaneously improved.

Additive Name 4-Acetyl-1-methylpyridinium *p*-toluene sulfonate (I)

Safety Although very limited safety information is available for this agent, in high exposure levels, it is mildly irritating to the respiratory system and skin.

Additive Structure

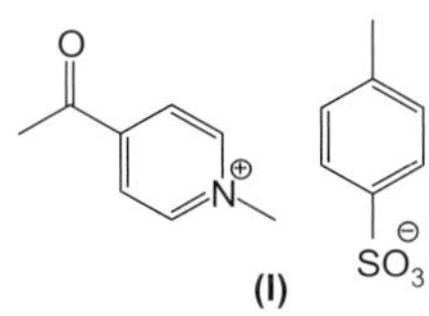

Additive Preparation

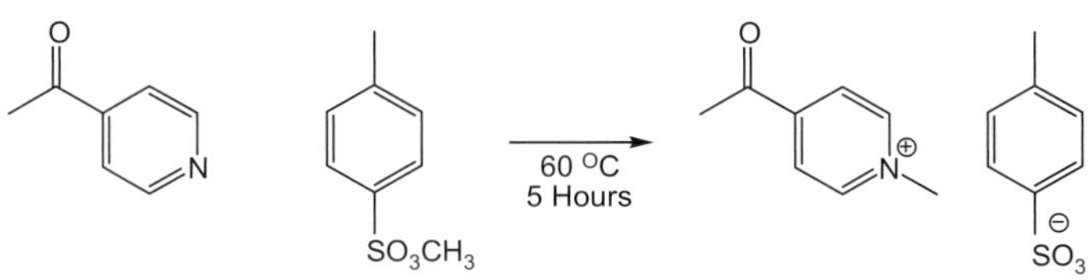

Additive Synthesis

1. *Preparation of 4-acetyl-1-methylpyridinium p-toluene sulfonate*

A glass reactor was charged with 30.0 g of 4-acetylpyridine, 55.8 g of *p*-toluene sulfonic acid methyl ester, and 500 ml of ethanol and then heated to reflux for 5 h. The solvent was removed under vacuum in a rotary evaporator, and the residue was dissolved in diethyl ether. After removing the diethyl ether phase, the product slowly crystallized, and the product was isolated in an 82.5% yield after drying.

^{1}H-NMR (400MHz, d_6-DMSO δ), 2.26 (s, 3H), 2.72 (s, 3H), 3.39 (s, 3H), 7.11 (d, 2H), 7.49 (d, 2H), 8.42 (d, 2H), 9.20 (d, 2H)

^{13}C-NMR (400MHz, d_6-DMSO δ), 20.8, 26.4; 48.1, 124.8, 125.3, 127.7, 138.9, 145.2, 146.5, 148.3;195.8.

Testing

Evaluation of the Lightening and Dyeing Power

Each strand of hair was measured colorimetrically before and after bleaching, and an L-value, a-value, and b-value were determined for each measurement. Dyeing power was determined from the Δ-E-value calculated from L-, a-, and b-values.

Lightening and dyeing properties of the experimental additive were determined by initially blending the additive into a into a blonding cream shown in Table 4 and then mixing it with a 1:1wt/wt ratio with the formulated developer dispersion given in Table 5. Lightening and dyeing power test results are provided in Table 6.

TABLE 4 Blonding cream formulations used in assessing the effectiveness of 4-acetyl 1-methylpyridinium *p*-toluene sulfonate

Component	Standard Formulation (wt%)	Experimental Formulation (wt%)
Hydrenol D	4.0	4.0
Lorol techn	2.4	2.4
Eumulgin B3	0.2	0.2
Eumulgin B2	0.2	0.2
Mergital CS 50 A	0.8	0.8
1,2-Propane diol	0.4	0.4
Xanthane FN	0.1	0.1
Crodafos CES	2.0	2.0
Amphoterge K-2	2.0	2.0
Merquat 281	3.0	3.0
Ammonium sulfate	2.0	2.0
Sodium sulfate	0.1	0.1
L-arginine	1.0	1.0
p-Toluenediamine sulfate	0.1	0.1
Resorcinol	0.1	0.1
HEDP 60%	0.2	0.2
Sodium silicate 40/42	0.5	0.5
Ammonia 25%	10.0	10.0
Experimental Additive	0	1
p-Toluene sulfonate acc. 1.1 Water	ad 100	ad 100

TABLE 5 Developing formulation used to assess the effectiveness of 4-acetyl-1-methylpyridinium *p*-toluene sulfonate in hair dying and lightening formulations

Component	Amount (wt%)
Sodium hydroxide 45%	0.73
Dipicolinic acid	0.10
Disodium pyrophosphate	0.03
HEDP® 60%	1.5
FAEOS®-Na C12-14 2EO 27%	2.0
Dow Corning® DB 110 A	0.07
Aculyn® 33A	2.5
Hydrogen peroxide 50%	20.00
Water	ad 100

Test Results

TABLE 6 Colorimetric measurements demonstrating the improved lightening and dyeing power of formulations additized with 4-acetyl-1-methylpyridinium *p*-toluene sulfonate. Higher values are preferred

Test	Unadditized Formulation	Additized Formulation
Lightening Effects	3.82	5.62
Dyeing Power	5.80	8.42

Advantages over Prior Art

Hair dyeing formulations containing the additive 4-acetyl-1-methylpyridinium *p*-toluene sulfonate are unreported. It is further unreported that hair formulations can simultaneously lighten and dye hair in reduced application times. Finally, hair formulations additized with 4-acetyl-1-methylpyridinium *p*-toluene sulfonate do not require ammonium hydroxide as a formulation component thereby reducing scalp and skin irritation on application.

Notes

1. In a subsequent investigation by Gross[1], 2-acetyl- (II), 2,6- (III), and 3,6-diacetyl-1-methylpyridinium acetate (IV) were also found to be effective as bleach activators when used in hair dyeing formulations.

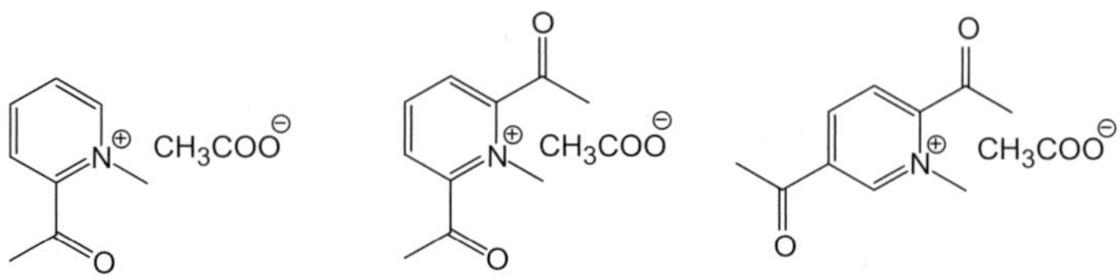

(II) (III) (IV)

2. Fadli[2] prepared hair dye formulations containing azo pyrazolinone methylpyridinium methylosulfates (IV) that were resistant to fading from shampoo washing or perspiration.

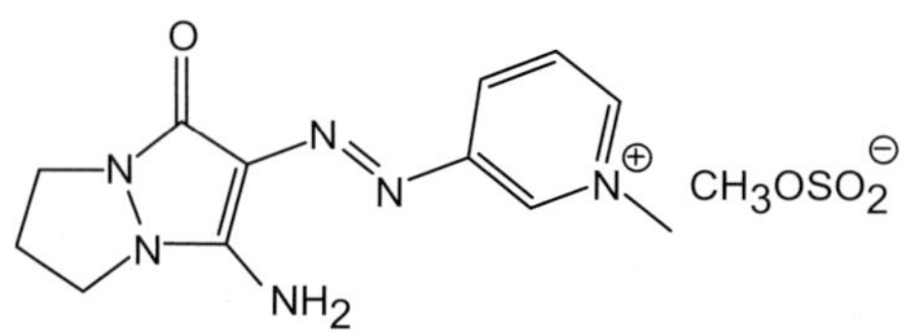

(IV)

3. Mono- (V) and diazomethane derivatives of pyrazolopyridine prepared by Fadli[3] were effective in simultaneously lightening and dyeing hair when used in dyeing formulations.

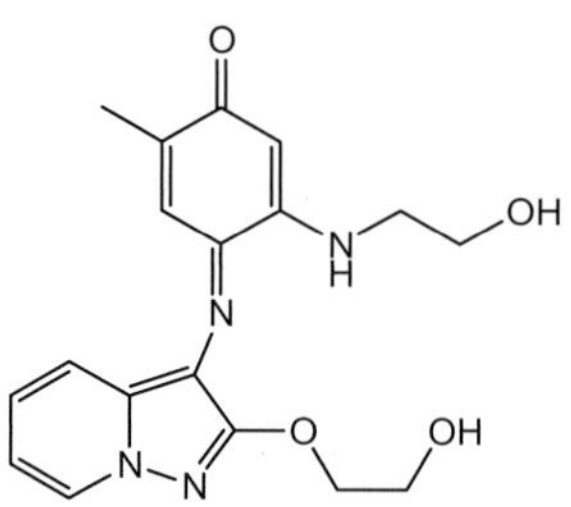

(V)

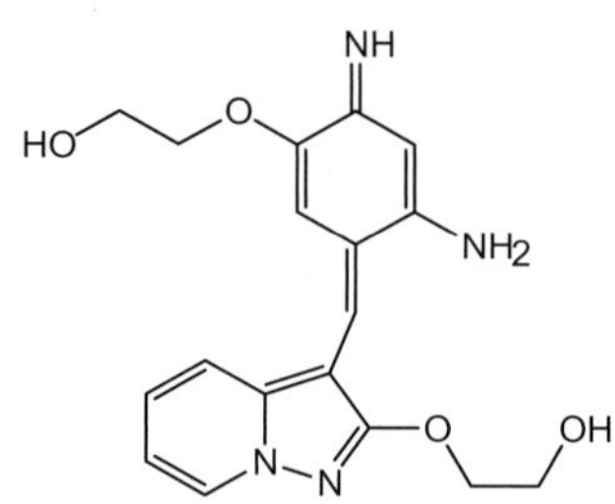

(VI)

4. Dye formulations containing a selected disulfide precursors (VII) were prepared by Greaves[4] that upon oxidation by hydrogen peroxide simultaneously both dyed and lightened hair.

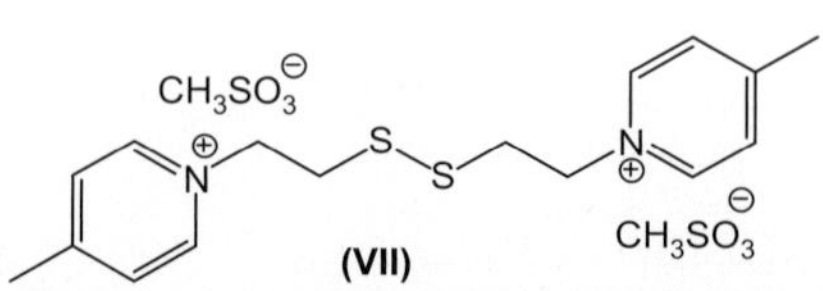

(VII)

REFERENCES

1. Wibke Gross et al, US. Patent 8,034,122 (October 11, 2011)
   Henkel AG & Co. KGaA (Duesseldorf, DE)
2. Aziz Fadli et al, US. Patent 8,034,124 (October 11, 2011)
   L'Oreal S.A. (Paris, FR)
3. Aziz Fadli et al, US. Patent 7,892,294 (February 11, 2011)
   L'Oreal S.A. (Paris, FR)
4. Andrew Greaves et al, US Patent 8,034,125 (October 11, 2011)
   L'Oreal S.A. (Paris, FR)

E. HAIR STYLING GEL FOR LUSTER AND STIFFNESS

Derivatized Cassia Tora Extract

Title:	Cationic polymers and fixative applications
US Patent:	8,003,585 (August 23, 2011)
Author:	Carole A. Lepilleur et al.
Assignee:	Lubrizol Advanced Materials, Inc. (Cleveland, OH)

Significance: The polysaccharide polygalactomanna obtained from Cassia obtusifolia of Cassia tora was solubilized in water by O-etherating with 2,3-epoxypropyl trimethyl chloride. When this material was modified to have a charge density as low as 1.8, it was used in hair styling formulations. The hair displayed desirable stiffness and improved luster under a variety of high testing conditions.

Additive Name Polygalactomannan-g-hydroxypropyltrimethyl ammonium chloride (I)

Safety Modified Cassia polymeric derivatives are considered innocuous.

Additive Structure

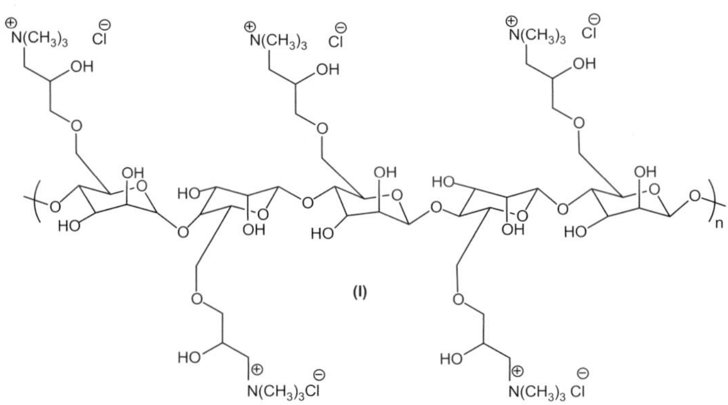

(I)

Additive Preparation

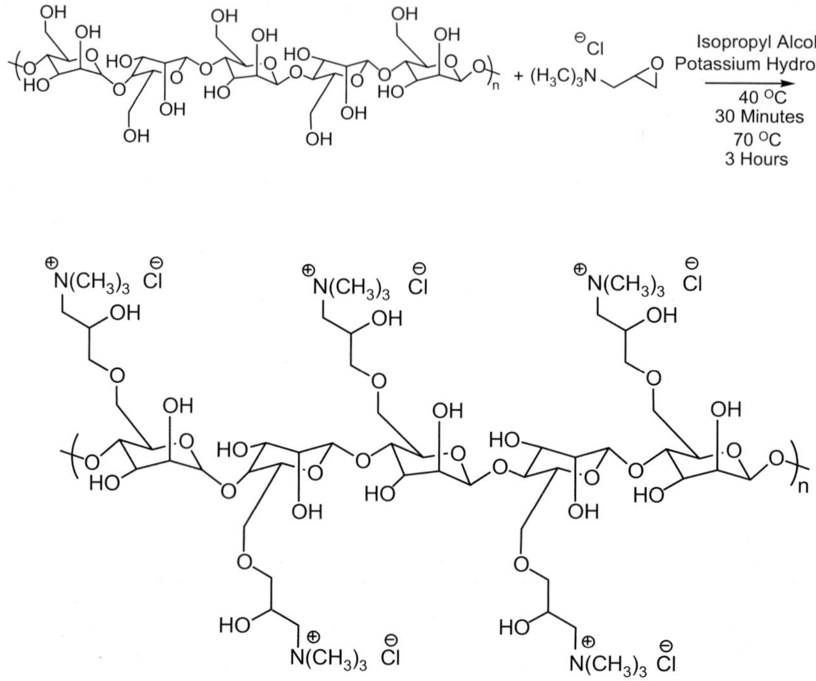

Additive Synthesis

1. *Preparation of polygalactomannan-g-hydroxypropyltrimethyl ammonium chloride*

A large reaction vessel was charged with 160 g of Cassia gum containing about 10% moisture, and 921 g of 44% isopropanol was dissolved in water and then treated with 4.5 g of potassium hydroxide. The mixture was then heated to 40 °C for 30 min until a slurry was formed and then treated with 92.8 g of 2,3-epoxypropyl trimethyl ammonium chloride. The reaction slurry was heated to 70 °C for 3 h and then cooled to 50 °C. It was then diluted with 380 g of 99% isopropanol and neutralized to a pH of about 7.0 using 50 wt% acetic acid. The mixture was filtered, washed once with 380 g of isopropanol, and then air dried overnight. Thereafter, it was oven dried at 100 °C for 4 h, and 179.3 of product was isolated having a charge density of 1.56 meq/g.

Testing

1. *Mechanical Stillness*

Aqueous fixative gels were made by blending the Step 1 product and guar gum (Novegum™. G888) and then treating European brown hair swatches with 0.8 g of the fixative. The coated hair swatches were then evaluated after exposure to varying humidity conditions and the hair stiffness assessed. Five replicates for each sample were prepared and retested, and the average Peak Force was recorded. Test results are provided in Table 7.

Test Results

TABLE 7 Summary of cationic charge densities and their effect on stiffness at 50% and 90% relative humidities. Cationic guars are provided as references

Sample	Cationic Charge Density (meq/g)	Relative Testing Humidity (%)	Work (N mm)	Peak Force (N)
Step 1 Additive	1.8	50	80.4	12.2
Additive-1	3	50	48.8	7.6
Step 1 Additive and Guar Gum	1.76	50	38.2	8.0
Cationic Guar	2.9	50	31.8	5.4
Cationic Guar	1.8	90	37.6	5.9
Step 1 Additive	3	90	60.7	9.0
Cationic Guar	1.76	90	28.8	4.2
Cationic Guar	2.9	90	31.1	4.7

Advantages over Prior Art

Hair fixatives typically consist of a judicious blend of hydrocolloid polymers, rheology modifiers, and auxiliary fixatives polymers and additives. The use of modified water-soluble Cassia extracts both simplifies this formulation process and generates a desirable consumer cosmetic product.

Notes

1. Leighton[1] prepared consumer hair care formations containing 1.4 wt% of hydroxypropyl starch phosphate (AMAZE®), which was effective in curl retention hair fixatives at high humidity.
2. Lepilleur[2] observed enhanced curl retention at high humidity when both polygalactomannan-g-2-hydroxy-3-(trimethyl ammonium)propyl chloride and guar gum were present in hair styling formulations.

3. Poly(ethylene glycol-ter-propylene oxide-ter-dimethylsiloxane) terpolymers containing itaconic acid ester termini (II) were prepared by Tamareselvy[3] and were effective as hair fixatives at high humidity when used in consumer personal care formulations.

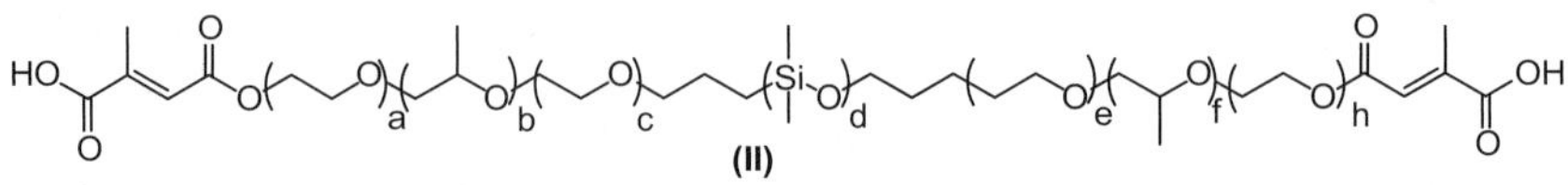

(II)

4. Hair fixative polyetheramines (III) consisting of the reaction product of 1,4-diamino-cyclohexane, polypropylene oxide and containing amine termini with adipic acid, repeat units (III) were prepared by Philbin[4] and used in consumer hair spray formulations.

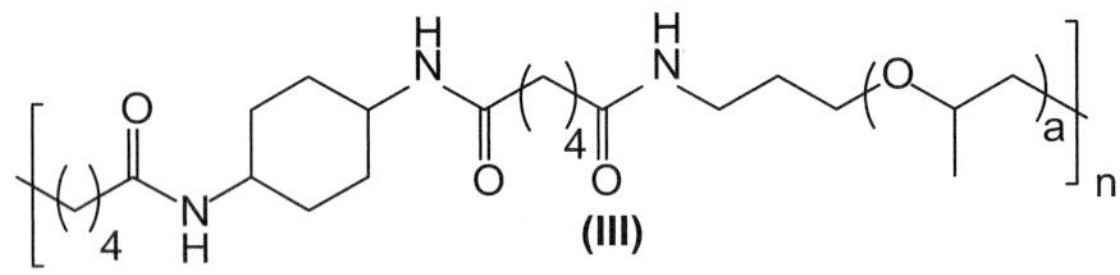

(III)

REFERENCES

1. John C.Leighton et al, US. Patent 7,972,589 (July 5, 2011) **Akzo Nobel N.V. (Arnhem, NL)**
2. Carole A. Lepilleur et al, US. Patent 7,759,296 (July 20, 2010) **Lubrizol Advanced Materials, Inc. (Cleveland, OH)**
3. Krishnan Tamareselvy et al, US. Patent 8,021,650 (September 20, 2011) **Lubrizol Advanced Materials, Inc. (Cleveland, OH)**
4. Michael Philbin et al, US. Patent 7,919,074 (April 5, 2011) **Akzo Nobel N.V. (Arnhem, NL)**

F. HAIR SHAMPOO FOAMING AGENT

Sodium Cocoyl α-Methyl and β-Methyl isothionates

Title:	Acylalkylisethionate esters and applications in consumer products
US Patent:	8,008,239 (August 30, 2011)
Author:	Prakasa R. Anantaneni et al.
Assignee:	Huntsman Petrochemical LLC (The Woodlands, TX)

Significance

Sodium cocoyl methyl- and ethylisothionates were prepared in quantitative yields by reacting the corresponding epoxide with sodium bisulfite and then postreacting the product mixture with hydrogenated C_8H_{17}–$C_{18}H_{35}$ carboxylic acids. Aqueous solutions or shampoo compositions additized with 0.5 wt% of the additives produced long-lasting creamy tight foams. Aqueous solutions containing 10 wt% of these additives were not irritating to either the skin or the eyes.

Additive Name

Sodium cocoyl α-methyl and β-methyl isethionate pair (I)
Sodium cocoyl α-ethyl and β-ethyl isethionate pair (II)

Safety

No toxicological data available, but these materials are believed to be non-toxic.

Additive Structures

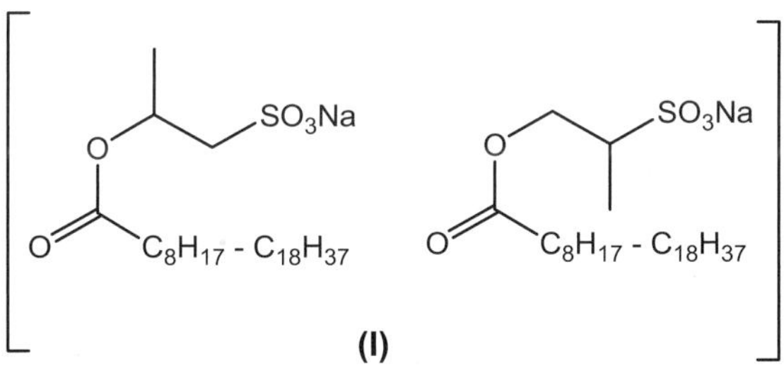

(I)

Additive Pair

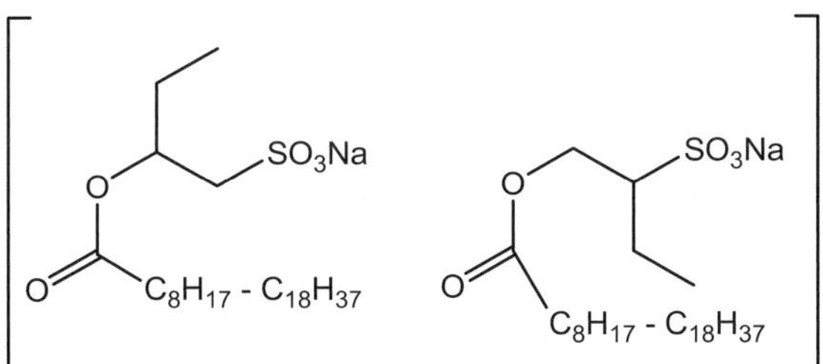

(II)

Additive Pair

Additive Preparation

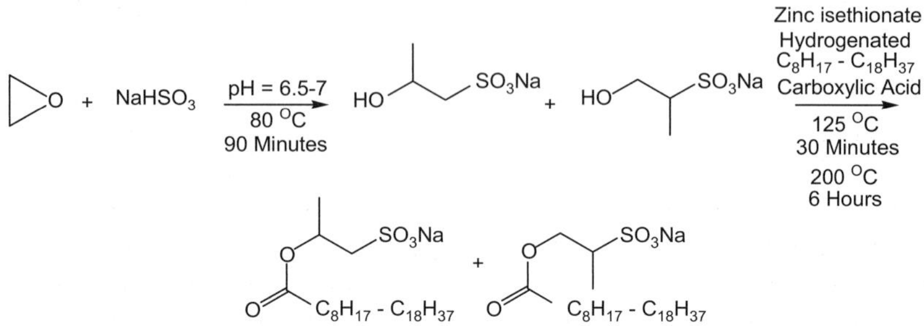

Additive Synthesis

1. *Preparation of a mixture of sodium α-methyl-2-hydroxy-ethane 1-sulfonate and β-methyl 2-hydroxy ethane-1-sulfonate*

A 3-gallon stainless steel autoclave reactor was charged with 9.40 pounds of 35% aqueous sodium bisulfite solution having a pH of between 6.5 and 7.0 and then purged with nitrogen to exclude air. The reactor was then heated to 70 °C and treated with 1.0 pound of propylene oxide and reacted at 60 psi for 90 min at 80 °C and then cooled, and the product mixture was isolated.

2. *Preparation of a mixture of sodium C_8H_{17}–$C_{18}H_{35}$ α-methyl and β-methyl isethionate*

A 500-ml round bottom flask equipped with mechanical stirrer, addition funnel, condenser, and thermocouples was charged with 212 g of hydrogenated C_8H_{17}–$C_{18}H_{35}$ carboxylic acid, 165 g of the Step 1 product mixture, and 5.00 g of zinc methyl isethionate and then heated to 125 °C for 30 min. The mixture was then heated for an additional 6 h at 200 °C at 10 mm Hg, and the product was isolated.

Testing

1. *Foaming Tests*

Foaming tests were performed using a 1-liter capped rotating measuring cylinder foam machine circulating at 30 rpm at an ambient temperature on samples containing 0.5 wt% of a selected additive. Foam heights in the graduated cylinder were measured at the start and after 10 min of rotation. Test results are provided in Table 8.

2. *Zein Irritation Score*

To evaluate the mildness of the products of the present invention, Zein scores were determined using 10 wt% of an aqueous solution of a selected sulfate and isethionate surfactant. Test results are provided in Table 9.

3. *Water Solubility*

Water solubility of experimental additives was determined by either adding molten or concentrates of a selected experimental additive to cold water containing betaines.

Test Results

1. *Foaming Testing*

TABLE 8 Foam test results indicating that while the reaction mixture comprising Additive 1 did not achieve the flash foam levels of the industry standard, sodium laureth sulfate, the foam generated a creamy appearance whether used alone or with other anionic surfactants

Sample	Ingredient (5 wt%)	Foam Height (0 min)	Foam Height (10 min)	Foam Appearance
1	80% Sodium Lauryl Sulfate 20% Cocamidopropyl Betaine	250	238	Open Foam
2	100% Sodium Lauryl Sulfate	200	170	Open Foam
3	80% Sodium Laureth 20% Cocamidopropyl Betaine	260	220	Open Foam
4	Sodium Cocoyl α-Methyl and β-Methyl Isethionate (Additive 1)	190	160	Creamy, Tight
5	80% Sodium Cocoyl α-Methyl and β-Methyl Isethionate and 20% Cocamidopropyl Betaine	190	180	Creamy, Tight

2. *Zein Irritation Score*

TABLE 9 Zein irritation test results indicating that the experimental isothionates are not skin or eye irritating at 10 wt% concentration levels

Ingredient	Zein Score
Sodium Lauryl Sulfate	527
Sodium Cocoyl Isethionate	254
Sodium Cocoyl Methyl Isethionate (Additive 1)	147
Sodium Cocoyl Ethyl Isethionate (Additive 2)	104
Sodium Lauroyl Isethionate	160
Sodium Lauroyl Methyl Isethionate	134
Sodium Lauroyl Ethyl Isethionate	187

3. *Water Solubility*

Aqueous solutions containing up to 17 wt% betaines dissolved up to 24 wt% of selected additives.

Advantages over Prior Art

Sodium C_8H_{17}–$C_{18}H_{35}$ α- and β-alkyl isothionates are anionic surfactants designed specifically to address the limited water solubility of sodium C_8H_{17}–$C_{18}H_{35}$ isothionates. Water soluble compositions of these experimental materials are nonirritating, display excellent foaming properties, and have been used successfully in shampoo, liquid soap, and body gel formulations.

Notes

1. Terada[1] prepared a two-component high foaming shampoo composition consisting of amodimethicone block polymer (III) and amino-modified polysiloxane-polyoxyalkylene terpolymer (IV).

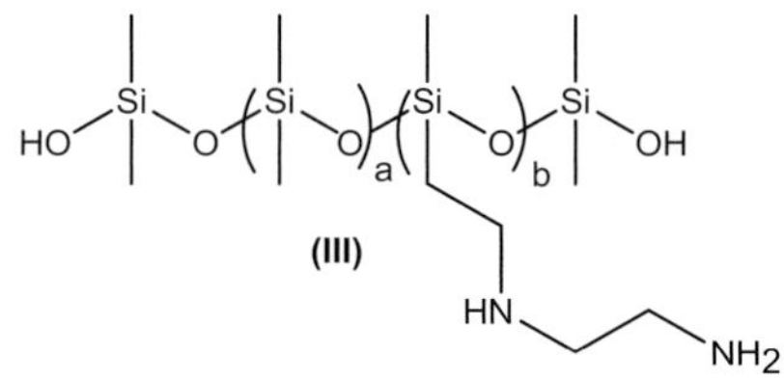

(III)

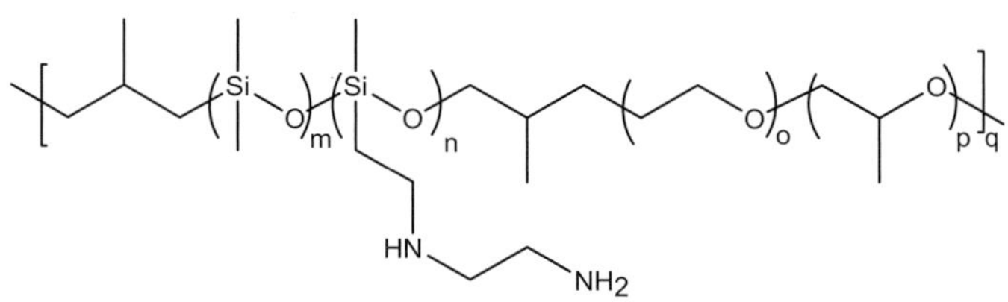

(IV)

2. Keller[2] prepared aerosol mousse formulations containing either cetyl trimethyl-ammonium chloride or ditallow dimethyl ammonium chloride with either cetyl and stearyl alcohols, which produced creamy and densely packed foams.
3. Gutierrez[3] prepared nonionic triglyceride surfactants alkoxylated with either ethylene or propylene oxide and then esterified these mixtures with capric acid (V) or coconut or palm oils, which quickly produced high form levels in water and were effective as detergent additives.

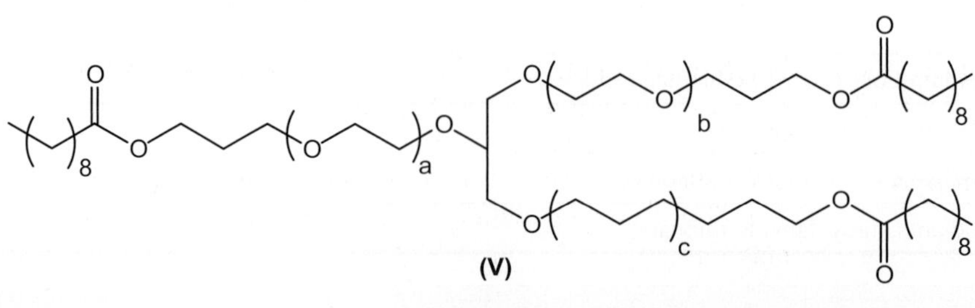

(V)

4. A three-component skin cleaning composition prepared by Uchiyama[4] consisting of sodium polyoxyethylene lauryl ether sulfate (VI) myristyl alcohol (VII), and cationic polydimethyldiallylammonium chloride (VIII) had high foam viscosities and creamy foam qualities when used in shampoo and soap formulations.

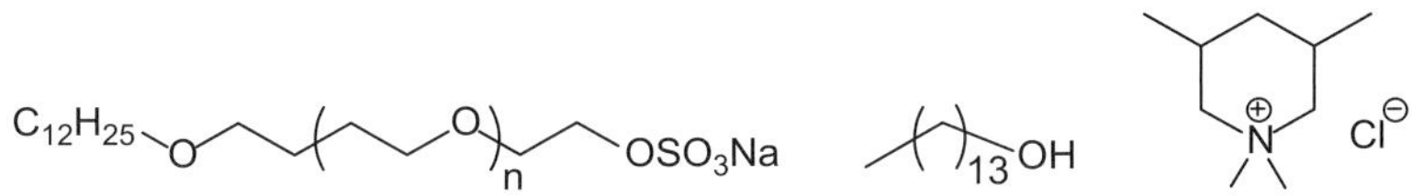

(VI) (VII) (VIII)

REFERENCES

1. Eiji Terada, US. Patent 8,003,085 (August 23, 2011)
   Kao Corporation (Tokyo, JP)
2. Walter Keller et al, US. Patent 8,017,106 (September 13, 2011)
   The Procter & Gamble Company (Cincinnati, OH)
3. Pey Gutierrez et al, US. Patent 7,989,411 (August 2, 2011)
   Kao Corporation, S.A. (Barcelona, ES)
4. Tomoko Uchiyama et al, US. Patent 7,964,545 (June 21, 2011)
   Kao Corporation (Tokyo, JP)

G. HAIR MOISTURIZING AGENT

Betaine Amphoteric Copolymers

Application:	Hair and Skin Moisturizing Agent
Title:	Polymer and cosmetic preparation
US Patent:	7,872,083 (January 18, 2011)
Author:	Tomohiko Kimura et al.
Assignee:	Shiseido Co., Ltd. (Chuo-ku, Tokyo, JP)
Significance:	Poly(dimethyldiallylammonium chloride-co-*N*,*N*-dimethylaminopropylacrylamide acetate betaine) was found to be effective as a facial cleansing agent and in providing the user with a moisturized feeling to the hair and skin when used in cosmetic formulations. Optimum cosmetic effects were achieved using this amphoteric polymer having an Mw of approximately 400,000 Da
Additive Name	Poly(dimethyldiallylammonium chloride-co-*N*,*N*-dimethylaminopropylacrylamide acetate betaine) (I)
Safety Concerns	Polymeric betaines are considered nontoxic since the acute oral male rat LD50 > 200 g/kg. In addition, these materials have negligible bioaccumulation, poor absorption, and rapid biodegradation.

Additive Structure

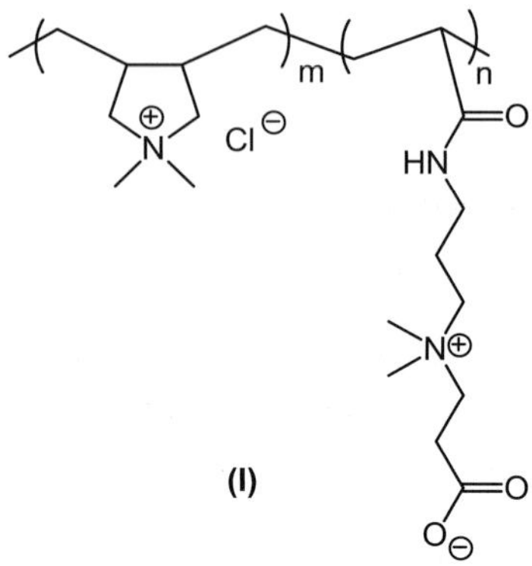

(I)

Additive Preparation

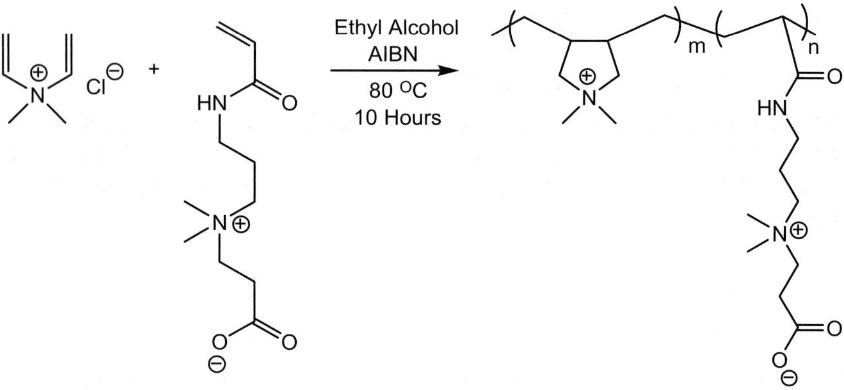

Additive Synthesis

1. *Preparation of poly(dimethyldiallylammonium chloride-co-N,N-dimethylaminopropylacrylamide acetate betaine)*

A five-necked flask attached to a reflux condenser, dropping funnel, thermometer, nitrogen-replacement glass tube, and a stirrer was charged with 64.6 g of dimethyldiallylammonium chloride, 128.5 g of *N,N*-dimethylaminopropylacrylamide acetate betaine, 100 g of ethanol, and 1.0 g of α,α′- azobisisovaleronitrile and then refluxed at 80 °C for 10 h. The mixture was then concentrated, and the product was isolated having an Mw of approximately 400,000 Da.

Testing

1. *Foam and Moisturizing Quality*

The foam quality of blends containing the experimental agent was evaluated using ten professional panelists according to the following evaluation criteria and points assigned.

5 points: creamy
4 points: slightly creamy
3 points: normal
2 points: less creamy
1 point: not creamy

Test results are provided in Table 10.

2. *Moisturizing Quality*

The moisturized feeling on the skin after cleansing with each facial cleanser was also evaluated by ten professional panelists according to the following evaluation criteria and points assigned:

5 points: moisturized
4 points: slightly moisturized
3 points: normal
2 points: less moisturized
1 point: not moisturized

Test results are provided in Table 10.

3. *Facial cleansers and shampoos*

The facial cleanser and shampooing properties of formulations were also evaluated using a 4:6 molar ratio of dimethyldiallylammonium chloride and *N,N*-dimethylaminopropylacrylamide acetate betaine, respectively, at varying molecular weights of the copolymer.

Test Results

TABLE 10 Effect of foam quality and moisturized feeling using skin ointments additized with 0.5 wt% of the Step 1 product and related amphoteric copolymer controls. Optimum facial cleanser and shampooing properties using a 4:6 molar ratio of dimethyldiallylammonium chloride and *N,N*-dimethylaminopropylacrylamide acetate betaine, respectively, was observed using poly(dimethyldiallylammonium chloride-co-*N,N*-dimethylaminopropylacrylamide acetate betaine) having an Mn of approximately 400,000 Da

Component	Treatment (wt%)	Treatment (wt%)	Treatment (wt%)	Treatment (wt%)	Treatment (wt%) (Unadditized Reference)
Step 1 Product	0.5	X	X	X	X
Polydimethyldiallylammonium chloride (Control)	X	0.5	X	X	X

(Continued)

TABLE 10 Effect of foam quality and moisturized feeling using skin ointments additized with 0.5 wt% of the Step 1 product and related amphoteric copolymer controls. Optimum facial cleanser and shampooing properties using a 4:6 molar ratio of dimethyldiallylammonium chloride and *N,N*-dimethylaminopropylacrylamide acetate betaine, respectively, was observed using poly(dimethyldiallylammonium chloride-co-*N,N*-dimethylaminopropylacrylamide acetate betaine) having an Mn of approximately 400,000 Da—*cont'd*

Component	Treatment (wt%)	Treatment (wt%)	Treatment (wt%)	Treatment (wt%)	Treatment (wt%) (Unadditized Reference)
Dimethyldiallylchloride acrylamide copolymer (Control)	X	X	0.5	X	X
Acrylic acid/dimethyldiallylammonium chloride/acrylamide copolymer (Control)	X	X	X	0.5	X
Glycerin	15.0	15.0	15.0	15.0	15.0
Sorbitol solution	20.0	20.0	20.0	20.0	20.0
Polyethylene glycol	5.0	5.0	5.0	5.0	5.0
Lauric acid	5.0	5.0	5.0	5.0	5.0
Myristic acid	12.0	12.0	12.0	12.0	12.0
Palmitic acid	5.0	5.0	5.0	5.0	5.0
Stearic acid	10.0	10.0	10.0	10.0	10.0
Potassium hydroxide	6.0	6.0	6.0	6.0	6.0
Ion-exchange water	Balance	Balance	Balance	Balance	Balance
Foam quality	4.6	2.5	3.0	3.0	2.0
Moisturized feeling	4.8	3.6	4.0	3.8	3.0

Advantages over Prior Art

The feeling of hair and skin after cleansing is an extremely important consumer concern. Most skin and hair cleansers contain surfactants that are high in cleansing power, which often lead to undesired defatting of the skin and hair. The current invention has devised a method using a cationic copolymer that is effective as a skin and hair cleaner but which limits skin and hair defatting to provide the consumer with a moisturized and comfortable feeling.

Notes

1. Ampholyte terpolymers (II) prepared by Melby[1] were used in skin care and shampoo formulations in personal care products to retain and attract moisture, soften skin, and to reduce skin irritation caused by cleansing detergents.

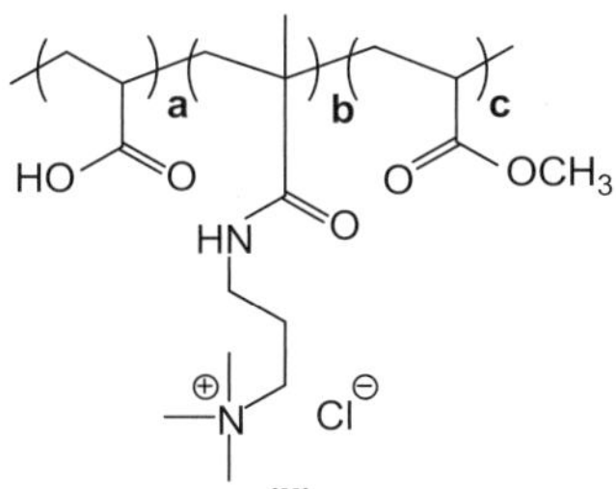

(II)

2. A cationic-neutral diblock copolymer consisting of poly(methylsulfate[2-(acryloyloxy)ethyl]-trimethylammonium)-b-(2-dimethylaminoethyl acrylate) (III) was prepared by Bayouzet[2] and was used in skin and hair shampoo compositions as a moisturizing and soothing agent.

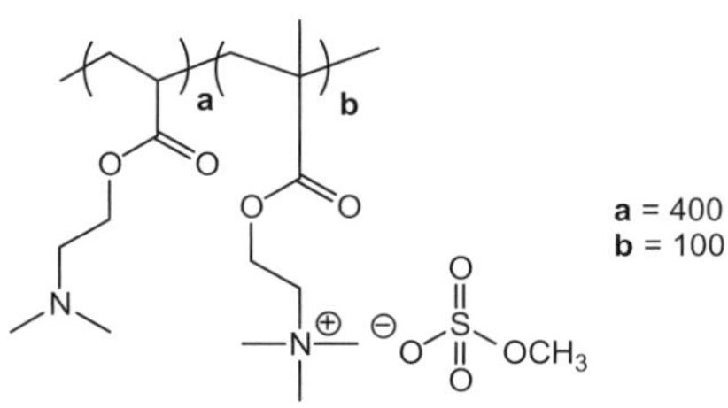

(III)

3. Skin care formulations were prepared by Tanabe[3] containing the sugar α,α-trehalose (IV) was effective as anti-inflammatory, antibacterial, and moisturizing agent.

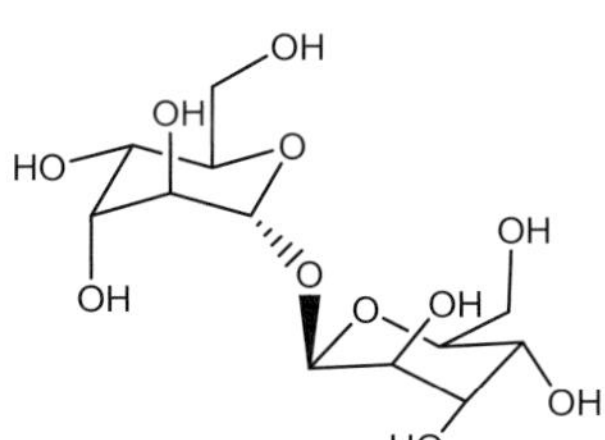

(IV)

4. Landa[4] observed an unusual moisturizing effect of skin care lotions additized with a 1:1 mixture of hydroxyethylurea (V) and ammonium lactate.

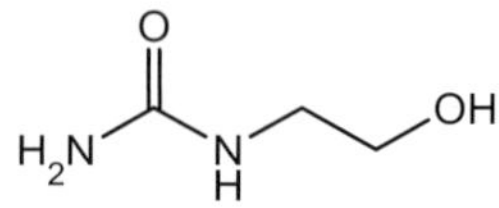

(V)

REFERENCES

1. Allan L. Melby et al, US Patent RE42,025 (January 11, 2011)
Nalco Company (Naperville, IL)
2. Bruno Bavouzet et al, US Patent 7,846,423 (December 10, 2010)
Rhodia Chimie (Billancourt Cedex, FR)
3. Fujimi Tanabe et al, US Patent 7,854,922 (December 21, 2010)
Kabushiki Kaisha Hayashibara Seibutsu Kagaku Kenkyujo (Okayama, JP)
4. Peter A. Landa, US Patent 7,816,310 (October 19, 2010)
Akzo Nobel N.V. (Arnhem, NL)

H. LOTIONS TO REDUCE FACIAL SKIN PORE SIZE

Keratinocytes

Title:	Skin pore minimizers and skin elasticity improvers
US Patent:	7,968,607 (June 28, 2011)
Author:	Takafumi Nishijima et al.
Assignee:	Koa Corporation (Tokyo, JP)
Significance:	A series of fatty acid phosphate and sulfate ester salts were prepared that reduced facial pore size when used in cosmetic formulations. After 8 weeks of treatments at concentrations as low as 1 μM, the keratinocyte contracting effect was as high as 73.5%, and pore size reduction of up to 30% was observed.
Additives	1-Isostearylglycerol-3-sulfate sodium salt (I) 1-Octadecylglycerol-3-phosphate disodium salt (II) 1-Isostearylglycerol-3-phosphate disodium salt (III) 1-β-O-dodecylglucopyranoside-6-sulfate sodium salt (IV) 1-Dodecylpropanediol-3-sulfate sodium salt (V)
Safety	Additives are surfactants and are considered nontoxic.

Additive Structures

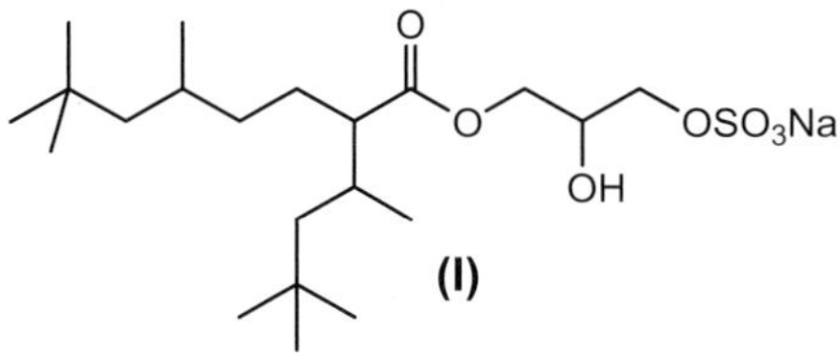

(I)

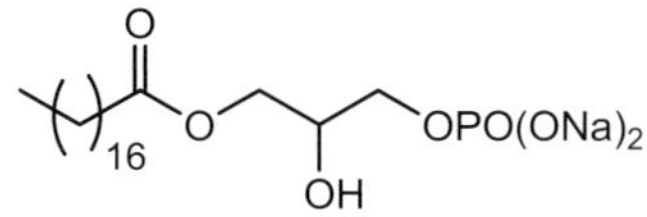

(II)

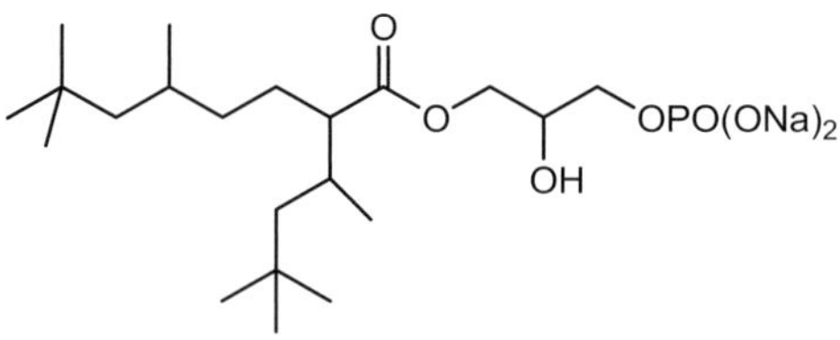

(III)

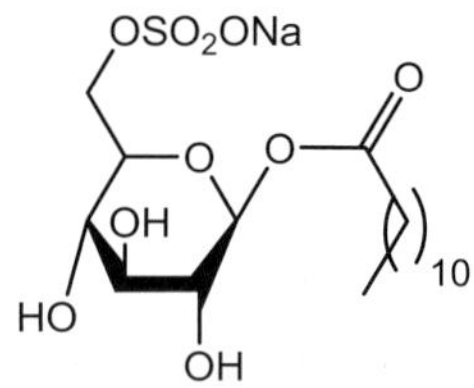

(IV)

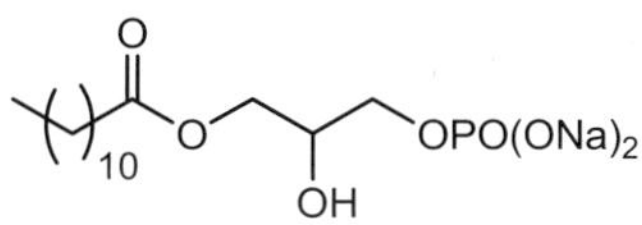

(V)

Additive Preparation

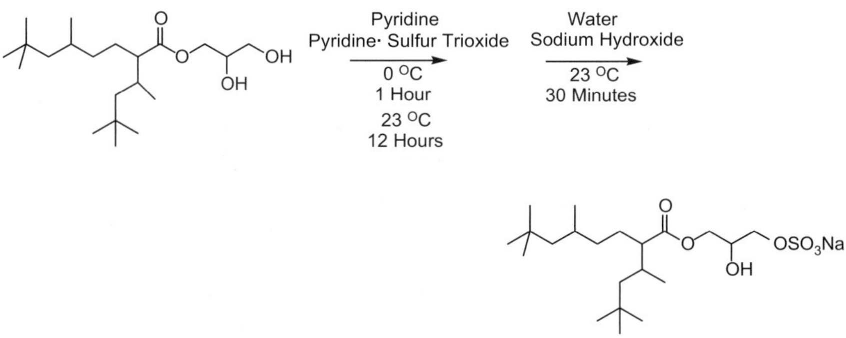

Additive Synthesis

1. Preparation of 1-isostearylglycerol-3-sulfate sodium salt

A 100-ml round bottom flask was charged with 10 g of isostearyl glyceryl ether dissolved in pyridine and then cooled with ice water and treated with 4.6 g of sulfur trioxide–pyridine complex. The mixture was then stirred for 1 h under ice cooling conditions and then at an ambient temperature for 12 h. It was then concentrated, and the residue was treated with purified water containing 1.16 g of sodium hydroxide, lyophilized, and the product was isolated.

Testing

1. Keratinocyte Contracting Effect

Type I collagen, MCDB 153 medium, and purified water were mixed under ice cooling, and 500 μl was added to each well in a 24-well plate and then warmed to 37 °C in an incubator to effect gelation. Using the MCDB 153 medium, 1-ml aliquots of keratinocytes were seeded to 2×10^4 cells/cm^2 onto collagen gels and then incubated for 24 h and peeled off with a pipette tip. The gels were then treated with a selected experimental agent at various concentrations, and collagen contractions were measured after 1 h. Test results are provided in Table 11.

2. Pore Reduction Testing

The ear skin pores of three 2.5-kg New Zealand male white rabbits were treated with a 2% solution of a selected experimental agent dissolved in ethyl alcohol. In each case, one ear was treated with 150 μl of ethyl alcohol alone, and the opposite ear was treated with 150 μl of ethyl alcohol containing 2% of a selected experimental agent. Each ear was treated twice daily. After 8 weeks of initiation of the application, the skin was collected, and the size of skin pores was measured by image analysis using a videomicroscope while exercising care to avoid drying. Test results are provided are provided below.

Test Results

1. Keratinocyte Contracting Effect

TABLE 11 Keratinocyte contracting effect of selected experimental agents

Ingredient	Concentration (μM)	Contraction (%)
Additive 2	50	77.0
Additive 3	50	62.3
Additive 4	50	90.0
Additive 5	10	88.0

2. Pore Reduction Testing

Videomicroscope studies indicated that experimental Additive 2 reduced pore size by 30%.

Advantages over Prior Art

The predominant method for reducing pore size is through keratotic plug removers. Their use, however, is problematic since the pore removal may shrink facial pores so that the facial features become even more conspicuous. The current invention provides an alternative to this method since in all formulations containing these experimental additives, statistically significant ketatinocyte contraction and pore size reduction were observed.

NOTES

1. Zimmerman[1] prepared a multicomponent cosmetic compositions designed to minimize or prevent the signs of aging, improve the esthetic appearance of skin, and to promote recovery from environmental stresses using extracts from rosemary, Centella, Echinacea, or Alpinia and a DNA repair enzyme.
2. A topical cosmetic formulation prepared by Faryniarz[2] containing 16.5% copper malonate and approximately 12.4% zinc malonate was reported to be effective in treating wrinkles and fine lines associated with age-related skin conditions.
3. Oddos[3] used a mixture of retinol (VI) and β-aminoisobutyric acid (VII) in skin care formulations to minimize facial wrinkles and fine lines, which induced a significant synergistic increase in cell proliferation.

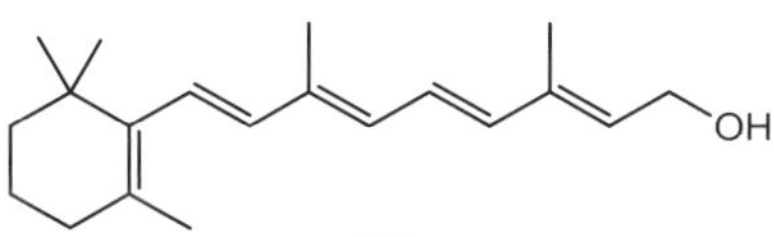

(VI)

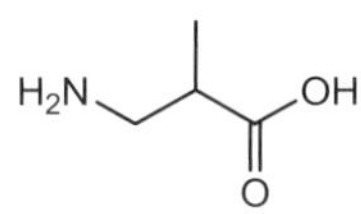

(VII)

4. *N*-(Hydroxyalkyl)sulfonamides (VIII) prepared by Marat[4] were used in nontherapeutic skin care formulations as a moisturizer and to reduce the appearance of facial fine lines and wrinkles.

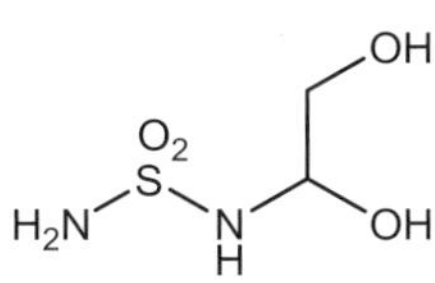

(VIII)

REFERENCES

1. Amy C. Zimmerman et al, US Patent 7,959,953 (June 14, 2011) **Access Business Group International LLC (Ada, MI)**
2. Joseph R. Faryniarz et al, US Patent 7,927,614 (April 19, 2011) **JR Chem, LLC (Key West, FL)**
3. Thierry Oddos et al, US Patent 7,928,142 (April 19, 2011) **Johnson & Johnson Consumer France SAS (FR)**
4. Xavier Marat et al, US Patent 7,887,824 (February 15, 2011) **L'Oreal (Paris, FR)**

I. LOTION SKIN ADDITIVES TO REDUCE FAT CELLS

Hydroxycitric Acid-2-Palmitate Disodium

Title:	Hydroxycitric acid derivatives and skin external preparations containing the same
US Patent:	7,846,970 (December 7, 2010)
Author:	Harumi Kamachi et al.
Assignee:	Showa Denko K.K. (Tokyo, JP)
Significance:	Formations containing the additives hydroxycitric acid-2-palmitate disodium or calcium have been found to be effective in lowering the fat content in the skin and improving skin appearance by increasing fat metabolism in the dermis and epidermis.

Additive Name Hydroxycitric acid-2-palmitate disodium (I)
Hydroxycitric acid-2-palmitate calcium (II)

Safety Concerns These chemical agents are considered innocuous and nontoxic even when exposed to acute dosages.

Additive Structure

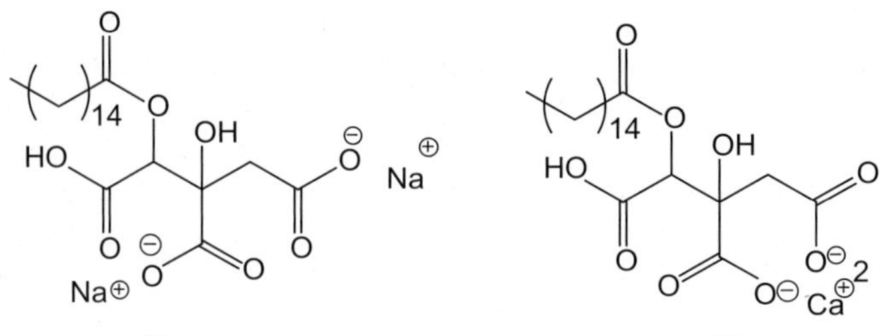

(I) (II)

Additive Preparation

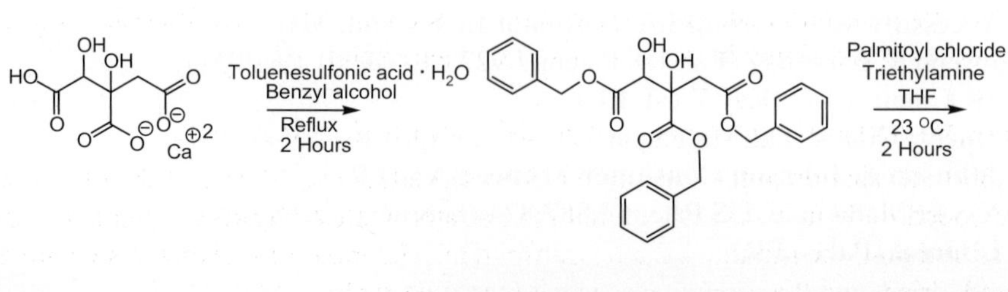

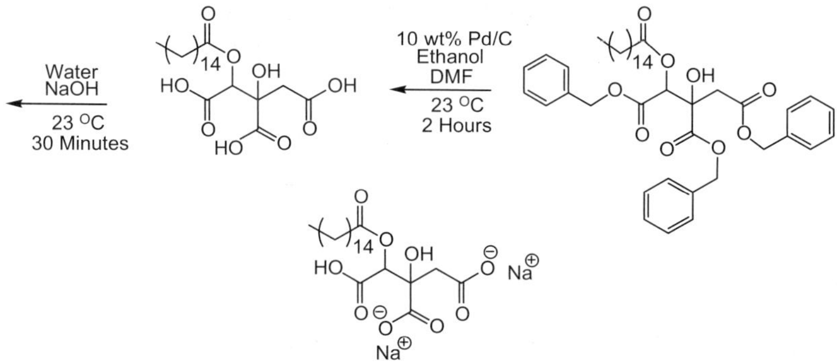

Additive Synthesis

1. Preparation of hydroxycitric acid tribenzyl ester

A 200-ml reactor was charged with 2.96g of calcium hydroxycitrate, 5.86g of toluenesulfonic acid monohydrate, 10g of benzyl alcohol, and 20ml of toluene and then refluxed 4h with azeotropic removal of water. The mixture was then cooled and treated with 50ml of ethyl acetate and slowly introduced into a 500-ml beaker containing 100ml of 5wt% aqueous sodium hydrogen carbonate. Insolubles were then filtered off, and the organic phase was isolated, washed with water, and then dried over anhydrous sodium sulfate. The liquid was concentrated, the residue purified by silica gel column chromatography using a 5:1 mixture of hexane/ethyl acetate, and the product isolated in a 40% percent yield as a white solid.

2. Preparation of hydroxycitric acid tribenzyl ester-2-palmitate

A 50-ml reaction flask was charged with 239mg of the Step 1 product, 5ml of THF, 165mg of palmitoyl chloride, and a solution of 61mg of triethylamine dissolved in 2ml of THF. The mixture was then cooled with ice and then stirred for 30min at this temperature for 3h. The reaction flask was then treated with 100ml of ethyl acetate and 50ml of water, and the organic layer was isolated. The organic phase was then washed with water, dried using anhydrous sodium sulfate, and concentrated The residue was purified using silica gel column chromatography using a 10:1 mixture of hexane/ethyl acetate, and the product was isolated in a 92% yield as a white solid.

3. Preparation of hydroxycitric acid-2-palmitate

A 50-ml reaction flask was charged with 300mg of the Step 2 product, 5ml apiece of ethanol and DMF, and 40mg of 10wt% palladium on carbon, and the catalytic reduction was performed for 2h. The catalyst was then filtered off, and the reaction mixture was concentrated. The residue was precipitated in hexane, filtered, dried, and the product was isolated in an 84% yield as a white solid.

4. Preparation of hydroxycitric acid-2-palmitate sodium salt

In a 500-ml reaction kettle, 10g of the Step 4 product and 200ml of distilled water were added to form a suspension and then treated with 2.15g of sodium hydroxide and stirred until the mixture became transparent. The mixture was then concentrated, filtered, dried, and the product was isolated in 99% yield.

Testing

1. Skin Permeation Test

A 2-cm^2 square piece of minipig skin was placed in the Netwell on assay rings on which a silicon sealant had been applied fixed to the epidermis of the skin. The skin culture medium was added to a multiwell plate, and the Netwell was set so that the lower portion of the skin would be soaked. A solution of a selected experimental agent was then slowly poured into the opening of the assay ring, and the units was sealed with multiplate seals and allowed to stand in an incubator at 37 °C under 5% carbon dioxide. After 4 h, the skin pieces were recovered from the Netwell rings and rinsed with distilled water. The assay rings were then detached and the central portion that had contacted with the experimental solution punched out using an 8-mm biopsy punch. The excised skin pieces were transferred to 1.5-ml tubes, and 1 ml of 0.5% trypsin solution was added to each tube and left undisturbed for 36 h at 4 °C. The trypsin solution was then removed, and the skin was washed thrice with distilled water and then heated in an incubator at 37 °C to release the epidermis. The epidermis was then homogenized and analyzed for hydroxycitric acid and hydroxycitric acid-2-palmitate using LC/MS. Test results are provided in Table 12.

2. Triglyceride Accumulation Test

Mouse 3T3-F442A cells were disseminated on a 96-well microplate and were incubated at 37 °C over a period of 2 weeks using a DMEM culture medium containing 10 wt% FCS. After differentiation of cells into adipocytes had occurred, the culture medium was replaced with one that contained 100 μM of a selected experimental agent, and incubation was continued for an additional 4 days. The control was incubated using a culture medium that contained no test articles. Thereafter, the cells were washed thrice with Dulbecco's PBS (−), and the triglyceride accumulated in the cells was determined by oil red staining. A summary of the accumulated triglyceride content is provided in Table 13.

Test Results

1. Skin Permeation Test

TABLE 12 **Permeation of underivatized hydroxycitric acid and hydroxycitric acid-2-palmitate into the dermis and epidermis of pig skin. Higher absorbance of hydroxycitric acid-2-palmitate suggests their effectiveness in facilitating the breakdown of accumulated fat**

Sample Analyzed	Substance to be Determined	Determined Value
Epidermis-1	Hydroxycitric acid	0.4
Epidermis-2	Hydroxycitric acid-2-palmitate (Step 1 Additive, disodium)	3.4
Dermis-1	Hydroxycitric acid	0.2
Dermis-2	Hydroxycitric acid-2-palmitate (Step 1 Additive, disodium)	3.8

2. *Triglyceride Accumulation Test*

TABLE 13 Effect of selected experimental agents in lowering the triglyceride content in mouse cells. These results suggest that hydroxycitric acid-2-palmitate salts are effective in both inhibiting the synthesis and accumulation of fat

Sample	Accumulated Triglycerides (Relative to Control)
Hydroxycitric acid-2-palmitate	100
Hydroxycitric acid-2-palmitate calcium salt (Step 2 Additive)	98
Hydroxycitric acid-2-palmitate sodium salt (Step 1 Additive)	52
Hydroxycitric acid monomethyl ester-2-palmitate sodium salt (Step 1 Additive)	84

Advantages over Prior Art

The invention addresses the need for a chemical agent that has high skin and percutaneous absorption properties needed to increase fat metabolism in skin. The nontoxic properties of hydroxycitric acid-2-palmitate suggest its use in consumer skin care formulations.

Notes

1. Tricreatine Hydroxycitrate (III) was prepared by Heuer[1] and used as a dietary supplemental to reduce adiposity, suppressing appetite and to improve muscle tone.

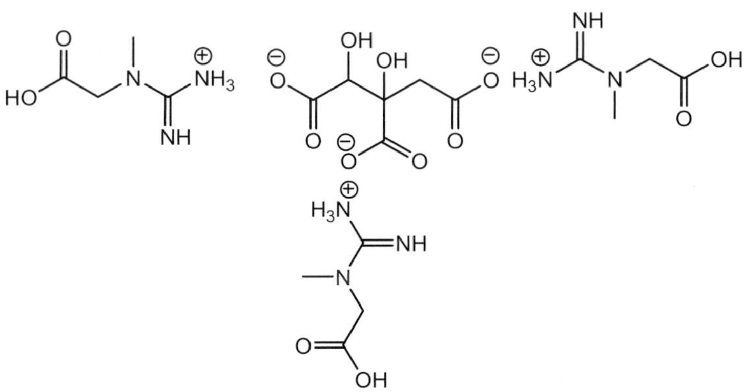

(III)

2. Topical application of ascorbyl esters (IV) prepared by Gupta[2] were effective in improving the skin appearance resulting from collagen loss, loss of skin pliability, puffiness, loss of skin suppleness and skin wrinkles.

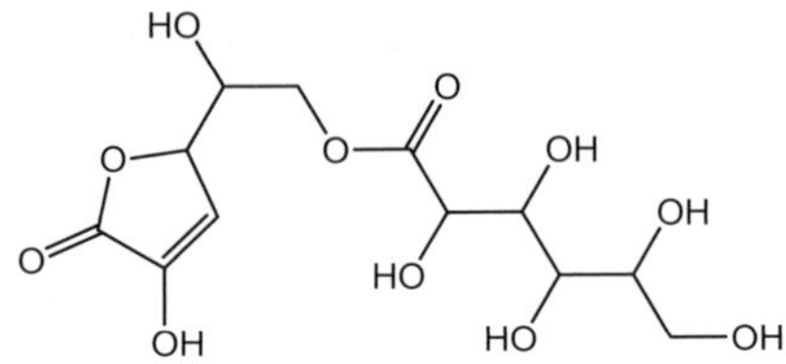

(IV)

3. Emollients containing furan alkenyls (V) prepared by Piccirilli[3] were effective in reducing flabby or gelatinous facial skin associated with diabetic conditions.

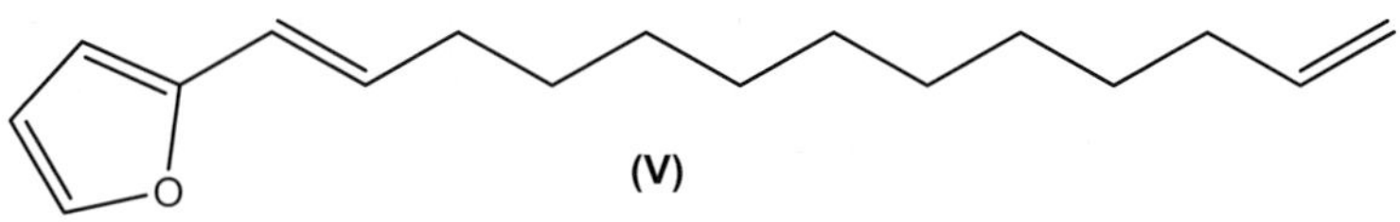

(V)

4. Skin lotions and creams containing 4-(3′,4′-dihydroxyphenyl)-butane-2-one (VI) or 4-(3′,4′-dihydroxyphenyl)-butane-2-ol (VII) were prepared by Yokota[4] and had high lipolytic activity and were useful in reducing fatty content in skin.

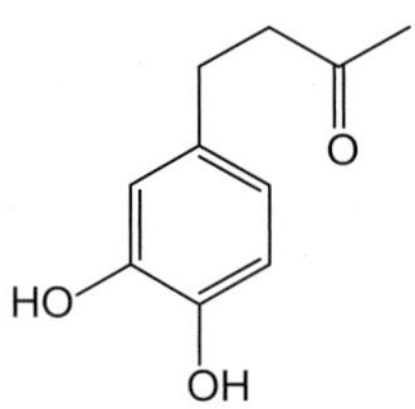

(VI)

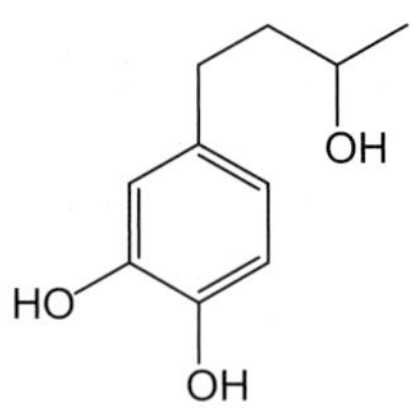

(VII)

REFERENCES

1. Marvin Heuer et al, US Patent 7,772,428 (August 10, 2010)
Northern Innovations and Formulations (Oakville, ON, CA)
2. Shyam K Gupta, US Patent 7,842,723 (November 30, 2010)
Bioderm Research (Scottsdale, AZ)
3. Antoine Piccirilli et al, US Patent 7,589,121 (September 15, 2009)
Laboratoires Expanscience (Courbevoie, FR)
4. Tomohiro Yokota et al, US Patent 7,025,985 (April 11, 2006)
Kanebo, Ltd. (Tokyo, JP)

J. LOTION TO REDUCE FINE LINES

PEG-12 Glyceryl Dimyristate

Application:	Reduction of Fine Lines
Title:	Compositions and methods useful for the reduction of fine lines and wrinkles
US Patent:	7,785,623 (August 31, 2010)
Author:	Brian C. Keller
Assignee:	National Institute of Health (Washington, DC)
Significance:	Two methods for preparing an agent for step over-the-counter consumer product designed to reduce fine lines and wrinkles is provided. Initially, the afflicted skin is treated with warm polyethylene glycol dioleate or PEG-12 glyceryl dimyristate in a lipid vesicle containing polyethylene glycol dioleate or PEG-12 glyceryl dimyristate to induce liposome formation. In the second phase, an electrical voltage from a small battery is applied directly to the facial ointment, which causes fine lines and wrinkles to diminish.

Additive Name	Polyethylene glyceryl dioleate (I)
CAS NUMBER	9009-91-0
Safety Concerns	Reports of the product being contaminated with 1,4-dioxane. Overall, however, the purified agent has shown very limited evidence of organ toxicity.

Additive Name	Polyethylene (PEG-12) glycol dimyristate (II)
CAS NUMBER	9004-89-1
Safety Concerns	Reports of the product being contaminated with 1,4-dioxane. Overall, however, the purified reagent has shown very limited evidence of organ toxicity.

Additive Structures

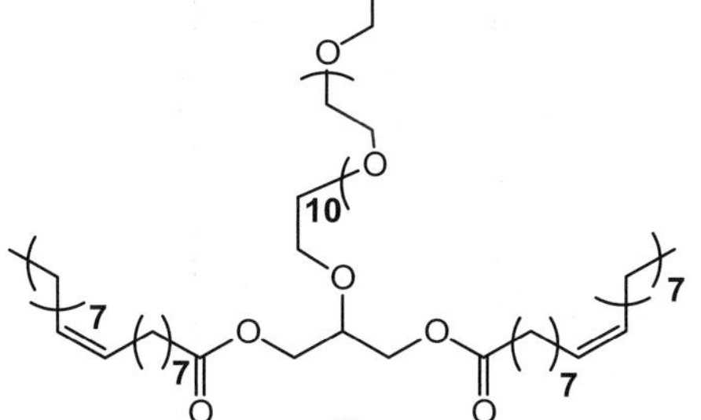

(I)

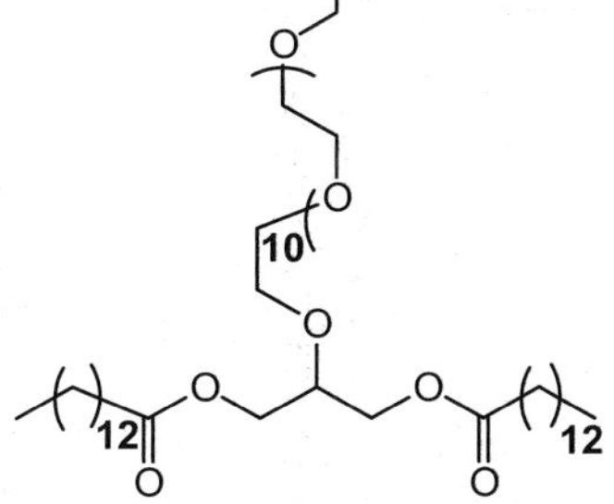

(II)

Additive Preparation Commercial reagent grade supplies of both additive components can be obtained from the following supplier.

Polyethylene Glyceryl Dioleate and Glyceryl Dimyristate

A & E Conock (Perfumery & Cosmetics) Ltd.
Alderhold Mill House
Fordingbridge
Hampshire SP6 1PU
England
44-(1425)-653367

Additive Treatment Formulation

TABLE 14 Summary of a typical gel formation containing polyethylene glyceryl dioleate designed to diminish facial fine lines and wrinkles

Ingredient	Ingredient Concentration (wt%)
Purified water	91.7
Polyethylene Glyceryl Dioleate	5.0
Stearylamine	0.5
Cholesterol	0.1
Phenoxyethanol	1.0
Imidazolidinyl urea	0.3
Triethanolamine	0.3
Methylparaben	0.25
Carbomer 940	0.25
Tocopheryl Acetate (Vitamin E)	0.1
Propylparaben	0.05
Retinyl Palmitate (Vitamin A)	0.45

Optimum delivery of cosmetic ingredients

The optimum method for applying the ointment is by warming the additive and then shaking or vortexing the cream and applying with a plastic applicator to spread the lipid bilayer on the skin. The afflicted area is then treated with a voltage in the range of about 0.5–10 volts from a small battery for a duration of about 1–10 min Table 14.

Advantages over Prior Art

The objective is to prepare an inexpensive over-the-counter kit that can be easily used by consumers. Existing therapies designed to reduce or eliminate fine lines and wrinkles are often invasive, cumbersome, and expensive. These methods may include laser resurfacing, chemical peels, dermabrasion, and cold ablation; implants comprising silicone, microlipoinjection, gore-tex, and collagen and muscle relaxants such as botulinum. Cosmetic surgery has also been used to minimize wrinkles.

Notes

1. In an earlier investigation of Kost,[1] ultrasound was used to enhance transdermal delivery of cosmetic and medicaments through the skin.
2. Keller[2] observed that diacylglycerol PEG lipid A liposome spontaneously formed lipid packing parameters when added to an aqueous solution and was effective in reducing fine lines and wrinkles.
3. Onyuksel[3] developed a method for preparing biologically active liposome agents that were effective in reducing facial lines and wrinkles. The method consisted of using vasoactive intestinal peptides and stabilized liposomes from of egg yolk phosphatidylcholine, egg yolk phosphatidylglycerol, cholesterol, and polyethylene glycol having an Mn of approximately 1900 Da linked to distearoyl- or phosphatidyl ethanolamine.
4. Gagnon[4] prepared an emulsion for promoting neutrophil survival in the treatment of neutropenia arising as a side effect in patients receiving chemotherapy and radiation therapy. The emulsion composition contained medium-chain fatty acids, mono- or diglycerides, or medium-chain triglycerides and had improved survival in vitro assays of neutrophil apoptosis.
5. Oddos[5] prepared a series of cosmetic additives (III)–(V) having retinoid-like properties, which were used in cosmetic compositions to diminish facial fine lines.

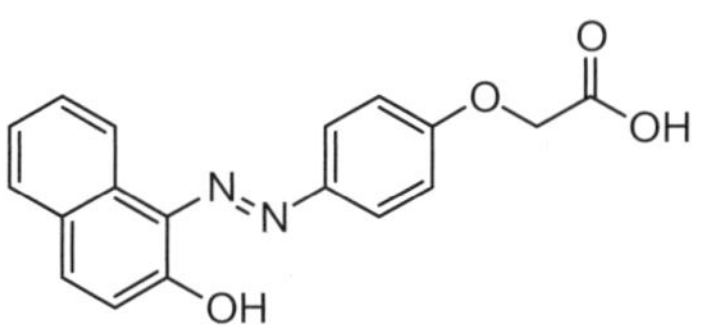

(III)

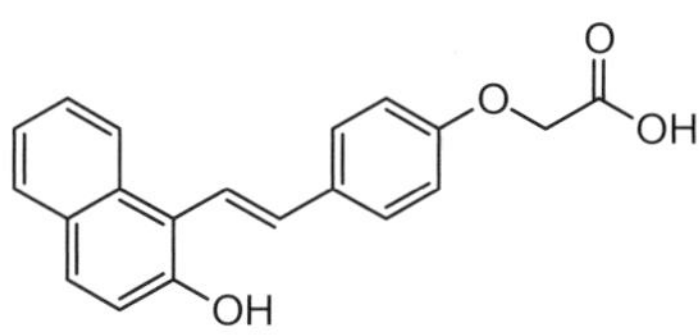

(IV)

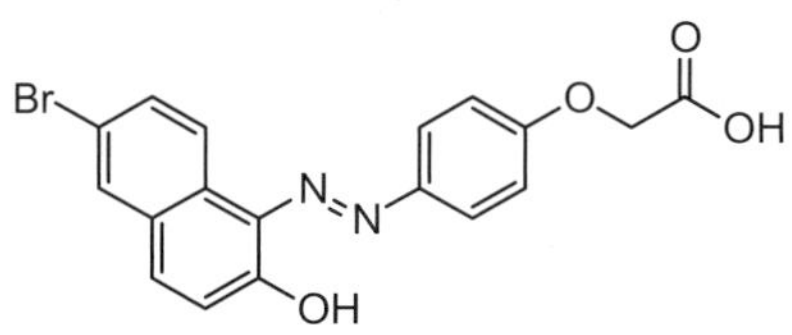

(V)

REFERENCES

1. Joseph Kost et al, US Patent Application 20010056255 (December 21, 2001) **National Institute of Health (Washington, DC)**
2. Brian Charles Keller, US Patent 6,030,930 (August 26, 2003) **National Institute of Health (Washington, DC)**

3. Hayat Onyuksel et al, US Patent 6,348,214 (February 19, 2002)
The Board of Trustees of the Illinois (Urbana, IL)
4. Lyne Gagnon et al, US Patent 7,745,488 (June 29, 2010)
Prometic Biosciences, Inc. (Quebec, CA)
5. Thierry Oddos et al, US Patent 7,803,966 (September 21, 2010)
Johnson & Johnson Consumer France, S.A.S. (France)

K. LOTION TO TREAT EXTREMELY DRY SKIN

N-Alkylsilylsulfamides

Application:	Skin Moisturizing Additive in Cosmetic Compositions
Title:	Sulfamide derivatives and cosmetic use thereof
US:	7,887,824 (February 15, 2011)
Author:	Xavier Marat et al.
Assignee:	L'Oreal (Paris, FR)
Significance:	Defatted strateum corneum was treated with a 3% solution of a selected sulfonamide in water across the transepidermal barrier so that the moisture loss was reduced to approximately ≤1% in a 24-h time period. Favorable moisturizing were also observed in skin care cream formulations additized with 3% - 6% of these experimental agents.
Additive Names	*N*-[Bis(trimethylsilyl)methyl]sulfamide (I) *N,N″*-Bis(2,3-dihydroxy-propyl)sulfamide (II) *N*-[Bis(trimethylsilyl)methyl]-N″-(2- hydroxyethyl)sulfamide (III) *N*-(4-Hydroxybutyl)sulfamide (IV) *N*-[Trimethylsilylmethyl]sulfamide (V) *N,N″*-bis(hydroxyethyl)sulfamide (VI) *N,N″*-Bis(hydroxypropyl)sulfamide (VII) *N,N″*-Bis[(2,2-dimethyl-1,3-dioxolan-4-yl)methyl]sulfamide (VIII) 2-[(Aminosulfonyl)amino]ethyl sulfamate (IX) 3-[(Aminosulfonyl)amino]propyl sulfamate (X)
Safety Concerns	Although essentially nontoxic, these materials are considered potentially very mild respiratory, eye, and skin irritants when used in cosmetic formulations. When formulating with these materials, respiratory and eye protection should be worn. After contact with skin, wash immediately with water.

Additive Structures

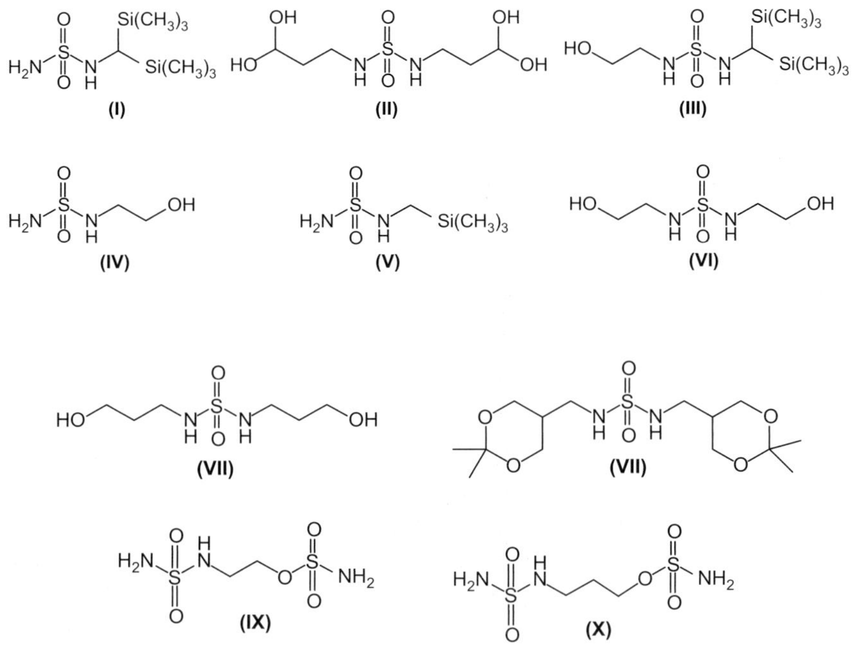

Additive Preparation

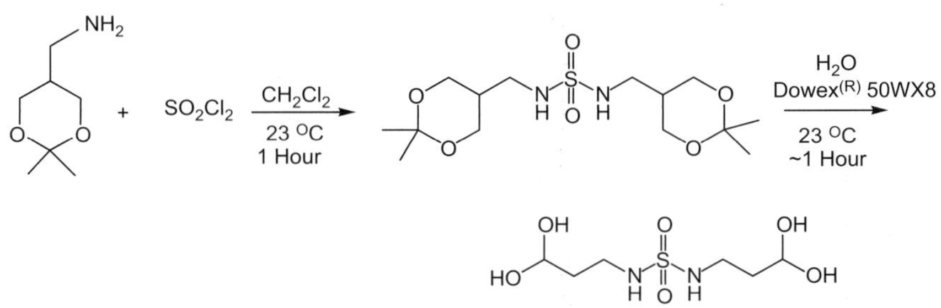

Additive Synthesis

1. *Preparation of N,N″-bis[(2,2-dimethyl-1,3-dioxolan-4-yl)-methyl]sulfamide*

A solution containing 1.55 ml of methamine dioxolane dissolved in 25 ml of methylene chloride containing 1.84 ml of triethylamine at 0 °C and then treated with the dropwise addition of 482 µl of sulfuryl chloride and slowly warmed to 20 °C. The mixture was then diluted with methylene chloride, washed with water and then with saturated sodium chloride solution. The organic phase was dried over sodium sulfate and then concentrated under reduced pressure, and 1.6 g of the product was isolated as a white solid.

2. Preparation of N,N″-(2,3-dihydroxypropyl)sulfamide

The Step 1 product was hydrolyzed by suspending it in water containing Dowex® 50WX8 acidic resin at 23 °C for approximately 1 h while monitoring the reaction progress using thin-layer chromatography. When the hydrolysis was complete, the mixture was filtered, and the filtrate was concentrated under vacuum to provide a vitrified oil. The oil was purified by chromatography using a mixture of diethyl ether and methanol, and 1.16 g of the product was isolated as a white solid having an MP = 80–81 °C.

Testing

1. Total Water Loss

Defatted stratum corneum was initially stored in a sealed container at 30 °C for 24 h and 40% relative humidity and then removed and coated with a 3% solution of a selected experimental agent. Samples were then removed from the container, and water loss measurements were determined using the Evaporimeter EP1 (Servomed, Inc.) after 2 and 24 h. Test results are summarized in Table 15.

2. Moisturizing Effects

Skin care creams were prepared using the formulation provided in Table 16.

TABLE 15 Skin care cream formulation used to evaluate the moisturizing effects containing selected experimental agents

Component	Amount (%)
Additive 1	3.0-6.0
Glyceryl monostearate	0.8
Cetyl alcohol	2.0
Stearyl alcohol	5.0
20 OE (polyoxyethylene stearate)	3.0
Carbopol 941 (crosslinked acrylic acid)	0.3
Caprylic/capric triglycerides	12.0
Preserving agents	qs
Water	qs 100.0%

Test Results

1. *Total Water Loss*

TABLE 16 Effects on water retention using defatted stratum corneum coated with a 3% solution of selected experimental agents. All the measurements were performed using an Evaporimeter EP1

Additive	Two-Hour Water Loss (%)	Twenty-four Water Loss (%)
Additive-I	0	0
Additive-III	4.6	1.0
Additive-IV	1.0	1.0
Control-1 (3% Urea in Water)	64.3	48.3
Control-2 (3% Glycerol in Water)	41.2	33.1

2. *Moisturizing Effects*

The moisturizing effects on human subjects was perceived as favorable for formulations additized with experimental Additives I, II, and IV.

Advantages over Prior Art

It is desirable to maintain the integrity of the stratum corneum since it provides protection against mechanical attack and the passage of chemical products and microorganisms. Sulfonamides described in the current invention limit water loss across the stratum corneum and thereby enhance the barrier elasticity and are extremely effective as moisturizers. In moisturizing formulations containing either glycol or urea, the stratum corneum barrier becomes distorted or compromised resulting in very high water trans-epidermal water loss and secondary afflictions associated with the compromised protective barrier.

Notes

1. Palumbo[1] developed a two-component additive for use in cosmetic skin care formulations consisting of glycerol triacetate (XI) and benzyl dimethyl ammonium chloride (XII), which when applied to the skin did not diminish or compromise the strateum corneum protective layer.

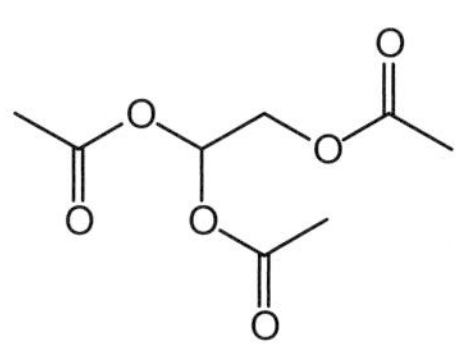

(XI)

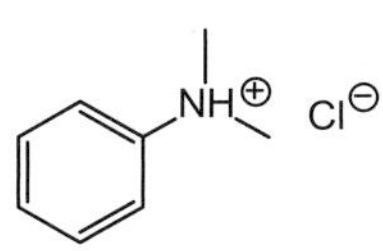

(XII)

2. Piccirilli[2] used 2-alkylfurans provided in Table 17, which were obtained from unsaponifiable avocado matter in cosmetic skin care cream formulations as both a moisturizer and for treating cellulite.

TABLE 17 2-Alkylfuran mixture obtained from unsaponifiable avocado matter. Skin care formulations were additized with 2 wt% of the mixture

2-Alkylfuran	Composition (%)
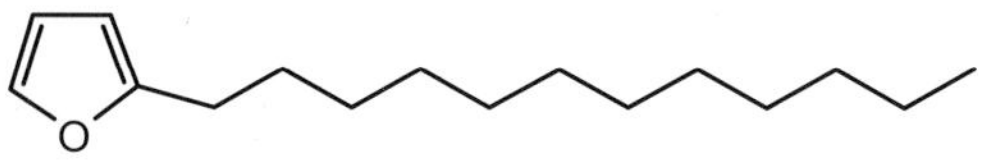	3–12
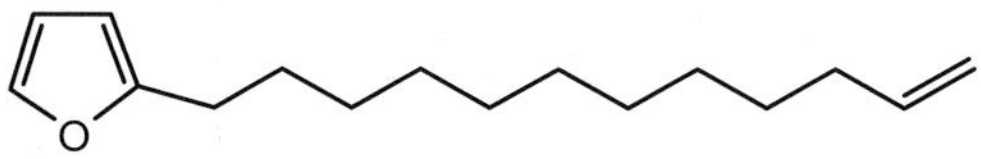	1–8
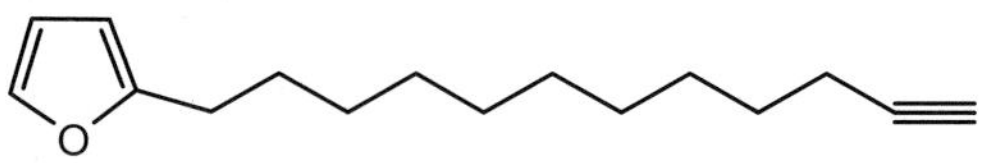	1–5
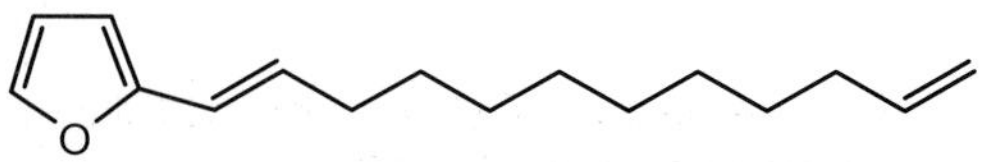	1–6
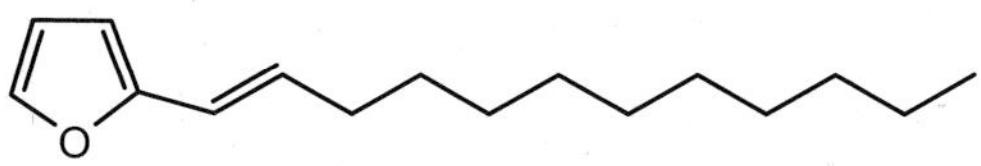	5–20
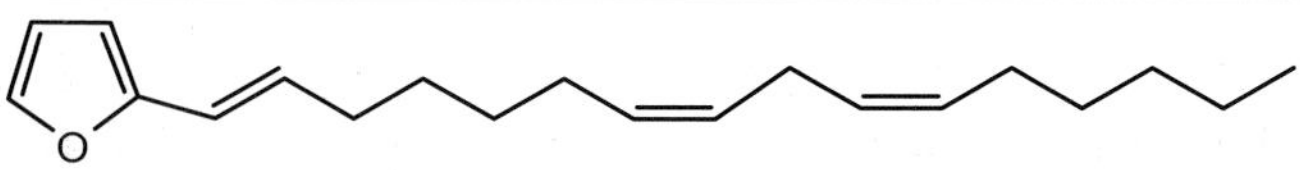	10–30
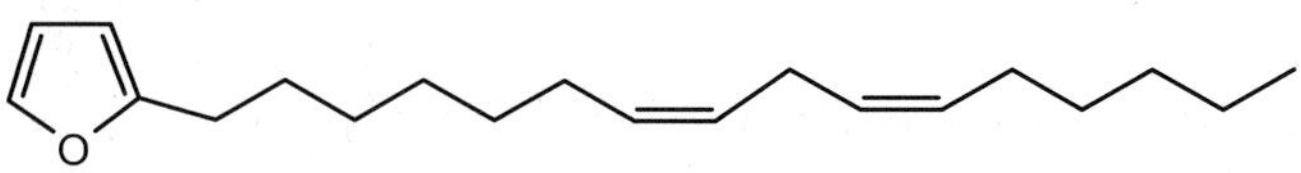	35–75

3. Skin care formulations additized with *N*-(2-hydroxyethyl)urea (XIII) and poly(2-acrylamido-2-methylpropanesulfonic acid) (XIV) were prepared by Moreno[3] and used in cosmetics as moisturizers and for treating mild skin and mucous membrane irritations.

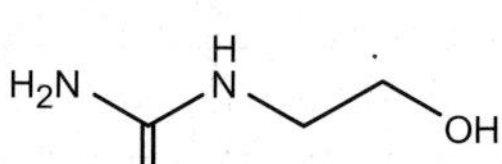

(XIII)

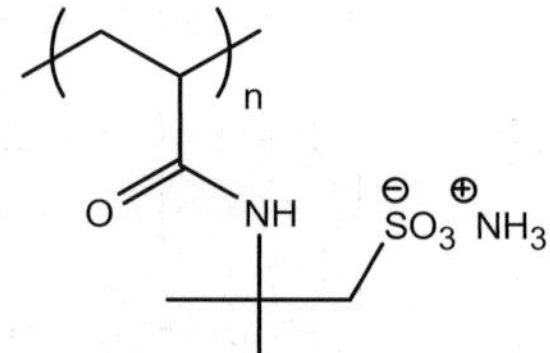

(XIV)

4. Thum[4] prepared ester-modified organopolysiloxanes (XV) emulsifying agents, which when used in cosmetic skin care formulations were effective as moisturizer sealants.

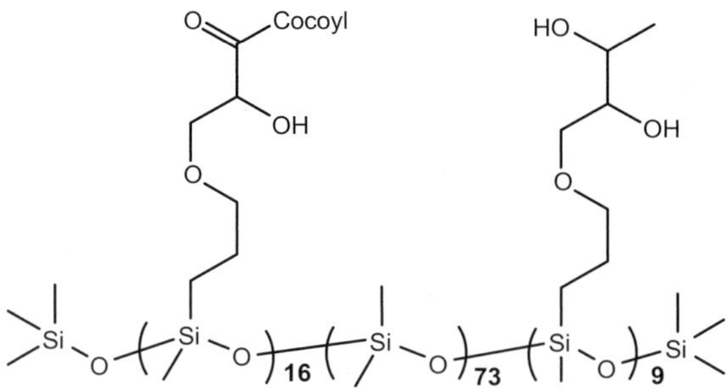

(XV)

REFERENCES

1. Gianfranco Palumbo et al, US Patent 6,855,326 (February 15, 2005) **The Procter & Gamble Company (Cincinnati, OH)**
2. Antoine Piccirilli et al, US Patent 7,872,043 (January 18, 2011) **Laboratories Expanscience (Courbevoie, FR)**
3. Angeles Fonolla Moreno, US Patent 7,615,586 (November 10, 2009) **L'Oreal (Paris, FR)**
4. Oliver Thum et al, US Patent 7,855, 265 (December 21, 2010) **Evonik Goldschmidt GmbH (Essen, DE)**

L. LOTION WITH HIGH EMULSION STABILITY AND LOW CRITICAL AGGREGATION CONCENTRATION

Hyaluronic acid Esters

Title:	Hyaluronic acid derivatives Novozymes Biopolymer A/S (Bagsvaerd, DK)
US Patent:	7,993,678 (August 9, 2011)
Author:	Kristoffer Tommeraas et al.
Assignee:	Novozymes Biopolymer A/S (Bagsvaerd, DK)
Significance:	Hyaluronic acid-g-2-alken-1-yl-succinic acid derivatives were prepared in a single step by condensing hyaluronic acid with 2-alken-1-yl-succinic anhydride. Additives derived from this process improved the emulsifier stability in cosmetics and skin care lotions and had critical aggregation concentrations of 0.002 to as low as 0.0001 mg/ml.

Additives Hyaluronic acid-g-2-octen-1-yl-succinic acid (I)
Hyaluronic acid-g-2-nonen-1-yl-succinic acid (II)
Hyaluronic acid-g-2-dodecen-1-yl-succinic acid (III)
Hyaluronic acid-g-phenylsuccinic acid (IV)

Safety Hyaluronic acid is found in many tissues of the body, such as skin, cartilage, and the vitreous humor and is considered to be nontoxic. Amphiphilic octenyl, nonen, dodecen, and phenylsuccinic anhydride–modified hyaluronic acid derivatives have previously been made and used in cosmetic formulations and are also considered nontoxic.

Additive Structures

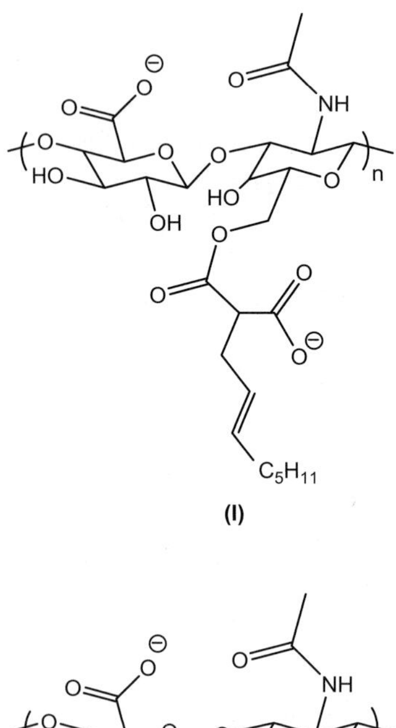

(I)

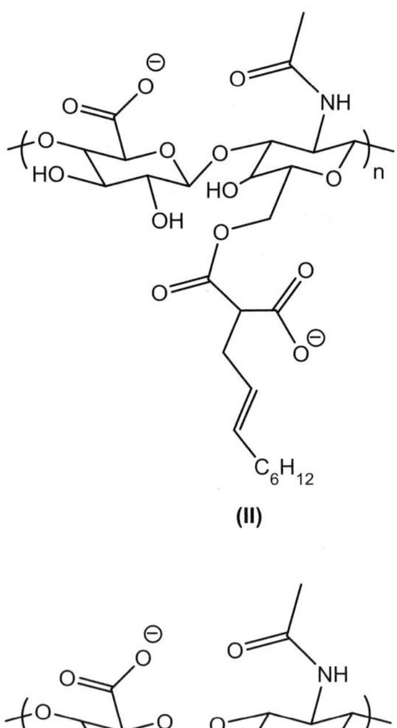

(II)

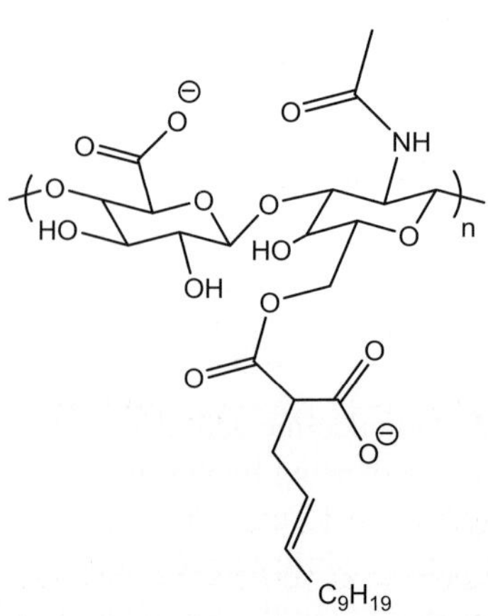

(III)

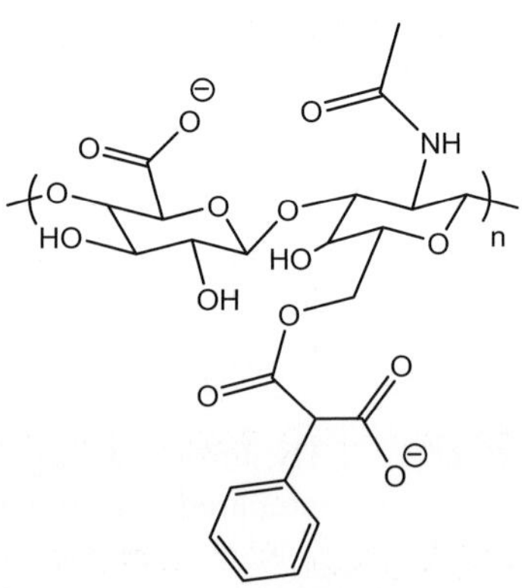

(IV)

Additive Preparation

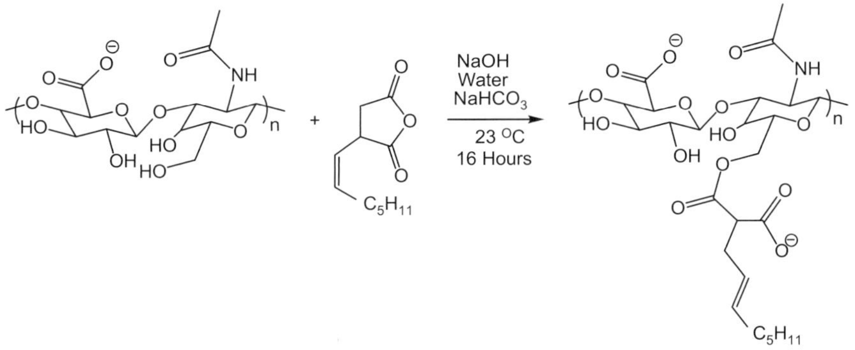

Additive Synthesis

1. Preparation of hyaluronic acid-g-2-octen-1-yl-succinic acid

A reactor was charged with 1.42 g of hyaluronic acid and 200 ml of Milli-Q® water and then stirred overnight to dissolve the contents. The pH of the solution was then adjusted to 9 using 4 M sodium hydroxide and then treated with 1 ml of 2-octen-1-yl-succinic anhydride. After the mixture was stirred for 16 h at 23 °C, it was further treated with 20 ml of saturated sodium bicarbonate, and the pH was adjusted to 6.8 with 1 M hydrochloric acid. The product was isolated by precipitation in four volumes of 96% ethanol followed by centrifugation at 3000 rpm for 15 min at 4 °C and then washed with 96% of ethanol before redissolving in Milli-Q® water and freeze drying.

Testing

1. Emulsion Stability

The four experimental additives and unmodified hyaluronic acid were formulated with three cosmetic oils including mineral oil, diethylhexyl carbonate, and ethylhexyl palmitate according to the following recipe:

1. Six milliliters of oil was added to 14 ml of an aqueous solution of 0.1 M sodium chloride followed by the addition of 0.29% of a selected hyaluronic acid-g-2-alken-1-ylsuccinic acid.
2. The solution was then mixed under strong shearing conditions for 25 s.
3. The emulsions were then left at an ambient temperature in the dark for 8 weeks while periodically inspecting after 24 h and 8 weeks for emulsion stability.

2. Critical Aggregation Concentration Emulsion Stability

The critical aggregation concentration of an experimental derivative was determined by calorimetry using an isothermal titration calorimeter. In this experiment, 0.294 ml of a selected experimentally additive was dissolved in 15 mg/ml of distilled water and then titrated using 1.4615 ml of distilled water in the calorimeter sample cell. A solution of 2 ml of the additive was then used to prepare a standard sample of 15 mg/ml and was then injected every 300 s, and the enthalpy variations were calculated over time. The change in enthalpies were then plotted as a function of the concentration of the additive in the sample cell, and the critical aggregation concentration of the sample was determined. Each experiment was repeated thrice, and the average critical aggregation concentration was reported. Test results for Additive 1 are provided in Table 18.

Test Results

1. Emulsion Stability

The four experimental additives showed increased emulsion stability compared with unmodified hyaluronic acid. This demonstrates that hyaluronic acid-g-2-alken-1-yl-succinic acid can be used as an emulsifier in cosmetics, particularly skin care lotions since hyaluronic acid-g-2-alken-1-yl-succinic acids bind stronger to the skin compared with unmodified hyaluronic acid.

2. Critical Aggregation Concentration

TABLE 18 Critical aggregation concentration test results for Additive 1, hyaluronic acid-g-2-octen-yl-succinic acid

Solution	Phosphate Buffer (M)	Critical Aggregation Concentration (mg/ml)
1	0.15	0.0001
2	0.50	0002
3	1.00	0006
4	1.50	0.001

Advantages over Prior Art

There are several distinct advantages of the current additive. First, additives of the current invention bind more strongly to the skin than do other heteropolysaccharides such as glucosaminoglycans. Moreover, because of the lower molecular weights of the current additives, greater penetration into the skin occurs more efficiently. Second, glucosaminoglycan lotion derivatives require several steps to be prepared in varying yields and require purification. Finally, related heteropolysaccharides have found only limited cosmetic applications because of their high viscosity. The additive of the current invention, by contrast, has a reduced viscosity, which makes it amenable to a wide variety of cosmetic, lotion, and drug delivery applications.

Notes

1. Boupat[1] prepared the additives poly(butyl acrylate-co-acrylic acid)-b-poly(methyl methacrylate-co-acrylic acid) (V) and poly(methyl methacrylate-co-methacrylic acid)-b-poly(butyl acrylate-co-methacrylic acid)-b-poly(methyl methacrylate-co-methacrylic acid) (VI), which were used in skin, lipstick, and hair formulations as moisture enhancers.

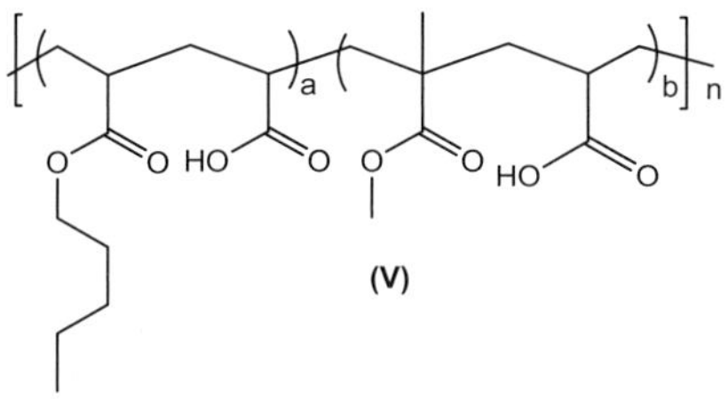

(V)

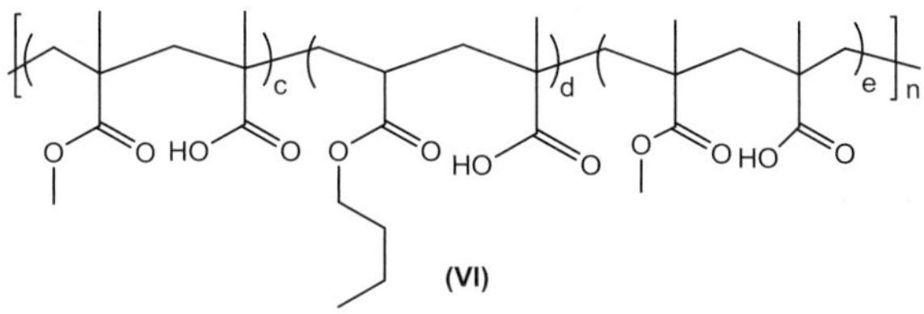

(VI)

2. Li[2] prepared self-emulsifying polyacrylate siloxane copolymer networks, which were soluble in water and prepared from the reaction of an organopolysiloxane (VII) with acrylic acid containing ascorbic acid and hydrogen peroxide, which was used in hair and skin care formulations as a moisturizer.

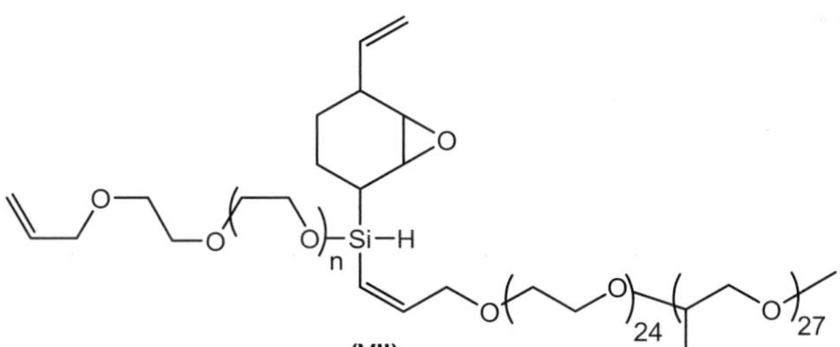

(VII)

3. Nguyen-Kim[1] prepared the water absorbent additive, poly(*t*-butyl methacrylate-ter-acrylic acid-ter-dimethylaminopropyl-methacrylamide) (VIII), which was used in cosmetic skin care formulations as a moisturizing agent.

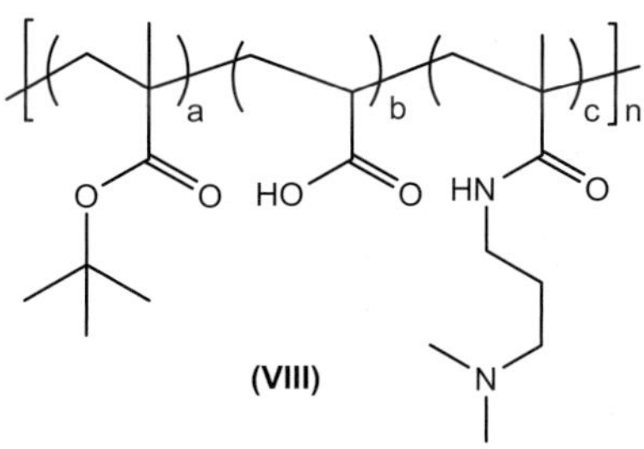

(VIII)

4. Yamato[4] prepared the additives α-N, γ-*N*-di-lauroyl-L-arginine (IX) and α-*N*, γ-*N*-di-palmitoyl-L-arginine, which were used in cosmetic skin care compositions to enhance the moisturizing properties of the formulations without imparting tackiness.

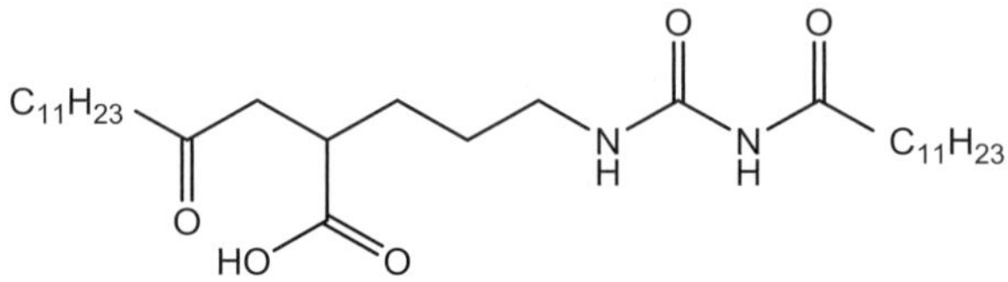

(IX)

REFERENCES

1. Nicolas Passade Boupat et al, 7,951,888 (May 31, 2011)
   L'Oreal S.A. (Paris, FR) Arkema France (Puteaux, FR)
2. Ning Lu et al, 7,833,541 (November 16, 2010)
   Momentive Performance Materials Inc. (Albany, NY)
3. Son Nguyen-Kim et al 7,829,070 (November 9, 2010)
   BASF SE (Ludwigshafen, DE)
4. Naoya Yamato et al, 7,608,250 (October 22, 2009)
   Ajinomoto Co., Inc. (Tokyo, JP)

M. SUNSCREENS FOR UV-A AND UV-B EXPOSURE

γ-Cyclodextrin Inclusion Complexes of Flavonoids

Title:	Flavonoid complexes
US Patent:	7,943,662 (May 17, 2011)
Author:	Christophe Carola et al.
Assignee:	Merck Patent Gesellschaft mit beschrankter Haftung (Darmstadt, DE)
Significance:	Oxidatively stable cosmetic formulations containing flavonoid derivatives encapsulated in hydroxypropyl-γ-cyclodextrin were prepared and found to be effective as sunscreen additives against UV-A and UV-B sunlight exposure in W/O and O/W lotion compositions. Lotions containing these chemical additives were stable and did not undergo color change or loss of activity after formulation.

Additives

7,4′-Dihydroxyflavone-hydroxypropyl-γ-cyclodextrin inclusion complex (I)
6,3′,4′-Trihydroxyflavone-hydroxypropyl-γ-cyclodextrin inclusion complex (II)
5,6,7-Trihydroxyflavone-hydroxypropyl-γ-cyclodextrin inclusion complex (III)
4′-Methoxy-6-Hydroxyflavone-hydroxypropyl-γ-cyclodextrin inclusion complex (IV)
7,8-Dihydroxy-3′,4′-dimethoxyflavone-hydroxypropyl-γ-cyclodextrin inclusion complex (V)
6,7-Dihydroxyflavone-hydroxypropyl-γ-cyclodextrin inclusion complex (VI)

Safety

Both flavone and γ-cyclodextrin derivatives are naturally occurring materials that are currently used as additives in cosmetics, pharmaceutical and food formulations, and do not pose a health hazard.

Additive Structures

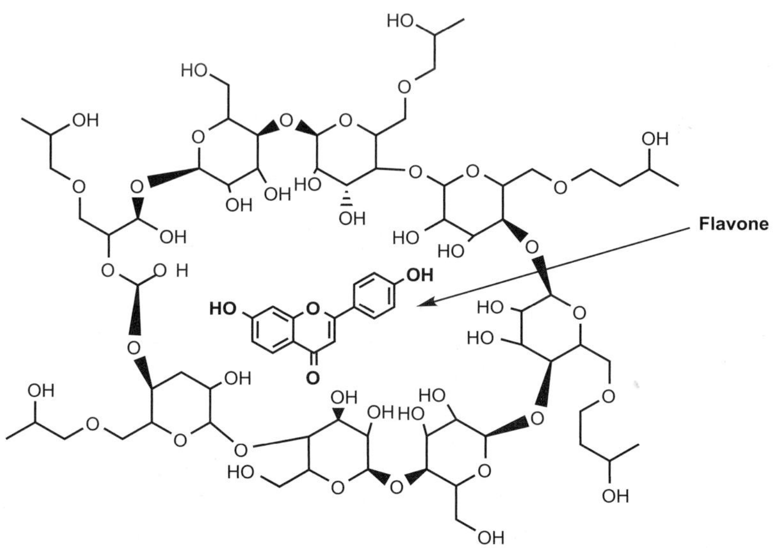

7,4'-Dihydroxyflavone-γ-cyclodextrin inclusion complex

(I)

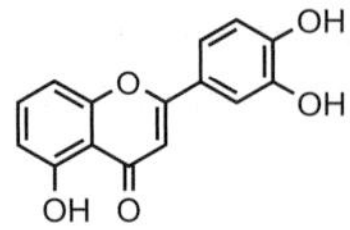

(II)

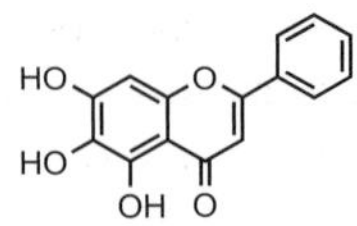

(III)

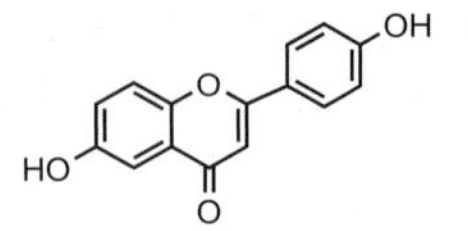

(IV)

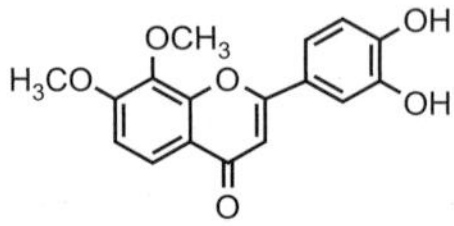

(V)

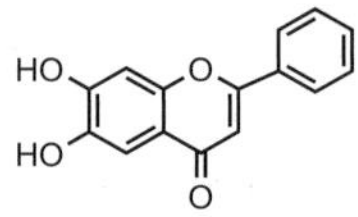

(VI)

Other flavones used in inclusion complexes

Additive Preparation

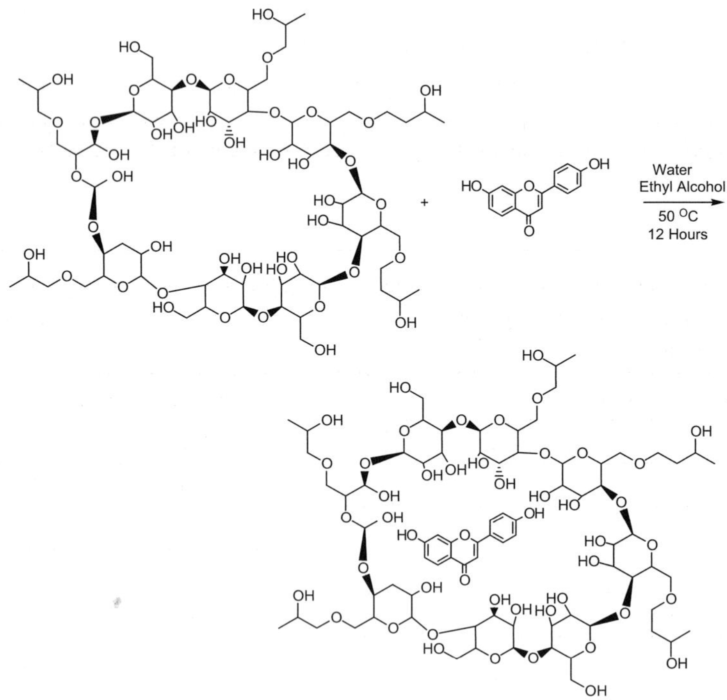

Additive Synthesis

1. *Preparation of 7,4′-dihydroxyflavone-hydroxypropyl-γ-cyclodextrin inclusion complex*

A 250-ml round bottom flask was charged with 3.1 g of hydroxypropyl-γ-cyclodextrin dissolved in 70 ml of water and then heated to 50 °C and treated with the dropwise addition 0.3 g of 7,4′-dihydroxyflavone dissolved in 60 ml of ethanol. The solution was heated for an additional 12 h and then concentrated to dryness at 40 °C at 200 mbar, and the product was isolated as a solid.

Testing

1. Antioxidancy

The antioxidancy of experimental agents was determined using the 2,2-diphenyl-1-picrylhydrazyl assay. In this experiment, 2,2-diphenyl-1-picrylhydrazyl has a strong free radical absorption band at 515 nm and a dark violet color. In the presence of free-radical scavengers, the free-radical electron associated with 2,2-diphenyl-1-picrylhydrazyl becomes paired, and both its absorption and color disappear. The antioxidancy of the tested material is then reported as the concentration of the experimental agent needed to deactivate 50% of 2,2-diphenyl-1-picrylhydrazyl free radical, E_{50}. Antioxidancy test results are provided below.

2. Sunscreen (W/O) Lotion for Skin Application

Sunscreen lotion formulations were prepared containing selected experimental flavone-γ-hydroxypropyl-cyclodextrin inclusion addition complexes and are provided in Table 19. Antioxidancy using these formulations are provided below.

TABLE 19 **W/O skin lotion additized with a selected experimental agent designed to provide photoprotection against UV rays having a wavelength of 280–400 nm**

Ingredient	Composition (wt%)
Polyglyceryl 2-Dipolyhydroxystearate	5.0
Beeswax	0.5
Zinc Stearate	0.5
Hexyl Laurate	9.0
Cetyl Isononanoate	6.0
Shea Butter	0.5
d,l-α-Tocopherol Acetate	1.0
Additive #4 Inclusion Complex	0.5
Glycerin	5.0
Magnesium sulfate Heptahydrate	1.0
Preservatives	q.s.
Water	to 100

Test Results

All experimental additives were effective as antioxidants or UV filters when evaluated using the 2,2-diphenyl-1-picrylhydrazyl free-radical assay or when used in skin care formulations. E_{50} test results were not provided.

Advantages over Prior Art

The inclusion of flavone derivatives into a hydroxypropyl-γ-cyclodextrin complex addresses several concerns. The first is the need for preparing sunscreen products that are both oxidatively stable and have long-term solubility properties in either O/W or W/O formulations. The second objective of this study is to replace flavonoid quercetin-containing sunscreen additives with a more photoactive flavone derivative. Finally, an unanticipated effect of encapsulating flavone derivatives into a hydroxypropyl-γ-cyclodextrin inclusion complex is their use as oral medicaments delivery agents.

Notes

1. A versatile and high yielding method for preparing flavones (VII) and (VIII) used lithiated 2-hydroxy-acetophenone and a selected benzoyl chloride and is described by Buchholz.[1]

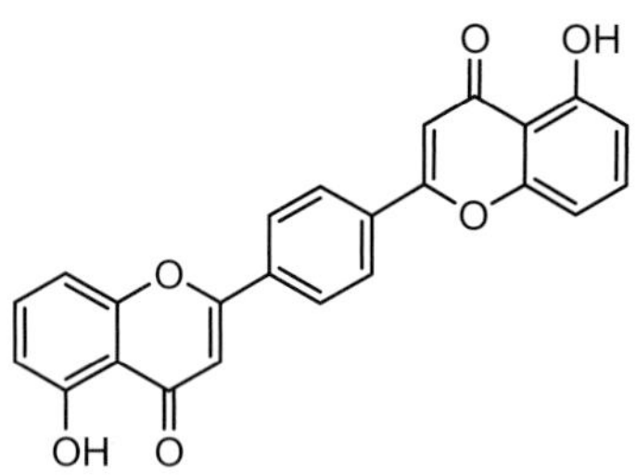

(VII)

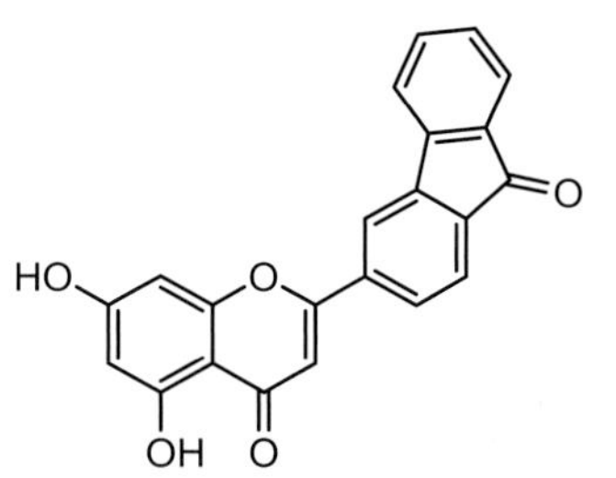

(VIII)

2. Polonka[2] prepared an opaque cosmetic formulation consisting of a polyamide resin having particle sizes of 100–1500 nm and containing octylmethoxycinnamate (IX) or 2-hydroxy-4-methoxy-benzophenone (X) (Benzophenone-3) as the sunscreen agent.

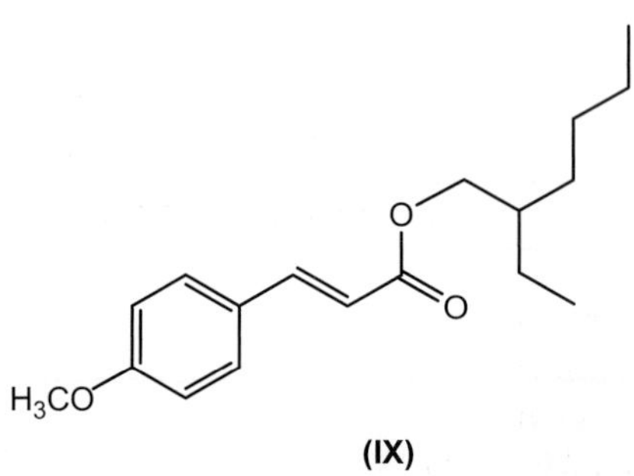

(IX)

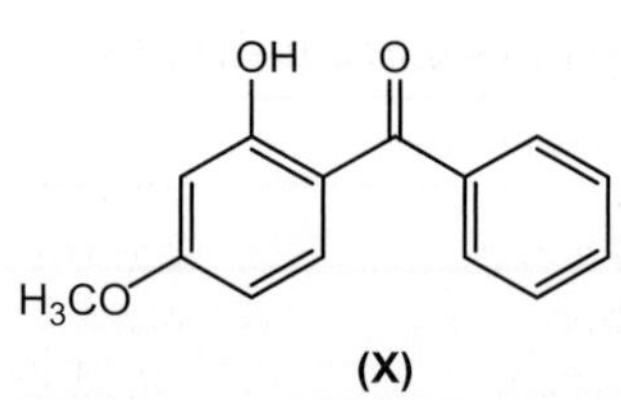

(X)

3. When the flavonoid quercetin (XI) was oxidized to 2-ethoxy-2-(3,4-dihydroxyphenyl)-3,3,5,7-tetrahydroxychroman-4-one (XII) by Carola[3] using oxygen in ethyl alcohol in the presence copper(II) chloride, the oxidized flavonoid had a higher antioxidative activity than its precursor did.

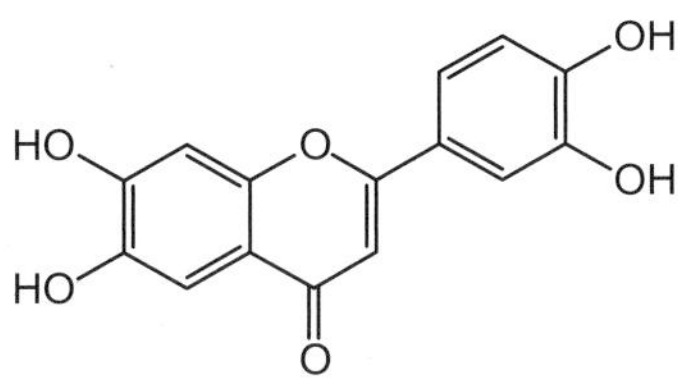

(XI)

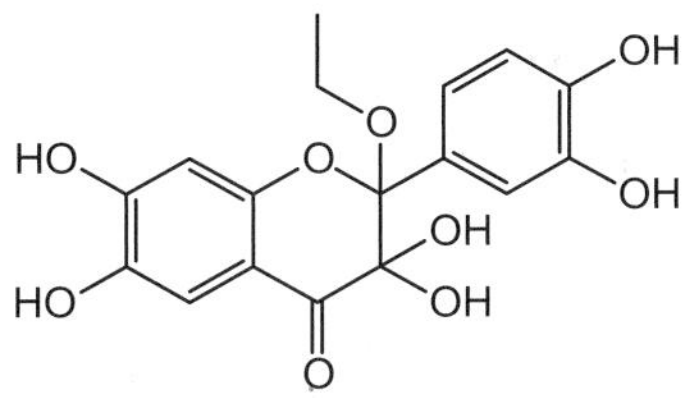

(XII)

4. 2-Oxadiazolechromone derivatives (XIII) prepared by Mujica-Fernaud[4] were used in skin care formulations to provide protection for skin proteins from environmental and age-related skin degeneration stresses.

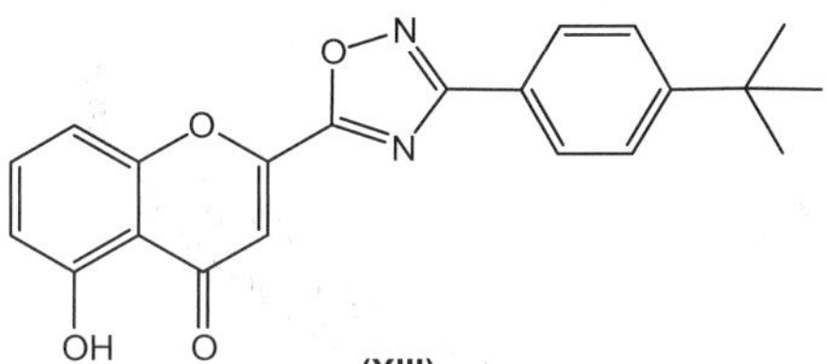

(XIII)

REFERENCES

1. Herwig Buchholz et al, US Patent 7,009,062 (March 7, 2006) **Merck Patent GmbH (Darmstadt, DE)**
2. Jack Polonka et al, US Patent 7,914,772 (March 29, 2011) **Conopco, Inc. (Englewood Cliffs, NJ)**
3. Christophe Carola et al, US Patent 7,867,993 (January 11, 2011) **Merck Patent GmbH (Darmstadt, DE)**
4. Teresa Mujica-Fernaud et al, US Patent 7,772,265 (August 10, 2010) **Merck Patent GmbH (Darmstadt, DE)**

N. SUNSCREENS FOR BOTH UV-A AND UV-B PROTECTION

Polymers Containing Both Benzamalonate and Aminobenzophenone

Title:	Polysiloxane sunscreens
US Patent:	7,906,108 (March 15, 2011)
Author:	Katja Berg-Schultz et al.
Assignee:	DSM IP Assets B.V. (Heerlen, NL)

Significance: A polymeric broad spectrum sunscreen additive designed to enhance both UV-A and UV-B protection was prepared by grafting a selected benzamalonate and an aminobenzophenone in varying ratios onto polydimethylsiloxane. Additives derived from this process were robust UV-A and UV-B absorbers that were easily incorporated in cosmetic lotion formulations or gel applicators.

Additive Name Polysiloxane-g-[2-(4-diethylamino-2-hydroxybenzoyl)benzoic acid-prop-2-ynyl]-g-[2-(prop-2-ynyloxy-benzylidene) diethyl malonate] (I).

Safety No safety data currently exist for polysiloxanes containing grafted UV-A and UV-B sunscreen monomers. In an earlier investigation by Bernascini,[1] however, a polysiloxane containing one of the monomers of the current invention was prepared as shown below. No safety concerns or alerts have been reported from their use.

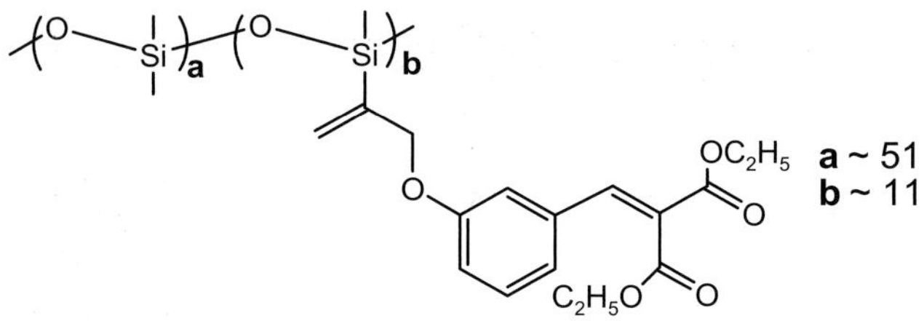

Prior Art

Additive Structure

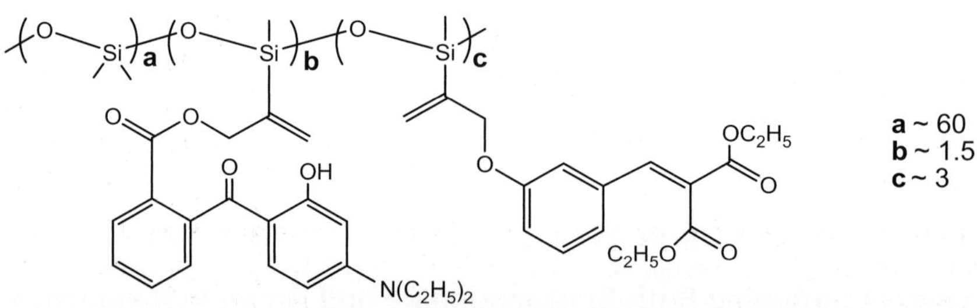

(I)

Current Additive

Additive Preparation

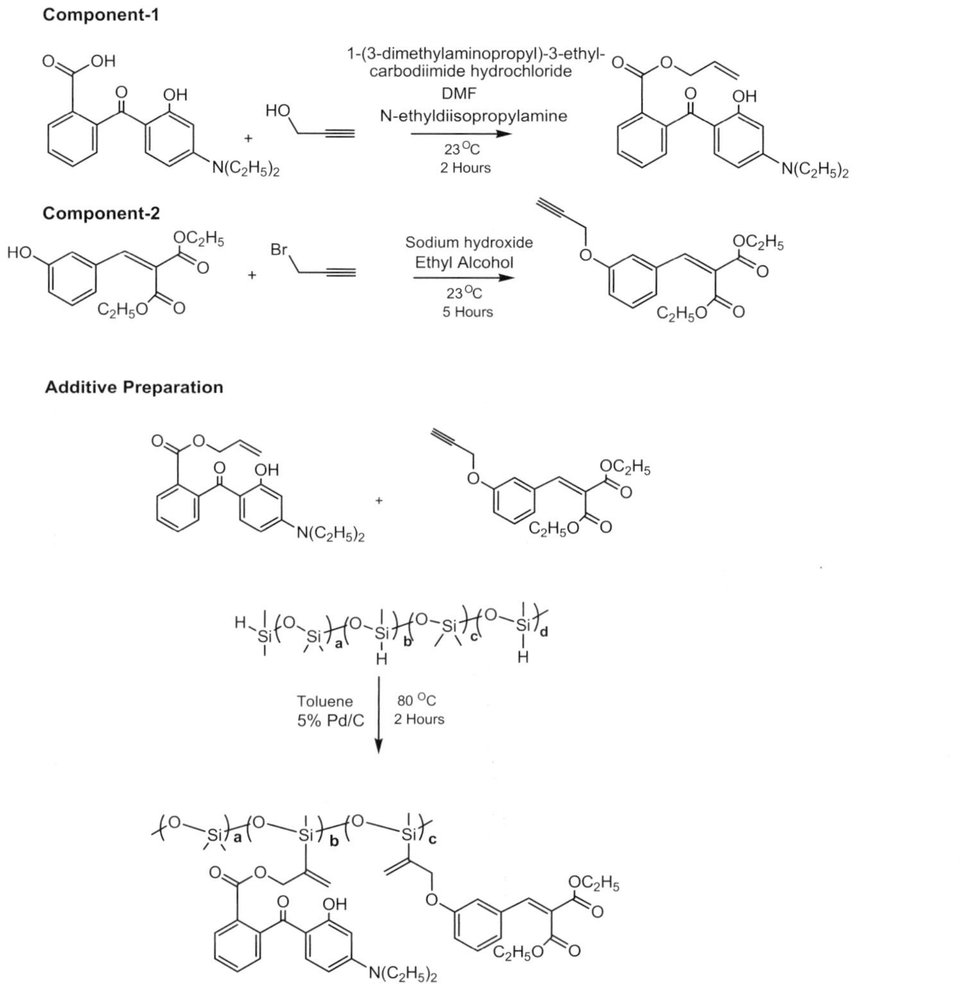

Additive Synthesis (COMPONENT-1)

1. Preparation of 2-(4-diethylamino-2-hydroxybenzoyl)benzoic acid prop-2-ynyl ester

A reactor was charged with a mixture consisting of 200 mmol of 2-(4-diethylamino-2-hydroxybenzoyl)benzoic acid, 500 mmol of propargylic alcohol, 220 mmol of 1-(3-dimethylaminopropyl)-3-ethylcarbodiimide hydrochloride, 500 mmol *N*-ethyldiisopropylamine, and 600 ml of DMF and then stirred at 23 °C for 2 h. Thereafter, the mixture was treated with 1000 ml of ethyl acetate, and the solution was extracted with 1000 ml of water. The organic layer was isolated, dried using sodium sulfate, and then concentrated. The product was purified using flash chromatography with *n*-hexane/ethyl acetate, 2:1, and the product was isolated in an 18% yield.

2. *Preparation of 2-(4-prop-2-ynyloxy-benzylidene)-malonic acid diethyl ester (COMPONENT-2)*

A reactor was charged with a mixture of 200 mmol of 2-(hydroxybenzylidene) diethyl malonate, 200 mmol of propargylic bromide, 50 mmol of sodium ethoxide, and 500 ml of ethyl alcohol and then stirred at 23 °C for 5 h. The mixture was then neutralized with 0.5 M hydrochloric, treated with 200 ml of water, and extracted with 500 ml of ethyl acetate. After concentrating the mixture, it was purified using flash chromatography with *n*-hexane/ethyl acetate/THF, 2:1:0.5, and the product was isolated a 41% yield.

3. *Preparation of polysiloxane graft polymer*

A reactor was charged with 2.20 mmol of the Step 1 product, 2.95 mmol of the Step 2 product, 5.0 g of polysiloxane AE®-151 (Wacker-Chemie) having a 4.15 mmol eq. SiH and 20 ml toluene under an inert atmosphere and then heated to 80 °C. The mixture was then treated with a catalytic amount of 5% platinum on carbon and heated for 24 h. The reaction mixture was then treated with activated charcoal, stirred for 1 h at an ambient temperature, and then filtered through Celite®. The solution was concentrated, and the crude product was purified by preparative GPC. The product was isolated as an oil having a UV-A (EtOH) absorbance of 354 nm ($\varepsilon = 50$) and a UV-B (EtOH) of 312 nm ($\varepsilon = 129$).

Testing

Testing of polysiloxane graft polymer in either lotion formulations or in gel applicators was not reported by the author.

Advantages over Prior Art

While there is no shortage of sunscreen formulations containing aminobenzophenone (Parcol®-1781), cinnamates (Parcol®-MCX), benzamalonates, or triazones (UVASORB®-HEB), polymers containing both these agents as grafts components of polymers are unreported. Sunscreen formulations containing both the polymeric sunscreens poly(dimethylsiloxane-g-aminobenzophenone-g-benzamalonate) prepared in this invention were both robust UV-A and UV-B absorbers, which were prepared in moderate yields in a single-step apiece reactions. Stable w/o lotion formulations containing up to 20% water and gel applicators were prepared, which provided waterproof protection. In addition, the synthetic method used to prepare these graft polymers can also be used to incorporate agents to provide protection against shorter UV wavelengths (UV-C).

Notes

1. A sunscreen formulation prepared by Polonka[2] consisted of polyamide particles derived from reacting 1,6-diaminohexane with 1,4-cyclohexane dicarboxylic acid dispersed octylmethoxycinnamate and benzophenone in W/O or O/W emulsion formulations.

2. A series of 1,4-dihydropyridine ionic derivatives (II) were prepared by Berg-Schultz[3] and were used in sunscreen to provide protection against UV-A.

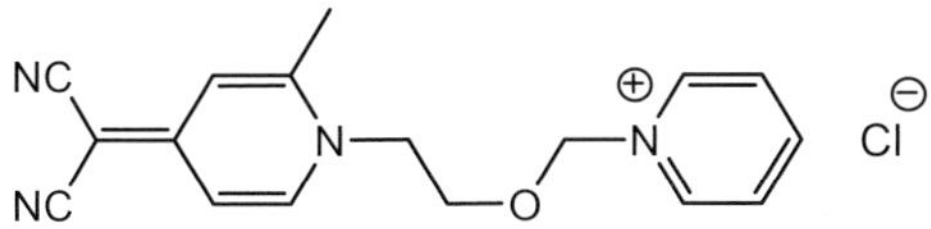

(II)

3. Merocyanine derivatives (III) prepared by Wagner[4] were used as daily sunscreen protection agents against UV-A and UV-B in lotions and cosmetic formulations.

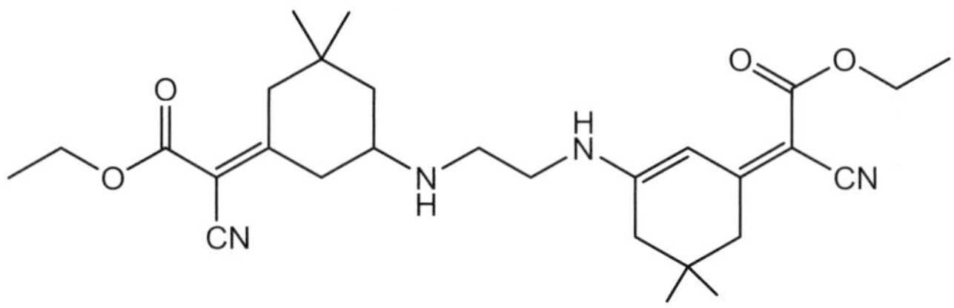

(III)

4. *N*-(2,3-Dihydroxypropyl)cinnamides (IV) prepared by Carola[5] were effective in protecting the skin and the hair against aging, dry skin, wrinkling, or pigment defects, and from exposure to UV-A and UV-B radiation.

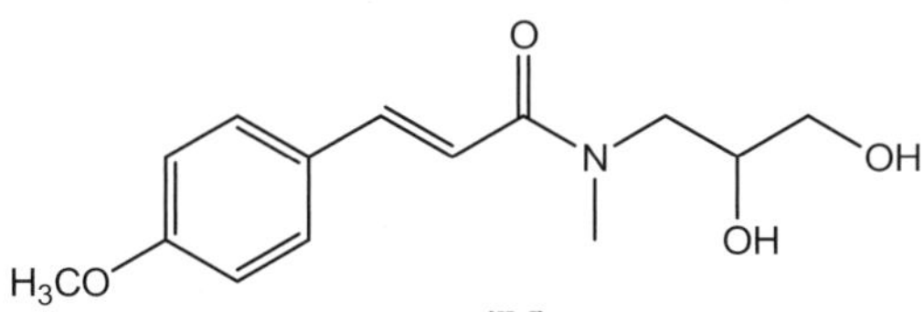

(IV)

REFERENCES

1. Pierclaudio Bernasconi et al, US Patent 6,193,959 (February 27, 2001) **Roche Vitamines Inc. (Parsippany, NJ)**
2. Jack Polonka et al, US Patent 7,914,772 (March 29, 2011) **Conopco, Inc. (Englewood Cliffs, NJ)**
3. Katja Berg-Schultz et al US Patent 7,897,779 (March 1, 2011) **DSM IP Assets B.V. (Heerlen, NL)**
4. Barbara Wagner et al et al, US Patent 7,893,292 (February 22, 2011) **BASF SE (Ludwigshafen, DE)**
5. Christophe Carola et al et al, US Patent 7,863,478 (January 4, 2011) **Merck Patent GmbH (Darmstadt, DE)**

O. SUNSCREENS FOR BOTH UV-A AND UV-B PROTECTION

Polymethacrylate Compositions Effective as UV-A and UV-B Absorbers

Title:	Ultraviolet absorber composition and resin composition
US Patent:	8,044,123 (October 25, 2011)
Author:	Ichiro Amasaki et al.
Assignee:	Fujifilm Corporation (Minato-Ku, Tokyo, JP)

Significance: Triphenyl-s-triazines effective as ultraviolet absorbers in the 340- to 400-nm wavelength range were prepared in quantitative yields in a three-step process. In the first step, salicylamide was reacted with a benzoyl chloride derivative to form an imide. This intermediate was then hydrolyzed to a benzooxazine-4-one and then cyclized to the triazine using phenylamidine. UV-A protection for polymethylmethacrylate blends additized with 0.4 wt% was at 92–97%.

Additive Names 2-(4-Carboxymethoxy)phenyl-4-(2-hydroxy-4-methoxyphenyl)-6-phenyl-s-triazine (I)
2-(4-Carboxy-(2-ethylhexoxy)phenyl-4-(2-hydroxy-4-methoxyphenyl)-6-phenyl-s-triazine (II)
2-(4-Carboxymethoxy)phenyl-4-(2-hydroxy-4-chlorophenyl)-6-phenyl-s-triazine,(III)
2-(4-Carboxymethoxy)phenyl-4-(2-hydroxy-3-naphthyl)-6-phenyl-s-triazine (IV)

Safety s-Triazine derivatives were determined to be skin and eye irritants in rabbits but did not behave as skin sensitizers in humans. Excessive and chronic exposure to these agents resulted in a hunched posture, dyspnea, and reduced locomotor activity in rabbits although full recovery was observed in 7 days in rabbits. The LD50 of the current additives was found to be >2000 mg/kg for related triaromatic s-triazine UV-absorbing agents.

Additive Structures

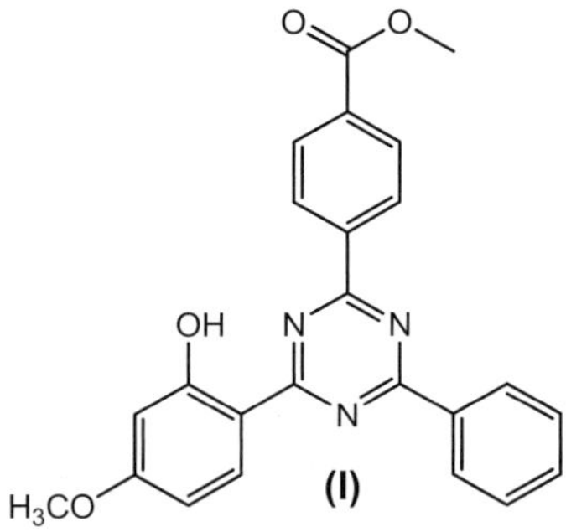

(I)

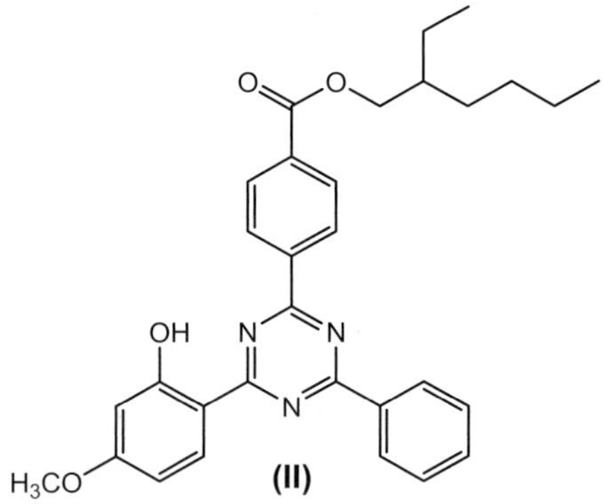

(II)

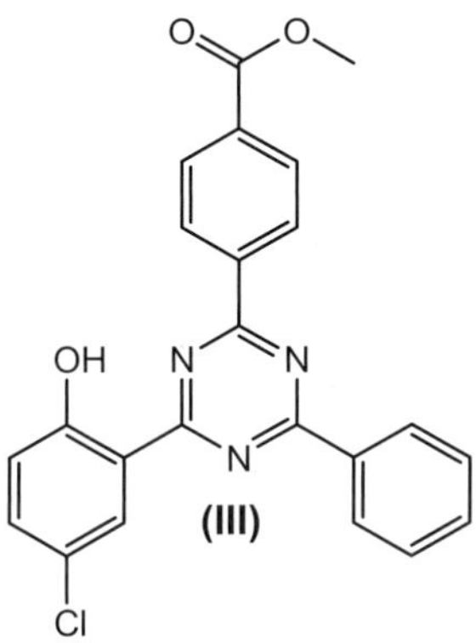

(III)

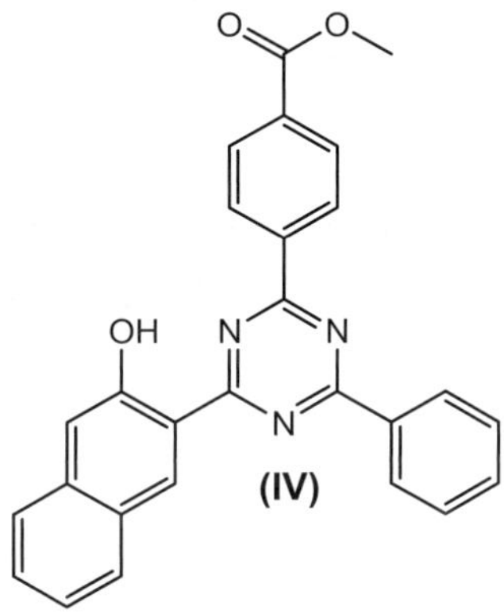

(IV)

Additive Preparation

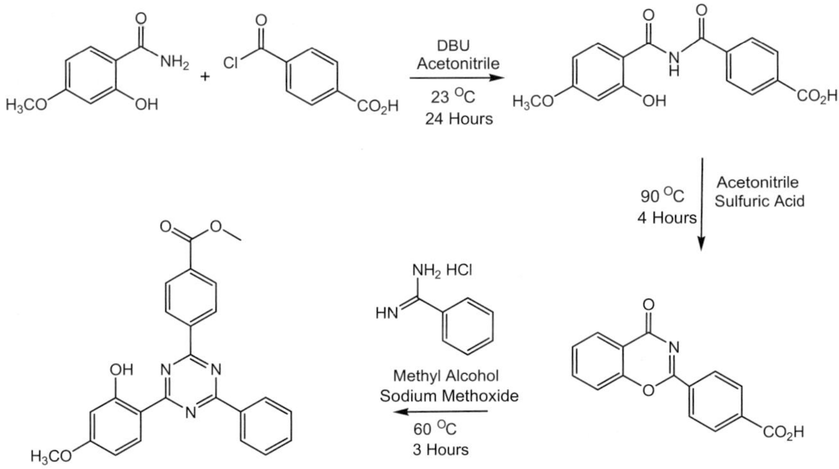

Additive Synthesis

1. Preparation of (2-hydroxy-4-methoxy)phenyl-(4′carboxymethoxy) phenylimide

A round-bottom flask was charged with 20.0 g of 4-methoxysalicylamide, 80 ml of acetonitrile, and 36.4 g of 1,8-diazabicyclo[5.4.0]undec-7-ene and then treated with 23.8 g of methyl 4-(chloroformyl)-benzoate. The mixture was stirred at an ambient temperature for 24 h and then treated with 100 ml of water and 20 ml of hydrochloric acid. A solid precipitate that was formed was collected by filtration and then washed with water, and the product was isolated in a 91% yield.

2. Preparation of 6-methoxy-4H-3,1-benzoxazine-4-one

A reactor was charged with 20.0 g of the Step 1 product, 200 ml of acetonitrile, and 8.9 g of 95% sulfuric acid and was then heated to 90 °C for 4 h. The mixture was then added to 80 ml of triethylamine and then cooled to an ambient temperature. A solid formed was collected by filtration and then washed, and the product was isolated in a 90% yield.

3. Preparation of 2-(4-carboxymethoxy)phenyl-4-(2-hydroxy-4-methoxyphenyl)-6-phenyl-s-triazine

A round bottom containing 2.8 g of benzamidine hydrochloride was treated with 100 ml of methanol, 3.4 g of a 28% solution of sodium methoxide dissolved in methanol, and 5.0 g of the Step 2 product and then stirred at 60 °C for 3 h. The mixture was then cooled to an ambient temperature and treated with 0.2 ml of hydrochloric acid. The precipitate which was formed collected by filtration and washed successively with water and methanol, and the product was isolated in a 92% yield.

Additive Testing

Ultraviolet Absorbance Testing Using Additives Encapsulated in Polymethylacrylate Resin

Polymethylacrylate resin was dissolved in methylene chloride to obtain a 22 wt% solution and then blended with 0.4 wt% of a selected additive. The UV absorber composition was then exposed to a wavelengths at 340–400 nm for 400 h. The retention of the each UV absorber was calculated at 400 nm according to the following formula:

$$\text{Retention (\%)} = 100\frac{100 - \text{transmittance after irradiation}}{100 - \text{transmittance before irradiation}}$$

Test results are provided in Table 20.

Test Results

TABLE 20 Efficiencies of selected s-triphenyl-s-triazines blended in polymethylmethacrylate additized with 0.4 wt% of experimental ultraviolet UV-A and UV-B absorbers

Additive	Retention (%)
I	93
II	93
III	97
IV	92
Tinuvin®-1277FF (Note 1)	33

Advantages over Prior Art

There are a very limited number of chemical additives that have ultraviolet shielding properties in the 340- 400-nm wavelength range. Encapsulating the desired additive in a polymethacrylate matrix to remain affixed to the skin is unreported. Although the current ultraviolet shielding additives require three synthetic steps to prepare, yields are essentially quantitative, and triazine derivatives are effective in the range of 340–400 nm from 92% to 97%.

Notes

1. Tinuvin® 1577 FF (2-(4,6-diphenyl-1,3,5-s-triazine (V) is an ultraviolet absorber, which is added to powdered or granular polymeric resins in treatment levels from 0.1 wt% to 8%.

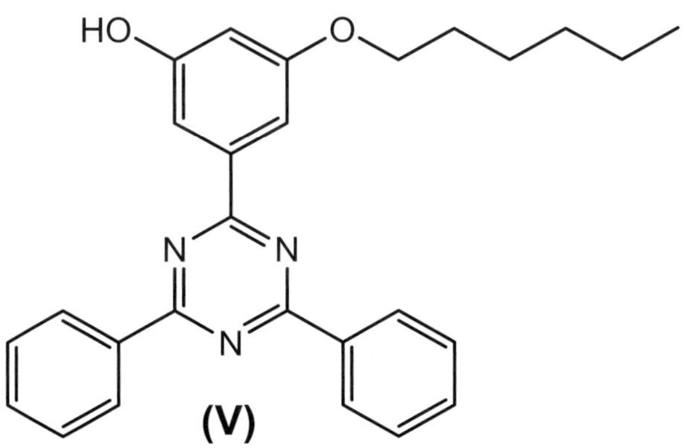

(V)

2. Heterocyclic derivatives (VI) and (VII) prepared by Amasaki[1] were effective ultraviolet light absorbing agents in the UV-A range.

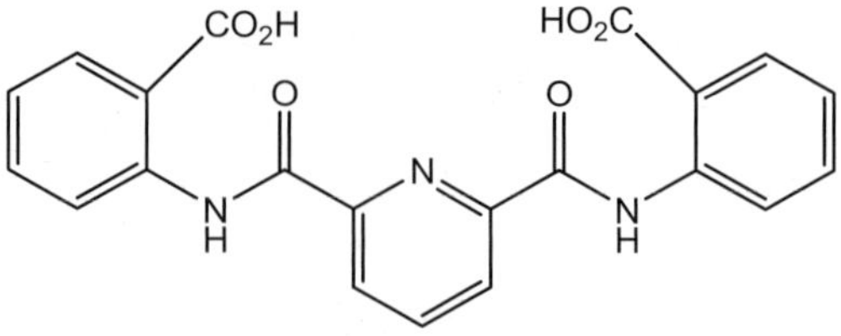

(VI)

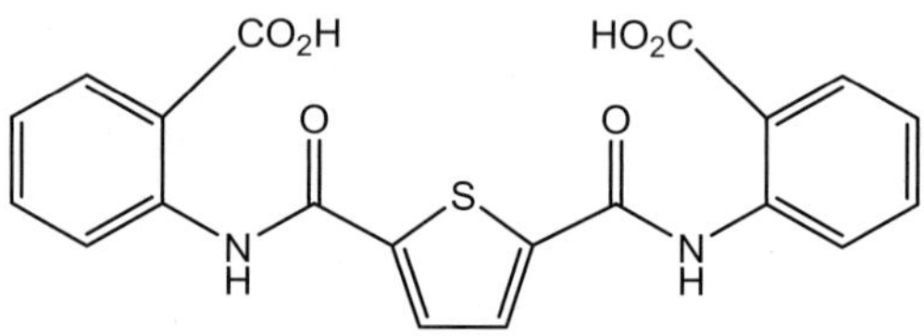

(VII)

3. Anino (VIII) and amide (IX) triphenyl-s-triazines prepared by Reinehr[2] were effective as ultraviolet absorbers and used in cosmetic formulations.

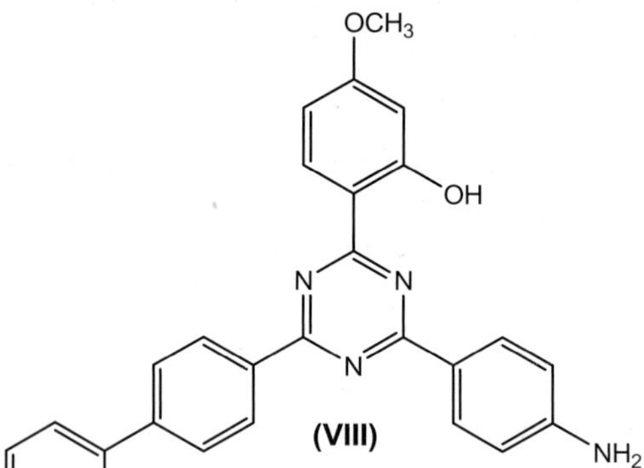

(VIII)

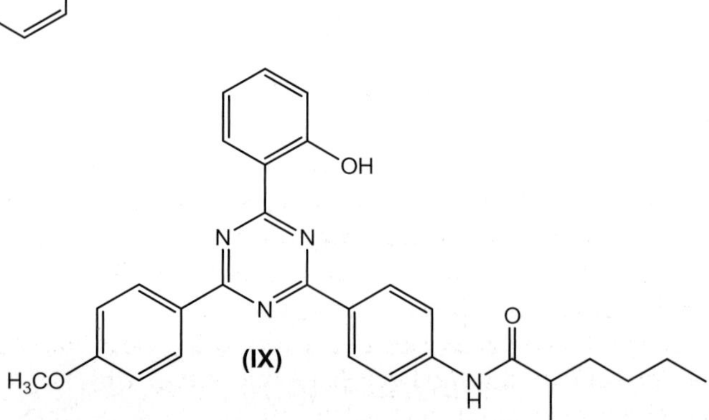

(IX)

4. Solanky[3] modified existing benzotriazole ultraviolet absorbers, which were effective as both antiozonants (X) and antioxidants and used in the protection of rubber.

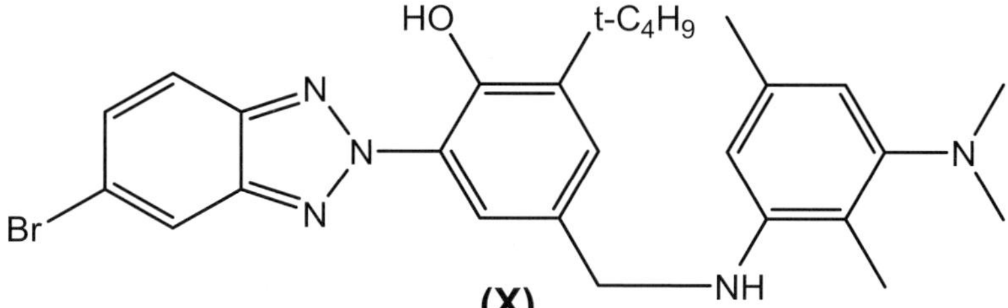

(X)

5. Gupta[4] prepared ultraviolet light absorbers consisting of phenylether-substituted hydroxyphenyl triazine (XI), which were used as UV-A and UV-B skin protectors in polymer blends.

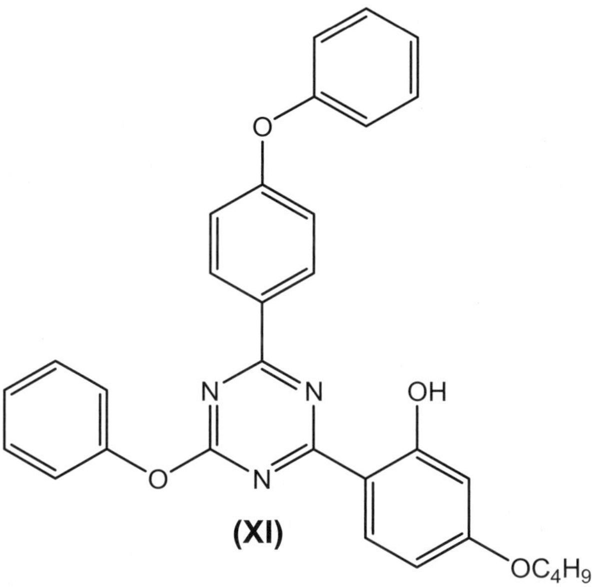

(XI)

REFERENCES

1. Ichiro Amasaki et al, US. Patent Application 20100004439 (January 7, 2010) **Fujifilm Corporation (Minato-Ku, Tokyo, JP)**
2. Dieter Reinehr et al, US. Patent 7,476,699 (January 13, 2009) **Ciba Specialty Chemicals Corp. (Tarrytown, NY)**
3. Shailendra Singh Solanky et al, US. Patent 7,211,675 (May 1, 2007) **Council of Scientific and Industrial Research (New Delhi, IN)**
4. Ram Baboo Gupta et al, US. Patent 6,855,269 (February 15, 2005) **Cytec Technology Corp. (Wilmington, DE)**

P. SUNSCREENS FOR BOTH UV-A AND UV-B PROTECTION

Oxidized Flavonoid Derivatives

Title:	Preparation containing oxidized flavonoid derivatives
US Patent:	7,867,993 (January 11, 2011)
Author:	Christophe Carola et al.
Assignee:	Merck Patent GmbH (Darmstadt, DE)

Significance: Oxidation of the quercetin produces a flavonoid derivative whose antioxidant potential is greater than that of quercetin. The enhanced antioxidant, 2-alkoxy-2-(3,4-dihydroxyphenol)-3,3,5,7-tetrahydroxychroman -4-one was prepared by oxidizing Quercetin and had a considerably higher antioxidancy than that of the predecessor and is readily soluble in water.

Additive Names 2-Ethoxy-2-(3,4-dihydroxyphenol)-3,3,5,7-tetrahydroxychroman-4-one (I)
2-Butoxy-2-(3,4-dihydroxyphenol)-3,3,5,7-tetrahydroxychroman-4-one (II)

Safety Concerns Flavonoid antioxidants are used as food stabilizers, additives, and nutritional supplements and are considered nontoxic.

Additive Structures

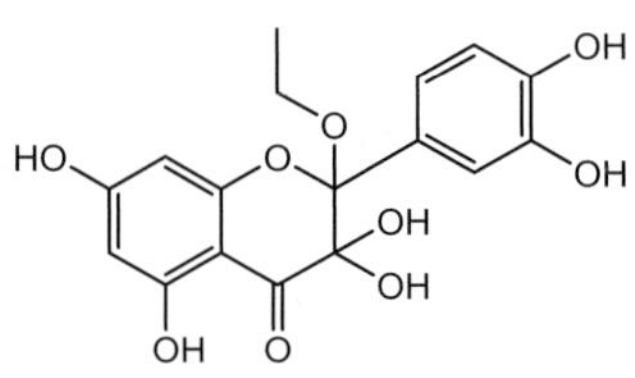

(I)

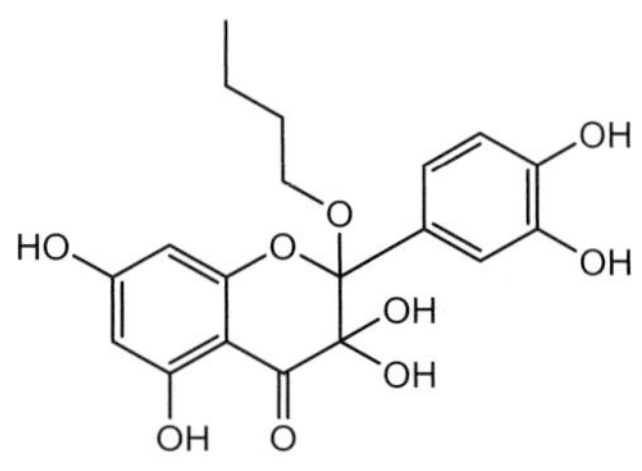

(II)

Additive Preparation

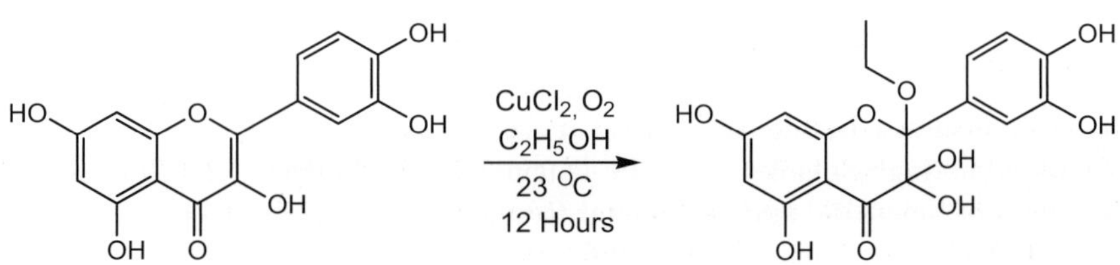

Additive Synthesis

1. Preparation of 2-ethoxy-2-(3,4-dihydroxyphenyl)-3,3,5,7-tetrahydroxychroman-4-one

A reactor was charged with 3.0 g of Quercetin dissolved in 320 ml of ethanol and then treated with 0.7 g of copper(II) chloride and stirred vigorously for 12 h under an oxygen atmosphere at 23 °C. The reaction solution was then concentrated and treated with 70 ml of 10% hydrochloric acid, and 2.6 g of the product was isolated as a beige solid.

Testing

Determination of the Antioxidative Action

The antioxidative action was investigated using the 2,2-diphenyl-1-picrylhydrazyl assay. In this test, the strong absorption band at 515 nm of 2,2-diphenyl-1-picrylhydrazyl free-radical diminishes as the electron becomes paired by a selected experimental agent. To quantify the antioxidative effect of the antioxidant, EC_{50} was determined, which is the concentration of the experimental agent, which deactivates 50% of 2,2-diphenyl-1-picrylhydrazyl free radical. The lower the EC_{50} value, the higher the capacity of the experimental agent to scavenge free radicals. Test results are provided in Table 21.

Test Results

For 2-butoxy-2-(3,4-dihydroxyphenyl)-3,3,5,7-tetrahydroxychroman-4-one, an EC_{50} value was obtained and compared with an EC_{50} value for Quercetin. Surprisingly, Additive 2 had a higher antioxidative activity than did Quercetin, which is regarded as a very strong antioxidant. Test results are provided in Table 21.

TABLE 21 Evidence of the enhanced antioxidancy of Additive 2 over Quercetin using the 2,2-diphenyl-1-picrylhydrazyl assay. Lower EC_{50} values are preferred

Antioxidant	EC_{50}
Quercetin	0.89
2-Butoxy-2-(3,4-dihydroxyphenol)-3,3,5,7-tetrahydroxychroman-4-one (Additive-2)	0.08

Advantages over Prior Art

Oxidized flavonoid derivatives of the current invention make these additives especially suitable as potent sunscreens. In addition toxicity for these and related chemical agents have not been reported and their water solubility enables then to be used in a wide variety of formulations for consumer products.

Notes

1. 2-Oxadiazolechromone derivatives (III) antioxidants prepared by Mujica-Fernaud[1] were used in skin cream formulations to minimize the effects of photo-aging and to protect the skin from stress associated with photoaging.

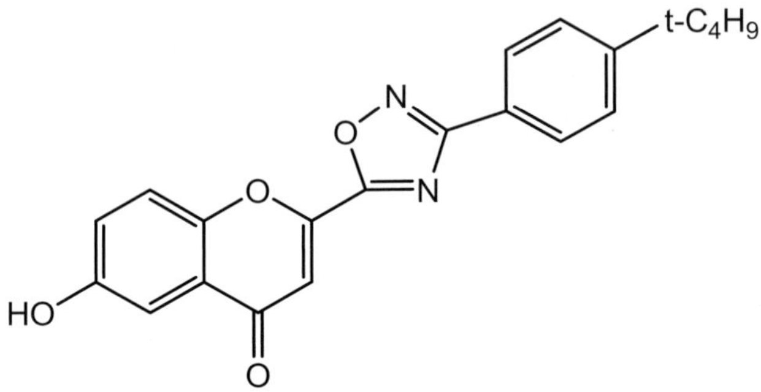

(III)

2. Buchholz[2] extracted the sulfonic acid monoester of a flavonoid derivative (IV) from Sidastrum micranthum, which was then used in cosmetic preparations for protection against photoaging, oxidative stress, inflammation, and skin irritation.

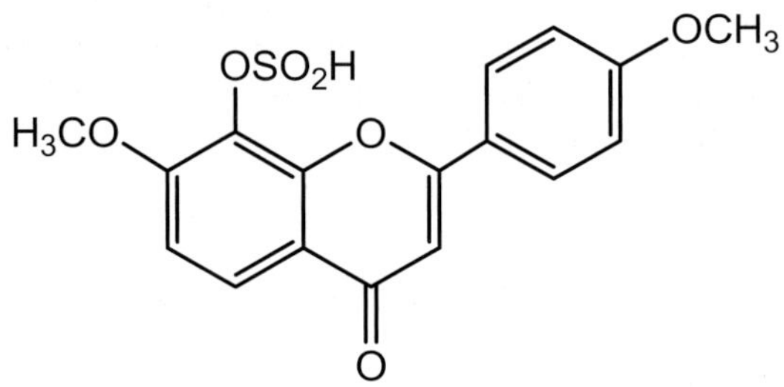

(IV)

3. UV-A light-protection agent using polyester additives (V) was prepared by Bonda[3] and used in tanning oil formulations.

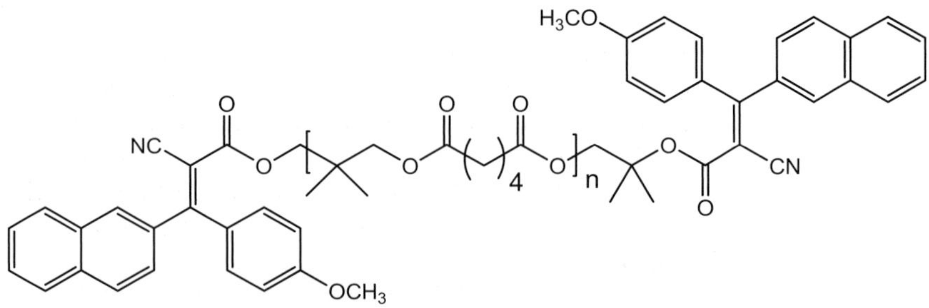

(V)

4. 5,6-Diphenyl-1,2,4-triazinic dimeric derivatives (VI) prepared by Picoul[4] were effective as UV-A and UV-B blocking agents in sun protection formulations.

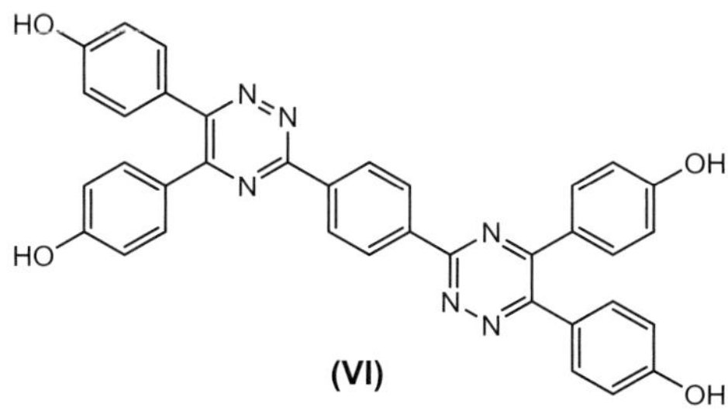

(VI)

REFERENCES

1. Teresa Mujica-Fernaud et al, US Patent 7,772,265 (August 10, 2010)
   Merck Patent GmbH (Darmstadt, DE)
2. Herwig Buchholz et al, US Patent 7,588,783 (September 15, 2009)
   Merck Patent GmbH (Darmstadt, DE)
3. Craig A. Bonda et al, US Patent 7,799,317 (September 21, 2010)
   Hallstar Innovations Corp. (Chicago, IL)
4. Willy Picoul et al, US Patent 7,816,520 (October 19, 2010)
   Pierre Fabre Dermo-Cosmetique (Boulogne-Billancourt, FR) Centre National de la Recherche Scientifique (CNRS) (Paris, FR) Universite Claude Bernard Lyon 1 (Villeurbanne, FR)

Q. NONIONIC SURFACTANTS FOR SHAMPOOS AND SOAPS

1,2,3-Trihydroxy-4-oxa-tetradecane

Title:	Glyceryl ether compounds and their use
US Patent:	8,026,378 (September 27, 2011)
Author:	Sergey Selifonov
Assignee:	Segetis, Inc. (Golden Valley, MN)

Significance: 4-Oxa-tetradecyl and 4-oxa-doeicosyl derivatives containing internal and terminal triols were prepared by reacting 1,2-decylylepoxide and 1,2-eicosyl epoxide with solketal and then hydrolyzing the intermediate acetal. These materials were determined to be suitable as nonionic surfactants in personal care products, particularly soaps and as emulsifiers in cosmetic formulations.

Additive Names 1,2,3-Trihydroxy-4-oxa-tetradecane (I)
1,2,Dihydroxy-5-hydroxymethyl-4-oxa-tridecane (II)
1,2,6-Trihydroxy-4-oxa-doeicosane (III))
1,2,Dihydroxy-5-hydroxymethyl-4-oxa-uneicosane (IV)

Safety Safety concerns associated with hydroxyethers prepared from glycerol and fatty chains or epoxyalkanes in the absence of an organic solvent are unreported particularly when used in soap and in cosmetic formulations as surfactants.

Additive Structures

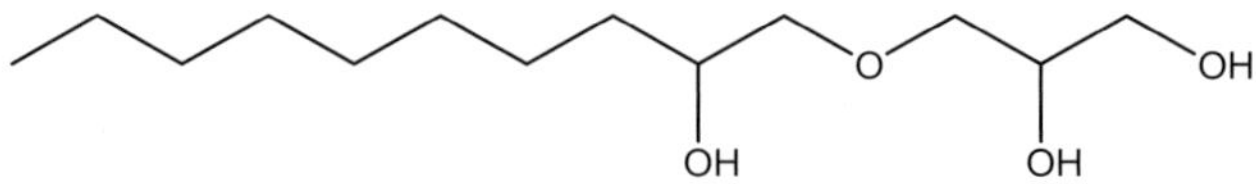

(I)

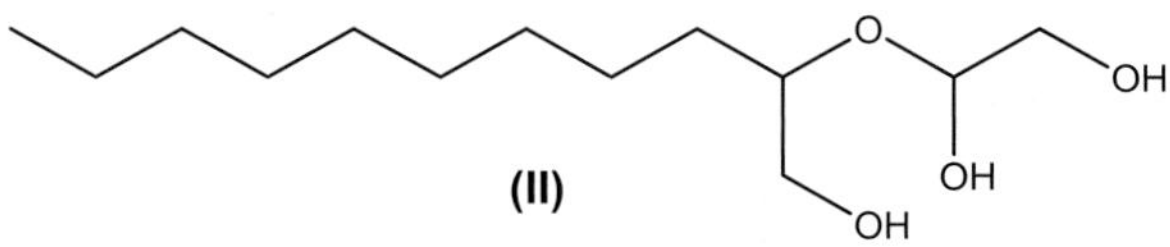

(II)

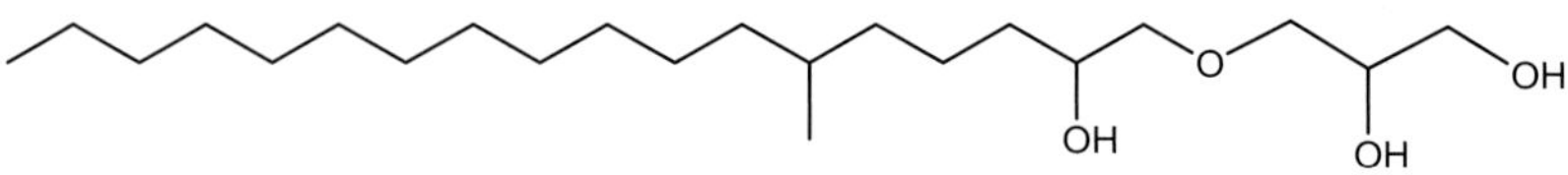

(III)

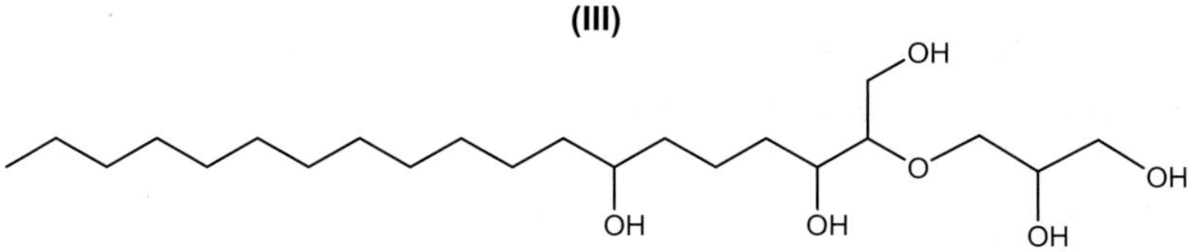

(IV)

Additive Preparation

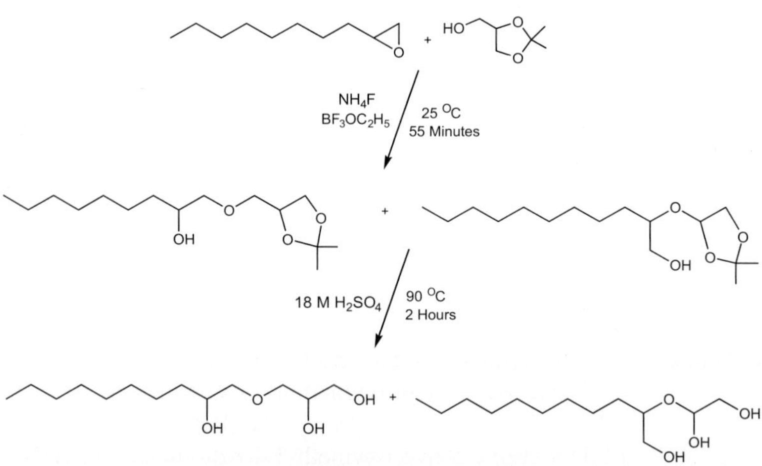

Additive Synthesis

1. *Preparation of 4-hydroxy-2-oxa-1-(5-(2,2,-dimethyl-1,3-dioxolanyl)) dodecane and 3-hydroxdymethyl-2-oxa-1-(5-(2,2,-dimethyl-1,3-dioxolanyl))undecane*

A reactor was charged with 5 g of 94% pure decene-1,2-oxide dissolved in 15 ml of solketal and then stirred at an ambient and treated with 0.05 ml of boron trifluoride diethyl etherate. The reaction mixture was then stirred for an additional 20 min, when a reaction exotherm occurred and increased the temperature to 55 °C. The reaction mixture was then treated with 0.5 g of ammonium fluoride, and the suspension was stirred for 2 h. After filtering and concentrating the mixture under reduced pressure, 7.8 g of the product mixture was isolated.

2. *Preparation of 1,2,3-trihydroxy-4-oxa-tetradecane and 1,2-dihydroxy-5-hydroxymethyl-4-oxa-tridecane*

A small reaction vessel was treated with 1 g of the Step 1 product and 0.01 g of sulfuric acid and then heated to 90 °C for 2 h. After cooling to an ambient temperature, the mixture was neutralized with 0.1 g calcium carbonate and then filtered, concentrated under reduced pressure, and 0.6 g of the product mixture isolated was viscous oily–waxy opalescent with a pearl-like appearance.

Additive Testing

Testing data are not supplied by the author.

Advantages over Prior Art

The current invention is distinguished from other nonionic surfactants and emulsifiers by their favorable economics and versatility. First, epoxides and polyepoxides are available from inexpensive precursors, including α-olefins and linseed and soybean oils. These materials can be easily converted into di-, tri-, and polyalcohols by acid hydrolysis and can be used in personal care, cleaning, and laundry formulations as well as in industrial applications.

Notes

1. The author also provided an example of the reaction of nonterminal epoxy (V), which was subsequently converted into internal alcohols and (VI) and (VII) by reacting with solketal and then postreacting with sulfuric acid as illustrated in Eq. 1.

Eq. 1.

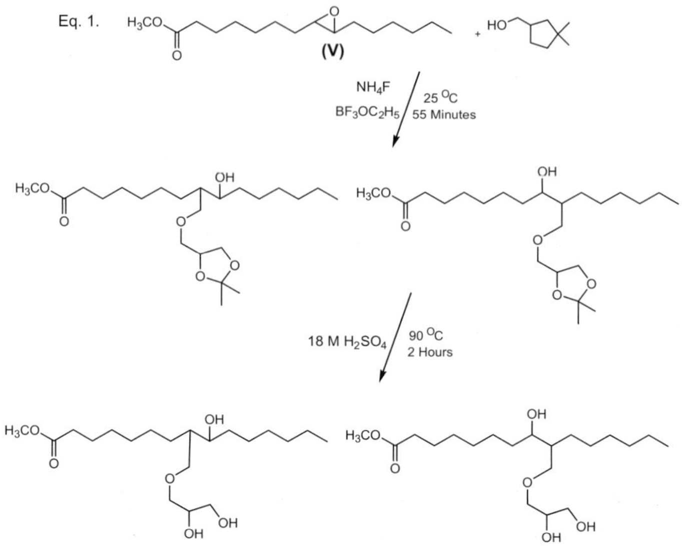

(VI) (VII)

2. Suppes[1] epoxidized soybean oil (VII) using chemoenzymatic NOVOZYME® 435 and then hydrolyzed the intermediate to the polyalcohol using either *Candida rugosa, Burkholderia cepacia*, or *Penicillium roquefortii*, which was subsequently used in forming rigid polyurethanes when reacted with aromatic diisocyanates.

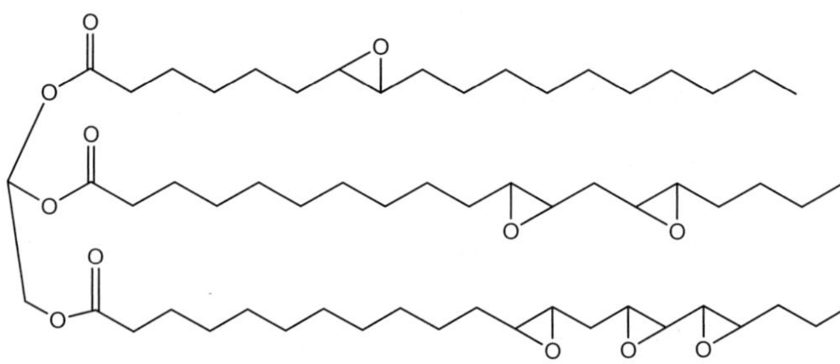

(VII)

3. Kurihara[2] prepared plasticizers for polylactic acid resins, which consisted of a mixture of epoxidized soybean oil, trioctyl phosphate, poly(adipic acid co-1,3-butylene glycol), processed oil, and naphthenic oil, which lowered the Tg of the resin by at least −23 °C.
4. Neigel[3] prepared alkyl glyceryl sulfonates (VIII) by the condensation of selected alkyl glyceryl epoxides with a mixture of sodium sulfite and sodium bisulfate, which were effective as topical antimicrobials.

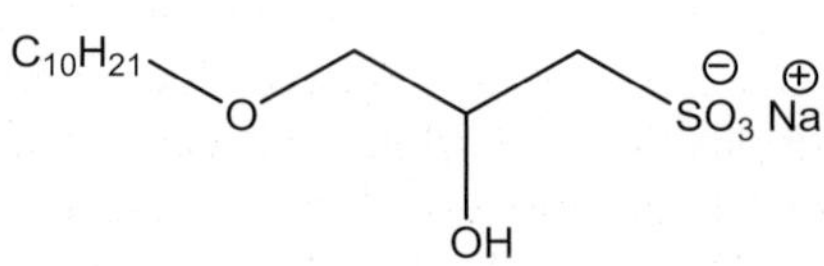

(VIII)

5. Su[4] prepared 3-trimethoxysilylpropyl-carbamic acid oxiranylmethyl esters (IX) which were designed to enhance curing rates in adhesive composition formulations.

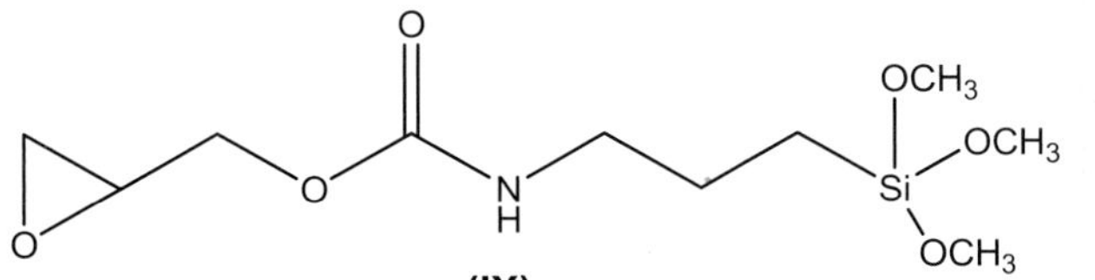

(IX)

REFERENCES

1. Galen Suppes et al, US. Patent 7,696,370 (April 13, 2010) **The Curators of the University of Missouri (Columbia, MO)**
2. Toyoaki Kurihara et al, US. Patent 8,022,139 (September 20, 2011) **Kuraray Co., Ltd. (Kurashiki-shi, JP)**
3. Dennis Neigel et al, US. Patent 8,053,596 (November 8, 2011) **Third Stream Bioscience, Inc. (Ft. Lauderdale, FL)**
4. Shiu-Chin H. Su et al, US. Patent 7,989,651 (August 2, 2011) **Momentive Performance Materials Inc. (Albany, NY)**

R. TOOTHPASTE AND MOUTHWASH

Menthanecarboxamide

Title:	Menthanecarboxamide derivatives having cooling properties
US Patent:	7,868,004 (January 11, 2011)
Author:	Lucienne Cole et al.
Assignee:	Givaudan S.A. (Vernier, CH)

Significance: Heterocyclic amide derivatives of menthane were found to be effective in providing a strong and long lasting cooling effect when used in toothpaste and mouthwash formulations. Menthanecarboxamide derivatives were prepared in a single step by the condensation of a heterocyclic amine with an optically inactive menthane acid chloride.

Additive Names *N*-Benzooxazol-4-yl 3-*p*-menthanecarboxamide (I)
N-Benzooxazol-6-yl 3-*p*-menthanecarboxamide (II)
N-Benzothiazol-6-yl menthanecarboxamide (III)
N-4-([1,2,4]Triazol-1-yl)-phenyl-3-*p*-menthanecarboxamide (IV)
N-4-(Oxazol-5-yl)-phenyl-3-*p*-menthanecarboxamide (V)
N-3-(Oxazol-5-yl)-phenyl-3-*p*-menthanecarboxamide (VI)

Safety Concerns Skin and oral formulations containing menthol or menthane derivatives are considered not to be toxic.

Additive Structures

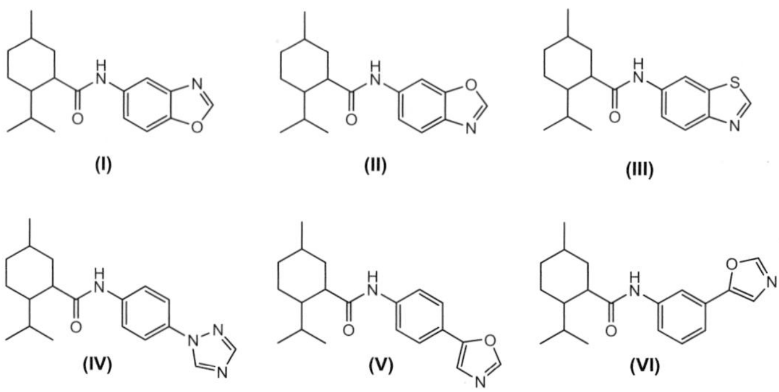

Additive Preparation

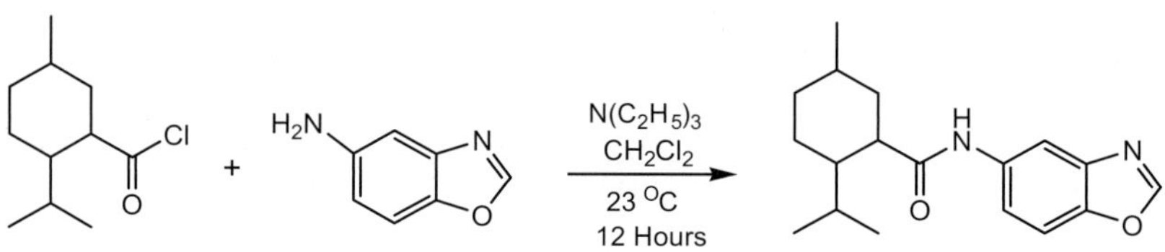

Additive Synthesis

1. Preparation of benzooxazol-4-yl-3-p-menthanecarboxamide

A flask was charged with 0.5 g of benzooxazol-4-yl-amine, 0.7 ml of triethylamine, and 10 ml of methylene chloride and then cooled to 0 °C and treated with the dropwise addition of 0.71 g of 3-*p*-menthanecarboxyl chloride over 5 min. The reaction mixture was then stirred overnight and then treated with 10 ml of water, and the layers were separated. The aqueous layer was then washed thrice with 5 ml of ethyl acetate, and the combined extracts were washed with 5 ml of water. The organic layer was dried using Na_2SO_4 and then concentrated under vacuum, and 0.46 g of the product was isolated.

Testing

1. Mouthwash Formulations

The feeling of coolness in mouthwash formulations for selected experimental agents was evaluated by professional panelists using the formulation given in Table 22 and 23. Ingredients were mixed and 30 ml put in the mouth, swished around, gargled, and spat out.

TABLE 22 Mouthwash formulation used to evaluate the coolness effect of *N*-benzooxazol-yl 3-*p*-menthanecarboxamide (I). Selected formulations evoked the feeling of coolness in mouthwash formulations as evaluated by panelists

Component	Treatment
Alcohol 95%	177 ml
Sorbitol 70%	290 g
Experimental Agent 1 (as a 1% solution in alcohol)	10 ml
Peppermint oil	0.30 g
Terpeneless	0.30 g
Methyl salicylate	0.640 g
Eucalyptol	0.922 g
Thymol	0.639 g
Benzoic acid	1.500 g
Pluronic F127	5.000 g
Sodium Saccharin	0.600 g
Sodium Citrate	0.300 g
Citric Acid	0.100 g
Water	q.s. 1 l

TABLE 23 Toothpaste formulation used to evaluate the coolness effect of *N*-3-(oxazol-5-yl)-phenyl-3-*p*-menthanecarboxamide, Additive (VI). Experimental Additives II, III, and V also performed exceptionally well

Component	Treatment
Opaque tooth gel	98.000 g
Experimental Agent 6	1.500 g (2% solution in PG)
Peppermint oil, Terpeneless	0.500 g

Testing Results

When used in mouthwash or toothpaste formulations, panelists reported an intense cooling sensation in all areas of the mouth and experienced a cooling perception that lasted for several hours.

Advantages over Prior Art

While *N*-substituted-*p*-menthane carboxamides that impart a cooling sensation to the skin of the mucous membrane are well known, their cooling sensation is moderate and fleeting. The identification, synthesis, and formulation of heterocyclic amide derivatives with enhanced and sustaining coolness are unreported.

Notes

1. Erman[1] prepared the physiological coolant *N*-[[5-methyl-2-(1-methylethyl) cyclohexyl]-carbonyl]glycine (VII), which was used in consumer oral product compositions.

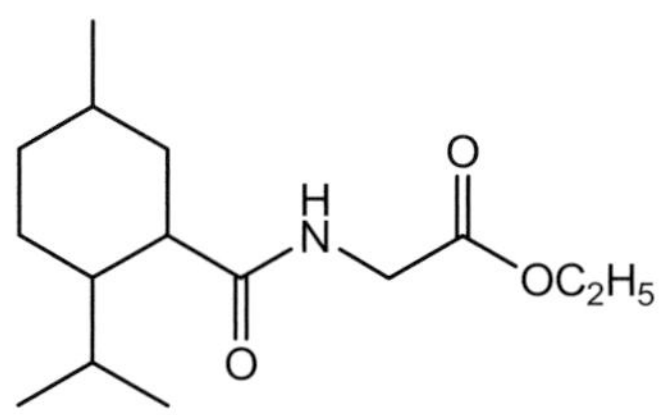

(VII)

2. Galopin[2] prepared *N*-(4-cyanomethylphenyl) *p*-menthanecarboxamide (VIII), which when used to additize toothpaste at 2 ppm provided a cooling effect that lasted for 91 min.

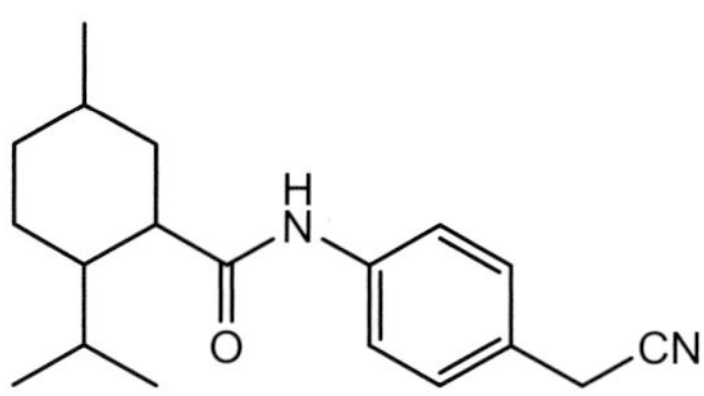

(VIII)

3. Galopin[3] determined that toothpaste additized with a mixture of menthol carboxamide and menthyl lactate, 1:11, respectively, had cooling properties superior to those of toothpastes individually additized with these agents.
4. Furrer[4] prepared *N*-(2-pyridin-2-ylethyl)-*p*-menthanecarboxamide[(1R,2S,5R)-2-isopropyl-5-methyl-*N*-(2-(pyridin-2-yl)ethyl)cyclohexanecarboxamide] (IX) and related pyridine derivatives, which were effective as cooling additives in toothpaste compositions.

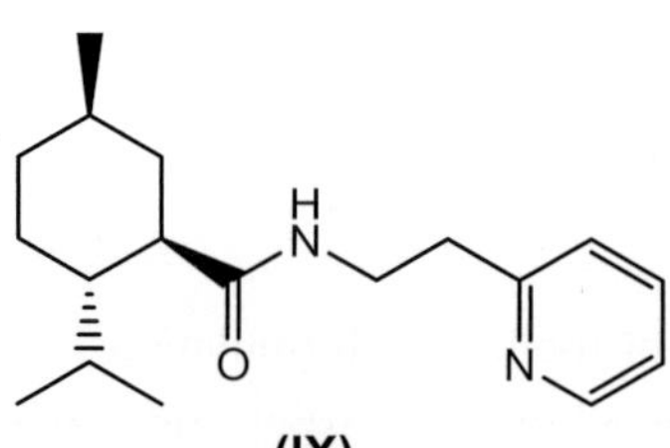

(IX)

REFERENCES

1. Mark B. Erman et al, US Patent 7,189,760 (March 13, 2007)
   Millennium Specialty Chemicals (Houston, TX)
2. Christophe C Galopin et al, US Patent 7,414,152 (August 19, 2008)
   Givaudan, SA (Vernier, CH)
3. Christophe C Galopin et al, US Patent Application 20060051301 (March 9, 2006)
   Givaudan, SA (Vernier, CH)
4. Stefan Michael Furrer et al, US Patent 7,880,011 (February 1, 2010)
   Givaudan, SA (Vernier, CH)

Chapter 10

Dental Adhesives

A. DENTAL ADHESIVE

Polymerizable 3,4-di-O-Methacryloyl-D-Mannitol

Title:	Polymerizable monomer, polymerizable composition, and dental material
US Patent:	7,968,615 (June 28, 2011)
Author:	Takahiro Sekiguchi
Assignee:	Kuraray Medical Inc. (Kurashiki-shi, JP)
Significance:	Dental adhesive formulations additized with 7% 3,4-di-O-methacryloyl-D-mannitol had a significantly improved adhesion to bovine dentin than did formulations containing an equivalent amount of glycerol dimethacrylate. The additive as well as mono- and di-saccharide diester analogs are readily prepared by condensing methacrylic acid or methacryloyl chloride with polyhydric alcohols.
Additive	3,4-Di-O-methacryloyl-D-mannitol (I)
Safety	3,4-Di-O-methacryloyl-D-mannitol is considered to be a polymerization hazard and should be protected from light and free-radical generators. Toxicity data are very limited.

Next Generation of International Chemical Additives. DOI: http://dx.doi.org/10.1016/B978-0-444-53788-1.00010-3

Additive Structure

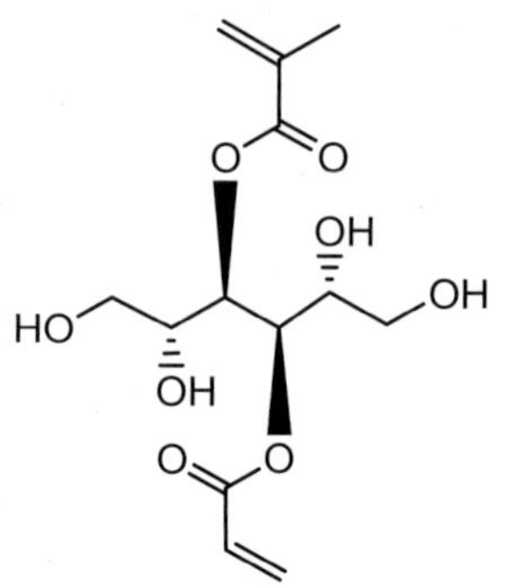

(I)

Additive Preparation

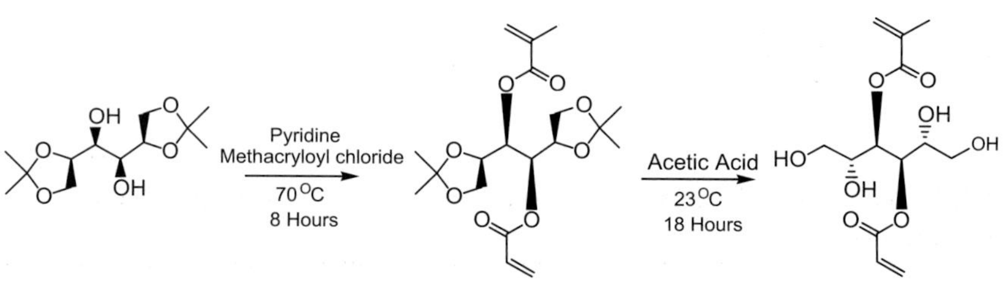

Additive Synthesis

1. Preparation of 1,2:5,6-di-O-isopropylidene-3,4-di-O-methacryloyl-D-mannitol

A 2-liter flask containing 700 ml of pyridine was treated with 65 g of 1,2:5,6-di-O-isopropylidene-D-mannitol and then cooled to 0 °C using an ice bath and treated with the dropwise addition of 60 g of methacryloyl chloride over 1 h. The mixture was then heated to 70 °C for 8 h and then cooled to an ambient temperature, and the contents were poured into a beaker containing 1 liter of ice water. The reaction mixture was then extracted five times using 1500 ml of diethyl ether and then concentrated, and an oily material was isolated. The crude product was purified using silica gel column chromatography with 7:3 hexane:diethyl ether, recrystallized in hexane, and the product was isolated in a 37% yield.

^{1}H-NMR (400 MHz, $CDCl_3$)	δ 1.31 (s, 6H), 1.36 (s, 6H), 1.96 (s, 6H), 3.85–3.96 (m, 4H), 4.21–4.27 (m, 2H), 5.43 (dd, 2H), 5.64 (s, 2H), 6.15 (s, 2H)
^{13}C-NMR (400 MHz, $CDCl_3$)	δ 18.2, 25.1, 26.3, 65.5, 71.6, 74.7, 109.3, 126.6, 135.6, 166.0

2. *Preparation of 3,4-di-O-methacryloyl-D-mannitol*

A 2-liter round-bottom flask was charged with 540 ml of concentrated acetic acid, 180 ml of water, and 18 g of the Step 1 product and then stirred for 18 h at an ambient temperature and concentrated. The oily material obtained was purified using silica gel column chromatography using ethyl acetate, and the product was isolated in a 60% yield.

^{1}H-NMR (400 MHz, $CDCl_3$)	δ 1.84 (s, 6H), 3.39 (dd, 2H), 3.51 (dd, 2H), 3.59–3.66 (m, 2H), 5.28 (d, 2H), 5.56 (s, 2H), 6.03 (s, 2H)
^{13}C-NMR (400 MHz, $CDCl_3$)	δ 18.4, 64.2, 71.6, 73.1, 126.8, 137.4, 167.9

Testing

A. Bonding Strength

Bonding strength of the experimental agent was evaluated using the formulations provided in Table 1. Test results are provided in Table 3.

TABLE 1 Dental formulations used to evaluate the bonding strength of the Step 2 additive

Component	Formulation #1 (g)	Formulation #1 (g)	Formulation #1 (g)	Comparison (g)
Additive 1	40	20	40	0
Glycerol dimethacrylate	0	0	0	40
2-Hydroxyethylmethacrylate	0	20	0	0
10-Methacryloyloxydecyl dihydrogen phosphate	15	15	15	15
2,4,6-Trimethylbenzoyldiphenyl-phosphine oxide	0.5	0.5	0	0.5
Camphor quinone	0	0	0.4	0
N,N-Dimethylaminobenzoic acid *n*-butoxyethyl ester	0	0	0.2	0
Distilled water	20	20	20	20
Ethyl alcohol	20	20	20	20

B. Bonding Strength to Bovine Teeth Dentin

The ability of the experimental additive to bond to the labial surface of ground bovine mandibular incisor was determined using the formulations provided in Table 2. To evaluate bond durability of samples, bonding strength of each sample was measured after 4000 thermal cycles were performed and when the sample had been immersed in water at 4 °C after being warmed to 60 °C for 1 min. Test results are provided in Table 4.

TABLE 2 Dental formulations used to evaluate the bonding strength to bovine teeth dentin using the Step 2 Additive

Component	Formulation #1 (g)	Formulation #1 (g)	Formulation #1 (g)	Comparison (g)
Additive 2	30	15	15	0
Glycerol dimethacrylate	0	0	0	30
2-Hydroxyethylmeth-acrylate	0	15	15	0
10-Methacryloy-loxydecyl dihydrogen phosphate	10	10	10	10
Bis glycerol methacrylate	45	45	45	45
2,4,6-Trimethylbenzo-yldiphenyl-phosphine oxide	2	2	0	2
Camphorquinone	0	0	1	0
N,*N*-dimethylaminoben-zoic acid *n*-butoxyethyl ester	0	0	1	0
Inorganic filler R972 (manufactured by Japan Aerosil Inc.)	5	5	5	5
Distilled water	20	20	20	20
Ethyl alcohol	20	20	20	20

Testing Results

A. Bonding Strength

TABLE 3 Results of bonding strength testing in dental formulations containing the Step 2 product indicating that optimum bonding strength was obtained using a treatment level of 22 wt% of 3,4-di-O-methacryloyl-D-mannitol

Bonding Strength to Dentine	Formulation #1 (MPa)	Formulation #1 (MPa)	Formulation #1 (MPa)	Comparison (MPa)
After 6000 thermal cycles	20.9	25.4	20.1	12.3

B. Bonding Strength to Bovine Teeth Dentin

TABLE 4 Results of bonding strength in dental formulations containing the Step 2 additive indicating that optimum bonding strength to bovine dentin was obtained using a treatment level of 7 wt%

Bonding Strength to Bovine Dentine	Example 1 (MPa)	Example 2 (MPa)	Example 3 (MPa)	Comparison (MPa)
After 24 h	12.1	16.1	15.1	8.2
After thermal cycle load	14.6	18.7	17.9	7.1

Advantages over Prior Art

Although polymerizable polyhydric alcohol mono-, di-, and tri-ester mixtures have been reported and used in dental adhesive packages, their dentine bonding strength is considerably less effective than that of the current additive. 3,4-Di-O-methacryloyl-D-mannitol is particularly unique since optimum decalcification and penetration into the collagen layer of dentil occur in adhesive formulations containing lower amounts of the additive than at higher treatment levels.

Notes

1. Kuboe[1] developed a method of increasing the bonding strength in dental compositions by surface treating 1.0–5.0 μm of irregularly shaped inorganic particles from 8-methacryloyloxyoctyl trimethoxysilane (II) and then copolymerizing with 2,2-bis[4-methacryloyl-oxypolyethoxyphenyl]-propane (III).

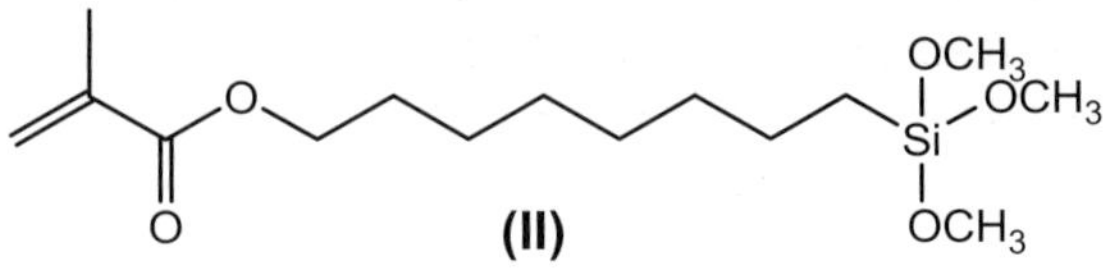

(II)

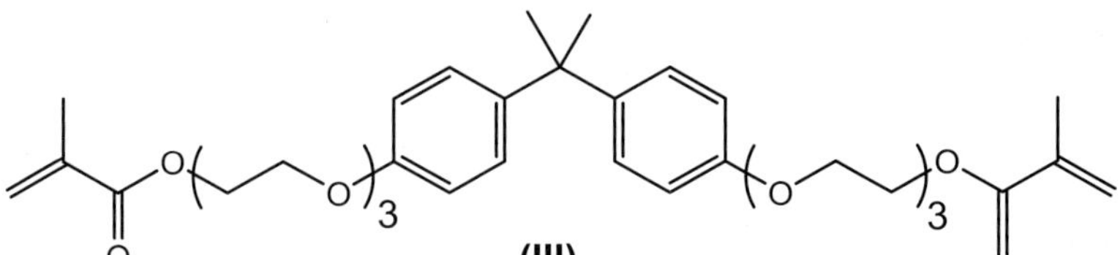

(III)

2. Moszner[2] prepared N,O-functionalized acrylic acid hydroxamide monophosphoric acid esters (IV), which were effective as crosslinkers for enamel and dentine adhesives.

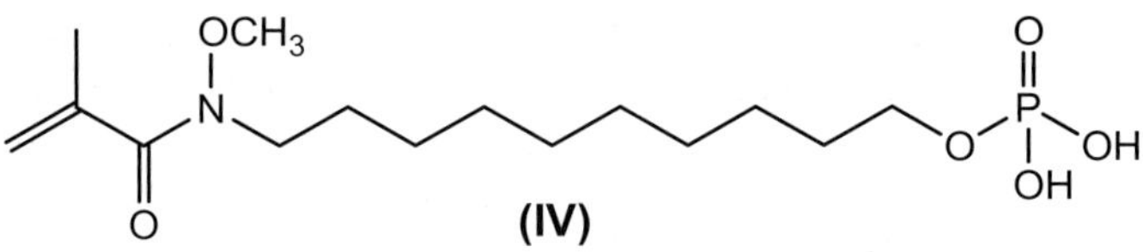

(IV)

3. Nakatsuka[3] prepared fluorocarbon-containing monophosphoric acid esters (V), which exhibited good bonding strength and high adhesive durability when used in dental adhesive formulations.

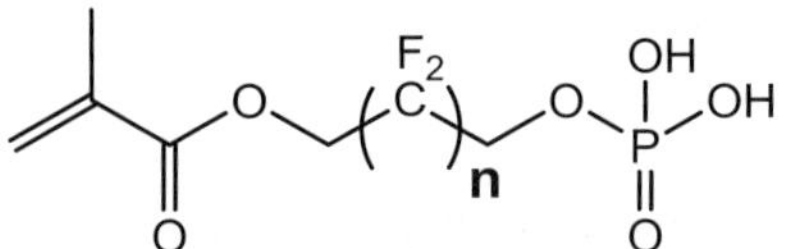

n = 2 - 12

(V)

4. Moszner[4] prepared high shearing strength dentine adhesives of a least 15.4 MPa consisting of dimethacrylamide derivatives (VI) in formulations containing glycerine dimethacrylate, 2-hydroxyethyl methacrylate and 2-[3-(dihydroxyphosphoryl)-2-oxa-butyl]-acrylic acid ethyl ester.

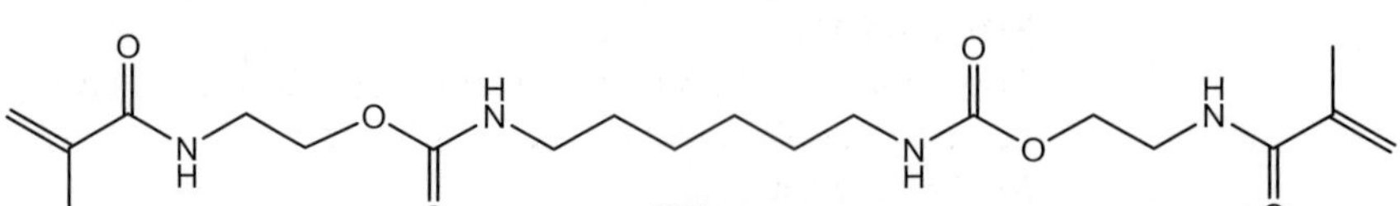

(VI)

REFERENCES

1. Yoshiko Kuboe et al, US Patent 7,951,851 (May 31, 2011)
   Kuraray Medical Inc. (Kurashiki-shi, JP)
2. Norbert Moszner et al, US Patent 7,932,303 (April 26, 2011)
   Ivoclar Vivadent AG (Schaan, LI)
3. Kazumitsu Nakatsuka et al, US Patent 7,655,722 (February 2, 2010)
   Kuraray Medical Inc. (Kurashiki-shi, JP)
4. Norbert Moszner et al, US Patent 6,953,832 (October 11, 2005)
   Ivoclar Vivadent AG (Schaan, LI)

B. DENTAL CROSSLINKING AGENTS

Polymerizable Azinidine Esters

Title:	Composition containing aziridino groups, method of production and use thereof
US Patent:	8,017,666 (August 30, 2011)
Author:	Peter Bissinger et al.
Assignee:	3 M Innovative Properties Company (Saint Paul, MN)
Significance:	Branched prepolyethers containing three or six 2-methyl-aziridino-crosslinking sites were prepared by esterification of the corresponding polyol with acrylic acid followed by the Michael addition of 2-methyl-aziridine. When used in dental formulations, the cured composite exhibited a Shore Hardness A of at least 50, high tensile strength, and an elongation at break of at least 200%.

Additive Name

3-(2-Methyl-aziridin-1-yl)-propionic acid 2,2-bis-[3-(2-methyl-aziridin-1-yl)-propionyloxy-methyl]-butyl ester (I) **(Crosslinker 1)**
MW = 467 Da

Tris(2-methylaziridinopropionate)-di(ε-caprolactone)-cyanuric adduct (II) **(Crosslinker 2)**
MW = 1709 Da

Penta(2-methylaziridinopropionate)-dipentaerythritol adduct (III) **(Crosslinker 3)**
MW = 796.0 Da

Tris(2-methylaziridinopropionate)-oligo(ethylene oxide)-trimethylolpropane adduct (IV) **(Crosslinker 4)**
MW = 600 Da

Tris(2-methylaziridinopropionate)-oligo(ethylene oxide)-pentaerythritol adduct (V) **(Crosslinker 5)**
MW = 1193 Da

Safety Polyaziridines are moderately toxic by inhalation or skin contact and are very mild skin irritants. When heated to decomposition, polyaziridines emit nitric oxide vapors.

Additive Structures

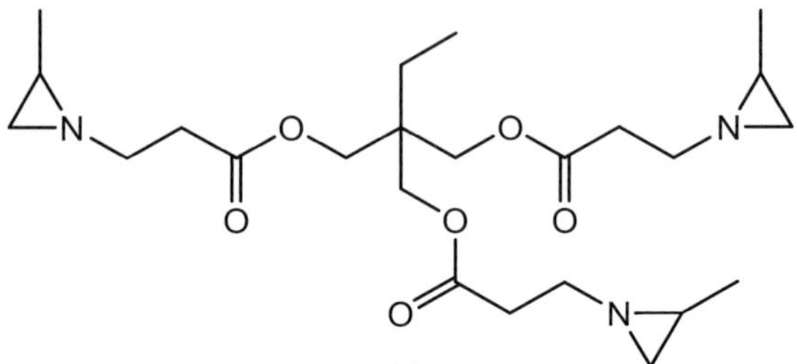

(I)

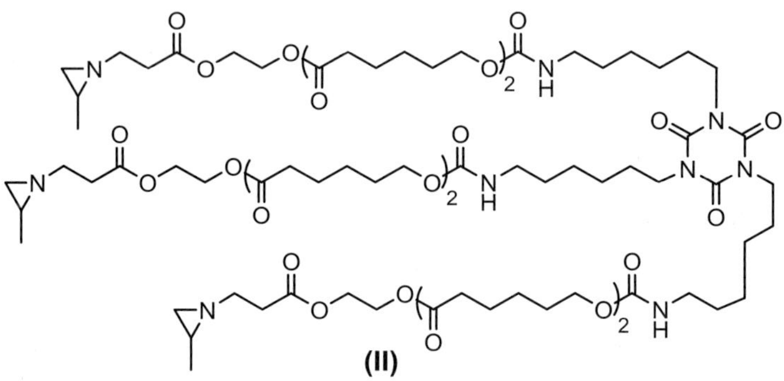

(II)

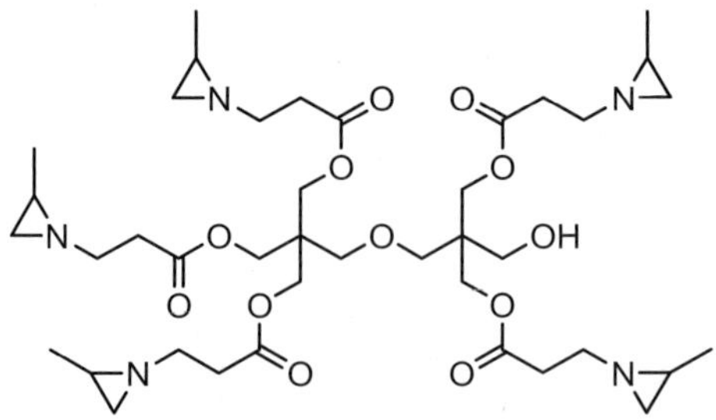

(III)

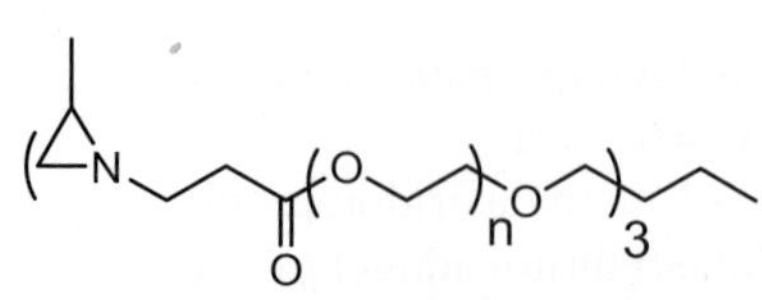

(IV)

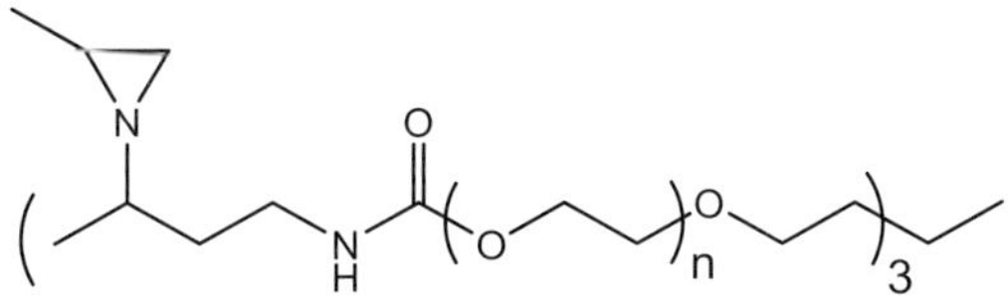

(V)

Additive Preparation

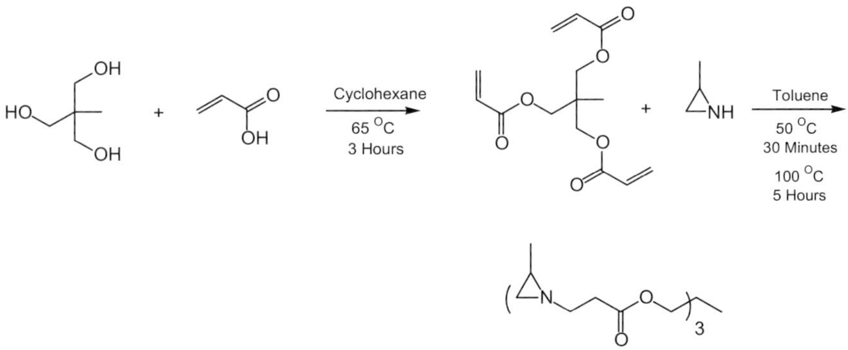

Additive Synthesis

1. Preparation of trimethylolpropane tris(acrylic acid ester) according to the method of Nowak[1]

A reactor equipped with a condenser and a Dean–Stark trap was charged with a stoichiometric amount of trimethylolpropane and acrylic acid and then diluted with 100 ml of cyclohexane and refluxed until no further water was collected. After purification, the product was isolated as a clear colorless liquid.

2. Preparation of trimethylolpropane tris(3-(2-methylaziridino-propionate))

A reactor was charged with the Step 1 product dissolved in 25–50 ml of toluene and then treated with the dropwise addition of a stoichiometric amount of 2-methylaziridine and heated to 50 °C for 30 min. The mixture was then further heated to 100 °C for 5 h while monitoring the disappearance of the double bond, using IR spectroscopy. Thereafter, the reaction mixture was cooled, and the product was isolated as a colorless fluid and used without purification.

Testing

Crosslinking Formulation

Crosslinking reactions were conducted using a base paste of 3.5% of the Step 2 product, 15.7% trisacylic triglyceride, 10.4% dibenzyl toluene, 1%

poly(ethylene oxide-co-propylene oxide) as surfactant, and 6.5% diatomaceous earth and then blended with a catalyst paste consisting of 19.3% sulfonium salt tetrafluoroborate, 40.5% acetyl tributyl citrate, 3.5% poly(ethylene oxide-co-propylene oxide) as surfactant, 24.1% highly dispersed silica, and 12.1% diatomaceous earth.

Mechanical Testing

Selected aziridine crosslinking agents were blended with the base paste so that the aziridine concentration in each blend was 2.5×10^{-2} mole. Cured sample specimens were then evaluated for Shore Hardness A, according to DIN 53505 and tensile strength and elongation at break according to DIN 50504. Test results are provided in Table 1.

Test Results

TABLE 1 Mechanical testing results for cured mixtures consisting of base paste additized with an experimental having an aziridine concentration of 2.5×10^{-2} mole

Crosslinker	Shore Hardness A (8 min)	Shore Hardness A (24 h)	Tensile Strength (MPa)	Elongation at Break (%)
Additive 1	31	54	2.35 ± 0.08	256 ± 14
Additive 2	27	52	1.96 ± 0.22	230 ± 39
Additive 3	26	52	2.68 ± 0.22	214 ± 25
Additive 4	27	52	2.28 ± 0.11	266 ± 20
Additive 5	39	54	1.99 ± 0.23	200 ± 37

Advantages over Prior Art

The inclusion of tensile strength properties of dental compositions as an indicator as to whether the prosthesis will rupture is not reported. The current additives also report enhanced speed of cure with enhanced physical properties over existing both linear and branched prepolysilicones containing pendant aziridine moieties and mono- and diaziridine-functionalized prepolyethers.

Notes

1. Klettke[1] prepared oligo- and polysiloxanes containing pendant aziridine functions (VI), which were used in sealing, molding, and in dental compositions as cross-linking agents.

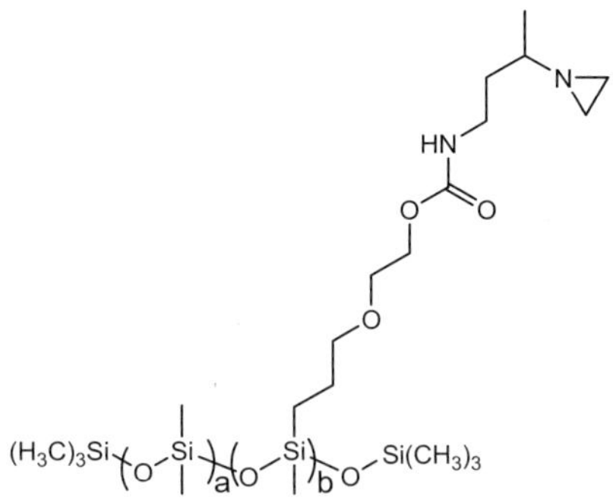

(VI)

2. Poly(ethylene glycol-co-propylene oxide) containing aziridine termini (VII) were prepared by Klettke[2] and used as crosslinking agents in dental compositions.

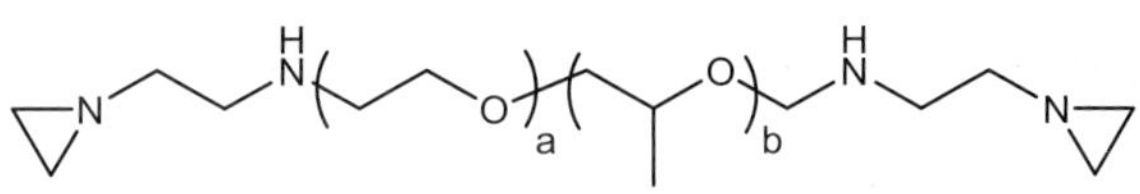

(VII)

3. Klettk[3] prepared a three-component dental composition consisting of poly(ethylene glucol)-co-tetramethylene oxide) (VII) having an Mn of approximately 6000 Da containing aziridine termini, *p*-toluene sulfonic acid-*N*-ethyl amide (VIII), and (S-lauryl-S-ethyl-sulfonium)-propionic acid-(2-ethyl-hexyl)-ester fluoroborate (IX), as the initiator, which had a setting time of ≤3 min at 23 °C.

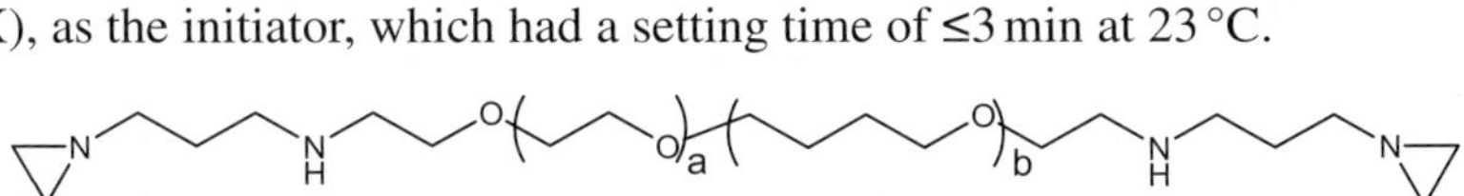

(VII)

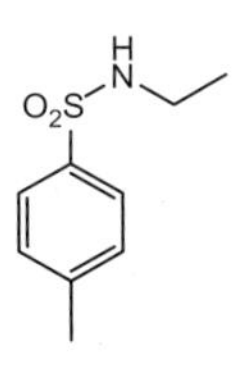

(VIII)

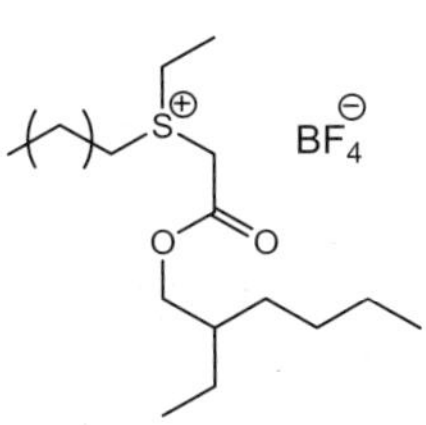

(IX)

4. Angeletakis[4] prepared a dental setting composition consisting of the ring-opening metathesis polymerization of a diester polyether containing norbornene carboxylate termini (X), polyethylene glycol urethane (XI), and 1,3-bis-(2,4,6-trimethylphenyl)-2-(imidazolidinylidene) dichloro(o-isopropoxyphenylmethylene) ruthenium catalyst (XII).

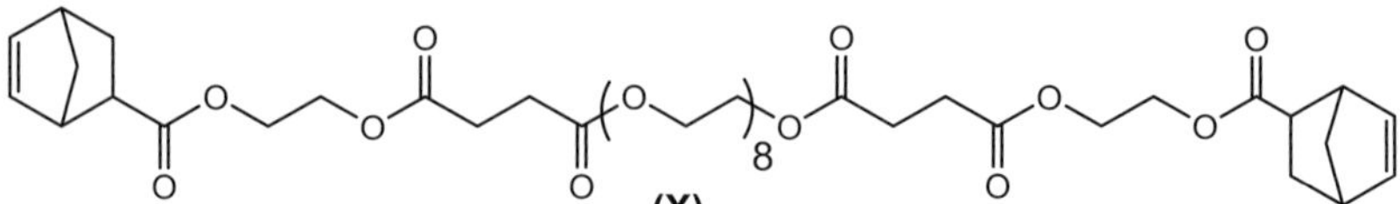

(X)

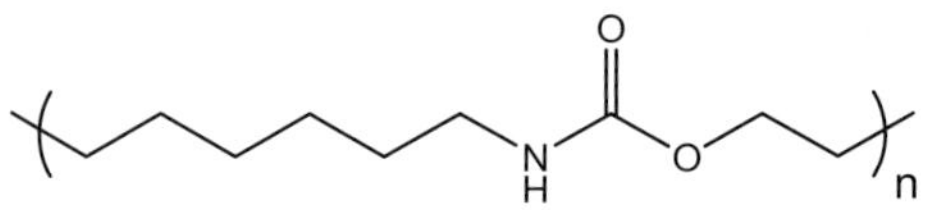

(XI)

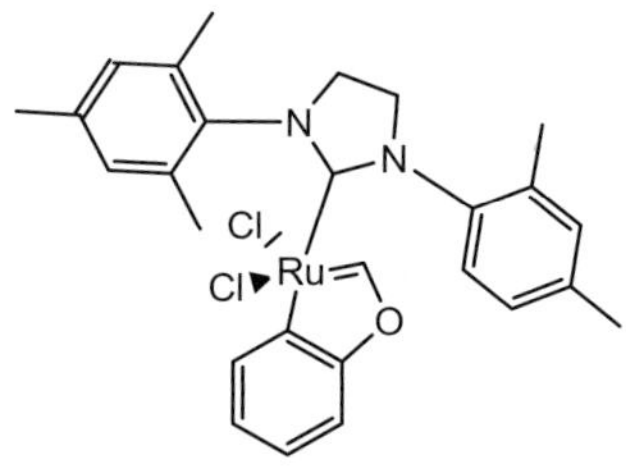

(XII)

REFERENCES

1. Thomas Klettke et al, US Patent 8,022,113 (September 20, 2011) **3M Innovative Properties Company (Saint Paul, MN)**
2. Thomas Klettke et al, US Patent 7,923,485 (April 12, 2011) **3M Innovative Properties Company (Saint Paul, MN)**
3. Thomas Klettke et al, US Patent **7,838,572** (November 23, 2011) **3M Innovative Properties Company (Saint Paul, MN)**
4. Christos Angeletakis, US Patent 7,625,551 (December 1, 2009) **Kerr Corporation (Orange, CA)**

Chapter 11

Disinfectants

A. DISINFECTANT ADDITIVE

Chlorine Dioxide–Glucose Clathrate

Title:	Clathrate of chlorine dioxide
US Patent:	7,842,195 (November 30, 2010)
Author:	Joseph Callerame
Assignee:	Joseph Callerame (Havasu City, AZ)

Significance: A stable solid glucose–chlorine dioxide clathrate was prepared by the absorption of nitrogen dioxide on glucose. The clathrate can be used to provide either a slow discharge chlorine dioxide over a period of months or it can be rapidly released by heating the solid to ≥37 °C.

Additive Name Chlorine dioxide–glucose clathrate (I)

Safety Concerns Information on the human health effects from exposure to chlorine dioxide clathrates is limited. The FDA has received several reports of health injuries from consumers using similar chlorine dioxide microbiocides including severe nausea, vomiting, and life-threatening low blood pressure from dehydration.

Additive Structure

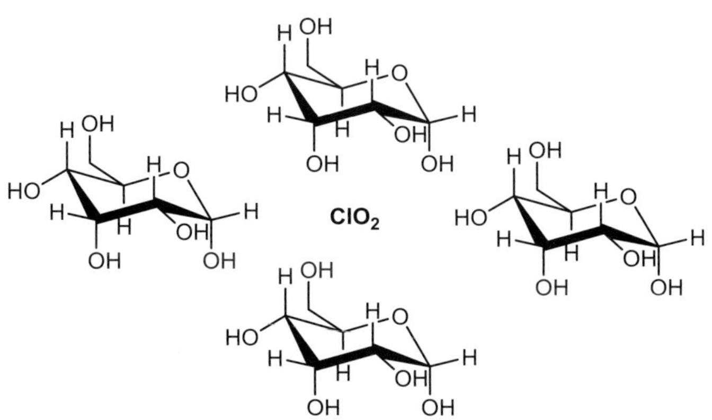

(I)

Next Generation of International Chemical Additives. DOI: http://dx.doi.org/10.1016/B978-0-444-53788-1.00011-5

Additive Preparation

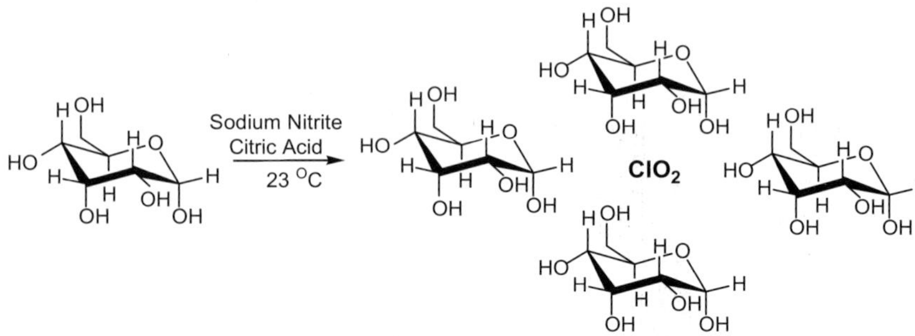

Additive Synthesis

1. Preparation of chlorine dioxide–glucose clathrate

A mortar and pestle containing 1.0 g of ground citric acid and 2.0 g of ground glucose was treated with sodium chlorite. Chlorine dioxide was then generated and dispersed into the solid blend over a period of 30 min, and yellow crystals were isolated. The clathrate began slowly releasing chlorine dioxide once the water content of the clathrate exceeded about 0.5 wt%. At temperatures >37 °C, the chlorine dioxide–glucose clathrate melted and completely released chlorine dioxide.

Stability Testing

The stability of the solid chlorine dioxide–glucose clathrate was evaluated by spectrophotometry by measuring the wavelength peak at 343 nm against a control sample. After 30 days, the clathrate continued to test positive for chlorine dioxide.

Advantages over Prior Art

Although silica- and clay-chlorine dioxide clathrates have previously been prepared, neither method releases chloride dioxide slowly. Glucose–chlorine dioxide can be selectively designed to expel chlorine dioxide very slowly or quickly as well as under percussion or grinding. These valuable delivery properties suggest broad applications in either consumer or industrial health-oriented products.

Notes

1. In an earlier investigation by Callerame,[1] high chlorine dioxide-containing chlorine dioxide–glucose clathrates were prepared by irradiating mixtures of sodium chlorite, citric acid, and glucose with ultraviolet light.
2. β-Cyclodextrin–chlorine dioxide clathrates were previously prepared by Beck[2] and used in solid pharmaceutical compositions as preservatives. In a subsequent investigation, Beck[3] prepared liquid compositions containing α- and β-cyclodextrin-chlorine dioxide clathrates, which were used in oral administration to humans or animals.

3. α-Cyclodextrin–chlorine dioxide clathrates containing >12 wt% chlorine dioxide were prepared by Harrison[4] and used as disinfectants.
4. Conant[5] determined that calcium lactate hydrate was effective in forming solid clathrates with hydrogen, methane, oxygen, or carbon dioxide.

REFERENCES

1. Joseph Callerame et al, US Patent 7,326,381 (February 5, 2008)
   Joseph Callerame (Havasu City, AZ)
2. Gary J. Beck et al, US Patent 6,358,935 (March 19, 2002)
   llergan Sales, Inc. (Irvine, CA)
3. Gary. Beck et al, US Patent 6,723,353 (April 20, 2004)
   Allergan Sales, Inc. (Irvine, CA)
4. Ken Harrison et al, US Patent Application 20090054375 (February 26, 2009)
   Ken Harrison (Madison, VA)
5. Lawrence D. Conant, US Patent 7,553,996 (June 30, 2009)
   Lawrence D. Conant; (Clinton, MA)

Chapter 12

Dispersants

A. NANOTUBE DISPERSANTS

Pyridinium-based Ionic Compounds

Title:	Ionic organic compound
US Patent:	7,858,799 (December 28, 2010)
Author:	Masaru Yoshida et al.
Assignee:	National Institute of Advanced Industrial Science and Technology (Tokyo, JP)
Significance:	Single-walled nanotube dispersions that were stable in water for over 6 months were prepared by blending with 1 wt% of a selected polymeric pyridine-containing ionic liquid followed by ultrasounding the mixture for 1 h. The ionic solvents were prepared in quantitative yield and in a single step, which did not require purification before using.

Additive Names

Poly(4-(4-benzyl)amidopyridinium) hexafluorophosphate (I)
Poly(4-(4-benzyl)amido-2,6-dichloropyridinium) hexafluorophosphate (II)
Poly(4-pentylamidopyridinium) hexafluorophosphate (III)
Poly(4-(*N*-benzyl)-N′-pyridinium urea) hexafluorophosphate (IV)

Safety Concerns

The toxicity of poly(4-(4-benzyl)amidopyridinium) hexafluorophosphate or polyvinylpyridinium hexafluorophosphates is known to cause moderate skin and eye irritation. Ingestion may cause nausea, cardiac disturbances, or central nervous system effects.

Next Generation of International Chemical Additives. DOI: http://dx.doi.org/10.1016/B978-0-444-53788-1.00012-7

Additive Structures

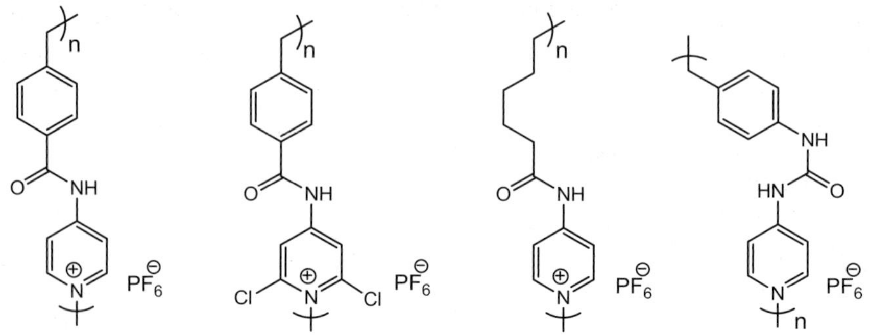

(I) (II) (III) (IV)

Additive Preparation

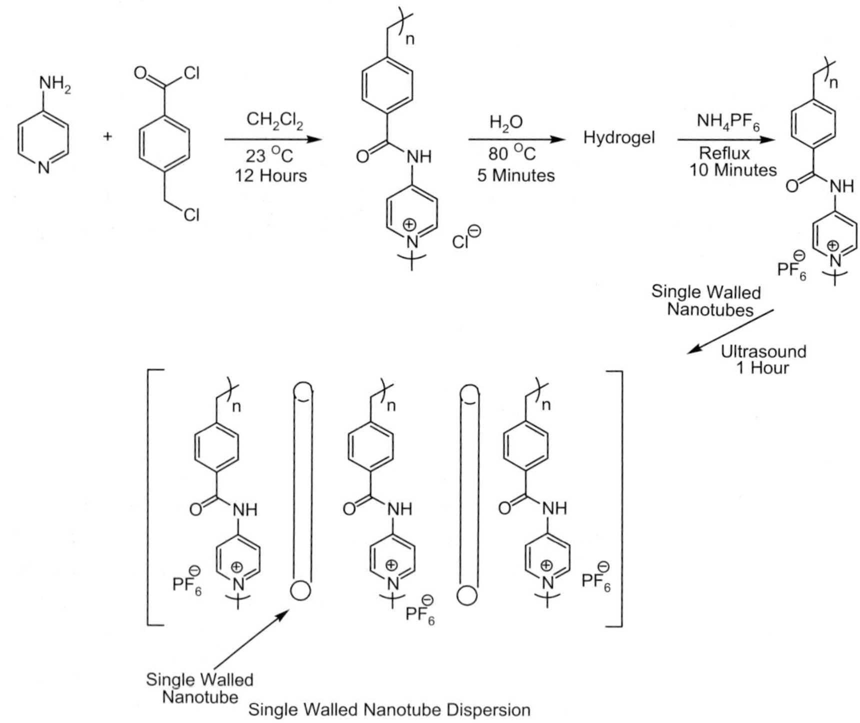

Additive Synthesis

1. Preparation of poly(4-(4-benzyl)amidopyridinium) chloride

At ambient temperature, a round bottom flask was charged with 4.27 g of 4-aminopyridine and 8.34 g of 4-(chloromethyl)benzoic acid chloride dissolved in 100 ml of anhydrous methylene chloride and then treated with 6.95 ml of triethylamine and stirred overnight. The reaction mixture was then filtered, and the product was isolated as a white precipitate in an 85% yield after filtration.

2. *Preparation of poly(4-(4-benzyl)amidopyridinium) chloride hydrogel*

A 1 wt% aqueous solution of the Step 1 product was mixed and heated to 80 °C to give a colorless solution. The solution was then slowly cooled to an ambient temperature for about 5 min whereupon a stable translucent hydrogel was formed. The critical gelation concentration was determined to be 7.5 g/l at an ambient temperature in neutral water.

3. *Preparation of poly(4-(4-benzyl)amidopyridinium) hexafluorophosphate*

A reaction kettle was charged with 860 mg of the Step 2 product and 200 ml of water and then heated until all the solids dissolved. The solution was then treated with 20 ml of an aqueous solution of 625 mg of ammonium hexafluorophosphate and then heated under reflux for 10 min. Thereafter, the solution was hot filtered, and the polymer was isolated in quantitative yield.

4. *Preparation of single-walled carbon nanotubes dispersed in liquid poly(4-(4-benzyl)amidopyridinium) hexafluorophosphate*

A 5-ml solution of the Step 3 product was treated with 0.5 mg of single-walled carbon nanotubes, and the mixture was ultrasonically irradiated for 1 h at 130 W and 35 kHz, and a black solution was formed without the appearance of precipitates. After 6 months, the single-walled carbon nanotube solution remained unchanged. Near infrared/UV spectra indicated the presence of sharp absorptions characteristic of individually dispersed single-walled carbon nanotubes in the wavelength range of 400–1600 nm when deuterated water was used.

Testing

No additional testing information was provided by the author.

Advantages over Prior Art

Although composites of single-walled nanotubes and polymeric solvents have been previously prepared, several distinct synthetic advantages are apparent in the current invention. First, the synthesis of the polymeric substrate is simple, requires inexpensive reagents and does not require purification before being used in subsequent blends. In addition, the synthesis quantitatively produces the target polymer. Finally, using small compact monomers to prepare the ionic substrate optimizes the ionic conductivity of the polymer.

Notes

1. In an earlier investigation by Zheng,[1] aqueous dispersions of single-walled nanotubes were prepared using peptide nucleic acid molecules such as single-stranded DNA and RNA dissolved in solutions containing tris(2-aminoethyl)amine, *N*-(2-hydroxyethyl)piperazine-N′-(2-ethanesulfonic) acid or 2-(*N*-morpholino)-ethanesulfonic acid.
2. Caroll[2] prepared polymer nanotube composite dispersions by the emulsion copolymerization of vinylidene difluoride containing hexafluoropropylene nanotubes.

3. A composite material was prepared by Asgari,[3] which consisted of a solgel covalently bonded to the nanoparticle, and this was then blended with a low molecular polysiloxane (DYNASIL®-40) and then gelled using 2-aminoethyl-3-aminopropyltrimethoxysilane (Dynasylan®).
4. Ford[4] solubilized carbon nanotubes in water by heating nanotubes containing 0.7–0.8 wt% of carboxylic acid groups with urea to 160 °C to form the ammonium salt.

REFERENCES

1. Ming Zheng et al, US Patent 7,588,941 (September 15, 2009)
   E.I. du Pont de Nemours and Company (Wilmington, DE)
2. David Carroll et al, US Patent 7,834,077 (November 16, 2010)
   Clemson University Research Foundation (Clemson, SC)
3. Soheil Asgari, US Patent 7,780,875 (August 24, 2010)
   Cinvention AG (Wiesbaden, DE)
4. William E. Ford et al, US Patent 7,854,914 (December 21, 2010)
   Sony Deutschland GmbH (Cologne, DE)

B. CEMENT DISPERSANTS

Methacrylic Acid Terpolymers Containing Polyethylene Glycol

Title:	Polymer and cement admixture using the same
US Patent:	7,851,576 (December 14, 2010)
Author:	Tsutomu Yuasa et al.
Assignee:	Nippon Shokubai Co., Ltd. (Osaka, JP)
Significance:	A reaction mixture of methoxypolyethylene glycol methacrylic acid ester and methacrylic acid initiated with the macroinitiator poly(oxyethylene-co-diazo) generated a terpolymer containing randomly incorporated polyethylene glycol. When this terpolymer was compounded with cement, sand, and water, the prehardened cement had an enhanced fluidity and dispersibility before setting.

Additive Name Poly(ethylene glycol-ter-methacrylic acid-ter-methoxypolyethylene glycol methacrylic acid ester) (I)

Safety Concerns While there are no toxicological data available for this additive, users are warned that polymethacrylic acid is an irritant and that gloves, chemical goggles, or a mask should be used.

Additive Structure

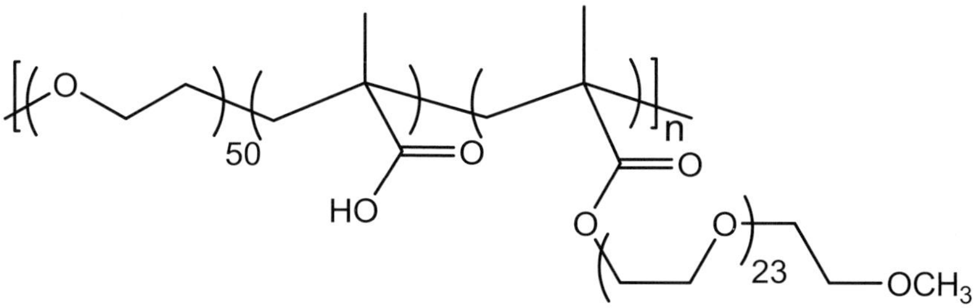

(I)

Additive Preparation

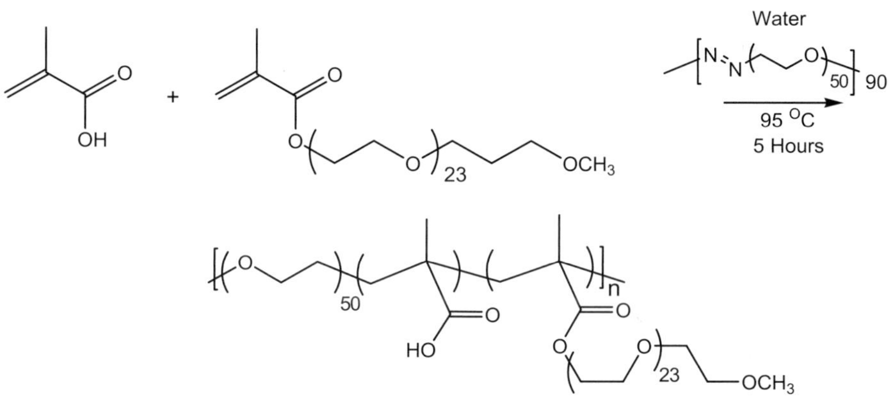

Additive Synthesis

1. Preparation of poly(ethylene glycol-ter-methacrylic acid-ter-methoxypolyethylene glycol methacrylic acid ester)

A glass reaction vessel equipped with a thermometer, stirrer, addition funnel, nitrogen inlet tube, and reflux condenser was charged with 100.0 g of water, then heated to 95 °C, and treated with 70.1 g of methoxypolyethylene glycol methacrylic acid ester having an ethylene oxide repeat unit of 23 and 9.9 g of methacrylic acid dissolved in 120.0 g of water. The polymerization was then initiated by the dropwise addition of the polyethylene glycol diazomacroinitiator having an Mn of approximately 4000 Da (VPE®-0401) dissolved in 80.0 g of water and added incrementally over a period of 3 h. The mixture was then heated for an additional 2 h at 95 °C, then concentrated, and the polymer was isolated. Physical properties of selected dispersants are provided in Table 1.

TABLE 1 Physical properties for copolymers prepared by using various ratios of methoxypolyethylene glycol methacrylic acid ester and methacrylic acid using the macroinitiators VPE®-0401, VPE®-0201, or VPE®-0601

Example	Monomers	Macroinitiator	Chain Transfer Agent	Mw (Da)
1	1) Methoxypolyethylene glycol methacrylic acid ester 2) Methacrylic acid	VPE-0401®*	None	52,300
2	1) Methoxypolyethylene glycol methacrylic acid ester 2) Methacrylic acid	VPE-0201®**	None	91,600
3	1) Methoxypolyethylene glycol methacrylic acid ester 2) Methacrylic acid	VPE-0601®***	None	85,600
Control-1	1) Methoxypolyethylene glycol methacrylic acid ester 2) Methacrylic acid 3) 30% Sodium hydroxide	V-50®****	None	24,200
Control-2	1) Methoxypolyethylene glycol methacrylic acid ester 2) Methacrylic acid	Ammonium persulfate	3-Mercapto-propionone	24,000

Poly(oxyethylene-co-diazo) having an Mn of approximately 23,500 Da with polyethylene oxide having an Mn of approximately 2000 Da (dp ~ 50) in the repeat unit and with a polymer repeat unit of 90.

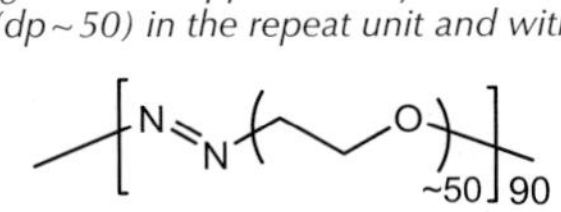

***Poly(oxyethylene-co-diazo) having an Mn of approximately 33,500 Da with polyethylene oxide having an Mn of approximately 2000 Da in the repeat unit and with a polymer repeat unit of 45.*
****Poly(oxyethylene-co-diazo) having an Mn of approximately 31,000 Da with polyethylene oxide having an Mn of approximately 6000 Da in the repeat unit and with a polymer repeat unit of 90.*
*****2,2′-Azobis(2-methylpropioneamidine) dihydrochloric acid salt*

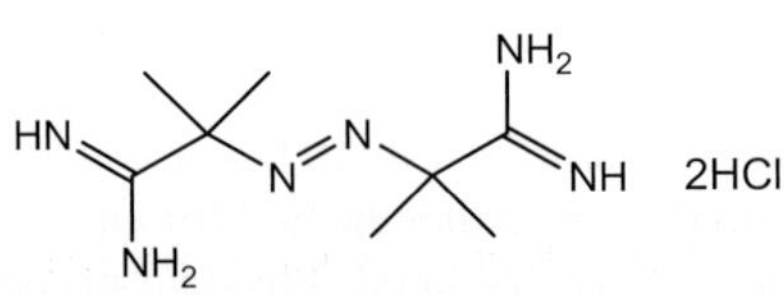

Testing

15. Hit Flow Rate

An experimental sample consisting of 550 g of Taiheiyo Normal Portland Cement, a 220 g of solution containing a selected experimental additive at varying concentrations, and the antifoaming agent MA404® were mixed in a Hobert-type mortar mixer for 30 s. During the mixing procedure, 1350 g of standard sand was added, and the mixture was further blended for an additional 30 s. The mortar was then hit 15 times with a stick, and after 15 s, the diameter of the hole was measured to determine the effect of the additive on cement dispersibility. Test results are provided in Table 2.

Test Results

TABLE 2 Effect on experimental additives on cement dispersibility using the 15 Hit Flow Rate Test. Higher values are indicative of enhanced cement dispersibility and stability

Example	Polymer Content in Cement Mixture (wt%)	15 Hit Flow Rate (mm)
1	0.09	229
2	0.10	239
3	0.09	240
Control-1	0.12	217
Control-2	0.18	223

Advantages over Prior Art

The novel method of randomly introducing polyethylene glycol into the terpolymer containing methoxypolyethylene glycol methacrylic acid ester and methacrylic acid using the macroinitiator poly(oxyethylene-co-diazo) has produced a cement dispersant able to form very fine small aggregates of prehardened cement compositions. Previously prepared cement dispersants containing block polyethylene glycol and derivatized with (meth)acrylate or styrene appear to lack the required structural segments or orientations needed to become better adsorbed onto cement particles.

Notes

1. Kraus[1] prepared a block copolymer effective as a cement dispersant, (II), using the atom transfer radical of the macroinitiator bromoisobutyric ester of polyethylene glycol monomethyl ether having an Mn of approximately 500 Da with *t*-butylmethacrylate.

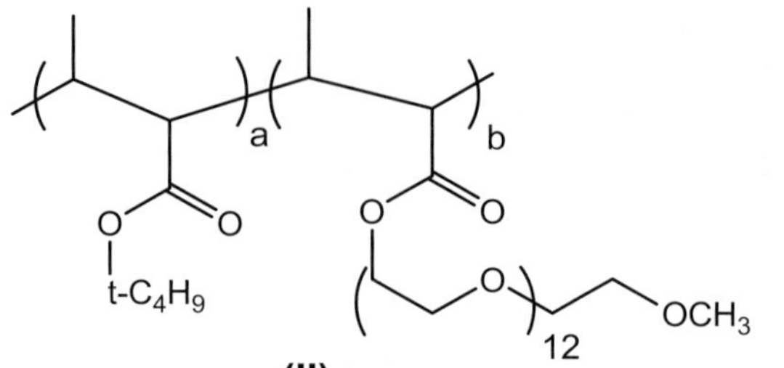

(II)

2. Bichler[2] prepared a yellowish cement dispersant copolymer (III) consisting of polyethylene glycol monovinylether having an ethylene glycol component with an Mn of approximately 5800 Da and acrylic acid.

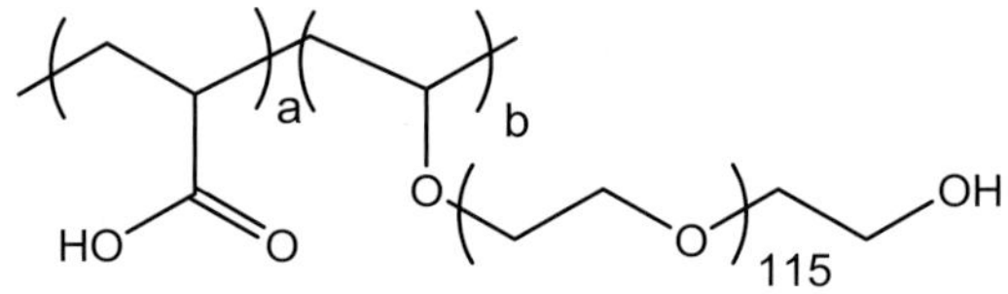

(III)

3. Kraus[3] prepared a cement dispersant copolymer (IV) containing poly(ethylene glycol) monomethyl ether methacrylate having a glycol component with an Mn of approximately 500 dalton functionalized with ethylene glycol methacrylate phosphate, which was also effective as an aqueous solid suspension stabilizer.

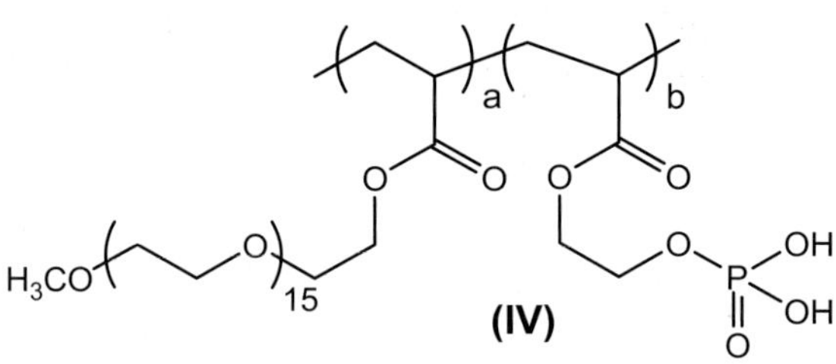

(IV)

4. A fluid loss control additive and cement dispersant consisting of humic acid salt containing grafted 2-acrylamido-2-methylpropanesulfonic acid salt, acrylamide, acrylic acid salt, and diallyldimethylammonium chloride was prepared by Lewis[4] and used in cement compositions to improve flowability.

REFERENCES

1. Alexander Kraus et al, US Patent 7,425,596 (September 16, 2008)
Goldschmidt GmbH (DE)
Construction Research & Technology GmbH (DE)
2. Manfred Bichler et al, US Patent 7,855,260 (December 21, 2010)
BASF Construction Polymers GmbH (Trostberg, DE)

3. Alexander Kraus et al, US Patent 7,842,766 (November 30, 2010)
Construction Research & Technology GmbH (DE)
4. Sam Lewis et al, US Patent 7,842,652 (November 30, 2010)
Halliburton Energy Services, Inc. (Duncan, OK)

C. PIGMENT DISPERSANTS

Poly(ethylene oxide)- or Poly(ethylene oxide-b-propylene-b-polybutylene oxide)

Title:	Dispersions containing alkoxylates of alicyclic polycyclic compounds
US Patent:	7,868,074 (January 11, 2011)
Author:	Lisa Marie Fine et al.
Assignee:	Ethox Chemicals, LLC (Greenville, SC)
Significance:	Surface active dispersants containing poly(ethylene oxide)- or poly(ethylene oxide-b-propylene-b-polybutylene oxide) have been found to be effective in lowering the interfacial tension of coloration pigments in both aqueous and nonaqueous media. Optimum pigment dispersion was obtained using cyclodecyloxopolyethylene oxide containing 70 moles ethylene oxide at a treatment level of 2.0 wt%.
Additive Name	Cyclodecyloxy-poly(ethylene oxide) (I) Adamantanoxy(ethylene oxide) (II) Terpineoxypoly(ethylene oxide) (III) Cyclodecyloxy-poly(ethylene oxide-ter-propylene oxide-ter-butylene oxide) (IV)
Safety Concerns	While polyethylene oxide and poly(ethylene oxide-co-propylene oxide-co-butylenes-oxide are considered to be safe overall, the presence of residual initiator raises health concerns. When handling these materials, contact lenses must not be worn. In the case of contact with the eyes, immediately flush eyes with cool water. In the case of skin contact, gently wash the contaminated area with soap and water.

Additive Structures

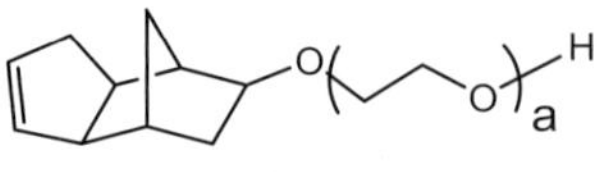

(I)

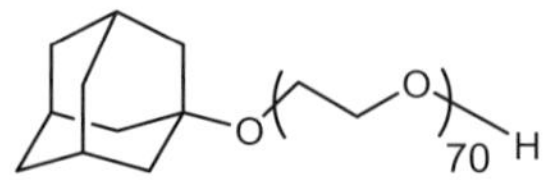

(II)

a=50,70,100,120

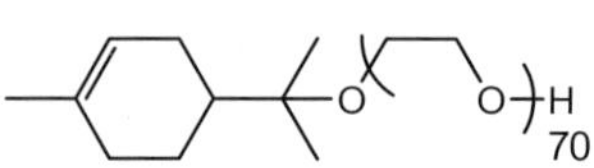

(III)

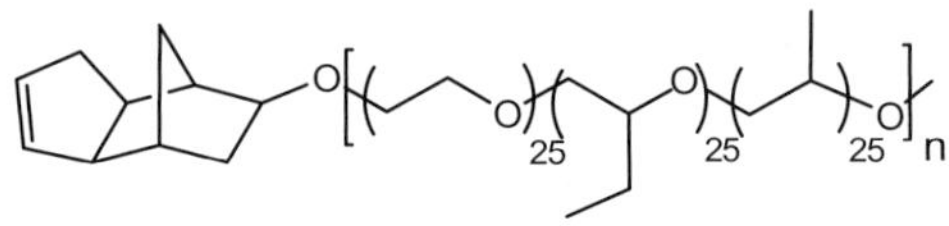

(IV)

Additive Preparation

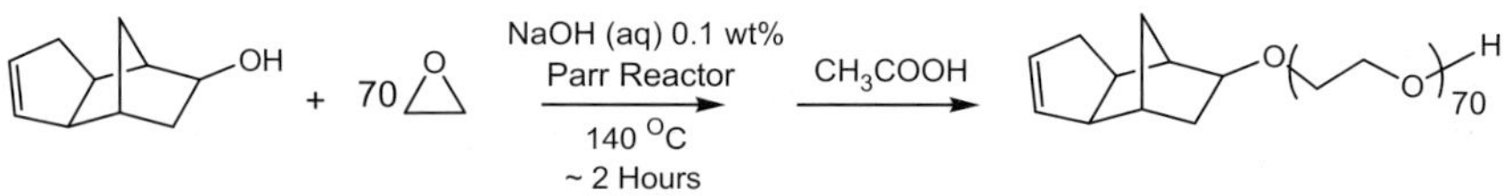

Additive Synthesis

1. Preparation of cyclodecyloxy-poly(ethylene oxide)

A Parr pressure reactor was charged with one equivalent of cyclodecyl alcohol and 0.1% potassium hydroxide and then purged with nitrogen and vacuum stripped for 30 min at 105 °C. The vessel pressure was then reduced, and ethylene oxide was slowly introduced to initiate the reaction. Thereafter, an additional 70 equivalents of ethylene oxide was introduced at 140 °C, and the reaction was continued until all ethylene oxide had been consumed. The reactor was then vacuum stripped, and the crude material was neutralized with acetic acid and the product isolated from the reactor.

Testing

1. Viscosity

In a typical process, pigments having an average particle diameter of 10–35 nm and a DBP of 60–150 ml/100 g were blended with a selected experimental dispersant containing 25–40 wt% solvent. The stability of the mixture was then ascertained by the shear rate in revolutions per minute needed to break the suspension. Test results for Phthalo Blue and Lithol Rubine dispersions are summarized in Tables 3 and 4, respectively.

Testing Results

1. Viscosity Testing for Phthalo Blue Dispersions

TABLE 3 Solution viscosities of Phthalo Blue dispersions stabilized after 48 h with cyclodecyloxy-poly(ethylene oxide) having an ethylene oxide content of 70 moles

RPM	Solution Viscosity (Control)	Solution Viscosity using Cyclodecyloxy-poly $(EO)_{70}$ (0.5%)	Solution Viscosity using Cyclodecyloxy-poly$(EO)_{70}$ (1.0%)	Solution Viscosity using Cyclodecyloxy-poly$(EO)_{70}$ (1.5%)	Solution Viscosity using Cyclodecyloxy-poly $(EO)_{70}$ (2.0%)
1.5	92160	2190	1210	2310	940
3	57410	1930	1030	2080	810
6	34940	1710	900	1840	720
12	22250	1440	800	1600	660
30	11990	1120	660	1320	580
60	7040	920	580	1120	520

2. Viscosity Testing for Lithol Rubine Dispersions

TABLE 4 Solution viscosities for Lithol Rubine dispersions stabilized after 24 h with cyclodecyloxy-poly(ethylene oxide) having an ethylene oxide content of 70 moles

RPM	Solution Viscosity (Control)	Solution Viscosity using Cyclodecyloxy-poly $(EO)_{70}$ (0.5%)	Solution Viscosity using Cyclodecyloxy-poly $(EO)_{70}$ (1.0%)	Solution Viscosity using Cyclodecyloxy-poly$(EO)_{70}$ (1.5%)	Solution Viscosity using Cyclodecyloxy-poly $(EO)_{70}$ (2.0%)
1.5	92460	7470	4370	11010	14070
3	56870	5770	3680	8370	10590
6	35410	4340	2920	6250	7820
12	22400	3200	2280	4600	5710
30	12180	2120	1630	4570	3720
60	7450	1520	1220	3060	–

Advantages over Prior Art

Although colorant pigment dispersants are well reported, there are very few instances of organic and inorganic pigment dispersants that can be used in both aqueous and nonaqueous media as described in the current invention. In addition, these additives are easily blended with pigments without causing flocculation to provide dispersions having very high flowability.

Notes

1. Styrenic polyethers (V) prepared by Weipert[1] were effective in dispersing insoluble fine powders in nonaqueous liquids and provided long-term stability without the formation of a hard cake or precipitates.

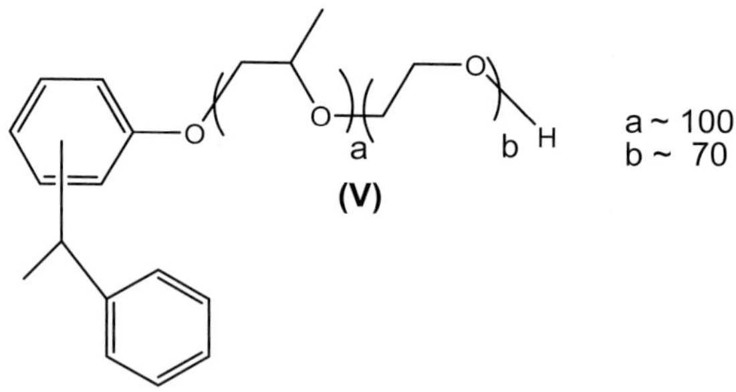

(V)

2. Waki[2] prepared an aqueous pigment dispersion for Yellow 5G pigment using poly(styrene-b-methylstyrene-b-acrylic acid) having an Mn of approximately 8000 Da and with a Tg = 75 °C blended glycerin and isopropyl alcohol. Fastgen Blue pigment TGR was also dissolved in poly(styrene-b-acrylic acid) having an Mn of approximately 12,000 Da with a Tg = 70 °C also blended with glycerin and isopropyl alcohol.
3. Weber[3] prepared diazo pigment dispersants that had a high affinity for those groups having diazo-containing dyes (IV) and (V).

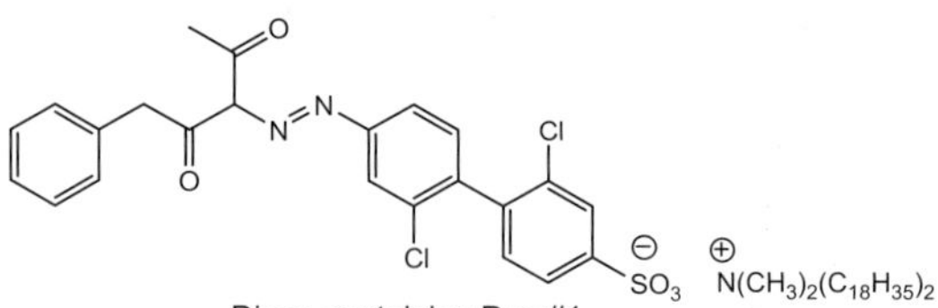

Diazo-containing Dye #1

(VI)

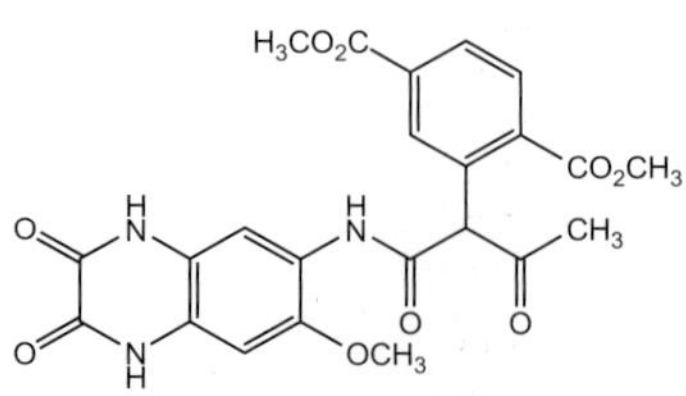

Diazo-containing Dye #2

(VII)

4. Hoogmartens[4] formed stable nonaqueous inkjet inks for diketopyrrolo-pyrrole pigments using polyethylene glycol dimethyl ether having an Mn of approximately ≥250 Da as the dispersing solvent.

REFERENCES

1. Paul David Weipert et al, US Patent 7,271,211 (September 18, 2007) **Ethox Chemicals, LLC (Greenville, SC)**
2. Minoru Waki et al, US Patent 7,858,676 (December 28, 2010) **Seiko Epson Corporation (Tokyo, JP) Mikuni Shikiso Kabushiki Kaisha (Himeji-shi, JP)**
3. Joachim Weber et al, US Patent 7,855,041 (December 21, 2010) **Clariant Produkte (Deutschland) GmbH (Frankfurt, DE)**
4. Ivan Hoogmartens, US Patent 7,854,799 (December 21, 2010) **Agfa Graphics NV (Mortsel, BE)**

Chapter 13

Drilling Additives

A. NATURAL GAS HYDRATE INHIBITORS

N-Ethoxylated Acid Ester Ammonium Salts

Title:	Additives for inhibiting the formation of gas hydrates
US Patent:	8,034,748 (October 11, 2011)
Author:	Uwe Dahlmann et al.
Assignee:	Clariant Produkte (Deutschland) GmbH (Frankfurt, DE)
Significance:	*N*-Ethoxylated acid ester ammonium salts were prepared by coupling *N*-Ethoxylated linear or cyclic amines with chloroacetic acid followed by esterification of the intermediate with branched or linear aliphatic alcohols. Autoclave testing with natural gas indicated that additives prepared from this process were effective as natural gas hydrate inhibitors since temperature and pressure changes were significantly lowered and torque minimized.
Additive Names	*N*-[(α–Diethylene glycol)-2-ethylhexylacetate]-cyclohexyl ammonium chloride (I) *N*-[(α–Diethylene glycol)-2-ethylhexylacetate]-1-butyl ammonium chloride (II) *N*-[(α–Decaethylene glycol)-dodecylacetate]-2-butyl ammonium chloride (III) *N*-[(α–Diethylene glycol)acetate]-1-butyl ammonium chloride (IV) *N*-[(α–Hexaethylene glycol)- 2-ethylhexylacetate]-1-butyl ammonium chloride (V)

Next Generation of International Chemical Additives. DOI: http://dx.doi.org/10.1016/B978-0-444-53788-1.00013-9

Safety

While little safety information is available on these additives, products are considered to be mixtures since purification of reaction products in both synthetic steps was limited. Particular concern is the presence of unreacted 2-ethylhexanol, chloro- and acetic acids, and sodium acetate. All these materials are considered to be moderate skin irritants and moderate to severe eye irritants.

Additive Structures

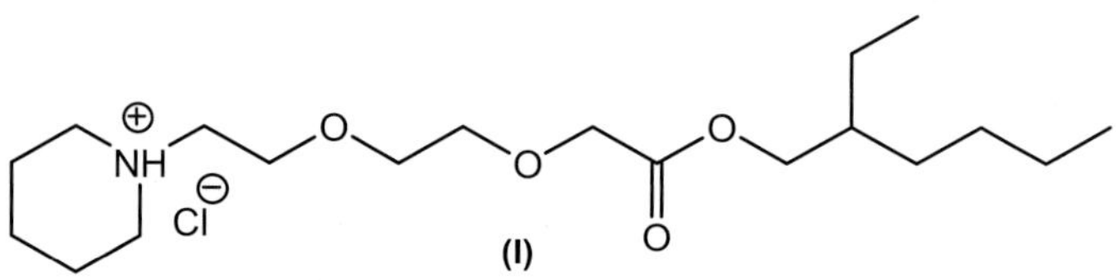

(I)

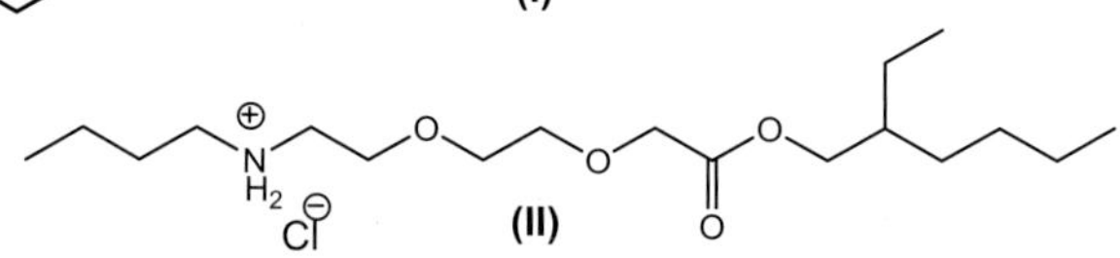

(II)

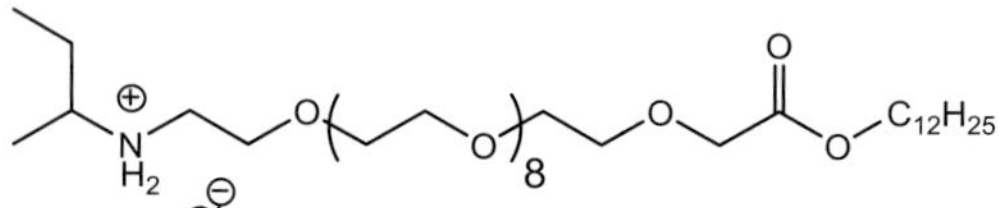

(III)

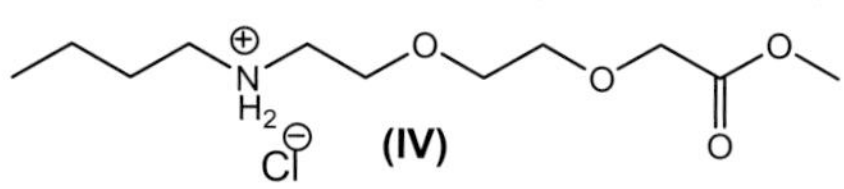

(IV)

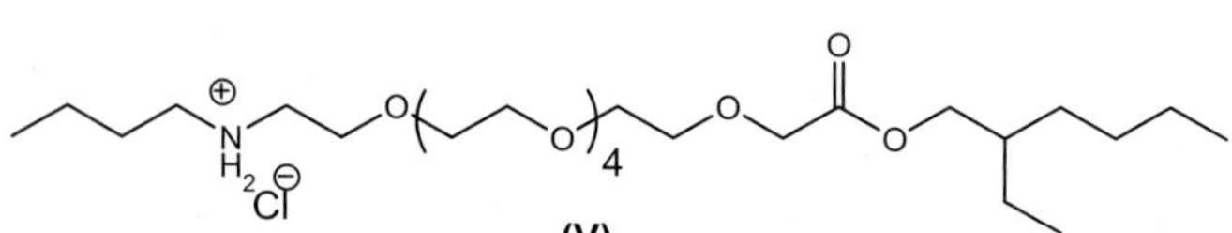

(V)

Additive Preparation

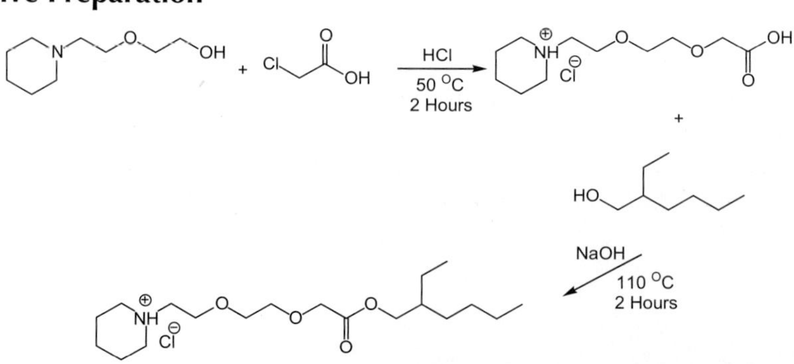

Additive Synthesis

1. Preparation of N-(α-diethylene glycolacetic acid)cyclohexyl ammonium chloride (Note 1)

A stirred apparatus was charged with 1 mole of *N*-diethylene glycol cyclohexyl amine and then heated to 40 °C and treated with 1.2 moles of sodium chloroacetate. The reaction temperature was then increased to 50 °C for 30 min, and then the contents were treated with the incremental addition of 1.5 moles of sodium hydroxide microprills at such a rate that the temperature did not exceed 55 °C. The mixture was then heated for an additional 2 h at 70 °C and then slowly treated with 10% hydrochloric acid until a pH of <3 was obtained. The mixture was then heated to 95 °C and then transferred to a heated stirred apparatus having a bottom outlet where the phases were separated after 15 min at 107 °C. The top phase was isolated, and residue water was removed by distillation under reduced pressure.

2. Preparation of N-[(α-diethylene glycol)-2-ethylhexylacetate]cyclohexyl ammonium chloride

An equimolar mixture consisting of the Step 1 product and 2-ethylhexyl alcohol containing a catalytic amount of sodium hydroxide was heated to 110 °C for 2 h and then cooled, and the product was isolated without purification.

Testing

Gas Hydrate Formation Inhibition Testing

To investigate the inhibiting action of selected additives, an autoclave containing a stirrer, temperature control, and pressure and torque sensor and having an internal volume of 450 ml was charged with natural gas.

At a starting temperature of 17.5 °C, the autoclave was cooled to 2 °C and then stirred for 18 h and heated to 17.5 °C. The initial pressure decrease corresponded to the thermal compression of the gas and was then recorded. When the formation of gas hydrate nuclei occurred during this cooling period, the pressure fell and an increase in both torque and temperature was observed . In the absence of the experimental inhibitors, further growth and agglomeration of the hydrate nuclei rapidly lead to a further increase in torque. When the mixture was heated, the gas hydrates decomposed, and the starting state of the experimental series was reestablished.

The effectiveness of the inhibiting action of the experimental additives was the time from the attainment of the minimum temperature of 2 °C to the first gas absorption, T_{ind}, or the time needed to increase the torque, T. Long induction times or agglomeration times are indicative of the effectiveness of the inhibitor. Torque measured in the autoclave also served as an indicator for the agglomeration of the hydrate crystals. The pressure drop, Δp, provides a semiquantitative method for measuring the amount of hydrate crystals formed. Test results are provided in Table 1.

Test Results

TABLE 1 Diminished pressure, temperature, and torque in natural gas mixtures resulting from reduced gas hydrate formation and increased agglomeration times for natural gas additized with a selected experimental additive

Example	Experimental Additive	Pressure Drop Δp (bar)	Temperature Rise ΔT (K)	Torque M_{max} (Ncm)
Blank	None	>40	>8	15.9
1	(I)	15.3	0.7	0.4
2	(II)	10.1	0.1	0.2
3	(III)	10.9	0.2	0.3
4	(IV)	10.3	0.1	0.1
5	(V)	9.4	0	0

Advantages over Prior Art

Chemical additives effective as both natural gas hydrate inhibitors and antiagglomeration agents that do not require coadditives, particularly anticorrosion materials, are unreported. Moreover, *N*-ethoxylated acid ester ammonium salts of the current investigation were effective at treatment levels of ≤5000 ppm, which is considerably less than the treatment of ≥2% for existing natural gas hydrate inhibitors.

Notes

1. Alkoxylated amines and derivatives are available from
   Huntsman Corporation
   10003 Woodloch Forest Drive
   The Woodlands, TX 77380

   or may prepared according to the method of Cuscurida.[1]
2. Dahlmann[2] modified dodecenyl anhydride with quaternized fatty acid alcohols (VI), which were effective as gas hydrate inhibitors.

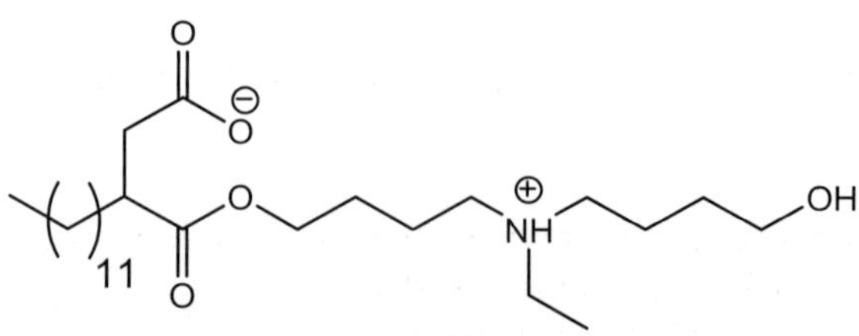

(VI)

3. 3,6,9-Trioxaundecanedioic acid quaternized acid diesters (VII) prepared by Dahlmann[3] were effective as agglomeration and gas hydrate inhibitors.

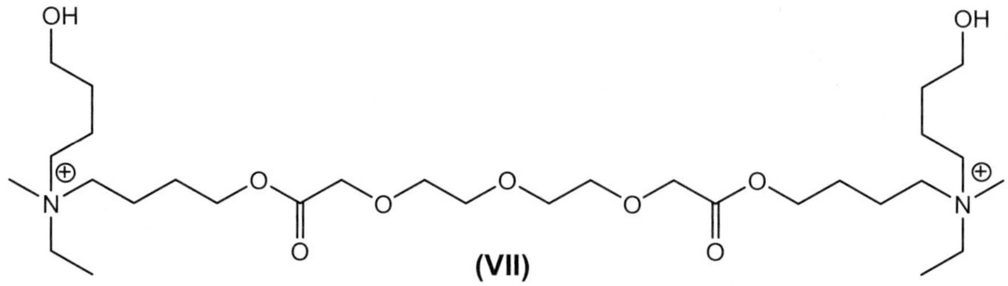

(VII)

4. Pakulski[4] determined that the reaction product of equimolar amounts of polyetheramine having a molecular weight of 320 Da and dodecyl-benzenesulfonic acid (VIII) were effective as gas hydrate inhibitors.

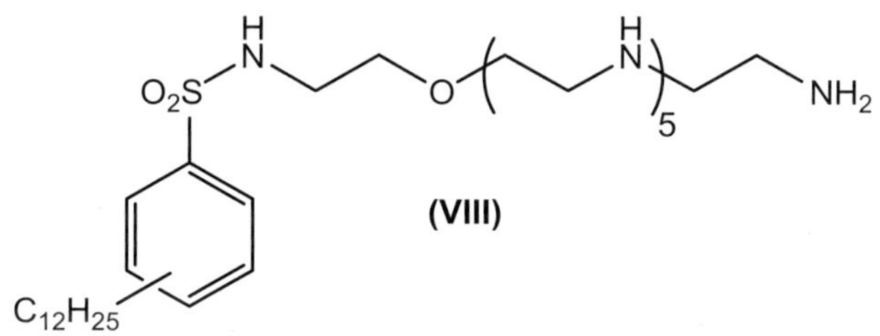

(VIII)

REFERENCES

1. Michael Cuscurida et al, US. Patent 4,612,335 (September 16, 1986) **Texaco, Inc. (White Plains, NY)**
2. Uwe Dahlmann et al, US. Patent 7,435,845 (October 14, 2008) **Clariant Produkte (Deutschland) GmbH (Sulzbach, DE)**
3. Uwe Dahlmann et al, US. Patent 7,348,451 (March 25, 2008) **Clariant Produkte (Deutschland) GmbH (Sulzbach, DE)**
4. Marek Pakulski et al, US. Patent 7,968,500 (June 28, 2011) **Baker Hughes Incorporated (Houston, TX)**

B. THIXOTROPIC AGENTS

Polyamides of Dodecyl Diamine and Hexatriacontanyl Dicarboxylic Acid

Title:	Supramolecular polymer-containing bituminous composition
US Patent:	8,076,398 (December 13, 2011)
Author:	Juan Antonio Gonzalez Leon et al.
Assignee:	Ceca S.A. (Colombes, FR)

Significance: High molecular weight polyamides were prepared by condensing hexatriacontanyl dicarboxylic acid with decyl diamine, pentacontanyl diamine, and poly(propylene oxide) diamine. When bitumen was blended with 5 wt% of these thixotropic polyamides, both the storage modulus and loss modulus at 40 °C, 60 °C, and 80 °C improved.

Additive Name Poly(pentacontanyl diamine-co-hexatriacontanyl dicarboxylic acid)di-1-(2-aminocarbonylethyl)imidazolin-2-one [Nylon 50/36-di-1-(2-aminocarbonylethyl)imidazolin-2-one] (I)
Poly(dodecyl diamine-co-tetratriacontanyl dicarboxylic acid) di-1-(2-aminocarbonylethyl)imidazolin-2-one [Nylon 10/36-di-1-(2-aminocarbonylethyl)imidazolin-2-one] (II)
Poly[(polypropylene oxide diamine)-co-tetratriacontanyl dicarboxylic acid]di-1-(2-aminocarbonylethyl)imidazolin-2-one [Nylon polyether/36-di-1-(2-aminocarbonylethyl)imidazolin-2-one] (III)

Safety If polycondensation products had been prepared using equimolar amounts of diamine and diacid, then the products would be considered to be Nylon analogs and innocuous. Since the industrial diacid, tetratriacontanyl dicarboxylic acid [Pripol 1040®] contains up to 23 wt% trimer acid, the polycondensate should be considered a very mild skin irritant.

Additive Structures

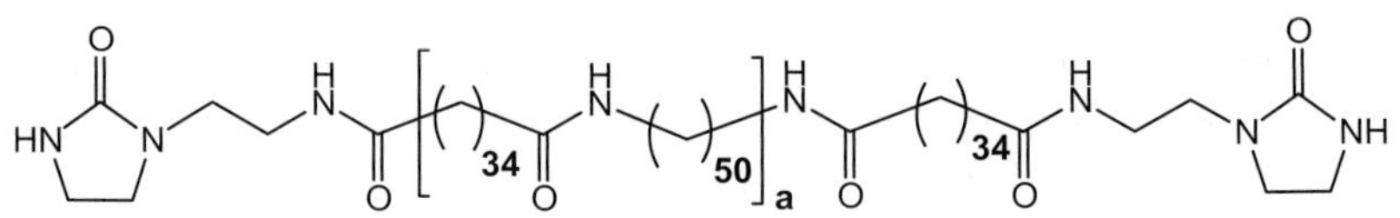

(I)

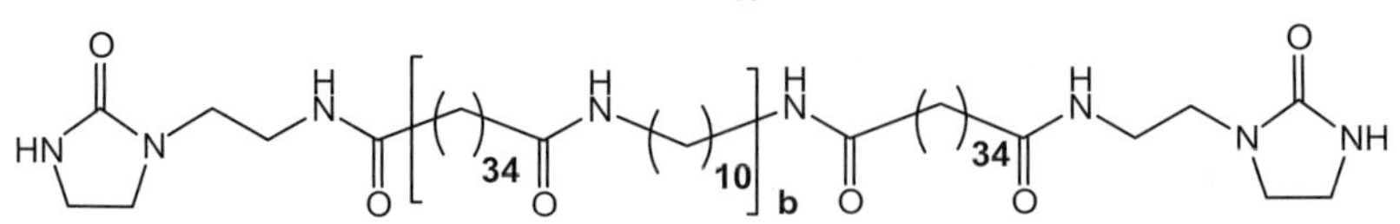

(II)

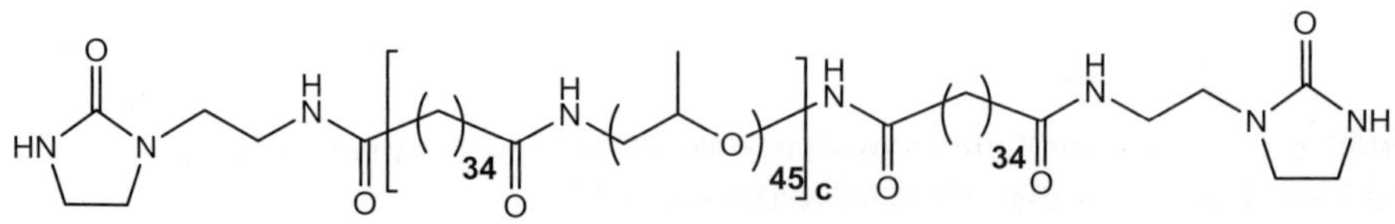

(III)

Additive Preparation

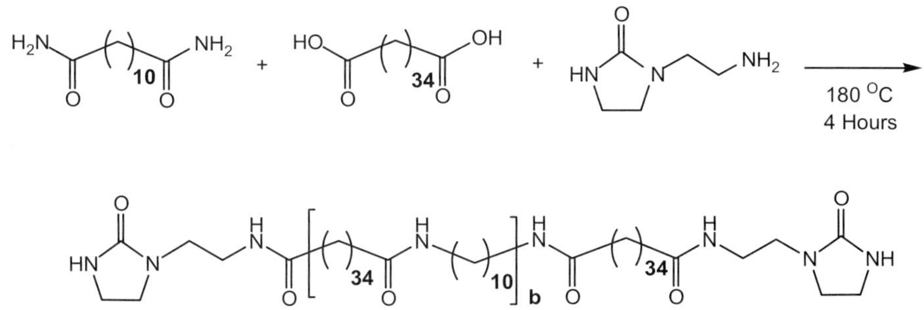

Additive Synthesis

1. *Preparation of Nylon 10/34 containing 2-(aminocarbonylethyl) imidazolin-2-one termini*

A 500-ml glass reactor equipped with a variable speed stirrer motor, inlets for introducing reactants and nitrogen, a temperature probe, reflux condenser, and a Dean–Stark trap was charged with 63 g of tetratriacontanyl dicarboxylic acid [Pripol 1040®] and then heated to 180 °C and treated with 16.8 g of pentacontanyl diamine [Versamine 551®]. After 1 h, the mixture was treated with the dropwise addition of 18.8 g of 1-(2-aminoethyl)imidazolin-2-one and then heated for an additional 3 h until no additional water was collected. The material was then poured from the reaction flask, and the product was isolated in the molten state.

Testing

Experimental compositions consisted of a mixture of five weight parts of a selected polycondensate blended with 100 weight parts of bitumen having a penetration of 35/50. Blends were then heated to 160 °C with rapid stirring for 15 min to obtain a homogeneous mixture and immediately evaluated.

1. *Storage Modulus and Loss Modulus*

Both storage modulus, G′, and loss modulus, G″, of experimental compositions were measured using a dynamical shear rheometer at 1.6 Hz at 40 °C, 60 °C, and 80 °C. Test results are provided in Table 2.

2. *Complex Viscosity*

The complex viscosity of selected polycondensates was measured using a dynamical shear rheometer at 1.6 Hz at 160 °C and 120 °C. In addition, the complex viscosity of poly(ethylene-co-vinyl acrylate) (Evathane 20/20®) was also determined as a comparison. Test results are provided in Table 3.

Testing Results

1. *Storage Modulus and Loss Modulus*

TABLE 2 Effect of the Step 2 product on the storage modulus, G′, and loss modulus, G″, for bitumen. In all cases, bitumen additized with experimental agents showed enhanced storage modulus and loss of modulus properties

Sample	G′ (MPa) 40 °C	G′ (MPa) 60 °C	G′ (MPa) 80 °C	G″ (MPa) 40 °C	G″ (MPa) 60 °C	G″ (MPa) 80 °C
35/50 Bitumen (Unadditized)	6.06×10^{-2}	6.16×10^{-4}	8.99×10^{-6}	2.46×10^{-1}	8.01×10^{-3}	5.51×10^{-4}
35/50 Bitumen + 5 wt% Additive 2	1.11×10^{-1}	2.31×10^{-3}	8.29×10^{-5}	2.83×10^{-1}	1.22×10^{-2}	9.46×10^{-4}

2. *Complex Viscosity*

TABLE 3 Complex viscosities for bitumen blends additized with Additive 1 and Additive 3. The complex viscosity of poly(ethylene-co-vinyl acrylate) was also used since it is often used to modify bitumen. Lower values are preferred

Sample	Viscosity (Pa· s) 140 °C	Viscosity (Pa· s) 120 °C
Poly(ethylene-co-vinyl acrylate) (Control)	45.1	126
Additive 1 (Nylon 50/34)	0.70	4.49
Additive 3 (Nylon polyether/34)	6.75	37.6

Advantages over Prior Art

Although paraffins or polyphosphoric acid derivatives are currently used to increase the stiffness of bitumen at high service temperatures to avoid rutting, the flexibility of additized blends diminishes at low operating temperatures. Incorporating 2–6 wt% of polystyrene, polybutadiene, or poly(ethylene-co-vinyl acetate) into bitumen blends improves properties, although this process requires long mixing times under vigorous agitation to achieve a good dispersion. The current investigation describes three oligomeric additives that form macromolecules by intermolecular hydrogen bonding while quickly dispersing into the bitumen. Bitumen-additive aggregates have improved storage modulus and loss modulus, which increases the range of their commercial application.

Notes

1. Additional supramolecular polymers effective as bitumen thixotropic additives were prepared by Gonzalez Leon[1] using oleic diacids (Empol® 1008), branched polyether/polyester diacids (Empol® 1080), and oleic triacids (Unidyme® 1080).
2. Mixtures of oleic diacids (Empol® 1008) and oleic triacids (Unidyme® 1080) were used by Grimaldi[2] terminated with 2-amino-2-imidazolidinone (IV), 1-(2-[(2-aminoethyl)amino]ethyl)-2-imidazolidinone (V), and 1-(2-{2-[(2-aminoethylamino)ethyl]amino}ethyl)-imidazolidin-2-one (VI), to prepare semicrystalline polyamides, which were effective as bitumen thixotropic additives.

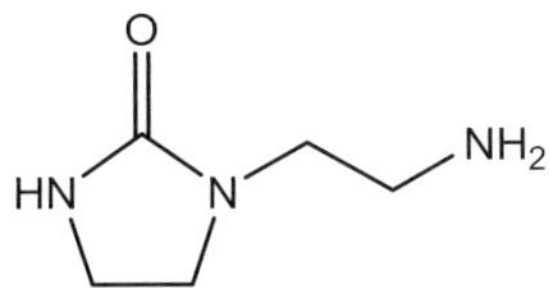

(IV)

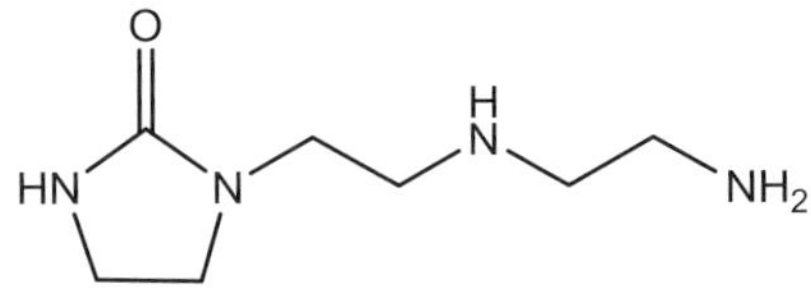

(V)

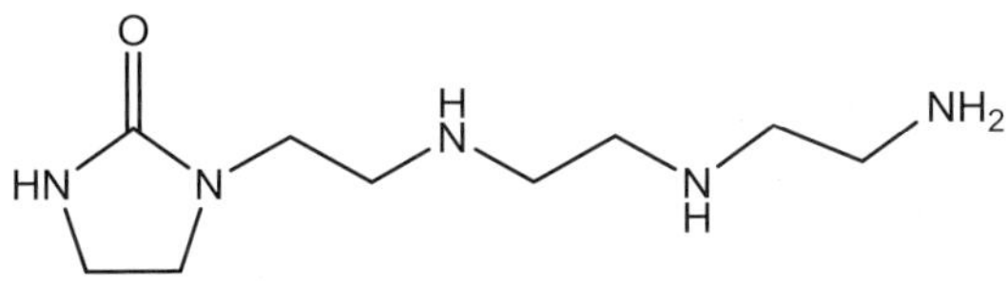

(VI)

3. Barreto[3] prepared a four-component aqueous bitumen thixotropic emulsion additive mixture consisting of the reaction product of poly(ethylene oxide-co-propylene oxide), dodecyl-benzenesulfonic acid, tallow amine, the salt of dodecylbenzenesulfonic acid, and cyclohexyl amine.
4. Tournilhac[4] prepared bitumen stabilizers by condensing fatty acid amides of tetraethylene diamie diamides (VII) with urea to generate imidazolidin-2-one (VIII) termini as illustrated in

Eq. 1.

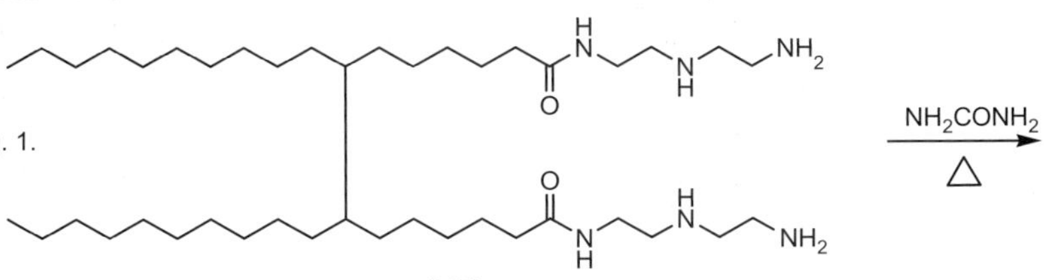

(VII)

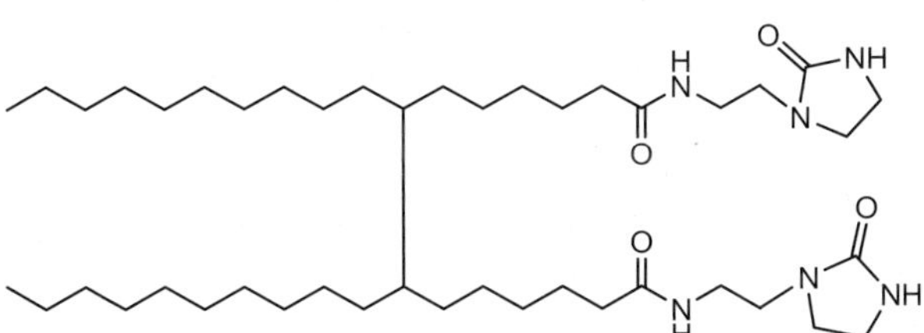

(VIII)

REFERENCES

1. Juan Antonio Gonzalez Leon et al, US Patent Application 20110009533 (January 13, 2011)
Ceca S.A. (Colombes, FR)
2. Sandra Grimaldi et al, US. Patent Application 20100135940 (June 3, 2010)
Ceca S.A. (Colombes, FR)
3 Gilles Barreto et al, US. Patent 7,951,858 (May 31, 2011)
Ceca S.A. (Colombes, FR)
4. Francois-Genes Tournilhac et al, US. Patent 7,750,113 (July 6, 2010)
Arkema France (Colombes, FR)

C. TUNNEL BORING DISPERSANTS EFFECTIVE AS ANTIWEAR, FRICTION MODIFIER, AND RHEOLOGY MODIFIER

Poly(α-hydroxymethyl)octaethylene carbomate

Title:	Glycerol polycarbonate, organic compositions containing the same, and method for obtaining the said compositions
US Patent:	7,928,182 (April 19, 2011)
Author:	Nguyen T. Dinh et al.
Assignee:	INRA (Institut National de Recherche Agronomique) (Paris, France) INPT (Institut National Polytechnique de Toulouse) (BP, France)
Significance:	Poly(α-hydroxymethyl)oxyethylene-co-carbonate was prepared in an 84% yield by reacting glycerol and urea under elevated pressure and temperature using zinc sulfate as the reaction catalyst. This composition was effective as both an antiwear and antifriction additive and as a drilling mud additive for tunnel boring.
Additive	Poly(α-hydroxymethyl)oxyethylene-co-carbonate (I)
Safety	The polymer is considered nontoxic to humans and animals. It is also biodegradable and not considered an environmental hazard.

Additive Structure

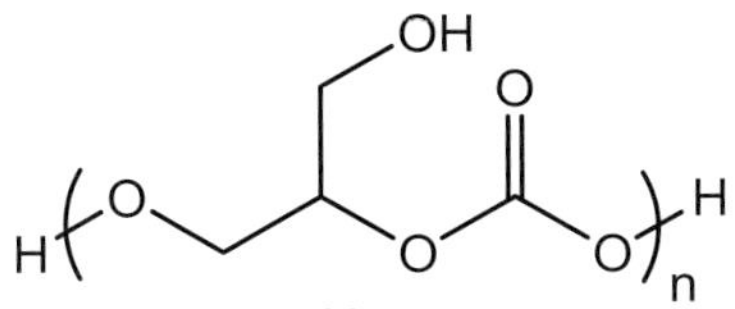

(I)

Additive Preparation

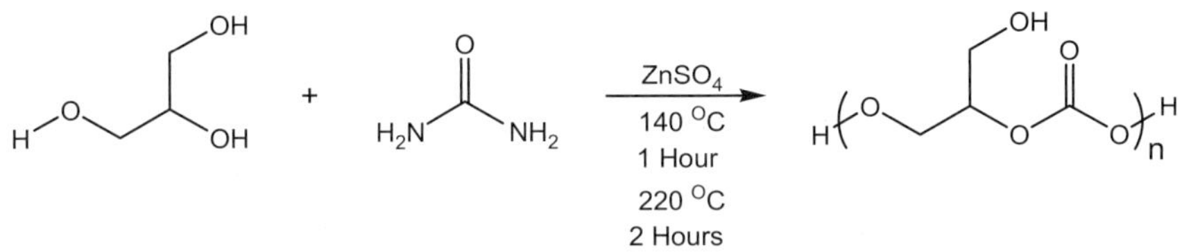

Additive Synthesis

1. *Preparation of poly(α-hydroxymethyl)oxyethylene-co-carbonate*

A reaction chamber was charged with 120 g of glycerol, 82 g of urea, and 2.6 g of zinc sulfate and then heated to 140 °C for 1 h while maintaining a set pressure of 4.0×10^3 Pa. The reaction temperature was then increased to 220 °C for 2 h at a pressure of 3.0×10^3 Pa and then cooled. The product mixture consisted of a liquid having a $[\eta] = 270$ cSt @ 40 °C comprising 84% of the product and 14% of unidentified reaction coproducts.

Testing

1. *Anti-wear*

The Four Ball Wear Test was performed on test samples according to the testing protocol described in ASTM D4172. In this test, wear was determined under a constant load of 40 kg while rotating on three other balls or by using the weld load of the balls under increasing load. Test results are provided in Table 4.

2. *Anti-friction Property Testing*

Antifriction testing was determined using the Cameron-Plint tribometer having a steel cylinder-on-flat configuration 100 C6 in reciprocal movement with a frequency of 6 Hz under a sliding speed of 10 cm/s and a maximum hertz load of 845 MPa. Test results are provided in Table 5.

3. *Drilling Mud Additive for Tunnel Borers or Petroleum Development*

Drilling muds were prepared containing 30 g/l of bentonite clay in water additized with 0.5% of a selected experimental agent, and the rheology of the mixture was evaluated. Test results are provided in Table 6.

Test Results

1. Four Ball Wear Testing

TABLE 4 Results of Four Ball Wear testing for experimental additives prepared by reacting 120 g of glycerol with 82 g of urea and 2.6 g zinc sulfate as the reaction catalyst

Additive	Additive Reaction Temperature (°C)	Solvent	Scar Diameter (mm)	Weld Load (kg)
2	180	As is	0.74	300
3	200	As is	0.40	300
1	160	As is	0.95	330
3	200	Glycerol Carbonate	0.87	230
1	160	Glycerol Carbonate	0.85	200
4	220	Glycerol Carbonate	0.85	260
4	220	Water	0.56	200
4	220	Water	0.57	170
HFC (ISO46)	–	Water	0.65	100
HFC (ISO46)	–	Plant Esters	0.60	150

Lower scar diameters are preferred.

2. Friction Testing

TABLE 5 Results of friction testing using the Cameron–Plint tribometer. Lower wear values are preferred

Additive	Additive Reaction Temperature (°C)	Friction Coefficient	Wear (μm)
3	200	0.08	~2
4	220	0.07	~2
HFDU ISO46 (Control)	–	0.14	20

3. *Rhelogy Testing*

TABLE 6 Effect of Additive 4 on the rheology of bentonite clay in water. A high flow threshold accompanied by the lowest possible dynamic viscosity at 1021 s^{-1} is preferred since this creates a seal that can be easily pumped in both directions

Mud Blend	Viscosity ($85\,s^{-1}$)	Viscosity ($170\,s^{-1}$)	Viscosity ($340\,s^{-1}$)	Viscosity ($511\,s^{-1}$)	Viscosity ($1021\,s^{-1}$)	Yield Point (Pa)
30 g/l Mud (Control)	5.4	5.4	5.25	5.3	6.75	Low, non measurable
30 g/l Mud and 0.5% Glycerol Carbonate	36.6	23.1	15.3	12.4	9.65	2.64
30 g/l Mud and 0.5% Additive 4	10.2	8.7	7.35	6.8	7.55	Low, non measurable

Advantages over Prior Art

The novelty of this invention is in using urea to form poly(α-hydroxymethyl)oxyethylene-co-carbonate. The current method does not require the use of expensive dialkyl carbonates as a reagent nor does it require the removal of alcohol produced from this process. Finally, oligomeric molecular weights of poly(α-hydroxymethyl)oxyethylene-co-carbonate produced using glycerol and urea are consistent with oligomer molecular weights prepared using dialkyl carbonates.

Notes

1. Acylated analogs of poly(α-hydroxymethyl)oxyethylene-co-α-hydroxymethyl) ethylene carbonate (II) were prepared by Dinh[1] and used as engine and industrial lubricants, grease, and hydraulic fluids additives.

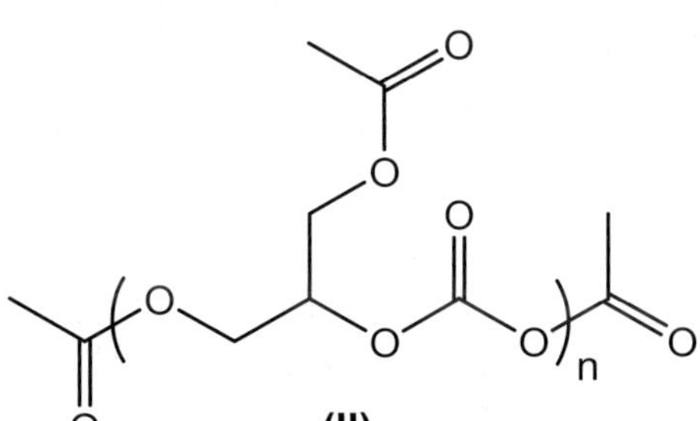

(II)

2. Polyglycerols esterified with oleic acid (III) were prepared by Yokota[2] and used in oil formulations for cutting and grinding.

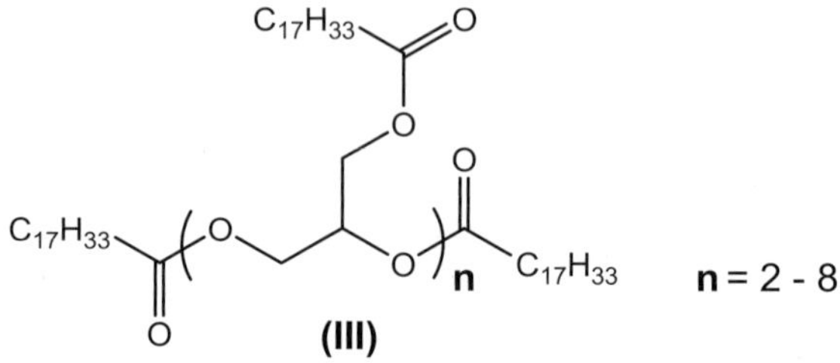

(III)

3. Dimethylsiloxane-modified polyglycerols (IV) were prepared by Miyanaga[3] and used in nonionic fabric and used as a surface modification agent.

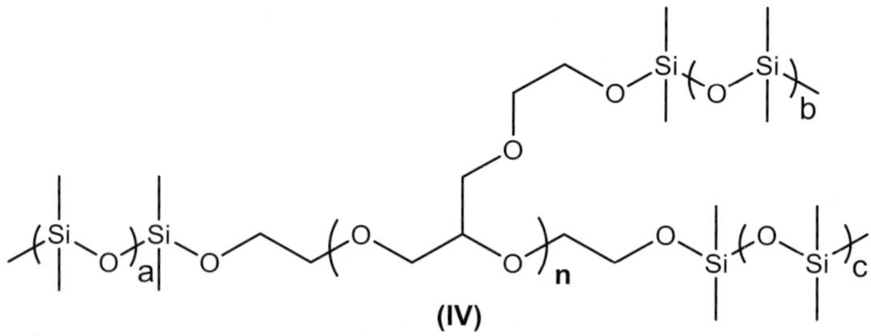

(IV)

4. Krull[4] prepared diesel lubricity additives consisting of partially esterified glycerol esters using soya oil fatty acid, tall oil fatty acid, tallow fatty acid, and oleic acid.

REFERENCES

1. Truong Dinh et al, US Patent 7,928,183 (April 19, 2011)
   INRA (Institut National de Recherche Agronomique) (Paris, FR)
   INPT (Institut National Polytechnique de Toulouse) (BP, FR)
2. Hideo Yokota et al, US Patent 7,838,472 (November 23, 2010)
   Nippon Oil Corporation (Tokyo, JP)
3. Seiichi Miyanaga, US Patent 7,655,744 (February 2, 2010)
   KAO Corporation (Tokyo, JP)
4. Matthias Krull, US Patent 7,815,696 (October 19, 2010)
   Clariant Produkte (Deutschland) GmbH (Sulzbach, DE)

Chapter 14

Drug Delivery

A. DRUG DELIVERY BY NANOPARTICLES

Poly(γ-Glutamic Acid)-Chitosan Nanoparticles

Title:	Nanoparticles for protein drug delivery
US:	7,897,585 (March 1, 2011)
Author:	Hsing-Wen Sung et al.
Assignee:	GP Medical, Inc. (Newport Beach, CA)
Significance:	A method for the oral delivery of a wide variety of pH-sensitive biologically active agents was demonstrated using nanoparticles having an outer chitosan skin and γ-polyglutamic acid core. Nanoparticles spontaneously formed when the two reagents were mixed using the ionic-gelation method. When the nanoparticle composite was blended with insulin, the delivery of this biologically active agent was either at 0 42 or 0.83 mg/ml over a 10-day period.
Additive Name	Poly(γ-glutamic acid)-chitosan nanoparticles (I)
Safety Concerns	None

Additive Structure

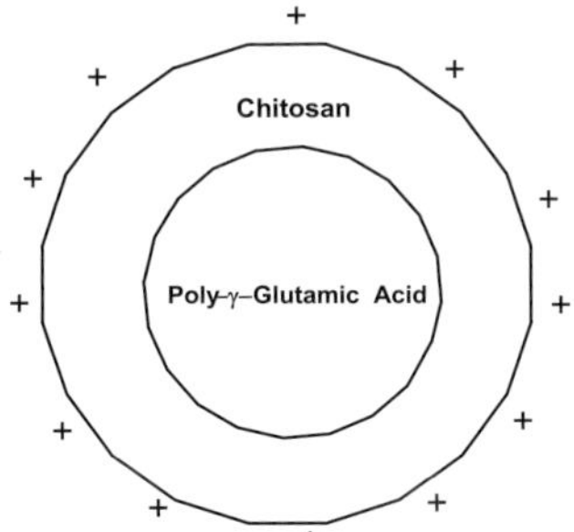

Poly-γ-Glutamic Acid—Chitosan Nanoparticle

(I)

Next Generation of International Chemical Additives. DOI: http://dx.doi.org/10.1016/B978-0-444-53788-1.00014-0

Additive Preparation

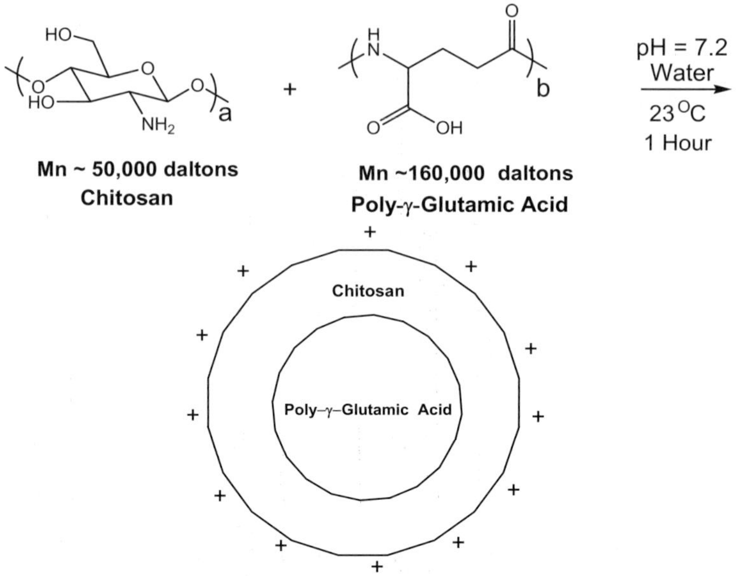

Additive Synthesis

1. Preparation of poly(γ-glutamic acid)-chitosan nanoparticles

Nanoparticles were prepared by the addition of a 2 ml solution of γ-PGA having an Mn of approximately 160,000 Da with a pH = 7.4 to either a 6 or 10-ml solution of chitosan having an Mn of approximately 50,000 Da at pH 6.0. The mixture was immediately stirred at an ambient temperature and nanoparticles that formed immediately were then collected by ultracentrifugation at 38,000 rpm after approximately 1 h. The supernatants were then discarded, and nanoparticles were purified by suspending in deionized water and isolated by the mild ionic-gelation method. These nanoparticles displayed characteristics of spheroidal configurations with a particle size distribution of between 50 and 400 nm and having a positively charged surface with a narrow polydispersity index. FT-IR indicated that peaks observed at 1563 and 1615 cm^{-1} were indicative of an $^{+}NH_3$ deformation in the presence of COO^-. The presence of a new peak at 1586 cm^{-1} was characteristic of $-NH_3$ deformations on chitosan at 1563 cm^{-1}. These observations were attributed to electrostatic interaction between the negatively charged carboxylic acid salts ($-COO^-$) on γ-glutamic acid and the positively charged amino groups ($-NH_3^+$) present on chitosan.

2. Effect of concentrations of γ-glutamic acid and chitosan on particle size of poly(γ-glutamic acid)-chitosan nanoparticles

Poly(γ-glutamic acid)-chitosan nanoparticles were also prepared using different reagent concentrations. A summary of nanoparticle size as a function of reagent concentrations used in their preparation is provided in Table 1.

TABLE 1 Effect on poly(γ-glutamic acid)-chitosan nanoparticle size as a function of reagent concentrations. All nanoparticles were prepared using the Step 1 reaction procedure

Treatment Level Poly(γ-glutamic acid) (%)	Chitosan (0.01%) (nm)	Chitosan (0.05%) (nm)	Chitosan (0.10%) (nm)	Chitosan (0.15%) (nm)	Chitosan (0.20%) (nm)
0.01	79.0	103.1	96.7	103.6	140.5
0.5	154.4	120.8	144.5	106.2	165.4

Testing

1. *Insulin loading properties using poly(γ-glutamic acid)-chitosan nanoparticles*

Fluorescence-labeled γ-glutamic acid was added to a chitosan solution to prepare fluorescence-labeled poly(γ-glutamic acid)-chitosan nanoparticles. Modified nanoparticles were then used in in vivo animal studies using confocal laser scanning microscopy to assess bioactivity. Insulin-loaded poly(γ-glutamic acid)-chitosan nanoparticles were then prepared using the ionic-gelation method with the modification that insulin was then thoroughly mixed. Physical properties of poly(γ-glutamic acid)-chitosan nanoparticles containing adsorbed insulin are summarized in Table 2.

2. *Insulin delivery effect of poly(γ-glutamic acid)-chitosan nanoparticles containing adsorbed insulin*

Experimental results are summarized below.

3. *pH stability of Poly(γ-glutamic acid)-chitosan nanoparticles modified with ethylenediamine tetraacetic acid*

The chitosan surface component of poly(γ-glutamic acid)-chitosan nanoparticles salt of ethylenediamine tetraacetic acid (II) was also prepared to determine the effect on the spherical biostability. Test results are provided in Table 3.

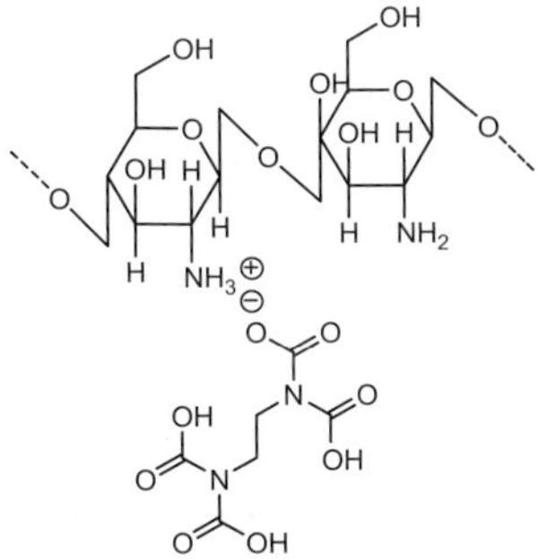

(II)

Test Results

1. *Insulin loading*

TABLE 2 The effect on poly(γ-glutamic acid)-chitosan nanoparticle physical properties after insulin loading

z	Mean Particle Size (nm)	Polydispersity Index	Zeta Potential
0	145.6	0.14	+32.11
0.42	185.1	0.31	+29.91
0.83	198.4	0.30	+27.83

2. *Insulin delivery*

When the particle size of insulin-loaded poly(γ-glutamic acid)-chitosan nanoparticles was approximately 170 nm, the average insulin delivery over a 40 day period was 0.083 mg/ml.

3. *pH Stability of modified γ-glutamic acid-chitosan nanoparticles*

The physical properties of poly(γ-glutamic acid)-chitosan nanoparticles modified with ethylenediamine tetraacetic acid are provided in Table 3.

TABLE 3 Effect of varying pH on poly(γ-glutamic acid)-chitosan nanoparticles modified with ethylenediamine tetraacetic acid suggesting the optimum pH drug delivery range was between pH 2.5 and 6

Nanoparticle pH Environment	Mean Particle Size (nm)	Zeta Potential (mV)	Polydispersity Index
1.2	N/A	N/A	1
2.0	N/A	N/A	1
2.5	113.3	38.6	0.14
6.0	104.1	36.2	0.11
6.6	245.6	12.9	0.17
7.0	N/A	N/A	1
7.4	N/A	N/A	1

Advantages over Prior Art

Polymeric biodegradable nanoparticles consisting of poly(ε-caprolactam)-lactide were previously prepared, but all efforts to utilize them as drug delivery agents were ineffective because of their hydrophobic characteristics. Composite nanoparticles containing γ-glutamic acid as the core and chitosan as the surface had an average zeta potentials of 27 mV. As a consequence of this unusual polarity, these composites can be successfully used as drug delivery agents for insulin, heparin, exenatide, pramlintide, calcitonin, teripapatide, vancomycin, and piperacillin.

Notes

1. Heppei[1] used chitosan-based transport systems for overcoming the blood–brain barrier and encapsulated nicotine, cerebrolysin, piracetam, phosphatidyl serine, and acetyl-L-carnitine to treat neurodenerative diseases.
2. In a subsequent investigation by Sung,[2] poly(γ-glutamic acid)-chitosan nanoparticle composites were used as transdermal drug delivery agents for rDNA-derived insulin glargine in animal subjects.
3. Polyanhydrides of amoxicillin (II) and cephalexin (III) were prepared by Giroux[3] and were successfully used as drug delivery agents in animals.

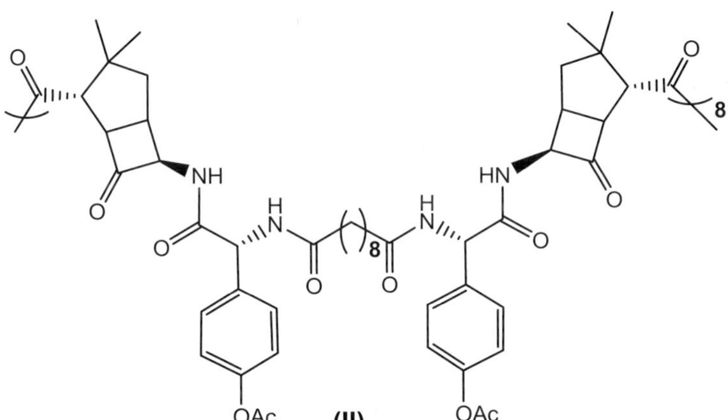

(II)

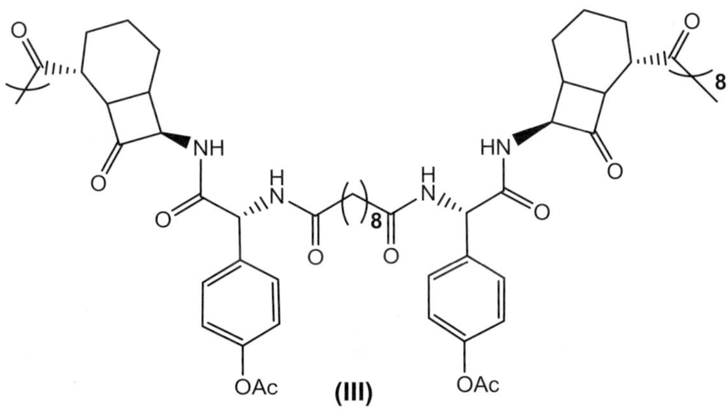

(III)

4. Transferrin-functionalized with both adamantane and polyethyleneglycol (IV) were prepared by Pun[4] and used as a delivery agent for antineoplastic agents.

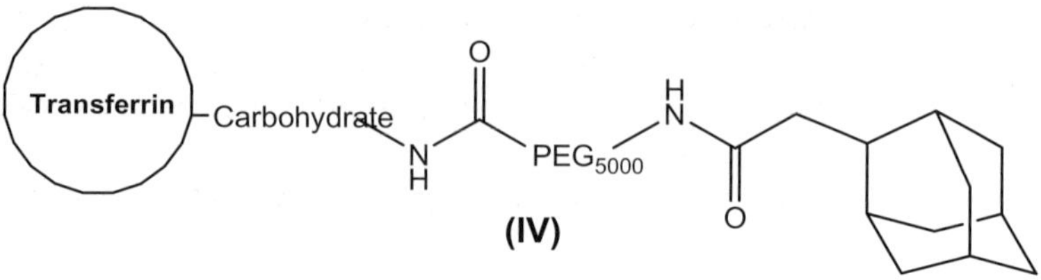

(IV)

REFERENCES

1. Katja Heppe et al, US Patent Application 20060051423 (March 9, 2006) **Bayer Aktiengesellschaft (Leverkusen, DE)**
2. Hsing-Wen Sung et al, US Patent 7,901,711 (March 8, 2011) **GP Medical, Inc. (Newport Beach, CA)**
3. Karen J. Giroux, US Patent Application 20060013851 (January 19, 2006) **Celgene Corporation (Summit, NJ)**
4. Suzie Hwa Pun et al, US Patent 7,807,198 (October 5, 2010) **California Institute of Technology (Pasadena, CA)** **Calando Pharmaceuticals, Inc. (Pasadena, CA)**

B. INULIN POLYKETAL AS A DRUG DELIVERY AGENT

Tetracyclic Delivery Agent

Title:	Biodegradable polyketal polymers and methods for their formation and use
US:	Patent: 7,951,898 (May 31, 2011)
Author:	Mikhail I. Papisov
Assignee:	The General Hospital Corporation (Boston, MA)
Significance:	Inulin was converted into a well-defined polyketal by oxidative cleavage with periodic acid followed by reduction of the intermediate using sodium borohydride. The polyketal and its succinyl conjugate were both nontoxic and biocompatible. Both agents are suitable as drug delivery additives for tetracyclic medicaments, particularly doxycycline and doxorubicin.
Additives	Poly(1-hydroxymethyl-1-(2-hydroxy-1-hydroxymethyl-ethoxy)-ethylene oxide) (I)
Safety	In initial studies, no toxicity in rats was observed after repeated saline intravenous injections containing either additive or the diethylenetriamine pentaacetic dicycloanhydride acid conjugate.

Additive Structure

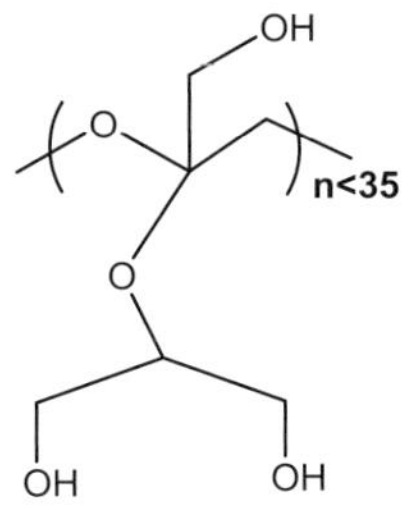

(I)

Additive Preparation

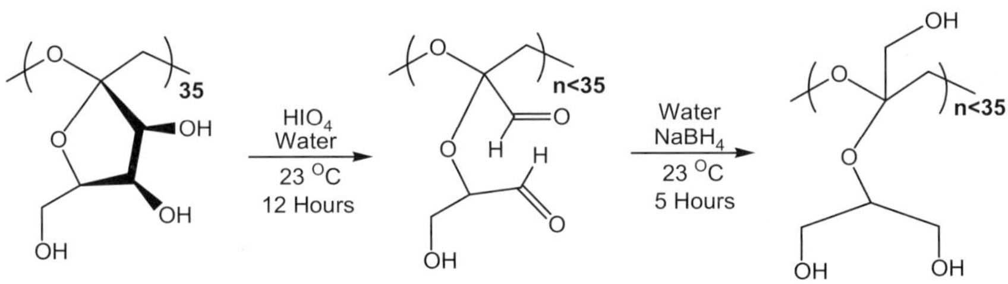

Additive Synthesis

1. Preparation of inulin polyaldehyde

A small reactor was charged with 0.5 g of inulin dissolved in 10 ml of water at 60 °C and then cooled to an ambient temperature and mixed with a suspension of 0.8 g sodium periodate dissolved in 2 ml of water. After overnight incubation, the reaction mixture was filtered and HPLC (High Performance Liquid Chromatography) indicated a lower molecular weight distribution as a result of oxidative degradation.

2. Preparation of poly(1-hydroxymethyl-1-(2-hydroxy-1-hydroxymethyl-ethoxy)-ethylene oxide)

The Step 1 product was cooled to 0 °C and then slowly poured into a cold solution of 1 g of sodium borohydride dissolved in 2 ml of water while stirring and periodically cooled to prevent the reaction temperature from rising >20 °C. After 5 h, the reaction was quenched and HPLC indicated that the molecular weight distribution remained unchanged from the Step 1 product. The product was then isolated by gel chromatography using Sephadex® G-25, and the product was isolated after lyophilization. Product structural conformation was performed using spectrometric methods (Note 1).

Testing

Acute Toxicity

The Step 2 product was dissolved in 0.9% saline at 100 mg/ml and injected intravenously into 6 mice at 100-mg/kg body weight. The animals were observed for 30 days after the injection. All the animals survived, and none showed any detectable signs of toxicity.

Advantages over Prior Art

Polyketals are a new class of biocompatible/biodegradable drug delivery additives derived from either dextran or inulin polysaccharides. Conceivably, other polyketal drug delivery agents can be prepared differently or related polysaccharides. Preliminary findings indicate that taxol and apyrase conjugates with succinylated polyketals are clinically more effective than when administrating these agents using this method.

Notes

1. Spectrometric analysis also detected minor structural irregularities (II) resulting from the incomplete oxidation degradation of Inulin.

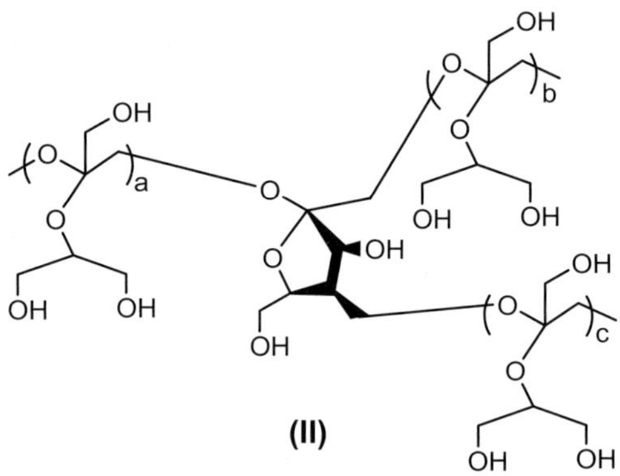

(II)

2. Poly(1-hydroxymethylethylene hydroxymethylformal) (III) was prepared by Papisov[1] by the oxidative cleavage and reduction of dextran and then used as a drug delivery agent for taxol (IV).

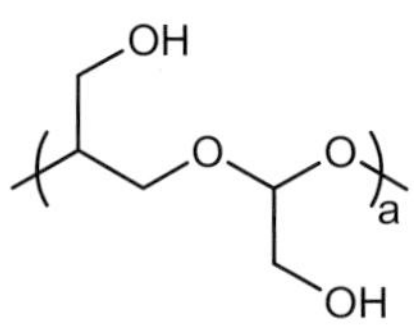

(III)

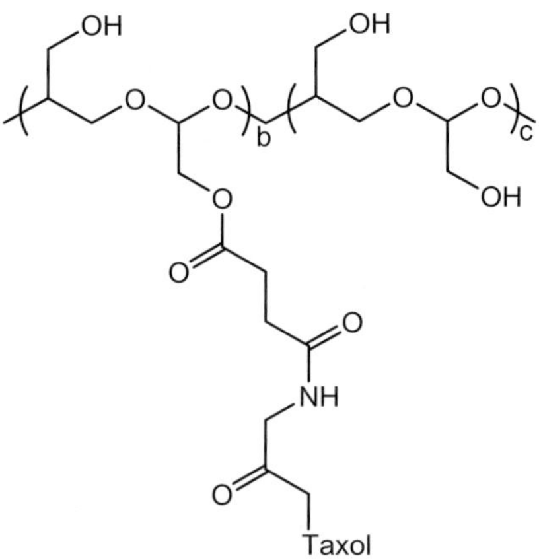

(IV)

3. Elmaleh[2] prepared the conjugate of succinylated poly(1-hydroxymethylethylene hydroxymethylformal) (V) with Apyrase and determined that conjugated apyrase exhibited a stronger enzymatic activity in vitro relative to unconjugated Apyrase.

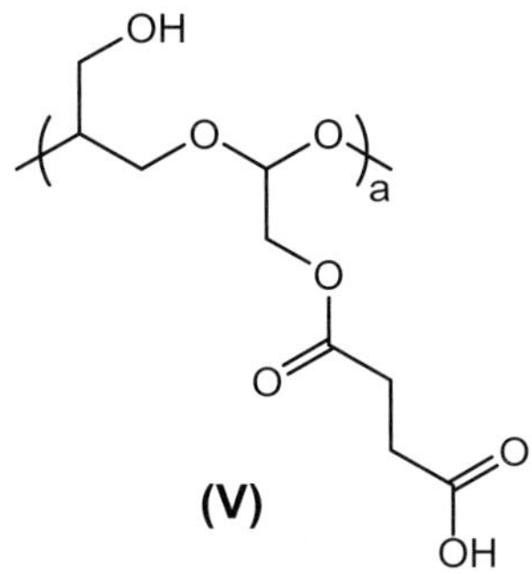

(V)

4. Bae[3] prepared pH-sensitive polymeric micelles consisting of a copolymer containing polyhistidine and polyethylene oxide (VI), which spontaneously formed micellars and was then used as a drug delivery agent for doxycycline and doxorubicin.

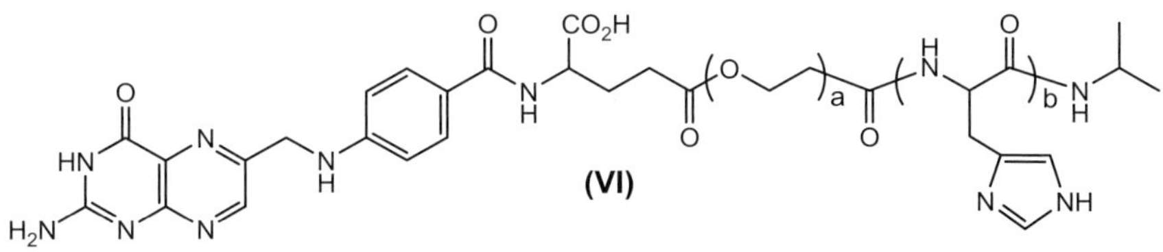

(VI)

5. O'Hagen[4] prepared a mucosal vaccine drug delivery device consisting of polyhyaluronic acid (VII), which was capable of containing up to a 40-wt% antigen load.

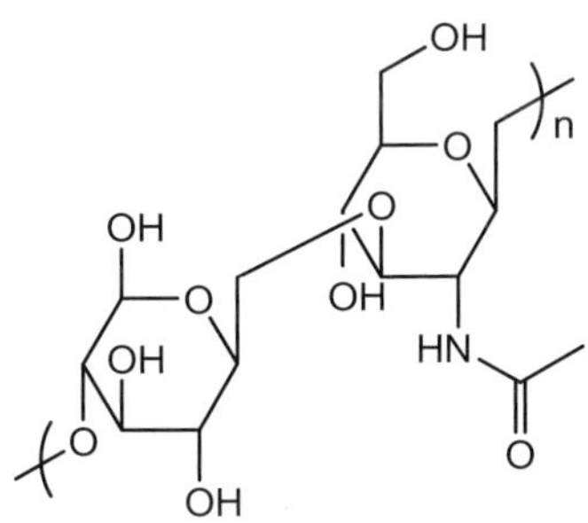

(VII)

REFERENCES

1. Mikhail I. Papisov et al, US Patent Application 20070190018 (August 16, 2007) **The General Hospital Corporation (Boston, MA)**
2. David R. Elmaleh et al, US Patent 7,785,618 (August 31, 2010) **The General Hospital Corporation (Boston, MA)**
3. You Han Bae et al, US Patent 7,951,846 (May 31, 2011) **University of Utah Research Foundation (Salt Lake City, UT)**
4. Derek O'Hagan et al, US Patent 7,947,309 (May 24, 2011) **Fidia Farmaceutici S.p.A (Abano Terme Padua, IT)**

Chapter 15

Electronic Additives

A. LOW DIELECTRIC CONSTANT AND LOW WATER ABSORPTION ELECTRONIC ADDITIVES

Polymeric Fluorine-containing Dicarboxylic Acids

Title:	Fluorine-containing dicarboxylic acids and their novel polymer compounds
US:	Patent 8,003,749 (August 23, 2011)
Author:	Satoru Narizuka et al.
Assignee:	Central Glass Company, Limited (Ube-shi, JP)
Significance:	Polyamides and polyesters having an Mn of at least 25,000 Da were prepared by reacting either bisphenol A diamine or bisphenol A with 1,1,1,3,3,3-hexafluoro-2-hydroxy-2-propane benzene-dicarboxylic acid. All fluorinated polymers had lower dielectric constants and decreased water absorption compared with their non-fluorinated counterparts.
Additive Name	Poly(4-(1,1,1,3,3,3-hexafluoro-2-hydroxy-2-propane)-1,3-benzene-dicarboxylic bisphenol A)polyester (I), poly(4-(1,1,1,3,3,3-hexafluoro-2-hydroxy-2-propane)-1,3-benzene-dicarboxylic bisphenol A)polyamide (II), poly(3-(1,1,1,3,3,3-hexafluoro-2-hydroxy-2-propane)-1,3-benzene-dicarboxylic bisphenol A)polyamide (III), poly(3-(1,1,1,3,3,3-hexafluoro-2-hydroxy-2-propane)-1,3-benzene-dicarboxylic hexafluoromethylolbisphenol A)polyamide (IV)
Safety	Additives of the current invention are considered extremely toxic. They are potentially fatal if ingested. They are also considered severe eye and respiratory system irritants. Skin contact can cause moderate-to-severe irritation. Combustion of perfluorinated aromatic diacids, diamides, or diesters will release hydrogen fluoride.

Next Generation of International Chemical Additives. DOI: http://dx.doi.org/10.1016/B978-0-444-53788-1.00015-2

Additive Structures

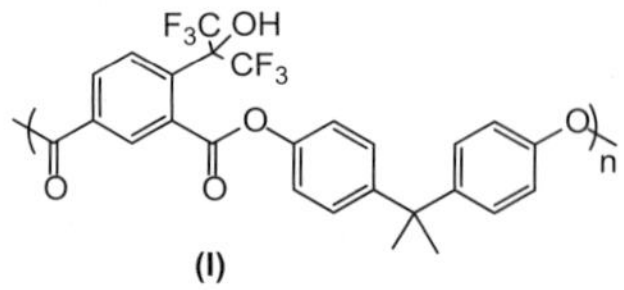

(I)

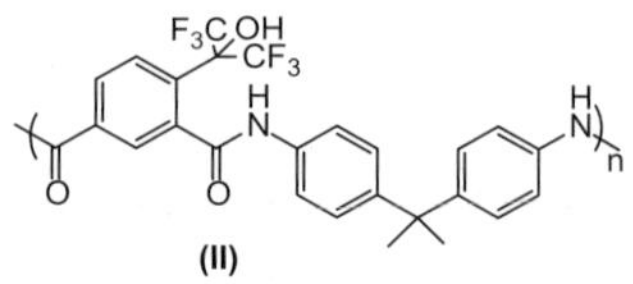

(II)

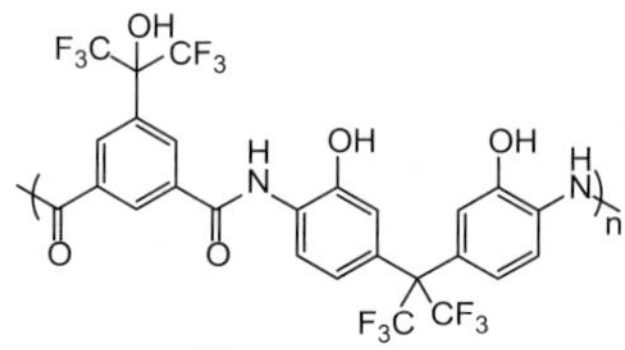

(III)

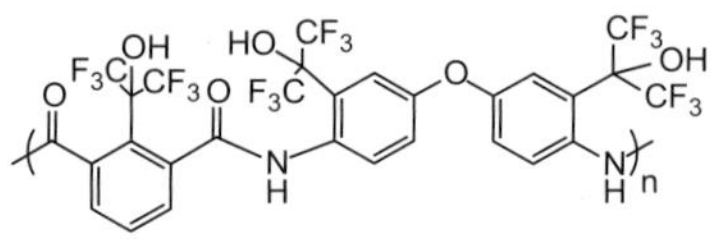

(IV)

Additive Preparation

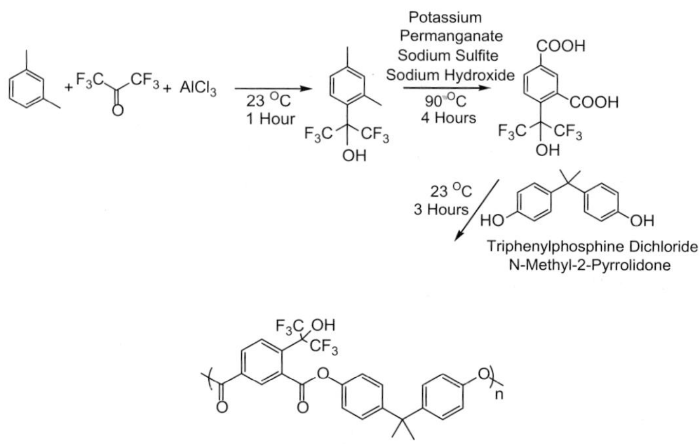

Additive Synthesis

1. *Preparation of 4-(1,1,1,3,3,3-hexafluoro-2-hydroxy-2-propane)-1,3-dimethyl benzene*

A 1-liter reactor was charged with 100.0g of *m*-xylene and 6.3g of aluminum chloride, and the inside reactor temperature was reduced to 10°C. The mixture was then treated with 164.2g of hexafluoroacetone, and the mixture was stirred at 23°C for 1h followed by the addition of 100ml of 10% hydrochloric acid. The aqueous layer was then extracted twice with 40ml of chloroform and then dried using anhydrous magnesium sulfate and filtered. After vacuum distillation, the product was obtained in an 84% yield having a purity of 95%.

^{1}H-NMR (d_6-DMSO, δ) 8.37 (s, 1H), 7.34 (d, J=8.0Hz, 1H), 7.12–7.06 (m, 2H), 2.52 (s, 3H), 2.26 (s, 3H).
^{19}F-NMR (d_6-DMSO, δ) −72.1 (s, 6F, CF_3).

2. *Preparation of 4-(1,1,1,3,3,3-hexafluoro-2-hydroxy-2-propane)-1,3-benzene-dicarboxylic acid*

A 300-ml reactor was charged with 10.0 g of the Step 1 product and 150 ml of 0.15 M sodium hydroxide and was then heated to 85 °C and treated with the incremental addition of 26.1 g of potassium permanganate over 1 h. Thereafter, the mixture was heated for 4 h and then treated with 11 ml of concentrated hydrochloric acid and sodium sulfite. The mixture was then extracted with 200 ml of diisopropyl ether, and the aqueous layer was further extracted twice with 50 ml of diisopropyl ether, and combined organic layers dried using magnesium sulfate. The mixture was then filtered, concentrated, and a pale yellow solid was isolated. The solid was purified by dissolving in 30 ml of toluene containing 4 ml of acetonitrile and then by heating to reflux. Upon cooling, 6.1 g of product was obtained having a purity of 99.5%.

^{1}H-NMR (d_6-DMSO, δ) 8.64 (dd, J=8.0 and 1.2 Hz, 1H), 8.61 (s, 1H), 8.28 (d, J=8.0 Hz, 1H).
^{13}C-NMR (d_6-DMSO, δ) 165.83 (s), 165.34 (s), 139.80 (s), 138.45 (s), 137.56 (s), 128.62 (s), 125.90 (s), 125.77 (s), 121.15 (q, J=284.7 Hz), 81.67 (sept, J=32.9 Hz).
^{19}F-NMR (d_6-DMSO, δ) -73.7 (s, 6F, CF_3).

3. *Preparation of bisphenol A perfluoropolyester*

A three-necked flask was charged with 2.00 g of the Step 2 product, 1.37 g of bisphenol A, 4.19 g of triphenylphosphine dichloride, and 12.0 g of *N*-methyl-2-pyrrolidone and then stirred at an ambient temperature for 3 h. The solution was then poured into 30 ml of methanol, and a precipitate was formed, which was isolated by filtration. After vacuum drying at 80 °C, 2.84 g of the polyester was isolated having an Mw of 25,000 Da. A summary of the physical properties of related polymers is provided in Table 1.

TABLE 1 Polyamides and polyester prepared by reacting 4-(1,1,1,3,3,3-hexafluoro-2-hydroxy-2-propane-1,3-benzene-dicarboxylic acid and bisphenol A diamine or bisphenol A

Polymer	Structure	Mn (Polydispersity) (Da)
1 (Step 3 Product)	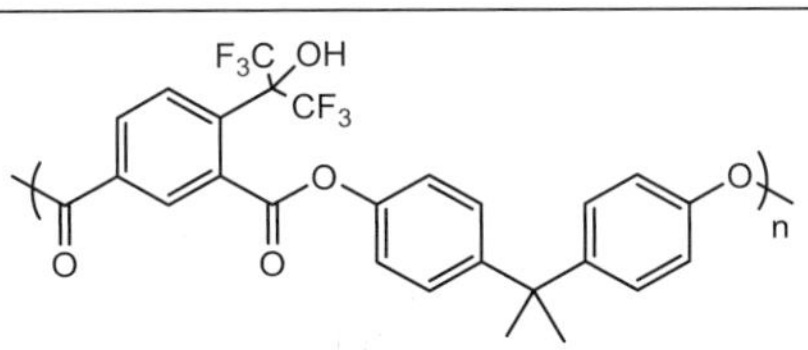	25,000 (2.34)
2	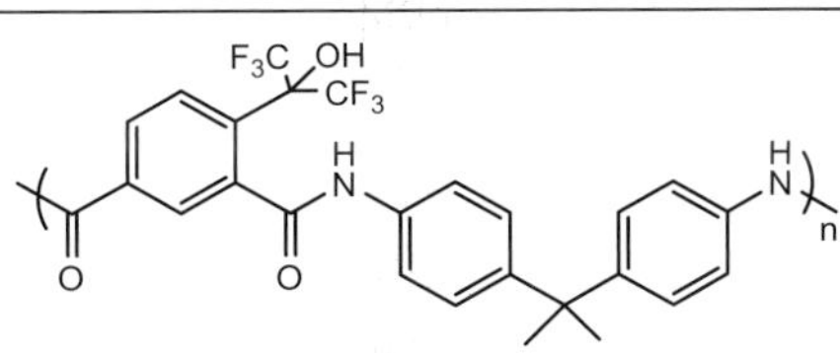	31,000 (2.52)

(Continued)

TABLE 1 Polyamides and polyester prepared by reacting 4-(1,1,1,3,3,3-hexafluoro-2-hydroxy-2-propane-1,3-benzene-dicarboxylic acid and bisphenol A diamine or bisphenol A—cont'd

Polymer	Structure	Mn (Polydispersity) (Da)
3	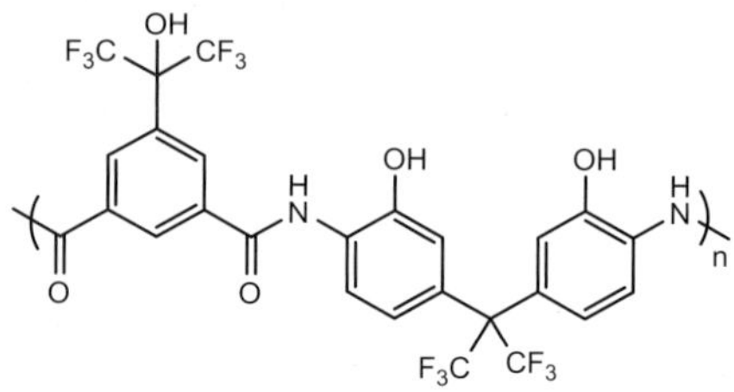	37,000 (2.17)
5	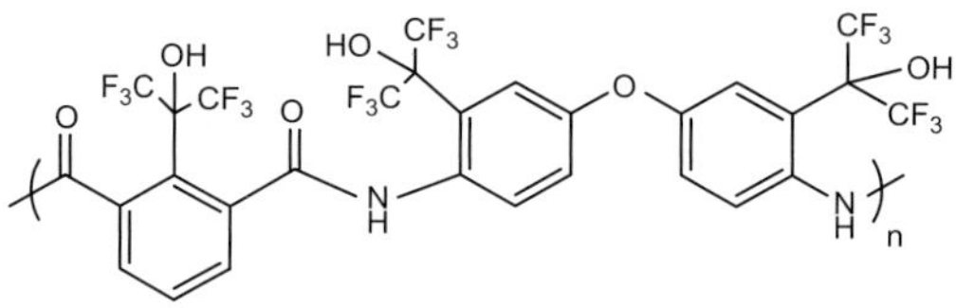	32,000 (2.45)

Testing

Dielectric Constant and Water Absorption Testing

A solution was prepared by dissolving 1.00g of a selected polymer in 4.00g of *N,N*-dimethyl-formamide and then filtered and applied onto a glass substrate by spin coating. The material was then heated to 80°C for 30min, 150°C for 30min, and then to 250°C for 1h. The transparent film was then removed from the glass substrate, and dielectric constant and water absorption properties were determined. Test results are provided in Table 2.

TABLE 2 Test results of dielectric constant and water absorption properties for polymers containing selected ,1,1,3,3,3-hexafluoro-2-hydroxy-2-propanebenzene 1,3-benzene-dicarboxylic acid. In all cases, fluorinated polymers outperformed nonfluorinated polymeric analogs

Polymer	Dielectric Constant	Water Absorption (%)
Polymer 1	3.10	2.00
Fluorine-free Polymer 1 Analog (Control)	3.40	2.40
Polymer 2	3.40	3.30
Fluorine-free Polymer 2 Analog (Control)	3.60	4.00
Polymer 3	2.70	2.60
Fluorine-free Polymer 3 Analog (Control)	3.20	3.20
Polymer 5	2.50	2.42
Fluorine-free Polymer 5 Analog (Control)	3.00	3.05

Test Results

Advantages over Prior Art

The current technology addresses three concerns associated with photoresist additives. First, materials produced according to these methods must be heat resistant. Second, the presence of the perfluoroalcohol or phenols in these materials must result in materials having excellent adhesion to both resins and fibers. Finally, the presence of cross-linkable alcohols and phenols makes these materials easily convertible into thermosets.

Notes

1. Selected 1,1,1,3,3,3-hexafluoro-2-hydroxy-2-propanebenzene diamines (IV) were prepared by Saegusa[1] and then subsequently used to prepare polyaromatic polyamides (V) for photoresist compositions.

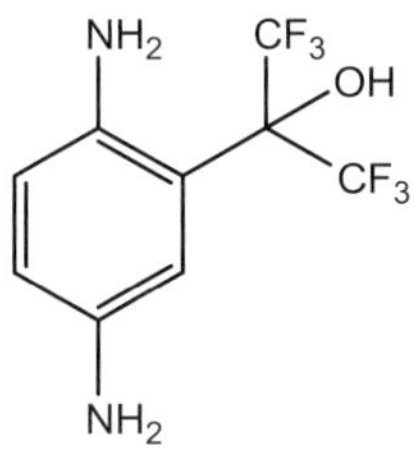

(IV)

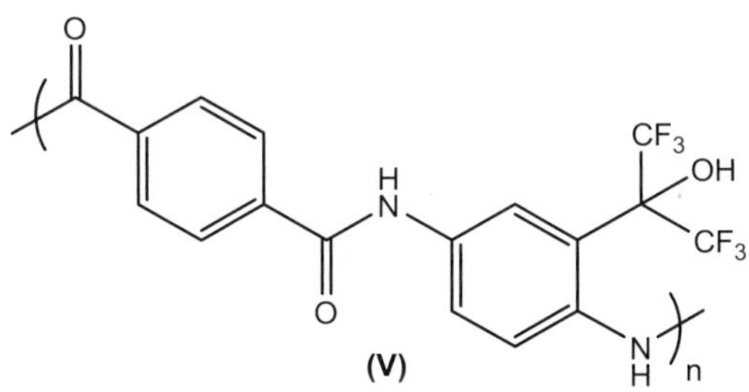

(V)

2. Compositions containing triphenysulfonium poly(acrylic acid-g-3,3,4,4-tetrafluorobutyl-4-sulfonic acid) (VI) were prepared by Nagai[2] and were effective as radiation-sensitive resins while not behaving as acid generators.

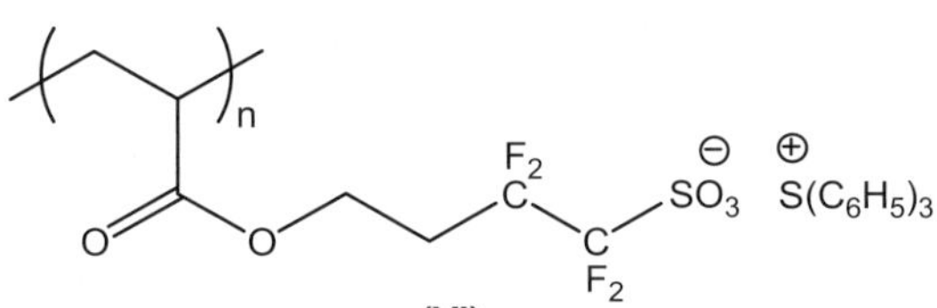

(VI)

3. Goldfarb[3] prepared polyperfluoromethacrylate esters (VII) and (VIII), which were effective as ultra low postexposure bake photoresist materials.

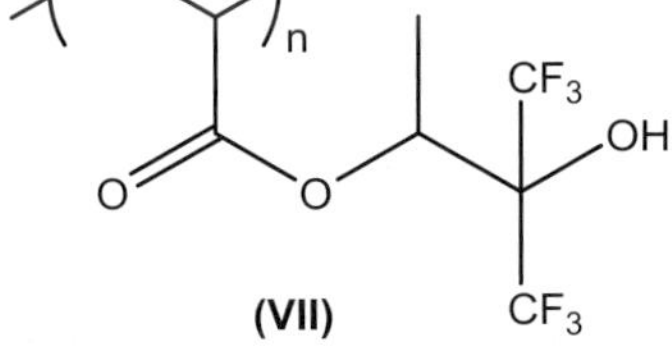

(VII)

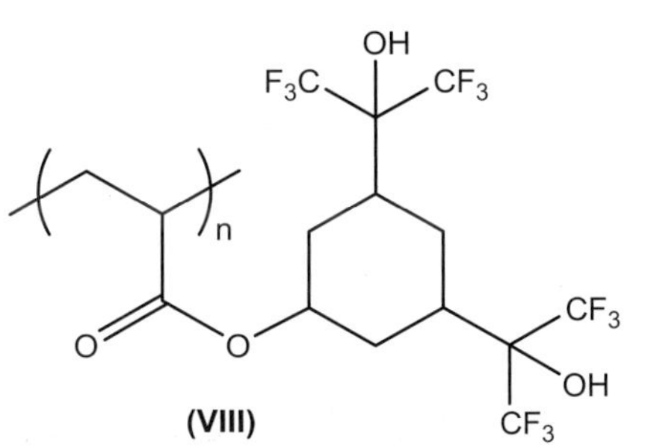

(VIII)

4. Naoyoshi[4] prepared polymerizable perfluoroadamantyl methylmethacrylate esters (IX) and (X), which were used in photoresist compositions to improve dry etching resistance in liquid immersion exposure methods while not reducing adhesion to a mold.

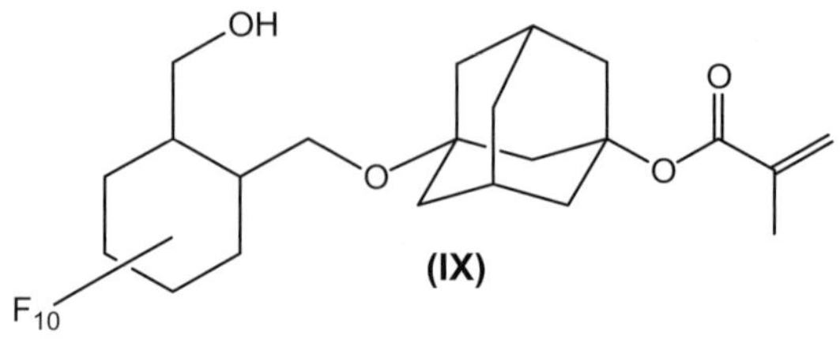

(IX)

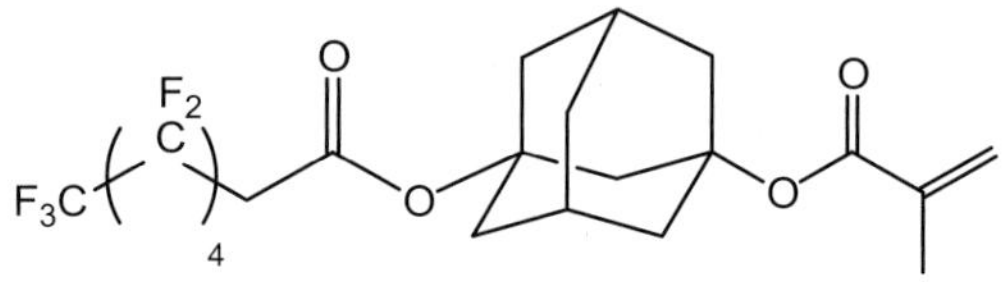

(X)

REFERENCES

1. Hiroshi Saegusa et al, US. Patent 7,932,348 (April 26, 2011)
   Central Glass Company, Limited (Ube-shi, JP)
2. Tomoki Nagai et al, US. Patent 8,026,039 (September 27, 2011)
   JSR Corporation (Tokyo, JP)
3. Leonardo Dario Goldfarb et al, US. Patent 8,017,303 (September 13, 2011)
   International Business Machines Corporation (Armonk, NY)
4. Naoyoshi Hatakeyama et al, US. Patent Application 20100022730 (January 28, 2010)
   Idemmitsu Kosan Co., Ltd. (Chiyoda-ku, JP)

Chapter 16

Extreme Pressure Additives

A. INCREASING STEEL TAPPING EFFICIENCY

Nitrated Products of Unsaturated Fatty Acids

Application:	Extreme Pressure Additives
Title:	Nitrated extreme pressure additives
US Patent:	7,960,323 (June 14, 2011)
Author:	Duong N. Nguyen
Assignee:	Dover Chemical Corporation (Dover, OH)
Significance:	Sulfur-, chlorine-, and phosphorous-free high performance lubricating oil additives were prepared by nitrating soybean oil, polyisobutylene, oleic acid or oleic acid esters, and nonylphenol. When used in lubricating oil formulations for steel and stainless steel tapping, nitrated agents outperformed all existing commercially available materials.

Additives

Nitrated soybean oil (I)
Nitrated esters of trimethylolpropane and oleic acid (II)
Nitrated polyisobutylene (III)
Nitrated oleic acid (IV)
Nitrated nonylphenol (V)

Safety

The nitration products of unsaturated fatty acids formed have been demonstrated to exert strong biological actions in endothelial cells and monocytes/macrophages.

In addition, there is strong evidence that unsaturated fatty nitroalkenes inhibit VSMC proliferation through activation of the Keap1/Nrf2 signaling pathway.

Next Generation of International Chemical Additives. DOI: http://dx.doi.org/10.1016/B978-0-444-53788-1.00016-4

Additive Structures

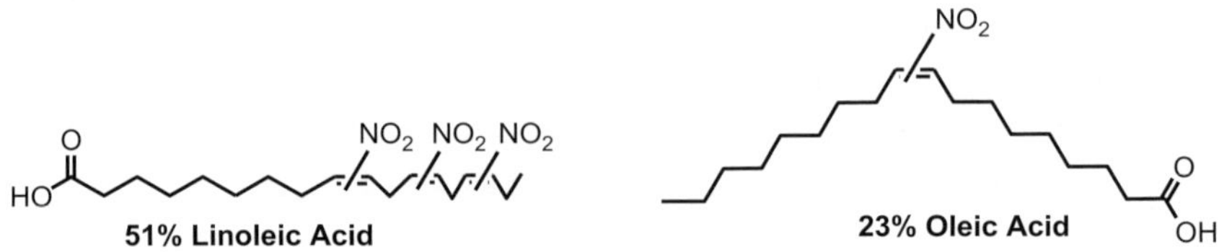

Major Soybean Oil Components

(I)

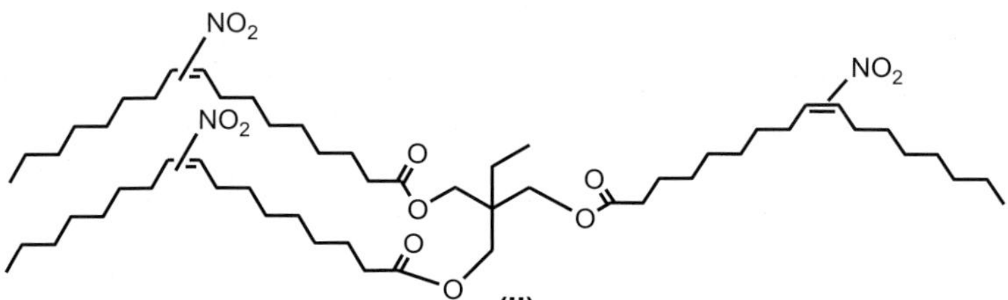

(II)

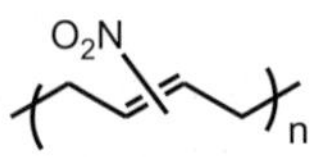

(III)

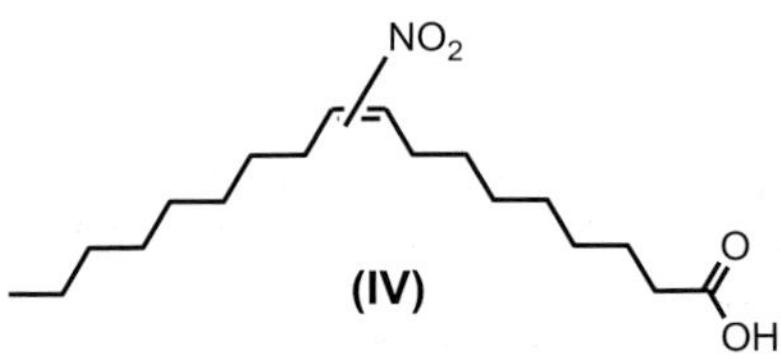

(IV)

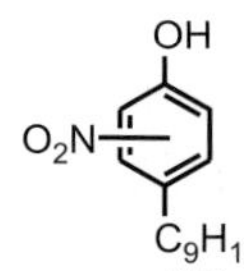

(V)

Additive Preparation

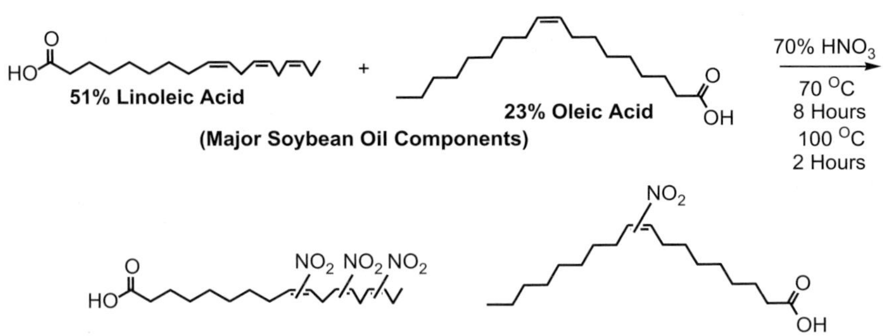

Additive Synthesis

1. Preparation of nitrated soybean oil

A three-necked round-bottom flask equipped with a thermometer, stirring bar, Dean-Stark trap, gas sparger, and a cer was charged with 89.0 g of refined soybean oil and

16.1 g of 70% nitric acid and then heated to 70 °C for 8 h. The temperature of the reaction was slowly increased to 100 °C for 2 h and then cooled to 80 °C. When the reaction was completed, the mixture was air blown to remove essentially all entrapped water. After cooling, 96.5 g of the product was isolated as a dark brown fluid having a nitro content of up to 18 wt%. FTIR confirmed the presence of the nitro group at 1551 cm^{-1}.

Testing

Experimental additives were evaluated using the laboratory-scaled Microtap® II G8 (Microtap USA, Inc.). In this test, holes were predrilled on the steel or stainless steel work-pieces. A series of tappings was then made through these predrilled holes at 300 rpm for stainless steel parts and at 450 rpm for regular steel parts. While the tap bit penetrated a 15-mm depth, all torques data points that were generated by friction were automatically recorded by the Microtap computer software, and an average value was calculated. Test results are provided in Tables 1 and 2.

Test Results

TABLE 1 Microtap tapping efficiency using lubricating fluid additized with 10% selected nitrated experimental extreme pressure additives conducted at with a depth of on stainless steel

Extreme Pressure Additive Sample	Steel Tapping Efficiency (%)
No additive (Standard)	100
Chlorinated Olefin (57% Cl) (Dow Chemical)	122
Sulfurized Fat (18% S) (Dow Chemical)	133
Sulfurized Olefin (40% S) (Dow Chemical)	123
2-Ethylhexyl Nitrate	124
Nitropropane	129
Nitro Soy	133
Nitro Synthetic Ester	130
Nitro Oleic Acid	127
Nitro PIB	126
Nitro Nonylphenol	123

TABLE 2 Microtap tapping efficiency using lubricating fluid additized with 10% selected nitrated experimental extreme pressure additives conducted at with a depth of on steel

Extreme Pressure Additive Sample	Stainless Steel Tapping Efficiency (%)
No additive (Standard)	Tap bit seized
Chlorinated Olefin (57% Cl) (Dow Chemical)	100
Sulfurized Fat (18% S) (Dow Chemical)	Tap bit seized
Sulfurized Olefin (40% S) (Dow Chemical)	Tap bit seized
2-Ethylhexyl Nitrate	Tap bit seized
Nitropropane	98
Nitro Soy	105
Nitro Synthetic Ester	100
Nitro Oleic Acid	100
Nitro PIB	99
Nitro Nonylphenol	100

Advantages over Prior Art

Nitrated high-performance lubricating additives have several distinct advantages over existing agents. First, they can be used in drilling applications for steel and stainless steel or alloys containing chromium, titanium, or nickel. Second, their high temperature performance range exceeds all additives containing phosphorous, chlorine, or sulfur. Finally, nitrated lubricating additives are readily prepared in high yields using 30–70% nitric acid and inexpensive oils.

Notes

1. Oil-soluble molybdenum derivatives derived from hydroxyethyl-substituted Mannich bases (VI) were prepared by Gatto,[1] which were effective in forming metal boundary layers in internal combustion engines.

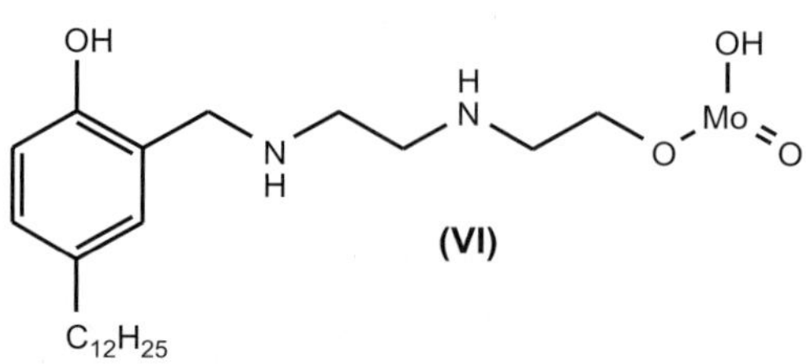

(VI)

2. Haiping[2] prepared nanofluids containing sulfonated nanotubes that were used in nanolubricants and nanogreases to improve the viscosity and load-bearing properties of each formulation.
3. Silicate drilling fluid formulations were prepared by Wu[3] containing a carrier oil and one or more extreme pressure additives consisting of sulfurized lecithins, sulfurized vegetable oil, sulfurized lard oil, or sulfurized vegetable esters and were designed to stabilize shale formations and improve the performance in silicate drilling fluids.
4. Yokota[4] prepared a 10-component high performance oil composition designed for cutting, grinding, and sliding surface consisting of:
   Triester of neopentylglycol and oleic acid
   Triester of trimethylolpropane and oleic acid
   Tetraester of trimethylolpropane and acid mixture of n-hexanic acid/n-octanic acid/n-decanic acid
   Tetraester of pentaerythritol and n-octanic acid
   Commercially available rapeseed oil
   Monoester of glycerol and oleic acid
   Sulfide ester (sulfur content: 9.9% by mass, inactive)
   Polysulfide (sulfur content: 36% by mass, active)
   Tricresylphosphate
   Di-2-ethylhexylphosphate

REFERENCES

1. Vincent J. Gatto et al, US Patent 7,960,321 (June 14, 2011)
   Afton Chemical Corporation (Richmond, VA)
2. Hong Haiping et al, US Patent 7,871,533 (January 18, 2011)
   South Dakota School of Mines and Technology (Rapid City, SD)
3. Anming Wu et al, US Patent 7,842,651 (November 30, 2010)
   Chengdu Cationic Chemistry Company, Inc. (Sichuan, CN)
4. Hideo Yokota et al, US Patent 7,838,472 (November 23, 2010)
   Nippon Oil Corporation (Tokyo, JP)

Chapter 17

Films

A. HEAT-CURABLE POLYIMIDE SILICONE RESINS

Polyimide Silicone Resins

Title:	Heat-curable polyimide silicone resin composition
US:	Patent: 7.981, 976 (July 12, 2011)
Author:	Yoshinori Yoneda et al.
Assignee:	Shin-Etsu Chemical Co., Ltd. (Tokyo, JP)
Significance:	A heat-curable aromatic polyimide silicone resin containing up to 12 vinyl appendages per repeat unit was prepared and then free radically cured using either *t*-butylperoxy-2-ethyl carbonate or 1,6-bis(*t*-butyl-carbonyl)hexane. Polyimide silicone coatings were completely solvent resistant and had excellent metal adhesive properties.
Additives	Polyimide silicone additive-1 (I) Polyimide silicone additive-2 (II) Polyimide silicone additive-3 (III) Polyimide silicone additive-4 (IV)
Safety	Aromatic polyimide resins are slight skin irritants, but they are not skin sensitizers. Effects of a single exposure by inhalation may include discomfort and difficulty in respiration after 4 h and the LC50 was15,600 mg/m^3 in rats. Effects of repeated oral exposure include reduced food consumption and reduced rate of weight gain. Human overexposure may include skin irritation or rash while inhalation may irritate the upper respiratory passages resulting in coughing and discomfort.

Next Generation of International Chemical Additives. DOI: http://dx.doi.org/10.1016/B978-0-444-53788-1.00017-6

Additive Structures

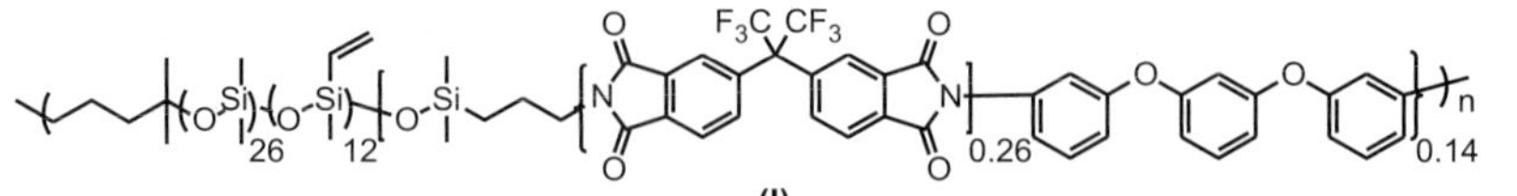

(I)

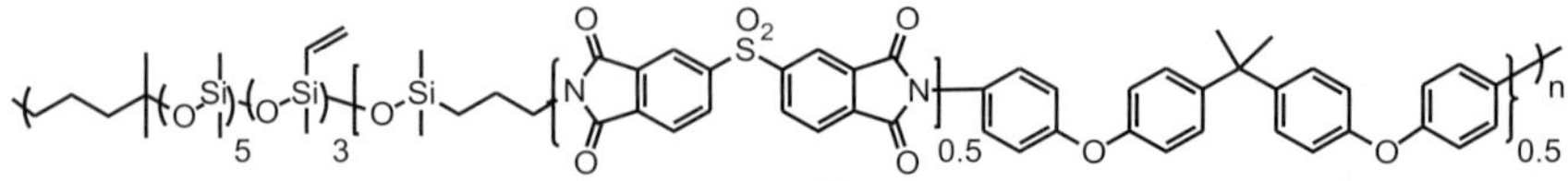

(II)

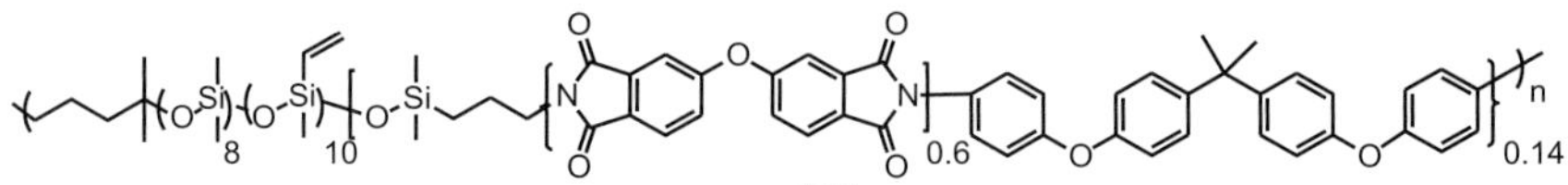

(III)

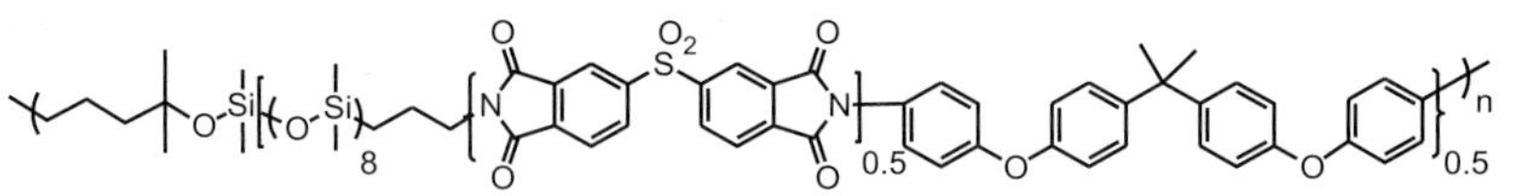

(IV)

Additive Preparation

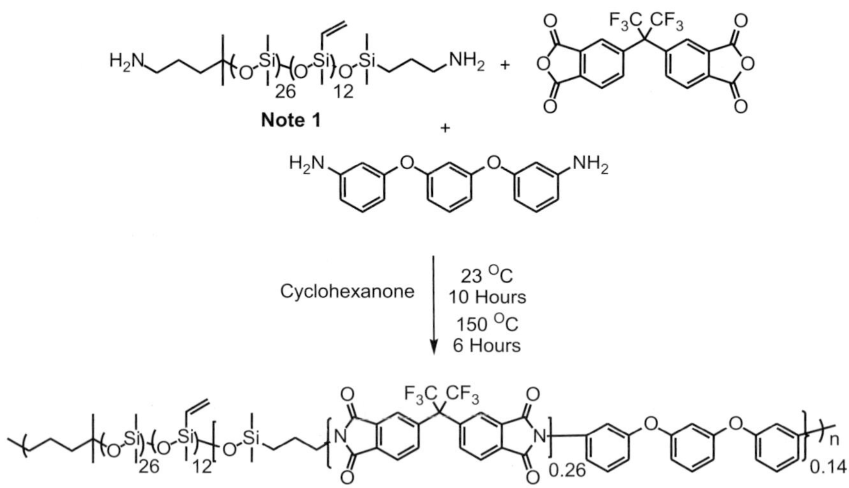

Additive Synthesis

1. Preparation of polyimide silicone additive

A flask provided with a stirrer, thermometer, and nitrogen purge equipment was charged with 88.8 g of 4,4′-hexafluoropropylidene bisphthalic acid anhydride and 500 g of cyclohexanone and then treated with the dropwise addition of 192.2 g of poly(dimethylsiloxane)-b-poly(methylvinylsiloxane)-bis(dimethylsiloxypropyldiamine) (Note 1) and 40.9 g of 1,3-bis(3-aminophenoxy)-benzene dissolved in 200 g of cyclohexanone. The reaction mixture was then stirred at 23 °C for 10 h and then treated with 70 g of xylene and further heated to 150 °C for 6 h. The solution was then cooled to 23 °C, and the product was isolated by precipitating in methyl alcohol, and the dried product had an Mn = 42, 200 Da.

Testing

1. *Preparation of polymer compositions*

Experimental conditions used to prepare polymer compositions using selected free radical initiators are provided in Table 1.

TABLE 1 Experimental conditions used to prepare cured polyimide silicone additives

Polyimide Silicone	Curing Agent	Curing Agent (g/100 g Resin)	Solvent	Final Curing Temperature (°C)
Additive-1	*t*-Butylperoxy-2-ethylhexyl carbonate	1	Methyl ethyl ketone	180
Additive-2	*t*-Butylperoxy-2-ethylhexyl carbonate	5	Methyl ethyl ketone	150
Additive-3	1,6-Bis(*t*-butyl-peroxycarbonyloxy) hexane	5	Methyl ethyl ketone	100
Additive-4	1,6-Bis(*t*-butyl- per-oxycarbonyloxy) hexane	3	Butyl acetate	150

2. *Solvent resistance*

A composition was prepared and applied to a plate coated with fluororesin to form a 0.1-mm-thick coating after drying and was then heated to 80 °C for 30 min and then further heated to 150 °C for 1 h. The cured film was then soaked in *N*-methyl-2-pyrrolidone for 30 min and examined. Test results are provided in Table 2.

3. *Adhesion strength*

Selected polymer compositions were applied onto a copper substrate and then heated at 80 °C for 30 min and cured by heating to 150 °C for 6 h. The heat resistance of the adhesion was measured after keeping the copper substrate attached to the cured film in a dryer at 150 °C for 360 h. The strength of the adhesive at elevated temperatures and humidity were measured after keeping the copper substrate attached to the cured

TABLE 2 Heat, solvent resistance, and adhesion strength testing for cured experimental polyimide silicone additives

Polyimide Silicone	Resistance to *N*-methyl-2-pyrrolidone	Heat Resistance of Additive	Heat and Humidity Resistance of Additive
Additive-1	Slightly swollen	100/100	100/100
Additive-2	No damage	100/100	100/100
Additive-3	No damage	100/100	100/100
Additive-4	No damage	100/100	100/100
Comparative	Dissolved	85/50	70/100

TABLE 3 Mechanical testing results for cured polyimide silicone additives

Polyimide Silicone	Tensile Elongation (%)	Tensile Strength (Mpa)
Additive-2	320	18
Additive-3	420	26
Additive-4	400	10

film in a saturated water vapor at 120 °C at 2 atmosphere pressure for 168 h. In the heat and humidity and resistance testing, the number of patches per 100 patches that did not peel from a substrate was measured, where a value of 100/100 indicates that no patch peeled off while 0/100 indicates that all 100 patches peeled off. Test results are provided in Table 2.

4. Mechanical strength

Tensile elongation and tensile strength of cured polymer composites were evaluated according to JIS K7113. Testing results are provided in Table 3.

Testing Results

1. Solvent Resistance
2. Adhesion Strength Testing Results
3. Mechanical Strength

Advantages over Prior Art

The current method addresses a need to quickly and quantitatively convert polyamic acid silicones into polyimide silicones for potential consumer markets. Current methods for this conversion can typically require up to 5 h of prolonged heating with up to 4% residue of amic acid remaining.

Notes

1. The preparation of the reagent, poly(dimethylsiloxane)-b-poly(methylvinylsiloxane)-bis(dimethylsiloxypropyldiamine) (V) is described by Koji.[1]

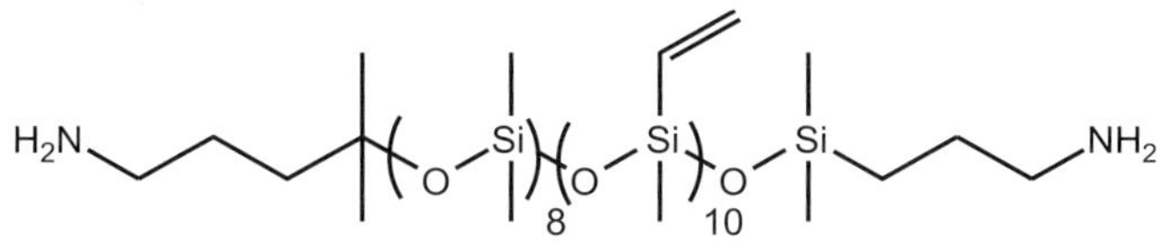

(V)

2. Yoshinori[2] previously prepared free radical polymerizable polyimide silicone resins containing mixed phenolics (VI), which were effective as an additives in heat-curable resin coating compositions.

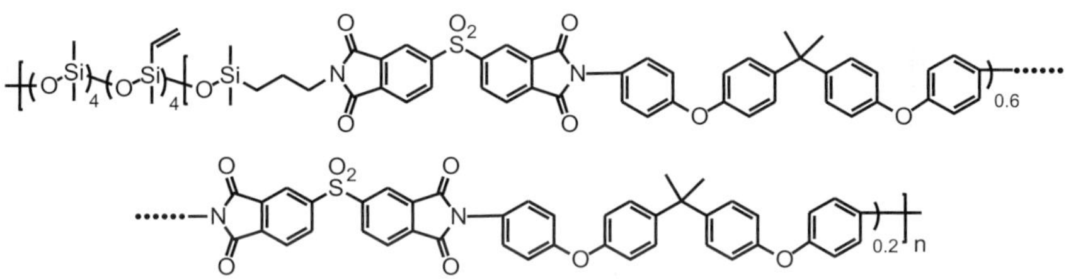

(VI)

3. Irisawa[3] prepared photopolymerizable polyimide silicone compositions consisting of difunctional acrylate aromatic esters (VII), which were effective as film additives and for forming liquid crystalline phases.

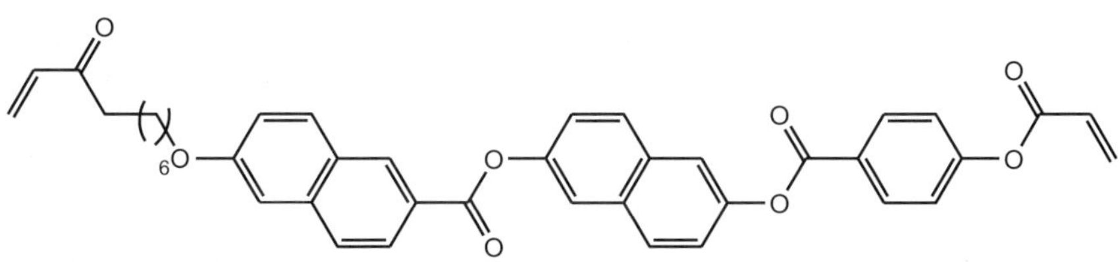

(VII)

4. Taniguchi[4] photopolymerized polyimide silicone resins (VIII) on silicone wafers using bis(α-isonitrosopropiophenone oxime)isophthal followed by curing at 170 °C to obtain films that resisted shrinkage and had good elongation at breakage.

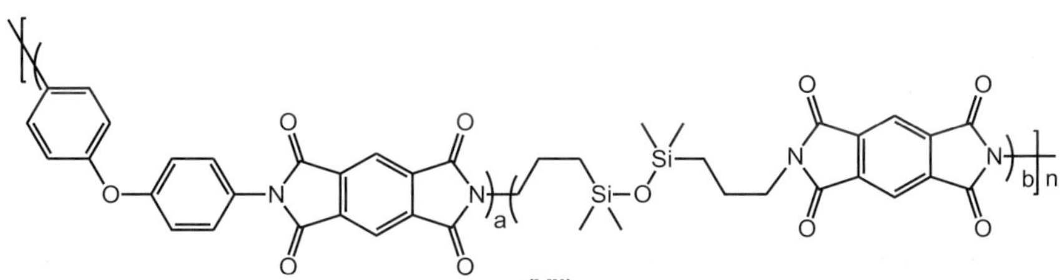

(VII)

5. Musa[5] prepared a reactive intermediate diimide (VIII) that when copolymerized with adipic acid at 145 °C was used as an adhesive additive in coatings and encapsulants on metal, glass, and plastic surfaces.

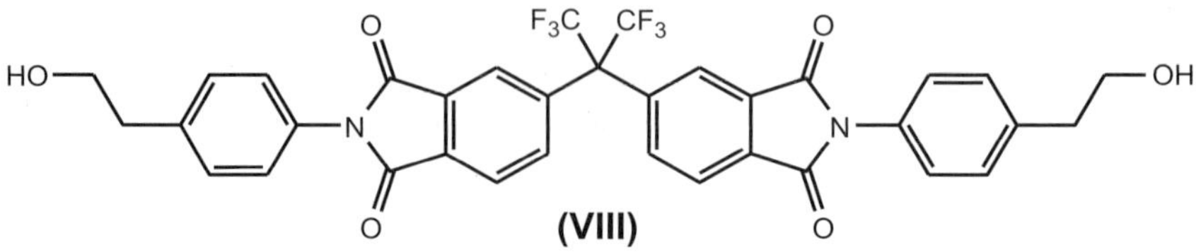

(VIII)

REFERENCES

1. Koji Okada et al, US Patent 7,141,614 (November 28, 2006) **Kaneka Corporation (Osaka, JP)**
2. Yoneda Yoshinori et al, US Patent 7,834,100 (November 16, 2010) **Kaneka Corporation (Osaka, JP)**
3. Masatomi Irisawa et al, US Patent 7,981,320 (July 13, 2011) **Adeka Corporation (Tokyo, JP)**
4. Ryuichiro Taniguchi et al, US Patent 7,977,400 (July 12, 2011) **Toray Industries, Inc.** (Tokyo, **JP**)
5. Osama M. Musa et al, US Patent 7,928,245 (April 11, 2011) **Henkel AG & Co. KGaA (Duesseldorf, DE)**

B. LOW VOLATILE ORGANIC FILMS

Derivatized 1,8-dioxa-4-azanaphthalenes

Title:	Tetrahydro-1,8-dioxa-4a-aza-naphthalenes in coating applications
US Patent:	7,812,173 (October 12, 2010)
Author:	Neville Everton Drysdale et al.
Assignee:	E.I. du Pont de Nemours and Company (Wilmington, DE)
Significance:	A single-step process for preparing tetrahydro-1,8-dioxa-4a-aza-naphthalene derivatives in high yields is described. Upon contact with water, tetrahydro-1,8-dioxa-4a-aza-naphthalenes are converted into acyclic diols and are reacted with diisocyanates to form polyurethanes. The polyurethanes are then used in low volatile organic emission compositions for preparing vary hard and clear coatings.

Additive Name 3,3,6,6,8a-Pentamethyl-tetrahydro-1,8-dioxa-4a-aza-naphthalene (I)

Safety Concerns Tetrahydro-1,8-dioxa-4a-methyl-aza-naphthalene is readily hydrolyzed in trace amounts of moisture generating di-(3-hydroxy-2,2-dimethylpropyl)-methyl amide. which is a moderate skin irritant.

Additive Name 3,3,6,6-Tetramethyl-8a-butyl-tetrahydro-1,8-dioxa-4a-aza-naphthalene (II)

Safety Concerns Tetrahydro-1,8-dioxa-4a-butyl-aza-naphthalene is readily hydrolyzed in trace amounts of moisture generating di-(3-hydroxy-2,2-dimethylpropyl)butyl amide, which is a moderate skin irritant.

Additive Structure

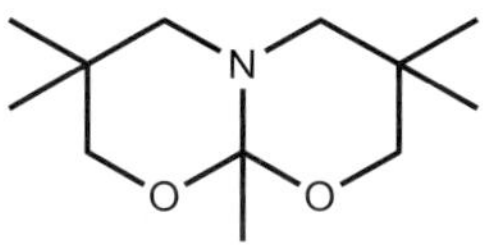

(I)

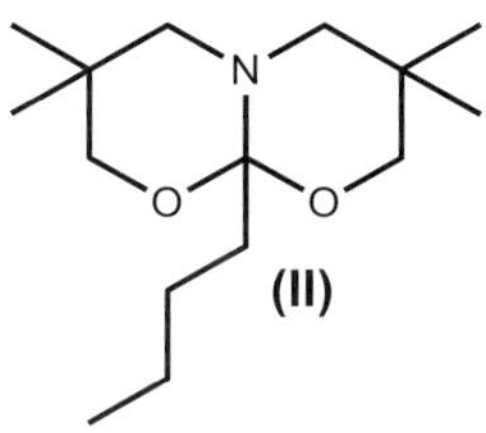

(II)

Additive Preparation

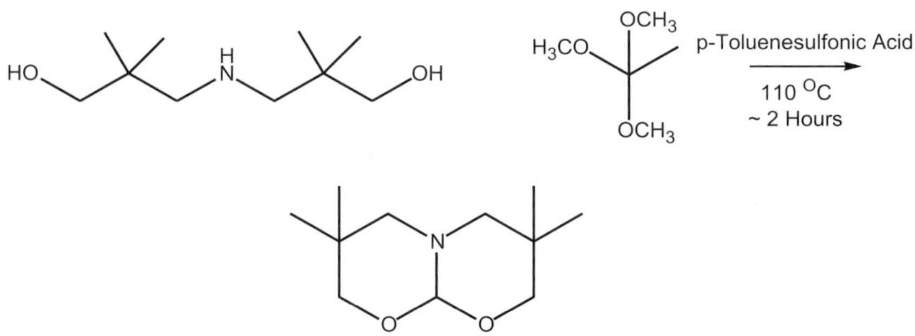

Film Coating

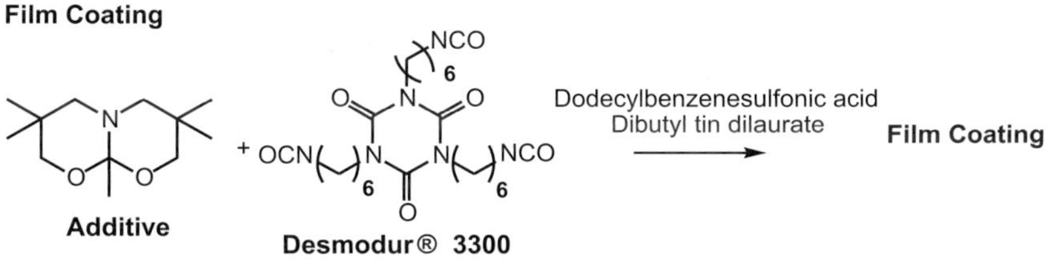

Additive Synthesis

1. Preparation of 3,3,6,6,8a-pentamethyl-tetrahydro-1,8-dioxa-4a-aza-naphthtalene

A reactor was charged with 128 g of 3-(3-hydroxy-2,2-dimethyl-propylamino)-2,2-dimethyl-propan-1-ol, 162 g of triethyl orthoacetate, and 2.0 g of toluene sulfonic acid and then heated to reflux until the theoretical amount of methanol had been collected. The reaction mixture was then cooled to an ambient temperature, and treated with triethylamine to neutralize the acid. The product was isolated by fractional vacuum distillations and isolated in a 90.5% yield as a clear liquid with a boiling point of 101–103 °C at 0.44 torr.

2. Additive Formulation in Coating Compositions

A glass vial was charged with 3.20 g of the Step 1 product and 7.02 g Desmodur® 3300 and then treated with 0.10 g of dibutyl tin dilaurate solution dissolved in 5 ml of propylene glycol monomethyl ether acetate and 0.10 g dodecylbenzenesulfonic acid solution dissolved in 3 g of propylene glycol monomethyl ether acetate. The solution was poured onto glass plates and allowed to dry overnight at an ambient temperature resulting in the formation of a very hard and clear coating. The remaining solution was stored in the sealed vial and remained flowable for an extended period of time.

Advantages over Prior Art

There are several distinct advantages to utilizing tetrahydro-1,8-dioxa-4a-aza-naphthalenes in preparing coating compositions. The additive is readily prepared in a high yield and can be used in low volatile organic emission coating formulations. In addition, coating compositions cure quickly and at relatively low temperature. Finally, sealed formulations appear to have an infinite shelf life.

Notes

1. Hydrolysis of tetrahydro-1,8-dioxa-4a-aza-naphthalenes occurs in water to generate the in situ diol (III).

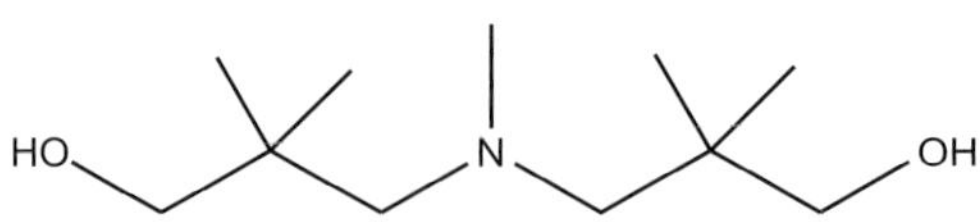

(III)

Hydrolysis Product

2. In an earlier investigation by Drysdale[1], coating compositions containing 2,6-dioxa-7-aza-bicyclo[2.2.2]octane derivatives (IV) were prepared and then hydrolyzed to the corresponding linear diol (V) and converted into the polyurethane by reacting with Desmodur® 3300.

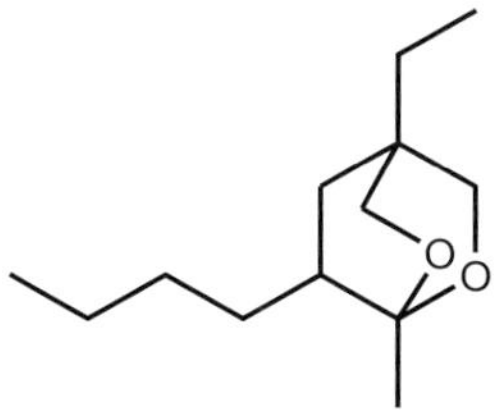

(IV)

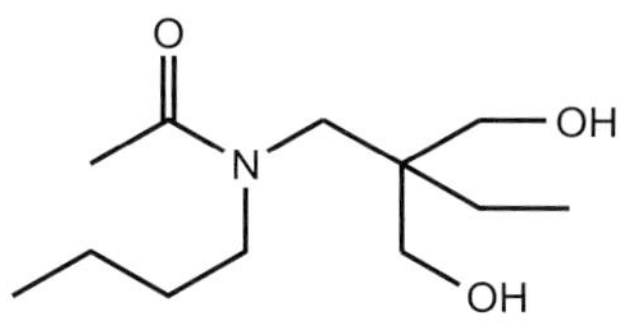

(V)

3. Polymeric crosslinkable compositions containing tetrahydro-1,8-dioxa-4a-aza-naphthalenes (VI) were prepared by Adelman[2] and used in the preparation of film coatings.

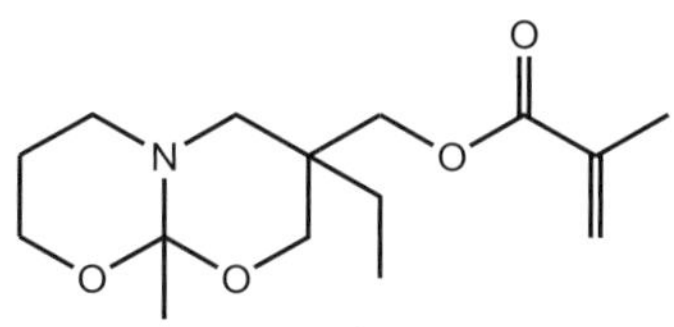

(VI)

4. Methacrylate tetrahydro-1,8-dioxa-4a-aza-naphthalenes (VII) were prepared by Drysdale[3] and used in coating compositions. In addition, Drysdale[4] prepared ultraviolet curable coating compositions containing methacrylate tetrahydro-1,8-dioxa-4a-aza-naphthalenes, which were used in coating formulations for preparing surfaces having hardness and clarity.

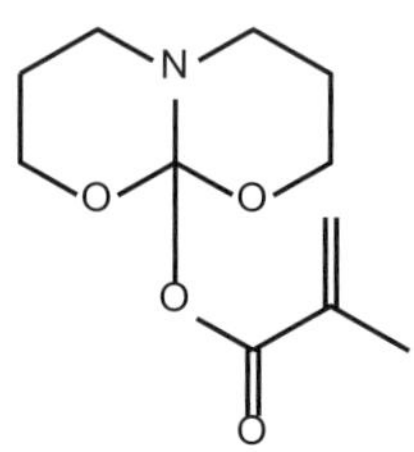

(VII)

REFERENCES

1. Neville Everton Drysdale, US Patent Application 20070049752 (March 1, 2007) **E.I. du Pont de Nemours and Company (Wilmington, DE)**
2. Douglas J. Adelman et al, US Patent Application 20050074615 (April 7, 2005) **E.I. du Pont de Nemours and Company (Wilmington, DE)**
3. Neville Everton Drysdale, US Patent 7,504,459 (March 17, 2009) **E.I. du Pont de Nemours and Company (Wilmington, DE)**
4. Neville Everton Drysdale, US Patent 7,608,645 (October 9, 2009) **E.I. du Pont de Nemours and Company (Wilmington, DE)**

C. NONADHESIVE SURFACE ADDITIVE TREATMENT

Bisphenol A Oligo- and Polycarbonates

Title:	Oligo- and polycarbonates terminated with silicon containing groups as surface modifiers
US Patent:	8,013,047 (September 6, 2011)
Author:	Fabien Jean Brand et al.
Assignee:	BASF SE Ludwigshafen (DE)
Significance:	Nonadhesive surface active agents, bisphenol A oligo- and polycarbonate containing decyl(dimethylsiloxane), were prepared by condensing bisphenol A oligo- and polycarbonate with decyl(dimethylsiloxane) and bis(2,4-dinitrophenyl) carbonate. Polycarbonate blends containing from 0.1 to 10 wt% of the nonadhesive additive had high surface energies, diminished surface adhesion of microbes, and algae and enhanced grease and dirt repellency.

Additive Names

Polysiloxane terminated bisphenol A oligocarbonate, a = 10 (I)
Polysiloxane terminated bisphenol A oligocarbonate, a = 13 (II)
Polysiloxane terminated bisphenol A oligocarbonate, a = 19 (III)
Polysiloxane terminated bisphenol A oligocarbonate, a = 27 (IV)
Polysiloxane terminated bisphenol A oligocarbonate, a = 37 (V)

Safety

There has been no report of toxicity associated with either bisphenol A oligo- and polycarbonates (Lexan®) or those containing polysiloxane termini.

Additive Structures

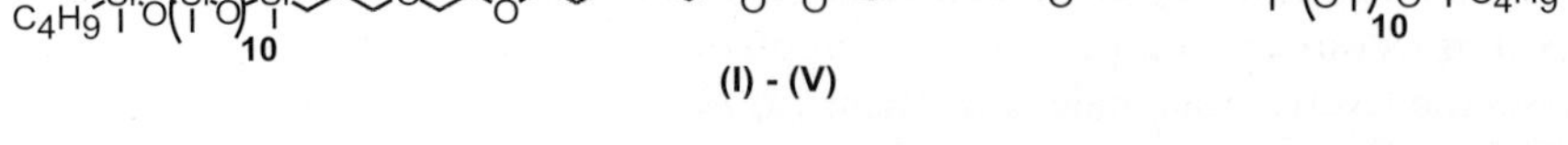

(I) - (V)

Additive Preparation

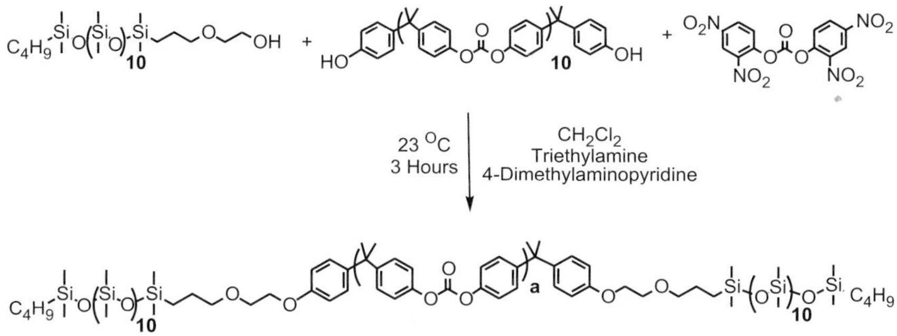

Additive Synthesis

1. Preparation of polysiloxane-terminated bisphenol A oligocarbonate

At an ambient temperature, a solution of 25 g of α-butyl-ϖ-[3-(2′-hydro-oxyethoxy) propyl]-polydimethylsiloxane having an Mn of approximately 878 Da and 11.22 g of bis(2,4-dinitrophenyl) carbonate dissolved in 200 ml of 1,2-dichloroethane was added dropwise to a solution of 4.32 ml of triethylamine dissolved in 50 ml of 1,2-dichloroethane and then stirred for 1 h. At ambient temperature, this solution was then slowly added to 200 ml of 1,2-dichloroethane containing 4.32 ml of triethylamine, 0.2 g of 4-dimethylaminopyridine, and 34.96 g of a bisphenol A oligocarbonate having an Mn of approximately 2700 Da and then stirred for 2 h. The mixture was then concentrated to 200 ml, and the product was precipitated by pouring into 2 l of methanol. The precipitate was filtered and then washed repeatedly with methanol until the solid became colorless. After drying under vacuum at 60 °C for 24 h, 51.6 g of the product was isolated.

Testing

1. Contact angle of polycarbonate blends

Polymer blends were prepared by blending the polycarbonate powder Makrolon® 3108 FBL and up to 10 wt% of a selected experimental additive and then mixed in a turbo mixer and extruded at 280 °C using a twin-screw extruder containing a flat die. In this way, polycarbonate tapes were prepared having a width of 20 mm and a thickness of 1 mm. Contact angles of the blends were then measured using a Dataphysics OCA 30 contact angle device by the sessile drop method and water as measuring liquid. The results are summarized in Table 4.

2. Reduced adhesion of algae on to polycarbonate blends

Adhesion of microbes to the polymer were measured using polymer blends containing the polycarbonate powder Makrolon® 3108 FBL® and up to 5 wt% of a selected experimental additive. The method entailed immersing the blend into an aquarium containing *Hematococcus pluvialis* and *Klebsormidium flaccidum*. Each microbe adhesion test was conducted for 4 weeks at 20 °C under 8 h of darkness and using a light intensity of 15,000–20,000 lux. Adhesion test results are provided in Table 5.

Test Results

1. Contact Angle of Polycarbonate Blends

TABLE 4 Contact angles of polycarbonate blends consisting of Makrolon® 3108 FBL and selected additives. Contact angle testing results strongly indicate the presence of a hydrophobic surface

Additive	Amount of Additive in Makrolon® 3108 FBL Polycarbonate Blend (wt%)	Contact Angle of Blend (degrees)
None	None	85.7
1	0.5	100.3
3	1.0	102.5
4	1.0	102.5
4	2.0	103.2
5	1.0	103.0
5	2.0	103.9
5	3.0	105.5
5	4.0	106.1
5	5.0	106.9

TABLE 5 Effect of algae adhesion on polycarbonate blends consisting of 5% Makrolon® 3108 FBL containing 3% of a selected polysiloxane terminated bisphenol A experimental oligocarbonate

Material	Surface Assessment After 12 Days	Surface Assessment After 4 Weeks
Glass	Slight Growth	Heavy Growth
Unadditized Makrolon® 3108 FBL	Moderate Growth	Moderate Growth
Makrolon® 3108 FBL Containing 3 wt% Additive-3	Slight Growth	Slight Growth
Makrolon® 3108 FBL Containing 3 wt% Additive-5	Slight Growth	Slight Growth

Advantages over Prior Art

Bisphenol A oligo- and polycarbonates terminated with decyl(dimethylsiloxane) are thermally stable additives that when melt blended with polycarbonates, polyester, or polyketones lower the surface energy of the thermoplastic resin. Although polyacrylic acid esters of poly(dimethylsiloxane) are also surface active, blends prepared using this additive cannot be melt processed with similar resins. Finally, while dendrimeric polycarbonates terminated with poly(dimethylsiloxane) can be injection molded with a variety of resins at elevated temperatures, they are relatively expensive to prepare.

Notes

1. Brunner[1] prepared fluorocarbon-terminated oligo- and polycarbonates (VI) additives that were used as nonadhesive surfaces.

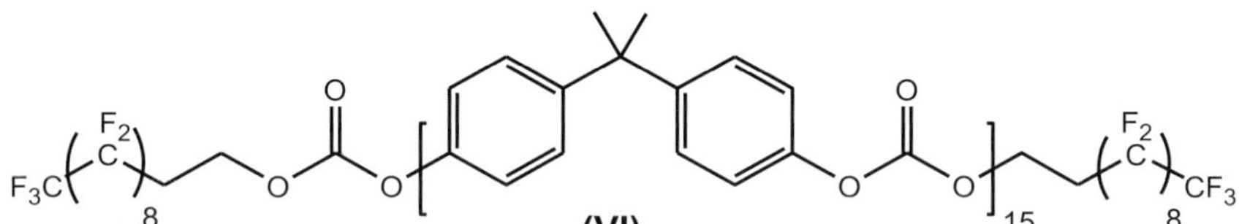

(VI)

2. Clark[2] polymerized perfluorourethane derivatives containing pendent siloxane groups (VII), which were used as durable printing ink repellants.

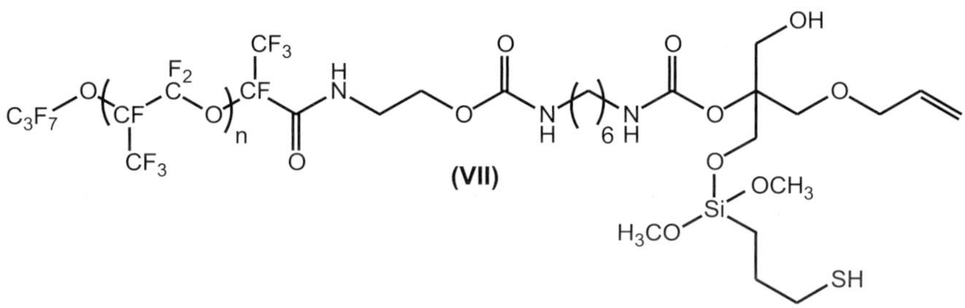

(VII)

3. The perfluorourethane siloxane surface treatment oligomer (VIII), which was designed to diminish dirt retention was prepared by Juhue,[3] and this was also found to be effective as an oil repellant because of the oleophobic character of the substrate.

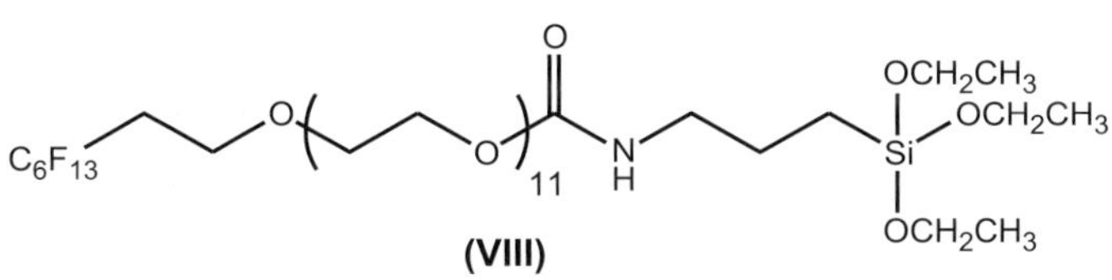

(VIII)

4. Low surface energy polyorganosiloxane polyoxamide copolymers (IX) prepared by Sherman[4] were used as coatings to both impede microbial adhesion and as antistatic additives.

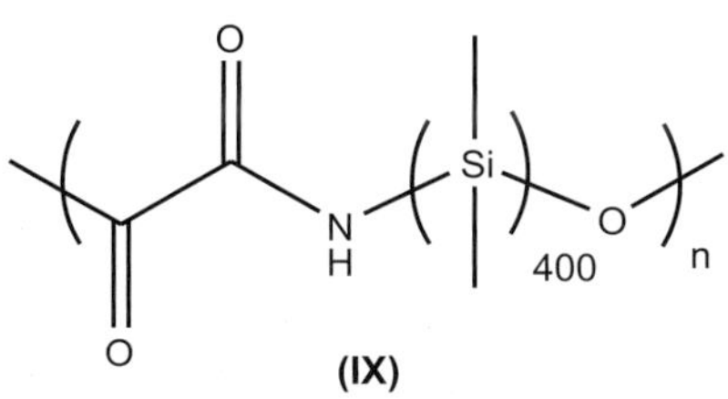

(IX)

REFERENCES

1. Martin Brunner et al, US. Patent 7,790,807 (September 7, 2010)
   Ciba Specialty Chemicals Corp. (Tarrytown, NY)
2. Gregory Clark, US. Patent 8,002,886 (August 23, 2011)
   3M Innovative Properties Company (St. Paul, MN)
3. Didier Juhue et al, US. Patent 7,276,618 (October 2, 2007)
   E. I. du Pont de Nemours and Company (Wilmington, DE)
4. Audrey A. Sherman et al, US. Patent 8,013,098 (September 6, 2011)
   3M Innovative Properties Company (St. Paul, MN)

Chapter 18

Fire Retardants

A. FLAME RETARDANT

Bisphenol A Phosphates

Title:	Sterically hindered phenolic phosphonates and polycarbonate resin composition using the same
US:	Patent 7,919,548 (April 5, 2011)
Author:	Beom Jun Joo et al.
Assignee:	Cheil Industries Inc. (Gumi-si, KR)
Significance:	The first and second average combustion times of bisphenol A-based polycarbonates were improved or became noncombustible by additizing with 3–5 wt% of sterically hindered halogen-free triaromatic phosphate flame retardants. Polycarbonates additized with phenolic phosphates had improved processability and maintained high notched impact strength.
Additive	Di-(4-*t*-butyphenyl)phenylphosphonate (I) Di-(2,4,6-trimethylphenyl)phenylphosphonate (II) Di-(2,6-dimethylphenyl)phenylphosphonate (III)
Safety	Although the toxicological properties of these substances have not been fully investigated, aromatic phosphonates are skin, respiratory tract, and eye irritants. Ingestion may cause digestive tract irritation resulting in nausea and vomiting.

Next Generation of International Chemical Additives. DOI: http://dx.doi.org/10.1016/B978-0-444-53788-1.00018-8

Additive Structures

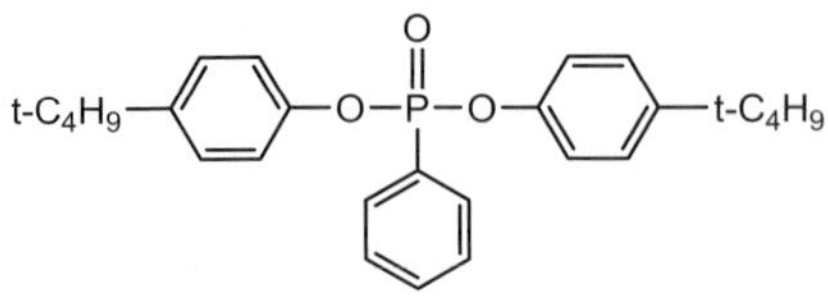

(I)

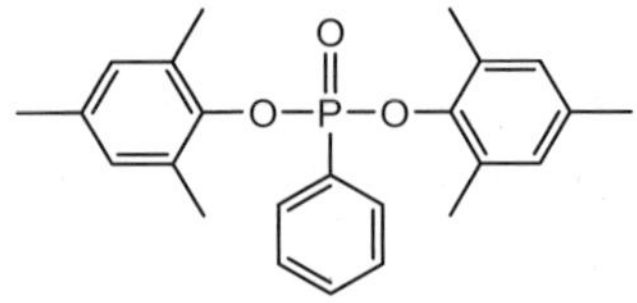

(II)

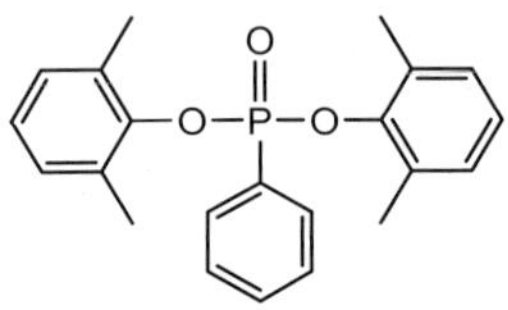

(III)

Additive Preparation

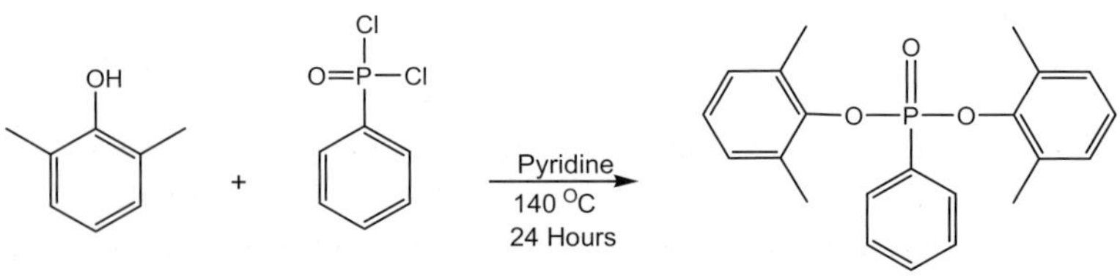

Additive Synthesis

1. Preparation of di-(2,6-dimethylphenyl)phenylphosphonate

A reactor was charged with 50 g of 2,6-dimethyl phenol, 43 g of phenyl phosphonic dichloride, and 400 ml of pyridine and then heated to 140 °C for 24 h The mixture was then cooled to an ambient temperature and pyridine was distilled off under reduced pressure until a residue was isolated. The residue was treated with 500 ml of distilled water and then stirred for 1 h and dissolved in pyridine chloric acid, and the salt was isolated. The mixture was then filtered, and the solid was dried for 24 h, and 69 g of the product was isolated as a white solid in 95% purity.

Testing

Flame Retardancy Testing

Pellets of selected experimental additives were blended with bisphenol A using polycarbonates having an Mw of approximately 25,000 Da were prepared by mixing the two components in a blender and then extruding through a twin-screw extruder at temperature range of 200–280 °C. The pellets were then dried at 80 °C for 2 h and then molded into test specimens for flame retardancy testing in an injection molding machine operating between 180 °C and 280 °C with a mold temperature of 40–80 °C. Flame retardancy was then determined by measuring the average combustion time using specimens having a thickness of about 1/8'' and a size of 125 mm by 13 mm in accordance with UL 94 VB. Test results are summarized in Table 1.

Test Results

TABLE 1 Flame retardancy test results for a polycarbonate resin additized with experimental phenolic phosphonates

Component	Test-1	Test-2	Test-3	Test-4	Test-4
Polycarbonate Resin (Control)	100	100	100	100	100
Polycarbonate Resin and Additive-I (wt%)	3	5	X	X	X
Polycarbonate Resin and Additive-II (wt%)	X	X	3	5	X
Polycarbonate Resin and Additive-III (wt%)	X	X	X	X	5
First Combustion Time (sec)	0	0	0	0	0.5
Second Combustion Time (sec)	9.3	5.3	3.3	0	8.8

Advantages over Prior Art

Nonsterically hindered di and triaromatic phosphate ester flame retardants are well known and widely used as retardants in resins and thermosets. Effective additive treatment levels using existing agents, however, typically exceed 9.5 wt%, which can compromise the resin mechanical properties. By limiting the effective flame retardancy treatment levels to ≤5 wt% using sterically hindered phosphates of the current invention, the resin mechanical and physical properties remain largely unchanged.

Notes

1. Freitag[1] prepared branched polyphosphates, which were effective as flame retarding agents by condensing various ratios of diphenylmethylphosphonate (IV), bisphenol A (V), 1,1,1-tris(trihydroxy)ethane (VI), and tetraphenylphosphonium phenolate (VII).

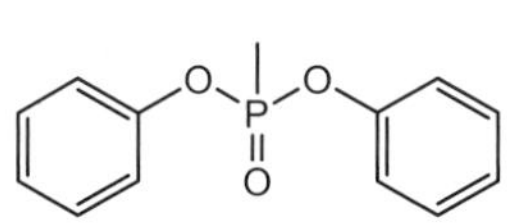

(IV)

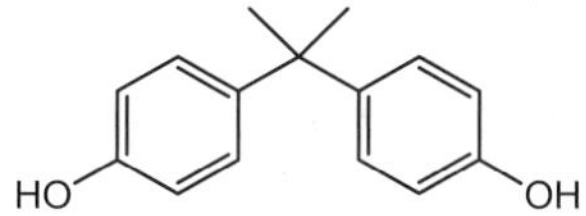

(V)

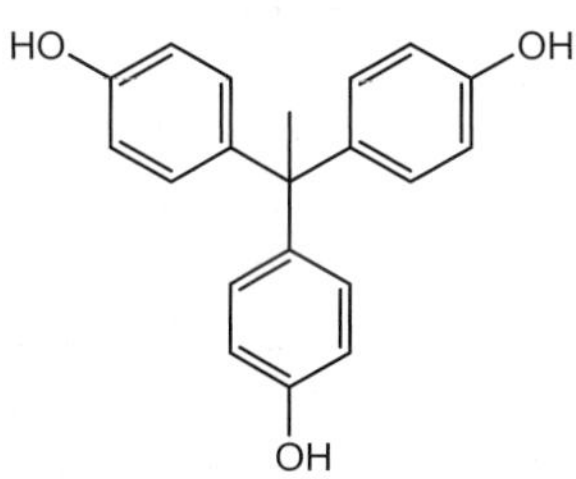

(VI)

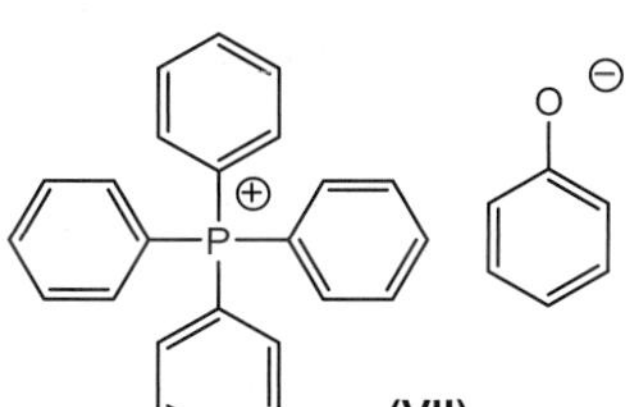

(VII)

2. Yao[2] determined that the aluminum salt of di-isobutylphosphinic acid was a particularly effective flame-retardant additive for nylon resins.
3. Hayata[3] additized bisphenol A polycarbonate having an [η] of approximately 14,500 Da with polydimethylsiloxanes having an [η] of approximately 17,000 Da with cenospheres having a particle size range of 150–300 μm and then used them as flame retardants.
4. Block polycarbonates additized with both methylphosphonic acid diphenylester were prepared by Freitag[4] as flame retardants in bisphenol A-based polycarbonates resins (VIII).

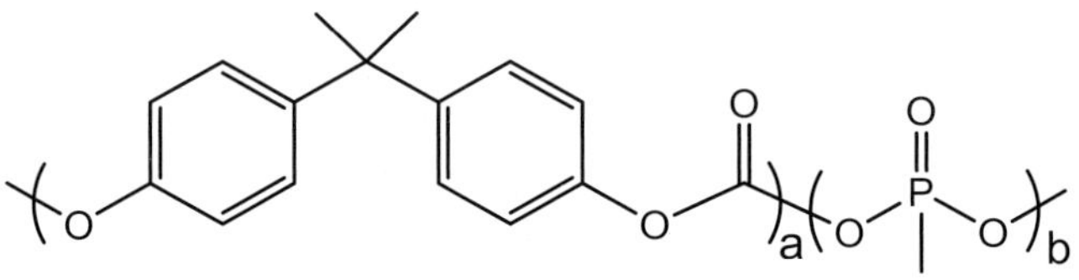

(VIII)

REFERENCES

1. Dieter Freitag et al, US Patent 7,816,486 (October 19, 2010)
FRX Polymers, Inc. (Chelmsford, MA)
2. Qiang Yao et al, US Patent 7,807,737 (October 5, 2010)
ICL-IP America Inc. (Ardsley, NY)
3. Yusuke Hayata et al, US Patent 7,888,409 (February 15, 2011)
Idemitsu Kosan Co., Ltd. (Tokyo, JP)
4. Dieter Freitag et al, US Patent 7,838,604 (November 23, 2010)
FRX Polymers, Inc. (Chelmsford, MA)

B. FLAME RETARDANT

Novolak Flame-Retardant Resins

Title:	Epoxy resin, blocked polyisocyanate or polymers with polyhydroxystyrenes having a novolak structure
US:	7,897,702 (March 1, 2011)
Author:	Wen-Chiung Su et al.
Assignee:	Chung-Shan Institute of Science and Technology, Armaments Bureau, Ministry of National Defense (Taoyuan County, TW)

Significance: A resin consisting of the reaction product of the flame retardant, 9,10-dihydro-9-oxa-10-phosphaphenanthrene-3,3′3″-tri(dimethyl-2,4-dioxoazetine) with a diepoxide and 4,4′-diaminodiphenylmethane exhibited char yields >32%. In addition, the protected resin exhibited excellent storage and loss modulus properties.

Additive Name 9,10-Dihydro-9 oxa-10-phosphaphenanthrene-3,3′3″-tri(dimethyl-2,4-dioxoazetine) (I)

Safety Concerns There is very limited safety information on the compound, 9,10-dihydro-9 oxa-10-phosphaphenanthrene-3,3′3″-tri(dimethyl-2,4-dioxoazetine).

The precursor, 9,10-dihydro-9-oxa-10-phosphaphenanthrene is considered a nontoxic environmentally friendly agent. When used as a flame retardant in linear or hyperbranched polymers there have been no reports suggesting toxicity human or environmental contamination resulting from its use.

Additive Structure

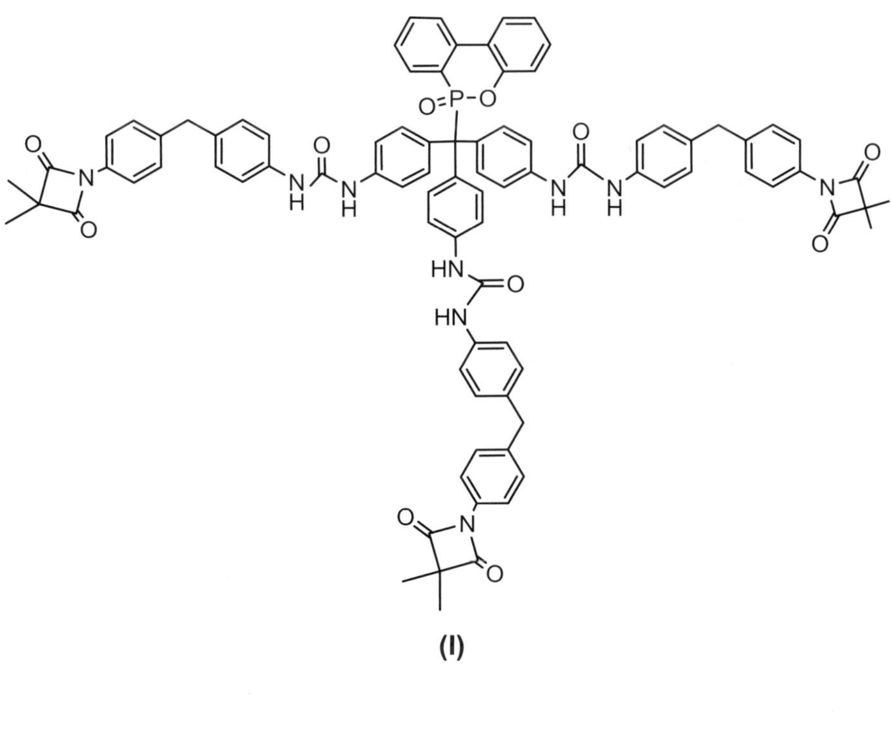

(I)

Additive Preparation

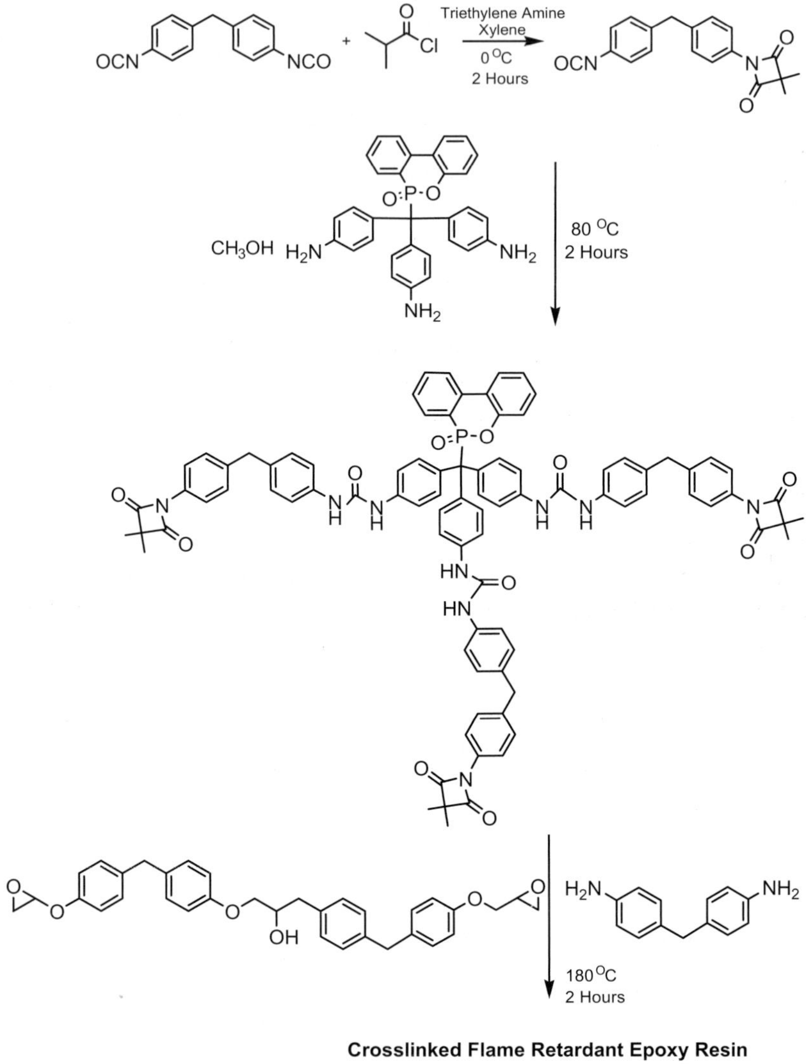

Additive Synthesis

1. *Preparation of 4-isocyanate-4-(3,3-dimethyl-2,4-dioxo-azetidino)-diphenylmethane*

A round-bottom flask was charged with equivalent amounts methylene-di-*p*-phenyl diisocyanate and isobutyryl chloride dissolved in xylene, and then the solution was heated for several hours. Thereafter, an equivalent of triethylamine was added, and the temperature was lowered to 0 °C and stirred until an orange solid was formed. The solution was then filtered, recrystallized in cyclohexane, and the product was isolated in a 45% yield as an orange solid.

2. *Preparation of 9,10-dihydro-9 oxa-10-phosphaphenanthrene-3,3′3″-tri(dimethyl-2,4-dioxoazetine)*

A reactor was charged with 3 equivalents of the Step 1 product and 1 equivalent of 9,10-dihydro-9-oxa-10-phosphaphenanthrene 10-Oxide (DOPO) dissolved in DMF and then stirred for several hours at 80 °C. Thereafter, the mixture was precipitated by pouring into a solution of a 1:1 water/methanol mixture, and the solid was isolated and then Soxhlet extracted using methyl alcohol. The product was isolated in 95% purity as a purple solid having a molecular weight of 504 Da with an MP = 329 °C.

3. *Preparation of crosslinked flame-retardant epoxy resin*

A container was charged with the epoxy resin BE® 187(Nan Ya Plastic Corp) having an epoxide equivalent of 187 g/eq, the Step 2 product, and 4,4′-diaminophenylmethane in a 1.3/0.3/1.0 molar ratio and then heated to 180 °C for 2 h. The container was then cooled, and a white solid was isolated.

Testing

A. *Thermal Properties*

Thermal properties of crosslinked epoxy resins prepared and then heated for 2 h using different reagent ratios were evaluated using DSC. Test results are provided in Table 2.

B. *Mechanical Properties*

Mechanical properties of crosslinked epoxy resins prepared with a curing time of 2 h using different reagent ratios were obtained by DMA. The results are provided in Table 3.

Test Results

A. *Thermal Property Test Results*

TABLE 2 DSC thermal test results for samples prepared using different ratios of Additive-1, BE® 187 epoxide and 4,4′-diaminophenylmethane. Crosslinked epoxy resin samples containing the experimental flame retardant consistently provided higher char yields over the reference sample

Sample	Diamine/Epoxide/ Additive (molar ratio)	Tg (°C)	Td (°C)	Char Yield (%)
1	1.0/1.0/0 (Reference)	None Detected	327	26.2
2	1.1/1.0/0.1	None Detected	318	32.1
3	1.3/1.0/0.3	200	301	31.7
4	1.5/1.0/0.5	None Detected	306	33.0

B. Mechanical Property Test Results

TABLE 3 DMA mechanical test results for samples prepared using different ratios of Additive-1, BE® 187 epoxide and 4,4′-diaminophenylmethane. The mechanical improvements were consistently observed using samples cured with Additive-1

Sample	Diamine/Epoxide/ Additive (molar ratio)	Storage Modulus at 100 °C (MPa)	Loss Modulus (MPa)	Tan δ (°C)	Tan δ_{max}
1	1.0/1.0/0 (Reference)	0.67	0.732	134.4	0.408
2	1.1/1.0/0.1	1.64	1.56	186	0.33
3	1.3/1.0/0.3	1.81	1.6	175.2 221	0.327 0.256
4	1.5/1.0/0.5	2.13	1.67	164.7 242	0.326 0.234

Advantages over Prior Art

Resins exhibiting both high char yields without compromising mechanical properties, particularly storage and loss modulus properties, are infrequently reported. This is due in part to the limited miscibility of the flame retardant and the release ease with which the flame retardant migrates to the surface of the resin. To address these two concerns, aromatic diamines were used in these resin formulations to limit migration and by optimizing the crosslink density of these flame retardants.

Notes

1. Additional bisphenylacylphosphine oxide derivatives (III)–(V) were previously prepared by Su[1] and used in fire retardant formulations.

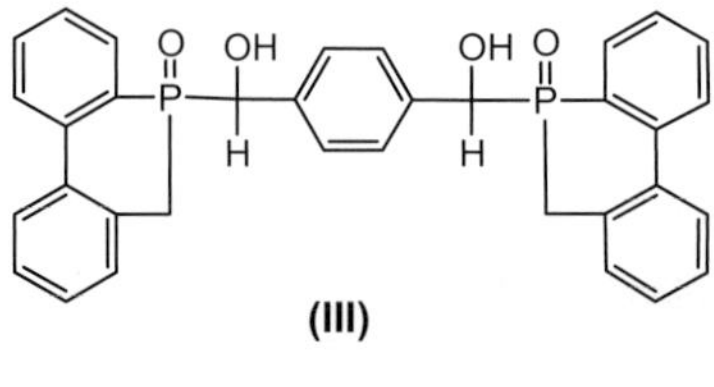

(III)

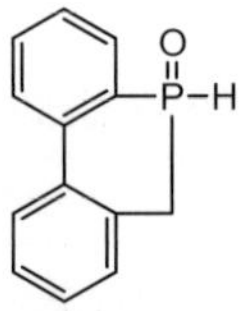

(IV)

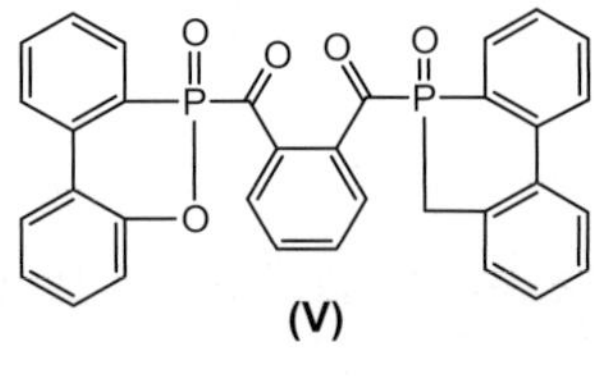

(V)

2. Lin[2] prepared a series of fire resistant phosphorus-based oxazine derivatives (VI), which had a high glass transition temperature, high modulus, low moist absorption rate, excellent electrical characteristics, and high char yield.

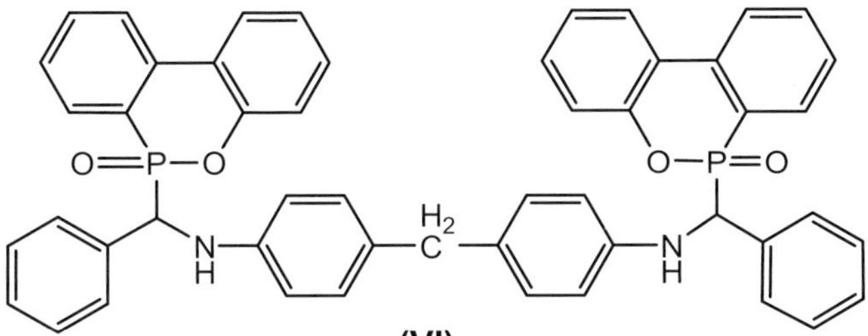

(VI)

3. Pan[3] prepared flame retardant polymer electrolyte compositions with hyper-branched dendrimer-like structures consisting of the condensation product of *N,N′*-4,4′-diphenylmethane-bismaleimide, propylene carbonate, and barbituric acid, which were then postreacted with lithium hexafluorophosphorus.
4 Emrick[4] prepared the polymeric flame retardant, poly(4,4-bishydroxydeoxybenzoin-co-phenylphorphorus dichloride) (VII) by interfacial polymerization, which had char yields between 45% and 57%.

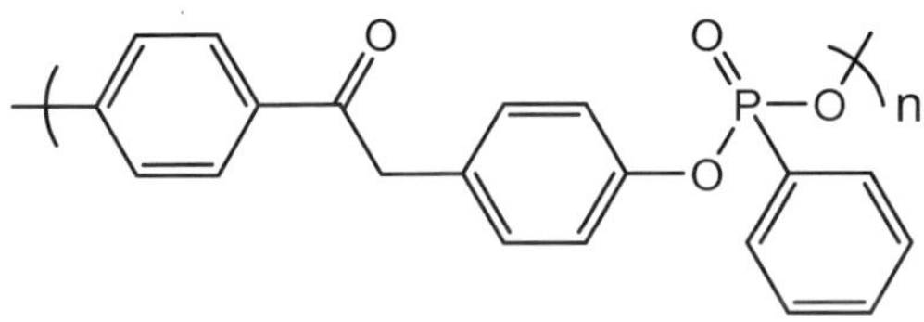

(VII)

REFERENCES

1. Wen-Chiung Su et al, US Patent 7,825,274 (November 2, 2010)
Chung Shan Institute of Science and Technology (Taoyuan County, TW)
2. Ching-Hsuan Lin et al, US Patent 7,566,780 (July 18, 2009)
Chang Chun Plastics Co., Ltd. (Hsinchu County, TW) National Chung Hsing University (Taichung, TW)
3. Jing-Pin Pan et al, US Patent 7,875,204 (January 25, 2010)
Industrial Technology Research Institute (Hsinchu, TW)
4. Todd Emrick et al, US Patent 7,863,400 (January 4, 2011)
University of Massachusetts (Boston, MA)

C. FLAME RETARDANT

Potassium 1,1,2,3,3,3-hexafluorpropanesulfonate

Title:	Flame retardant polycarbonate compositions
US:	Patent 7,947,770 (May 24, 2011)
Author:	Christopher P. Junk et al.

Assignee:	E. I. du Pont de Nemours and Company (Wilmington, DE)
Significance:	Resin compositions additized at low levels with potassium perfluoroalkyl- or perfluoroalkylimidazolium sulfonates salts had improved flame retardancy and antidrip properties of molten burning aromatic polycarbonate resins. Compositions containing these additives had a V-0 blend classification rating for all experimental additives.
Additives	Tetra-*n*-butylphosphonium 1,1,2,3,3,3-hexafluoropropanesulfonate (I) Potassium 1,1,2,3,3,3-hexafluorpropanesulfonate (II) Potassium 1,1,2-trifluoro-2-(perfluoroethoxy)ethanesulfonate (III) 1-Butyl-3-methylimidazolium-1,1,2,2 tetrafluoroethane sulfonate (IV)
Safety	Potassium perfluoroethane and propane sulfonic acid salts are moderate to severe eye, skin, and respiratory system irritants. Tetra-*n*-butylphosphonium 1,1,2,3,3,3-hexafluoropropane and 1-butyl-3-methylimidazolium-1,1,2,2 tetrafluoroethane sulfonate are considered very mild eye, skin, and respiratory system irritants

Additive Structures

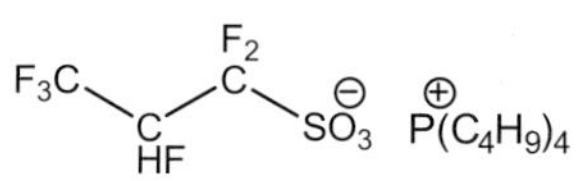

(I)

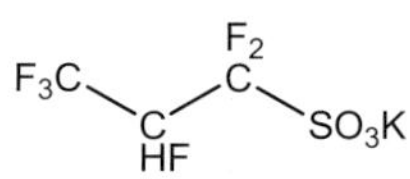

(II)

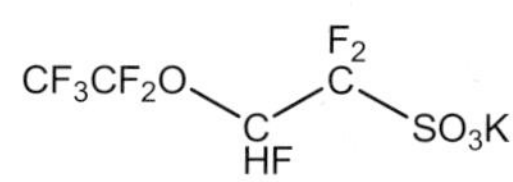

(III)

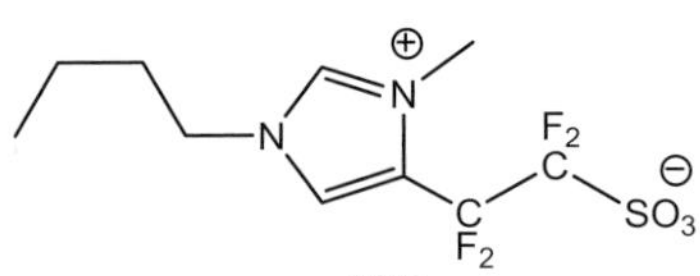

(IV)

Additive Preparation

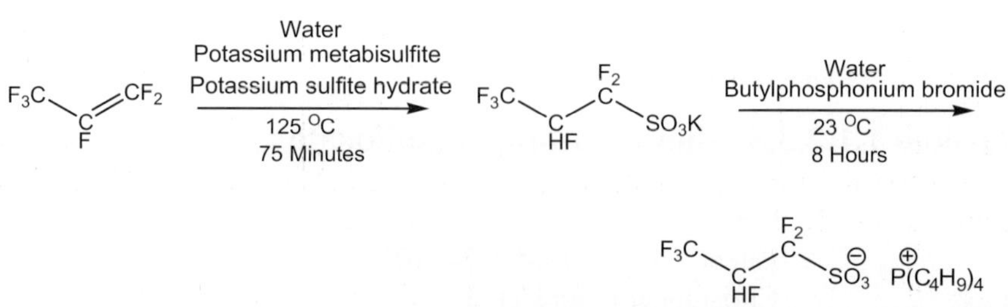

Additive Synthesis

1. *Preparation of potassium 1,1,2,3,3,3-hexafluoropropanesulfonate according to the method of Harmer*[1]

A one-gallon Hastelloy reaction vessel was charged with a solution of 130 g of 95% potassium sulfite hydrate, 448 g of 99% potassium metabisulfite, and 1300 ml of deionized water and then cooled to −35 °C and evacuated to 83 kPa and purged with nitrogen. The vessel was then treated with 550 g of hexafluoropropylene and heated to 125 °C for 75 min and then vented and cooled to 25 °C. After filtration, the product was isolated as a white powder in an 80% yield.

Water by Karl-Fisher titration: 0.15 wt%

Analysis Calculated for $C_3H_{0.4}SF_6$; Found. $C_3H_{0.4}N_{0.1}SF_6$

2. *Preparation of tetra-n-butylphosphonium 1,1,2,3,3,3-hexafluoropropanesulfonate*

A 4-liter reaction flask was charged with 2800 ml of deionized water and 551.2 g of the Step 1 product at an ambient temperature and then treated with 75 wt% of an aqueous solution containing 923.0 g of tetra-*n*-butylphosphonium bromide. The mixture was stirred for 8 h at an ambient temperature, filtered, and a crude white product isolated. The white solid was then stirred in 1000 ml of saturated sodium carbonate for 30 min, refiltered, and further washed thrice with 1000-ml portions of deionized water. After isolating the solid by filtration, it was dried at 70 °C @ 100 Torr for 18 h, and the product was isolated in an 89% yield isolated having a melting point of 74 °C.

Water by Karl-Fisher titration: 671 ppm

^{19}F-NMR (CD_2Cl_2) δ −74.9 (m, 3F); −115.3, −122.8 (ABq, J = 264 Hz, 2F); −210.1 (dm, 1F)

^{1}H-NMR (CD_2Cl_2) δ 1.0 (t, J = 7.3 Hz, 12H); 1.5 (m, 16H); 2.2 (m, 8H); 5.4 (dm, JFH = 54 Hz, 1H). Analysis for $C_{19}H_{0.37}F_6O_{0.3}$

Calculated: % C, 46.5: H, 7.6; Found C, 46.9: H, 8.0: N, <0.1.

Testing

Flame Retardancy of Polycarbonate Resins

Experimental additives were evaluated according to the standard flame retardancy test procedure described in UL-94. In this test, blended specimens were evaluated in the vertical position using a regulated flame placed onto a test specimen for 10 s, and the time was measured until the flame self extinguished, which was recorded as t1. Once the flame was extinguished, the regulated flame was again placed onto the test specimen for 10 s, and the second burn time was recorded as t2. Test results are provided in Table 4.

Test Results

TABLE 4 Flame retardancy testing for polycarbonate Lexan® HF1110 additized with experimental additives. All values represent the average of five measurements

Flame Retardant	t1 (s)	t2 (s)	Flaming Drip	Blend Classification
None	9.5	3.2	Yes	V-2
Additive-1	1.0	2.4	No	V-0
Additive-2	2.4	9.0	No	V-0
Additive-3	4.3	2.4	No	V-0
Additive-4	0.9	0.8	No	V-0

Advantages over Prior Art

The prevailing flame retardants used in polycarbonate resins are bromine based, but concerns over suspected toxicity have resulted in a need for cost-effective replacements. Although potassium perfluorobutane sulfonic acid has previously been used in polysiloxane resins to improve both the flame retardancy and self-fire extinguishing of the composition; the additive was expensive to prepare and difficult to compound. In addition, existing phosphonate-based polycarbonate flame retardants require very high treatment levels that can potentially impact the mechanical properties of the resin. Moreover, these additives require expensive multisynthetic processes to prepare. The current method has addressed all of these concerns by producing a cost-effective highly effective polycarbonate resin flame retardant.

Notes

1. Costanzi[2] used halogen-free flame retardants aluminium- (V) and calcium hypophosphite (VI) and triphenyl phosphate (VII) in recycled polycarbonate materials.

$Al(H_2PO_2)_3$
(V)

$Ca(H_2PO_2)_2$
(VI)

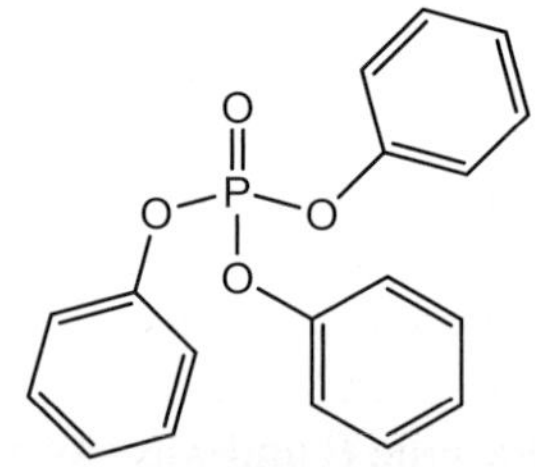

(VII)

2. Kim[3] prepared flame-retardant polyarylate resins by copolymerizing a three-component additive mixture consisting of 4,4′-dihydroxyphenyl-acetylene, phthaloyl chloride, and bishydroxyphenol (VIII).

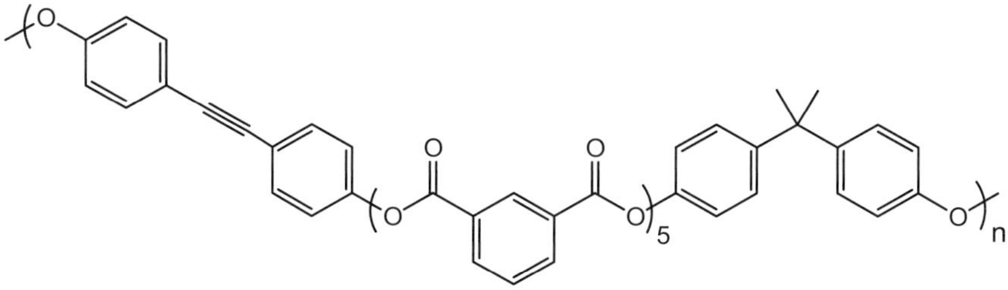

(VIII)

3. Hong[4] prepared the aluminum salt of ethyl carboxyethyl methylphosphinate (IX) as a flame retardant for high impact polystyrene and polyphenylene ether resins.

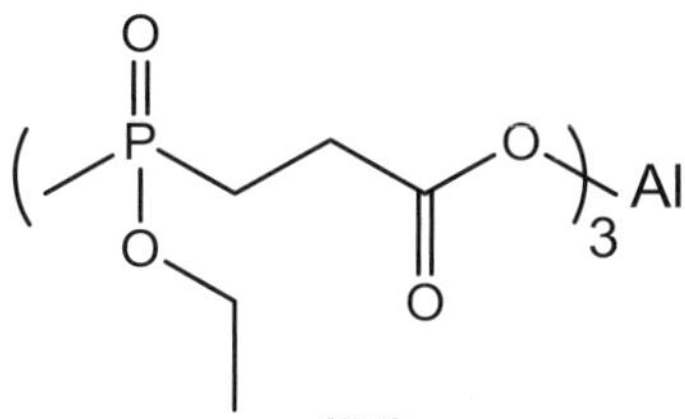

(IX)

4. Levchik[5] prepared oligomeric phosphonate flame retardant compositions consisting of the reaction product of bisphenol diepoxide and oligomeric methyl di-butylphosphates (X).

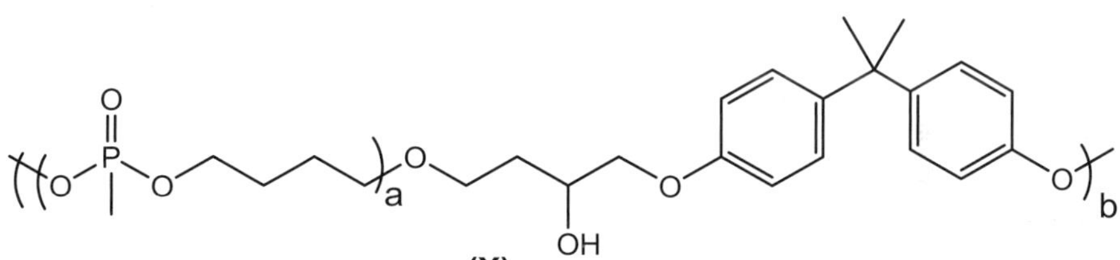

(X)

REFERENCES

1. Mark A. Harmer et al, US Patent Application 20060276670 (December 7, 2006)
E. I. du Pont de Nemours and Company (Wilmington, DE)
2. Silvestro Costanzi et al, US Patent 7,608,654 (October 22, 2009)
Italmatch Chemicals S.P.A. (Genova, IT)
3. Kyoung Mahn Kim, US Patent 7,943,722 (May 17, 2011)
Korea Research Institute of Chemical Technology (Daejeon, KR)
4. Sang Hyun Hong et al, US Patent 7,939,588 (November 10, 2011)
Cheil Industries Inc (Gyeongsangbuk-do, KR)
5. Sergei Levchik et al, US Patent 7,910,665 (March 22, 2011)
ICL-IP America Inc. (Ardsley, NY)

D. FLAME RETARDANTS FOR ENGINEERED PLASTICS

Polytetrabromobisphenol A

Application:
Title: Bromine Compounds Ltd. (Beer Sheva, IL)
US Patent: 8,067,497 (November 29, 2011)
Author: Yoav Bar-Yaakov et al.
Assignee: Bromine Compounds Ltd. (Beer Sheva, IL)

Significance: A high molecular weight flame retardant was prepared by condensing the oligometric reaction product of tetrabromobisphenol A and epichlordrin with tetrabromobisphenol A and then quantitatively capping with tribromophenol. When compounded with poly(butylene terphthalate), the high performance resin had improved processing properties and enhanced high impact strength.

Additive Name Poly[(tetrabromobisphenol A-co-oxy-2-hydroxypropyl)-co-tetrabromobisphenol A]di-tribromophenol

Safety Perbrominated phenols and bisphenols are considered skin irritants and harmful if absorbed through the skin. They are also considered severe eye irritants. Inhalation causes respiratory tract irritation. In addition, residual monomers present in the additive are very toxic to aquatic life with long-lasting effects.

Additive Structure

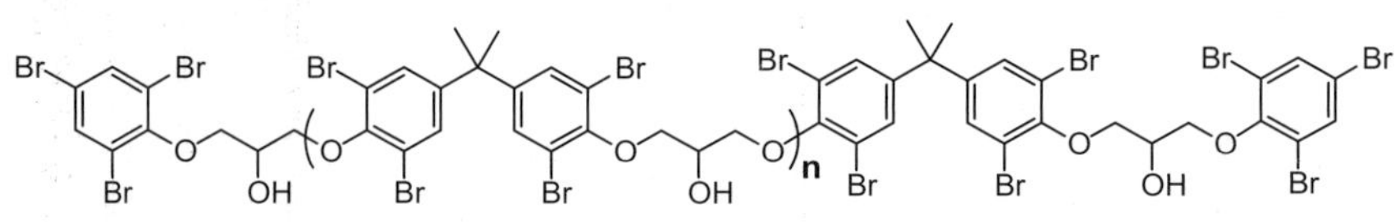

(I)

Additive Preparation

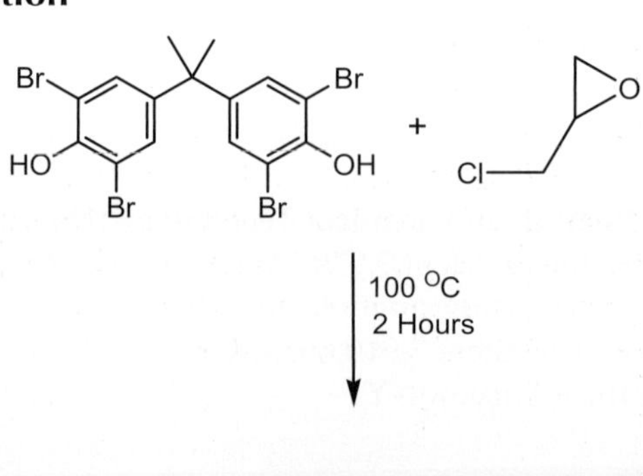

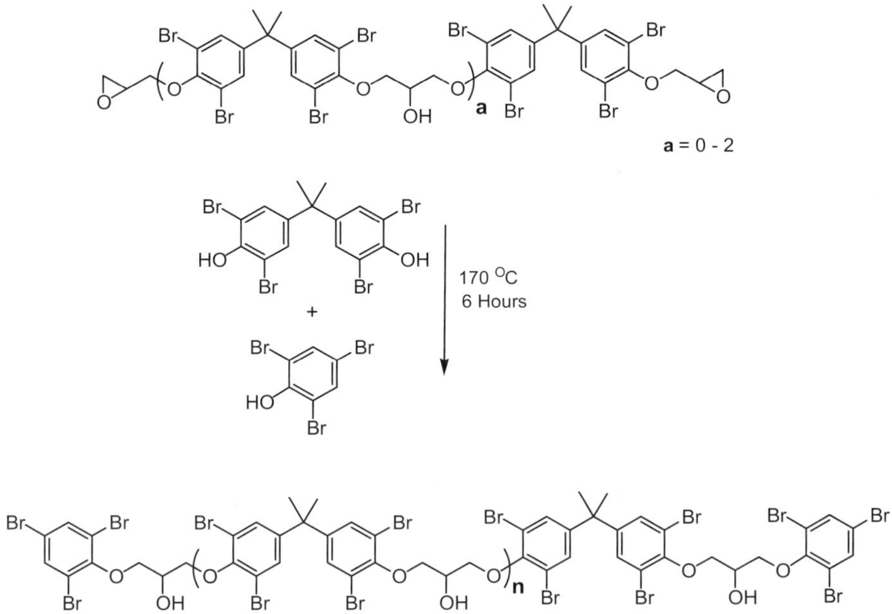

Additive Synthesis

1. *Preparation of oligo(tetrabromobisphenol A-co-oxy-2-hydroxypropyl) diglycidylether*

A 1-liter glass reactor equipped with a heating mantle, thermometer, stirrer, and nitrogen inlet was charged with 664 g of tetrabromobisphenol A and a stoichometric amount of epichlorohydrin and then heated at 100 °C for 2 h, and the product was isolated.

MP = 80 °C

Epoxy equivalent molecular weight of the intermediate = 477 g/equivalent

Volatile content = 21 ppm

2. *Preparation of poly[(tetrabromobisphenol A-co-oxy-2-hydroxypropyl)-co-tetrabromo-bisphenol A]di-tribromophenol*

A nitrogen-purged reaction vessel was charged with the Step 1 product and then heated to 150 °C and treated with 323.6 g of tetrabromobisphenol, 52 g of tribromophenol, and 0.03 g of tetrabutyl phosphonium bromide. The kettle temperature was then reduced to 100 °C but rose exothermally to 170 °C in 45 min. The mixture was then poured into an aluminum tray and placed in an oven heated to 170 °C for 6 h and then cooled to an ambient temperature, and the resin was used without purification. The physical properties of the resin are provided below.

Weight average molecular weight = 16, 050 Da
Number average molecular weight = 10,100 Da
Softening point = 187 °C
Volatile content = 13 ppm
Bromine content = 53.1 wt%
Epoxy equivalent molecular weight 27,000 g/equivalent
Acid value = 0.07 mg KOH/g
Epoxy end groups = 18.5 mol%

Testing

Compounding was performed in a Berstorff corotating twin-screw extruder type ZR-25 with a ratio of the length over the diameter of the screws equal to 32. Twin-screw processing temperatures are provided below:

Zone 1	Zone 2	Zone 3	Zone 4	Zone 5	Zone 6	Zone 7	Zone 8	Die
180 °C	220 °C	240 °C	250 °C	260 °C	260 °C	265 °C	270 °C	280 °C

Screw Speed = 230rpm
Feed Rate = 10Kg/hr

The blend composition used to evaluate the experimental flame retardant is provided in Table 5 while flame retardancy testing is provided in Table 6.

TABLE 5 Blend composition using the Step 2 product and a low molecular weight Step 2 product analog having an Mn of approximately 2000 Da

Component	Step 2 Flame Retardant	Low Molecular Step 2 Flame Retardant Analog
Poly(butylene Terephthalate)	52.3	52.3
Glass fiber pbt 1a 1 h (Owens Corning)	30	30
Flame retardant	13.5	13.5
Antimony trioxide master batch concentrate (M0112-Kafrit)	42	42

Test Results

TABLE 6 Test results involving formulations containing the Step 2 product and a low molecular weight Step 2 analog having an Mn of approximately 2000 Da. While both formulations have good mechanical and electrical properties with excellent flammability reduction, the Step 2 product has better flow in the mold and higher impact strength

Variable	Step 2 Product	Low Molecular Step 2 Flame Retardant Analog
Rating UL-94 at 1.6 mm	V-0	V-0
Rating UL-94 at 0.8 mm	V-0	V-0
Comparative tracking index (V)	300 325	300 325
Spiral flow (inch)	41.2	30.0
HDT, ASTM D646 @ 1820 kPa (°C.)	244.1	243.3
Notched Izod Impact, ASTM D2244 (J/m)	99.4	83.3
Tensile strength at yield, ASTM D638 (MPa)	**Step 2 Flame Retardant**	**Low Molecular Step 2 Flame Retardant Analog**
Tensile strength at break (MPa)	172.8	178.5
Elongation at break (%)	2.1	2.6
Tensile modulus (MPa)	11408	10292

Advantages over Prior Art

The enhanced flame retardancy and improved thermal properties of the current additive are due almost exclusively to the quantitative ring opening of the terminal epoxides using tribromophenol. Previously attempts to ring-open epoxy-terminated brominated polystyrene bisphenol A resulted in 60–50% endcapping.

Notes

1. Luther[1] improved the thermal stability and flame retardancy of thermoplastic polyester resins including polyethylene terephthalate, polybutylene terephthalate, polytrimethylene terephthalate, and polycyclohexylene terephthalate by blending with brominated polystyrene having an Mw from 3000 Da to about 10,000 Da with a polydispersity of between 1 and 4.

2. Kiuchi[2] prepared a flame-retardant epoxy resin composition by kneading together a blend of epoxy resin (II) and epoxy resin (III), fused flake silica powder, carnauba wax, and triphenylphosphine for 5 min at 100 °C.

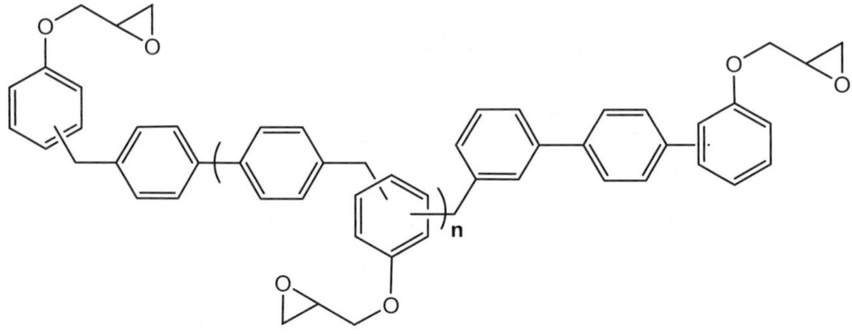

(II)

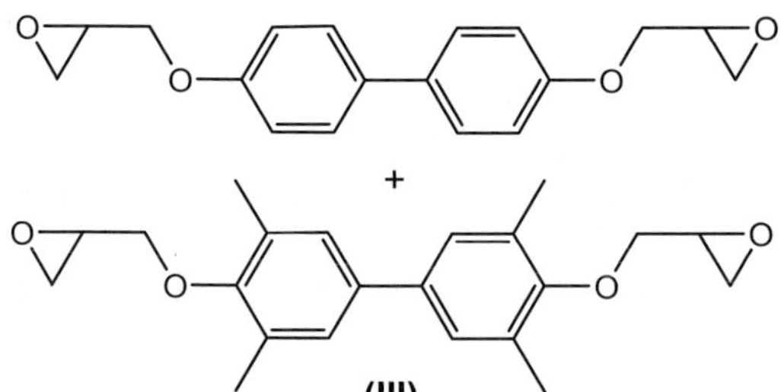

(III)

3. Tozzi-Spadoni[3] prepared flame-retardant viscose fibers and textiles by dosing an aqueous solution of cellulose xanthogenate with tetrabromobisphenol A or selected bis-(2-thiono-1,3,2-dioxaphosphorinanyl) oxides (IV).

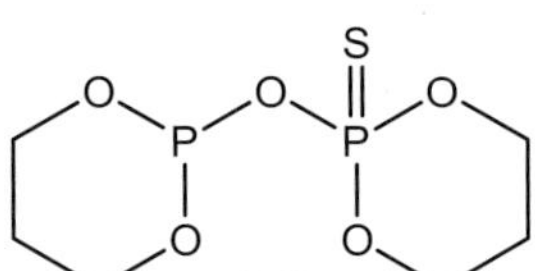

(IV)

4. Harada[4] additized poly(2,6-dimethylphenylene ether) with the flame-retardant magnesium hydroxide, which enabled the thermoplastic resin to remain flexible for subsequent use in electrical wires.

REFERENCES

1 Douglas W Luther, US Patent 8,067,088 (November 29, 2011)
Albemarle Corporation (Richmond, VA)

2 Yukihiro Kiuchi et al, US. Patent 7,799,852 (September 11, 2010)
NEC Corporation (Tokyo, JP) Sumitomo Bakelite Company Limited (Tokyo, JP)

3. Francesco Tozzi-Spadoni et al, US Patent 8,070,996 (December 6, 2011)
Lenzing Aktiengesellschaft (Lenzing, AT)

4. Tamotsu Harada et al, US. Patent 8,062,747 (November 22, 2011)
Sabic Innovative Plastics IP B.V. (NL)

E. HALOGEN-FREE RESIN FLAME RETARDANT

4,4-Dimethyl-*t*-Butyl-P-Oxide Dioxaphorinane

Title:	Nonhalogen flameproof resin composition
US Patent:	7,956,110 (June 7, 2011)
Author:	Min Soo Lee et al.
Assignee:	Cheil Industries Inc. (Gumi-si, KR)

Significance: The flame retardancy, mechanical properties, and melt index behavior of polystyrene-polyphenylene oxide blends were dramatically improved after molding at elevated temperatures with 5 wt% dioxaphorinane P-oxide flame retardants. These agents are readily prepared in two steps in a very high purity using inexpensive reagents.

Additives

4,4-Dimethyl-*t*-butyl-P-oxide dioxaphorinane (I)
4,4-Dimethylphenyl-P-oxide dioxaphorinane (II)
4,4-Dimethyl-methyl-P-oxide dioxaphorinane phosphorinate (III)
4,4-Dimethyl-butyl-P-oxide dioxaphorinane phosphorinate (IV)

Safety

P-oxide dioxaphorinanes are skin and eye irritants. In addition, they have been shown to be moderate to severe respiratory irritants. Working in enclosed work spaces requires that the operator wear an adequate respirator. Clothing contaminated with powder from these agents must be adequately cleaned before reuse.

Additive Structures

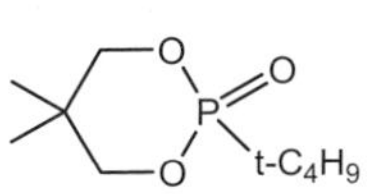

(I)

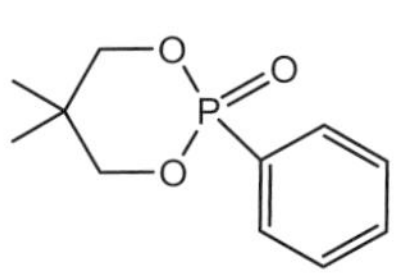

(II)

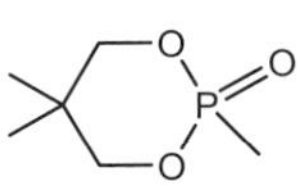

(III)

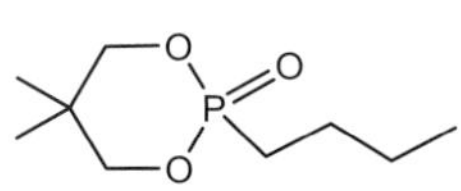

(IV)

Additive Preparation

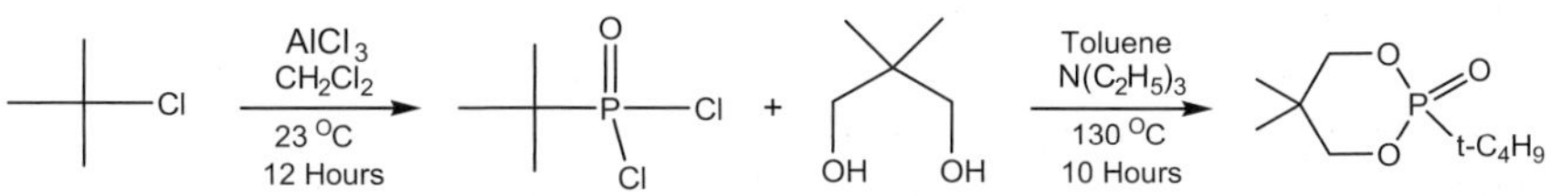

Additive Synthesis

1. *Preparation of t-butyl phosphonic dichloride*

A reactor was charged with 130 g of aluminum chloride dissolved in 200 ml of dichloromethane and then treated with the dropwise addition of 93 g of *t*-butyl chloride with stirring at 23°C. The mixture was then stirred for 12 h at an ambient temperature and then poured into 200 ml of chloroform and treated with 400 ml of dilute hydrochloric acid. The organic layer was then isolated, concentrated under reduced pressure, and the product was isolated and used without further purification.

2. *Preparation of 4,4-dimethyl-t-butyl-P-oxide dioxaphorinane*

A reactor was charged with the entire Step 1 product and then treated with one equivalent of 2,2-dimethyl-1,3-propanediol, two equivalents of triethylamine, and 200 ml of toluene and then refluxed for 10 h. The cooled mixture was then treated with 100 ml of ethyl acetate and 200 ml of water and stirred until all the solids dissolved. The organic layer was isolated, concentrated, and the product isolated in a 50% yield in 99% or higher purity.

Testing and Treatment Levels

Blending compositions and additive treatment levels are provided in Table 7.

TABLE 7 Experimental blends used in determining fire retardancy, impact strength, and melt index of resins additized with the Step product a that of a competitor

Component	Blend Composition 1 (g)	Blend Composition D2 (g)	Blend Composition 3 (g)	Blend Composition 4 (g)	Blend Composition 5 (g)
Polystyrene	85	85	85	90	85
Polyphenylene oxide	15	15	15	10	15
Step 2 Product	20	10	5	20	X
Competitor Additive*	X	X	X	X	20

**Tetra-2,6-dimethyl phenyl resorcinol diphosphate (Daihachi Chemical Industry Co.).*

A. *Fire Retardancy*

Components were mixed and then extruded at 200–280 °C using a conventional twin-screw extruder and isolated as pellets. The pellets were then dried at 80 °C for 2 h and then molded into test specimens for evaluating flame retardancy by injection molding between 180–280 °C with a mold temperature of between 40 °C and 80 °C Flame retardancy was then measured in accordance with UL 94 VB using samples with a 1/8'' thickness. Flame retardancy test results are provided in Table 2.

B. Impact Strength

Impact strength was measured in accordance with ASTM D 256 using samples with a 1/8'' thickness. IZOD test results are provided in Table 6.

C. Melt Index

The melt index was measured at 200 °C under a load of 5 kg according to ASTM D1238. Melt index test results provided in Table 8.

Test Results

TABLE 8 Favorable flame retardancy, impact strength, and melt index test results for polystyrene-polyphenylene blends additized with the Step 2 product, 4,4-dimethyl-2,6-dioxa-*t*-butylphosphorinate

Test	Blend Composition 1	Blend Composition 2	Blend Composition 3	Blend Composition 4	Blend Composition 5
UL94 Flame Retardancy	V-0	V-1	V-1	V-1	Fail
IZOD (kgf cm/cm)	8.8	8.1	7.0	7.5	7.0
Melt Index	8.2	8.1	8.0	8.4	7.9

Advantages over Prior Art

P-oxide dioxaphorinane derivatives of the current invention were extremely effective for when used in polystyrene-polyphenylene blends. In all cases, polymers additized with these agents had enhanced flame retardancy, impact strength, and melt index improvements at treatment levels as low as 5 wt%. Finally, P-oxide dioxaphorinanes can be readily prepared, are easily blended, and can be prepared in purities >99%.

Notes

1. Additional P-oxide dioxaphorinane derivatives (V) were prepared by Ku[1] and used as flame retardants in poly(butadiene-co-styrene-acrylnitrile).

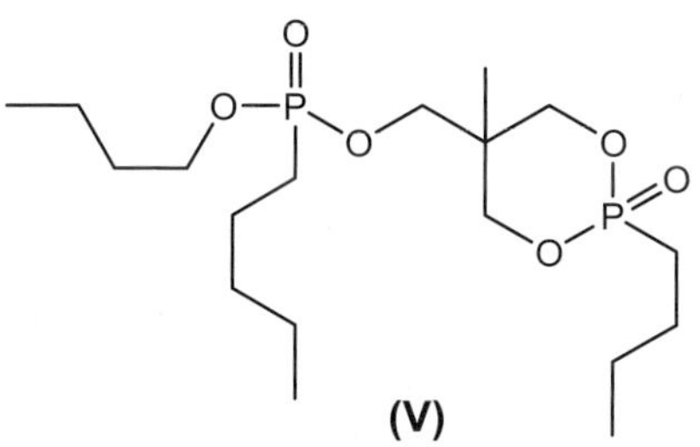

(V)

2. Byrd[2] prepared oligomeric polysiliconated phosphoramides (VI), which were used as fire retardants in clothing and cloth.

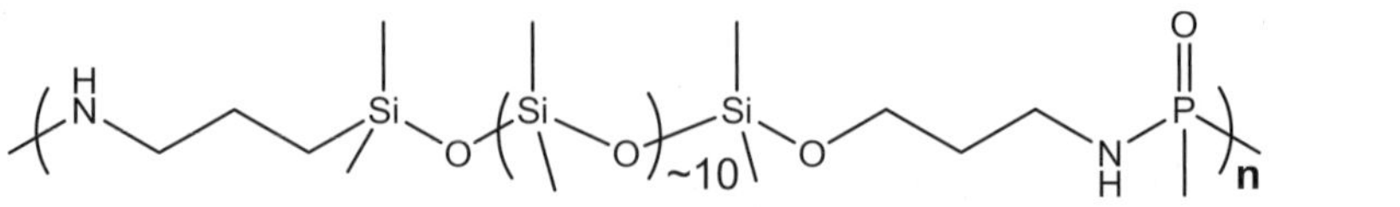

(VI)

3. Junk[3] prepared and used both potassium 1,1,2,2 tetrafluoroethane sulfonate (VII) and potassium 1,1,2,3,3,3-hexafluoropropanesulfonate (VIII), as a flame retardant for Lexan®-1110 aromatic polycarbonate.

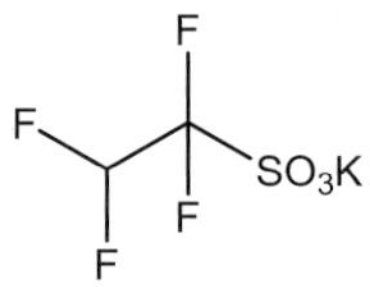

(VII)

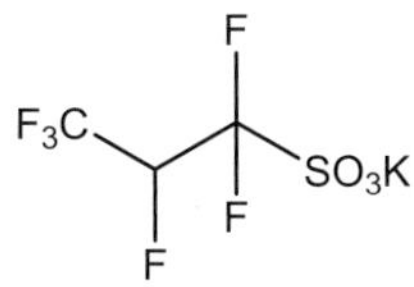

(VIII)

4. Hong[4] prepared versatile flame retardants consisting of carboxyethyl phosphinate acid ester derivatives (IX), which were used as flame retardants in polyethylene, polypropylene, polystyrene acrylonitrile–butadiene–styrene terpolymer, high impact polystyrene, acrylonitrile–styrene–acrylate terpolymer, and methylmethacrylate–butadiene–styrene terpolymer resins.

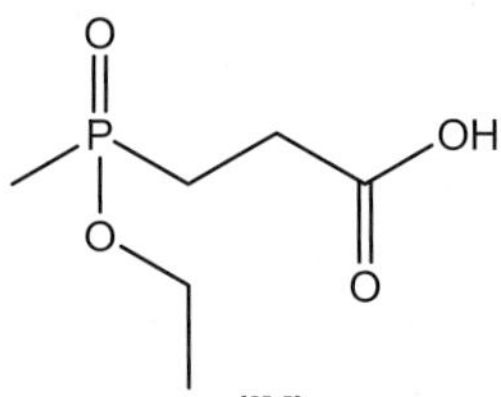

(IX)

REFERENCES

1. Jeong Hwan Ku et al, US Patent Application 20070032579 (February 8, 2007)
Cheil Industries Inc. (Gumi-si, KR)
2. Norman R. Byrd et al, US Patent 7,956,149 (June 7, 2011)
The Boeing Company (Chicago, IL)
3. Christopher P. Junk et al, US Patent 7,947,770 (March 24, 2011)
E. I. du Pont de Nemours and Company (Wilmington, DE)
4. Sang Hyun Hong et al, US Patent 7,939,588 (May 10, 2011)
Cheil Industries Inc (Gyeongsangbuk-do, KR)

Chapter 19

Flame Suppressants

A. PYROTECHNICALLY FLAME SUPPRESSANTS

Potassium Bromate

Application:	***Flame Suppressant***
Title:	Flame suppressant aerosol generant
US Patent:	7,906,034 (March 15, 2011)
Author:	Philip L. Posson et al.
Assignee:	Goodrich Corporation (Charlotte, NC)
Significance:	A pyrotechnically generated aerosol effective as a flame suppressant was prepared by the low-temperature thermal decomposition of potassium bromate. The aerosol contained potassium and bromine free radicals, which flame suppressed combusting phenolic resins in confined spaces without self-igniting.
Additive Composition	Sodium bromate (oxidant) (I) Potassium cyanurate dehydrate (fuel) (II) Polyvinyl alcohol (binder) (III) Glycerol (plasticizer) (IV)
Safety	The flame suppressant composition was prepared as small solid plugs under a pressing force of 50,000 pounds. The material is easily ignited from −30 °C to an ambient temperature in the presence of sparks or other ignition sources and should be considered a severe fire hazard. The decomposition burn rate of the flame suppressant is from approximately five to about 60 s per inch.

Next Generation of International Chemical Additives. DOI: http://dx.doi.org/10.1016/B978-0-444-53788-1.00019-X

Additive Components

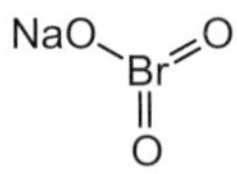

(I)

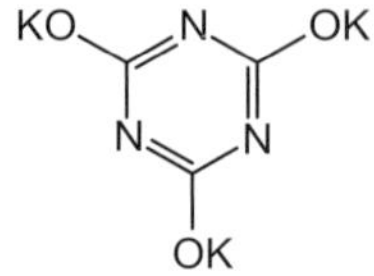

(II)

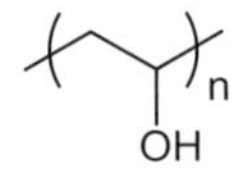

(III)

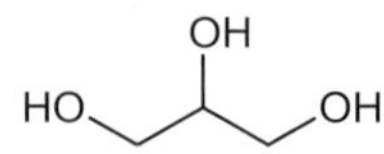

(IV)

Additive Composition

1. Preparation of pyrotechnic aerosol fire suppressant composition

A glass flask was charged with 165 g of commercial-grade cyanuric acid dehydrate, 92 g of anhydrous potassium carbonate powder, and 75 ml of water, which generated a thick slurry of carbon dioxide. The mixture was then heated to 100 °C and then cooled to an ambient temperature, and excess liquid was decanted. The mixture was then mixed with 260 g of ground potassium bromate, 1.5% polyvinyl alcohol binder, and 1.5% glycerol. After blending the mixture, it was dried and pressed into cylinders having a diameter of about 1.1" each weighing 50 g by using a pressing force of 50,000 pounds.

Testing

Forty-seven cylinders were arranged symmetrically on a laminated phenolic fabric circular plate, which was 7 mm thick and 280 mm wide and where the Step 1 cylinders were attached to the bottom of the circular plate. The outer rim of the plate was then raised to 13 mm to form a 25-mm-wide lip while a second plate was attached above the cylinders forming a 13-mm-wide annular vent around the circumference of the disc-shaped container. The inner fused ends and the center cylinder were primed with a pyrotechnic slurry. The assembly was then ignited using pull-wire igniters.

Testing Results

Whether flame suppressant testing was conducted at −30 F or at an ambient temperature, the Step 1 composition burned for <30 s while emitting a thick flame-suppressive aerosol with no visible flame. The phenolic fabric discs in both cases were discolored but were not consumed by the burning of the flame suppressant composition.

Advantages over Prior Art

Chemical flame suppressants are designed to supply free radicals to quench combustion-generated radicals. The aerosol delivery of these combustion deactivating free radicals is novel. The composition is particularly advantageous when used in confined spaces since the free fall velocity of the free radicals in an aerosol is lower than the air current velocity.

Notes

1. Fire suppressant compositions containing PB_3, $POBr_3$, $SOBr_2$, BBr_3, PBr_5, NOBr, and BrF were prepared by Huntington,[1] which had an ozone depletion potential of <0.1%.
2. A flame-suppressive aerosol device was prepared by Morton[2] and consisted of initially nebulizing boron trichloride liquid and sodium hydroxide solution, which upon mixing formed an aerosol with droplet diameters between 1 and 10 μm to deactivate free radicals.
3. Stevens[3] prepared a power-operated gas generating system containing nitrocellulose as the gas generator and molybdenum dioxide/copper oxide catalyst as the flame suppressant for use in confined spaces.
4. A fire suppressant aerosol containing 1,1,1,trifluoro-3-chloro-propene alone or in conjunction with trans-1,2-dichloroethylene was injected into a tube or nebulizer to form an aerosol of droplets, which deactivated free radicals.

REFERENCES

1. John H.Huntington et al, US Patent RE41,558 (August 24, 2010) **Eclipse Aerospace, Inc. (Charleston, SC)**
2. David A. V. Morton, US Patent 5,938,969 (August 17, 1999) **AEA Technology plc (Didcot, GB)**
3. Bruce A. Stevens et al US Patent Application 20100275808 (November 4, 2010) **TK Holdings, Inc. (Armada, MI)**
4. Rajiv R. Singh et al, US Patent Application 20090305876 (December 10, 2009) **Honeywell International, Inc. (Morristown, NJ)**

Chapter 20

Food Additives

A. IRON SUPPLEMENT

Poly(Iron Hydroxide-co-Saccharides)

Application:	Food Additive
Title:	Synthesis of high molecular weight iron-saccharidic complexes
US Patent:	7,964,568 (June 21, 2011)
Author:	Robert A. Beck et al.
Assignee:	Chromaceutical Advanced Technologies, Inc. (Hopkinton, MA)
Significance:	Polymer aggregates having weight average molecular weights >1.5×10^6 Da were prepared by mixing an aqueous suspension of ferric hydroxide in a mixtures of glucose, sodium gluconate, or sucrose in the presence of base. Weight average molecular weights >3.0×10^6 Da were obtained at pH = 9.0 using either sodium hydroxide or sodium carbonate and refluxing the mixture for 2 h.
Additives	Poly(ferric hydroxide-co-erythorbic acid) (I) Poly(ferric hydroxide-co-gluconate) (II) Poly(ferric hydroxide-co-glucose) (III) Poly(ferric hydroxide-co-sucrose) (IV) Poly(ferric hydroxide-co-gluconate-co-sucrose) (V)
Safety	Parenteral iron-saccharidic complexes are inherently safe and are designed to be used as an iron delivery system in the treatment of anemia and related pathologies.

Next Generation of International Chemical Additives. DOI: http://dx.doi.org/10.1016/B978-0-444-53788-1.00020-6

Additive Structures

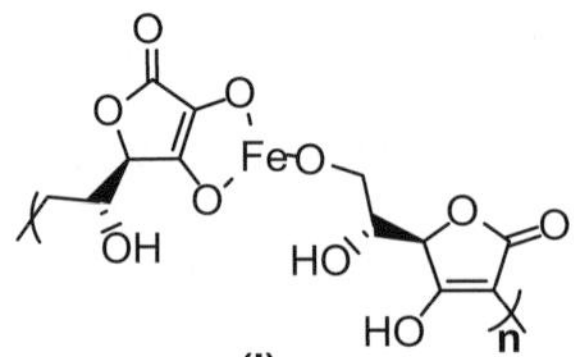

(I)

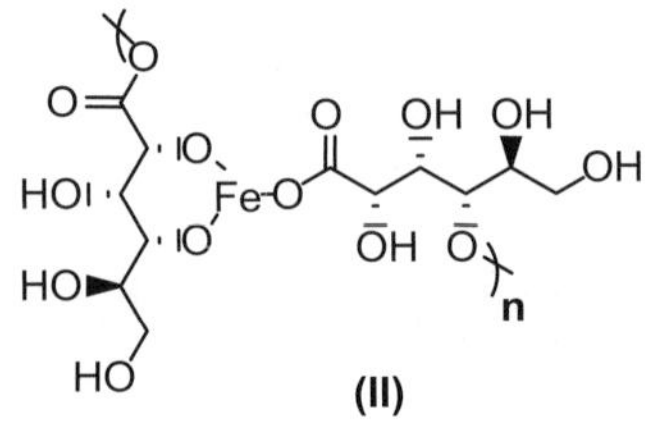

(II)

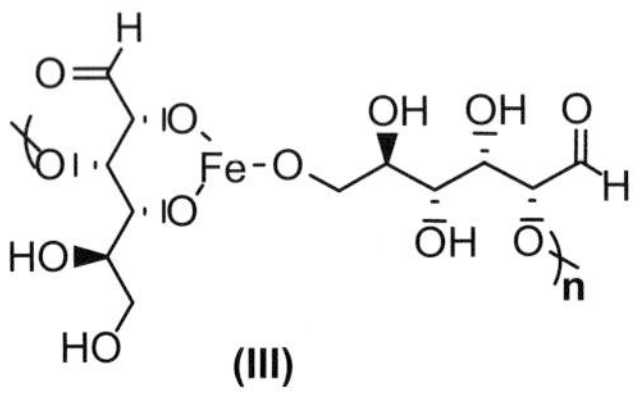

(III)

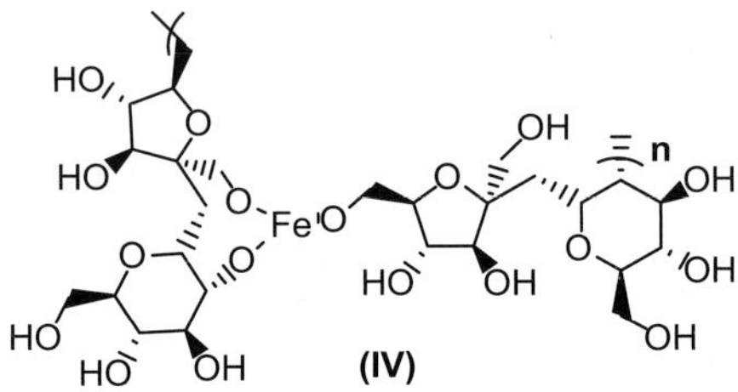

(IV)

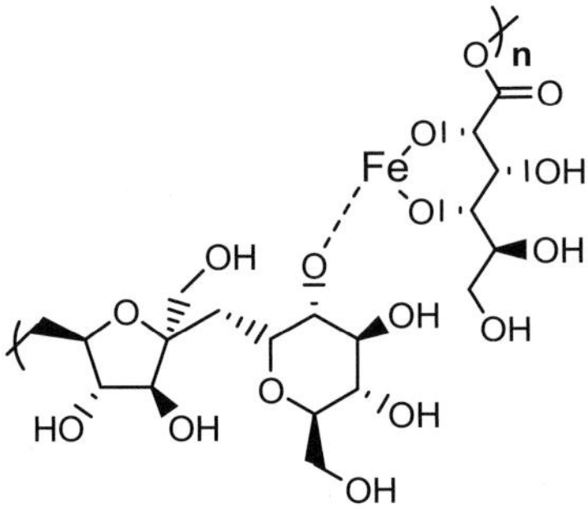

(V)

Additive Preparation

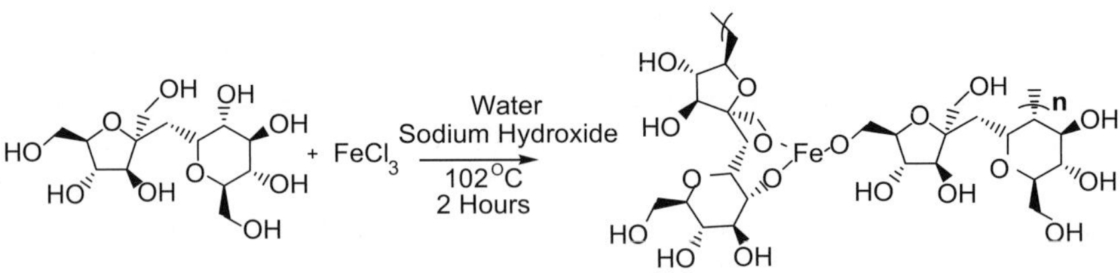

Additive Synthesis

1. Preparation of poly(ferric hydroxide-co-sucrose)

A round-bottom flask was charged with 200 ml of 0.3699 M solution of ferric chloride hexahydrate at 20 °C and then treated with sucrose to produce a final carbohydrate concentration of 0.877 M. After sucrose had completely dissolved, the mixture was treated with sufficient 8.0 M sodium hydroxide to give the solution a pH = 11.5 and then heated to 102 °C for 2 h. After cooling, the product was isolated and had an $M_w = 1.57 \times 10^6$ Da with a polydispersity of 1.32. Reaction conditions and physical properties for other saccharide/ferric hydroxide polymers are provided in Table 1.

TABLE 1 Summary of the effects of reaction time, temperature, and base selection on the physical properties of iron polysaccharide polymer mixing speeds and reagent stoichiometries were identical in all studies

Base	Saccharide	Reaction Temperature (°C)	Reaction Time (h)	Reaction pH	Iron–Polymer Mw (1.00×10^6 Da)
Na_2CO_3	Sodium Gluconate	100	2	9	1.5
Na_2CO_3	Sodium Gluconate	20	16	6	No product generated
Na_2CO_3	Sodium Gluconate	52	24	7	2.0
Na_2CO_3	Glucose	102	1.5	9	3.0
NaOH	Glucose	102	1.5	9	3.6
NaOH and Na_2CO_3	Glucose	102	1.5	9	0.61
Na_2CO_3	Sodium Gluconate/Sucrose	102	2	6	0.59

Testing

No testing results were supplied.

Advantages over Prior Art

Ferric ions are necessary for hematopoiesis. To ensure a slow, consistent, and prolonged delivery of ferric ions for at least a 24-hour period after a single dosage, the ferric-saccharide polymeric aggregate must be very large. This requirement has been achieved in the current invention. The current ferric ion delivery system, Ferrlecit®, consists of a solution of sodium ferric gluconate aggregates having an M_w of approximately 3.1×10^5 Da and also contains 20 wt% excess sucrose. In addition, therapeutic regimen involving Ferrlecit® requires multiple daily dosages to ensure that clinically significant ferric ion levels are maintained.

Notes

1. Beck[1] used Ferrlecit®, an iron-saccharidic consisting of sodium ferric gluconate complex dissolved in sucrose and having an Mw of approximately 310,000 Da, as a substitute for iron as a method for preparing purified hematinic iron-saccharadic complexes.
2. Groman[2] prepared colloids having mean volume diameters of ≤30 nm consisting of dextrans or polyether glycols having Mns from between 4000 and 10,000 Da coated with iron oxide, which were used as MRI-contrasting agents.

3. Helenek[3] prepared iron–saccharide complexes containing carboxymaltose, mannitol, polymaltose gluconate, and sorbitol carbohydrates, which were used in the treatment of iron deficiency anemia.
4. Lehmann[4] developed a method for treating disturbances in iron metabolism by using iron (III) complexed with the glycoprotein hormone erythropoietin.

REFERENCES

1. Robert A Beck et al, US Patent 7,883,897 (February 8, 2011)
   Chromaceutical Advanced Technologies, Inc. (Hopkinton, MA)
2. Ernest V. Groman et al, US Patent 7,871,597 (January 18, 2011)
   AMAG Pharmaceuticals, Inc. (Lexington, MA)
3. Mary Jane Helenek et al, US Patent 7,754,702 (July 13, 2010)
   Luitpold Pharmaceuticals, Inc. (Shirley, NY)
4. Paul Lehmann, US Patent 6,710,028 (March 23, 2004)
   Roche Diagnostics GmbH (Mannheim, DE)

B. FOOD ADDITIVE TO COUNTER BITTER TASTE

hT2R14 Antagonists

Title:	Compounds that inhibit (block) *bitter* taste in composition and use thereof
US:	Patent 8,076,491 (December 13, 2011)
Author:	Donald S. Karanewsky et al.
Assignee:	Senomyx, Inc. (San Diego, CA)
Significance:	Flavor modulators effective as hT2R8 antagonists were prepared in a moderate yield in five steps. When blended with bitter coffee containing hT2R14 antagonists, the perception of bitterness in the beverage was eliminated in coffee additized with between 30 and 10 μM of the experimental agents.
Additive Names	*N*-(1-((3,5-Dimethylisoxazol-4-yl)methyl)-1H-pyrazol-4-yl)-2-phenylpropanamide (I) *N*-(1-((3,5-Dimethylisoxazol-4-yl)methyl)-1H-pyrazol-4-yl)-3-methoxybenzamide (II) *N*-(1-((3,5-Dimethylisoxazol-4-yl)methyl)-1H-pyrazol-4-yl) benzo[d][1,3]diox-ole-5-carboxamide (III) *N*-(1-((3,5-Dimethylisoxazol-4-yl)methyl)-1H-pyrazol-4-yl)-2,5-dimethoxybenzamide (IV) *N*-(1-((3,5-Dimethylisoxazol-4-yl)methyl)-1H-pyrazol-4-yl)-1-phenylcyclopro-panecarboxamide (V) *N*-(1-((3,5-Dimethylisoxazol-4-yl)methyl)-1H-pyrazol-4-yl)-2-methoxybenzamide (VII)

Safety Chemical agents derived from the current investigation are designed to be used as chemical food additives and are believed to nontoxic.

Additive Structures

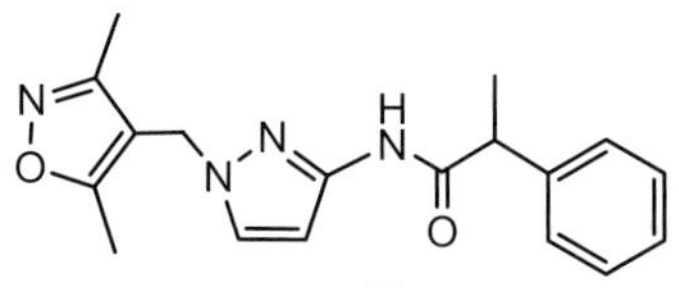

(I)

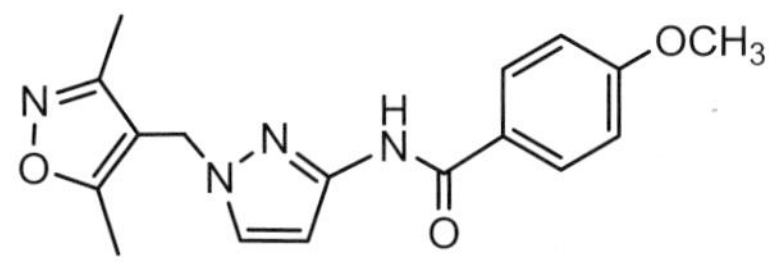

(II)

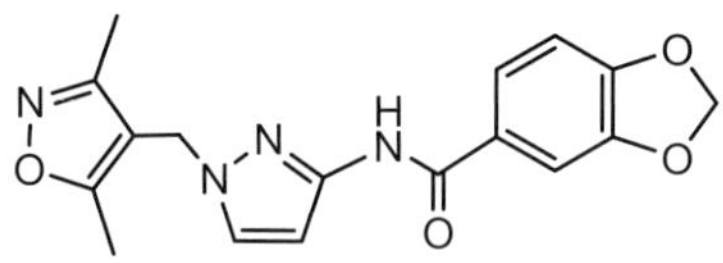

(III)

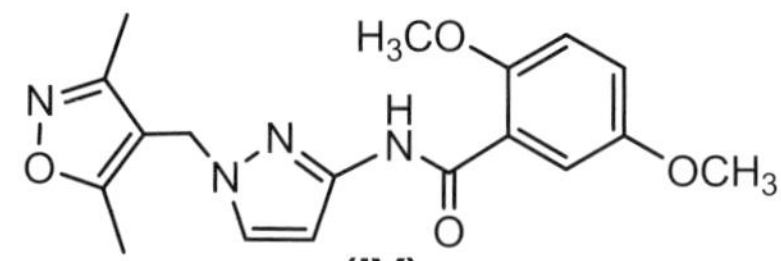

(IV)

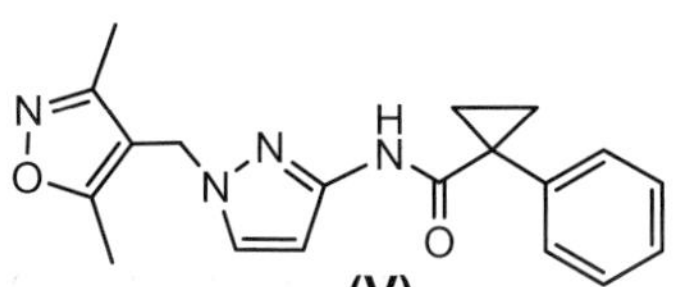

(V)

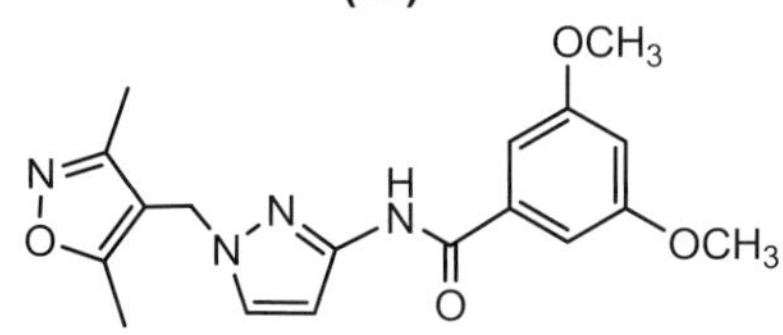

(VI)

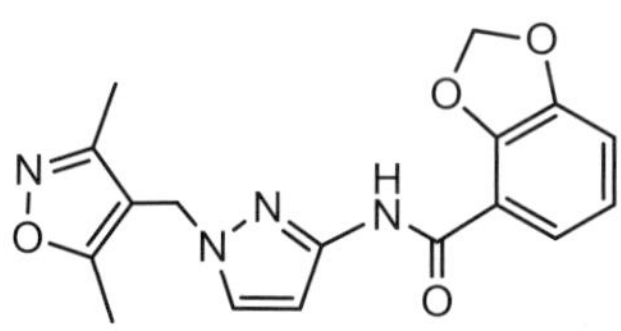

(VII)

Additive Preparation

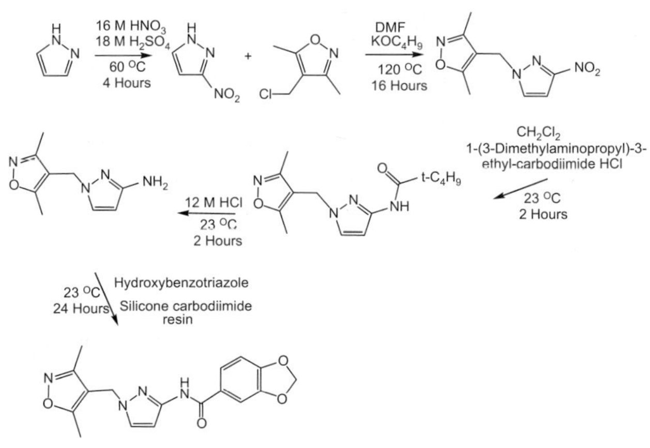

Additive Synthesis

1. *Preparation of 4-nitro-1H-pyrazole*

A flask containing 100 ml of 18 M sulfuric acid was treated portionwise with 10 g of pyrazole while maintaining the internal reaction temperature <50 °C. The mixture was then treated with the dropwise addition of 10 ml of 16 M nitric acid while again maintaining the internal reaction temperature <50 °C. The mixture was then heated to 60 °C and stirred for 4 h, and the pH was adjusted to 10 using 150 ml of 18 M aqueous sodium hydroxide solution. The product precipitated from the solution and was isolated as a white solid after it was collected by filtration. The solid was washed with water, dried under high vacuum, and the product was isolated in a 42% yield as a white solid.

^{13}C-NMR (100 MHz, d-$CDCl_3$ δ) 126.4, 137.0

2. *Preparation of 3,5-dimethyl-4-((4-nitro-1H-pyrazol-1-yl)methyl)isoxazole*

A small round bottom was charged with 3.8 g of the Step 1 product and 80 ml of *N,N*-dimethylformamide and then cooled to 0 °C and treated with 4.2 g of potassium *t*-butoxide. The mixture was stirred at an ambient temperature for 30 min and then treated with 5 g of 4-(chloromethyl)-3,5-dimethylisoxazole and refluxed for 16 h. After cooling to an ambient temperature, the mixture was precipitated in water and isolated by filtration and then washed in water and recollected. After drying under vacuum, the product was isolated in a 78% yield as a slightly yellow solid.

^{1}H-NMR (400 MHz, d-$CDCl_3$ δ) 2.23 (s, 3H), 2.46 (s, 3H), 5.08 (s, 2H), 8.02 (s, 1H), 8.08 (s, 1H)

3. *Preparation of t-butyl 1-((3,5-dimethylisoxazol-4-yl)methyl)-1H-pyrazol-4-ylcarbamate*

The entire Step 2 product was treated with 598 mg of 1-(3-dimethylaminopropyl)-3-ethylcarbodiimide hydrogen chloride and then stirred for 4 h and diluted with 5 ml of dichloromethane. The mixture was then washed twice with a saturated solution of 5 ml of sodium bicarbonate and 5 ml of a saturated solution of brine. The organic layer was collected, dried, filtered, and solvents removed under vacuum.

4. *Preparation of 1-((3,5-dimethylisoxazol-4-yl)methyl)-1H-pyrazol-4-amine hydrochloride*

A small flask was charged with 592 mg of the Step 3 product and 20 ml of 4 M hydrochloric acid dissolved in dioxane and then stirred for 2 h at an ambient temperature. The mixture was concentrated under reduced pressure and the residue taken up in a 1/1 mixture of 30 ml ethyl acetate/hexanes and concentrated. The solid was triturated with hexanes, collected by filtration, and the product was isolated in a 99% yield.

^{1}H-NMR (400 MHz, d-$CDCl_3$ δ) 2.11 (s, 3H), 2.38 (s, 3H), 5.16 (s, 2H), 7.51 (s, 1H), 8.03 (s, 1H), 10.27 (bs, 3H)

5. *Preparation of N-(1-((3,5-dimethylisoxazol-4-yl)methyl)-1H-pyrazol-4-yl)benzo[d]-[1,3]dioxole-5-carboxamide*

A mixture consisting of 8 mg of the Step 4 product dissolved in 200 μl dimethylformamide containing 7 mg of benzo[d][1,3]dioxole-5-carboxylic acid was treated with 70 mg of silicone carbodiimide resin and then loaded into a 1.2 ml 96 well Greiner plate followed by the addition of amine and acid. The reaction well mixture was then treated with 6 mg hydroxybenzotriazole dissolved in 100 μl dimethylformamide and then shaken overnight at an ambient temperature. The well mixture was then treated with 35 mg PS-Trisamine resin to remove excess carboxylic acid and hydroxybenzotriazole. It was then shaken overnight at an ambient temperature, treated with 200 μl of acetonitrile, and then shaken for about 1 min and the top clear solution transferred into a new plate. The extraction process was repeated twice, and the product was isolated in a 6% yield.

MS M+H calculated 341.1, found 341.2

Testing

1. *Bitter coffee taste test*

Taste tests were performed using coffee bitter fractions or bitter instant coffee containing hT2R8 and hT2R14 antagonists and evaluated by 4–5 taste panelists. Unadditized and additized coffee samples were given to the taste panelists who were required to identify the bitterness of the sample set within each pair. Testing results are provided in Tables 2 and 3, respectively.

2. *hT2R8 and hT2R14 screening assay*

Partially purified bitter fraction from coffee was used to screen the 25 human T2Rs in transiently transfected HEK cells using human embryonic kidney cells that stably express large T-cell antigen, and G15 protein was transiently transfected with an hT2R expression plasmid. Thereafter, a fluorescent-based assay was used to detect changes in calcium concentration in the transiently transfected cells. The coffee fraction had a high level of fluorescence, which interfered with the assay. To overcome the interference, a number of blue dyes were tested for ability to block the fluorescence from the coffee fraction. Screening assay results are provided in Table 4.

Test Results

1. Detection of bitterness in coffee fractions

TABLE 2 Detection of bitterness in coffee fractions additized with selected hT2R8 antagonists and 10 μM of hT2R14 antagonist prepared by Li* (VIII)

		Selected Bitterness		
Agent	Concentration (μM)	Without Antagonist Combination	With Antagonist Combination	*P* Value
Additive 1	30	32	0	<0.001
Additive 2	30	15	1	0.001

**hT2R14 prepared by Li.*

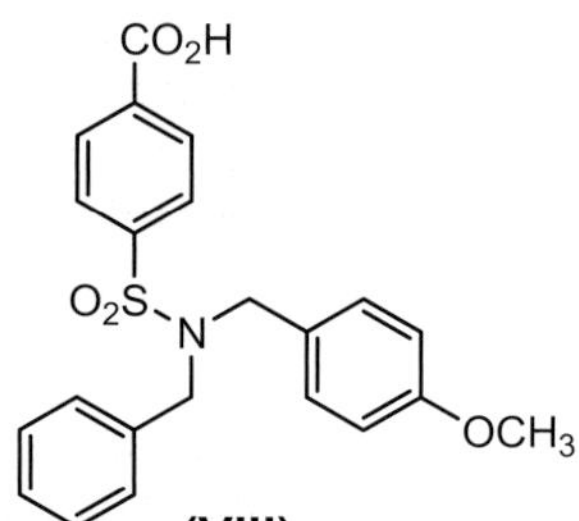

(VIII)

2. Detection of bitterness in instant coffee

TABLE 3 Detection of bitterness in two types of instant coffee additized with hT2R8 antagonists and 10 mM hT2R14 antagonist (VIII) described by Li*

			Selected Bitterness		
Instant Coffee	Agent	Concentration (μM)	Without Antagonist	With Antagonist	*P* Value
Medium	Additive 1	30	16	0	<0.001
Medium Dark	Additive 2	30	13	3	0.021

**hT2R14 prepared by Li.*

TABLE 4 List of chemical agents effective as hT2R5, -9, -13, -54, -67, and -75 antagonists determined by the screening assay

hT2R5, -9, -13, -54, -67- and -75 Antagonists	Effective Concentration (μM)
Picric Acid	0.05
Chloroquine pH 6.5	50
Picoline	10
Andrographolide	0.5
Ofloxacin	1
Strychnine	50
Oxyphenonium	1
Aristolochic Acid	2
Salicin	1
Denatoni	0.5
Strychnine	1
Ranitidine	5
Cinchonine	150
Aristolochic Acid	25
Caffeine	1
Andrographolide	100
Aristolochic Acid	1
Oleuropein	1
Andrographolide	5
Picric Acid	10

Advantages over Prior Art

This discovery has identified and synthesized at least seven unique food additives that inhibit the detection of bitter taste in humans by antagonizing hT2R8 receptors. Their application, however, is anticipated to be used as medicinal agents.

Notes

1 Additional hT2R14 (IX) and (X), and hT2R8 antagonists (XI) and (XII) are described by Li[1].

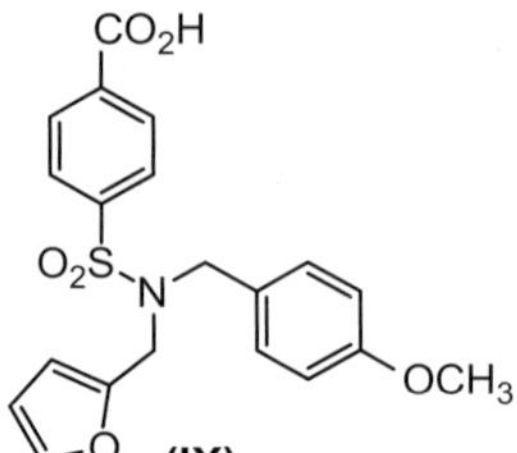

(IX)

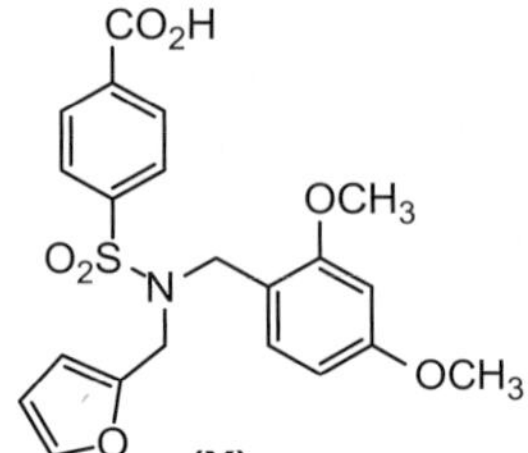

(X)

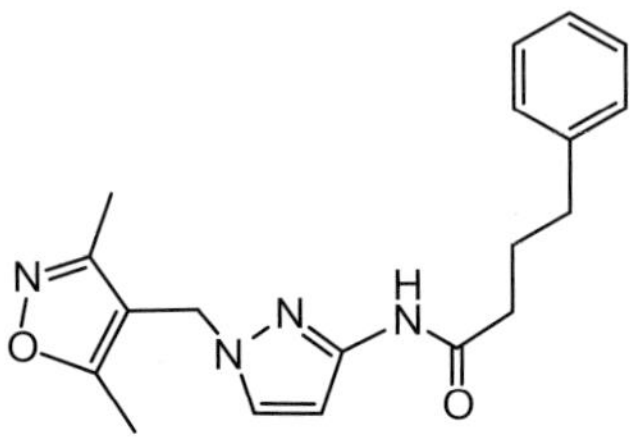

(XI)

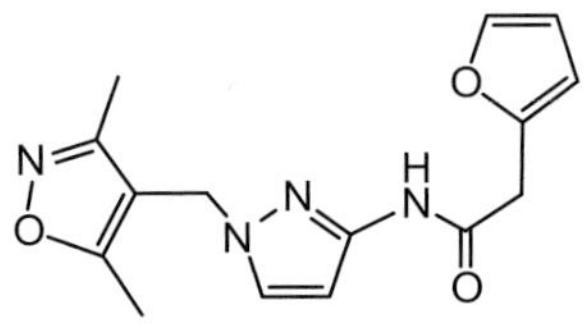

(XII)

2 Li[2] determined that acetaminophen (XIII) and denatomium (XIV) were effective as hT2R54 antagonists. Strychnine (XV) was effective as an hT2R8, -9, -10, -54-, and -75 antagonists while ranitidine (XVI) was effective as hT2R8 and -10 antagonists.

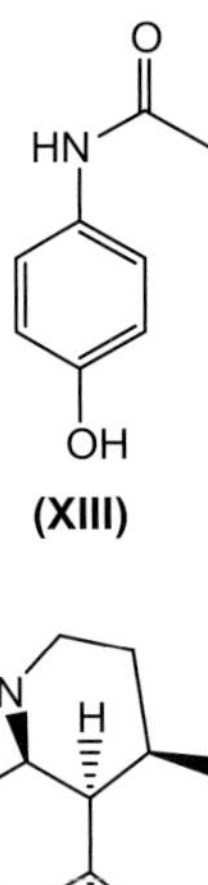

(XIII)

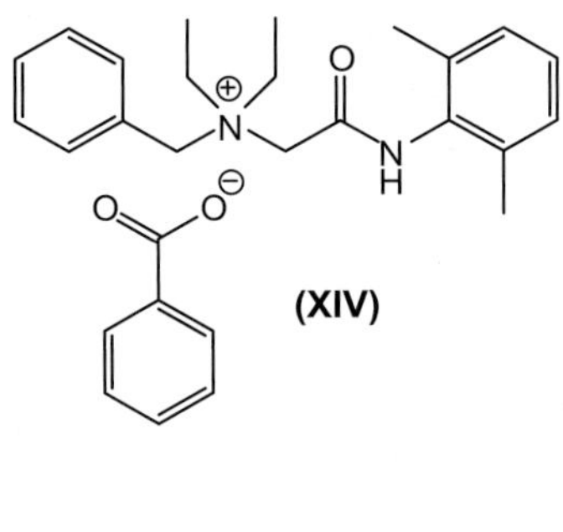

(XIV)

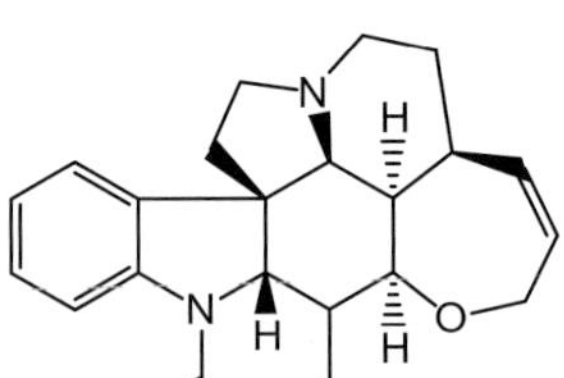

(XV)

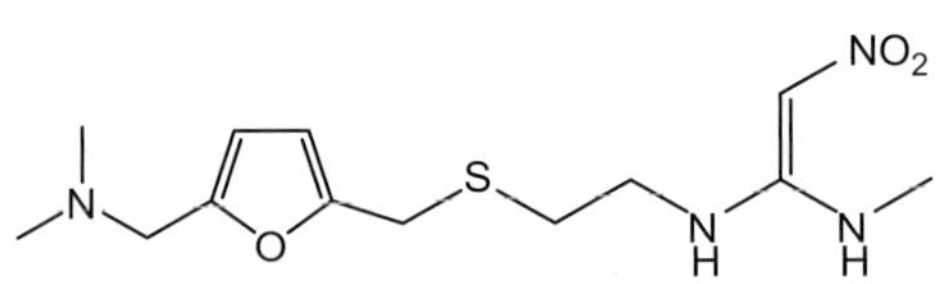

(XVI)

3 Alkyldienamides (XVII) and (XVIII) prepared by Dewis[3] were used as food additives to neutralize bitterness and umami characteristics of certain foods, respectively.

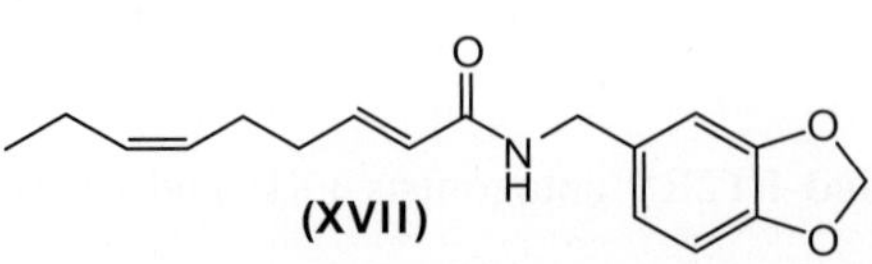

(XVII)

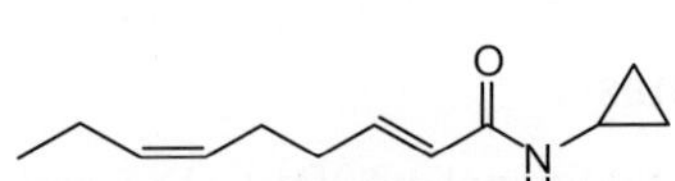

(XVIII)

4 Dang[4] determined that when azelastine (XIX) a second-generation selective histamine antagonist was used in nasal spay formulations, it minimizes or masked the unpleasant bitter taste associated with postnasal drip

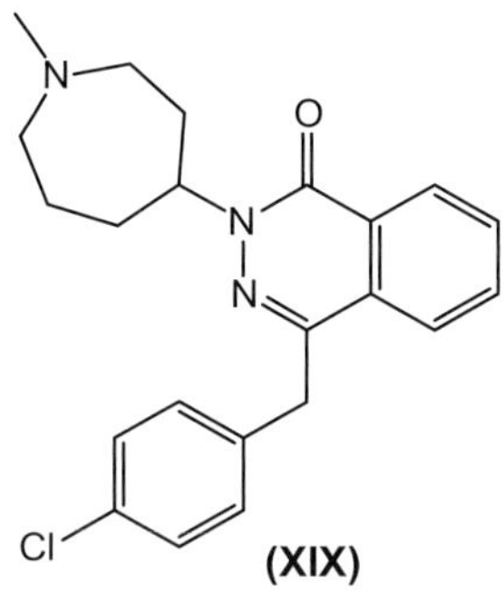

(XIX)

REFERENCES

1. Xiaodong Li et al, US 7,939,671 (May 10, 2011)
Senomyx, Inc. (San Diego, CA)

2. Xiaodong Li et al, US. Patent 7,407,765 (August 5, 2008)
Senomyx, Inc. (San Diego, CA)

3 Mark L. Dewis et al, US. Patent 8,075,937 (December 13, 2011)
International Flavors & Fragrances Inc. (New York, NY)

4. Phuong Grace Dang et al, US. Patent 8,071,073 (December 6, 2011)
MEDA Pharmaceuticals Inc. (Somerset, NJ)

Chapter 21

INK

A. INK AND INK-JET COLORANT DISPERSANTS

Grafted Aminophenylpolyacids

Title:	Modified colorants with aliphatic polyacid groups
US Patent:	8,021,471 (September 20, 2011)
Author:	Paul S. Palumbol
Assignee:	Cabot Corporation (Boston, MA)

Significance: A method for improving the dispersancy of carbon black colorants for inkjet printing applications is described. In this process, aromatic di and tricarboxylic acids were azografted onto the surface of carbon black using a diazonium grafting reaction involving an aromatic amine present in aromatic di and tricarboxylic acids. Once modified, carbon black colorants were used in inkjet formulation where the print was clear, distinct, and outer diameters considerably larger than formulations using unmodified carbon black colorants.

Additive Names

4-Aminophenylsuccinic acid (I)
3-(4-Aminophenyl)glutaric acid (II)
4-Aminohomophthalic acid (III)
3-(4-Aminophenyl)tricarballylic acid (IV)

Safety

Depending on the degree of exposure, aminophenyl di and tricarboxylic acids cause moderate to severe eye irritation. These agents are considered digestive tract irritants if consumed. Dust and particulates cause respiratory tract irritation.

Next Generation of International Chemical Additives. DOI: http://dx.doi.org/10.1016/B978-0-444-53788-1.00021-8

Additive Structures

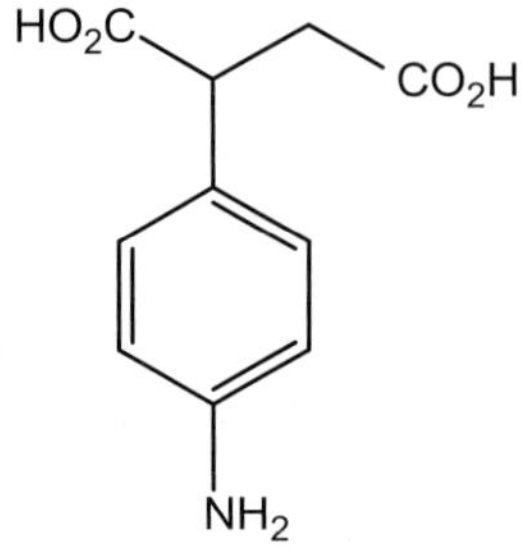

(I)

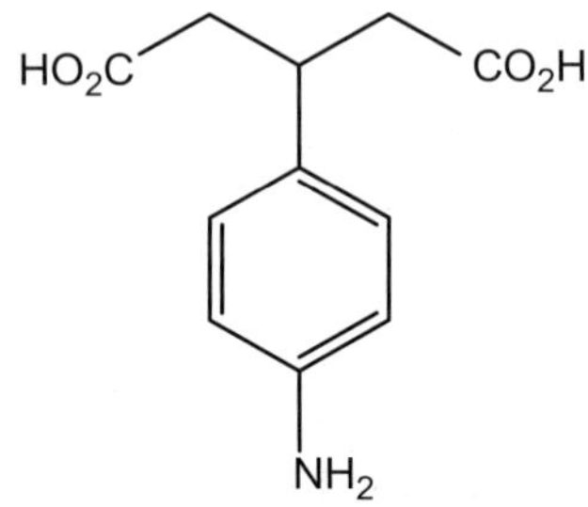

(II)

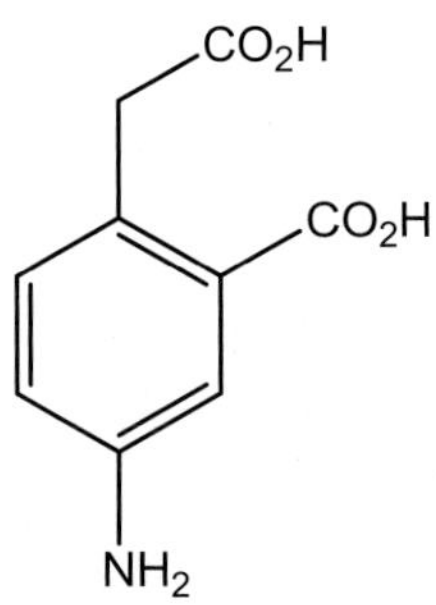

(III)

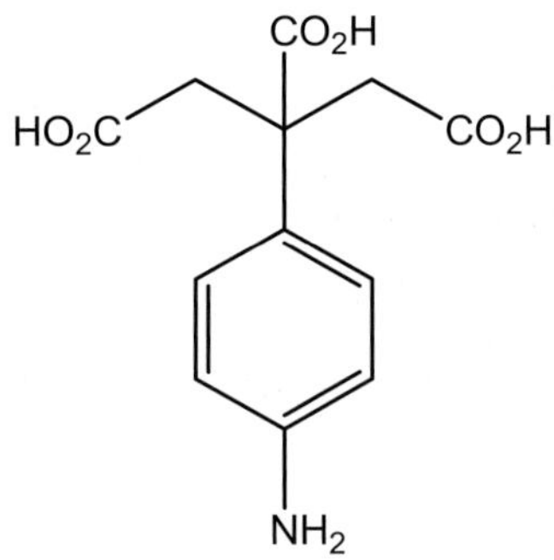

(IV)

Additive Preparation

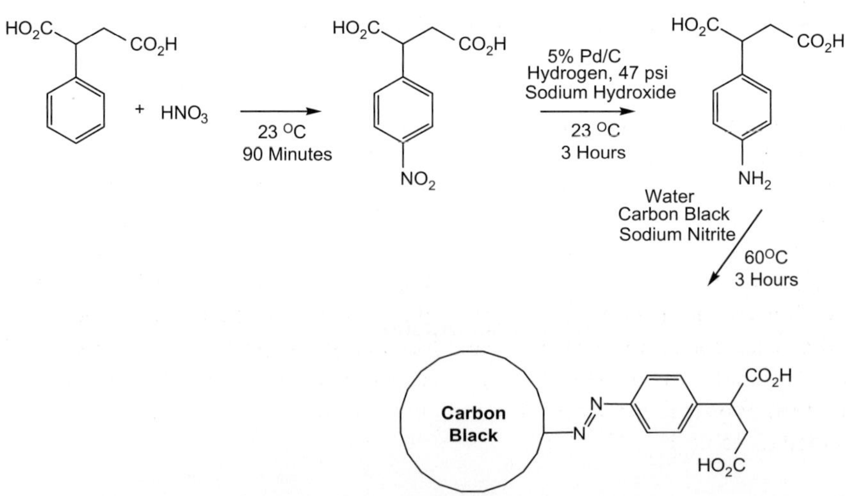

Additive Synthesis

1. Preparation of 4-nitrohomophthalic acid

A three-neck round-bottom flask containing 142 ml of fuming nitric acid having a density of 1.52 g/ml was treated portionwise with 45 g of homophthalic acid while being cooled in an ice bath. The addition of homophthalic acid was at such a rate that the reaction temperature was kept <22 °C. After the addition was completed, the bath was removed, and the mixture was slowly warmed to an ambient temperature and then stirred for 90 min, recooled, and poured into 500 ml of crushed ice. The yellow precipitate that formed was filtered and then washed with cold deionized water, and the product was isolated in a 60% yield.

^{1}H-NMR (d_6-DMSO, δ) 8.6 (s, 1H), 8.35 (d, J = 8 Hz, 1H), 7.67 (d, J = 8 Hz, 1H), 4.1 (s, 2H).

2. Preparation of 4-aminohomophthalic acid

A 500-ml Parr bottle was charged with 15.75 g of the Step 1 product, 150 ml of 1M NaOH, and 0.55 g of 5% Pd/C and then hydrogenated at 47 psi for 3 h. The solution was then filtered through Celite®, washed with a small amount of water, and acidified with hydrochloric acid to a pH of 4–5. Thereafter, the hydrogenated solution was stored in the refrigerator overnight, and the product was precipitated from solution and isolated. Second and third crops were also obtained by concentrating the mixture at an ambient temperature so that the overall yield was 81%.

^{1}H-NMR (d_6-DMSO, δ) 7.13 (s, 1H), 6.91 (d, J = 8 Hz, 1H), 6.64 (d, J = 8 Hz, 1H), 3.69 (s, 2H);

IR (KBr, cm^{-1}): 3320–3640 (br), 1700 (s), 1550 (s)

3. Preparation of carbon black-g-azohomophthalic acid

A 1-liter stainless steel beaker equipped with an overhead paddle stirrer was charged with 50 g of carbon black and 7.6 g of the Step 2 product and then suspended in 441 ml of deionized water. The mixture was then heated to 60 °C and then treated with the dropwise addition of 10.7 g of 20% aqueous sodium nitrite over several minutes. When the addition was completed, the mixture was heated for 3 h at 60 °C and then cooled to an ambient temperature, and the pH was adjusted to 9.25 using 1 M sodium hydroxide. The dispersion was diafiltered using a peristaltic diafiltration unit equipped with a Spectrum polysulfone 0.05 μ hollow fiber membrane with deionized water for a time equivalent to 5 diafiltration volumes. The final dispersion was adjusted to about 15% solids and then centrifuged at 4000 rpm for 10 min. After decanting from the settled pellet, the dispersion of modified carbon black was analyzed for percent solids, viscosity, surface tension, sodium content, and TGA. Physical properties for selected modified carbon black colorants are provided in Table 1.

TABLE 1 Physical properties of carbon black surface containing azografted aromatic di and tricarboxylic acids

Colorant	Solids (%)	pH	Viscosity (cP)	Surface Tension (dyne/cm)	Sodium Content (ppm)	TGA (%)
Carbon Black-g-Additive 1	12.2	9.3	3.4	73.8	7215	3.4
Carbon Black-g-Additive 1	12.1	9.2	3.4	73.1	10972	4.97
Carbon Black-g-Additive 4	12.1	8.8	2.72	73.7	7238	3.94
Carbon Black-g-Additive 4	11.7	8.5	2.06	72.8	11582	5.9
Carbon Black-g-Additive 2	15.2	7.7	3.56	72.6	5814	3.81
Carbon Black-g-Additive 2	10.1	7.5	2.2	72.3	7661	5.34
Carbon Black-g-Additive 3	11.9	8.4	2.44	73.0	9751	–

Testing

1. Colorant formulation

Modified carbon black dispersions were used to prepare a number of inkjet inks. All inks were prepared at a 4% pigment concentration in the formulations that are listed in Table 2.

TABLE 2 Ink formulations used to evaluate the dispersancy of carbon black containing azografted aromatic di- and tricarboxylic acids

Additive	Formulation I (%)	Formulation II (%)	Formulation III (%)
Surfynol 465	1	0.2	1
Triethyleneglycol	5	X	X
Monobutylether Glycerin	10	X	7
Diethyleneglycol	X	X	5
Trimethylolpropane	X	5	X
2-Pyrrolidinone	X	7	X
1,5-Pentanediol	X	7	X
Surface Modified Carbon Black Pigment	4	4	4
Deionized Water	Remainder	Remainder	Remainder

2. *Ink dispersancy testing*

Formulated inks were placed in ink cartridges, and printing was carried out using a commercial inkjet printer with six different paper types including Hammermill™ Copy Plus™ (HCP), Xerox™ 4024, HP Bright White™, and Epson Enhanced Matt™.

In these evaluations, a series of black squares were printed in triplicate. The prints were analyzed on the Image Expert using an X-Rite™ 938 Spectrodensitometer. From the triplicate data, an average optical density across the paper set was calculated. A summary of the test results is provided in Table 3.

Test Results

TABLE 3 Printing test results using pigments consisting of carbon black azografted with aromatic di and tricarboxylic acids on commercial printing paper. All modified carbon black pigments displayed larger outer diameters than from ink formations using Cab-O-Jet® unmodified carbon black

Grafted Polyacid On Carbon Black	Ink Formulation	Treatment Level (%)	HCP (OD)	Xerox-4024 (OD)	Optical Density HP Brightness White	Optical Density Epson Enhanced Matt	Average Optical Density
I	I	0.4	1.13	1.16	1.36	1.57	1.25
II	I	0.4	1.13	1.16	1.39	1.58	1.26
III	I	0.7	1.24	1.20	1.39	1.56	1.28
IV	I	0.7	1.14	1.22	1.36	1.41	1.28
I	II	0.4	1.32	1.37	1.54	1.65	1.45
IV	II	0.4	1.30	1.39	1.56	1.60	1.44
II	III	0.4	1.33	1.35	1.54	1.67	1.40
III	III	0.7	1.09	1.17	1.53	1.67	1.30

Advantages over Prior Art

Although the incorporation of dispersants onto the surface of carbon black by diazonium coupling in inkjet formulations has been previously reported, the optical density of the print varied considerably depending on the printing paper. Moreover, the optical density of print on recycled paper is unreported. The single step incorporation of aromatic di and tricarboxylic acid dispersants onto the carbon black surface addresses the shortcomings of previously reported methods.

Notes

1. Vasudevan[1] prepared polymerizable dye–monomer conjugates (V) for carbon black, titanium oxide, cobalt blue, chrome yellow, and iron oxide pigment particles to improve the dispersancy of the solid.

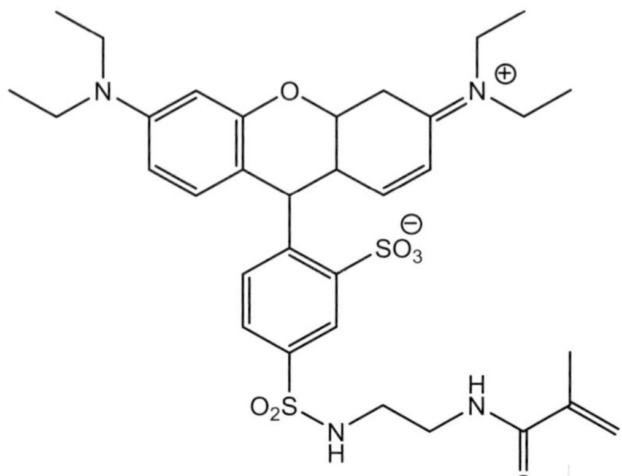

(V)

2. Lee[2] prepared self-dispersible coloring agents having long-term storage stability by grafting sulfonic acid derivatives onto the surface of carbon black (VI).

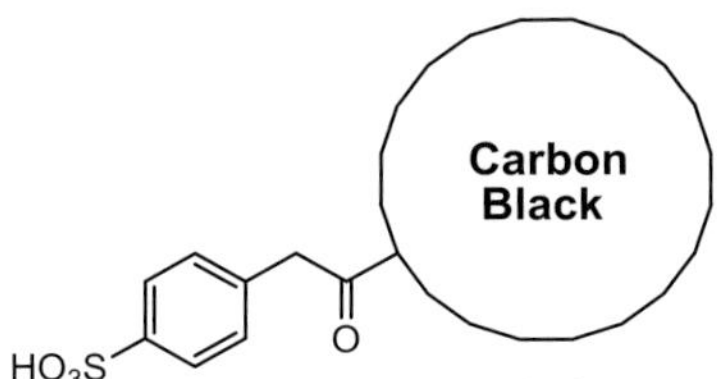

(VI)

3. Lee[3] prepared ink compositions containing self-dispersible metals using pyridine and bipyridine carboxylic acid or phenolic complexes (VII), and (VIII), respectively.

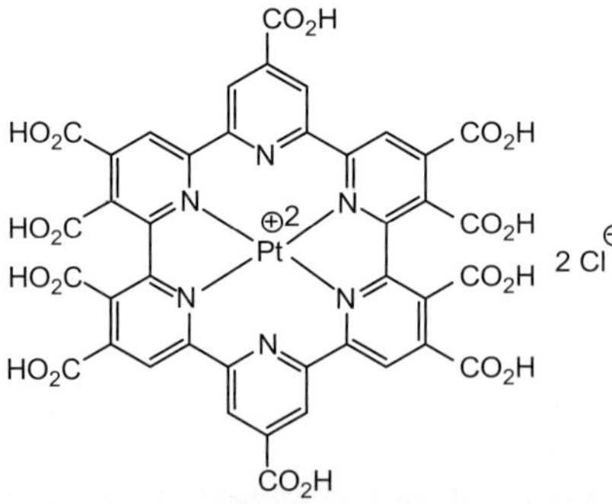

(VI)

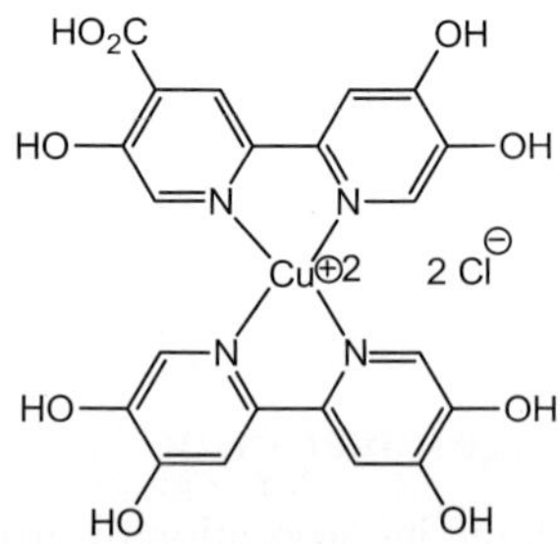

(VIII)

4. Sujeeth[4] dispersed carbon black by introducing an sulfanilic acid surface (IX) by initially reacting carbon black with sulfonyl chloride or chlorosulfonic acid and then treating the modified surface with aminocarboxylic acid esters and hydrolyzing the intermediate.

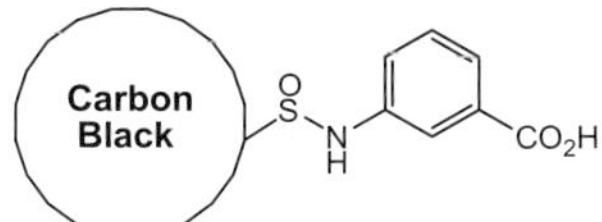

(IX)

REFERENCES

1. Sundar Vasudevan et al, US. Patent 7,868,059 (January 11, 2011)
   Hewlett-Packard Development Company, L.P. (Houston, TX)
2. Jong-in Lee et al, US. Patent 7,393,403 (July 1, 2008)
   Samsung Eletronics Co., Ltd. (Suwon-Si, KR)
3. Jong-in Lee et al, US. Patent 7,718,803 (May 18, 2010)
   Samsung Eletronics Co., Ltd. (Suwon-Si, KR)
4. Puthalath Sujeeth et al, US. Patent 7,964,033 (January 21, 2011)
   Sensient Colors LLC (St. Louis, MO)

B. JET INK STABILIZER

2,2,6,6-Tetramethylpiperidine Derivatives

Title: Water compatible sterically hindered alkoxyamines and hydroxy substituted alkoxyamines
US Patent: 7,947,832 (May 24, 2011)
Author: Mervin G. Wood et al.
Assignee: BASF SE (Ludwigshafen, DE)

Significance: There are a very limited number of inkjet additives that are designed to maintain a high jet printing ink density when using cyan, magenta, or yellow dyes after exposure of a 650–700 mg/m^2 ink sample to a Xenon lamp @ 461 W/m^2. After exposure to 4–8 weeks of inks additized with 2,2,6,6-tetramethylpiperidine derivatives, the prinking density of exposed ink remained essentially unchanged. Finally, 2,2,6,6-tetramethylpiperidine ink derivatives are readily prepared in moderately to good yield in two steps.

Additives 1-(2-Hydroxy-2-methylpropoxy)-2,2,6,6-tetramethyl-*N*-hydroxy-methyl-4-piperidine (I)
1-(2-Hydroxy-2-methylpropoxy)-2,2,6,6-tetramethyl-4-piperidine hydroxyl oxime (II)
1-(2-Hydroxy-2-methylpropoxy)-2,2,6,6-tetramethyl-4-piperidinium ethylenediamine acetic acid (III)
1-(2-Hydroxy-2-methylpropoxy)-2,2,6,6-tetramethyl-4-piperidinium acetic acid (IV)
1-(2-Hydroxy-2-methylpropoxy)-2,2,6,6-tetramethyl-4-piperidine-*N*-3-propylsulfate (V)
1-(2-Hydroxy-2-methylpropoxy)-2,2,6,6-tetramethyl-4-piperidinium-*N*-2-ethyl hydroxyl chloride (VI)

Safety 4-Amino-2,2,6,6-tetramethylpiperidine and oxime- and amide-derivatives are moderately toxic if ingested. When heated to decomposition, they emit toxic nitric oxide/dioxide vapors. Overall, chemical additives have an LD50 of 906 mg/kg in oral rat studies. No toxic effects have been reported other than lethal dose values.

Additive Structures

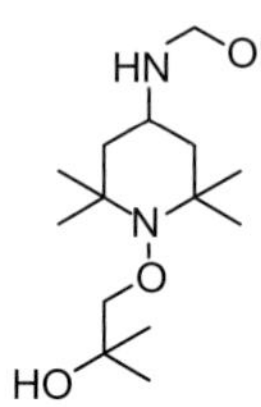

(I)

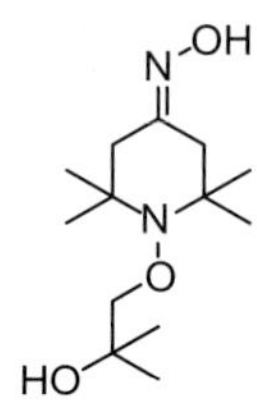

(II)

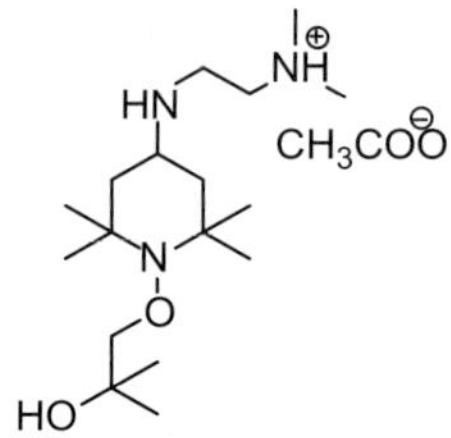

(III)

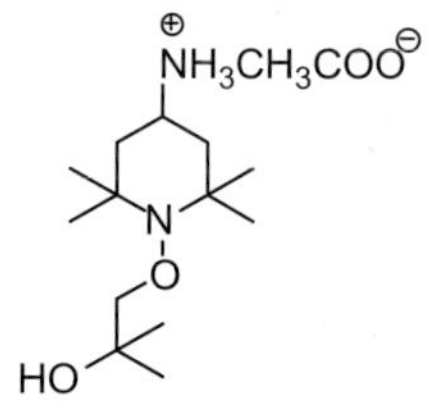

(IV)

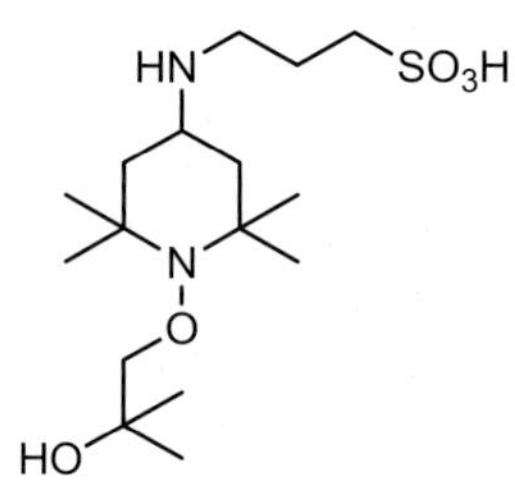

(V)

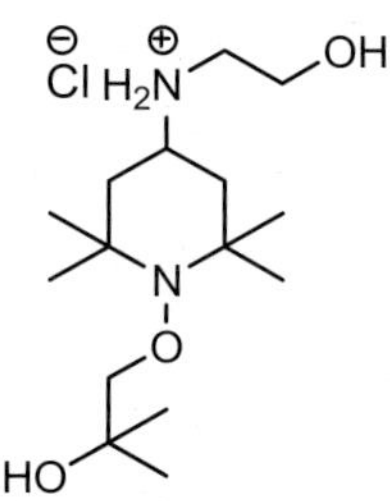

(VI)

Additive Preparation

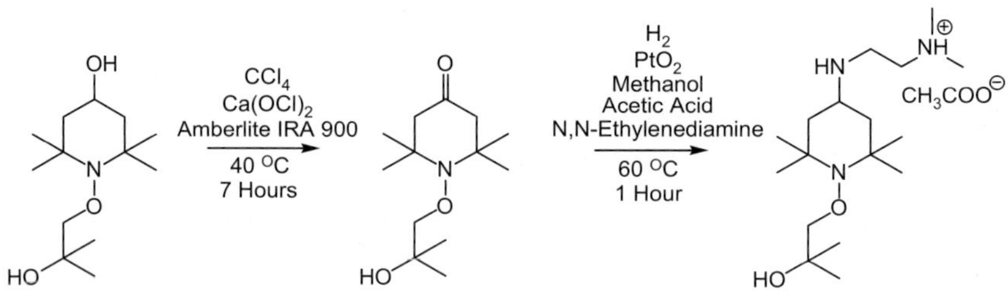

Additive Synthesis

1. *Preparation of 1-(2-hydroxy-2-methylpropoxy)-2,2,6,6-tetramethyl-4-piperidone*

A 1000-ml glass bottom reactor was charged with 42.5 g of 4-hydroxy-1-(2-hydroxy-2-methylpropoxy)-2,2,6,6-tetramethylpiperidine, 100 g of calcium hypochlorite, 25 g of Amberlite® IRA 900 ion exchange resin, and 600 ml of carbon tetrachloride and then heated 40 °C for 7 h. After stirring overnight at an ambient temperature, the slurry was filtered, the filtrate washed twice with water, and then dried over magnesium sulfate and refiltered. The solution was then distilled, and the product was isolated in an 88% yield as a white solid with a melting point of 58–61 °C.

2. *Preparation of 1-(2-hydroxy-2-methylpropoxy)-2,2,6,6-tetramethyl-4-(1-amino-3-dimethylammonium) acetate*

A reactor was charged with 5.28 g of the Step 1 product, 2.2 g of *N,N*-ethylenediamine, 0.5 g of platinum dioxide, 75 ml of absolute methanol, and then pressured up to 45 psig with hydrogen gas. The reactor was heated to 60 °C for 1 h and then vented, and the catalyst was removed by filtration. The reaction solvent was removed by distillation, and the concentrate was dissolved in 10 ml of acetonitrile and then titrated with 1.5 g of glacial acetic acid. Excess solvent and acetic acid were removed by distillation, and the residue was dried overnight in a vacuum oven. The product was obtained in a 93% yield as a white solid having a melting point of 72–74 °C.

Testing

Testing entailed preparation of an ink-jet formulation consisting of 2 g of a selected dye dissolved in 20 g of diethylene glycol containing 78 g of deionized water. The mixture was then treated with 0.15 g of a selected experimental ink-jet stabilizing additive dissolved in 2.85 g of ink. The mixture was filtered and then transferred into an Deskjet® 510 printer cartridge. An image was then printed onto plain paper and then dried at 50 °C under vacuum for 2 h. The image was irradiated behind a 5-mm-thick window glass using a Xenon lamp having an intensity of 461 W/m^2 at 43 °C and 50% relative humidity. The color density of each image was measured before and after exposure using a MacBeth® TR 924 densitometer. Test results are provided in Tables 4–6 for cyan, magenta, and yellow dyes, respectively.

Test Results

TABLE 4 **Print density for cyan dye placed in a Deskjet® 510 printer cartridge and then irradiating the printed image using a Xenon lamp. Lower numbers are preferred**

Experimental Stabilizer	Print Density of Cyan Dye After 4 Weeks
None	7.29
Additive VI	5.98
Additive IV	5.85
Additive II	5.29
Additive VI (Twice the usual concentration)	5.11
Additive IV (Twice the usual concentration)	4.95
Additive IV (Twice the usual concentration)	4.34

TABLE 5 Print density for magenta dye placed in a Deskjet® 510 printer cartridge and then irradiating the printed image using a Xenon lamp. Lower numbers are preferred

Experimental Stabilizer	Print Density of Magenta Dye After 4 Weeks
None	20.16
Additive III	16.62
Additive III (Twice the usual concentration)	14.14
Additive IV (Twice the usual concentration)	8.60

TABLE 6 Print density for the yellow dye placed in a Deskjet® 510 printer cartridge and then irradiating the printed image using a Xenon lamp. Lower numbers are preferred

Experimental Stabilizer	Print Density of Yellow Dye After 4 Weeks
None	4.68
Additive VI	3.15
Additive II	1.36

Advantages over Prior Art

There are a very limited number of jet ink additives that are designed to maintain high-density jet ink print quality using inks containing cyan, magenta, or yellow dyes when exposed to a Xenon lamp @ 461 W/m^2. The efficacy of 2,2,6,6-tetramethylpiperidine derivatives when used as ink additives as ink stabilizing agents after exposure to Xenon lamp-induced free radicals is most likely due to the stabilizing effects of the 2,2,6,6-tetramethyl-4-piperidine hindered amine.

Notes

1. Reinehr[1] prepared amino- and hydroxysubstituted triphenyl-s-triazines as jet-ink stabilizers (VII), which protected images from damage by light, oxygen, and heat

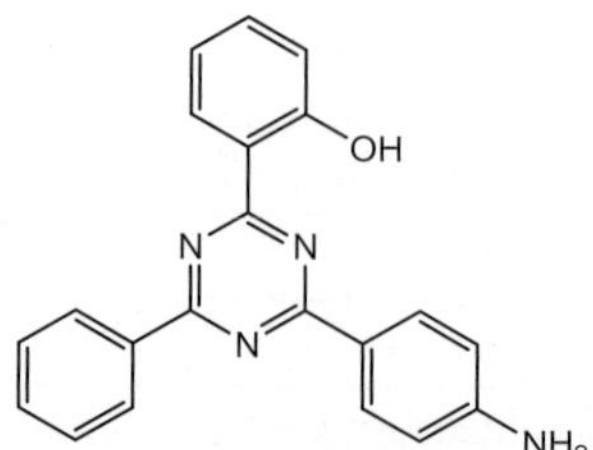

(VII)

2 Haase[2] incorporated hydroxyphenyl benzophenone (VIII), onto select dyes as a method of stabilizing them from the damage of UV radiation.

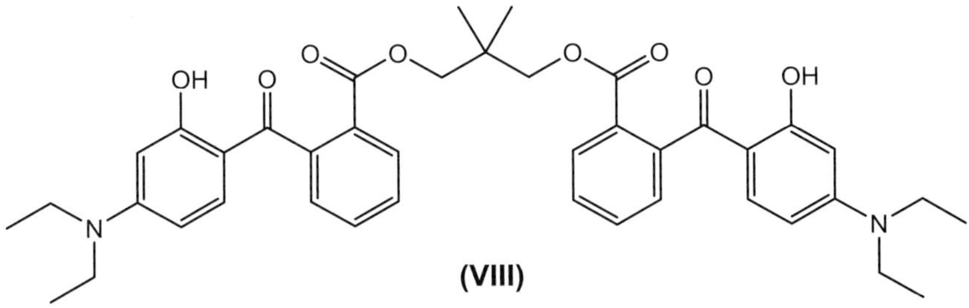

(VIII)

3. Yokota[3] prepared an ultraviolet curable resin composition consisting of the reaction product of Cyracure® UVR-6110 (IX) and a polyurethane prepolymer (X) that had excellent jet-ink coating properties but modest stabilizing effects.

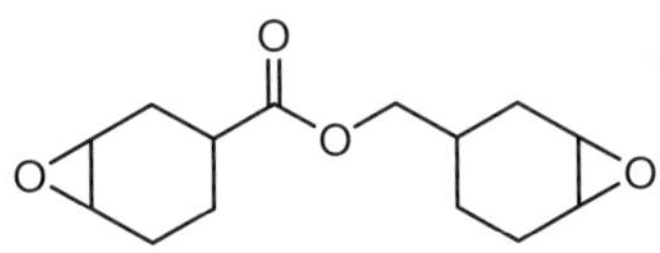

(IX)

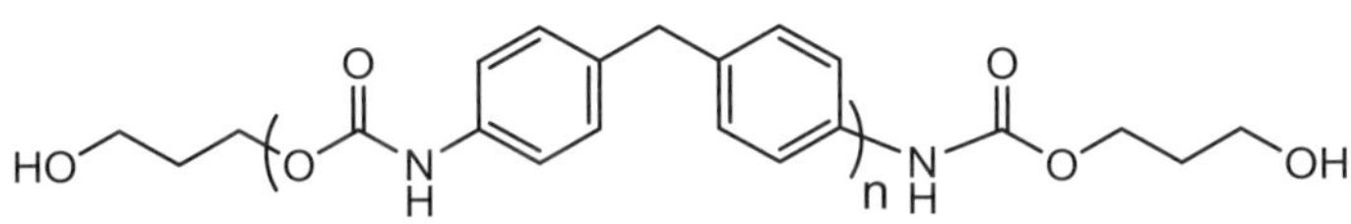

(X)

4. Amasaki[4] prepared heterocyclic ultraviolet absorbent derivatives (XI), which favorably improved the printing quality in jet-ink coatings.

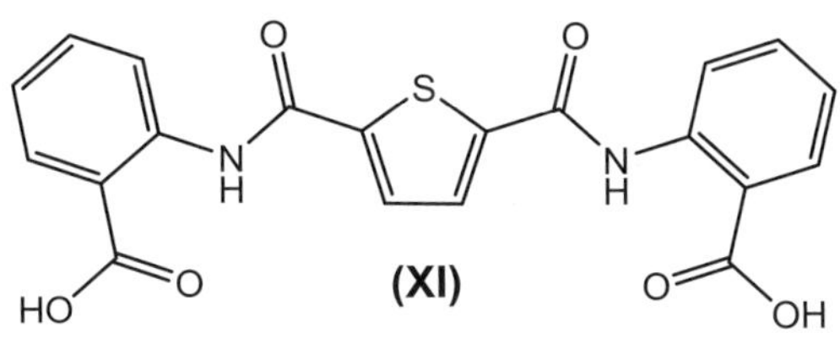

(XI)

REFERENCES

1. Dieter Reinehr et al, 7,476,699 (January 13, 2009)
   Ciba Specialty Chemicals Corp. (Tarrytown, NY)
2. Jurg Haase et al, 7,374,749 (May 20, 2008)
   Ciba Specialty Chemicals Corp. (Tarrytown, NY)
3. Kouichi Yokota et al, 7,977,401 (July 12, 2011)
   DIC Corporation (Tokyo, JP)
4. Ichiro Amasaki et al, 7,897,761 (March 1, 2011)
   Fujifilm Corporation (Minato-Ku, Tokyo, JP)

Chapter 22

Laundry

A. FABRIC WHITENERS

Metal Complexes As Hydrogen Peroxide Decomposition Agents

Title:	Use of metal complex compounds as oxidation catalysts
US Patent:	8,044,013 (October 25, 2011)
Author:	Gunther Schlingloff et al.
Assignee:	BASF SE (Ludwigshafen, DE)

Significance: A series of hydrogen peroxide decomposition catalysts were prepared by complexing manganese (II) dichloride with 4-substituted-[2,2′;6,6′]-terpyridines. When these agents were added to liquid detergents or laundry formulations containing hydrogen peroxide, fabric whiteness was notably improved when laundering at 40 °C.

Additive Names

1′H-[2,2′;6′,2″]Terpyridin-4′-one manganese (II) dichloride (I)
[2,2′;6′,2″]Terpyridin-4′-yl-hydrazine manganese (II) dichloride (II)
2-(Methyl-[2,2′;6′,2″]terpyridin-4′-yl-amino)-ethanol manganese (II) dichloride (III)
4′-Pyrrolidin-1-yl-[2,2′;6′,2″]terpyridine manganese (II) dichloride (IV)
2-[(2-Hydroxyethyl)-[2,2′;6′,2″]terpyridin-4′-yl-amino]-ethanol manganese (II) dichloride (V)
4′-(4-Methyl-piperazin-1-yl)-[2,2′;6′,2″]terpyridine manganese (II) dichloride (VI)
1,1-Dimethyl-4-[2,2′;6′,2″]terpyridin-4′-yl-piperazin-1-ium iodide manganese (II) dichloride (VII)
4′-Piperidin-1-yl-[2,2′;6′,2″]terpyridine manganese (II) dichloride (VIII)
4′-Morpholin-4-yl-[2,2′;6′,2″]terpyridine manganese(II) dichloride (IX)

Next Generation of International Chemical Additives. DOI: http://dx.doi.org/10.1016/B978-0-444-53788-1.00022-X

Safety

2,2′:6′,2″-Terpyridine-4′-substituted ligands are irritants to the eyes, respiratory system, and the skin and are considered harmful if swallowed. Risk of serious damage to the eyes has been reported using these reagents. Suitable protective clothing, gloves, and eye/face protection should be worn when formulating. If contact occurs with the eyes, rinse immediately with plenty of water and seek medical advice. Combustion of manganese complexes releases manganese, which is moderately toxic.

Additive Structures

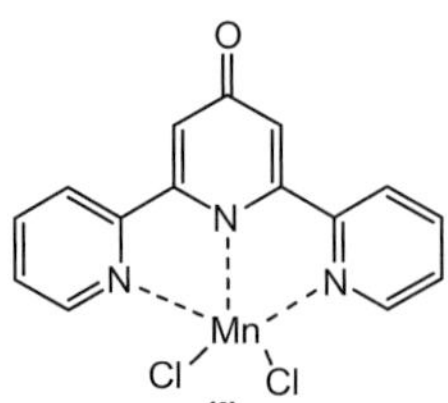
(I)

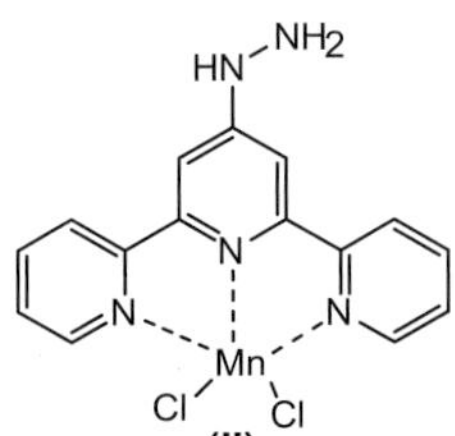
(II)

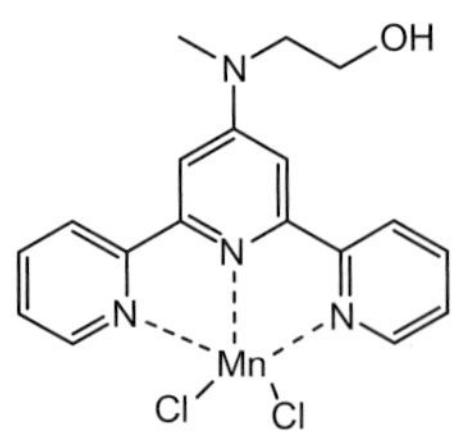
(III)

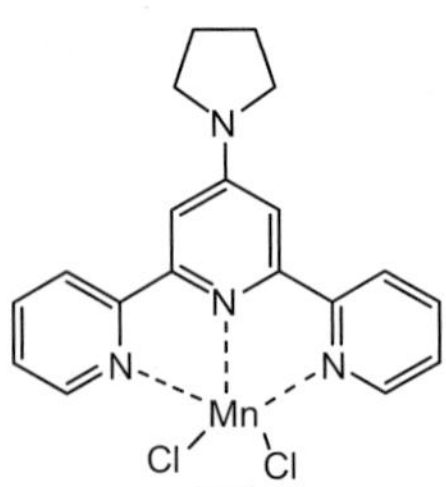
(IV)

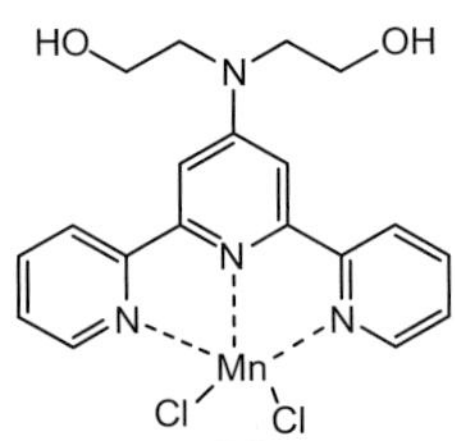
(V)

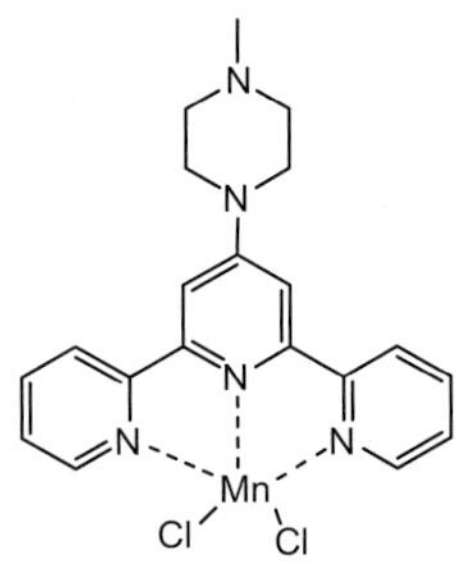
(VI)

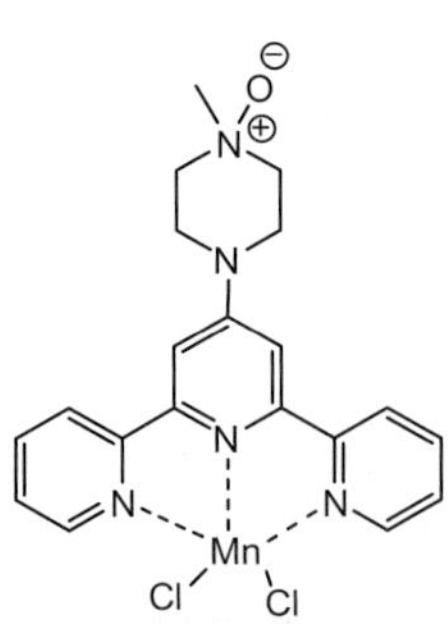
(VII)

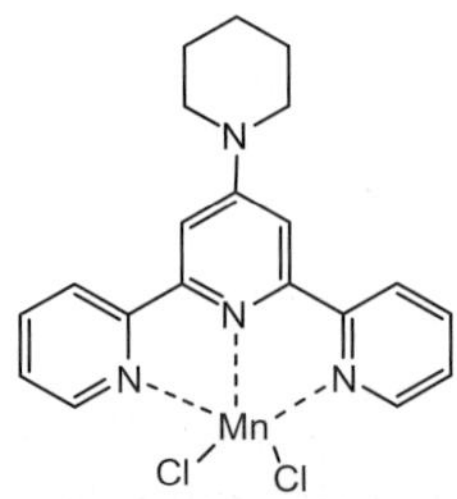
(VIII)

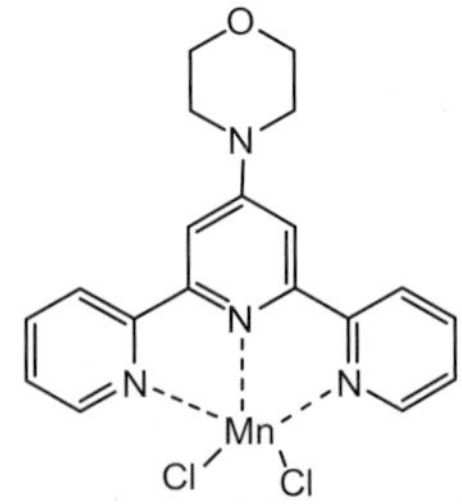
(IX)

Additive Preparation

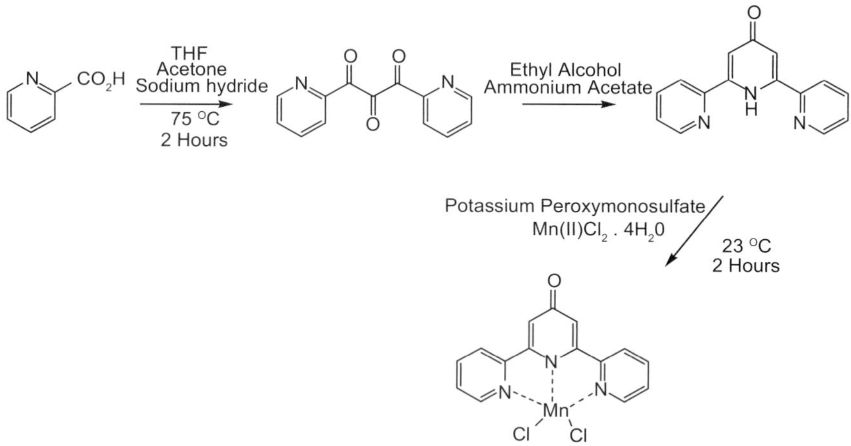

Additive Synthesis

1. *Preparation of 1,5-di-pyridin-2-yl-pentane-1,3,5-trione*

A solution of 22.7 g of pyridine-2-carboxylic acid ethyl ester and 3.6 ml of dry acetone was treated over a of 4-hour period with a suspension of 100 ml of tetrahydrofuran containing 6 g of sodium hydride and then refluxed for 2 h The mixture was then concentrated and precipitated in 200 ml of ice water containing 50% glacial acetic acid. The mixture was then filtered, and the product was isolated as a yellow solid.

IR (cm^{-1}): 2953 (s); 2923 (vs); 2854 (m); 1605 (m); 1560 (s); 1447 (w); 1433 (w); 1374 (m); 1280 (w); 786 (w).

2. *Preparation of 1′H-[2,2′;6′,2″]terpyridin-4′-one*

A mixture of 10 g of the Step 1 product and 20 g of ammonium acetate was refluxed in 250 ml of ethanol for 8 h and then concentrated to half volume. The mixture was then cooled to an ambient temperature, filtered, and the product was isolated as a white solid.

^{1}H-NMR (360 MHz, d_6-DMSO, δ): 7.40–7.50 (qm, 2H); 7.87 (s, 2H); 7.92–8.0 (tm, 2H); 8.57 (d, 2H, 7.7 Hz); 8.68 (d, 2H, J = 4.5 Hz), 10.9 (s, 1H)

3. *Preparation of 1′H-[2,2′;6′,2″]terpyridin-4′-one manganese (II) dichloride*

A solution consisting of 1.78 g of the Step 2 product and 1.75 g of manganese (II) chloride tetrahydrate dissolved in 35 ml of water was treated with the dropwise addition with 3.28 g of potassium peroxomonosulfate dissolved in 20 ml of water. The mixture was then stirred for 2 h at an ambient temperature and then filtered and washed with 25 ml of water. The material was dried for 12 h at 50 °C under high vacuum, and 2.05 g of an olive-green product was isolated.

IR (cm^{-1}): 3068 (m), 1613 (m), 1602 (m), 1587 (s), 1480 (m), 1099 (vs), 1053 (w), 1028 (s), 1011 (s), 788 (m)

Additive Testing

1. Bleaching Action in Washing Agents

A white tea-stained cotton fabric was treated with a mixture consisting of 80 ml of a commercial liquid washing detergent containing a methanol solution of a selected experimental agent, 50 μmol/l of lithium hydroxide, and 8.6 mmol/l of hydrogen peroxide and then washed in a steel beaker for 30 min at 40 °C The bleaching effects of the experimental agents were then evaluated by measuring the difference in lightness according to the CIE protocol. Stain removal was determined spectrophotometrically in comparison with values obtained without the addition of catalyst as a reference. Test results are provided in Table 1.

2. Cleaning Performance on Soiled Surfaces

A tea-stained cup containing 100 ml of a buffered solution consisting of 10 mM sodium carbonate having a pH = 10.0 and containing 1.1 mM of hydrogen peroxide with 7.3 μM of a selected experimental lightening agent was washed at 53 °C for 12 min and maintained at the final temperature for 23 min The fabric was then rinsed with water, air dried, and evaluated visually on a scale of 1 (no cleaning) to 10 (clean). A catalyst-free analog was also used as reference. Results are provided in Table 2.

Test Results

1. Bleaching action in washing agents

TABLE 1 Effectiveness of selected experimental agents in lightening tea-stained cotton fabrics when laundered in detergent containing hydrogen peroxide and an experimental hydrogen peroxide decomposition activator

Additive	Fabric Lightness Increase	Difference in Lightness from Unadditized With Fabric Lightness
(I)	5.0	4.4
(II)	5.0	5.5
(III)	5.2	5.3
(IV)	5.8	3.0
(V)	5.6	5.3
(VI)	5.0	6.2
(VII)	4.5	4.3
(VIII)	3.8	–

2. Cleaning performance on soiled surfaces

TABLE 2 Cleaning properties of a detergent containing hydrogen peroxide additized with 7.3 μM of a selected manganese complex peroxide activator

Additive	Visual Rating
(I)	5.0
(IV)	7.2
(V)	6.9
(IX)	5.2
Unadditized	3.8

Advantages over Prior Art

A peroxide activator designed for use in laundering formulations was previously prepared and consisted of manganous (II) 1,4,7-trimethyl-1,4,7-triazacyclononane. Although this agent was effective as a peroxide activator at low laundering temperatures, it caused considerable damage to both the fabric dye and fibers. The [2,2′;6,6′] Terpyridines of the current invention are effective as low temperature peroxide activators, are extremely versatile, and do not damage fabric dyes or fibers.

Notes

1. Murrer[1] used tris(2,2′-bipyridyl-4,4′-dicarboxylate)ruthenium (II) dichloride (X) as a sensitizer and nanocrystalline titanium dioxide sol to remove 4-chlorophenol and other halogenated pollutants on titanium dioxide surfaces.

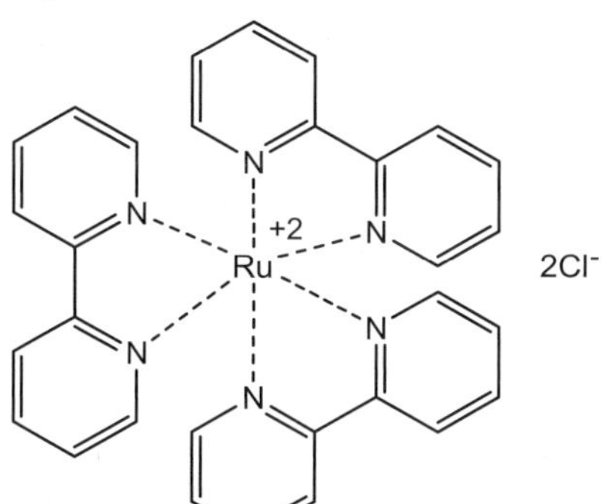

(X)

2. Schlingloff[2] prepared the peroxide activator, 4,4″-bis(4,4-dimethylpiperazinium-1-yl)-1′H-[2,2′;6′,2″]terpyridin-4′-one manganese(II) dichloride (XI), which was effective as a peroxide decomposition agent while having limited dye transfer properties.

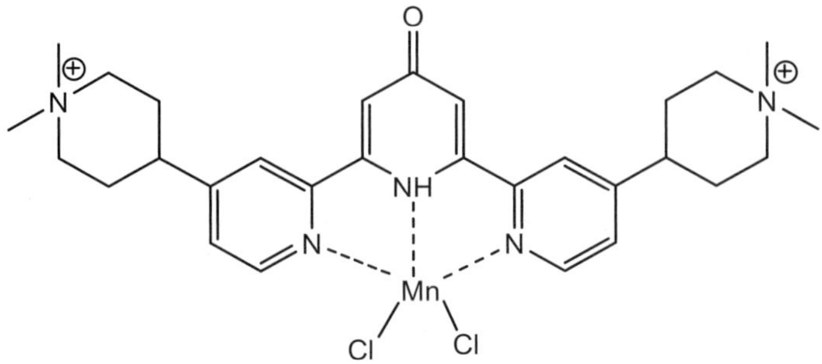

(XI)

3. Ford[3] used azo initiators (XII) and (XIII) to improve the bleaching performance and stain-removal properties in fabrics.

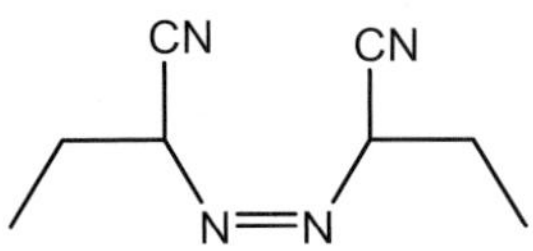

(XII)

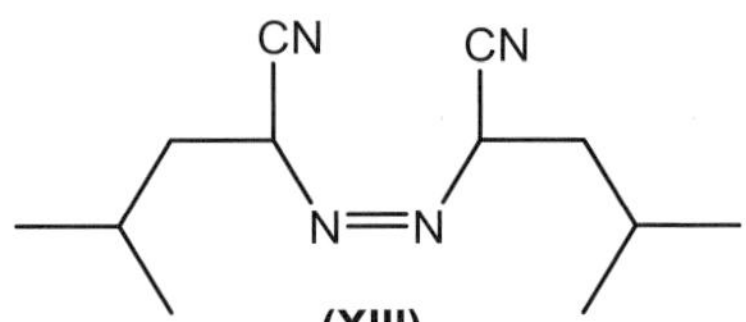

(XIII)

4. de Almeida[4] activated the decomposition of peroxides by adding both 1,4,7-trimethyl-1,4,7-triazacyclononane (XIV) and manganese(II) oxide and then laundering soiled fabrics at pH 10.9.

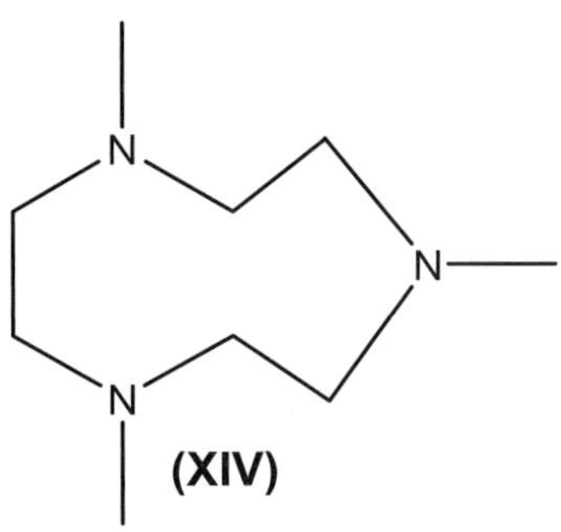

(XIV)

REFERENCES

1. Barry Anthony Murrer et al, US. Patent 7,351,294 (April 1, 2008)
Reckitt Benckiser (UK) Limited (Berkshire, GB)
2. Gunther Schlingloff et al, US. Patent 7,494,964 (February 24, 2009)
Ciba Specialty Chemicals Corp. (Tarrytown, NY)
3. Francis Cornelio Ford et al, US. Patent 7,713,920 (May 11, 2010
The Procter & Gamble Company (Cincinnati, OH)
4. Joaquim Manuel Henriques de Almeida et al, US. Patent 7,976,582 (July 12, 2011)
Conopco, Inc. (Englewood Cliffs, NJ)

B. NON-RINSE FABRIC SOFTENER

Fatty Acids Ester Amines

Title:	Nonrinse fabric softener
US Patent:	7,951,769 (May 31, 2011)
Author:	Maria Jose Bermejo Oses et al.
Assignee:	Kao Corporation, S.A. (Barcelona, ES)

Significance: A mixture containing quarternary tallow triethanolamine and esteramine salts was prepared by heating together tallow fatty acid and triethanolamine and then blending with hexadecyltrimethylammonium sulfate. Clothing washed with 1 wt% of this fabric softener formulation had less residual detergent after rinsing and improved and enhanced softness.

Additives (Note 1) Tallow triethanolamine esteramine mixture (I)
Hexadecyltrimethylammonium sulfate (II)

Safety Both hexyldecyldimethylammonium sulfate and tallow fatty acid/triethanolamine reaction products can cause eye burns and permanent tissue damage. In high concentrations, these agents are moderate respiratory tract irritants.

Additive Blend Components

Component 1

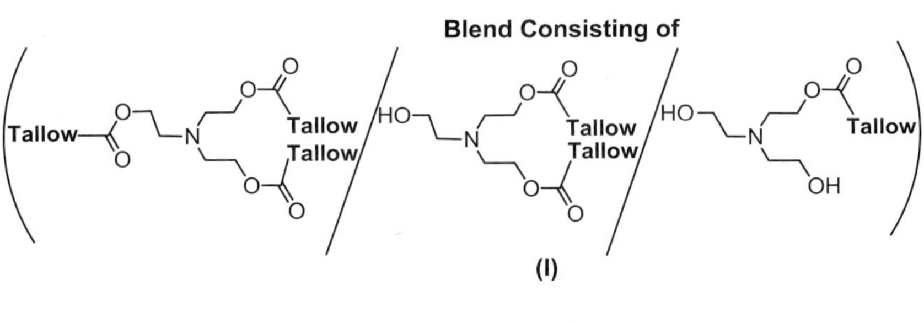

(I)

Component 2

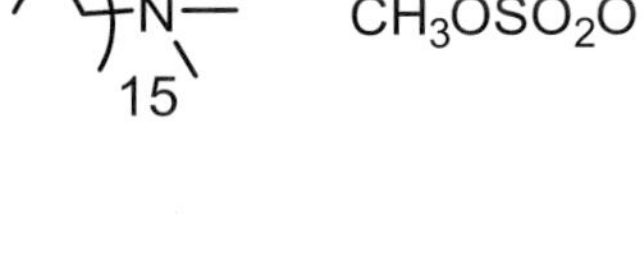

Additive Preparation

Component 1

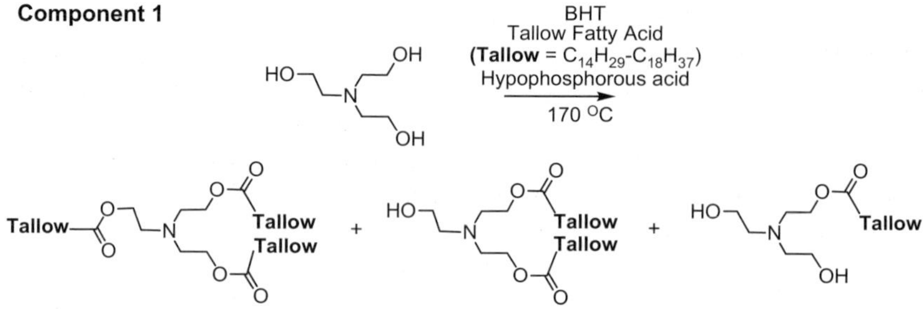

Component 2

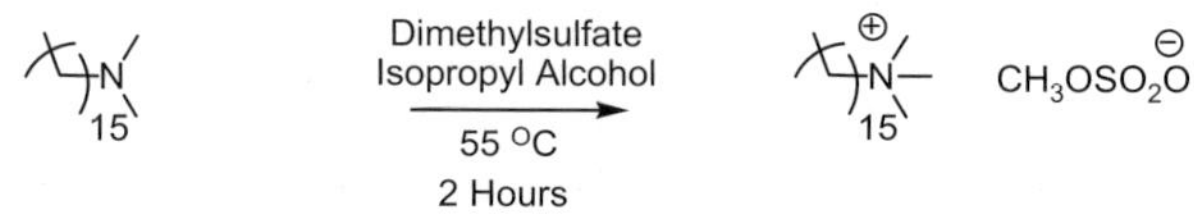

Additive Synthesis

1. *Preparation of tallow triethanolamine esteramine mixture (Component 1)*

A reactor was charged with 786 g of tallow fatty acid, 261 g of triethanolamine, 1 g of 50% hypophosphorous acid, 0.6 g of butylated hydroxytoluene and then heated with stirring to 170 °C until the acid value of the mixture was <5 mg KOH/g. After cooling, 995 g of product was isolated consisting of a mixture of mono, di, and triesteramines.

2. *Preparation of hexadecyltrimethylammonium sulfate (Component 2)*

A mixture consisting of 52 g of *N,N*-dimethylhexadecane, 143 ml of isopropyl alcohol, and 235 g of dimethylsulfate was heated with stirring to 55 °C for 2 h, and the product was isolated in quantitative yield.

Testing

Component Blending

Testing was conducted using a blend consisting of the Step 1 and Step 2 products. The components were mixed and heated for 1 h at 55 °C to prepare a 0.9 weight ratio of the tallow triethanolamine esteramine mixture and hexadecyltrimethylammonium sulfate, respectively. The fabric additive mixture was then dissolved in water by heating to 45 °C for 1 h. Defoaming silicone agents, perfume, dye, and preservative were then blended into the mixture to achieve a homogeneous composition.

1. White residue evaluation

A mixture consisting of 60 g of the commercial powdered detergent was stirred at an ambient temperature until complete dissolution. Cotton terry towels having a total weight of 300 g were added to the detergent solution and maintained under movement for 1 min The towels were manually squeezed and transferred to a softener rinse bath containing 45 g of the experimental composition dissolved in 3 l of water at an ambient temperature. Towels were maintained in the softening bath for 30 s and then removed and manually squeezed. Two minutes after the towels were removed from the softening bath, they were filtered through a cotton knit black fabric, and the residue of the filtration was visually assessed using a rating scale. Test results are summarized below.

2. Softness evaluation

The White Residue Evaluation procedure was again performed using cotton terry towels having a total weight of 1 kg and then manually washed at 20 °C for 5 min and manually squeezed. The cotton terry towels were then immersed in a softening bath containing 25 g of the experimental softening composition dissolved in 20 l of water for 3 min and then manually squeezed. After drying indoors at 20 °C, the softening effect was assessed by an expert panel of 20 testers by means of a rating scale. Test results are summarized below.

Test Results

1. White residue evaluation

Experimental results indicated that cotton terry towels rinsed with the experimental fabric softener composition consistently had greater reduction of white residues or undesirable flocs without affecting the softness.

2. Softness evaluation

Dried terry cotton terry towels rinsed with the fabric softener composition consistently received excellent softness ratings by the sensory test panelists.

Advantages over Prior Art

Fabric softeners typically consist of either one esteramine salt or a mixture of two esteramine salts derived from different substrates. The unusual synergism associated with the combination of a fatty acid triethanolamine esteramine mixture and quarternary fatty acid triethanolamine esteramine salts is unreported. The high adsorption properties of this novel mixture are also reflected in low residue detergent and enhancing softness in fabrics.

Notes

1. A likely third unreported component present in the invention is quarternary ester salt mixtures (III) of Component 1 generated in situ during the blending of Components 1 and 2.

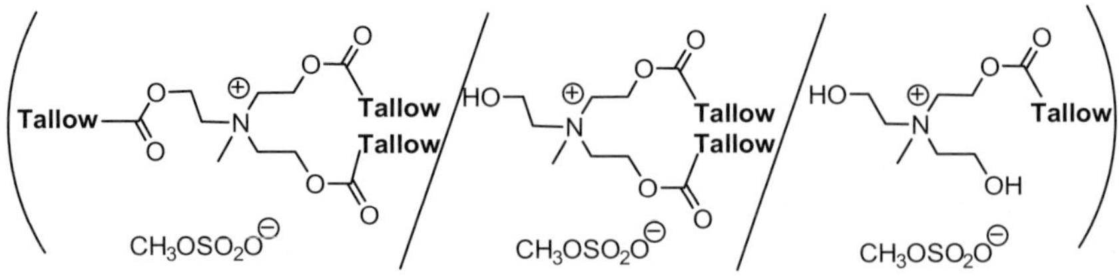

(III)

2. Declercq[1] prepared fabric compositions having reduced detergent residue by sonicating cationic fabric softeners at ≤170 nm.
3. Ferenz[2] prepared polysiloxanes containing quaternary ammonium groups (IV), which were effective as cationic fabric softeners that imparted softness to the dried fabric.

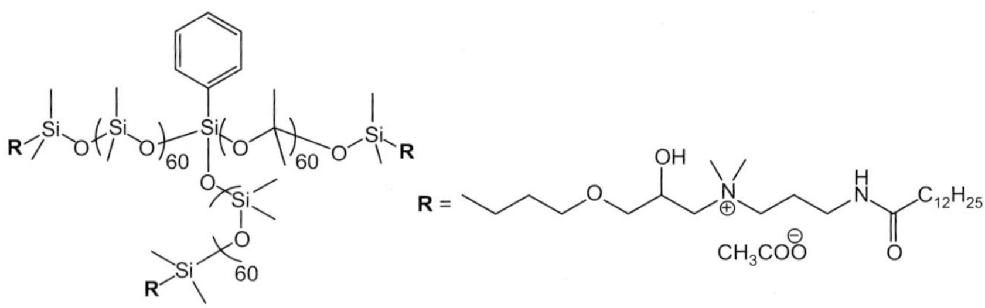

(IV)

4. Cowman[3] ethoxylated fatty acid esters containing triethanolamine (V), which were found to be effective as softening agents in laundering compositions.

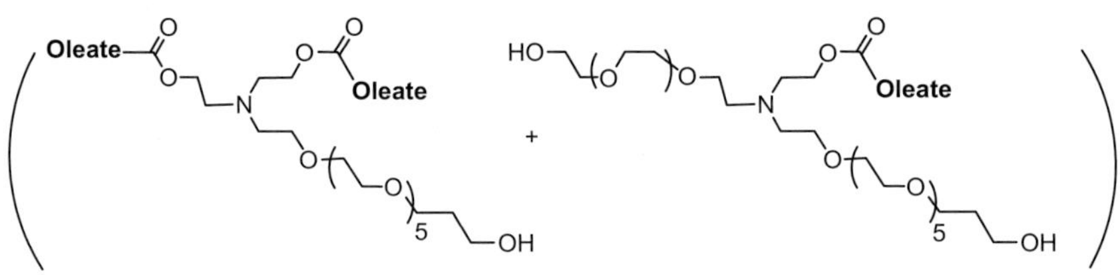

(V)

5. Harada[4] developed a method for quantifying the amount of residual detergent on fabrics by treating the residue with a solution of bis(2,4,5-trichloro-6-carbobutoxyphenyl oxalate) (VI) or bis(2,4,5-trichloro-6-carbopentoxyphenyl oxalate) and hydrogen peroxide and then measuring the fluorescence.

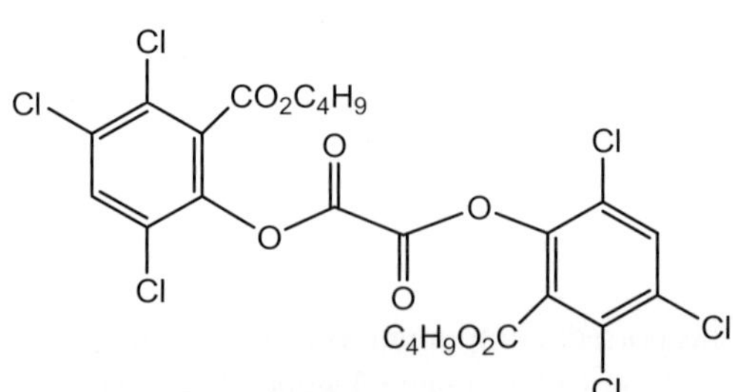

(VI)

REFERENCES

1. Marc Johan Declercq et al, US Patent 7,833,961 (November 16, 2010) **The Procter & Gamble Company (Cincinnati, OH)**
2. Michael Ferenz et al, US Patent 7,834,122 (November 16, 2010) **Goldschmidt GmbH (DE)**
3. John Stuart Cowman et al, US Patent Application 20070265469 (November 15, 2007) **Clariant Corporation (Charlotte, NC)**
4. Shiro Harada, US Patent 7,955,860 (June 7, 2011) **Lumica Corporation (Koga, JP)**

Chapter 23

Medical

A. CHAIN EXTENDERS FOR TISSUE ADHESIVES

Eight-arm Poly(ethylene oxide) Amides and *N*-Hydroxylsuccinimides

Title:	Tissue adhesives with modified elasticity
US Patent:	7,960,498 (June 14, 2011)
Author:	Henry Keith Chenault et al.
Assignee:	Actamax Surgical Materials, LLC (Berkeley, CA)

Significance: Tissue adhesives were prepared by condensing amine-terminated eight-arm poly(ethylene oxide) with either a single or multiple arm poly(ethylene oxide) carboxylic acid or *N*-hydroxylsuccinimide. These materials were made for biocompatibility and elastic adhesion by extruding with oxidized dextran at 120 °C. Elongation at break for selected extrudates was as high as 152%.

Additives

Eight-arm poly(ethylene oxide) amide (I)
Eight-arm poly(ethylene oxide) *N*-hydroxylsuccinimde (II)

Safety

Experimental linkers are biocompatible and biodegradable and are considered to be safe and nontoxic when used as implants or adhesives.

Next Generation of International Chemical Additives. DOI: http://dx.doi.org/10.1016/B978-0-444-53788-1.00023-1

Additive Structures

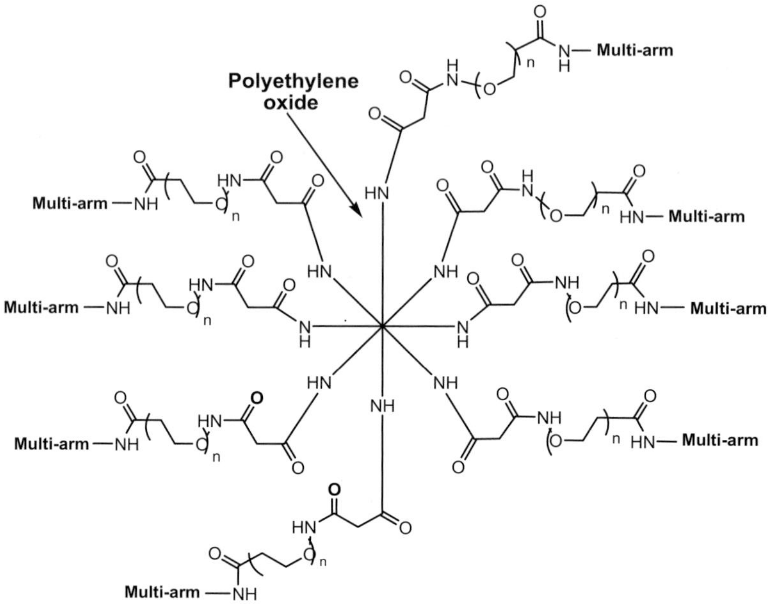

(I)

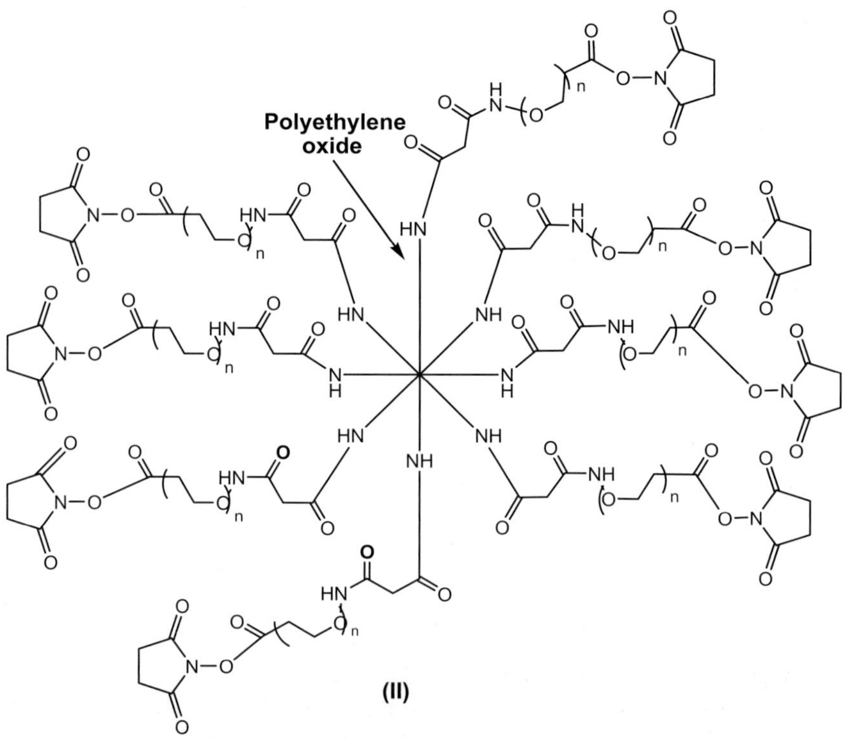

(II)

Additive Preparation

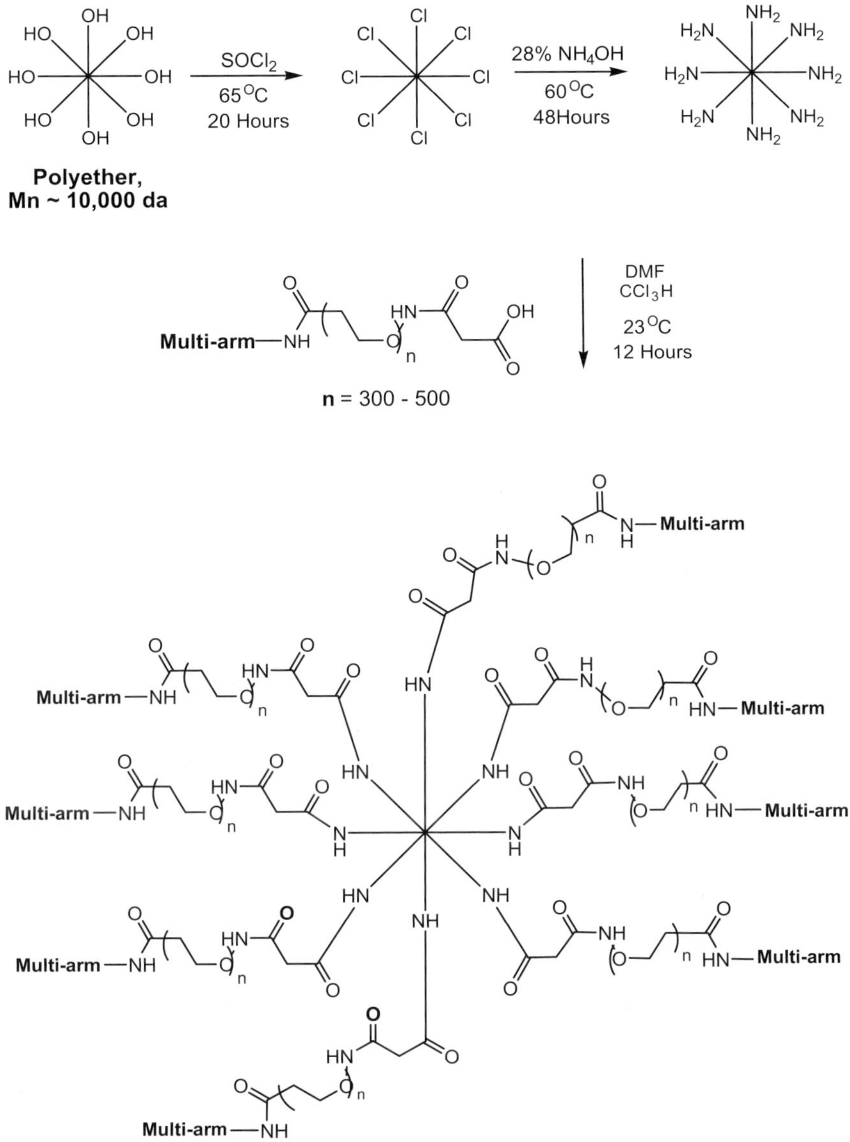

Additive Synthesis

1. Preparation of 8-arm polyether chloride

A reactor charged with 100 g of an 8-arm polyethylene oxide octadiol having an Mw of approximately 10,000 Da was dried by heating with stirring to 85 °C under a vacuum of 0.06 torr for 4–16 h. The dried material was then cooled to an ambient temperature and treated with 35 ml of thionyl chloride and the mixture was heated to 85 °C for 20 h. Excess thionyl chloride was then removed by vacuum distillation and the 8-arm polyether chloride isolated.

^{1}H-NMR (500 MHz, d_6-DMSO) δ 3.71-3.69 (m, 16H), 3.67-3.65 (m, 16H), 3.50 (s, ~ 800H).

2. *Preparation of 8-arm polyether amine*

A stainless steel vessel was charged with 100.9 g of the Step 1 product dissolved in 640 ml of 12 M ammonium hydroxide and then heated in a sealed vessel to 60 °C for 48 h generating 40 psig of pressure. The solution was then cooled to an ambient temperature and sparged for 2 h with dry nitrogen to drive off approximately 50–70 g of excess ammonium hydroxide. The solution was then passed through a column of Purolite® A-860 anion exchange resin, and the liquid was collected and concentrated under reduced pressure to about 200 g. The liquid was then frozen and lyophilized, and the product was isolated.

^{1}H-NMR (500 MHz, d_6-DMSO) δ 3.50 (m, ~800H), 3.34 (t, J = 5.8 Hz, 16H), 2.62 (t, J = 5.8 Hz, 16H).

3. *Preparation of polyethylene oxide N-hydroxylsuccinimide*

In a typical coupling experiment, 40 g of the Step 2 product was dissolved in 80 ml of DMF and 40 ml of methylene chloride and then treated with 1.22 g of polyethylene oxide *N*-hydroxylsuccinimide dissolved in 40 ml of methylene chloride. The mixture was then stirred overnight at an ambient temperature and then filtered, diluted with 300 ml of chloroform, and then washed thrice with 80 ml of saturated potassium carbonate. The combined aqueous washes were backextracted with 40 ml of chloroform, and combined chloroform layers were dried over magnesium sulfate. After filtering and concentrating at 40 °C @ 15 torr, the residue was treated with 35 ml water and reconcentrated at 40 °C @15 torr to remove trace amounts of dichloromethane and chloroform. The clarified solution was frozen and lyophilized, and the product was isolated in a 92% yield.

^{1}H-NMR (500 MHz, d_6-DMSO) δ 3.50 (m, ~ 880H), 3.35 (t, J = 5.8 Hz, 16H), 2.64 (t, J = 5.8 Hz, 16H), 1.03 (d, J = 6.0 Hz, 8.0H).

Testing

1. *In vitro swine uterine horn burst strength*

Double-barreled syringes were loaded in one barrel with a 25 wt% of the reference oxidized dextran dialdehyde and 25 wt% of the Step 3 product. A 0.25-inch circular hole was punched out of the wall of a 6- to 8-cm section of clean, fresh swine uterine horn, and the wound was sealed by applying approximately 0.4 ml of each adhesive. After the adhesive remained on the horn for 5 min, both ends of the uterine horn were clamped shut. A 14-gauge syringe needle connected to a plastic tubing of syringe pump equipped with a pressure meter was inserted through the uterine horn wall of the intestinal lumen. The section of the uterine horn was then submerged in a clear dish of water, and air was pumped using a syringe pump into the intestinal section at 11 ml/h and the pressure at which each sealed wound leaked was recorded. Test results are provided below.

2. *In vitro degradation of adhesive hydrogels*

Double-barreled syringes were loaded in one barrel with a 25 wt% of the reference oxidized dextran dialdehyde and 25 wt% of the Step 3 product. Adhesive mixtures were then extruded from each double-barreled syringe into a 10-mm diameter polycarbonate circular container to form a pellet. The pellets were then weighed and then placed individually in jars containing phosphate-buffered saline at pH = 7.4 and incubated at 37 °C on an orbital shaker set at 80 rpm. The pellets were monitored over time for degradation by visual inspection and by weight gain or loss. Test results are provided below.

3. *In Vitro Biocompatibility with NIH3T3 Mouse Fibroblasts*

Testing was done using NIH3T3 mouse fibroblast cell cultures according to ISO 10993-5:1999. The NIH3T3 mouse fibroblast cells were grown in Dulbecco's modified essential medium supplemented with 10% of fetal calf serum. In this test, a double-barreled syringe was loaded in one barrel with a 25 wt% of the reference oxidized dextran dialdehyde and 25 wt% of the Step 3 product. Adhesive hydrogel was then extruded from each double-barreled syringe into the well of a polystyrene culture plate such that about one-third of the well bottom was covered. The well was then sterilized under UV light and seeded with 50,000-100,000 NIH3T3 cells. After 48 h of incubation at 37 °C, the cell cultures were examined for cell rounding, confluent coating of the well bottom, growth up to the edges of the hydrogel, and overgrowth of the hydrogel. Test results are provided below.

Test Results

1. *In Vitro Swine Uterine Horn Burst Strength*

The Step 3 product exhibited a significantly higher intraintestinal burst pressure than dextran dialdehyde by at least 1 psi.

2. *In Vitro Degradation of Adhesive Hydrogels*

After 86 days of testing, oxidized dextran dialdehyde was completely hydrolyzed while >44 wt% of the Step 3 product remained.

3. *In Vitro Biocompatibility with NIH3T3 Mouse Fibroblasts*

In vitro biocompatibility of the Step 3 product on NIH3T3 mouse for cell rounding, confluent coating of the well bottom, growth up to the edges of the hydrogel and overgrowth of the hydrogel was consistently superior to that of oxidized dextran dialdehyde.

Advantages over Prior Art

Tissue adhesives prepared using oxidized dextran address a series of concerns that have limited hydrogel- and collagen-containing additives. First, tissues prepared according to the current invention are extremely flexible and have excellent elongation properties. As a result, polymeric agents are able to form intimate contact with the wound to minimize the risk of infection. Finally, they are nontoxic and are easily removed from the wound surface by wetting the substrate with water.

Notes

1. Other extensions used in the current invention are shown below.

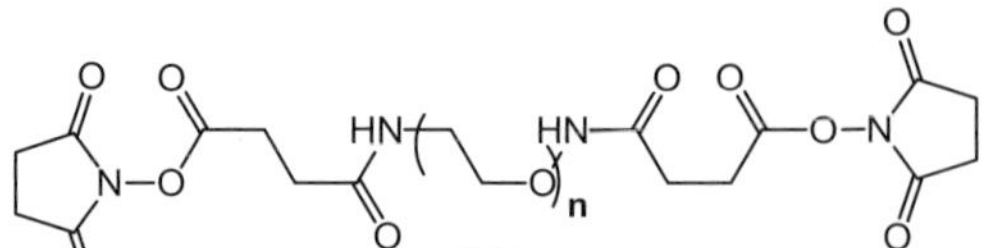

(III)

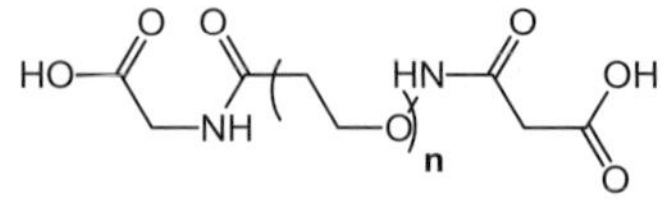

(IV)

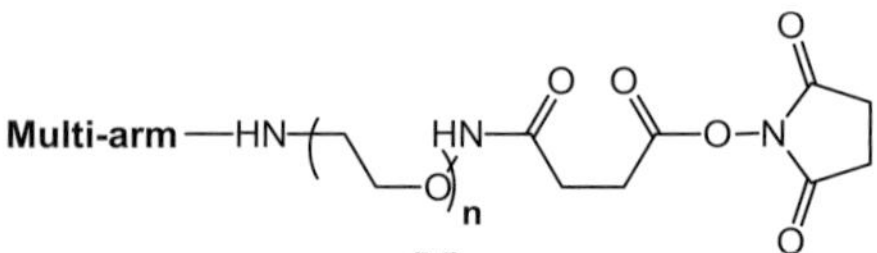

(V)

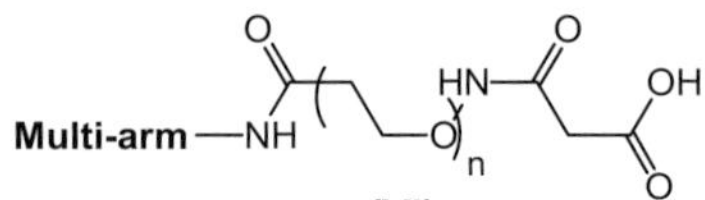

(VI)

n = 300 - 500

2. Greener[1] prepared a topical skin cream formulation consisting of the imide of poly(maleic anhydride-co-polyvinyl ether) (VII) functionalized with desferrioxamine (VIII) as a cream for wound covering which was designed to neutralize the detrimental effects of protease activity without significant dermal penetration.

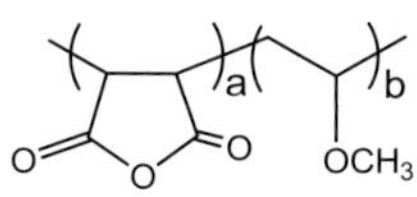

(VII)

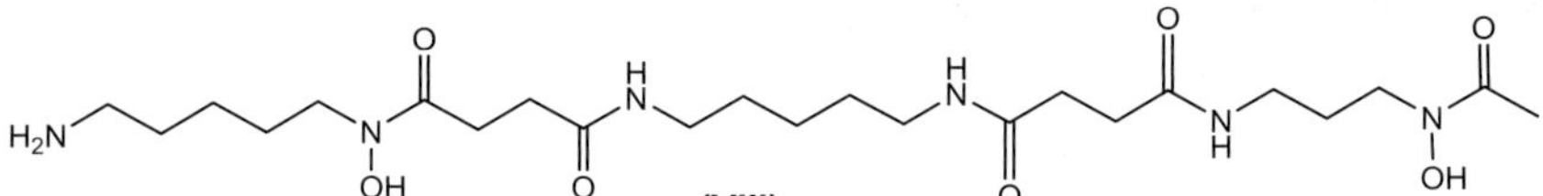

(VIII)

3. Fossum[2] prepared water-swellable crosslinked copolymers consisting or poly(acrylic acid-co-methylene bisacrylamide) (IX), which were effective as bandages to absorb secretions from wounds.

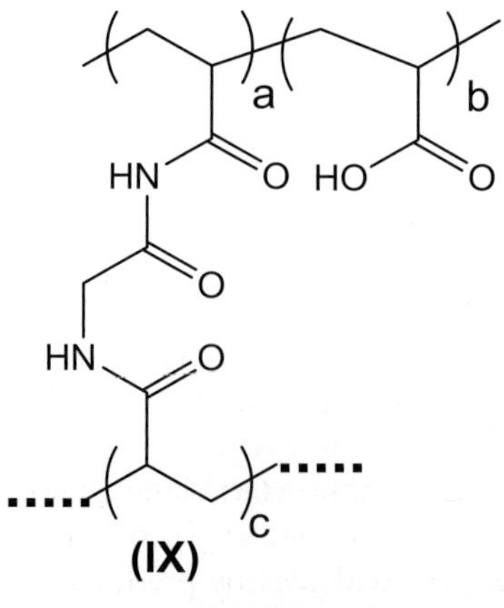

(IX)

4. Hadba[3] prepared a biocompatible polymer (X) suitable as a dressing for moderate to severe skin injuries by treating methyl poly(ethylene glycol) sequentially with hexamethylene diisocyanate, d-sorbitol, and lactic acid.

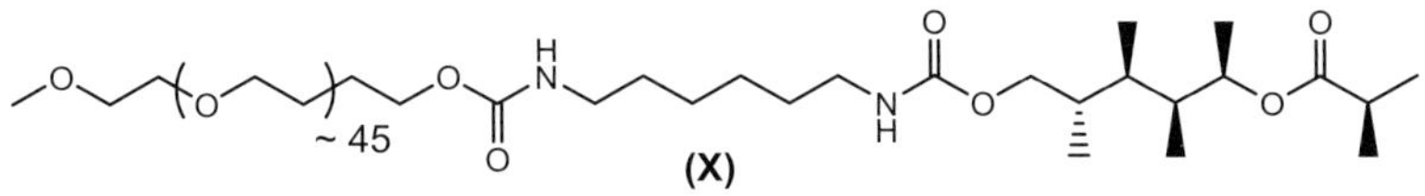

(X)

5. Skin care formulations prepared by Toujo[4] were additized with phosphorylcholine derivatives (XI), which were designed to improve moisture retention and adhesion to skin.

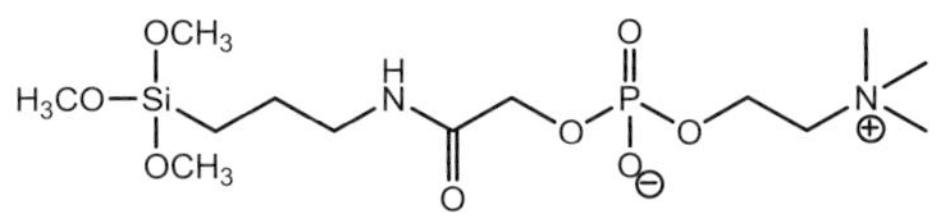

(XI)

REFERENCES

1. Bryan Greener, al, US Patent 7,955,594 (June 7, 2011)
   Smith & Nephew, PLC (London, GB)
2. Renae D. Fossum et al, US Patent 7,947,865 (May 24, 2011)
   The Procter & Gamble Company (Cincinnati, OH)
3. Ahmad R. Hadba et al, US Patent 7,858,078 (December 28, 2010)
   Tyco Healthcare Group LP (North Haven, CT)
4. Yousuke Toujo et al, US Patent 7,906,670 (March 15, 2011)
   Shiseido Co., Ltd. (Tokyo, JP)

B. CHAIN EXTENDERS FOR TISSUE ADHESIVES

Sixteen-arm Poly(ethylene oxide) Crosslinked With Dextran Dialdehyde

Title:	Tissue adhesives with modified elasticity
US Patent:	7,960,498 (June 14, 2011)
Author:	Henry Keith Chenault et al.
Assignee:	Actamax Surgical Materials, LLC (Berkeley, CA)

Significance: Polyamine adhesives were prepared by condensing 8-arm star poly(ethylene glycol) with linear poly(ethylene/propylene glycol) dimesylates and then postreacting the intermediate with dextran dialdehyde. When the product was used as adhesives, they had high burst levels in swine uterine horn and demonstrated excellent biocompatibility with NIH3T3 mouse fibroblasts.

Additive Name Poly(8-arm star poly(ethylene glycol) amine-co-(polyethylene glycol-b-polypropylene glycol-b-polyethylene glycol) dextran dialdehyde condensation product.

Safety Multiarm poly(ethylene glycol) amine crosslinkers and dextrtan dialdehyde condensation products are used as adhesives or implants and are biocompatible and considered innocuous.

Additive Structure

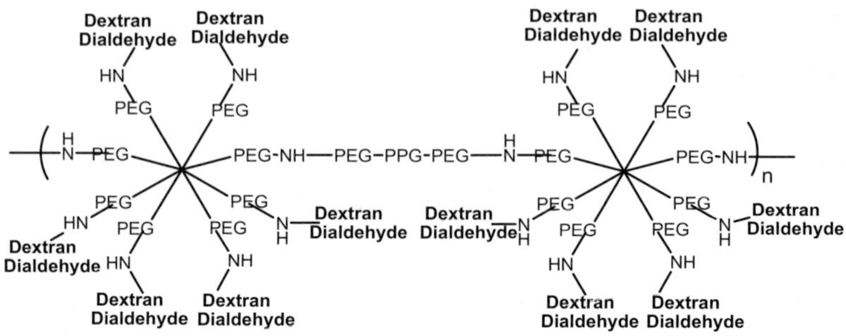

(I)

Additive Preparation

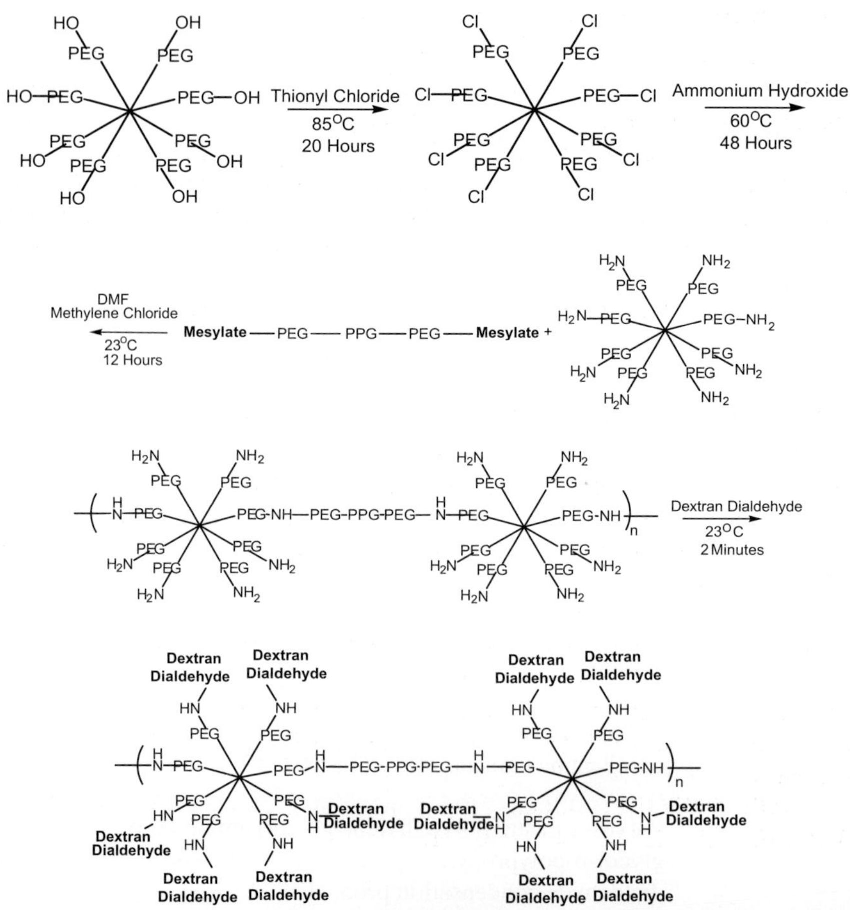

1. Preparation of 8-arm star poly(ethylene glycol) chloride

A dried reactor was charged with 100g of dried hydroxyl-terminated 8-arm star poly(ethylene glycol) having an Mn of approximately 10,000Da and then treated with excess thionyl chloride at an ambient temperature and heated to 85 °C for 20h. Unreacted thionyl chloride was then removed under high vacuum and 8-arm poly(ethylene glycol) chloride isolated.

^{1}H-NMR (500MHz, DMSO-d_6,) δ Cl: 3.71–3.69 (m, 16H), 3.67–3.65 (m, 16H), 3.50 (s ~0.800H).

2. Preparation of 8-arm star poly(ethylene glycol) amine

A metal reactor was charged with 100.9g of the Step 1 product dissolved in 640ml of 12M aqueous ammonium hydroxide and then heated in a sealed vessel to 60 °C for 48h. The solution was then cooled to an ambient temperature and sparged for 2h with dry nitrogen. The resulting product was then passed through an anion exchange resin, lyophilized, and the product was isolated.

^{1}H-NMR (500MHz, DMSO-d_6,) δ 3.50 (m ~t.800H), 3.34 (t, J=5.8Hz, 16H), 2.62 (t, J=5.8Hz, 16H).

3. Preparation of poly(ethylene glycol) dimesylate crosslinker

A sample of 20g of poly(propylene glycol) having an Mn of approximately 4000Da was dissolved in 140ml of methylene chloride and then treated with 2.9ml of triethylamine and 1.6ml of methanesulfonyl chloride. The mixture was then stirred at an ambient temperature, diluted with 200ml of chloroform and then washed thrice with 100ml of 1M potassium dihydrogen phosphate. Thereafter, the reaction product was then washed with 100ml of saturated potassium carbonate, dried with magnesium sulfate, filtered, concentrated, and 9.8g of the product was isolated.

^{1}H-NMR (500MHz, DMSO-d_6,) δ 4.29 (m, 4H), 3.66 (m, 4H), 3.50 (m ~.360H), 3.16 (s, 6H).

4. Preparation of 16-arm star poly(ethylene glycol) using poly(propylene glycol) dimesylate

A reactor was charged with 1.0g of the Step 2 product dissolved in 2ml of DMF containing 1ml of methylene chloride and then treated with 31mg of the Step 3 product dissolved in 1ml of dichloromethane. The mixture was then stirred overnight at an ambient temperature and then filtered, diluted with 10ml of chloroform, and washed thrice with 2ml of saturated potassium carbonate. The combined aqueous washes were backextracted with 1ml of chloroform and combined extracts were dried over magnesium sulfate, filtered, and concentrated. The residue was treated with 2g of water forming a cloudy solution and then reconcentrated to remove dichloromethane and chloroform. The clarified solution was then lyophilized, and 0.98g of the product was isolated.

^{1}H-NMR (500MHz, DMSO-d_6,) δ3.59 (m~960H), 3.46 (t, J=5.2Hz, 16H), 2.81 (t, J=5.2Hz, 14H), 2.76 (t, J=5.5Hz, 2H).

5. *Preparation of the condensation product of 8-arm star poly(ethylene glycol) star crosslinked with 16-arm star poly(ethyleneglycol) crosslinked with dextran dialdehyde*

A 25 wt % solution of dextran dialdehyde having an Mn of approximately 10,000 Da containing 48% aldehyde was loaded into one barrel of a double-barreled syringe containing a static mixing tip and 25 weight % solution of the Step 4 product loaded into the other barrel. Both reagents were simultaneously extruded onto a sheet of glass to form an elongated structure that was about 6 cm in length and 4–5 mm in diameter. The samples were cured at an ambient temperature and were then tested for elasticity after 1–2 min.

Testing

1. *In Vitro Swine Uterine Horn Burst Strength*

Double-barreled syringes were loaded with a 25 wt% of dextran dialdehyde and 25 wt% of the Step 4 product in the other barrel. A 0.25-inch circular hole was punched out of the wall of a 6- to 8-cm section of clean, fresh swine uterine horn, and the wound sealed by applying approximately 0.4 ml from each barrel. After the adhesive remained on the horn for 5 min, both ends uterine horn were clamped shut. A 14-gauge syringe needle connected to a plastic tubing of a syringe pump equipped with a pressure meter was then inserted through the uterine horn wall into the intestinal lumen. The section of the uterine horn was then submerged in a clear dish of water and air pumped by the syringe pump into the intestinal section at 11 ml/h. The pressure at which each sealed wound leaked was then recorded. Test results are given in Table 1.

2. *In Vitro Biocompatibility with NIH3T3 Mouse Fibroblasts*

Testing was done using NIH3T3 mouse fibroblast cell cultures according to ISO 10993-5:1999. The NIH3T3 mouse fibroblast cells were grown in Dulbecco's modified essential medium supplemented with 10% fetal calf serum.

Double-barreled syringes were loaded in with one barrel containing a 25 wt% of dextran dialdehyde and 25 wt% of the Step 4 product in the other barrel. The adhesive hydrogel was then extruded from each barrel of the syringe into the well of a polystyrene culture plate such that about one-third of the well bottom was covered. The well was then sterilized under UV light and seeded with 50,000–100,000 NIH3T3 cells. After 48 h of incubation at 37 °C, the cell cultures were examined for cell rounding, confluent coating of the well bottom, growth up to the edges of the hydrogel, and overgrowth of the hydrogel. Test results are provided in Table 2.

Test Results

1. *In Vitro Swine Uterine Horn Burst Strength*

TABLE 1 Effect of biocompatible adhesives prepared by the condensation of 8-arm star poly(ethylene glycol) with selected bifunctional poly(ethylene/propylene glycol) dimesylates and then postreacted with dextran dialdehyde on the bursting strength of swine uterine horn

Example	Poly(ethylene glycol) Dimesylate Chain Extender	8-Arm Star Polyethylene Polyamine (wt%)	Uterine Horn Burst Pressure (psi)
26	PEG-PPG-PEG (Pluronic® P104)	25	3.7
30	PEG-PPG-PEG (Pluronic® F88)	30	2.9
31	PEG-PPG-PEG (Pluronic® F108)	30	3.0
4	PEG-PPG-PEG (Pluronic® F88)	50	3.6
5	PEG-PPG-PEG (Pluronic® F108)	50	3.5

2. *In Vitro Biocompatibility with NIH3T3 Mouse Fibroblasts*

TABLE 2 In vitro biocompatibility testing of adhesives prepared by the reacting 8-arm star poly(ethylene glycol) amines with selected poly(ethylene/propylene glycol) dimesylates and then postreacted with dextran dialdehyde and exposing to NIH3T3 mouse fibroblasts

	Examples			
	Ex. 26	Ex. 26	Ex. 31	Ex. 31
Chain Extender Multiarm Dimesylate	PEG-PPG-PEG (Pluronic® P104)	PEG-PPG-PEG (Pluronic® P104)	PEG-PPG-PEG (Pluronic® F108)	PEG-PPG-PEG (Pluronic® F108
8-Arm Star PEG or Chain- Extended Polyamine Product (wt %)	30	50	30	50
Cell Rounding	No	No	No	No
Confluent Coating	Yes	Yes	Yes	Yes
Growth to the Edge	Yes	Yes	Yes	Yes

Advances over Prior Art

There are currently two classes of tissue adhesives. The first comprise the traditional adhesives such as polycyanoacrylate and fibrin-based agents. The second comprise the hydrogels particularly those based on collagen and chitosan. Polycyanoacrylates are limited to topical wound closure applications because of the release of toxic degradation products; fibrin-based adhesives have only moderate mechanical strength. Hydrogels either swell or dissolve too quickly and typically lack either the required adhesive or mechanical strength. Multiarm polyether crosslinked star polymers are biocompatible while possessing excellent tissue adhesion, which are mechanically sound.

Notes

1. Kodokian[1] prepared polysaccharide-based polymer tissue adhesives consisting of the reaction product of 3, 4, or 8-arm star poly(ethylene glycol) amines with dextran, chitin, starch, agar, cellulose, and hyaluronic acid dialdehydes having an Mn of approximately 30,000 Da.
2. Sawhney[2] prepared a biocompatible hydrogel effective as a tissue adhesive and tissue sealant consisting of the reaction product of a 4-arm star of succinimidyl glutarate polyethylene glycol) (II) having an Mn of approximately 10,000 Da with a 4-arm star poly(ethylene glycol) amine (III) having an Mn of approximately 2000 Da.

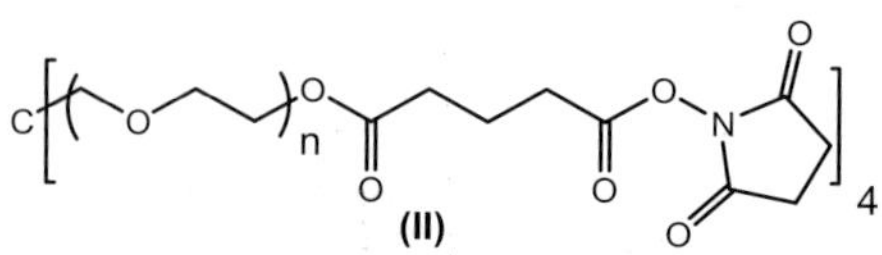
(II)

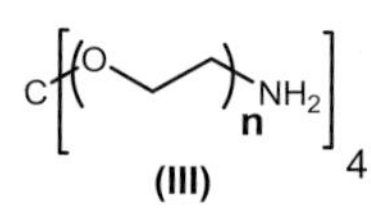
(III)

3. Hadba[3] prepared biocompatible tissue sealant and adhesives consisting of the reaction product of polyethylene oxide di(4,4′-diphenylmethane diisocyanate) crosslinked with 1,2,3-propylenetriamine as illustrated in Eq. 1.

Eq. 1.
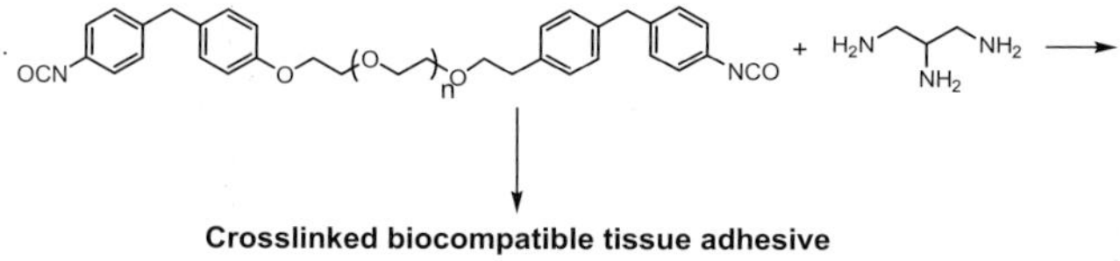

4. Biodegradable surgical compositions effective as bioadhesives (IV) were prepared by Halab[4] by initially condensing hexamethylene diisocyanate with monomethoxypolyethylene glycol and then postreacting with d-sorbitol and l-lactide.

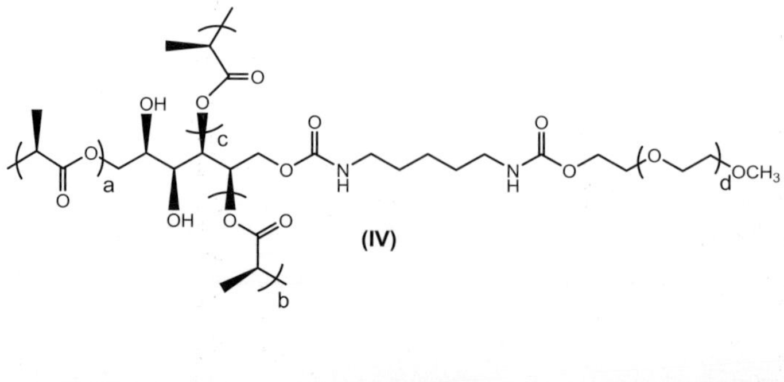
(IV)

REFERENCES

1. George K Kodokian et al, US. Patent Application 20060078536 (April 13, 2006) **E I Du Pont De Nemours and Company**
2. Amarpreet S.Sawhney et al, US. Patent 8,003,705 (August 23, 2011) **Incept LLC (Bedford, MA)**
3. Ahmad R. Hadba et al, US. Patent 7,998,466 (August 16, 2011) **Tyco Healthcare Group LP (North Haven, CT)**
4. Ahmad R. Hadba et al, US. Patent 7,947,263 (May 24, 2011) **Tyco Healthcare Group LP (North Haven, CT)**

C. POLYURETHANE STENT AND SCAFFOLD EXTENDERS

Aliphatic Hydroxy Carboxylic Acids

Title:	Chain extenders
US Patent:	7,964,696 (June 21, 2011)
Author:	Pathiraja A.Gunatillake
Assignee:	Polynovo Biomaterials Pty Limited (Victoria, AU)
Significance:	Biodegradable nonhydrogen bonding ('soft' segment) chain extenders were prepared by reacting a mono or dicarboxylic acid with an aliphatic diol. When these chain extending agents were incorporated into polyether containing medical devices, the molecular weight was reduced by at least 30% after refluxing in water for 24 h.
Additives	Hydroxyacetic acid 3-hydroxypropyl ester (I) 6-Hydroxyhexanoic acid 2-hydroxyethyl ester (II) 6-Hydroxyhexanoic acid 4-hydroxybutyl ester (III) Succinic acid bis-2-hydroxyethyl) ester (IV) trans-Buta-2-enedioic acid bis(2-hydroxyethyl) ester (V)
Safety	Experimental esters and esters contained in block copolymer implants as biodegradable porous scaffolds or stent components and are considered nontoxic.

Additive Structures

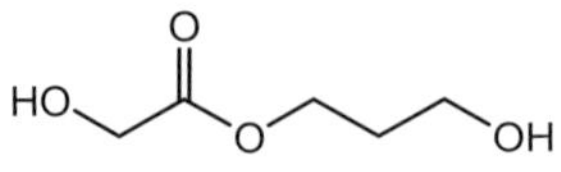

(I)

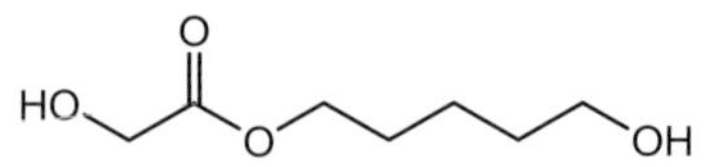

(II)

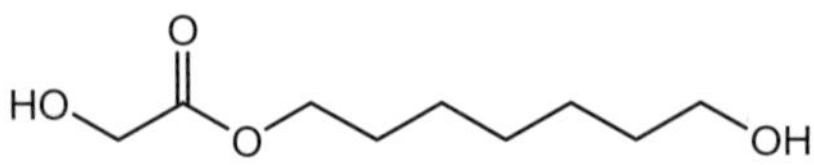

(III)

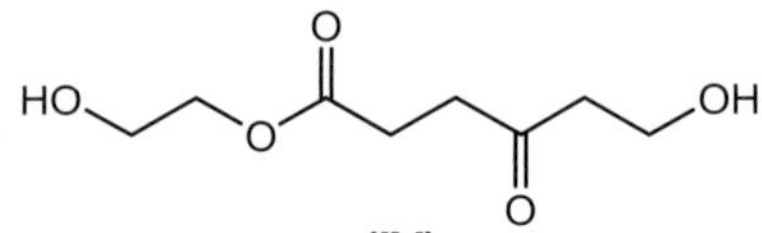

(IV)

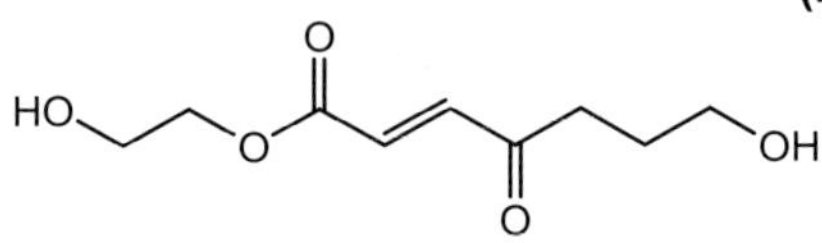

(V)

Additive Preparation

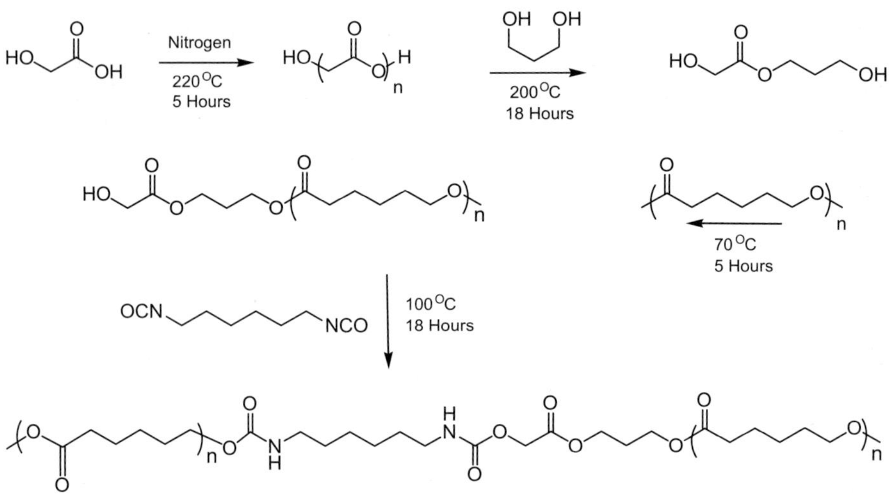

Additive Synthesis

1. Preparation of polyglycolic acid

A reaction vessel containing 56.7 g of glycolic acid was heated to 220 °C for 5 h to dehydrate, and the product was isolated as a white solid polymer.

2. Preparation of hydroxyacetic acid 3-hydroxypropyl ester

A large round-bottom flask charged with 283.6 g of 1,3-propane diol and 43 g of the Step 1 product was heated to 200 °C for 18 h. The transesterification diol was then purified by fractional distillation using a Kugelrohr reactor while heating to 50 °C at 0.01–0.001 torr to remove unreacted starting reagents. The fractionalization temperature was then increased to 70 °C at 0.01–0.001 torr and the product was isolated in a 53% yield.

3. *Preparation of polyurethanes using hydroxyacetic acid 3-hydroxypropyl ester as the chain extenders*

A mixture of 35.000 g of poly(ε-caprolactone) diol, 21.311 g of the Step 2 product, and 0.050 g of stannous octoate wasweighed in a 100 ml predried polypropylene beaker and then covered with aluminum foil and heated to 70 °C in a laboratory oven. The mixture was then treated incrementally with 43.689 g of predried hexamethylene diisocyanate until gelation occurred. Thereafter, the mixture was poured onto a Teflon® coated metal tray to cure at 100 °C for 18 h, and the product was isolated as a colorless solid.

Testing

Degradation testing

Testing was conducted using melt-pressed specimens according to ASTM F 1635. Polymers were melt pressed to 100–200 μm, and strips were then cut to 5 m^3 and 0.45 m^3. The buffer consisted of 0.1 M PBS of pH 7.4 and degradation testing was conducted at 37 °C. The solution/specimen ratio was between 100:1 and 300:1, and 0.1% sodium azide was added as an antimicrobial agent. Finally, all samples were placed in a 50 rpm agitated incubator. Test results are provided in Table 3.

Test Results

TABLE 3 Degradation testing results for biocompatible polyurethanes using experimental chain extenders. All polymeric agents were prepared using agents and procedures described in Step 3

Coupling Agent	Predegradation		Postdegradation	
	Mn (1×10^5)	Mw (1×10^5)	Mn (1×10^5)	Mw (1×10^5)
Additive-1	2.8	4.7	1.9	3.1
Additive-2	1.5	2.7	1.3	2.1
Additive-3	3.4	9.1	1.9	3.8
Additive-4	12.8	44.3	11.8	53.1
Additive-5	1.7	3.0	1.6	2.9

Advantages over Prior Art

The objective of this investigation was to prepare polymeric medical devices with predictable and systematic degradation rates. Esters and urethanes generated using aliphatic chain ester diols would be most susceptible to both hydrolytic and enzymatic degradation. Molecular fragments from this process would either be bioresorbed or discharged as waste. Finally, the low crystallinity of the experimental extenders would not be expected to negatively impact the biocompatibility of the medical device.

Notes

1. Fortune[1] used tetrafunctional polyethylene oxide succinimidyl carbonates (VI), as both chain extenders and crosslinking agents when preparing biocompatible biomaterials.

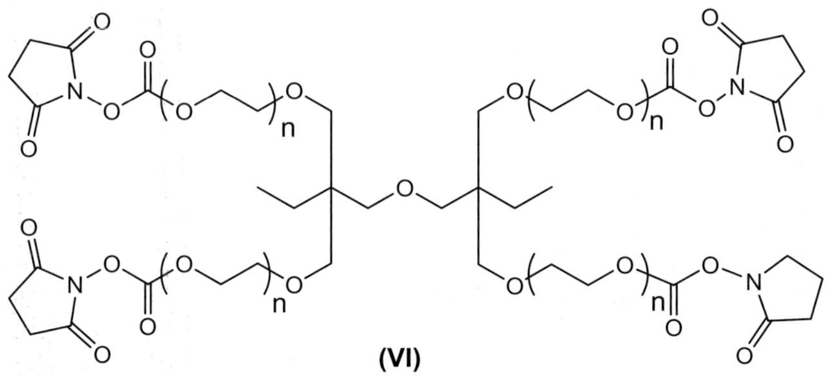

(VI)

2. 3,3′,5,5′-Tetraethyl-4,4′-diaminodiphenylmethane (VII) was prepared by Lee[2] and used as a chain extender in the reaction involving 4,4-methylenediphenyl diisocyanate (VIII) and Jeffamine® T-5000 polyetheramine (IX).

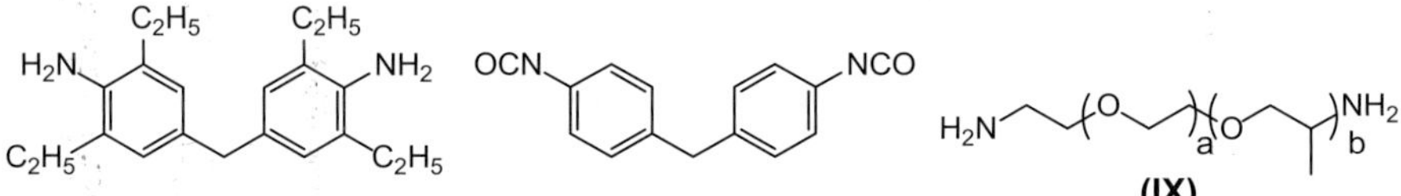

(VII) (VIII) (IX)

3. Polyhydroxydiamines (X) and (XI) were prepared by Swedo[3] and used as chain extenders which exhibited very low or no VOC.

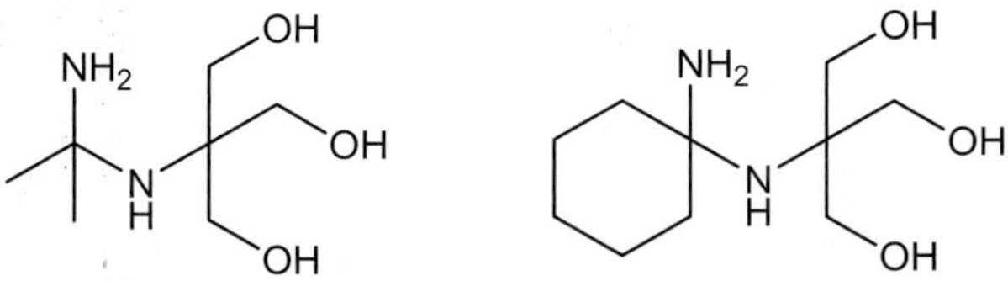

(X) (XI)

4. Awojulu[4] prepared polybutylacrylate end capped with glycidyl methacrylate (XII) for use as a chain extender for processing polylactic acid.

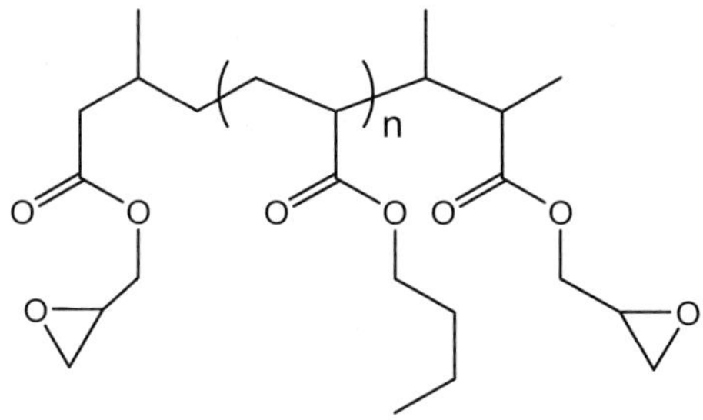

(XII)

REFERENCES

1. D. Fortune et al, US. Patent 7,727,547 (June 1, 2010)
Tissuemed Limited (Leeds, England)

2. J. Lee et al, US. Patent 7,964,695 (June 21, 2011)
Albemarle Corporation (Richmond, Va)

3. R.J. Swedo et al, US. Patent 5,102,972 (April 7, 1992)
Allied-Signal, Inc (Morris Township, NJ)

4. A.E. Awojulu et al, US. Patent 7,855,252 (December 2q1, 2010)
BASF Corporation (Wyandotte, MI)

Chapter 24

Ophthalmology

A. ANTIMICROBIAL AGENT FOR CONTACT LENS

Polyquaternium Polymers

Title:	Polyquaternium-1 synthesis methods and associated formulations
US Patent:	7,999,063 (August 16, 2011)
Author:	Zhi-Jian Yu
Assignee:	Abbott Medical Optics Inc. (Santa Ana, CA)
Significance:	The polyquaternium polymer, poly(*N*,*N*-dimethyl-1-aza-3-pentene)-di-(*N*,*N*-dimethyl-1-aza-6-*N*,*N*,*N*- triethanolamine-5-aza-3-hexene) chloride, was prepared by condensing 1,4-dichloro-2-butene and 1,4-bis-dimethylamine-butene and then postreacting with triethanolamine. When used as a disinfectant in contact eye formulations, optimum antibiotic effect against S. marcescenes, C. albicans, F. salani, S. aureus and P. aeruginosa was observed for polyquaternium polymers having Mns of at least ≥30,000 Da at treatment rates from 0.00001 wt% to 0.01 wt%.

Additive Poly(*N*,*N*-dimethyl-1-aza-3-pentene)-di-(*N*,*N*-dimethyl-1-aza-6-*N*,*N*,*N*-triethanolamine-5-aza-3-hexene) chloride (I)

Safety Although little safety data exist on poly(*N*,*N*-dimethyl-1-aza-3-pentene)-di-(*N*,*N*-dimethyl-1-aza-6-*N*,*N*,*N*-triethanolamine-5-aza-3-hexene) chloride, dimers of 1,4-bis-dimethylamino-2-butene and triethanolamine are suspected of being weakly teratogenic.

Next Generation of International Chemical Additives. DOI: http://dx.doi.org/10.1016/B978-0-444-53788-1.00024-3

Additive Structure

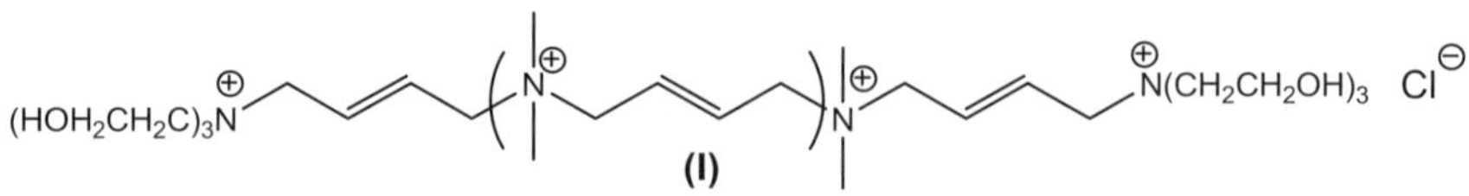

(I)

Additive Preparation

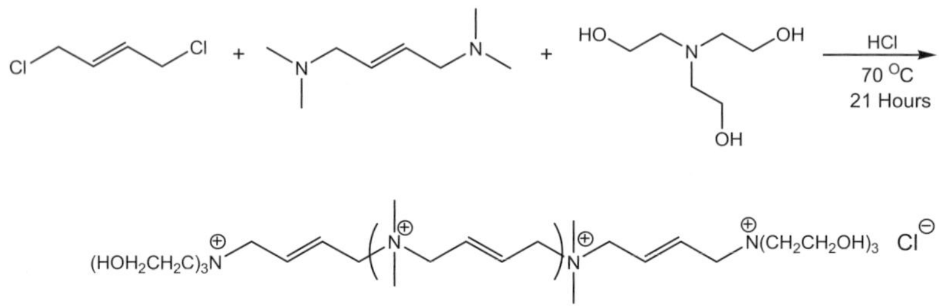

Additive Synthesis

1. *Preparation of poly(N,N-dimethyl-1-aza-3-pentene)-di-(N,N-dimethyl-1-aza-6-N,N,N-triethanolamine-5-aza-3-hexene) chloride*

A 100 ml round-bottom flask was charged with 10.14 g of 1,4-bis-dimethylamino-2-butene, 6.4 g of triethanolamine, 4.92 ml of 6 M HCl, and 18.8 g of water and then submerged into an ice water bath and treated with the dropwise addition of 9.8 g of 1,4-dichloro-2-butene under constant stirring. The ice-bath was removed after all of the 1,4-dichloro-2-butene was added, and the flask was submerged into a 25–40 °C water bath for 20 min. The water bath temperature was then increased to 70 °C, and the mixture was heated for 21 h. Reaction conditions and corresponding number average molecular weights are provided in Table 1.

TABLE 1 Reaction scoping and corresponding number average molecular weights for the preparation of poly(*N*,*N*-dimethyl-1-aza-3-pentene)-di-(*N*,*N*-dimethyl-1-aza-6-*N*,*N*,*N*-triethanolamine-5-aza-3-hexene) chloride

Reaction	Ratio 1,4-dichloro-2-butene/triethanolamine	Ratio 1,4-dichloro-2-butene/triethanol-amine/HCl	Reaction Time (h)	Reaction Temperature (°C)	Mn (k)
1	0.83	1/1.2/0.83	5	70	6.94
2	1.67	1/1.1/0.41	21	70	11.4
3	1.25	1/1.1/0.55	8	75	9.3

TABLE 1 Reaction scoping and corresponding number average molecular weights for the preparation of poly(*N*,*N*-dimethyl-1-aza-3-pentene)-di-(*N*,*N*-dimethyl-1-aza-6-*N*,*N*,*N*-triethanolamine-5-aza-3-hexene) chloride—*Cont'd*

Reaction	Ratio 1,4-dichloro-2-butene/ triethanolamine	Ratio 1,4-dichloro-2-butene/ triethanol-amine/HCl	Reaction Time (h)	Reaction Temperature (°C)	Mn (k)
4	0.83	1/1.2/0.83	8	60	7.7
5	1.11	1/1.15/0.62	6	60	10
6	5.0	1/1.1/1.1	8	75	26

Testing

Antimicrobial Activity

The antimicrobial activity of the experimental agents was assessed against *Serratia marcescens* 13880, *Candida albicans* 10231, *Psedomonas aeruginosa* 9027, *Staphylococcus aureus* 6538, and *Fusarium solani* 36031 using the contact lens formulation provided in Table 2. Antimicrobial test results are provided in Table 3.

TABLE 2 Contact lens formulation used to assess the antimicrobial activity of the experimental additive

Ingredient	Dosage (wt%)
Hydroxypropylmethylcellulose	0.20
Sodium Chloride	0.59
Potassium Chloride	0.14
2-Amino-2-hydroxymethyl-1,3-propanediol HCl	0.055
2-Amino-2-hydroxymethyl-1,3-propanediol	0.021
Taurine	0.05
Poloxamer 237	0.05
Edetate Disodium	0.01
Propylene Glycol	0.50
Purified Water	98.38

Test Results

TABLE 3 Antimicrobial test results suggesting that higher molecular weights of the experimental disinfectants are more effective against *S. marcescens* 13880, *C. albicans* 10231, and *F. solani* 3603

Mn Experimental Additive (Da)	Dosage (wt%)	Dosage (wt%)	Dosage (wt%)	Dosage (wt%)
34,000	0.0007	X	X	X
30,000	X	0.00015	X	X
7000	X	X	0.00015	X
6000	X	X	X	0.0007
	X	X	X	X
Microbial Log Drop After 6 h				
S. marcescens 13880	2.14	3.19	2.56	0.80
C. albicans 10231	2.97	3.46	3.35	2.39
F. solani 36031	3.63	4.17	3.87	1.93

Advantages over Prior Art

Polyquaternium derivatives have previously been prepared and used an antifungal additives, preserving ophthalmic compositions, and as disinfecting contact lens solutions against bacteria, algae, and fungi. A chief disadvantage of these methods is that the incorporation of the termini, triethanolamine has varied from as low as 10% to as high as 45%. The inability to incorporate quantitative amounts of triethanolamine on the polyquaternium termini has resulted in the diminished disinfectant activity of the additive. The current invention has devised a synthetic route where this concern in completely addressed.

Notes

1. In a subsequent investigation by Yu[1], poly(*N,N*-dimethyl-1-aza-3-pentene)-di-(*N,N*-dimethyl-1-aza-6-*N,N,N*-triethanolamine-5-aza-3-hexene) chloride was also found to be effective against S Aureus 6538 and P. aeruginosa 9027 as provided in Table 4.

TABLE 4 Antimicrobial test results suggesting that higher molecular weights of the experimental additive are more effective against *S. aureus* 6538 and *P. aeruginosa* 9027

Mn Experimental Additive (Da)	Dosage (wt%)	Dosage (wt%)	Dosage (wt%)
34,000	0.0007	X	X
30,000	X	0.00015	X
7000	X	X	0.00015
Microbial Log Drop After 6 h			
S. aureus 6538	2.89	2.99	3.322
P. aeruginosa 9027	>5.00	>5.00	>5.00

2. Nemoto[2] was successful in developing an additive package designed to decrease the terpenoid eye irritating content of xanthan gum in polyethylene terephthalate eye drop containers using the following formulation provided in Table 5.

TABLE 5 Four successful additive formulations prepared by Nemoto[2] designed to decrease terpenoid eye irritating content of xanthan gum present in polyethylene terephthalate eye drop containers

Component (g/100 ml)	Ex. 1	Ex. 2	Ex. 3	Ex. 4
Xanthan gum	0.2	0.5	0.7	0.5
Sodium chloride	0.9	0.9	0.9	0.9
l-Menthol	0.01	0.01	0.01	–
dl-Camphor	–	–	–	0.01
Purified water	e.q.	e.q.	e.q.	e.q.

3. Donley[3] developed a method for sterilizing and cleaning contact lenses by washing the lens thoroughly at approximately 50 °C in a steel chamber using 3 to 8 ppm of 3% hydrogen peroxide solution and then thoroughly rinsing the lenses with sterilized water.

4. Stopek[4] prepared antimicrobial furanone/phosphorylcholine copolymers (II), which were used in optical lenses to provide effective levels of antimicrobial activity for an extended period of time.

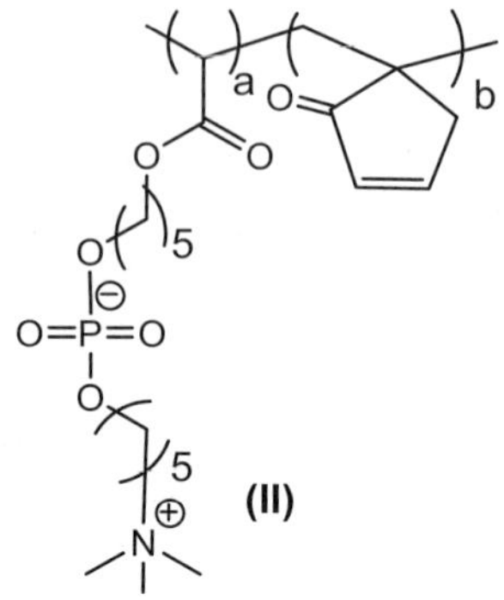

(II)

REFERENCES

1. Zhi-Jian Yu, 7,999,064 (August 16, 2011)
   Abbott Medical Optics Inc. (Santa Ana, CA)
2. Fukiko Nemoto et al, 7,998,942 (November 16, 2011)
   Senju Pharamaceutical Co., Ltd. (Osaka, JP)
3. Keith K. Donley, 7,998,405 (August 16, 2011)
   Keith K. Donley, (Port Aransas, TX)
4. Joshua B Stopek, 7,999,051 (August 16, 2011)
   Tyco Healthcare Group LP (North Haven, CT)

B. UV-ABSORBERS FOR OPHTHALMIC LENS MATERIALS

Polynitromethacrylates

Title:	UV-absorbers for ophthalmic lens materials
US Patent:	7,947,849 (May 24, 2011)
Author:	Walter R. Laredo
Assignee:	Alcon, Inc. (Hunenberg, CH)

Significance: Ophthalmic lens formulations containing blended or copolymerized nitropropenylbenzene derivatives have been found to be effective in blocking ultraviolet light in the 400- to450-nm range. Lens formulations copolymerized with 2-(2-methoxy-4-(2-nitroprop-1-enyl)phenoxy)ethyl methacrylate absorbed light in a broader and higher range than the industrial standard, 3-(3-*t*-butyl-4-hydroxy-5-(5-methoxy-2H-benzo[d][1,2,3]triazol-2-yl)phenoxy)propyl methacrylate.

Additive

5-(2-Nitroprop-1-enyl)benzo[d][1,3]dioxole (I)
1-Chloro-4-(2-nitroprop-1-enyl)benzene (II)
1,2-Dimethoxy-4-(2-nitroprop-1-enyl)benzene (III)
2-(2-Methoxy-4-(2-nitroprop-1-enyl)phenoxy)ethyl methacrylate (IV)

Safety

Although the toxicological properties of monomeric nitroprop-1-enylbenzene derivatives have not been fully investigated, they may cause eye and skin irritation. Ingestion of these agents may result in gastrointestinal irritation with nausea, vomiting, and diarrhea. No information is available about chronic exposure to these materials. The toxicological properties of copolymers containing nitroprop-1-enyl-benzene is unreported.

Additive Structures

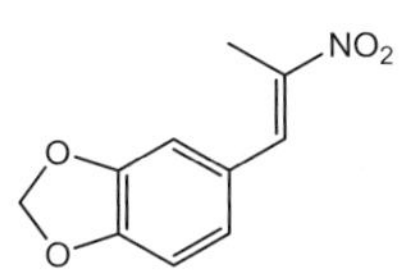

(I)

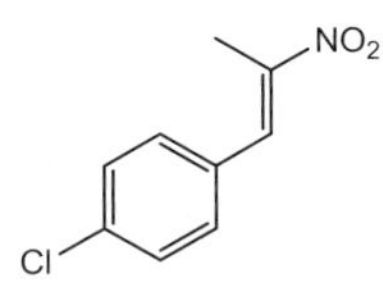

(II)

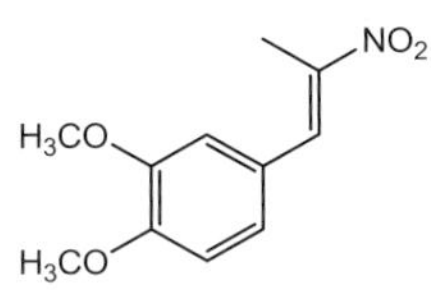

(III)

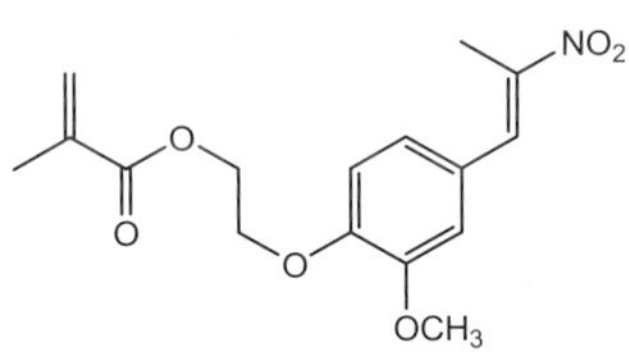

(IV)

Additive Preparation

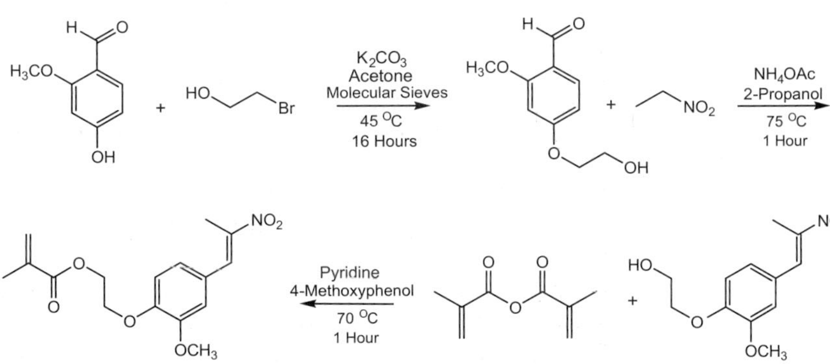

Additive Synthesis

1. Preparation of 4-(2-hydroxyethoxy)-3-methoxybenzaldehyde

A reaction flask was charged with 25.9 g of vanillin, 33.3 g of 2-bromoethanol, 35.2 g of potassium carbonate, and 20 g of molecular sieves and then treated with 600 ml of acetone and refluxed for 16 h. The mixture was then filtered to remove solids, and the solution was concentrated to yield a yellow oil. After recrystallization in ethyl alcohol at −20 °C, the product was isolated in a 40% yield as white crystals.

2. Preparation of 2-(2-methoxy-4-(2-nitroprop-1-enyl)phenoxy)ethanol

A 200-ml round-bottom flask was charged with 5.04 g of the Step 1 product, 9.80 g of ammonium acetate, 13.5 g of dry molecular sieves, 82 g of 2-propanol, and 83 g of nitroethane and then refluxed for 90 min. The reaction mixture was then poured into 200 ml of THF, filtered, and the solution was concentrated. After recrystallization in 2-propanol, the product was isolated in a 50% yield.

3. Preparation of 2-(2-methoxy-4-(2-nitroprop-1-enyl)phenoxy)ethyl methacrylate

A reaction flask was charged with 0.40 g of the Step 2 product dissolved in 10 ml of anhydrous pyridine, 20 mg of 4-methoxyphenol, and 0.54 g of 94% methacrylic anhydride and then heated to 70 °C for 1 h. The mixture was then treated with 100 ml of methylene chloride, and the organic layer was isolated and washed thrice with 500 ml 0.2 M hydrochloric acid, twice with 500 ml of 0.15 M sodium carbonate, and once with 500 ml of water. The organic layer was dried over anhydrous sodium sulfate, filtered, concentrated, and an oil was isolated. The oil was recrystallized in 5 ml of methyl alcohol at −20 °C and the product was isolated in a 40% yield as yellow needles.

Testing

Ultraviolet Absorbance Testing

The effectiveness of the experimental ultraviolet absorbers was determined by preparing 0.9-mm-thick lens containing selected experimental agents in ophthalmic lens formulations as illustrated in Tables 6–9. Absorbance testing results are provided in Table 10 and as illustrated in Tables 6–9. Absorbance test results are provided in Table 10.

TABLE 6 Ocular lens formulations containing Additive 1

Component	Lens Composition A (wt%)	Lens Composition B (wt%)
Additive 1	0.1	0.25
PEA	50.0	49.9
PEMA	44.0	43.9
Terg®15S30-MA	5.0	5.0

TABLE 6 Ocular lens formulations containing Additive 1—*Cont'd*

Component	Lens Composition A (wt%)	Lens Composition B (wt%)
BDDA	1.0	1.0
AIBN	0.5	0.5

2-Phenylethyl acrylate (**PEA**)
2-Phenylethyl methacrylate (**PEMA**)
Methacrylate derivative of Tergitol®15S30 (**Terg®15S30-MA**)
1,4-butanediol diacrylate (**BDDA**)
2,2′-azobis(2-methylpropionitrile) (**AIBN**)

TABLE 7 Ocular lens formulations containing Additive 2

Component	Lens Composition C (wt%)	Lens Composition D (wt%)
Additive 2	0.1	0.25
PEA	66.1	66.1
PEMA	30.4	30.4
Terg15S30-MA	5.0	5.0
BDDA	3.4	3.2
AIBN	0.5	0.5

2-Phenylethyl acrylate (**PEA**)
2-Phenylethyl methacrylate (**PEMA**)
Methacrylate derivative of Tergitol®15S30 (**Terg®15S30-MA**)
1,4-Butanediol diacrylate (**BDDA**)
2,2′-azobis(2-methylpropionitrile) (**AIBN**)

TABLE 8 Ocular lens formulations containing Additive 3

Component	Lens Composition E (wt%)	Lens Composition F (wt%)
Additive 3	0.1	0.25
PEA	66.1	66.0
PEMA	30.5	30.5
DEGEMA	3.4	3.2
Irgacure® 819	0.2	0.2

2-Phenylethyl acrylate (**PEA**)
2-Phenylethyl methacrylate (**PEMA**)
3-(3-t-Butyl-4-hydroxy-5-(5-methoxy-2H-benzo[d][1,2,3]-triazol-2-yl)phenoxy)propylmethacrylate (**Control**)
Diethylene glycol dimethacrylate (**DEGDMA**)
Irgacure® 819

TABLE 9 Ocular lens formulations containing Additive 4

Component	Lens Composition G (wt%)	Lens Composition H (wt%)	Lens Composition I (wt%)
Additive 4	0.75	1.5	0
Control	0	0	1.8
PEA	65.7	65.2	65.0
PEMA	30.3	30.1	30.0
BDDA	3.2	3.2	3.2
AIBN	0.5	0.5	0.5

2-Phenylethyl acrylate (**PEA**)
2-Phenylethyl methacrylate (**PEMA**)
1,4-Butanediol diacrylate (**BDDA**)
2,2′-azobis(2-methylpropionitrile) (**AIBN**)
3-(3-*t*-Butyl-4-hydroxy-5-(5-methoxy-2H-benzo[d][1,2,3]-triazol-2-yl)phenoxy)propylmethacrylate (**Control**)

Testing Results

TABLE 10 Effectiveness of experimental additives at blocking out ultraviolet in 0.9-mm-thick ophthalmic lenses

Lens Composition	Wavelength at 10% Transmission (nm)	Wavelength at 50% Transmission (nm)
Table 6 Composition A	409	424
Table 6 Composition B	421	432
Table 7 Composition C	361	373
Table 7 Composition D	372	391
Table 8 Composition E	383	436
Table 8 Composition E	423	446
Table 9 Composition F	433	433
Table 9 Composition G	438	448
Table 9 Composition H	425	433

Advantages over Prior Art

The use of nitropropenylbenzene derivatives as additives in ophthalmic lens formulations as ultraviolet filters is unreported. Nitropropenylbenzene derivatives are readily prepared from relatively inexpensive starting reagents, and these chemical additives do not appear to coalesce into isolated domains when blended or copolymerized in lens formulations.

Notes

1. The ultraviolet light absorbing additive 2,4-dihydroxy-5-(4-(2-hydroxyethyl) phenylazo)-benzophenone (V) was converted into the polymerizable 2-methacryloyloxyethyl urethane monomer (VI) by Satake[1] and used in ophthalmic lens formulations as an ultraviolet filtering agent.

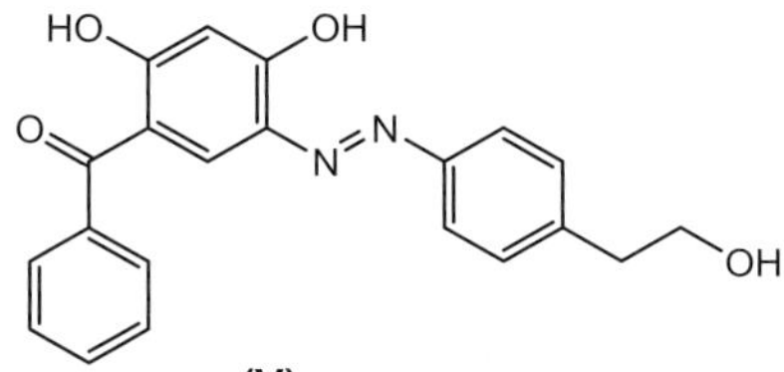

(V)

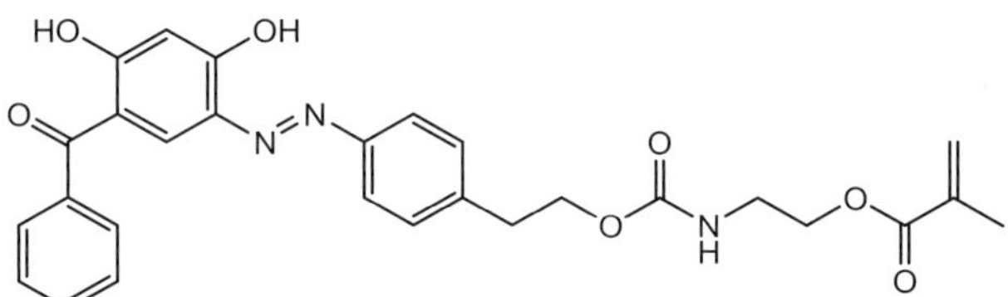

(VI)

2. 1,2,3-Triazole ultraviolet absorbing additives (VII) and (VIII) were prepared by Schlueter[2] and used in ophthalmic lens formulations as ultraviolet filters.

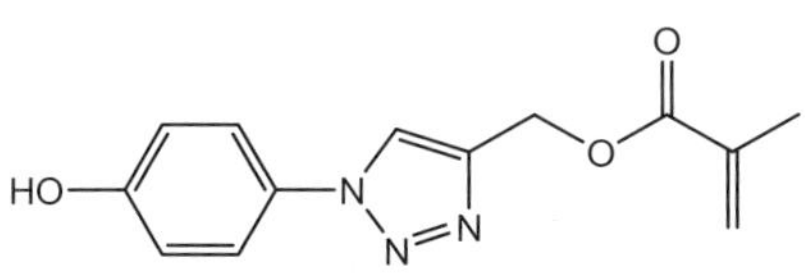

(VII)

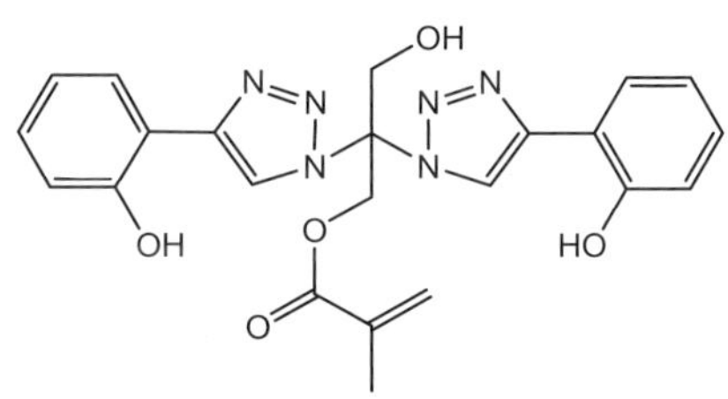

(VIII)

3. Swarup[3] prepared ophthalmic lens coating formulations effective as ultraviolet and oxygen barriers consisting of hyperbranched polyester alcohol containing photochromic indeno-fused naphthopyran (IX).

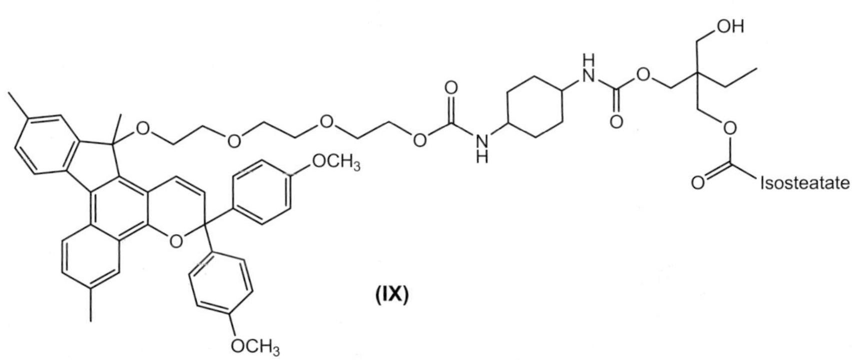

(IX)

4. Mentak[4] prepared ultraviolet, violet, and blue light absorbing monomers for use in ophthalmic lens formulations consisting of acrylamide and methacrylamides of 4-(4-nitrophenylazo)aniline (Orange Dye 3) (X) and 3-[*N*-ethyl-4-(4-nitrophenylazo)phenylamino]propionitrile (Orange Dye 25) (XI).

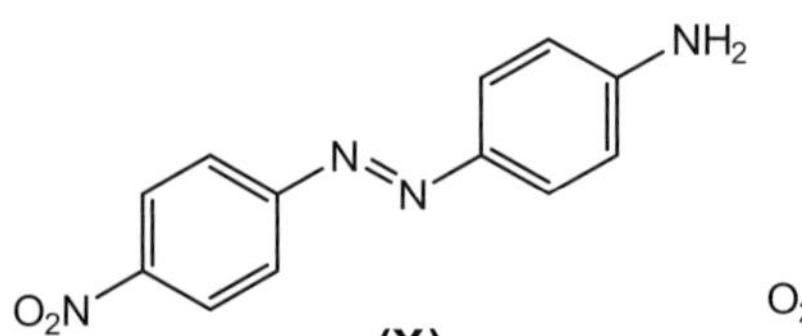

(X)

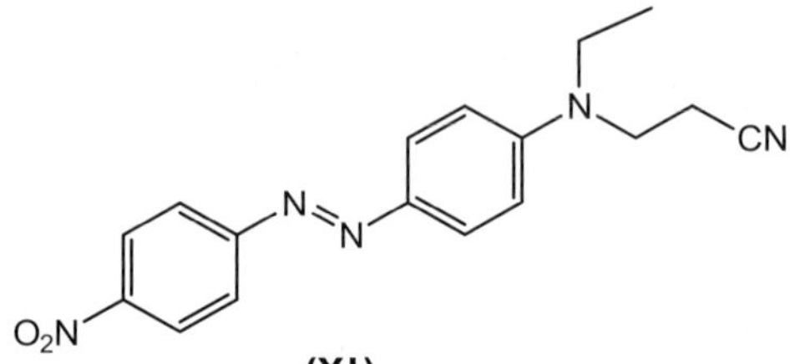

(XI)

REFERENCES

1. Kohsuke Satake et al, US Patent 7,662,937 (February 16, 2010) **Menicon Co., Ltd. (Aichi, JP)**
2. Douglas C. Schlueter, US Patent 7,909,458 (March 22, 2011) **Alcon, Inc. (Hunenberg, CH)**
3. Shanti Swarup et al, US Patent 7,907,346 (March 15, 2011) **Transitions Optical, Inc. (Pinellas Park, FL)**
4. Khalid Mentak, US Patent 7,842,367 (November 30, 2010 **Key Medical Technologies, Inc. (San Ramon, CA)**

Chapter 25

Paint Additives

A. CRACK RESISTANT PAINT

Hydrophobic Modified Polyrotaxanes

Title:	Hydrophobic modified polyrotaxane and crosslinked polyrotaxane
US Patent:	7,943,718 (May 17, 2011)
Author:	Kohzo Ito et al.
Assignee:	Nissan Motor Co., Ltd. (Tokyo, JP)

Significance: A paint additive with the ability to uniformly disperse tension along its axis has been prepared. The additive is a rotaxane containing polyethylene oxide as the axis and α-cyclodextrin as the rotator end capped with adamantine and then postreacted with propylene oxide and γ-caprolactone.

Additive Polyethylene oxide-α-cyclodextrin rotaxane-g-poly(propylene oxide-b-γ-caprolactone) (I)

Safety The additive is a new composition of matter. Safety and handling information is unavailable.

Additive Structure

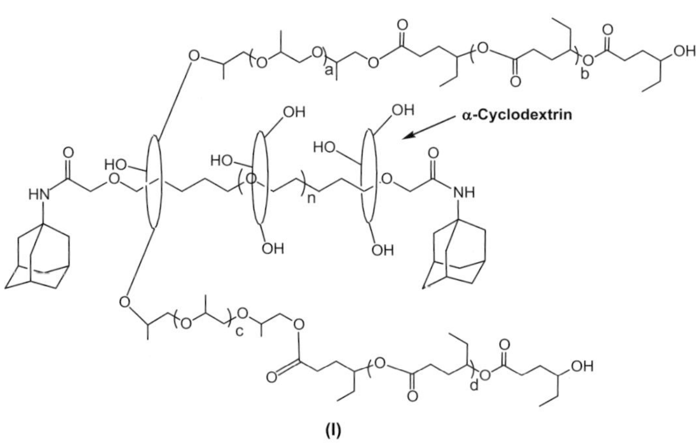

(I)

Next Generation of International Chemical Additives. DOI: http://dx.doi.org/10.1016/B978-0-444-53788-1.00025-5

Additive Preparation

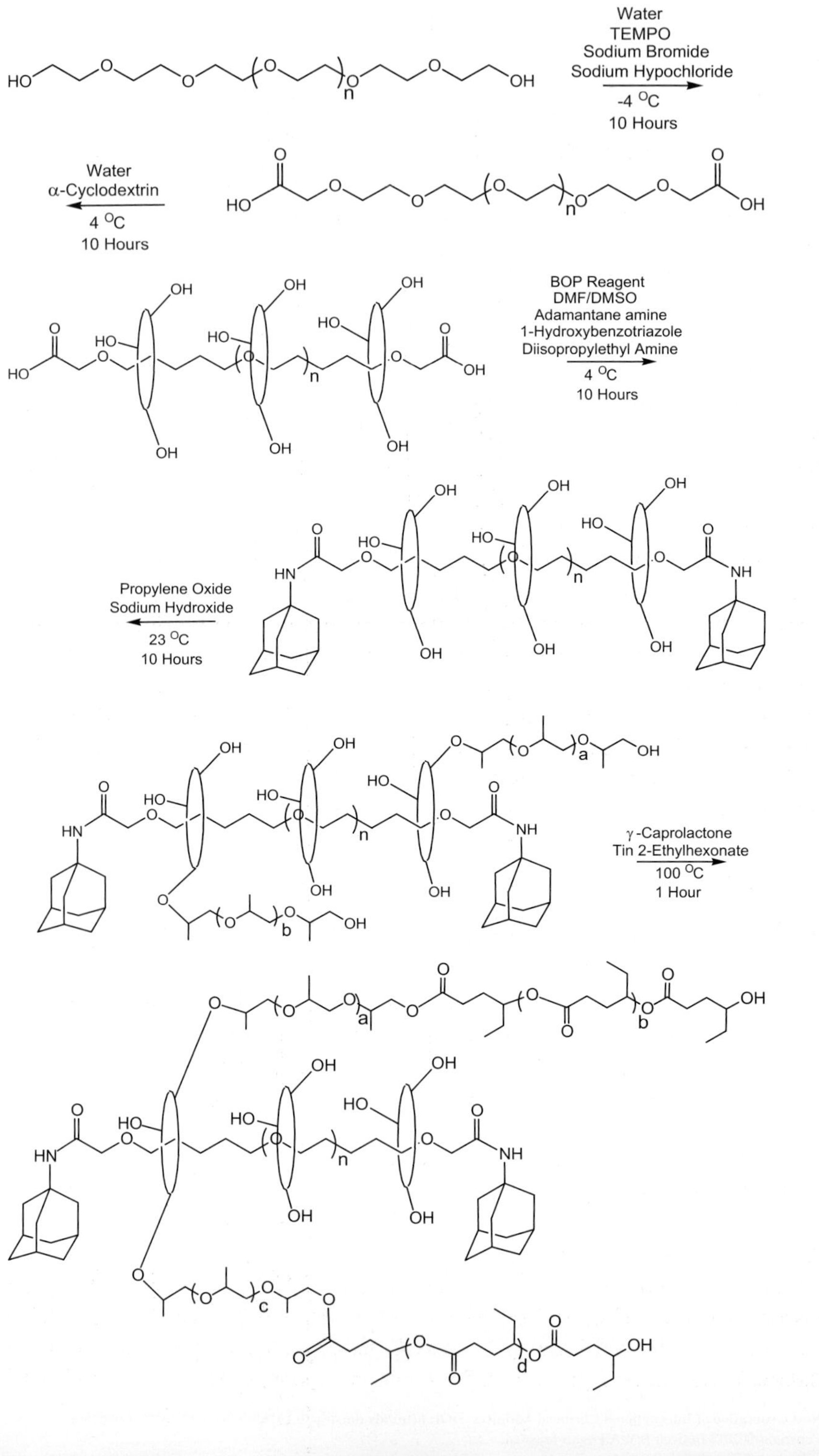

Additive Synthesis

1. Preparation of polyethylene glycol dicarboxylic acid

A reactor was charged with 10 g of polyethylene glycol having an Mn of approximately 5000 Da, 100 mg of 2,2,6,6-tetramethyl-1-piperidinyl-oxy radical, and 1 g of sodium bromide dissolved in 100 ml of water and then treated with 5 ml of 5% aqueous sodium hypochlorite. The mixture was stirred at an ambient temperature for 10 min and then treated with 5 ml of ethanol to terminate the reaction. The product was extracted thrice with 50 ml of methylene chloride and then concentrated by distillation. The mixture was then dissolved in warm ethanol and then cooled to −4 °C for 10 h, and the product was isolated after drying.

2. Preparation of inclusion complex of polyethylene glycol dicarboxylic acid and α-cyclodextrin

A pear-shaped flask was charged with 3 g of the Step 1 product and 12 g of α-cyclodextrin dissolved in 50 ml of water at 70 °C and then mixed and cooled to 4 °C for 10 h. The inclusion complex was precipitated in a cream state and was recovered.

3. Preparation of adamantane-capped inclusion complex of polyethylene glycol dicarboxylic acid and α-cyclodextrin

A pear-shaped flask was charged with 14 g of the Step 2 product dispersed in a solvent mixture consisting of dimethylformamide and dimethylsulfoxide, 75/25, and then treated with 3 g of benzotirazole-1-yl-oxy-tris(dimethylamino)phosphonium hexafluorophosphate, 1 g of 1-hydroxybenzotriazole, 1.4 g of adamantane amine, and 1.25 ml of diisopropylethyl amine dissolved in 10 ml of dimethylformamide. The mixture was then shaken at an ambient temperature and then cooled to 4 °C for 10 h and treated with 50 ml of a mixture of dimethylformamide and methanol, 1/1. The mixture was then centrifuged, and the precipitated product was isolated. The product was vacuum dried and then dissolved in 50 ml of dimethylsulfoxide, precipitated in 700 ml of water, and the purified product was isolated by centrifuging.

4. Preparation of hydroxylpropylated adamantane-capped inclusion complex of polyethylene glycol dicarboxylic acid and α-cyclodextrin

A reactor was charged with 500 mg of the Step 3 product dissolved in 50 ml of 1 M aqueous solution of sodium hydroxide and then treated with 3.83 g of propylene oxide. The mixture was then stirred for 10 h at an ambient temperature and then neutralized with 1 M aqueous solution of hydrochloric acid and the product was isolated after dialysis.

5. Preparation of block copolymer of caprolactone and hydroxylpropylated adamantane-capped inclusion complex of polyethylene glycol dicarboxylic acid and α-cyclodextrin

A reactor was charged with 500 mg of the Step 4 product, 10 ml of γ-caprolactone, and 0.2 ml of tin 2-ethylhexanoate and then heated to 100 °C for 1 h. The intermediate was dissolved in 50 ml of toluene and precipitated in 450 ml of hexane, dried, and the product was recovered. ^{1}H-NMR and GPC analyses indicated that α-cyclodextrin inclusion was 0.06 wt% and that the degree of γ-caprolactone incorporation was approximately 0.02 wt%.

Testing

Test data are not provided by the author.

Advantages over Prior Art

Paint additives containing rotaxanes are unreported in the patent literature. The ability of these materials to behave as slide ring or topological gels may explain their remarkable resistance to cracking and excellent mechanical strength. Hydrophilic/hydrophobic properties of these additives can be customized by postreacting the rotaxane with monomers of varying polarity.

Notes

1. Hydrophilic rotaxanes previously prepared by Ito[1] were used in cationic electrodeposition paint formulations and had excellent marring resistance, produced no cracking, and had excellent weatherability properties.
2. The preparation of the inclusion complex of polyethylene glycol and α-cyclodextrin capped with adamantane acetic acid (II) is described by Ito.[2]

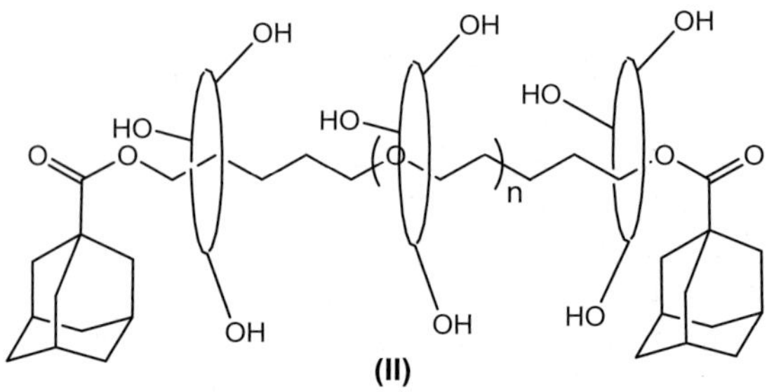

(II)

3. In an earlier investigation, Ito[3] prepared the hydroxylpropylated adamantane-capped inclusion complex of polyethylene glycol dicarboxylic acid and α-cyclodextrin postreacted with ε-caprolactone (III), which was used in curable solvent-based top coating material formulations.

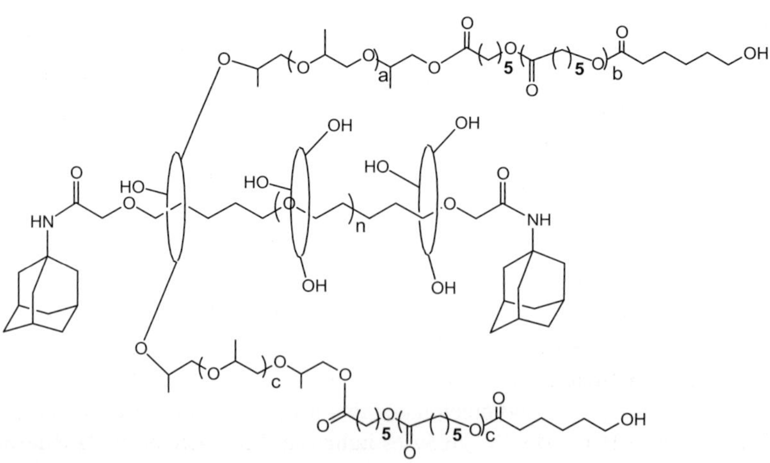

(III)

4. The adamantane ε-capped inclusion complex of polyethylene glycol dicarboxylic acid and α-cyclodextrin containing 30% methoxylation (IV) was prepared by Ito[4] using sodium methoxide and methyl iodide and used in paint formulations requiring unique viscoelastic properties.

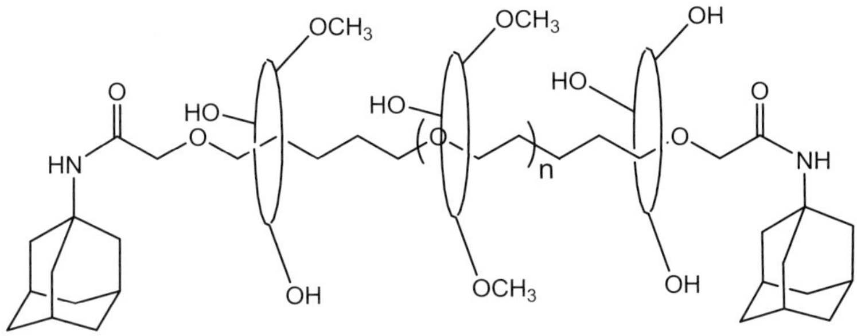

(IV)

REFERENCES

1. Kohzo Ito et al US Patent 7,893,155 (February 22, 2011) **Nissan Motor Co., Ltd. (Kanagawa, JP)**
2. Kohzo Ito et al US Patent 7,847,049 (December 7, 2010) **Nissan Motor Co., Ltd. (Kanagawa, JP)**
3. Kohzo Ito et al US Patent 7,799,867 (September 21, 2010) **Nissan Motor Co., Ltd. (Kanagawa, JP)**
4. Kohzo Ito et al, US Patent 7,622,527 (November 24, 2009) **Nissan Motor Co., Ltd. (Kanagawa, JP)**

B. CREEPAGE RESISTANT PAINT COMPOSITIONS

Oxiranyl Methyl Ester Derivatives

Title:	Epoxy silanes, processes for their manufacture and curable compositions containing the same
US Patent:	7,989,651 (August 2, 2011)
Author:	Shiu-Chin H.Su et al.
Assignee:	Momentive Performance Materials Inc. (Albany, NY)

Significance:	Water-soluble latex paint formulations additized with up to 2 wt% of epoxysilane derivatives dramatically reduced creeping behavior on freshly painted surfaces by up to 50%. Curing the painting formulation at 120 °C for 15 min completely eliminated creaking behavior.

Additives	3-Trimethoxysilylpropylcarbamic acid oxiranyl methyl ester (I) 3-Triethoxysilylpropylcarbamic acid oxiranyl methyl ester (II)

Safety Both reagents are acutely toxic if inhaled (Category 4). Both reagents cause severe skin and eye irritation (Category 2). High single target organ toxicity has been reported after a single exposure. Reagents must not come in contact with water, alcohols, or amines or a potentially violent hydrolysis reaction may occur.

Additive Structures

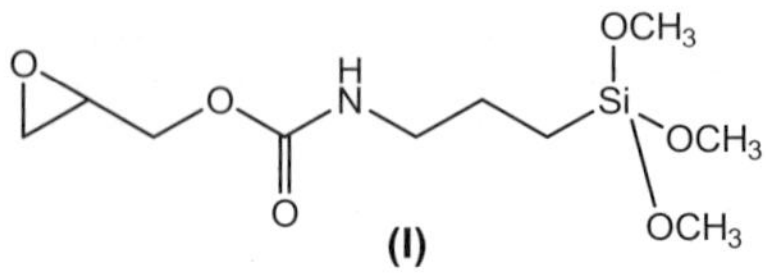

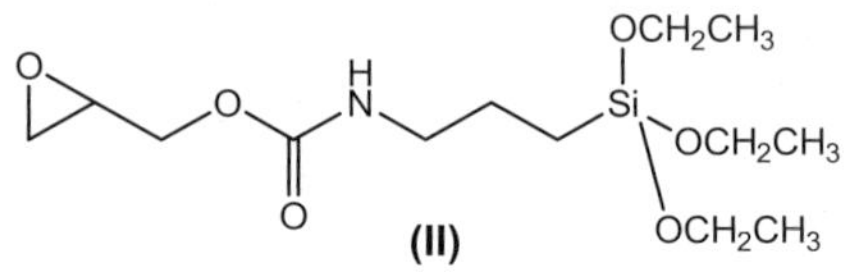

Additive Preparation

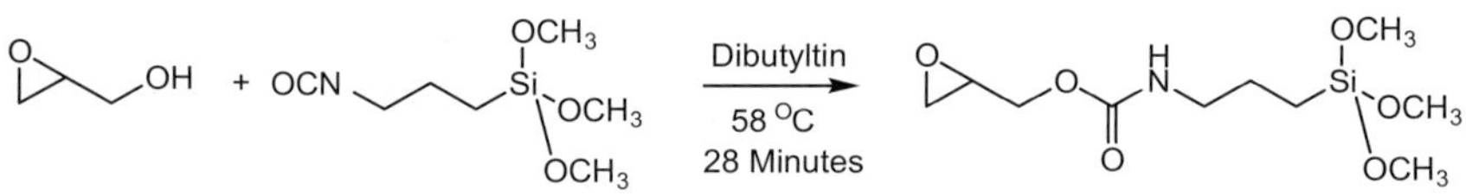

Additive Synthesis

1. Preparation of 3-trimethoxysilylpropylcarbamic acid oxiranyl methyl ester

At an ambient temperature, a separatory funnel charged with 7.4 g of glycidol was slowly added to a mixture of 107 mmol of γ-isocyanatopropyltrimethoxysilane containing 0.0295 g of dibutyltin. The temperature of the mixture was then increased to 53 °C over 25 min, and it was then cooled to an ambient temperature for 2 h. After isolation, the product was characterized by infrared analysis confirming the absence of the isocyanate peak at 2300 cm^{-1} while ^{13}C-NMR analysis confirmed the presence of the target molecule.

Testing

Paint Creepage Test

A 2% weight treatment solution of Experimental Additives 1 and 2 formulated with 50 mg of nonionic Triton® X-100, 1 g of silane, and 48.95 g of distilled water was used to coat degreased and unpolished test panels. Panels were initially treated with an immersion of zinc phosphate coating using Henkel Bonderite® 958 and then by a Parcolene® 60 rinsing with water. Panels to be coated with the experimental treatment additive compositions were then cured at 120 °C for 20 min before being painted. The paint consisted of a primerless polyester-based system from Sherwin Williams® and was applied by a drawdown technique where the dry film thickness was approximately 1.2 mil. The baking conditions were 15 min at 175 °C. Creepage was measured according to the ASTM D1654 protocol. Test results are provided in Table 1.

Results

Paint Creepage Test

TABLE 1 Paint creepage test results for latex formulations additized with 2 wt% of the experimental agents. Lower creep numbers are preferred

Panel Coating	Average Creeping Distance (mm)	Minimum Creeping Distance (mm)	Minimum Creeping Distance (mm)
Additive 1	2.4	1.5	3.3
Additive 2	2.1	0.8	3.3
Unmodified Surface (Control Panel)	2.2	1.5	3.0

Advantages over Prior Art

Epoxy silanes are well-established promoters, additives, and crosslinking agents. This investigation represents the first use of epoxy silanes in water-soluble latex paint formulations designed to control the creeping behavior. Paint containing crosslinkable epoxy silanes reduced creepage without inducing paint discoloration.

Notes

1. Lejeune[1] prepared a very low organic emission latex paint formulation having excellent physical properties by additizing the paint formulation with γ-glycido xypropyltrimethoxysilane (Silquest® A-187) (III) and 2-methyl-1,3-propanediol (IV) as the crosslinking agent.

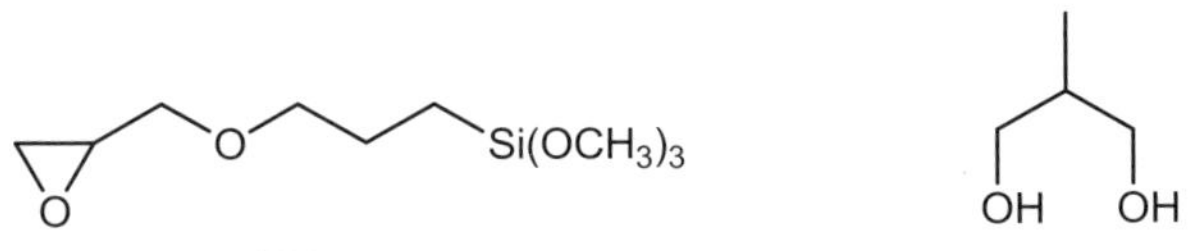

(III) **(IV)**

2. Lin[2] prepared a silicone-containing paint adhesion additive containing organopolysiloxanes (V) and hydrogenated polybutadiene diol (VI), which had excellent mechanical properties and improved adhesion of coatings of latex paints, adhesives, and other surface treatments.

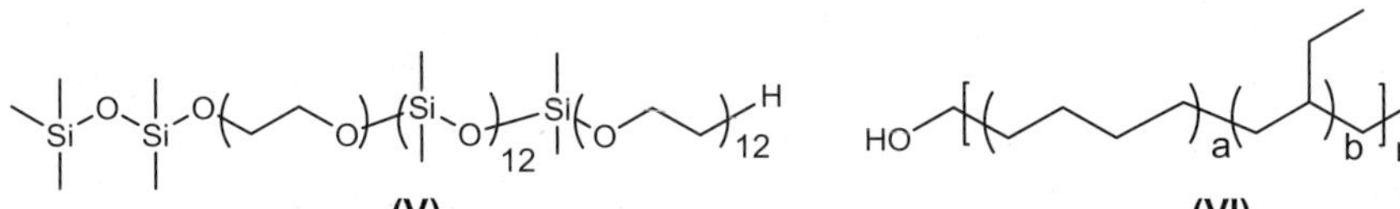

(V) **(VI)**

3. Brown[3] additized water-soluble latex paint with phosphoethylmethacrylate (VII) and phosphodi(ethyl methacrylate) (VIII) to improve the painted surface mechanical properties and to impart a slightly opaque color to the dried surface.

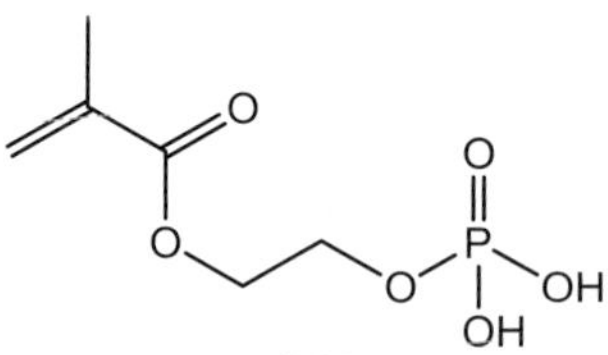

(VI)

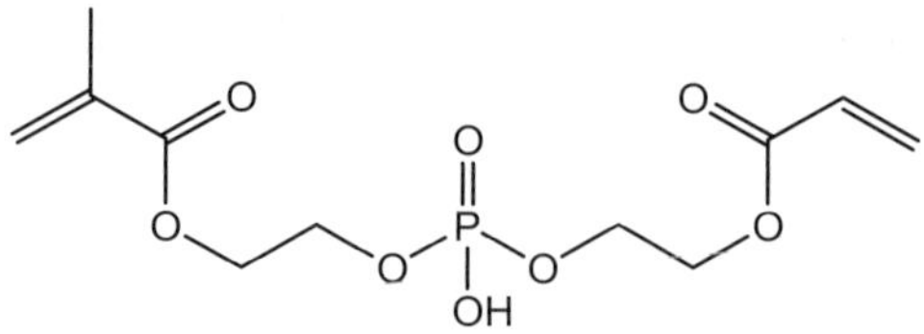

(VII)

REFERENCES

1. Alain Lejeune et al, 7,985,821 (July 26, 2011)
   Momentive Performance Materials Inc. (Albany, NY)
2. Zhiqing Lin et al, 7,569,645 (August 4, 2009)
   Momentive Performance Materials Inc. (Albany, NY)
3. Ward T. Brown et al, 7,179,531 (February 20, 2007)
   Rohm and Haas Company (Philadelphia, PA)

C. ELASTIC PAINT COMPOSITIONS

Polydimethylpolysiloxane

Title:	Highly elastomeric and paintable silicone compositions
US Patent:	7,994,261 (August 9, 2011)
Author:	Melvin Dale Beers et al,
Assignee:	Tremco, Incorporated (Beachwood, OH)
Significance:	Seven silicone-containing painting formulations were prepared that were extremely elastic, curable, and had paintable surfaces. The silicone formulations each contained an organopolysiloxane, a functionalized silicone crosslinker, and an organic polymer. The highly elastic, curable, and paintable silicone compositions have an elongation of at least 150% and were used as paintable sealants and caulks.
Additives	Silanol terminated dimethylpolysiloxane, Mn of approximately 10,000–50,000 Da (I) Silanol terminated dimethylpolysiloxane, Mn of approximately 5000–20,000 Da (II) Vinyltris-methylethylketoximosilane (III) Dimethyl bis-sec-butylaminosilane (IV) Hexamethyldisilazane (V) Aminopropyltriethoxysilane (VI) Dibutyltin diacetate (VII) Trimethoxysilylalkyl terminated poyurethane (VIII)

Safety

While no safety reports have been filed with any government agency, many of the latex paint formulations containing silylcrosslinking agents must be protected from moisture. Amino-containing monomers are considered to be moderate dermal and eye irritants. In addition, ketooximes can potentially form acetone and reactive amines. The dried latex paint, by contrast, is chemically inert since the potential for crosslinking reactions has been eliminated.

Latex Paint Additives

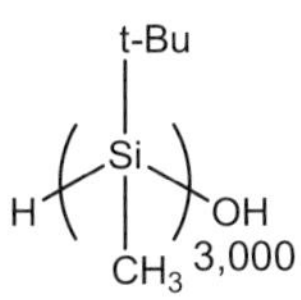

(I)

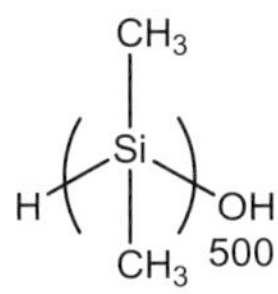

(II)

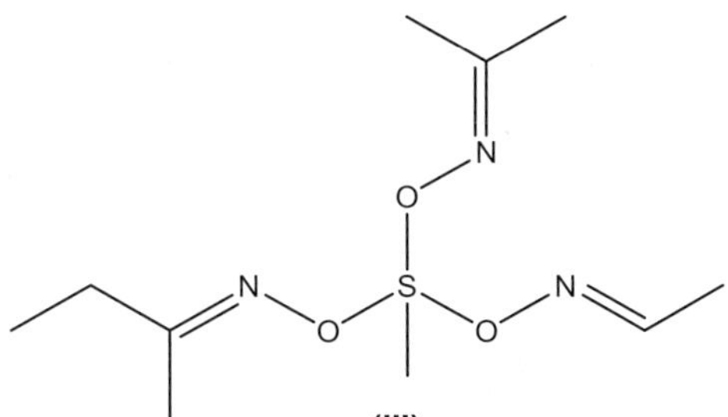

(III)

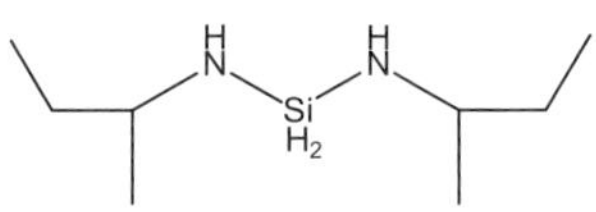

(IV)

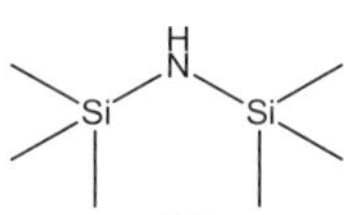

(V)

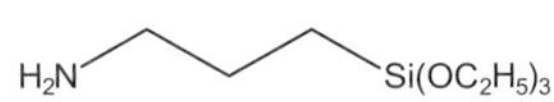

(VI)

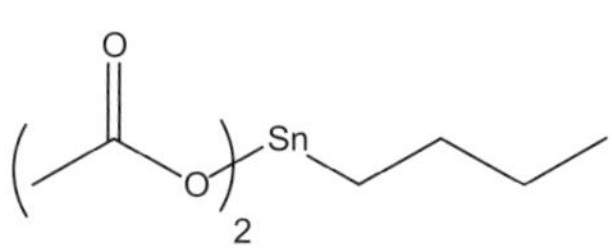

(VII)

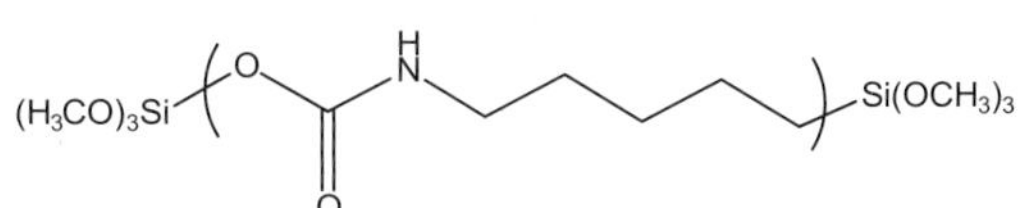

(VIII)

Painting Formulations

Seven unique painting formulations were prepared, which were designed to highlight a specific physical property of the dried paint coating. Formulations are provided below.

1. Painting Formulation #1

In this example, about 43% of an organic polymer (based on the total weight of organic and inorganic polymer) was used to prepare a medium modulus sealant with a Shore-A Hardness of 15. The organic polymer comprised a silyl terminated polyurethane. The latex formulation is provided in Table 2.

TABLE 2 Experimental latex paint formulation resulting in a smooth, "fish eyes"-free surface coating. This composition also produced surfaces having excellent weatherability and showed no evidence of surface cracks or change in Shore Hardness after >20,000 h of UV exposure in the Xenon weatherometer

Composition	Weight Percent
Silanol Terminated Dimethylpolysiloxane: Mn ~ 2400 Da	10.44
Silanol Terminated Dimethylpolysiloxane Polymer: Mn ~ 50,000 Da	23.43
Soft Acrylic Filler (Thixotrope)	19.38
Hydrophobic-Treated Precipitated Calcium Carbonate	5.86
Hexamethyldisilazane	0.35
Hydrophobic-Treated Ground Calcium Carbonate	2.34
Acrylic Functional Plasticizer	7.14
Dimethylbis-Secondary Butylaminosilane	0.64
Methyltris-Methylethylketoximosilane	3.99
A trimethoxysilylalkyl Terminated Polyurethane	25.56
Aminoethylaminopropyltrimethoxysilane	0.86
Dibutyltin dilaurate	0.01
Total	100.00

2. *Painting Formulation #2*

In this example, 29 wt% of the organic polymer was used to prepare a medium modulus sealant with a Shore-A Hardness of 15 with a fast curing silicone sealant having medium modulus properties. The latex paint formulation is provided in Table 3.

TABLE 3 Weight composition of paint formulation designed to provide a paintable fast curing silicone sealant having medium modulus properties combined with excellent adhesion properties on plastics, glass, and anodized aluminum

Composition	Weight Percent
Silanol Terminated Dimethylpolysiloxane Polymer: Mn ~ 50,000 Da	25.46
20,000 cps. Silanol Terminated Dimethylpolysiloxane Polymer: Mn ~ 20,000 Da	16.97
Hydrophobic-Treated Precipitated Calcium Carbonate	33.94
Hexamethyldisilazane	0.85

TABLE 3 **Weight composition of paint formulation designed to provide a paintable fast curing silicone sealant having medium modulus properties combined with excellent adhesion properties on plastics, glass, and anodized aluminum—*Cont'd***

Composition	Weight Percent
Dimethyl bis-secondary Butylaminosilane	2.25
Vinyltris-methylethylketoximosilane	1.70
Alkoxy Silyl Terminated Polyether	16.97
Aminoethylaminopropyltrimethoxysilane	1.80
Dibutyltin diacetate	0.06
Total	100.00

3. *Painting Formulation #3*

In this formulation, 24 wt% of the organic polymer was used to prepare a medium modulus sealant. The organic polymer comprised a silyl-terminated polyether polymer Table 4.

TABLE 4 **This formulation was used to prepare a medium modulus sealant using the silyl-terminated polyether component and had good adhesion to plastics, glass, and anodized aluminum**

Composition	Weight Percent
Silanol Terminated Dimethylpolysiloxane Polymer: Mn ~ 50,000 Da	3.83
Silanol Terminated dimethylpolysiloxane Polymer: Mn ~ 20,000 Da	34.51
2400 cps. Silanol Terminated dimethylpolysiloxane Polymer	4.02
Vinyltris-Methylethylketoximosilane	1.53
Dimethyl bis-Secondary butylaminosilane	2.037
Hexamethyldisilazane	0.84
Hydrophobic Treated Precipitated Calcium Carbonate	30.68
Soft Acrylic Filler	7.48
Alkoxy Silyl Terminated Polyether Polymer	13.42
Aminopropyltriethoxysilane	1.63
Dibutyltin diacetate	0.023
Total	100.00

4. *Painting Formulation #4*

In this example, 26 wt% of the organic polymer was used to prepare a medium modulus sealant. The organic polymer comprised an MDI-terminated polyurethane polymer. The formulation is provided in Table 5.

TABLE 5 In this formulation, an organic polymer was used to prepare a medium modulus sealant. The organic polymer comprised an MDI-terminated polyurethane polymer

Composition	Weight Percent
Silanol Terminated Dimethylpolysiloxane Polymer: Mn ~ 50,000 Da	26.12
Silanol Terminated Dimethylpolysiloxane Polymer: Mn ~ 20,000 Da	17.41
Vinyltris-methylethylketoximosilane	2.394
Dimethyl bis-secondary butylaminosilane	1.22
Hexamethyldisilazane	0.871
Hydrophobic Treated Precipitated Calcium Carbonate	34.83
MDI-Terminated Polyurethane Prepolymer	15.24
Aminopropyltriethoxysilane	1.85
Dibutyltin diacetate	0.065
Total	100.00

5. *Painting Formulation #5*

In this example, 31 wt% of the organic oligomer was used to prepare a silicone sealant. The organic oligomer comprised an oximo silane terminated polyether polyol. The oximo silane terminated polyether polyol was prepared by prereacting it with vinylmethylethylketoximosilane. The formulation is provided in Table 6.

TABLE 6 The above composition was painted after curing, yielding a smooth painted surface having no "fish eyes." The above composition passed the paint adhesion test and had excellent weatherability as indicated by the absence of surface cracks and no change in Shore-A Hardness after >20,000 h of UV exposure in the Xenon weatherometer

Composition	Weight Percent
Silanol Terminated Dimethylpolysiloxane Polymer: Mn ~ 50,000 Da	24.90
Silanol Terminated Dimethylpolysiloxane Polymer: Mn ~ 20,000 Da	16.60
Vinyltris-Methylethylketoximosilane	2.90
Oximo Silane Terminated Polyether Polyol	17.43
Dimethylbis-Secondary butylaminosilane	2.20

TABLE 6 The above composition was painted after curing, yielding a smooth painted surface having no "fish eyes." The above composition passed the paint adhesion test and had excellent weatherability as indicated by the absence of surface cracks and no change in Shore-A Hardness after >20,000 h of UV exposure in the Xenon weatherometer—*Cont'd*

Composition	Weight Percent
Hexamethyldisilazane	0.95
Hydrophobic treated precipitated Calcium Carbonate	33.20
Aminopropyltriethoxysilane	1.76
Dibutyltin diacetate	0.06
Total	100.00

6. *Painting Formulation #6*

In this example, 15 wt% of the organic polymer was used to prepare a paintable sealant. The sealant had a Shore-A Hardness of 40, accompanied by an elongation of 760%, and tensile strength of 220 psi. The organic polymer comprised an acrylic terpolymer and polyether polyol. The paint formulation is provided in Table 7.

TABLE 7 After curing, the above sealant was painted with acrylic latex paint. The painted surface was smooth and showed no "fish eyes." Three days after painting, the surface retained excellent paint adhesion test

Composition	Weight Percent
Silanol Terminated Dimethylpolysiloxane Polymer: Mn ~ 50,000 Da	24.42
Silanol Terminated Dimethylpolysiloxane Polymer: Mn ~ 20,000 Da	16.28
Hydrophobic Precipitated Calcium Carbonate	32.56
Ground Calcium Carbonate	7.79
Talc	0.90
Titanium Dioxide	0.26
Hexamethyldisilazane	0.81
Dimethyl bis-secondary butylaminosilane	2.16
Vinyltris-methylethylketoximosilane	1.63
Ethyl acrylate acrylonitrile Acrylic Acid Terpolymer	6.81
Polyether Polyol	0.20
Butyl Benzyl Phthalate	0.77

(Continued)

TABLE 7 After curing, the above sealant was painted with acrylic latex paint. The painted surface was smooth and showed no "fish eyes." Three days after painting, the surface retained excellent paint adhesion test—*Cont'd*

Composition	Weight Percent
Solvent	2.77
Ceramic Fiber	0.85
Aminoethylaminopropyltrimethoxysilane	1.73
Dibutyltin diacetate	0.06
Total	100.00

7. Painting Formulation #7

In this example, 9% of an organic polymer was used to prepare a paintable sealant that had a Shore-A Hardness of 43, elongation of 355%, and a tensile strength of 300 psi. The organic polymer comprised an isobutylene–isoprene copolymer and polybutene with a tackifier. The formulation is provided in Table 8.

TABLE 8 Latex paint formulation containing 9% organic polymer used to prepare a paintable sealant. It had a Shore-A Hardness of 43, accompanied by an elongation of 355%, and had a tensile strength of 300 psi

Composition	Weight Percent
50,000 cps. silanol terminated polydimethylsiloxane polymer: Mn ~ 50,000 Da	24.42
20,000 cps. silanol terminated polydimethylsiloxane polymer: Mn ~ 20,000 Da	16.28
Hydrophobic precipitated calcium carbonate	32.56
Ground calcium carbonate	11.14
Titanium dioxide	0.27
Magnesium carbonate	0.22
Crystalline silica	0.09
Hexamethyldisilazane	0.81
Dimethylbis-secondary butylaminosilane	2.16
Vinyltris-methylethylketoximosilane	1.63
Polybutene	2.81
Isobutylene isoprene copolymer	1.14
Hydrocarbon resin	0.24

TABLE 8 Latex paint formulation containing 9% organic polymer used to prepare a paintable sealant. It had a Shore-A Hardness of 43, accompanied by an elongation of 355%, and had a tensile strength of 300 psi—*Cont'd*

Composition	Weight Percent
Castor oil	0.24
Solvent	2.89
Aminoethylaminopropyltrimethoxysilane	1.73
Dibutyltin diacetate	0.06
Total	100.00

Testing

The curable silicone compositions described were tested according to the paint adhesion tape test described in the ASTM test method D3359. In this paint adhesion test, the sealant composition was affixed to a surface and painted in crosshatches and then the paint and sealant were cut through using a razor blade. A 0.5-inch wide by 4-inch long strip of clear Scotch® brand adhesive tape and was then firmly affixed across the precut crosshatches and pulled away at a 90° angle. If the paint remains intact on the surface of the sealant, it is considered to have passed the test.

Test Results

All seven paint formulations tested resulted in a smooth surface without the presence of "fish eyes" to mar the finish. In addition, all seven painting formulations passed the paint adhesion test while maintaining excellent weatherability as demonstrated by the absence of surface cracks and consistent Shore-A Hardness values.

Advantages over Prior Art

Seven versatile water-soluble latex paint formulations containing crosslinkable siloxane polymers were prepared that had excellent adherence to glass, plastic, or anodized aluminum surfaces. Selected coatings had Shore-A Hardness of at least 40 with an elongation of 355%, which is unsurpassed by existing painting formulations.

Notes

1. Yeats[1] prepared an ambient temperature curing coating painting formulation containing polysiloxane butyl acrylate, glycidyl methacrylate, methyl methacrylate, and di-tert butyl peroxide, which was resistant to UV-A or UV-B radiation under accelerated test conditions.

2. Ootake[2] incorporated the polyorganosiloxane (IX) in painting formulations to improve the weatherability properties and flexibility of the coating finish.

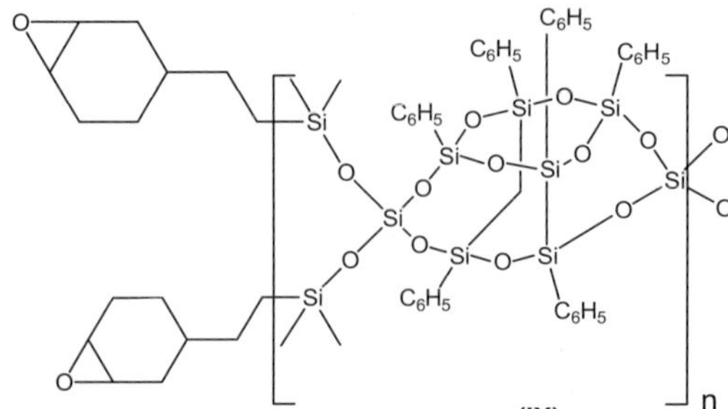

(IX)

3. In an earlier investigation by Ootake,[3] the organosiloxane additive (X) was incorporated into a painting formulation to enhance the flame retardancy, heat resistance, weatherability, light resistance, electric insulating property, surface hardness, and mechanical strength of the polymer finish.

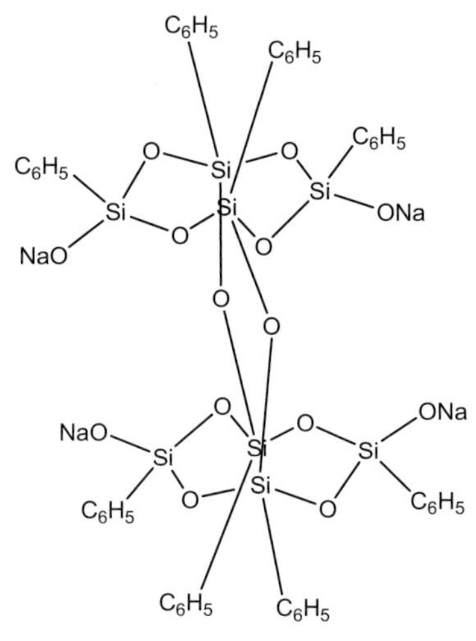

(X)

4. Lichtenhan[4] prepared nanoreinforced paint coatings with improved hydrophobicity, thermal stability, hardness, and durability using polyhedral oligomeric silsesquioxane (POSS) (IV) derivatives. The nanoscopic dimensions and organic inorganic hybrid polymer composition of POSS derivatives were useful for coating fillers derived from minerals, metals, glasses, and polymeric materials.

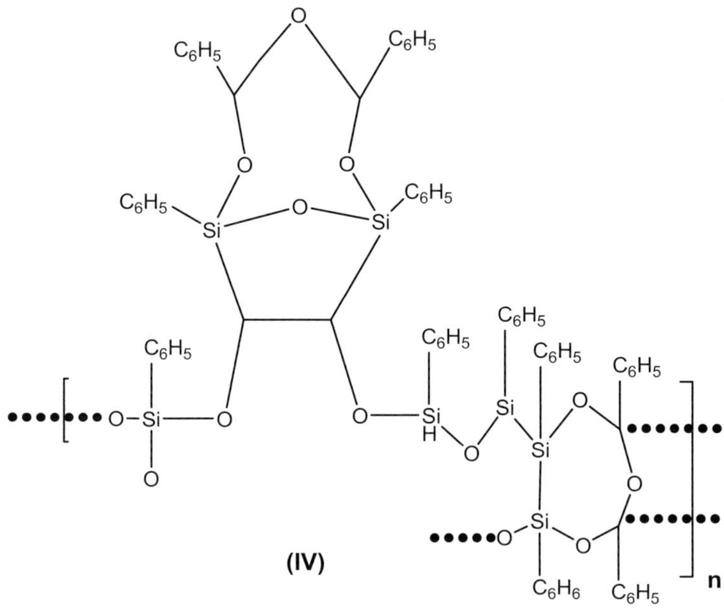

(IV)

REFERENCES

1. Keith Yeats et al, 7,026,400 (April 11, 2006)
   International Coatings Limited (London, GB)
2. Nobumasa Ootake et al, 7,939,617 (May 10, 2011)
   Chisso Corporation (Osaka, **JP**)
3. Nobumasa Ootake et al 7,524,917 (April 28 2009)
   Chisso Petrochemical Corporation (Tokyo, JP)
   Chisso Corporation (Osak, JP)
4. Joseph D.Lichtenhan et al. 7,638,195, (December, 29, 2009)
   Hybrid Plastics, Inc. (Hattiesburg, MS)

D. LOW-VISCOSITY PAINT COMPOSITIONS

Oligomeric Isocyanates

Title:	Method for the preparation of low-viscosity (poly) isocyanates
US Patent:	8,039,574 (October 18, 2011)
Author:	Eugenie Charriere et al.
Assignee:	Rhodia Chimie (Courbevoie Cedex, FR)

Significance: Two methods for preparing low-viscosity cyclic isocyanate-terminated oligomers are described. In the first method, hexamethylene diisocyanate was converted into 11% cyclic dimers and 51% cyclic trimers by heating to 140 °C for 2 h with a catalytic amount of hexamethyldisilazene. In the second method, hexamethylene diisocyanate and pentaerythritol were heated to 140 °C for up to 6 h producing isocyanate-terminated cyclic dicarbamate and tricarbamate in 20.2% and 33.2%, respectively. In both methods, the low-viscosity of the products makes them easily blendable in paint formulations.

Additive Names Dimeric hexamethylene diisocyanate (I)
Trimeric hexamethylene diisocyanate (II)

Safety Diisocyanate dimers, trimers, and oligomers are eye, skin, and upper respiratory system irritants. Repeated exposure may cause nervous system depression. Extreme overexposure may result in unconsciousness and possibly death. Prolonged or chronic overexposure to these agents has been associated with hematopoietic disorders.

Additive Structures

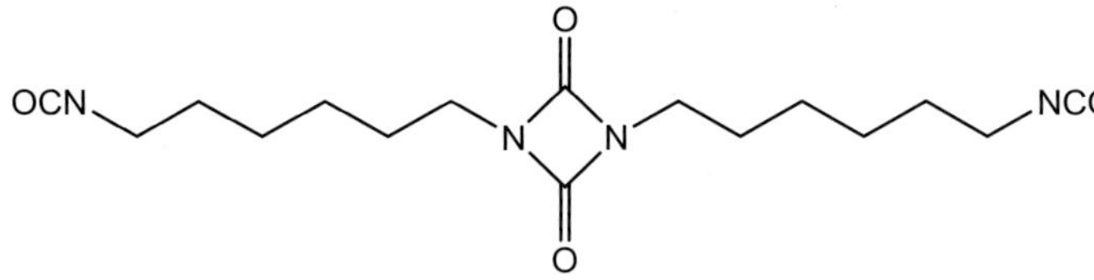

(I)

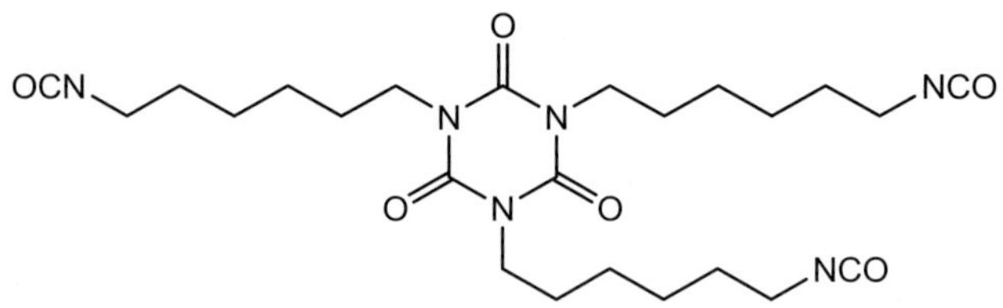

(II)

Additive Preparation

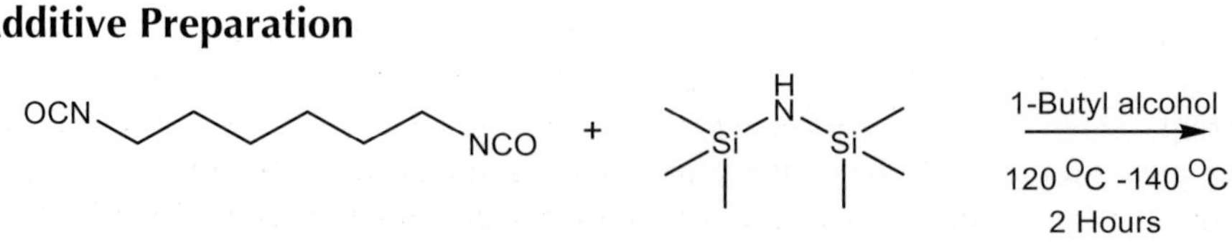

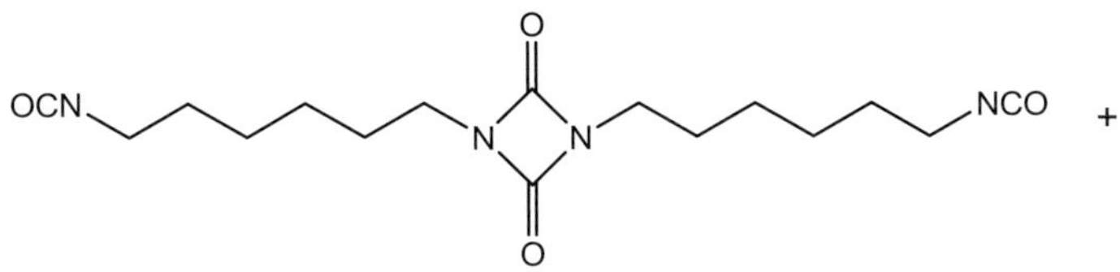

11%

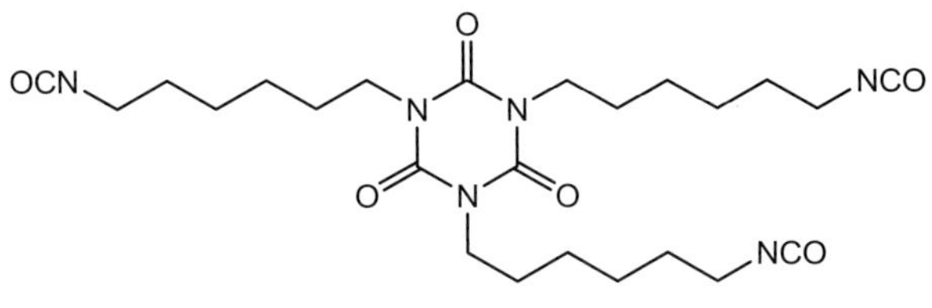

51%

Prepolymer Preparation

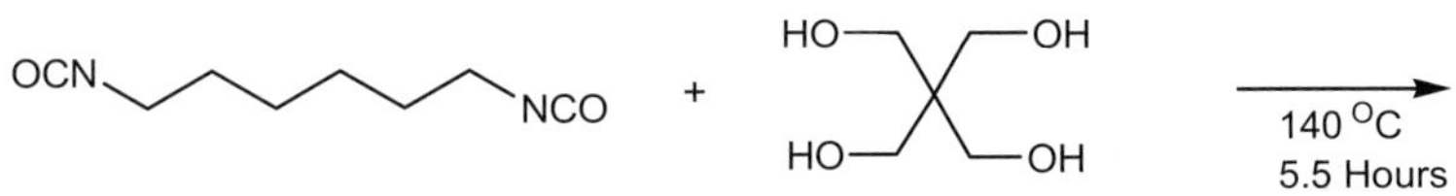

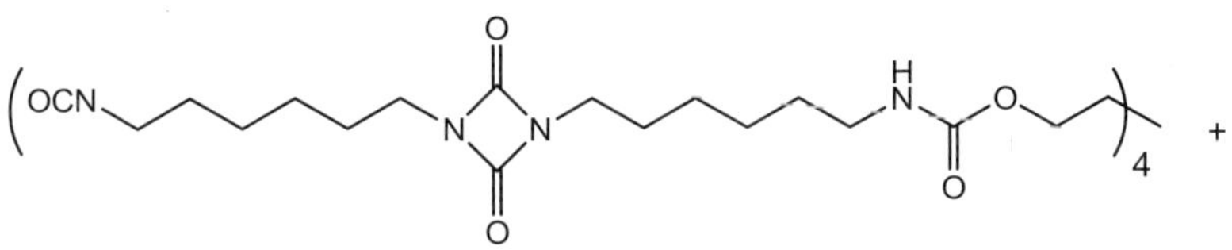

20.2%

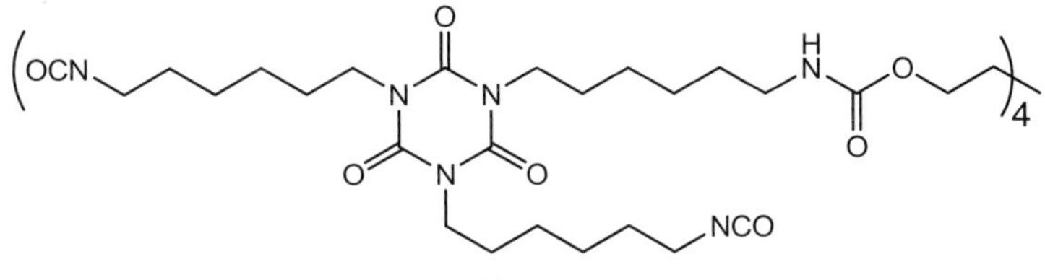

33.2%

Additive Synthesis

1. Preparation of hexamethylene diisocyanate dimer and trimer in the presence of hexamethyldisilazane

A reactor was charged with 1 kg of hexamethylene diisocyanate and 10 g of hexamethyldisilazane and then heated between 130 °C and 140 °C for 2 h with vigorous stirring. The mixture was then cooled to 80 °C and treated with 4.6 g of *n*-butanol and then stirred for an additional 20 min to deactivate the catalyst. The reaction mixture products were isolated by vacuum distillation and are summarized in Table 9.

TABLE 9 Summary of reaction products produced by heating hexamethylene diisocyanate to 130 °C for 2 h using hexamethyldisilazane as the reaction catalyst

Component	Amount (wt%)
Unreacted hexamethylene diisocyanate	0.24
Mono- and di-butyl hexamethylene carbamate	0.5
Dimeric hexamethylene diisocyanate	11.3
Trimeric hexamethylene diisocyanate	51.3
Bis-(trimeric hexamethylene diisocyanate)	25.2
Unidentified oligomers	11.46

2. *Preparation of pentaerythrityl carbamate prepolymer by the in situ dimerization and trimerization of hexamethylene diisocyanate in the absence of hexamethyldisilazane*

A reactor charged with 4942.3 g of hexamethylene diisocyanate and 100.1 g of pentaerythritol at 23 °C under an inert atmosphere was gradually heated to 140 °C for 2 h. After heating and stirring were continued for an additional 35 min, the mixture became translucent, and insoluble materials were observed. After an additional 3 h of heating and stirring, the mixture was cooled, and 3.7 g of insoluble mass was isolated by filtration. The mixture was then distilled at 160 °C at 0.5 to 1 mbar, and the reaction products were isolated as summarized in Table 10.

TABLE 10 Summary of the reaction products produced by heating hexamethylene diisocyanate and pentaerythritol to 140 °C for 5.5 h in the absence of hexamethyldisilazane

Component	Amount (wt%)
Unreacted hexamethylene diisocyanate	0.11
Dimeric hexamethylene carbamate	20.2
Trimeric hexamethylene carbamate	33.2
Pentameric hexamethylene carbamate	18.3
Oligomers/polymers	27.54
Biuret	0.2

Testing

Material testing data are not supplied by author.

Advantages over Prior Art

The method for preparing low-viscosity hexamethylene diisocyanate dimers and trimers using hexamethyldisilazane as the reaction catalyst is unreported. Moreover, the preparation of low-viscosity dimeric and trimeric isocyanate urethane prepolymers is also unreported. All of these intermediates are readily blendable into paint formulation to provide smooth coated surfaces. Another application for these materials is as components in adhesive formulations.

Notes

1. Bernard[1] condensed hexamethylene diisocyanate with aminopropyltriethoxysilane to form the corresponding biuret (III), which was used in adhesive and in paint formulations.

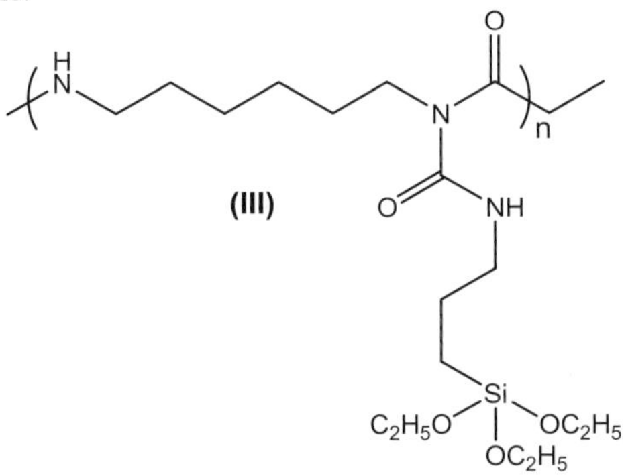

(III)

2. Bernard[2] prepared a polyallophanate (IV) containing up to 17 wt% terminal isocyanate by condensing hexamethylene diisocyanate with 1-butyl alcohol using dibutyltin dilaurate as the reaction catalyst.

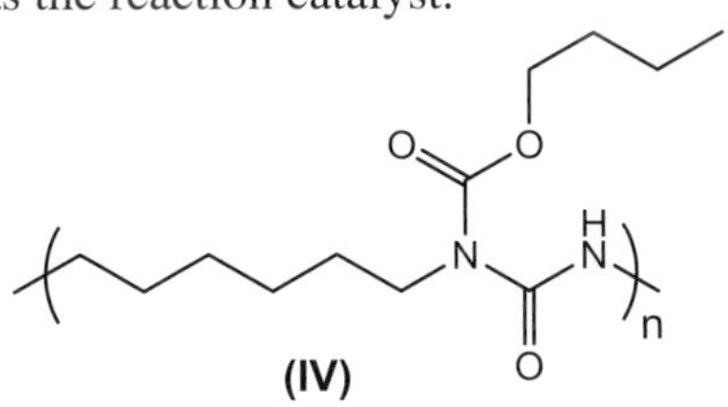

(IV)

3. Bernard[3] converted hexamethylene diisocyanate into an oligouretidinedione (V) containing up to 18.7% isocyanate content using tri-*n*-butylphosphine as catalyst and by heating the mixture to 60 °C for 10 h.

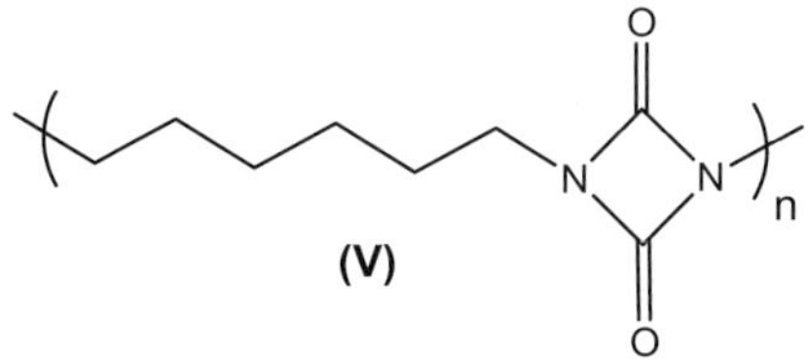

(V)

4. Bernard[4] reacted isophorone diisocyanate, lanthanum tris(2-methoxy-ethylene glycolate), and imidazole to generate equivalent amounts of isophorone oligomeric dimers (VI) and trimers (VII).

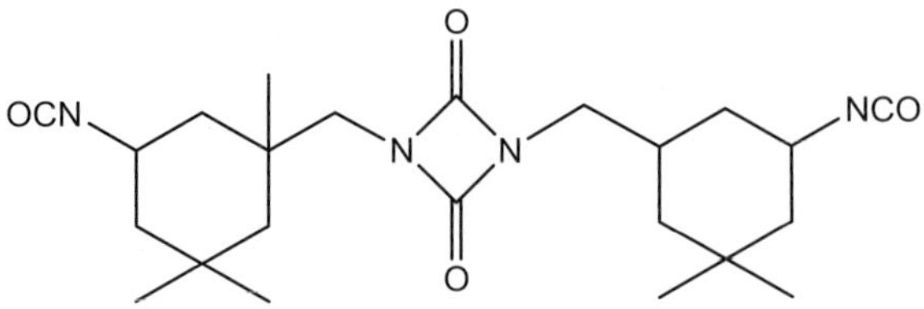

(VI)

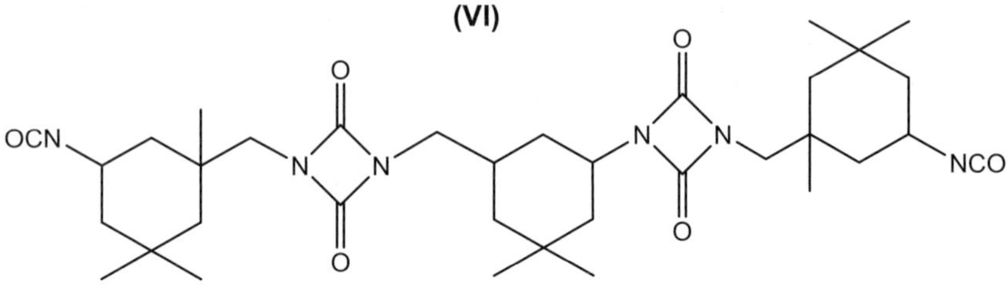

(VII)

REFERENCES

1. Jean-Marie Bernard et al, US. Patent 7,875,693 (January 25, 2011) **Perstorp France (Saint Priest, FR)**
2. Jean-Marie Bernard et al, US. Patent 7,635,742 (December 22, 2009) **Rhodia Chimie (Aubervilliers, FR)**
3. Jean-Marie Bernard et al, US. 7,163,995 (January 16, 2007) **Rhodia Chimie (Aubervilliers, FR)**
4. Jean-Marie Bernard et al, US. Patent 7,524,435 (April 28, 2009) **Rhodia Chimie (Aubervilliers, FR)**

E. RHEOLOGY CONTROL AGENTS

Lysine Diurea/Diurethane Methyl Esters

Title:	Rheology control agents
US Patent:	8,030,513 (October 4, 2011)
Author:	Christian Peter Lenges et al.
Assignee:	E. I. du Pont de Nemours and Company (Wilmington, DE)

Significance: Hydrophobically modified ureas and urethanes were prepared in quantitative yields by the condensation of lysine diisocyanate methyl ester or 1,6-hexanediisocyanate with aliphatic or cycloaliphatic amines or alcohols. Latex paint formulations containing high Tg acrylate or low molecular weight polyester resins were completely compatible with experimental agents. In all cases, the viscosity of the latex paint increased and gelling times decreased for blends additized with 5 wt% of these additives.

Additive Names *N,N′*-Diundecyl lysine diurea methyl ester (I)
N,N′-Dicyclohexyl lysine diurea/diurethane methyl ester (II)
N,N′-Diheptyl lysine diurea/diurethane methyl ester (III)
N,N′-Dipentyldiethylester-1,6-hexyldiurea, (IV)
N,N′-(1-Benzyl-2-methoxyethoxy)ethylester-1,6-hexyldiurea (V)

Safety Hydrophobically modified ureas and urethanes are nonirritating to either the eyes or the skin. Their oral LD50 in rats is >5 g/kg and is considered nontoxic.

Additive Structures

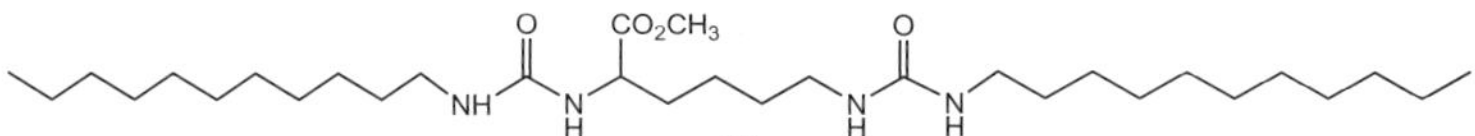

(I)

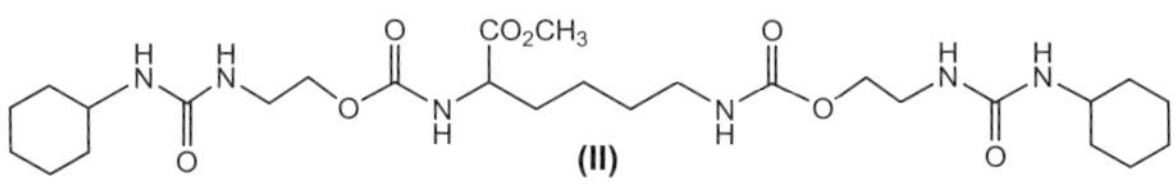

(II)

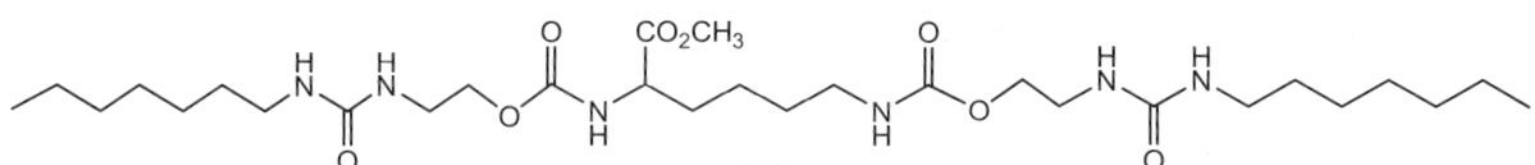

(III)

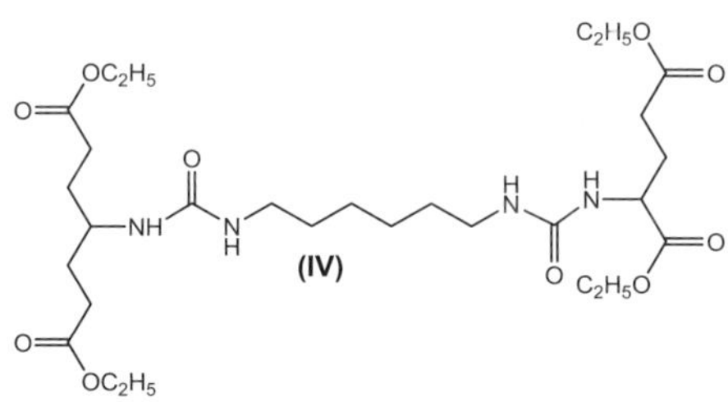

(IV)

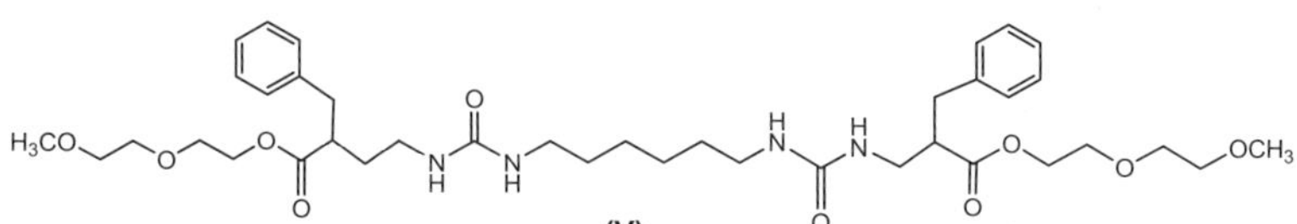

(V)

Additive Preparation

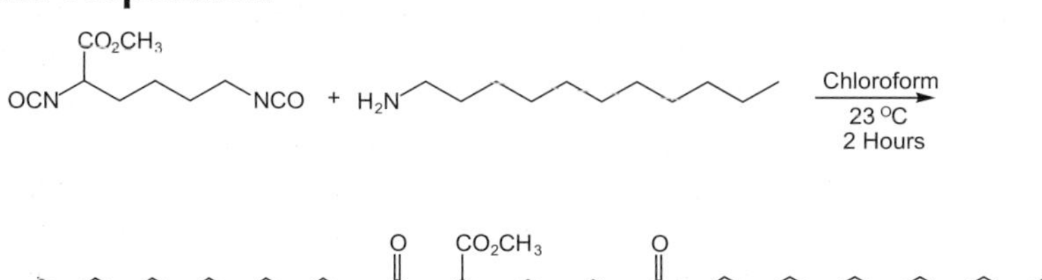

Additive Synthesis

1. Preparation of N,N′-diundecyl lysine diurea methyl ester

A round-bottom container with 37.05 g of undecyl amine dissolved in 700 ml of chloroform was treated with the dropwise addition of 23.18 g of lysine diisocyanate methyl ester and then vigorously stirred at an ambient temperature for 2 h. After infrared spectroscopic analysis indicated that the reaction was complete, the mixture was filtered, concentrated, and the product was isolated in an 89.8% yield.

Testing

1. Viscosity Testing

An acrylic or polyester-based coating resin was blended with 2 wt% of a selected rheology control agent dissolved in 80 wt% butyl acetate and 20 wt% butyl alcohol and was then aged for 1 h, and the viscosity was determined. Test results are provided in Table 11.

2. Gel Time

The gelation or viscosity increase the times for blends additized with 5 wt% of selected rheology agents was tested in waterborne-based coating compositions containing Aquacryl® 514. In this test, sealed containers of the blend were routinely inverted, and gel or viscosity increase times of the mixture were measured. Test results are provided in Table 12.

Test Results

1. Viscosity Testing

TABLE 11 **The effect of selected rheology reagents on the viscosity of coating resins dissolved in 80 wt% butyl acetate and 20 wt% butyl alcohol. Both high glass transition temperatures, Tg, of acrylate resins, and low number average molecular weights, Mn, polyester resins were used in this evaluation**

Rheology Agent	Resin	Viscosity 0.5 rpm (cps)	Viscosity 5 rpm (cps)	Viscosity 50 rpm (cps)	Viscosity 100 rpm (cps)	Viscosity 250 rpm (cps)
Additive 1	High Tg Acrylate Resin	28200	4560	668	496	416
Additive 2	High Tg Acrylate Resin	6400	1580	446	345	262
Additive 5	High Tg Acrylate Resin	28400	2240	616	497	409
Additive 2	Low Mn Polyester	44600	2000	720	553	358
Additive 5	Low Mn Polyester	11200	1440	340	270	236
Unadditized	Low Mn Polyester	60	60	68	65	63

2. Gel Time

TABLE 12 The effect of gelling times for fully formulated paint additized with 5 wt% selected rheology agents

Rheology Agent	Gel Time (hh:mm:ss)
Additive 2	3:00:00
Additive 3	12:00:00
Additive 4	00:10:00

Advantages over Prior Art

The objective of this investigation was to prepare rheological control agents compatible with low organic emission automotive spray paint finishes that have favorable weatherable properties and which contain resins. Hydroxyl functionalized monoamines are currently used as rheological control agents in automotive paint formulations, but they have marginally resin compatibility. While currently urea and urethane rheology control agents derived from toluene diisocyanate have good resin compatibility, their weatherability is poor, gel times are long, and additive dispersability is limited.

Notes

1. Additional rheology agents (VI) and (VII) were prepared by Lenges[1] and used in paint formulations.

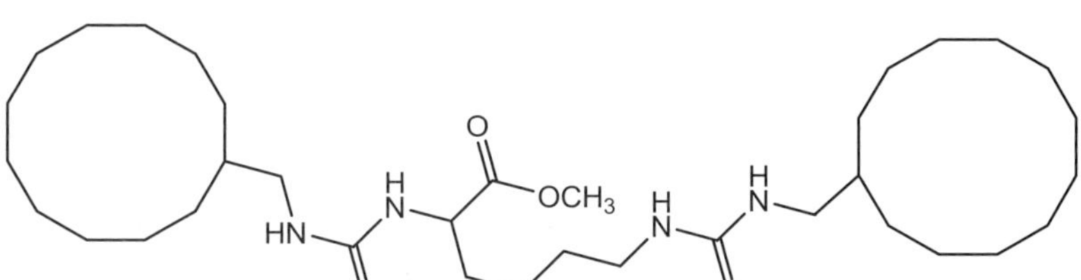

(VI)

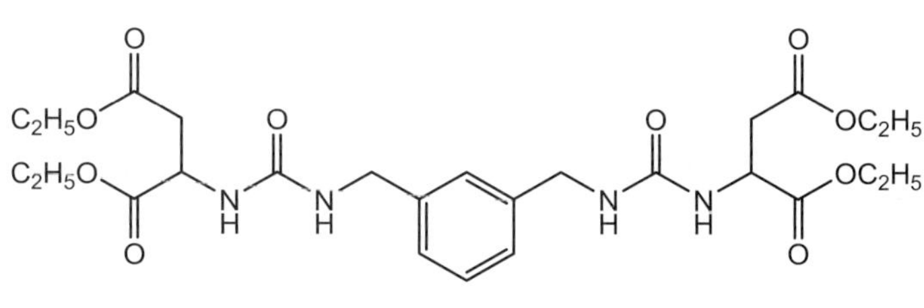

(VII)

2. The rheological control agent, *N*-2-ethylhexanoyl-l-glutamic acid dibutylamide (VIII) prepared by Yamato[2] was used as a gelling agent in the oil component of cosmetic formulations designed for use in dermal applications, particularly lipsticks.

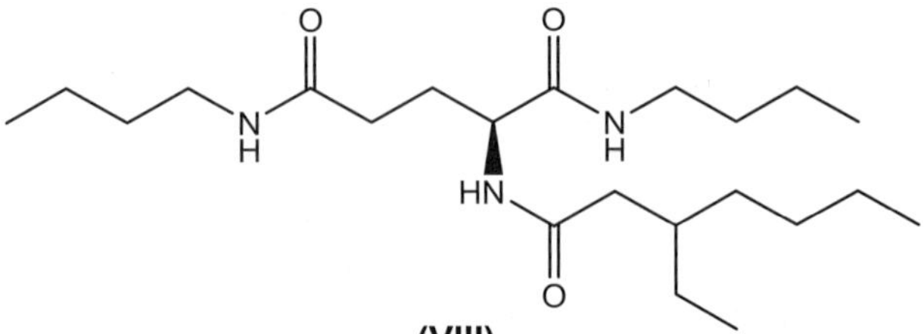

(VIII)

3. Yang[3] used polyalkoxylate derivatives (IX) and (X) as rheological additives for aqueous systems and emulsions in hydrophobically modified alkali soluble paint formulations.

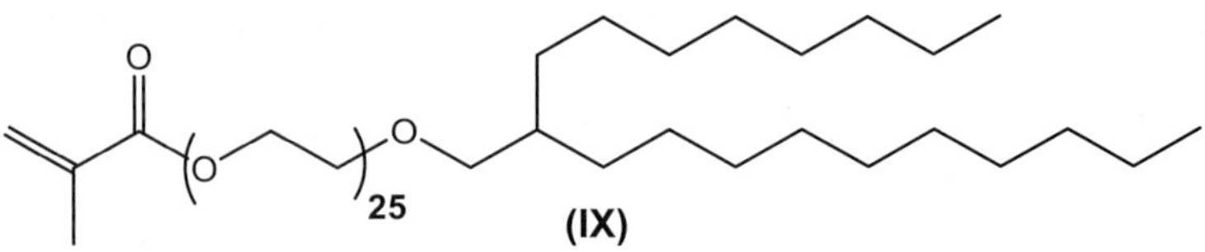

(IX)

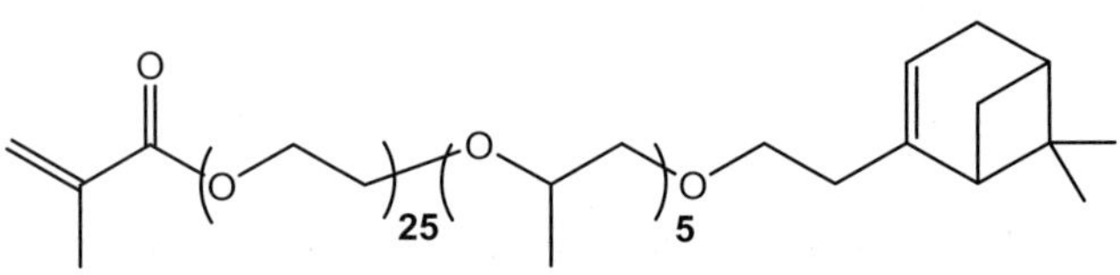

(X)

4. Haubennestel[4] prepared fatty acid polyamides (XI) as rheological agents, which were effective when used in epoxide-based coatings and unsaturated polyester resin formulations.

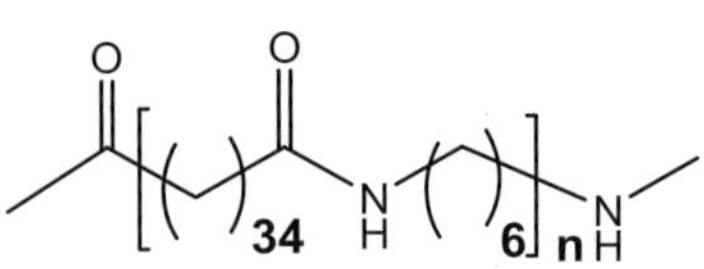

(XI)

REFERENCES

1. Christian Peter Lenges et al, US 7,741,510 (June 22, 2010)
E. I. du Pont de Nemours and Company (Wilmington, DE)
2. Naoya Yamato et al, US. Patent 7,244,419 (July 17, 2007)
Ajinomoto Co., Inc. (Tokyo, JP)
3 Hui Shirley Yang et al, US. Patent 8,071,674 (December 6, 2011)
Rhodia Operations (Aubervilliers, FR)
4. Karlheinz Haubennestel et al, US. Patent 7,834,126 (November 16, 2010)
BYK-Chemie GmbH (DE)

F. SURFACE ACTIVE AGENTS

Fluoroalkyl Phosphate Compositions

Title:	Fluoroalkyl phosphate compositions
US Patent:	7,951,975 (May 31, 2011)
Author:	Sheng Peng et al.
Assignee:	E.I. du Pont de Nemours and Company (Wilmington, DE)

Significance: Perfluoroalkylphosphate ammonium salts containing six or fewer perfluorinated carbons were found to be effective as surface active agents when used in paint formulations. Treatment levels as low as 0.001 wt% lowered paint composition surface tension by as much as 45%. Contact angles were reduced by as much as 300% in formulations containing 0.3 wt% additive.

Additives

1,1,1,2,2,3,3,4,4-Perfluoro-6-(diethanolammonium)phosphate hexane (I)
1,1,1,2,2,3,3,4,4-Perfluoro-6-(diammonium)phosphate hexane (II)
1,1,1,2,2,3,3,4,4,6,6-Perfluoro-8-(diammonium)phosphate octane (III)
1,1,1,2,2,3,3,4,4,6,6,8,8-Perfluoro-10-(diammonium)phosphate octane (IV)

Safety

Based on related products, eye contact with these agents may result in general discomfort, tearing, or blurring of vision. Other potential health effects may include nasal, throat, or lung irritation caused by spray or mist inhalation. Chronic inhalation may be toxic to the lungs followed by severe shortness of breath hours after the exposure.

Additive Structures

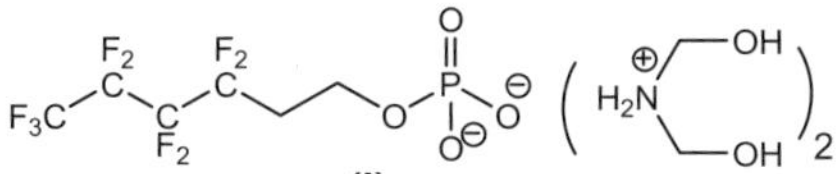

(I)

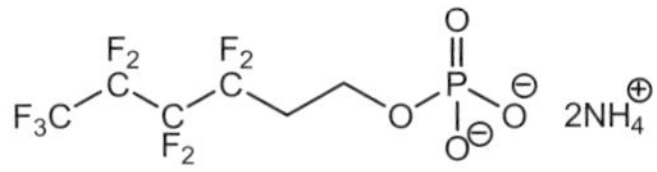

(II)

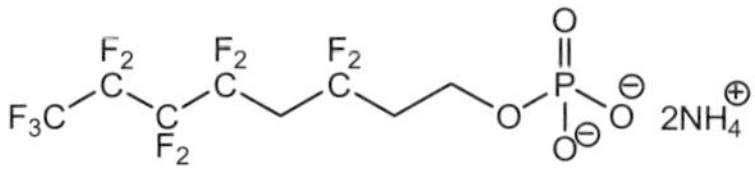

(III)

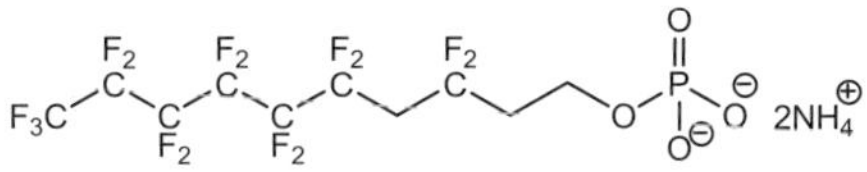

(IV)

Additive Preparation

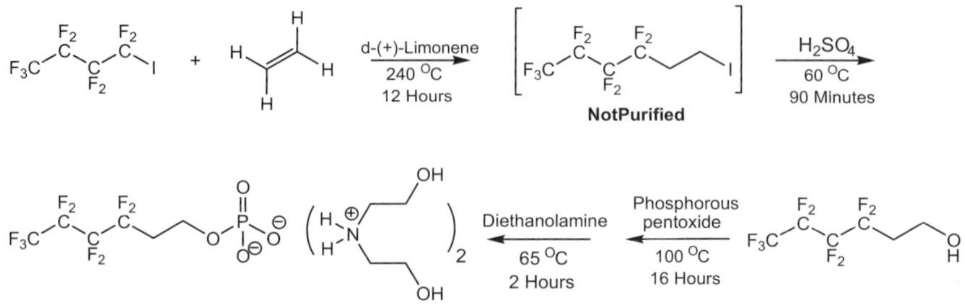

Additive Synthesis

1. Preparation of 1,1,1,2,2,3,3,4,4-perfluoro-6-hexanol

An autoclave containing 217 g of $C_4F_9CH_2CF_2I$ and 1 g of d-(+)-limonene was charged with 25 g of ethylene and then heated to 240 °C for 12 h. After vacuum distillation, 50 g of the intermediate was converted into the corresponding alcohol by washing with 70 ml of fuming sulfuric acid and then stirring at 60 °C for 90 min and quenching with ice-cold water containing 1.5 wt % sodium sulfite. The mixture was then heated to 95 °C for 30 min, and the bottom layer was isolated. The bottom layer was washed with 10 wt % aqueous sodium acetate, dried, and the product was isolated by distillation at 54 °C to 57 °C at 2 mmHg.

2. Preparation of 1,1,1,2,2,3,3,4,4,6,6-perfluoro-6-(diethanolammonium) phosphate hexane

A reactor was charged with 1.44 g of phosphorous pentoxide, and 10 g of the Step 1 product was then heated to 100 °C for 16 h. The reaction mixture was then treated with 30 ml of diethanolamine over 90 min and then stirred for 2 h at 65 °C, and the product was isolated.

Testing

1. Contact Angles

Contact angles of selected experimental agents were added to a semigloss latex paint formulation in an amount of 0.03 wt% by dry weight of the composition and contact angles were measured using hexadecane as the liquid. Test results are provided in Table 13.

TABLE 13 Contact angle test results for paint additized with 0.03 wt% of experimental additives using hexadecane. Higher values are preferred since they are indicative of greater dirt and soil repellency

Additive	Contact Angle (Degrees)
Control	28.1
Additive 1	76.8
Additive 2	67.5
Additive 3	62.0
Additive 4	64.2

2. *Surface Tension*

Surface tensions of selected experimental agents were measured using a Kruess Tensiometer, K11, Version 2.501 at varying concentrations, and test are results provided in Table 14.

TABLE 14 Surface tension test results for paint formulations additized with experimental additives. The surface tension of an unadditized paint coating was 72.3 dynes/cm

	Surface Tension (dynes/cm)					
Additive	Treat Rate (0.001 wt%)	Treat Rate (0.005 wt%)	Treat Rate (0.010 wt%)	Treat Rate (0.050 wt%)	Treat Rate (0.020 wt%)	Treat Rate (0.500 wt%)
Additive 1	58.5	45.1	35.3	35.7	32.0	31.5
Additive 2	65.6	55.4	53.5	41.4	32.6	29.6
Additive 3	38.0	21.9	18.8	17.7	17.3	17.2
Additive 4	51.4	39.9	35.6	26.8	22.7	21.8

Test Results

Advantages over Prior Art

It has previously been thought by Honda[1] that optimum parallel alignment of perfluoroalkyl chains occurs when at least eight perfluorinated carbons are present. The current discovery contests this earlier observation and suggests that high performance of perflorinated surfactants may be as effective as longer perfluoro chain analogs.

Notes

1. Miller[2] prepared an effective two component masonry sealing composition consisting of polymerizable perfluoromethacrylate ester (V) and perfluorophosphate ammonium salt (VI).

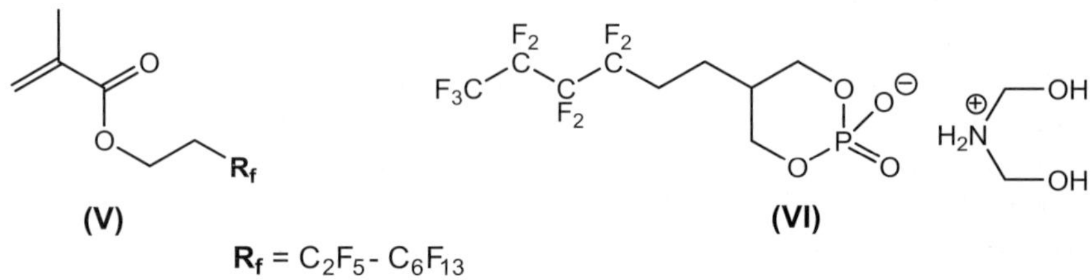

2. Perfluorosulfonate copolymers (VII) prepared by Savu[3] were effective in lowering the surface tension of water and used in polyurethane coating formulations.

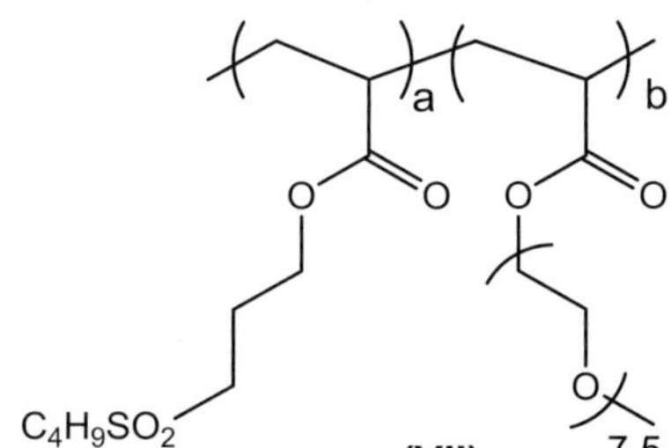

(VII)

3. Zullo[4] prepared water-based latex paint coatings additized with surface active potassium or ammonium 9-decenoic acid VIII, which resisted in-can spoilage and were effective as antimicrobial agents.

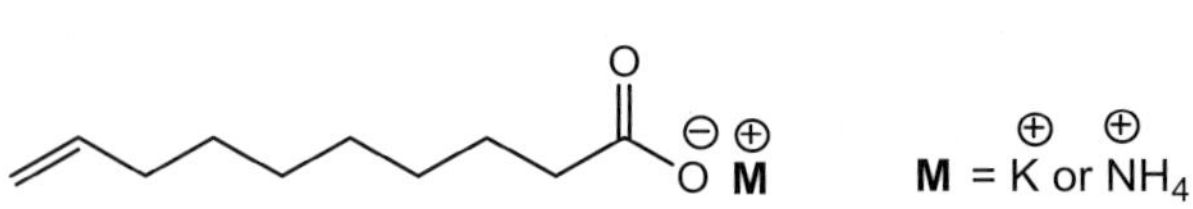

(VIII)

4. Peng[4] prepared short-chained perfluoroalkyl urethane additives (IX), which improved the wash durability and oil repellency of carpets.

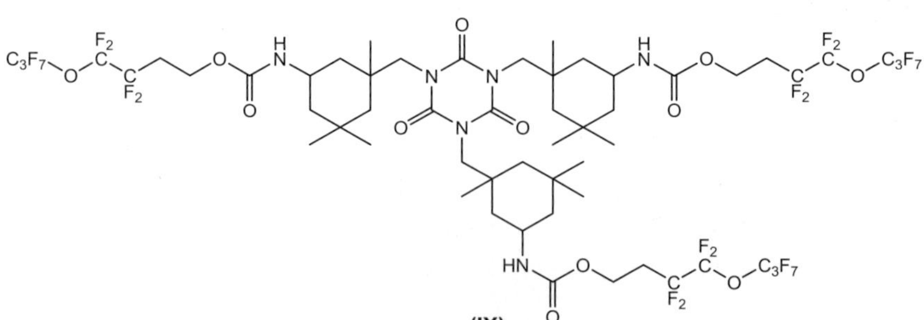

(IX)

REFERENCES

1. T. Honda et al, "Molecular aggregation structure and surface properties of poly(fluoroalkyl acrylate) thin films," Macromolecules, 2005, 38, 5699–5705
2. Chandra Tier Miller et al, US Patent 7,951,975 (May 31, 2011)
   E. I. du Pont de Nemours and Company (Wilmington, DE)
3. Patricia M. Savu et al, US Patent 7,662,896 (February 16, 2010)
   3M Innovative Properties Company (St. Paul, MN)
4. Jill Louise Zullo et al, US Patent 7,951,232 (May 31, 2011)
   Elevance Renewable Sciences, Inc. (Bolingbrook, IL)
5. Sheng Peng et al, US Patent 7,956,124 (June 7, 2011)
   E. I. du Pont de Nemours and Company (Wilmington, DE) OH

G. WATER RESISTANCE IN PAINT

Sodium Di(2-propylheptyl)sulfosuccinate

Title:	Salts of alkyl esters of sulfonated dicarboxylic acids and compositions containing the same
US Patent:	7,981,932 (July 12, 2011)
Author:	Joachim Meyer et al.
Assignee:	Cognis IP Management GmbH (Duesseldorf, DE)

Significance: The water-soluble latex additive, sodium di(2-propylheptyl)sulfosuccinate, was quantitatively prepared by condensing maleic anhydride with 2-propylheptanol and then postreacting with sodium disulfite. When used in paint formulations at 0.7 wt%, dried paint surfaces had excellent water resistance and diminished surface tensions.

Additive Sodium di(2-propylheptyl)sulfosuccinate (I)

Safety Harmful if swallowed. May be harmful if inhaled or absorbed through the skin. Skin and respiratory irritant. Severe eye irritant.

Additive Structure

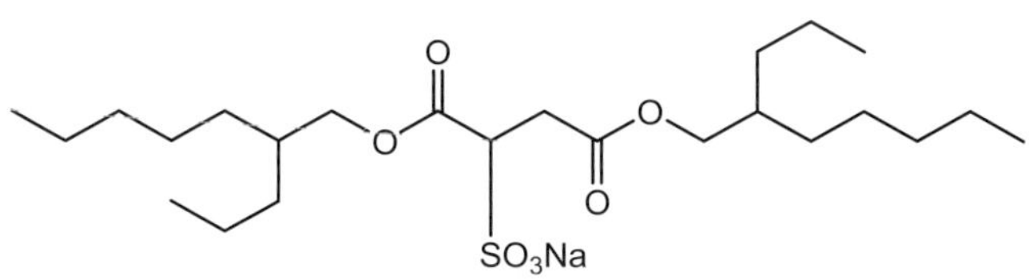

(I)

Additive Preparation

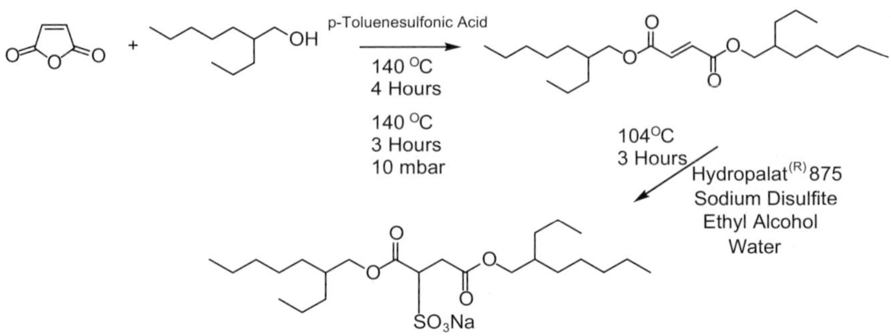

Additive Synthesis

1. Preparation of di-(2-propylheptyl)-maleate

A round-bottom flask was charged with 392 g of maleic anhydride, 1330 g of 2-propylheptanol, and 8.6 g of *p*-toluenesulfonic acid monohydrate and then slowly heated to 140 °C for 4 h. The product was then further heated for an additional 3 h while slowly reducing the pressure to 10 mbar, and the product was isolated having a residual acid value of 1.66 mg KOH/g.

2. Preparation of sodium di(2-propylheptyl)sulfosuccinate

A reactor was charged with 574 g of the Step 1 product, 40.0 g of Hydropalat® 875,142 g of sodium disulfite and 194 g of deionized water and then refluxed for 3 h. On cooling to ambient temperature, the product gelled and was liquefied by dissolving in 50 g of ethanol at 80 °C. The product had an acid value of 0.18 mg KOH/g; Epton 14.01%; dry residue 71.66% with a sodium sulfate content of 0.32%.

Testing

1. Dynamic Surface Tension Determination

Paints were prepared additized with 0.1 wt% of the experimental additive, and the dynamic surface tension was determined with a Kruss bubble tensiometer at various bubble frequencies. Test results are provided in Table 15.

2. Water Resistance of Paints

Water stability was determined to DIN 68861. The paint formulation used in this test is provided in Table 16.

TABLE 15 Paint formulation used to evaluate the water resistance of the experimental additive, sodium di(2-propylheptyl)sulfosuccinate

Component	Amount (g)
Basic paint	100
NeoPac® E-106	90.6
DSX® 1514	0.3
Butyl glycol	2.2
Methyl Diglycol	2.2
Dehydran® 1293,	1.0
Deionized Water	0.7
Experimental Additive	0.7 wt%

Test Results

1. Dynamic Surface Tension Testing

TABLE 16 Dynamic surface tension test results for latex paint additized with 0.1 wt% of the experimental additive sodium di(2-propylheptyl)sulfosuccinate and the comparison additive, di(2-ethylhexyl) maleate at various bubbling frequencies. Lower surface tension values are preferred

Bubble Frequency (Hz)	Comparison Additive (mN/m)	Experimental Additive (mN/m)
0.1	33.8	26.1
0.5	36.1	26.9
1.0	37.2	28.3
5.0	39.5	41.5
10.0	40.6	50.9

2. Water Stability

TABLE 17 Water stability test results of a dried painted surface containing the experimental additive and a comparative. Lower water resistance values are preferred

Example	Paint Formulation	Water Resistance of Dried Paint
Comparative	Paint Additized with Comparison Additive, di(2-ethylhexyl) maleate	3
Experimental	Paint Additized with Experimental Additive, sodium di(2-propylheptyl) sulfosuccinate	1

Advantages over Prior Art

The predominant water-soluble paint surfactants are di-octyl- and di(ethylhexyl) sulfo-succinates. Neither of these agents appears to be especially effective in imparting water resistance to latex paint. In addition, the presence of unreacted alcohol imparts an unpleasant odor to paint formulations additized with these agents. The use of sodium di(2-propylheptyl)sulfosuccinate eliminates both these concerns and improves the dynamic surface tension of the painted surface Table 17.

Notes

1. Bao[1] incorporated dioctyl sodium sulfosuccinate (II) into water-soluble latex paint formulations containing poly(isodecyl methacrylate), which reduced drag reducer by 19% at 2 ppm.

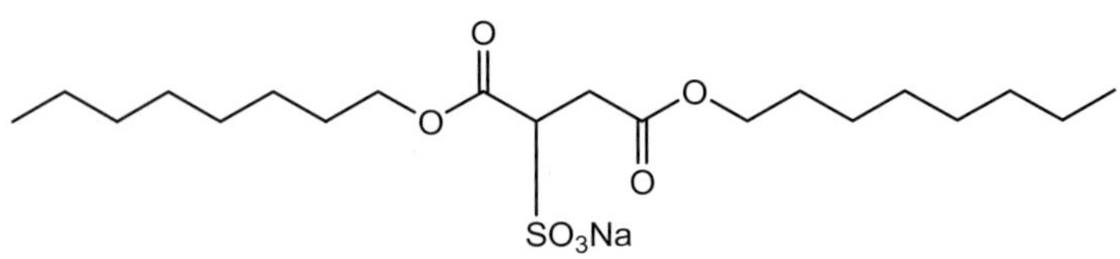

(II)

2. Vyorykka[2] prepared water-soluble latex paints containing acrylic acid, styrene, or oleic acid, which improved contact angle, normalized stain size, rolling angle, and elasticity. In this process, the monomer was initially polymerized and then polymeric particles were mechanically dispersed into the paint after agglomeration.
3. Water-soluble latex paint formulations containing poly(ethyl acrylate-co-methacrylic acid-styrene) blended with difunctional bisphenol A-epichlorohydrin liquid epoxy resins (EPON® 828), bisphenol A, butyl cellosolve, and tri-*n*-butylamine were prepared by Schutte[3] with improved coating adherence properties when applied to medicament-contact surfaces.

4. Water-soluble latex paint containing poly(butyl acrylate-co-methyl methacrylate-co-methacrylic acid-co-acetoacetoxyethyl methacrylate-co-*N*-(2-methacryloyloxyethyl)ethylene urea) (ROHAMERE® 6852) (III) was prepared by Wu[4] and had excellent adhesion properties to surfaces at low and elevated temperatures.

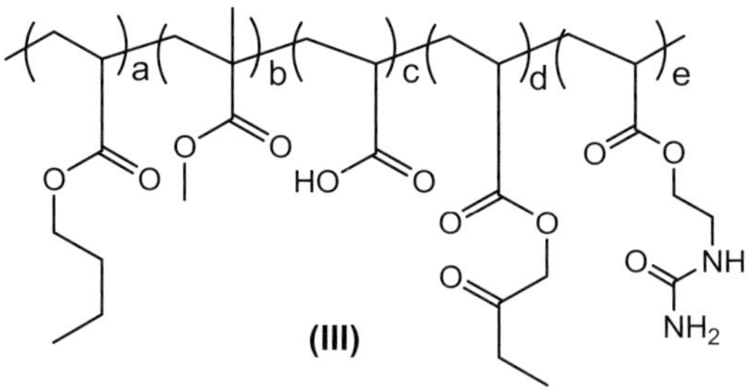

(III)

REFERENCES

1. Zhiyi Bao et al, US Patent Application 20110184121 (July 28, 2011) **ConocoPhillips Company (Houston, TX)**
2. Jouko T. Vyorykka et al, US Patent Application 20110021698 (January 27, 2011) **The Dow Chemical Company (Minneapolis, MN)**
3. Grant Schutte et al, US Patent Application 20100282248 (November 11, 2010) **Valspar Sourcing, Inc. (Minneapolis, MN)**
4. Wenjun Wu et al, US Patent Application 20110152439 (June 23, 2011) **Ankema Inc. (Philadelphia, PA)**

Chapter 26

Photography

A. RADIATION CURABLE FILMS

Triazine Carbamate Acrylate

Title:	Radiation curable 1,3,5-triazine carbamates and 1,3,5-triazine ureas containing vinyl groups, methacryloyl groups, or acryloyl groups
US Patent:	8,044,197 (October 25, 2011)
Author:	Yvonne Heischkel et al.
Assignee:	BASF SE (Ludwigshafen, DE)
Significance:	The radiation curable additive of 2,4,6-tris-(2-ethoxyacrylatocarbamoyl)-1,3,5-triazine was prepared by condensing 2,4,6-tris-(methylcarbamoyl)-1,3,5-triazine with 2-hydroxyethyl acrylate. This intermediate was then postreacted with 1,4-butanediol and poly(adipic acid-co-isophthalic acid-co-1,6-hexanediol)diol and could be thermally and photochemically cured. Films produced were clean, nonyellowing, with favorable pendulum hardness, and Erichsen cupping.

Additive Names

2,4,6-Tris-(2-ethoxyacrylatocarbamoyl)-1,3,5-triazine (I)
2,4,6-Tris-(2-ethoxyacrylate)-1,3,5-triazine (II)
2,4,6-Tris-(2-ethoxyacrylate)-isocyanuric acid (III)

Safety

Both cyanuric and isocyanuric acids are severe skin and eye irritants. The urethane-free additives are considered extremely toxic and photopolymerization hazards.

Next Generation of International Chemical Additives. DOI: http://dx.doi.org/10.1016/B978-0-444-53788-1.00026-7

Additive Structures

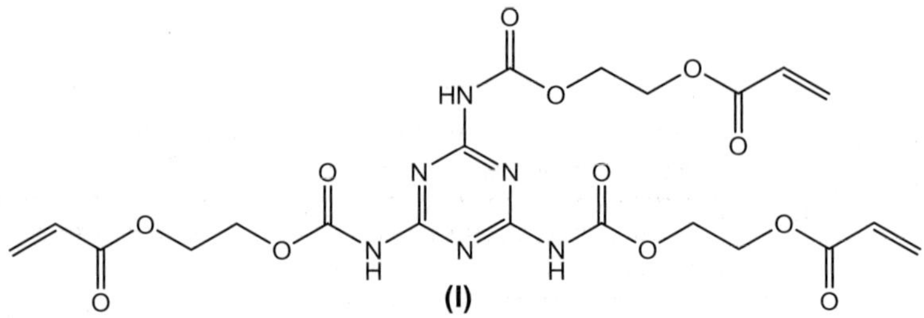

(I)

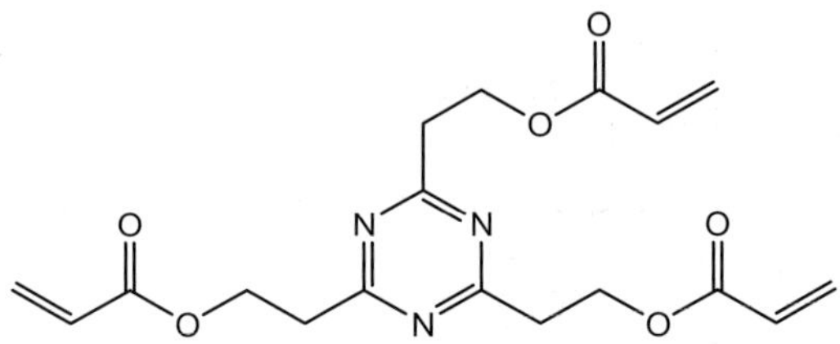

(II)

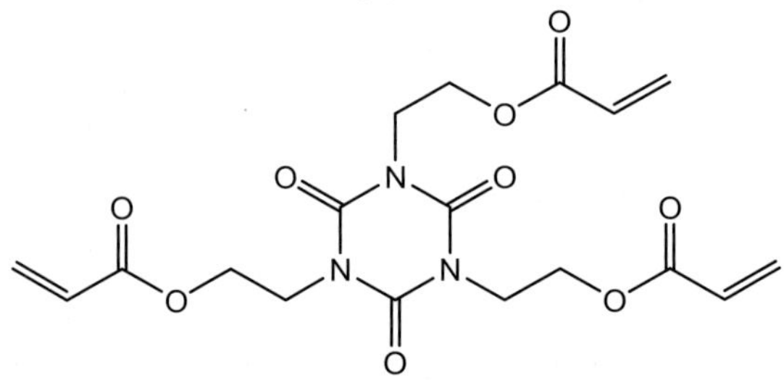

(III)

Additive Preparation

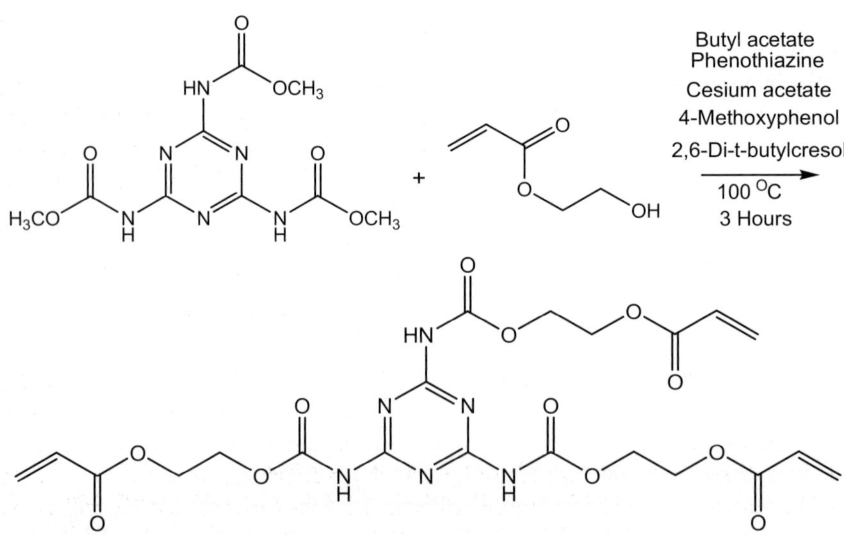

In situ Additive Preparation and Polymerization

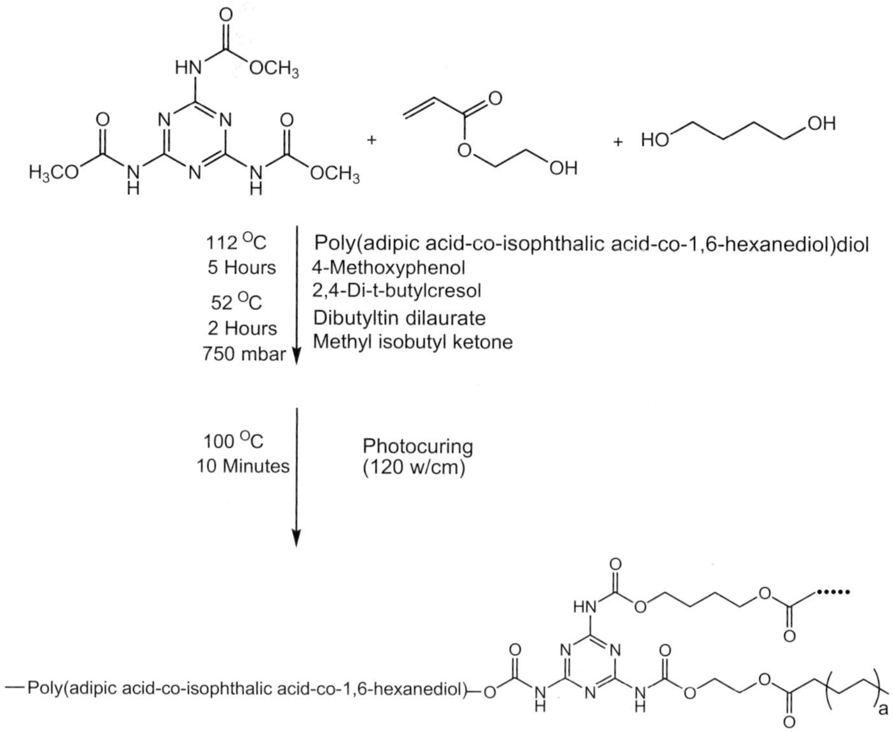

Additive Synthesis

1. *Preparation of 2,4,6-tris-(2-ethoxyacrylatocarbamoyl)-1,3,5-triazine*

A 250 ml four-necked reaction flask equipped with a distillation bridge, Liebig condenser, and stirrer was charged with 12.32 g of *n*-butyl acetate, 6.0 g of 2,4,6-tris-(methylcarbamoyl)-1,3,5-triazine (Note 1), 6.97 g of 2-hydroxyethyl acrylate, 12.5 mg of 4-methoxyphenol, 4 mg of 2,6-di-*t*-butyl-*p*-cresol, 0.3 mg of phenothiazine, and 0.96 mg of cesium acetate and then heated to 110 °C for 6 h. After methanol was distilled from the reaction vessel, the product was isolated and used without further purification.

2. *In situ preparation and polymerization of 2,4,6-tris-(2-ethoxyacrylatocarbamoyl)-1,3,5-triazine with poly(adipic acid-co-isophthalic acid-co-1,6-hexanediol)diol*

A suspension of 0.136 g of *p*-methoxyphenol, 0.045 g of di-*t*-butyl-*p*-cresol, 0.003 g of phenothiazine, 0.016 g of dibutyltin dilaurate, 2,4,6-trismethylcarbamoyl-1,3,5-triazine, poly(adipic acid-co-isophthalic acid-co-1,6-hexanediol)diol, and hydroxyethyl acrylate was dissolved in 30.0 ml of methyl isobutyl ketone and then stirred at 112 °C for 4 h. The reaction mixture was then distilled at 52 °C at 750 mbar for 2 h, and a clear resin solution was isolated.

Additive Testing

1. Pendulum hardness

Pendulum hardness was determined in accordance with DIN 53157 and is a measure of the hardness of the coating where results are reported in seconds until the pendulum comes to a standstill. In this test, high values denote high hardness. Films used for determining pendulum hardness were applied to glass substrate using a sharp blade so that the cured film thickness was 100 μm. Test results are provided in Table 1.

2. Erichsen cupping

Erichsen cupping testing was performed in accordance with DIN 53156. This test is a measure of the flexibility and elasticity of a substrate or coating where high testing values indicate high flexibility. Films used for determining Erichsen cupping were conducted on sheet metal using 50 μm coatings before curing.

Test Results

TABLE 1 Physical properties of cured materials containing the Additive 1 postreacted with poly(adipic acid-co-isophthalic acid-co-1,6-hexanediol)-diol. Optimum curing was obtained using a combination of both thermal and photochemical curing methods

Example	Additive I (moles)	Hydroxy-ethyl Acrylate (moles)	Polyester diol (moles)	Curing	Pendulum Hardness (sec)	Erichsen Cupping (mm)
1	0.4	1.2	0.2	Thermal	8	9.8
				Photochem	10	9.9
				Both	39	9.8
2	0.2	0.6	0.05	Thermal	17	9.8
				Photochem	21	9.8
				Both	76	9.4
3	0.1	0.3	–	Thermal	46	5.1
				Photochem	200	1.2
				Both	235	2.2
4	0.4	1.2	0.008	Thermal	126	5.1
				Photochem	153	1.1
				Both	235	3.8
5	0.1	0.3	6.7	Thermal	66	4.5
				Photochem	129	1.1
				Both	210	1.1

Advantages over Prior Art

Although radiation-activated s-triazines have previously been prepared, each method has inherent limitations. s-Triazines containing UV-activated trihalomethyl groups produced films that were gradually yellowed. Melamine modified with formaldehyde and s-triazines containing ethylenically unsaturated groups were sensitive to acids because of the presence of aminal or *N,O*-acetals. The present invention provides a method for producing a nonyellowing acid-insensitive film in high yield that is both thermally and photochemically curable.

Notes

1. 2,4,6-tris-(Methylcarbamoyl)-1,3,5-triazine was previously prepared by Wagner[1] by condensing 2,4,6-triisocyanato-1,3,5-triazine (IV), with methyl alcohol using tin(II) *n*-octanoate or tin(II) 2-ethylhexanoate as the reaction catalyst.

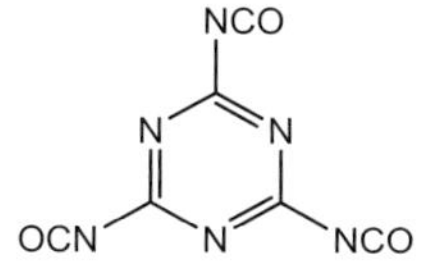

(IV)

2. Schneider[2] developed a more cost-effective method for preparing 2,4,6-tris-(alkyl-carbamoyl)-1,3,5-triazines, which entailed condensing melamine with ethylene carbonate and 1-butyl alcohol (Eq. 1).

Eq. 1.

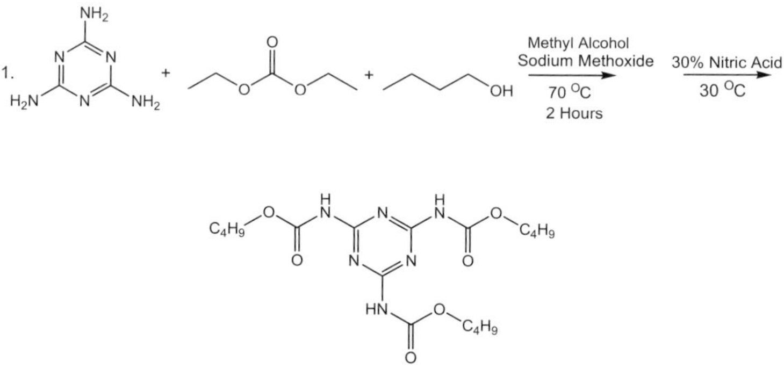

3. Bruchmann[3] prepared nonyellowing films having favorable pendulum hardness and Ericksen cupping properties by using hyperbranched polyesters consisting of (Case I), cyclohexane-1,2-dicarboxylic anhydride, pentaerythritol, and 1,4-cyclohexanediol or (Case II), adipic acid, pentaerythritol, and 1,4-cyclohexanedimethanol as illustrated below.

   Case I: Polyester hyperbranched polyester components:

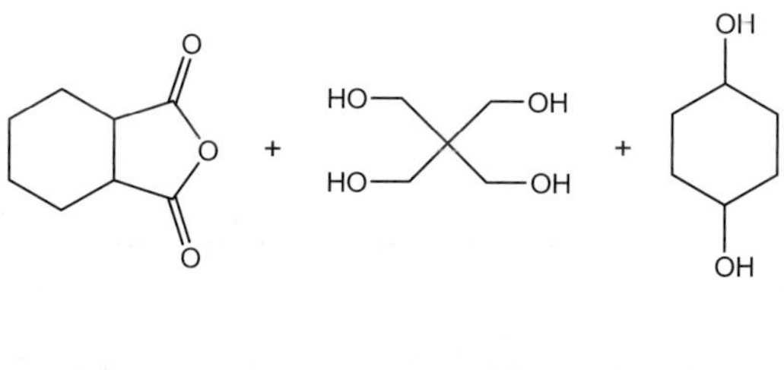

Case II: Polyester hyperbranched polyester components:

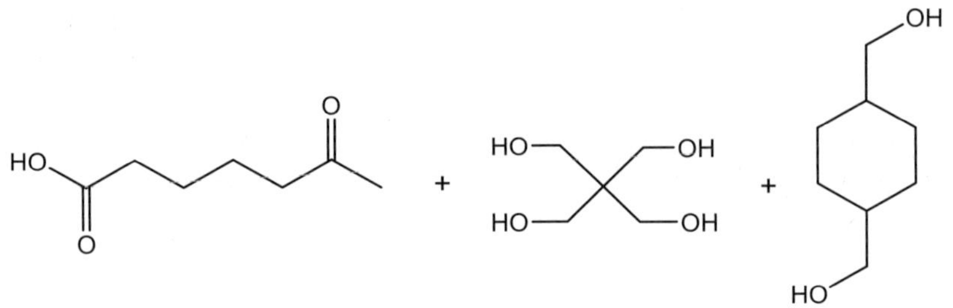

4. Heischkel[4] prepared UV-curable aqueous polyurethane dispersions consisting of the reaction product of polycaprolactonediol, 1,2-ethanediol, hydroxyethyl acrylate, dimethylolpropionic acid, isophorone diisocyanate, and tris(hexamethylenediisocyanato)isocyanurate.

REFERENCES

1. Eva Wagner et al, US. Patent 7,517,474 (April 14, 2009) **BASF Aktiengesellschaft (Ludwigshafen, DE)**
2. Joerg Schneider et al, US. Patent 7,507,818 (March 24, 2009) **BASF Aktiengesellschaft (Ludwigshafen, DE)**
3. Bernd Bruchmann et al, US. Patent 8,044,170 (October 25, 2011) **BASF Aktiengesellschaft (Ludwigshafen, DE)**
4. Yvonne Heischkel et al, US. Patent 8,048,937 (November 1, 2011) **BASF Aktiengesellschaft (Ludwigshafen, DE)**

Chapter 27

Polymers

A. ANTIHAZE AGENTS

Substituted Alditols

Title:	Substituted alditol compounds, compositions, and methods
US Patent:	7,888,454 (February 15, 2011)
Author:	Chunping Xie
Assignee:	Milliken & Company (Spartanburg, SC)

Significance: Substantial haze reduction was observed in poly(ethylene-co-propylene) and polypropylene resins when they were additized with dibenzylidene sorbitol nucleating agents. Optimum resin clarity was obtained using dibenzylidene sorbitol derivatives containing substituents in the aromatic ring.

Additive Names

Bis-1,3,2,4-(4′-ethylbenzylidene)-1-allyl sorbitol (I)
Bis-1,3,2,4-(5′,6′,7′,8′-tetrahydro-2-naphthaldehydebenzylidene)-1-allyl xylitol (II)
Bis-1,3:2,4-(3′,4′-dimethylbenzylidene)-1-propyl xylitol (III)
Bis-1,3,2,4-(3′-methyl-4′-fluoro-benzylidene)-1-propyl sorbitol (IV)
Bis-1,3,2-4-(3′,4′-dimethylbenzylidene)-1′-methyl-2′-propenyl sorbitol, (V)

Safety Concerns Slight skin irritation and slight-to-moderate eye irritation have been reported when using these clarifiers. Oral and dermal toxicities are unreported. Dust clouds are potentially explosive.

Next Generation of International Chemical Additives. DOI: http://dx.doi.org/10.1016/B978-0-444-53788-1.00027-9

Additive Structures

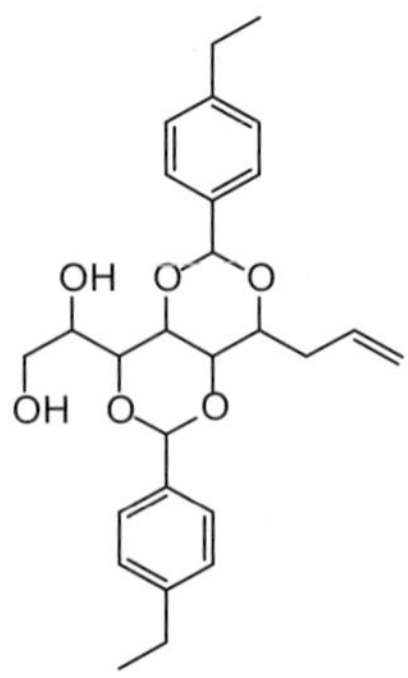

(I)

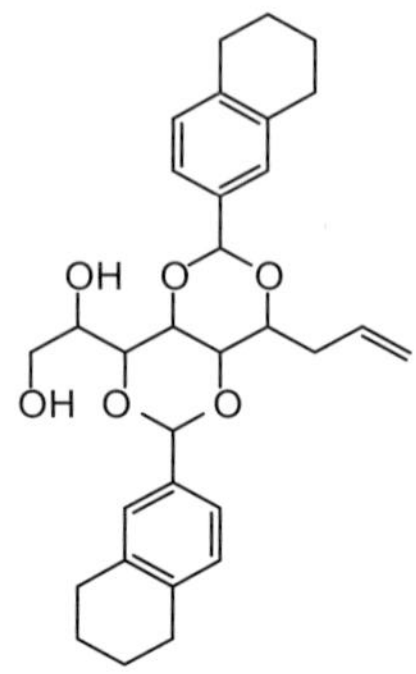

(II)

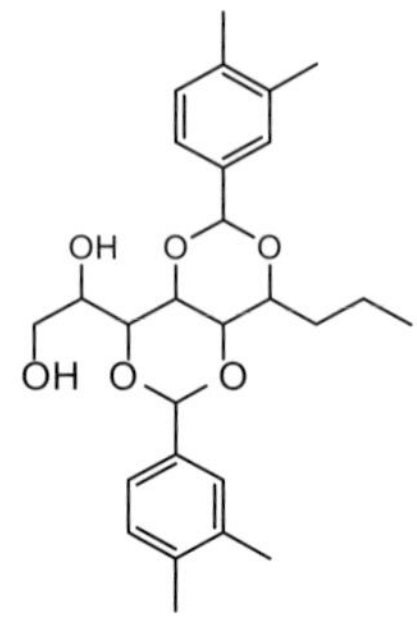

(III)

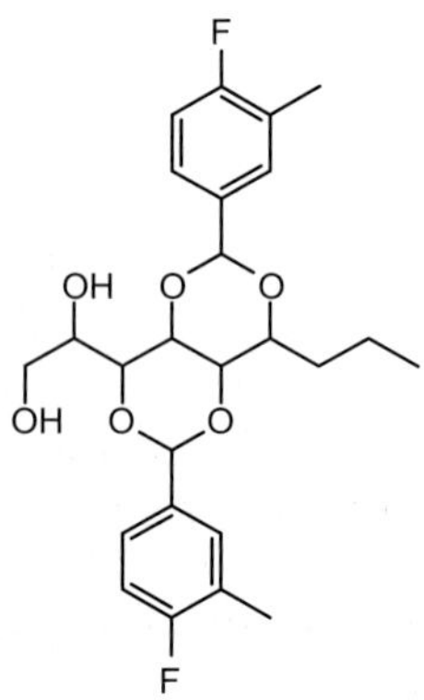

(IV)

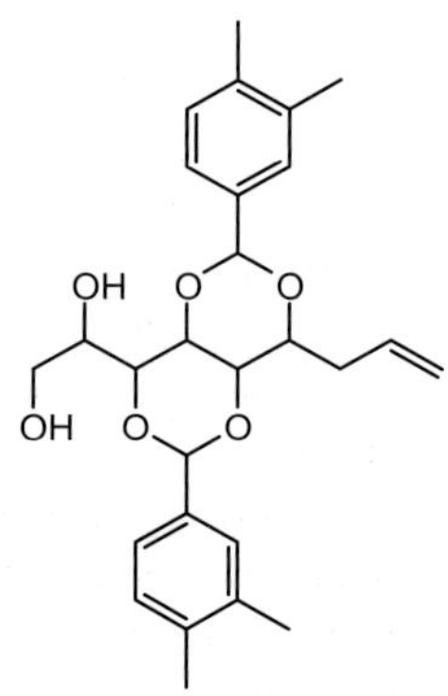

(V)

Additive Preparation

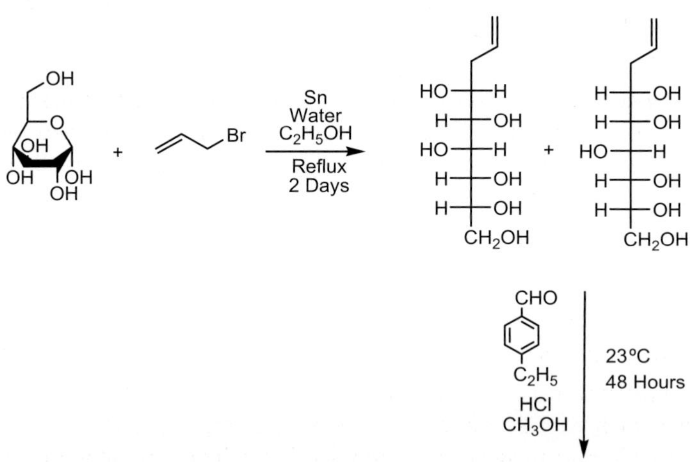

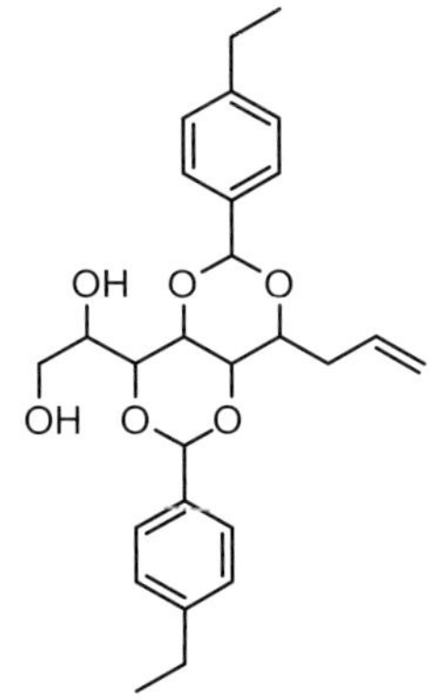

Additive Synthesis

1. Preparation of 1-allyl sorbitol

A 3-liter three-necked round-bottom flask equipped with a heating mantle, stirrer, nitrogen inlet, and condenser was charged with 900 ml of ethanol, 150 ml of water, 180 of d-glucose, 119 g of 100 mesh tin powder, and 121 g of allyl bromide. The mixture was stirred and heated to 60 °C when an unexpected significant exotherm accompanied by gas evolution occurred. The gray suspension was stirred at reflux for two days when it turned into an orange/brown color. The cooled reaction mixture was then treated with 200 ml of 5 M NaOH, filtered, and then decolorized with activated carbon. The solution was refiltered, concentrated, and 200 g of white syrup was isolated consisting of a 7:1 threo–erythro mixture isolated. The syrup was used in subsequent reactions without further purification.

2. Preparation of bis-1,3,2,4-(4′-ethylbenzylidene)-1-allyl sorbitol

A 2-liter reaction kettle, equipped with a stirrer and nitrogen inlet, was charged with 111 g of the Step 1 product mixture dissolved in 100 ml of 6 M HCl and 134 g of 4-ethylbenzaldehyde dissolved in 800 ml of methanol and then stirred for 48 h at an ambient temperature. A precipitate formed which was isolated by filtration and then washed with 250 ml of 1 M NaOH and suspended in water containing a small amount of NaOH. The suspension was boiled in water, refiltered, washed seven times with 500 ml of boiling water and then dried overnight. The powder was stirred in 500 ml of boiling cyclohexane, filtered, and washed twice with 250 ml of boiling cyclohexane. The white powder was dried in a vacuum oven, and 107 g of the product was isolated having a purity >99% with an m.p = 244–246 °C.

Testing

1. Haze Improvement

A mixture consisting of a selected experimental agent and 0.05 wt% of Irganox® 1010, 0.1 wt% of Irgafos® 168, and 0.08 wt% of calcium stearate was blended with either polypropylene or random poly(propylene-co-ethylene) containing 3% ethylene content and then dry blended in a mechanical mixer. The blend was then extruded through a single screw extruder at 240 °C and pelletized. The pellets were then converted into 1.27-mm-thick plaques by injection molding at 220 °C and evaluated. Haze testing results are provided in Table 1.

2. *Gel Formation*

Solid gel formation evaluation consisted of mixing a selected experimental agent in a specific organic solvent and then heating to between 77 and 149 °C under agitation for 5–120 min. The solution was then poured into a rigid gel stick mold, and gel formation properties were evaluated. Mold testing results are provided in Table 2.

Test Results

1. *Haze Improvement*

TABLE 1 Haze improvement in poly(propylene-co-ethylene) and polypropylene resins using selected sorbitols derivatized from aromatic acetals

Anti-Haze Additive	Polymer	Tc (°C)	Haze (%)
None	Poly(propylene-co-ethylene)	101.2	44.3
Millad® 3988*	Poly(propylene-co-ethylene)	113.6	7.2
Additive 2	Poly(propylene-co-ethylene)	115.1	4.9
Additive 10	Poly(propylene-co-ethylene)	109.6	11.3
Additive 13	Poly(propylene-co-ethylene)	109.5	8.4
Additive 14	Poly(propylene-co-ethylene)	111.5	5.6
Additive 15	Poly(propylene-co-ethylene)	107.0	8.8
None	Polypropylene	116.6	58.1
Millad® 3988	Polypropylene	124.0	11.7
Additive 2	Polypropylene	125.2	7.5

**3,4-Dimethylbenzylidene sorbitol.*

2. *Gel Formation*

TABLE 2 Gel formation testing properties of selected experimental agents in different organic solvents

Solvent	Experimental Additive	Gel Formation	Gel Character
1,2-Propanediol	Additive 2	Y	Hard
1,3-Propanediol	Additive 2	Y	Hard
2-Chlorotoluene	Additive 1	Y	Soft
Toluene	Additive 2	Y	Soft
Benzonitrile	Additive 1	Y	Soft
1,2-Propanediol	Additive 3	Y	Hard
2-Chlorotoluene	Additive 1	Y	Hard
Benzonitrile	Additive 2	Y	Hard
1,2-Propanediol	Additive 2	Y	Hard
1,3-Propanediol	Additive 2	Y	Hard
2-Chlorotoluene	Additive 2	Y	Soft
1,2-Propanediol	Additive 3	Y	Hard

Advantages over Prior Art

Additives of the current invention represent the next generation of sorbitol-based anti-hazing agents for polymer resins. In addition, the preparation of these agents is versatile and can used to produce a wide variety of analogs. Finally, sorbitol derivatives derived from this process are also effective in molding applications because of their favorable gelling properties.

Notes

1. To eliminate film haze, Zhang[1] additized thermoplastic materials used in medical solution pouches or meat packaging with poly(ethylene-co-vinyl alcohol) containing 27% and 32% ethylene.

2. Sisson[2] used the colorants Renol® Red 4 (VI) and Renol® Blue NE 51050340 (VII) to mask the haze caused by particles ranging in size from 380 to about 720 nm present in thermoplastic resins.

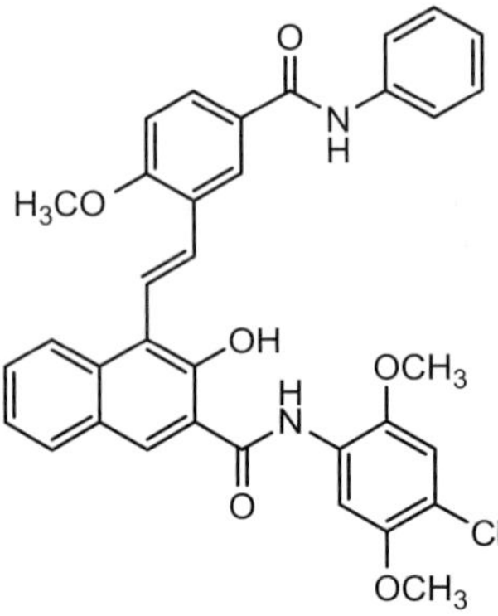

(VI)

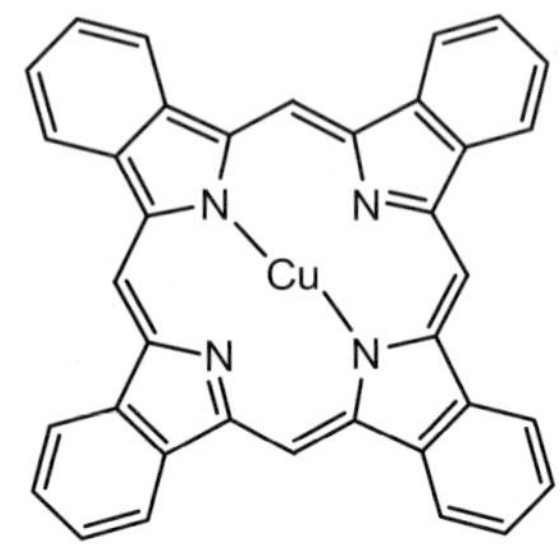

(VII)

3. Hannsen[3] reduced the haze in molded polymer articles consisting of either high-density polyethylene or polypropylene using the nucleating agent, disodium bicyclo[2.2.1]heptane-2,3-dicarboxylate (VIII).

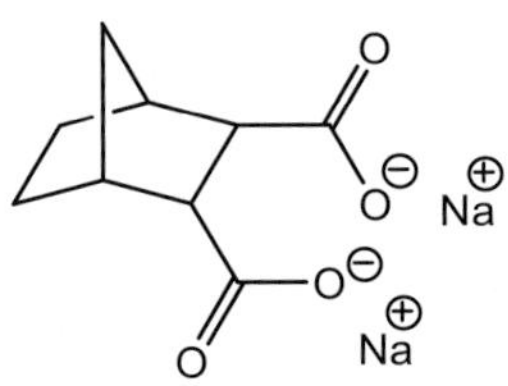

(VIII)

4. Yamamoto[4] improved the clarity of poly(acrylic acid-co-methyl methacrylate) resin by dehydration and dealcoholization of the resin by heating the material to between 250 and 330 °C and forming glutaric anhydride repeat units (IX).

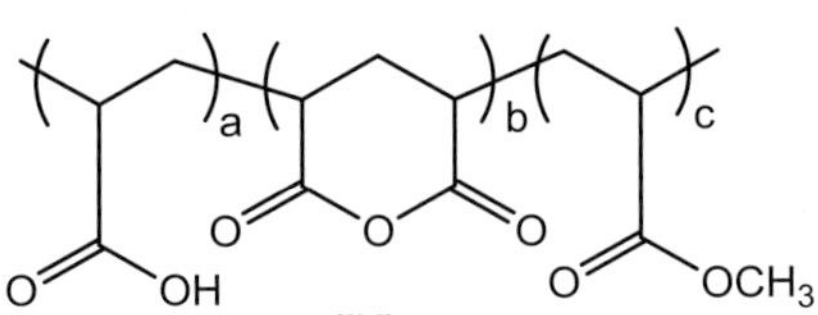

(IX)

REFERENCES

1. David D.Zhang, US Patent 7,834,089 (November 16, 2010)
E. I. du Pont de Nemours and Company (Wilmington, DE)
2. Edwin A. Sisson et al, US Patent 7,833,595 (November 16, 2010)
M&G USA Corporation (Apple Grove, WV)

3. Rob Hanssen et al, US Patent 7,786,203 (August 31, 2010)
Milliken & Company (Spartanburg, SC)
4. Daisuke Yamamoto et al, US Patent 7,888,456 (February 15, 2011)
Toray Industries, Inc. (Tokyo, JP)

B. ANTIOXIDANT ESTERS FOR THERMOPLASTICS

2,4-*t*-Pentyl Bisphenol A Carboxyethyl Acrylate

Title:	Bisphenol monoester compound
US Patent:	8,076,402 (December 13, 2011)
Author:	Ryoji Soma et al.
Assignee:	Sumitomo Chemical Company, Limited (Tokyo, JP)
Significance:	An analog of bisphenol A was prepared by condensing 2,4-di-*t*-pentylphenol with acetaldehyde and then converting it into a monoester by reacting with 2-carboxyethyl acrylate. When compounded with poly(ethylene-b-propylene) and heated to 180 °C in air for 2 h, the blend had an improved Degree of Discoloration over an existing commercial analog.
Additive Name	2,4-Di-*t*-pentyl-6-[1-(3,5-di-*t*-pentyl-2-hydroxyphenyl)ethyl]-2- carboxyethyl acrylate (I)
Safety	Although little safety information is available on the current material, it is virtually identical to the antioxidant Sumilizer GS® (II) manufactured by Sumitomo Chemical Company, Ltd. Sumilizer GS® has very low skin and oral toxicity and is a very mild eye irritant. The agent is nongenotoxic in either vitro gene cell mutation or chromosome aberration testing.

Additive Structure

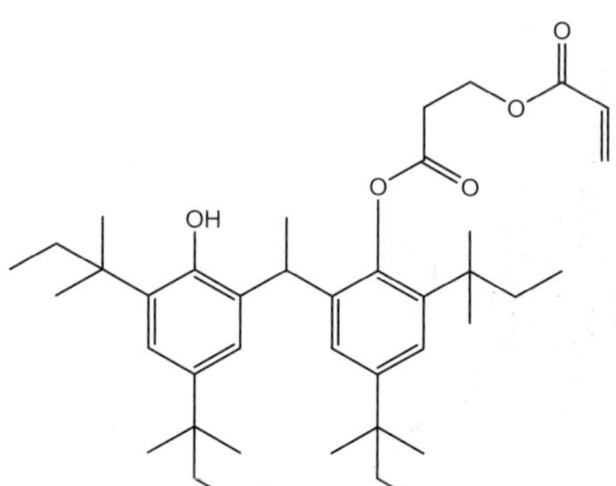

(I)

Additive Preparation

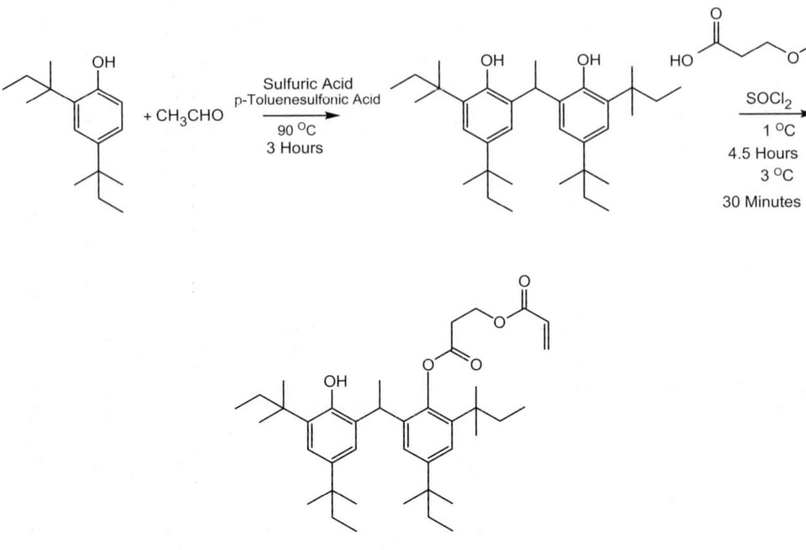

Additive Synthesis

1. *Preparation of 2,4-di-t-pentyl-6-[1-(3,5-di-t-pentyl-2-hydroxyphenyl) ethyl]*

A 250-ml round bottom container was charged with 117.5 g of 2,4-di-*t*-pentylphenol, 0.9 g of *p*-toluenesulfonic acid monohydrate, and 6.9 g of 78% sulfuric acid and then treated with the dropwise addition of 12.1 g of acetaldehyde and 28 g of xylene over 2 h at 35 °C. The mixture was further heated for 3 h at 90 °C and then cooled and further treated with 157 g of xylene, 59 g of water, and 11.7 g of 25% aqueous sodium hydroxide. After removing the aqueous phase, the organic phase was washed with water, azeotropically distilled, and 292 g of a pale yellow oil was isolated.

2. *Preparation of 2,4-di-t-pentyl-6-[1-(3,5-di-t-pentyl-2-hydroxyphenyl) ethyl]-2-carboxyethyl acrylate*

A reaction vessel was charged with the entire Step I product and 28 g of 50% aqueous potassium hydroxide and then heated to 98 °C for 3 h. After cooling to 50 °C, hexane was added to the vessel, and the mixture was azeotropically distilled to 106 °C and then cooled to 0 °C. A separate reaction vessel was charged with 35.3 g of 2-carboxyethyl acrylate and cooled to 1 °C and then treated with the dropwise addition of 26.7 g of thionyl chloride over 4.5 h. The contents of the second vessel were then added dropwise to the anhydrous Step 1 vessel over 1 h at 3 °C and stirred for 30 min. The mixture was then treated with 150 ml of water and vigorously stirred for 10 min at 30 °C. The organic layer was isolated and purified using the procedure described in Step, and 303 g of the product was isolated as a brown oil having a purity of 93%.

Testing

Antioxidancy

A blend consisting of 11.6 mg of the Step 2 product dissolved in 5 ml chloroform was added to 2 g of poly(ethylene-b-propylene) having an MI of 9–10 g/10 min at 230 °C under a weight of 2.16 kg and then stirred, and chloroform was vaporized off. Antioxidancy of the blend was then evaluated by heating the mixture in a gear oven for 2 h at 180 °C in air. Test results are provided in Table 3.

Test Results

TABLE 3 Oxidative stability of poly(ethylene-b-propylene) additized with 5.8 wt% of the Step 2 product and a comparative additive and then heated for 2 h at 180 °C in air. Lower values are preferred

Additive	Discoloration Degree (YI Value)	Discoloration Degree (Improvement Index) (%)
Step 2 Product	10.03	0
Comparative*	12.22	7.1

2,4-Di-t-pentyl-6-[1-(3,5-di-t-pentyl-2-hydroxyphenyl)ethyl]phenyl acrylate (II).

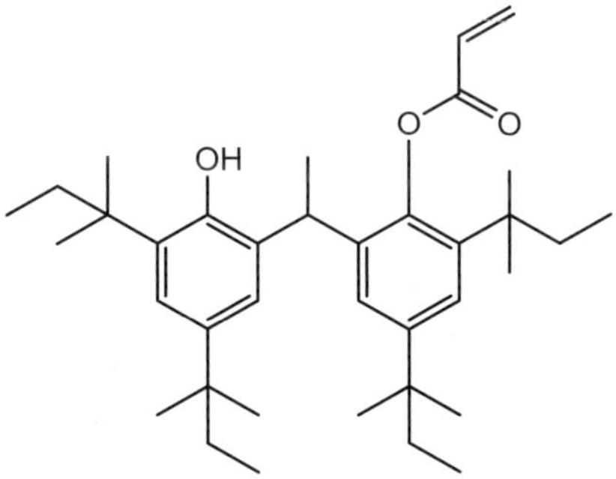

(II)

Advantages over Prior Art

The bisphenol of the current invention and Sumilizer GS® manufactured by Sumitomo Chemical Company, Ltd. are essentially identical. The improved antioxidancy of the carboxyethyl acrylate-functionalized reflects higher grafting levels to poly(ethylene-b- propylene) than when using the acrylated-functionalized analog. This suggests that the efficacy of an antioxidant reflects higher grafting efficancy so that antioxidancy of existing antioxidants can potentially be enhanced by judiciously extending chain lengths.

Notes

1. Additional bisphenol antioxidant monoester analogs (III) and (IV) were previously prepared by Kimura[1] and used as antioxidants in engineered materials.

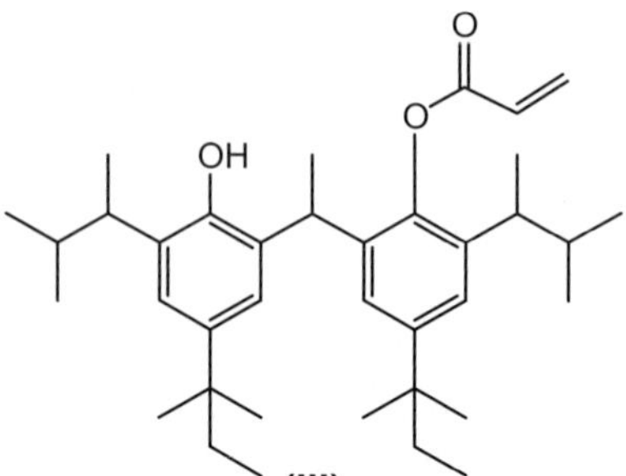

(III)

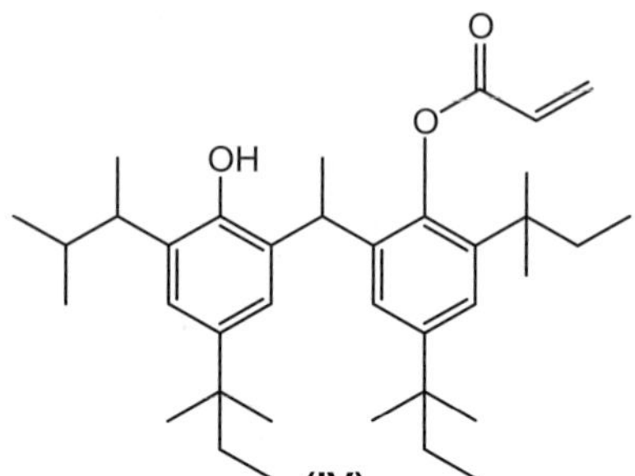

(IV)

2. The oligomeric antioxidant (V) prepared by Cholli[2] was used in thermoplastics, automotive additives, and cosmetic formulations.

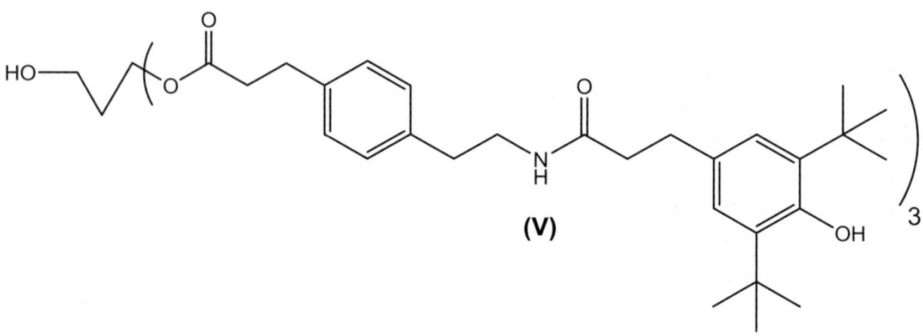

(V)

3. The polymeric phenolic antioxidant (VI) prepared by Cholli[3] was more effective at lower treatment rates than Irgonox® 1010 when used in processing polypropylene at elevated temperatures.

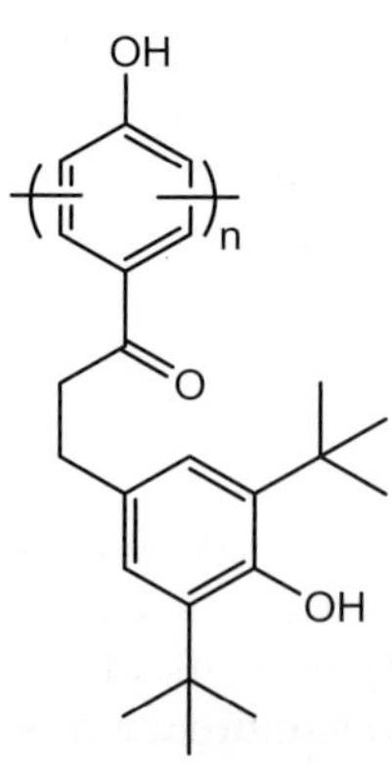

(VI)

4. Cholli[4] used Coprinus cinereus peroxidase and Arthromyces ramosus peroxidase to prepare high molecular weight aromatic antioxidants (VII) and (VIII) from the corresponding monomers.

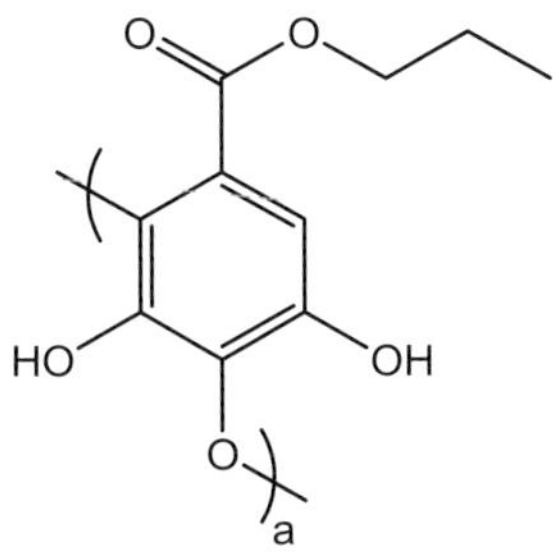

(VII)

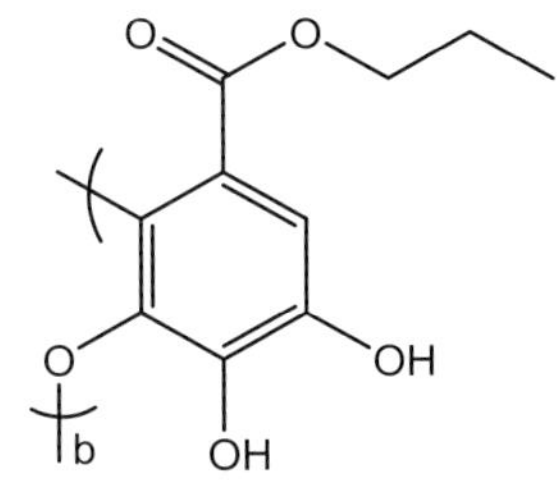

(VIII)

REFERENCES

1. Kenji Kimura et al, US Patent 7,977,418 (July 12, 2011)
   Sumitomo Chemical Company, Limited (Tokyo, JP)
2. Ashok L. Cholli et al, US. Patent 8,039,673 (October 18, 2011)
   Polnox Corporation (Lowell, MA)
3. Ashok L. Cholli et al, US. Patent 7,923,587 (April 12, 2011)
   Polnox Corporation (Lowell, MA)
4. Ashok L. Cholli et al, US. Patent 7,754,267 (July 13, 2011)
   The United States of America as represented by the Secretary of the Army (Washington, DC) University of Massachusetts Lowell (Lowell, MA)

C. ANTIOXIDANTS

Phenolic Phosphite Antioxidants

Title:	Macromolecular antioxidants based on sterically hindered phenols and phosphites
US Patent:	7,956,153 (June 7, 2011)
Author:	Ashok L. Cholli et al.
Assignee:	Polnox Corporation (Lowell, MA)
Significance:	A phenolic phosphite antioxidant was inexpensively prepared which had an oxidative induction time of 25 min longer than the Irganox® antioxidant product line. The additive is designed to be used in polymer blends to prevent material discoloration during high temperature extrusion or other high temperature processing.

Additive Tris[*N*-(4-hydroxyphenyl)-3-(2,6-di-*t*-butyl-4-hydroxyphenyl) propionamide]-phosphite (I)

Safety This additive is analogous to Irganox®-related antioxidants. Irganox®-based antioxidants can cause irritation to the eyes, skin, and to the respiratory system. Liver and thyroid are especially affected after repeated exposure to these and similar Irganox®-related antioxidant agents. The use of a respirator is strongly encouraged to avoid dust inhalation when working in enclosed spaces.

Additive Structure (Note 1)

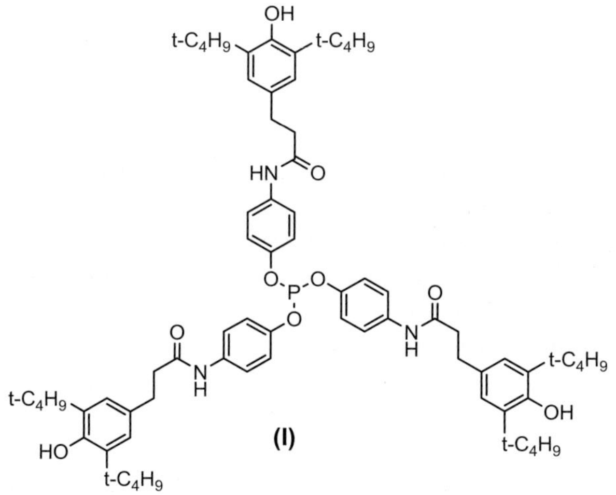

(I)

Additive Preparation

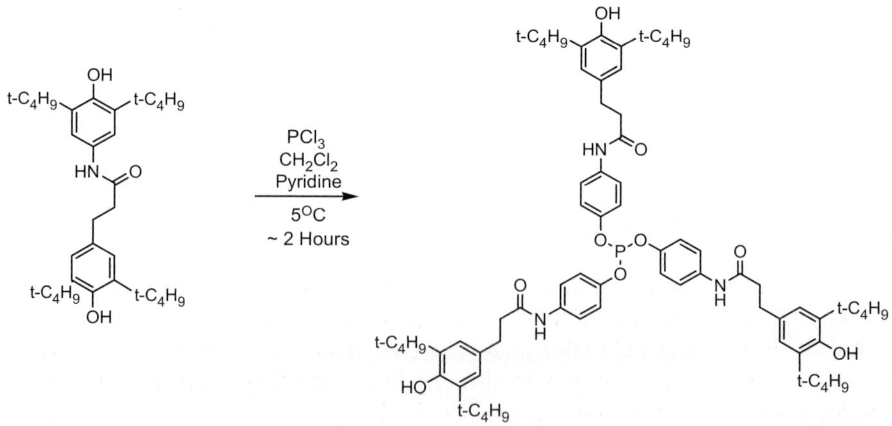

Additive Synthesis

1. *Preparation of tris[N-(4-hydroxyphenyl)-3-(2,6-di-t-butyl-4-hydroxy-phenyl)-propionamide]phosphite*

A 1-liter two-necked round-bottom flask equipped with a thermometer, dropping funnel, and a nitrogen inlet was charged with 100 g of *N*-(4-hydroxyphenyl)-3-(2,6-di-*t*-butyl-4-hydroxy phenyl) propionamide, 22 ml of pyridine dissolved in 200 ml of methylene chloride and then treated with 7.86 ml of phosphorous trichloride at 0–5 °C over a period of 30–40 min. The course of the reaction was monitored by TLC/HPLC and terminated when the reagents had been consumed. The mixture was then distilled to remove dichloromethane, and the residue was treated with 100 ml of methanol. A white crystalline precipitate which formed was isolated and washed several times with cold methanol. After drying, the product was isolated and had an MP = 224–245 °C.

Testing and Treatment Levels

Unadditized polypropylene powder was additized with 1000 ppm of the Step 1 product and then extruded through a single screw extruder and isolated in the form of pellets. Additized polypropylene pellets were then subjected to an accelerated oxidative stability test using differential scanning calorimetry according to the testing protocol of ASTM D3895. A reference sample additized with 1000 ppm of Irganox® 1010 was also prepared. Test Results are provided in Table 4.

Test Results

TABLE 4 Comparison of oxidative onset and termination of the Step 1 product and the reference Irganox® 1010 when blended in polypropylene at 1000 ppm and extruded at 230 °C

Antioxidant	Oxidation Onset (min)	Oxidation Completion (min)
Experimental Additive	12	37
Irganox® 1010* (Reference)	17	39

Tetrakis-(methylene-(3,5-di-t-butyl-1-hydrocinnamate))methane.

Advantages over Prior Art

Within the past 5 years, the Polnox Corporation has been engaged in an aggressive research program to develop antioxidants having both higher antioxidant activity and improved stability. These materials are likely replacements for the Irganox® antioxidant product line. Finally, with only modest modifications, these agents can be used as petroleum-based products and in processed foods.

Notes

1. A second antioxidant (II) was also prepared in this invention by the author, but its synthesis was not provided.

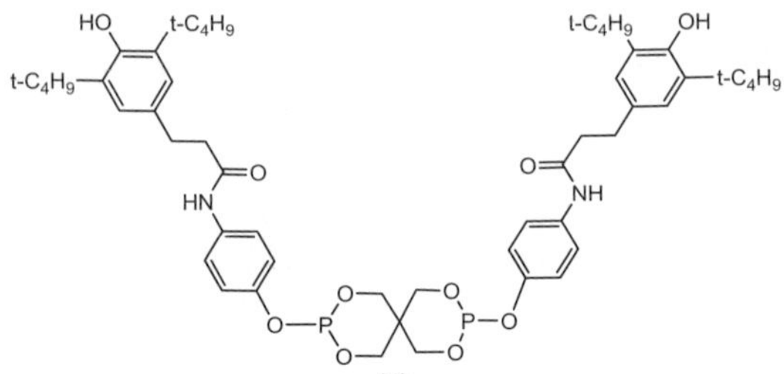

(II)

2. Phenolic-based antioxidants (III) having an Mn of approximately 40,000 Da were prepared by Cholli[1] by polymerization of phenolic esters using horse radish peroxidase.

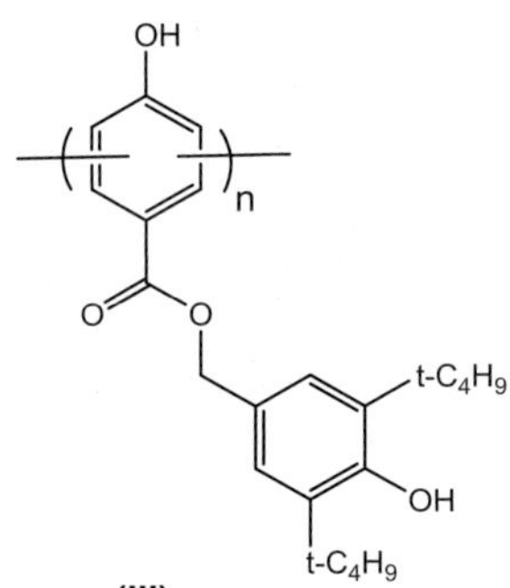

(III)

3. Cholli[2] prepared pentaerythritol-based tetraester antioxidants (IV), which were used in polymer blending formulation to minimize discoloration during extrusion processing.

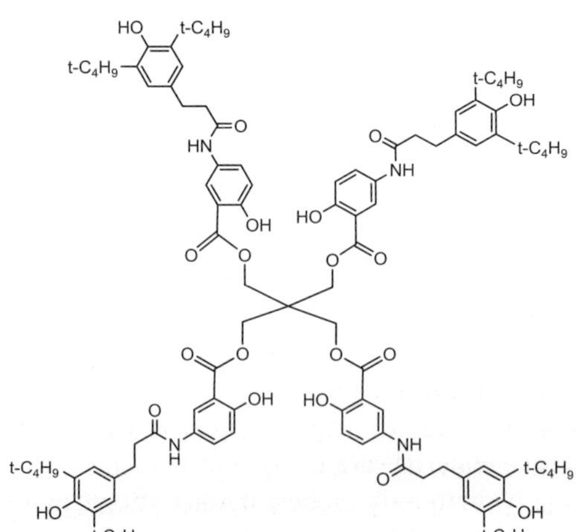

(IV)

4. Kumar[3] prepared nitrogen and hindered phenol-containing antioxidants (V), which had high oxidative induction times and were used in propylene blending formulations.

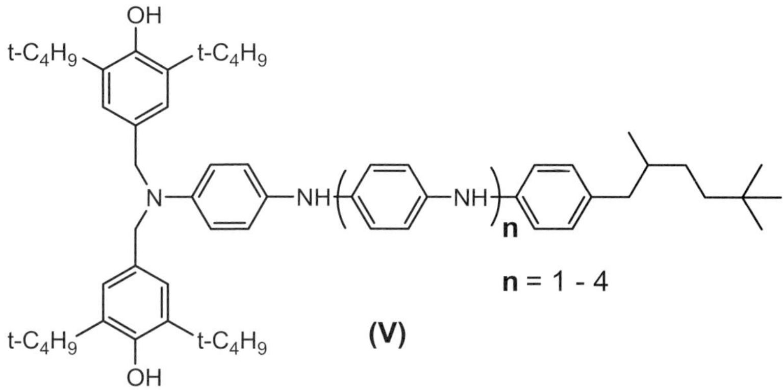

5. Rajesh[4] prepared gemini antioxidants (VI), which were blended in polypropylene, elastomers, and lubricant formulations.

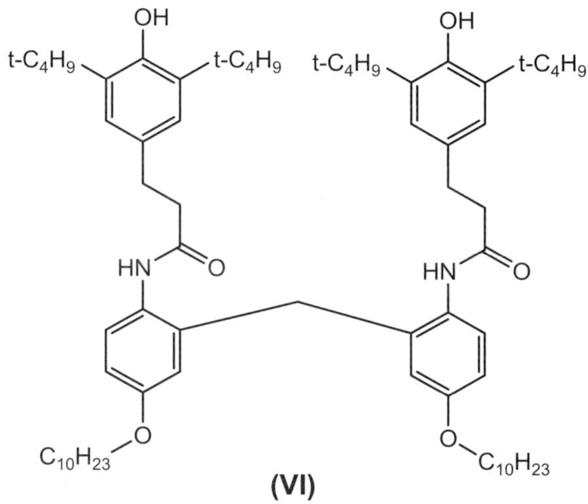

REFERENCES

1. Ashok L Cholli, US Patent 7,923,587 (April 12, 2011)
   Polnox Corporation (Lowell, MA)
2. Ashok L Cholli et al, US Patent Application 20090184294 (July 23, 2009)
   Polnox Corporation (Lowell, MA)
3. Rajesh Kumar et al, US Patent 7,799,948 (September 21, 2010)
   Polnox Corporation (Lowell, MA)
4. Rajesh Kumar et al, US Patent 7,705,185 (April 27, 2010)
   Polnox Corporation (Lowell, MA)

D. OXIDATIVELY AND ABRASIVE-RESISTANT POLYMERS

POSS-Containing High Performance Monomers

Title:	POSS enhanced compositions
US Patent:	8,058,380 (November 15, 2011)
Author:	Vandana Vij et al.
Assignee:	The United States of America as represented by the Secretary of the Air Force (Washington, DC)
Significance:	Polyimide resins consisting of heptaphenylbis(3- or 4-aminophenyl)-POSS methylsilane, -bis(4-aminophenyl) ether, or -pyromellitic dianhydride were prepared in two synthetic steps at elevated temperatures. These high performance polyaromatic materials were readily moldable into abrasive-resistant solids having excellent thermooxidative properties and good responses to mechanical stresses.
Additive Names	Heptaphenylbis(4-aminophenyl)-POSS methylsilane (para-POSS-dianiline monomer) (I) Heptaphenylbis(3-aminophenyl) POSS methylsilane (meta-POSS-dianiline monomer) (II)
Safety	POSS-containing additives and hybrid polymers are considered to be both eye and skin irritants. Inhalation of these agents can cause headaches, nausea, dizziness, and respiratory irritation. Ingestion can cause severe gastric system distress.

Additive Structures

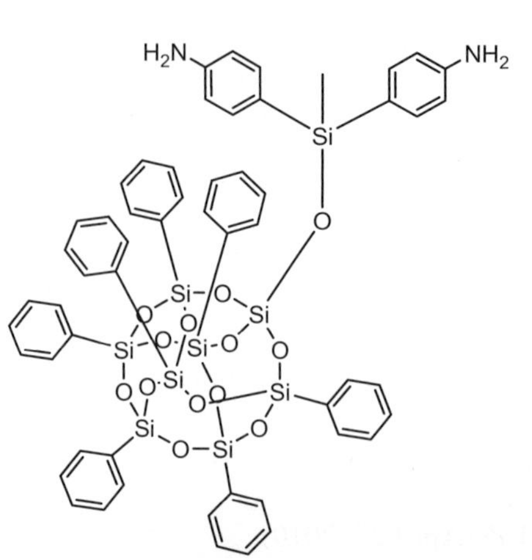

(I)

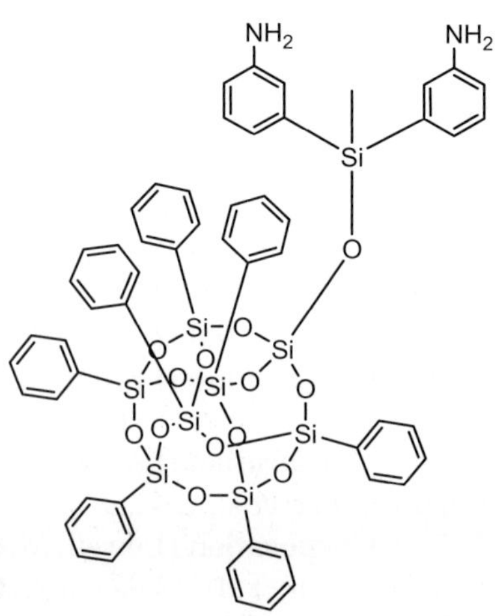

(II)

Additive Preparation

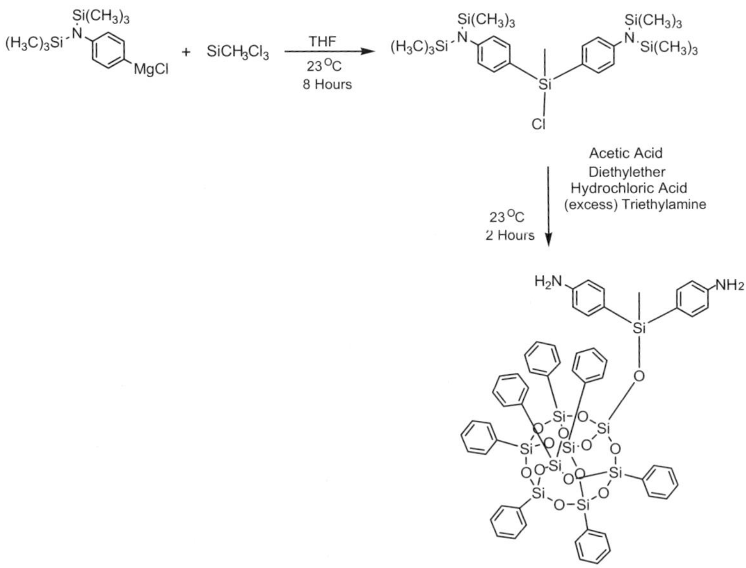

High Performance Composite Preparation

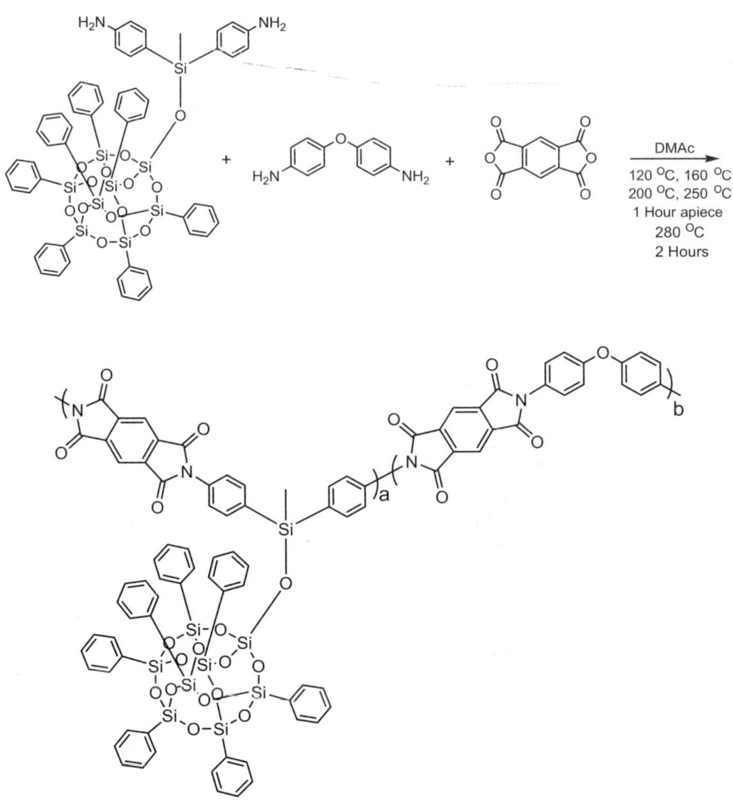

Additive Synthesis

1. *Preparation of di(4-N,N-bis(trimethylsilyl)aminophenylchloromethylsilane*

An addition funnel was charged with 30 ml of a 1 M solution of 4-[bis(trimethylsilyl) amino]phenyl magnesium chloride dissolved in 25 ml of anhydrous tetrahydrofuran and slowly added to a stirred mixture of 2.252 g of trichoromethylsilane dissolved in 15 ml of anhydrous tetrahydrofuran. The mixture was then stirred overnight at an ambient temperature and then concentrated and extracted with hexane. After filtering was completed, the light yellow colored filtrate was transferred to a distillation flask, and all volatiles were removed under vacuum. The residual viscous liquid was distilled using a Kugelrohr apparatus, and the product was isolated in a 90% yield as a colorless, viscous liquid, and isolated in a 92% yield after drying.

^{1}H-NMR ($CDCl_3$ δ) 0.08 ppm (36H, NSi CH3), 0.93 ppm (s, 3H, SiCH3), 6.94–7.36 ppm (8H).

^{29}Si-NMR ($CDCl_3$ δ) 4.83 ppm, 9.89 ppm

2. *Preparation of heptaphenylbis(4-aminophenyl) POSS methylsilane (meta-POSS-dianiline-monomer)*

A 30-ml solution of [bis(trimethylsilyl) amino]-phenylmagnesium chloride dissolved in 25 ml of THF was put into an addition funnel and slowly added to a stirrred mixture of trichlorodi-ethylsilane dissolved in 15 ml of anhydrous tetrahydrofuran and stirred overnight. The solvent was removed from the reaction mixture, and dry hexane was added to extract the product by filtration. The light yellow colored filtrate was transferred to a distillation flask, and all volatiles were removed under a vacuum. The residual was distilled using a Kugelrohr apparatus to give di(*N,N*-bis(trimethylsilyl)aminophenylchloromethylsilane as a colorless, viscous liquid in a 90% yield.

^{1}H-NMR ($CDCl_3$) δ 0.08 ppm (36H, NSi CH3), 0.93 ppm (s, 3H, SiCH3), 6.94–7.36 ppm (8H)

^{29}Si-NMR ($CDCl_3$ δ) −8.185, −77.95, -77.77, −109.251

3. *Preparation of poly(para-POSS-dianiline-co-pyromellitic dianhydride-co-4-oxydianiline) imide*

A reactor was charged with 5 wt% of para-POSS-dianiline monomer containing 1.9% cage and 0.901 g of bis(4-aminophenyl)ether dissolved in 4 ml of *N,N*-dimethylacrylamide. The mixture was stirred until both solids dissolved and then treated with 1.00 g of pyromellitic dianhydride dissolved in 4 ml of *N,N*-dimethylacrylamide. The mixture was heated until it became viscous and was then cast on a glass slide and placed into a vacuum oven at 80 °C for 4 h under a nitrogen purge. The glass slide was then placed in a nitrogen circulated vacuum oven at 120 °C, 160 °C, 200 °C, and 250 °C for 1 h a piece and then heated to 280 °C for 2 h to ensure complete imidization.

Testing

Test results are not provided by the authors.

Advantages over Prior Art

The high performance compositions containing POSS octaanilines are useful in preparing well-defined hybrid polymers. In addition, these agents can act as crosslinking agents and for preparing oxidatively resistant as well as noncrosslinkable terpolymers by judiciously altering the stoichiometry of reaction monomers. Reaction monomers particularly suitable include epoxides, esters, cyanate esters, and epoxides.

Notes

1. Basheer[1] prepared a rigid epoxy resin using rigid octaaminophenyl silsesquioxane (III) as a curing agent with tetraglycidyl 3-xylenediamine (IV).

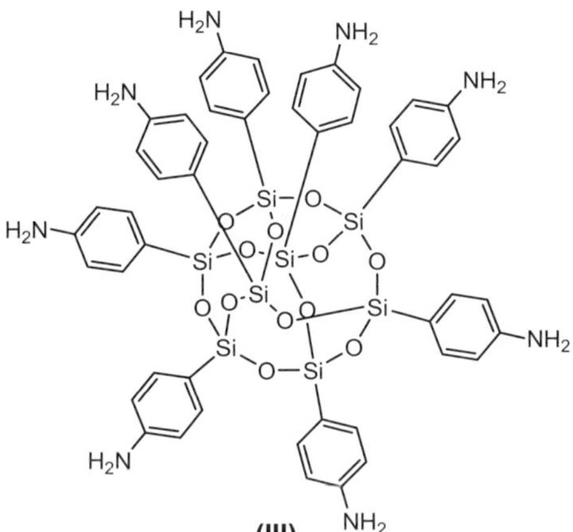

(III)

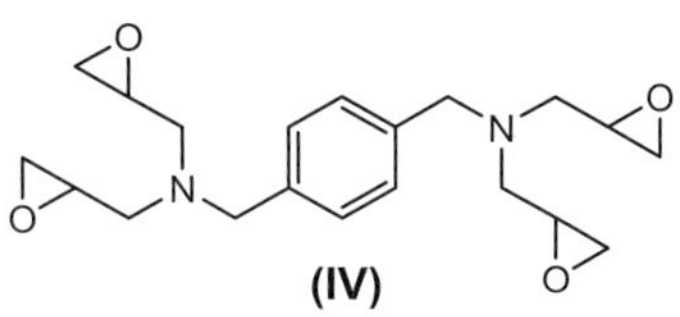

(IV)

2. Poe[2] prepared polymers consisting of polyimides containing grafted heptaphenyl silsesquioxane (V), which were used as high performance materials in the aerospace industry.

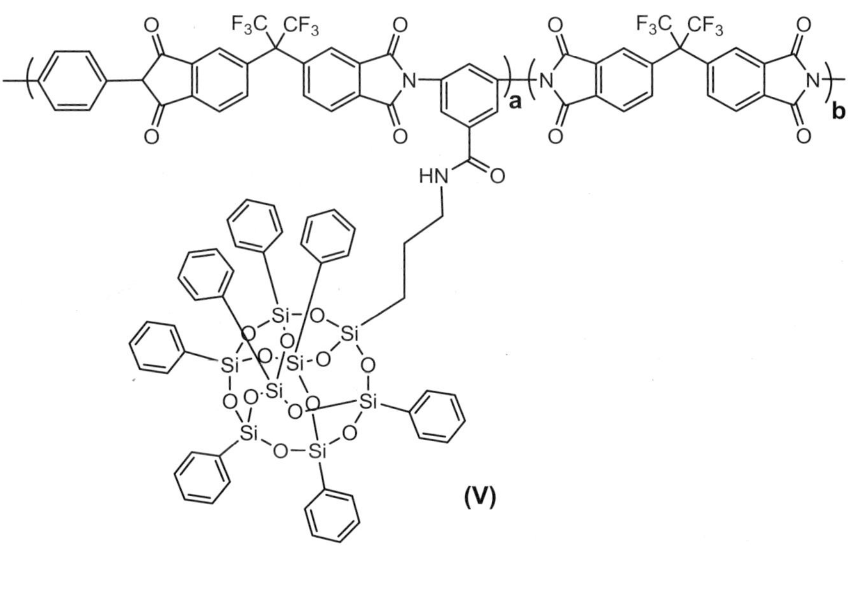

(V)

3. Chen[3] prepared polyimide–titanium hybrid materials (VI) containing oligomeric titanium oxide termini, which did not undergo macrophase separation and had decomposition temperatures >550 °C.

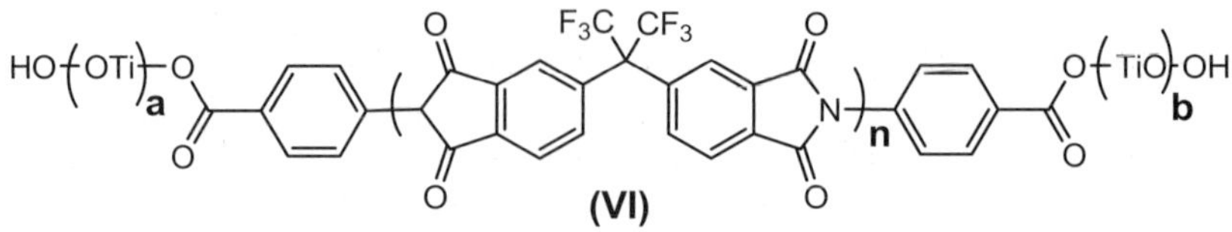

(VI)

4. Lichtenhan[4] prepared optically clear thermoset silsesquioxane resins by treating vinyl silsesquioxane with phenyltrisdimethlysiloxy silane and then cured them.

REFERENCES

1. Rafil Basheer et al, US. Patent 7,332,822 (February 19, 2008) **Delphi Technologies, Inc. (Troy, MI)**
2. Garrett Poe et al, US. Patent 7,619,042 (November 17, 2009) **NeXolve Corporation** (Huntsville, AL)
3. Wen-Chang Chen et al, US. Patent Application 20090092759 (April 9, 2009) **National Taiwan University (Taipei City, TW)**
4. Joseph D. Lichtenhan et al, US. Patent 7,553,904 (June 20, 2009) **Hybrid Plastics, Inc. (Hattiesburg, MS)**

E. BLENDS AND RECYCLING

2,3-Epoxypropyl-3,5-di-*t*-Butyl-4-Hydroxybenzene Esters

Title:	Functionalized polymers
US Patent:	7,807,757 (October 5, 2010)
Author:	Rudolf Pfaendner et al.
Assignee:	Ciba Specialty Chemicals Corp. (Tarrytown, NY)

Significance: A method for preparing stable polymer blends of dissimilar polarities for use in engineered materials or for recycling incompatible plastics is described. The two-step process entails initially functionalizing poly(styrene-co-ethylene-co-butadiene-co-styrene) or polyethylene-co-acrylic-co-acrylate) with either 2,3-epoxypropyl-3,5-di-*t*-butyl-4-hydroxybenzene propionate or 1,2,2,6,6-tetramethyl-4-(2,3-epoxypropyloxy) piperidine. Blends of incompatible Nylon-6 and low-density polyethylene were then prepared by additizing with up to 8 wt% of the functionalized polymer.

Additive Name 2,3-Epoxypropyl-3,5-di-*t*-butyl-4-hydroxybenzene propionate (I) containing up to 7 wt% unreacted 3,5-di-*t*-butyl-4-hydroxybenzene

Safety Concerns Very strong skin and respiratory irritant.

Additive Name 1,2,2,6,6-Pentamethyl-4-(2,3-epoxypropyloxy)piperidine (II) containing varying amounts of unreacted 1,2,2,6,6-pentamethyl-4-hydroxyl)piperidine

Safety Concerns Moderate skin and respiratory irritant.

Additive Structures

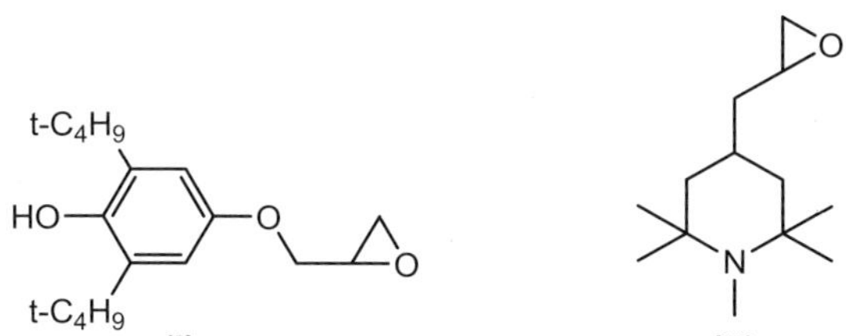

Polymer Modification

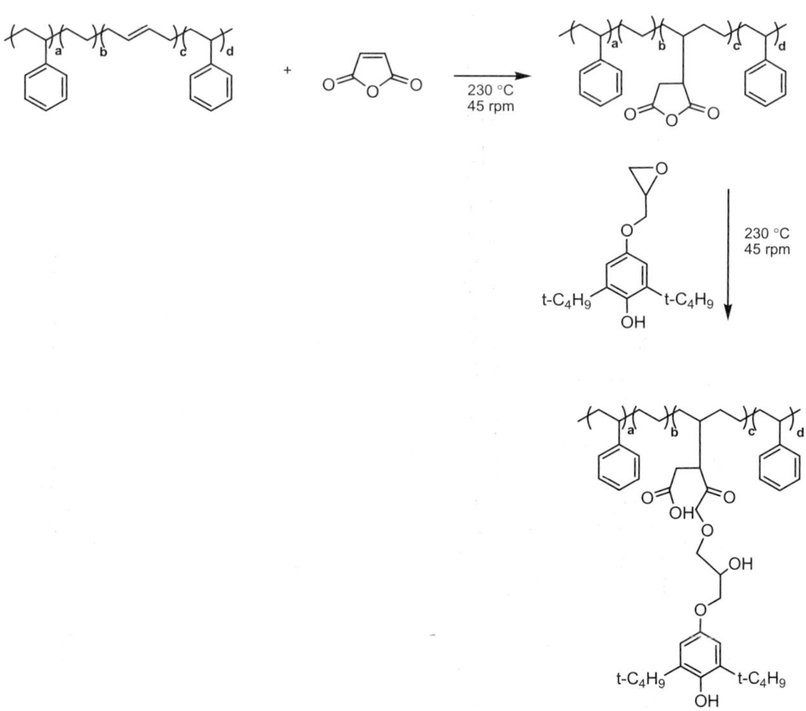

1. Preparation of modified poly(styrene-co-ethylene-co-butadiene-co-styrene)

A mixture of poly(styrene-co-ethylene-co-butadiene-co-styrene) was mixed with maleic anhydride and then extruded through a five-heating zone Haake TW 100 twin-screw extruder at 210–230 °C at 45 rpm. The mixture was then blended and reextruded with 2% of 2,3-epoxypropyl 3,5-bis(1,1-di-*t*-butyl)-4-hydroxybenzenepropionate. Gas chromatography analysis indicated that both additives were completely consumed.

Blending

Incompatible blends were prepared consisting of 70% Nylon-6 and high-density polyethylene and up to 8 wt% of the functionalized polymer. The components were mixed in a Haake TW 100 twin-screw extruder and extruded at 240 °C at 75 rpm.

Testing

Impact strength of blends was evaluated using DIN 53 448 at 10 °C in a circulating air oven as summarized in Table 5. In all cases, 30% Nylon 6 and low-density polyethylene were mixed and then blended with a selected functionalized polymer and extruded at 240 °C at 75 rpm, and the blend was evaluated. Test results are provided in Table 5.

Test Results

TABLE 5 Impact strength as a function of oven aging of blends consisting of 30 wt% Nylon-6 and low-density polyethylene additized with modified poly(styrene-co-ethylene-co-butadiene-co-styrene)

Polymer	Pendant Functionalizing Agent	Polymeric Additive Treatment in Blend (wt%)	Impact Strength (kJ/m^2) Initial	Impact Strength (kJ/m^2) 1000 h	Impact Strength (kJ/m^2) 2000 h
PSEBS*	Phenolic**	2	544	479	448
PSEBS	Piperidine‡	2	569	456	283
PEAAc†	Phenolic	2	663	583	538
PEAAc	Piperidine	4	702	651	503
PEAAc	Piperidine	6	668	670	533
Reference¶	Phenolic	2	571	458	311

**Poly(styrene-co-ethylene-co-butadiene-co-styrene).*
†Poly(ethylene-co-acrylic-co-acrylate).
¶Poly(styrene-b-butadiene-styrene)-g-succinic anhydride.
***2,3-Epoxypropyl-3,5-di-t-butyl-4-hydroxybenzene propionate.*
‡1,2,2,6,6-Pentamethyl-4-(2,3-epoxypropyloxy)piperidine.

Advantages over Prior Art

The invention provides a method for stabilizing and phase compatibilizing two or more plastics of dissimilar compositions. The method produces useful phase-stabilized polymer blends that have good mechanical properties. Both compatibilizers and stabilizers are easily prepared using phenol or piperidine derivatives.

Notes

1. Additional stabilizing agents useful for functionalizing polymers in recycling or blending applications were illustrated by Hoffman[1] who grafted pentaerythrityl ester of (3,5-di-*t*-butyl-4-hydroxyphenyl)propionic acid (III) tri-(2,4-di-*t*-butylphenyl)phosphite (IV), and octadecyl ester of (3,5-di-*t*-butyl-4-hydroxyphenyl) propionic acid (V) onto polymeric substrates.

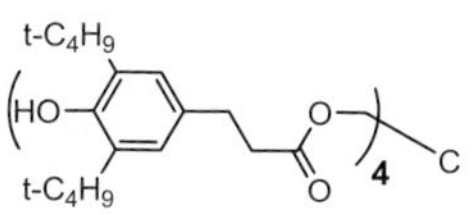

(III)

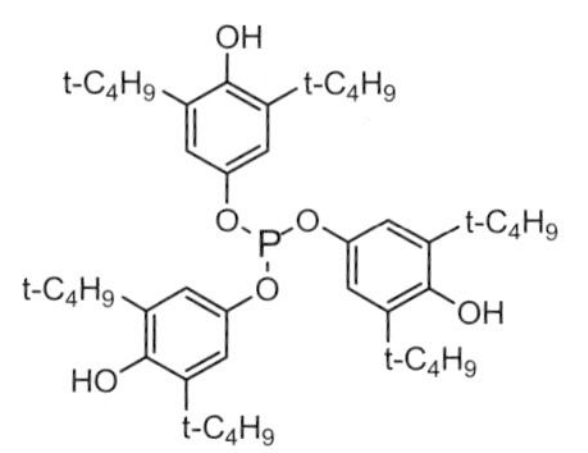

(IV)

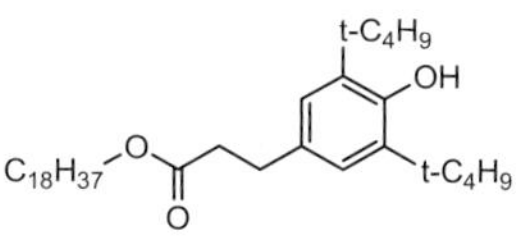

(V)

2. Vollenberg[2] recycled postconsumer polyethylene terephthalate by blending it with Hytrel® 4056 thermoplastic polyester elastomers and Valox® 315 polybutylene terephthalate and then processing the blend through a 27-mm Werner Pfeiderer twin screw extruder at 250 °C with a screw speed of approximately 450 rpm.
3. Renz[3] incorporated phenolic (VI) and piperidine/benzotriazole (VII) derivatives into poly(vinylalcohol)-co-poly(vinylamine) as a method for preparing oxidatively and photochemically stabilizing blends, which were compatible with methoxypolyethylene glycol poly(methacrylate) (Sartomer® CD 550).

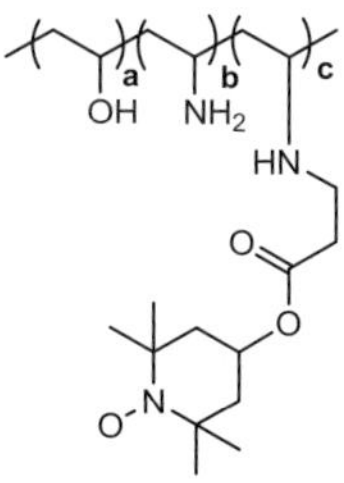

(VI)

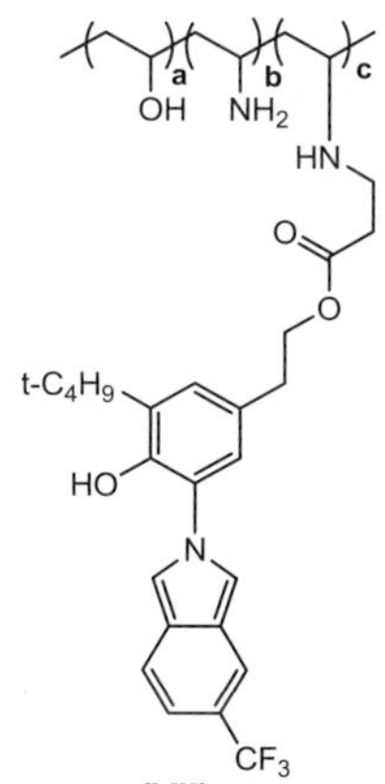

(VII)

REFERENCES

1. Kurt Hoffmann et al, US Patent 6,525,158 (February 25, 2003) **Ciba Specialty Chemicals Corporation** (Tarrytown, NY)
2. Peter H. Vollenberg et al, US Patent 7,799,838 (September 21, 2010) **Sabic Innovative Plastics IP B.V. (NL)**
3. Walter L. Renz et al, US Patent 7,572,843 (August 11, 2009) **Ciba Specialty Chemicals Corporation** (Tarrytown, NY)

F. COMPATIBILIZING AGENTS

Poly(Ethylene-co-Propylene) and poly(Methyl Methacrylate) Esters

Title:	Multibranched polymer, process for producing the same, and applications thereof
US Patent:	7,795,347 (September 14, 2010)
Author:	Hideyuki Kaneko et al.
Assignee:	Mitsui Chemicals, Inc. (Tokyo, JP)
Significance:	A method for preparing intimately mixed blends of hydrophobic and hydrophilic polymers using block, tri, or tetrabranched star-shaped polymers as compatibilizing agents has been developed. Block and multibranched polymer-blending additives each contain covalently attached poly(ethylene-co-propylene) and poly(methyl-methacrylate), which ensures a high degree of mixing in blends containing polymers of dissimilar polarities.

Additive Name	Poly(ethylene-co-propylene)
CAS Number	9010-79-1
Safety Concerns	Very slight skin irritant.

Additive Name	Methyl Methacrylate
CAS Number	80-62-6
Safety Concerns	Strong skin and eye irritant. Hazardous if ingested or inhaled.

Additive Structures

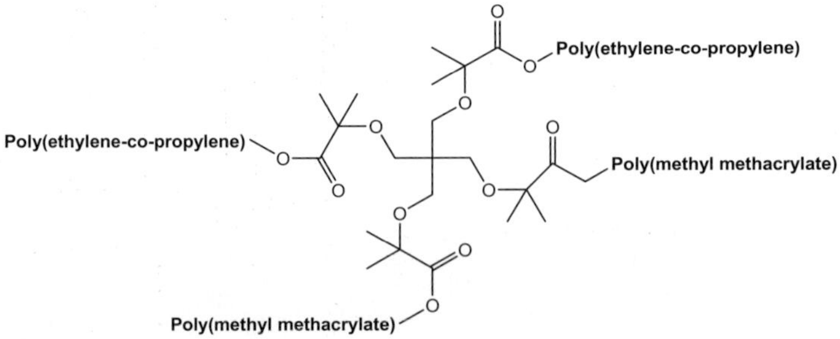

(I)

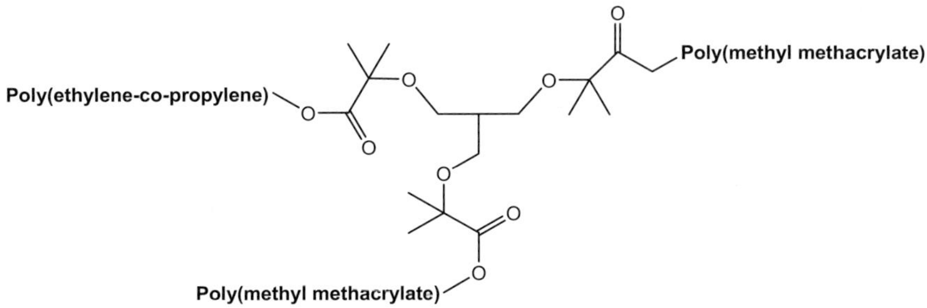

(II)

(III)

Step 1. Preparation of radical polymerization initiator

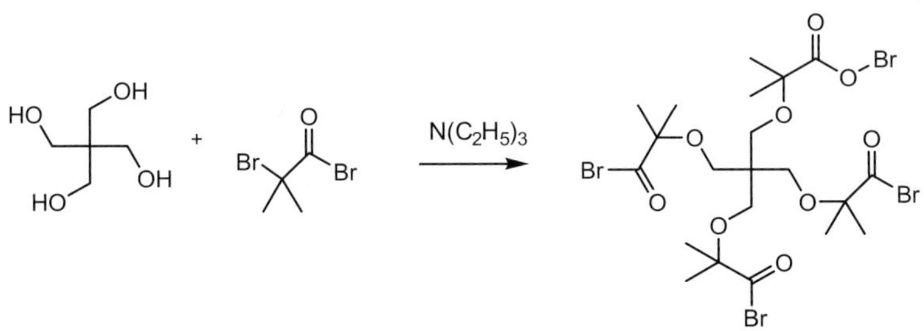

Radical Polymerization Initiator

Step 2. Preparation of star-shaped poly(methacrylate-g-ethylene-co-propylene) synthesis

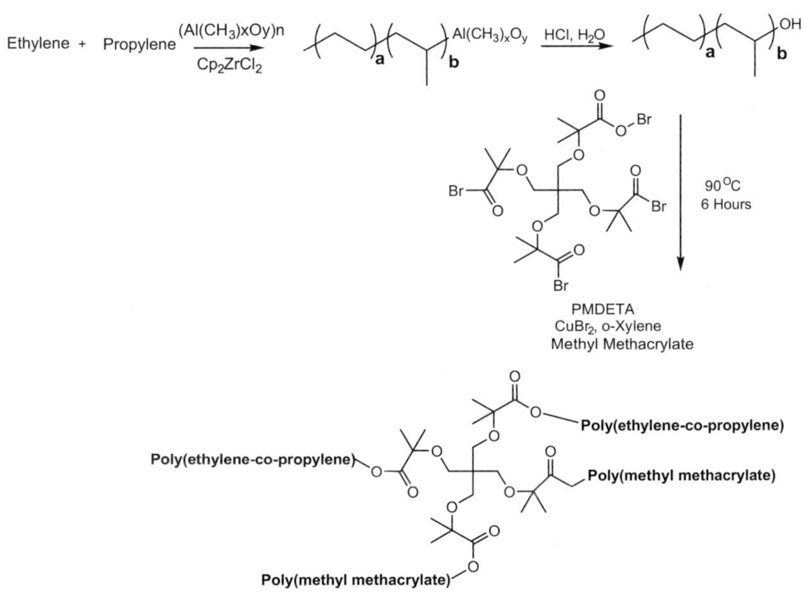

Additive Synthesis

Step 1. Preparation of poly(ethylene-co-propylene) containing an aluminum terminus

A 1-liter glass autoclave was charged with 800 ml of purified toluene purged with nitrogen and saturated with ethylene and 80 propylene the polymerized by treating with 1.12 ml of methyl aluminoxane dissolved in toluene and 0.02 mmol of dicyclopentadienyl zirconium dichloride at 50 °C. After 2 h, the polymerization was terminated by adding a small amount of isobutyl alcohol, and the solution was washed five times with 100 ml of 1 M aqueous hydrochloric acid solution and twice with 100 ml of water. The organic layer was then dried with magnesium sulfate, filtered, concentrated, and 118.7 g of a colorless oil was isolated. A 1-liter reactor was then charged with 100 g of the intermediate dissolved in 500 ml of toluene and treated with 50 ml of diisobutyl aluminum hydride and then heated to 110 °C for 6 h under stirring, and the product was isolated having an Mn = 430 Da, Mw = 1690 Da having a polydispersity of 4.0.

Step 2. Preparation of poly(ethylene-co-propylene) with a hydroxyl terminus

The Step 1 product was dissolved in 500 ml of toluene and then transferred to a separatory funnel and washed thrice with 300 ml of 1 M aqueous hydrochloric acid solution and then twice with 200 ml of water. The organic layer was then dried with magnesium sulfate, filtered, concentrated, and dried for 10 h under vacuum. The product was isolated as a yellow oil in 107.9 g having a hydroxyl group content of 2.8 mol%.

Step 3. Preparation of radical polymerization of a four-arm initiator

A 50-ml Shrenk tube was charged with 1.0 g of pentaerythritol, 10 ml of THF, and 8.2 ml of triethylamine and then treated with the dropwise addition of 7.26 ml of 2-bromoisobutyric acid bromide. The mixture was then stirred at an ambient temperature for 24 h, and the resulting dark brown slurry was washed seven times with 100 ml of 0.5 N hydrochloric acid solution and thrice with 100 ml of water. The organic layer was dried with magnesium sulfate, filtered, and the filtrate was concentrated to give 9.7 g of a dark brown solid. This crude product was dissolved in 10 ml of benzene and purified by column chromatography to give 6 g of a brown solid. The solid was then recrystallized from hexane, and 3.7 g of the product was isolated as white bar-shaped crystals.

Step 4. Preparation of star-shaped poly(methylmethacrylate)-co-(ethylene-co-propylene)

A 300-ml Schlenk tube was charged with 9.8 g of the Step 2 product dissolved in 3.5 ml of o-xylene, 0.21 ml of methyl methacrylate, 1.2 ml of a solution of copper bromide, and *N,N,N′,N″,N′*-pentamethyldiethylenetriamine (PMDETA) dissolved in o-xylene [0.03 mmol/ml in terms of Cu atoms, 0.06 mmol/ml in terms of PMDETA] and 0.2 ml of o-xylene solution containing 0.05 mmol/ml of the Step 3 four-arm linking agent (initiator) and then heated to 90 °C for 6 h. The reaction mixture was then poured into 400 ml of methanol to precipitate the product. The precipitated polymer was filtered with a glass filter, washed thrice with 10 ml of hexane, and thrice with 10 ml of methanol. The precipitate was dried for 10 h, and 0.22 g of the product was isolated having a methylmethacrylate/ethylene-co-propylene weight ratio of 95:5, respectively.

Additive Formulation

Two types of blends were prepared. The first consisted of blends containing 15% wt% of the Step 4 star-shaped product, poly(methylmethacrylate)-co-(ethylene-co-propylene) [blending agent]. The second blend consisted of a 1:1 wt/wt t mixture of poly(ethylene-co-propylene) having an Mn = 41,000 Da and poly(methylmethacrylate) having an Mn = 28,000 Da. Blends were prepared in o-xylene by stirring at 130 °C 1 h.

Testing

Blend compatibility was evaluated using transmission electron microscopy. Blends containing the star-shaped blending agent appeared uniform and finely dispersed. Blends not containing the experimental additive contained an islet PMMA phase separated as a macrophase in the order of several 10 nm in the EPR sea phase.

Advantages over Prior Art

Multiarm polymer compatiblizing agents prepared in this investigation can be routinely used to form blends of polyolefins such as polyethylene, polypropylene, and poly(ethylene-co-propylene) as well with more polar monomers such as vinyl acetate, methacrylate, or maleic anhydride. Intimately blended materials can be expected to have unique physical properties including printability, coating properties, adhesiveness, and compatibility with other polar polymers.

Notes

1. Kennedy[1] prepared the star copolymers (IV) containing polyisobutylene and polyacrylonitrile as a method for forming intimate blends of incompatible polyisobutylene and polyacrylonitrile.

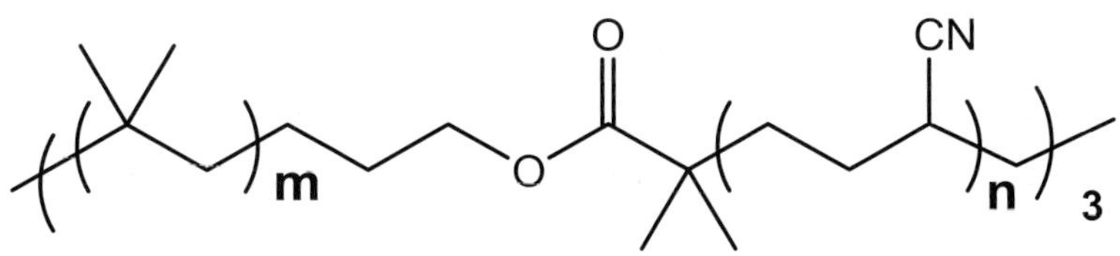

(IV)

2. A triblock polymer containing a polar sulfonated component (V) was prepared by Willis[2]. In this process, poly(4-*t*-butylstyrene-b-styrene-b-1,3-butadiene) was anionically prepared, hydrogenated to poly(4-*t*-butylcyclohexane-b-cyclohexane-b-1,3-butadiene), and then sulfonated with acetylsulfate. Materials prepared in this application were used in membranes.

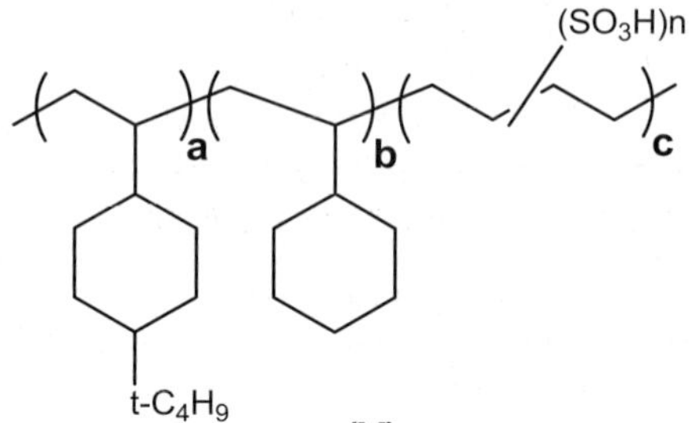

(V)

3. A method for orienting the lamellae in poly(styrene-b-methylmethacrylate) was developed by Cheng[3] by blending with poly(epoxydicyclopentadienylmethacrylate) (VI) in propylene glycol methyl ether acetate and γ-butyrolactone and then casting a 3% solution onto a silicon substrate and baking at 200 °C for 1 min.

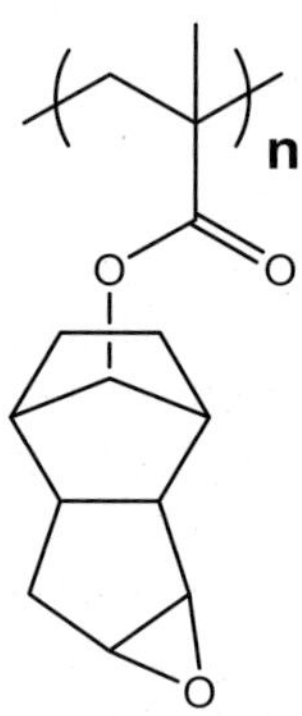

(VI)

4. A method of compatibilizing metals present in blending components was prepared by Faust[4] using organometallic-polyisomonoolefin block copolymers containing vinylferrocene (VII).

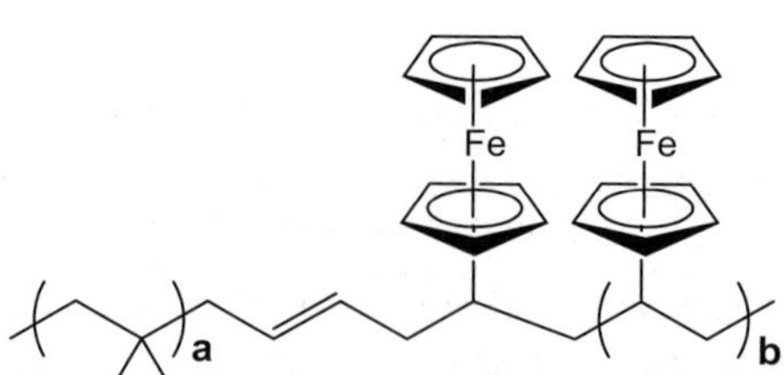

(VII)

REFERENCES

1. Joseph P. Kennedy et al, US Patent Application 20030236354 (December 25, 2003)
   Michigan State University (Ann Arbor, MI)
2. Carl L. Willis et al, US Patent 7,737,224 (June 15, 2010)
   Kraton Polymers U.S. LLC (Houston, TX)
3. Joy Cheng et al, US Patent 7,763,319 (July 27, 2010)
   International Business Machines Corporation (Armonk, NY)
4. Rudolf Faust et al, US Patent 7,786,217 (August 31, 2010)
   University of Massachusetts Lowell (Lowell, MA)

G. CRYSTALLINE INHIBITORS

Cyclobutanediol

Title:	Polyester compositions containing cyclobutanediol having a certain combination of inherent viscosity and moderate glass transition temperature and articles made there from
US Patent:	7,781,562 (August 24, 2010)
Author:	Emmett Dudley Crawford et al.
Assignee:	Eastman Chemical Company (Kingsport, TN)
Significance:	The commercially available polyester, poly(1,4-cyclohexylenedimethylene-terephthalate), crystallizes rapidly upon cooling from the melt making it difficult to process when using extrusion or injection molding. Although isophthalic acid and ethylene glycol have previously been used to slow down the poly(1,4-cyclohexylenedimethylene terephthalate) crystallization rate, materials containing these monomers exhibit deficiencies in glass transition temperature and impact strength. It has been determined that the incorporation of 2,2,4,4-tetramethyl-1,3-cyclobutanediol into poly(1,4-cyclohexylenedimethylene terephthalate) decreases the crystallization rate by at least 50%, increases the Tg, and lowers the polymer density.
Additive Name	2,2,4,4-Tetramethyl-1,3-cyclobutanediol
CAS Number	3010-96-6
Safety Concerns	The additive is an extremely flammable solid. The additive is irritating to the eyes, respiratory system, and the skin.

Additive Structure

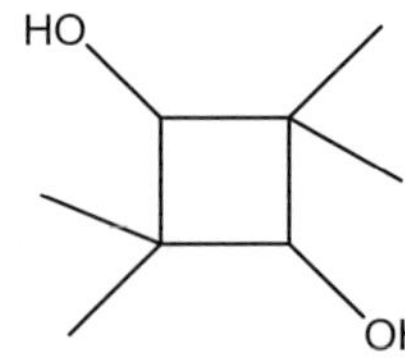

2,2,4,4-Tetramethyl-1,3-Cyclobutanediol
(Cis/Trans ~1:1)

Additive Source: Commercial reagent grade supplies of this additive component can be obtained from the following supplier:

2,2,4,4-Tetramethyl-1,3-cyclobutanediol
Suzhou Industrial Park Yacoo Chemical Reagent Co., Ltd.
Suite 209 A3 Building, 218 Xinghu Road, Suzhou Industrial Park
SuZhou, Jiangsu, China
86-512-62956016-888
The preparation of laboratory scale amounts is described by Theilheimer.[1]

Polymer Preparation

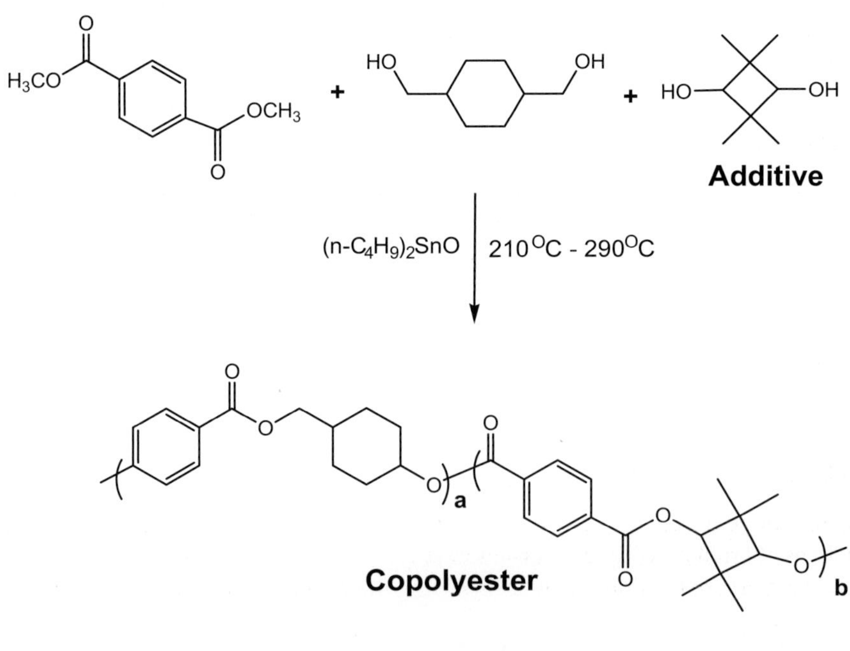

1. Method of incorporating additives into polymers

A mixture of 77.68 g of dimethyl terephthalate, 46.98 g 1,4-cyclohexanedimethanol 68.92 g, 21.33 g of 2,2,4,4-tetramethyl-1,3-cyclobutanediol, and 0.046 g of dibutyl tin oxide was placed in a 500-ml flask equipped with a nitrogen inlet, a metal stirrer, and a short distillation column. The flask was placed in a Wood's metal bath already heated to 210 °C and then stirred at 200 rpm throughout the experiment. The reaction temperature was then gradually increased to 260 °C over 30 min and then maintained at this temperature for 120 min. The temperature was then further increased to 290 °C for 30 min followed by the introduction of a vacuum until the pressure inside the flask reached 100 mm of Hg. The stirring speed was then reduced to 100 rpm while the pressure inside the flask was further reduced to 0.3 mm of Hg. The stirring speed was further reduced to 50 rpm while a pressure of 0.3 mm of Hg was maintained for 60 min to remove unreacted diols. The copolyester was colorless and was isolated as a white solid having a Tg of 103 °C and an inherent viscosity of 0.65 dl/g. Physical properties of the polyester as a function of reaction conditions are provided in Table 6.

TABLE 6 Physical properties of polyesters prepared by transesterification using dimethyl terephthalate with either ethylene glycol or 2,2,4,4-tetramethyl-1,3-cyclobutanediol. The objective of these experiments was to decrease copolymer crystallization half times

Diol	Comonomer (mol%)	η (dl/g)	Density (g/ml)	Tg (°C)	Tg (max) (°C)	Crystallization Times 150 °C	170 °C	200 °C
Ethylene Glycol	19.8	0.713	1.219	87.7	290	2.5 min	1.4 min	1.7 min
2,2,4,4-Tetramethyl-1,3-cyclobutanediol*	20.0	0.731	1.188	100.5	290	>180 min	23.3 min	25.2 min
2,2,4,4-Tetramethyl-1,3-cyclobutanediol*	40.1	0.53	1.172	122.0	260	>5 days	>5 days	>5 days

**1,4-Cyclohexanedimethanol constituted the comonomer.*

Additive Treatment Level

The partial replacement of 1,4-cyclohexanedimethanol by as little 12 mol% to as much as 41 mol% with 2,2,4,4-tetramethyl-1,3-cyclobutanediol was effective in decreasing crystallization rate half times and improving copolyester Tgs.

Advantages over Prior Art

Poly(1,4-cyclohexylenedimethylene terephthalate) is used to produce films, sheeting, and molded articles. While the polyester is prepared using terephthalic acid and 1,4-cyclohexanedimethanol, the material exhibits low glass transition temperature and short crystallization half times. When ethylene glycol was added as a termonomer, the crystallization half times marginally improved but ductile-to-brittle transition temperatures and even lower glass transition temperatures were also observed. Coreacting terephthalic acid or dimethyl terephthalate with 1,4-cyclohexanedimethanol and 2,2,4,4-tetramethyl-1,3-cyclobutanediol generates a high-performance plastic having a good dimensional stability, high heat resistance, good impact strength, and long crystallization half times.

Notes

1. Earlier preparations and physical properties of poly(1,4-cyclohexylenedimethylene terephthalate) containing 20 mol% of 5/95 cis/trans and 65/35 cis/trans 2,2,4,4-tetramethyl-1,3-cyclobutanediol are described by Hale.[2]
2. A summary of high notched Izod impact strength testing and glass transition trends for copolyester formulations consisting of varying molar ratios of dimethyl terephthalate, 1,4-cyclohexanedimethanol, 2,2,4,4-tetramethyl-1,3-cyclobutanediol is described by Crawford.[3]
3. Crawford[4] prepared copolyesters of terephthalic acid, 2,2,4,4-tetramethyl-1,3-cyclobutanediol, 1,4-cyclohexanedimethanol, and ethylene glycol, which had good flexural modulus, good clarity, and good color. These copolyesters were easily converted into articles and were used in thermoformed sheet and film applications.

REFERENCES

1. Werner Theilheimer. Synthetic Methods of Organic Chemistry. Volume 16, 1962, p. 29
2. Wesley Raymond Hale et al, US Patent 7,588,794 (November 30, 2006)
   Wesley Raymond Hale et al, US Patent Application 20060270773 (November 30, 2006)
   Eastman Chemical Company, Kingsport, TN
3. Emmett Dudley Crawford et al, US Patent 7,740,941 (June 22, 2010)
   Eastman Chemical Company, Kingsport, TN
4. Emmett Dudley Crawford et al, US Patent 7,737,246 (June 15, 2010)
   Eastman Chemical Company, Kingsport, TN

H. CURING AGENTS

Azlactones

Title:	Azlactone telechelic polymer
US Patent:	7,932,325 (April 26, 2011)
Author:	Kevin M. Lewandowski et al.
Assignee:	3 M Innovative Properties Company (St. Paul, MN)

Significance A curable diacrylate/acrylate polyester containing grafted acetoacetate telechelic appendages was prepared containing 4,4-dimethyl-2-ethyl-2-oxazolin-5-one termini using a base-catalyzed Michael addition reaction. The telechelic polymer had a controlled molecular weight and was readily crosslinkable whether the azlactone ring remained intact or was hydrolyzed to a reactive alanine derivative.

Additive 4,4-Dimethyl-2-ethyl-2-oxazolin-5-one-poly(1,6-hexanediol diacrylate-co-2-oxo-butyric acid ethyl acrylate ester)-4,4-dimethyl-2-ethyl-2-oxazolin-5-one (I)

Safety No safety information is available on the experimental additive of the current invention. The reagent 4,4-dimethyl-2-vinyl-2-oxazolin-5-one is considered, however, extremely flammable (EHU6f) and a mild irritant.

Additive Structure

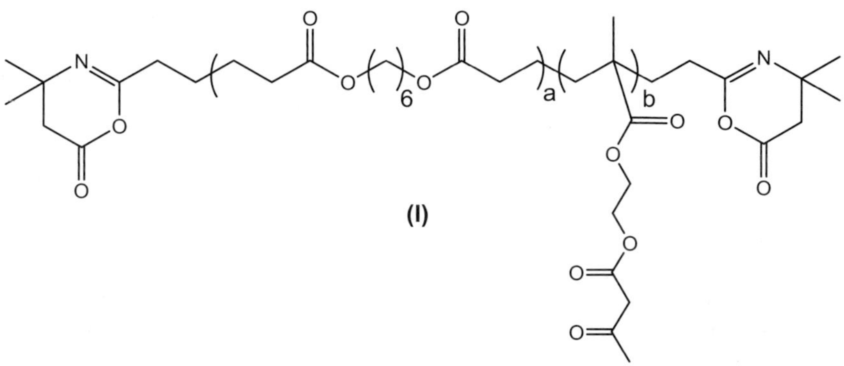

(I)

Additive Preparation

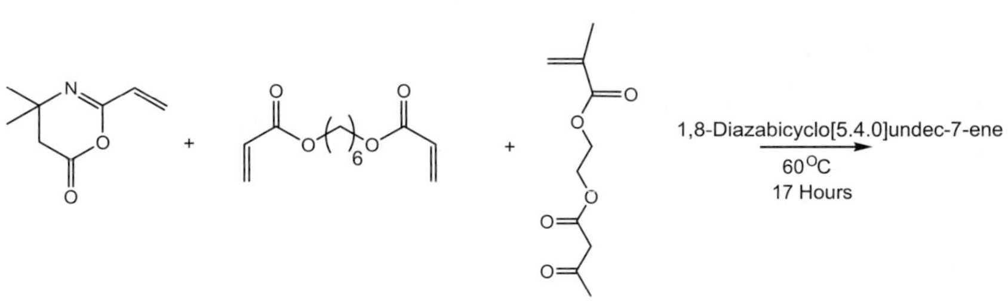

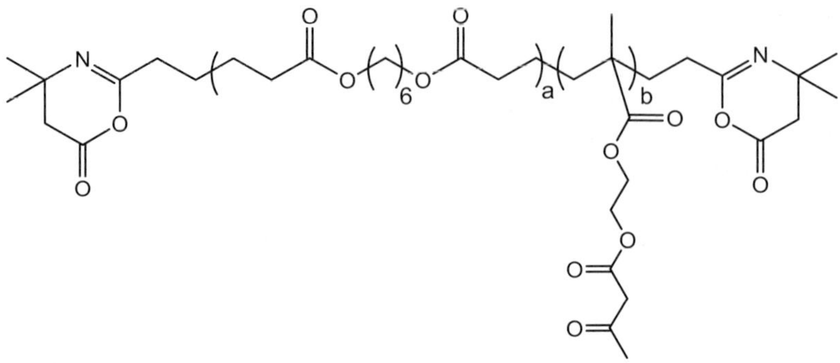

Additive Synthesis

1. Preparation of azlactone terminated telechelic methacrylate oligomers

In a typical preparation of azlactone terminated telechelic polymers, a reaction vessel was charged with reagents provided in Table 7 and then sealed and placed in a water shaker bath at 60 °C for 17 h. The compositions and molecular weights for each reaction were then determined by gel permeation chromatography and are summarized in Table 7.

TABLE 7 Reaction scoping and physical properties of azlactone-terminated telechelic polymers

Reaction	Acrylate Acac* (g)	HDDA† (g)	VDM¶ (g)	DBU** (g)	Di/Mono acrylate Ratio	Mw (1×10^3)	Mn (1×10^3)
1	2.5	2.64	0	0.025	60/40	72.1	4.7
2	2.5	2.38	0.17	0.025	90/10	20.9	3.4
3	2.5	2.11	0.32	0.025	80/20	7.6	2.2
4	2.5	1.85	0.49	0.025	70/30	1.3	0.8
5	2.5	1.59	0.65	0.025	60/40	1.3	0.7
6	2.5	1.32	0.81	0.025	50/50	1.4	0.8

**Acrylate Acac 2-Oxo-butyric acid ethyl acrylate ester acetoacetate.*
†HDDA 1,6-Hexanediol diacrylate.
¶VDM 4,4-Dimethyl-2-vinyl-2-oxazolin-5-one.
***DBU 1,8-Diazabicyclo[5.4.0]undec-7-ene.*

Testing

Testing for the azlactone telechelic polymer was not provided.

Advantages over Prior Art

The preparation of telechelic polymers with regulated and controlled molecular weights by using specified stoichiometries of Michael donor to Michael acceptor with the basic catalyst 1,8-diazabicyclo-[5.4.0]undec-7-ene is unreported. Polymer termini containing 4,4-dimethyl-2-vinyl-2-oxazolin-5-one are also a potential replacement for isocyanate-terminated telechelic polymers since hydrolysis of azlactones by moisture produces reactive carboxylic acid.

Notes

1. Ring-opened azlactone initiators (II) were prepared by Fansler[1] and used to make multiarm star polymers (III) for composite applications.

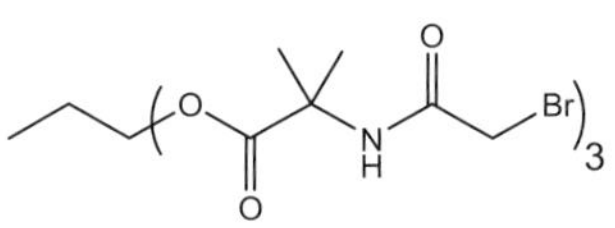

(II)

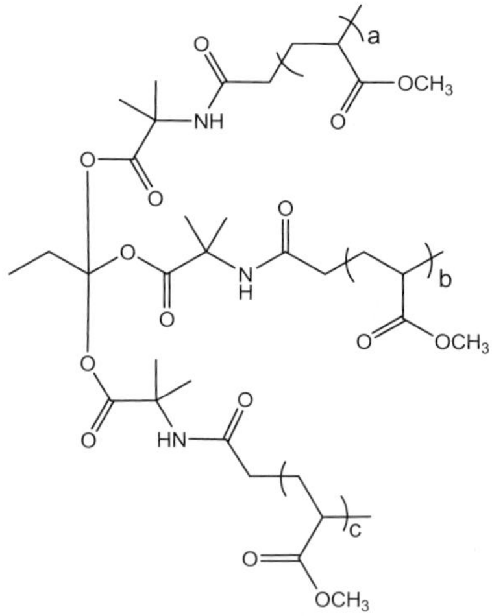

(III)

2. A series of multifunctional polybutadiene telechelic derivatives (IV) were prepared by David[2] using the thiol–ene reaction and used as intermediates in preparing liquid crystals.

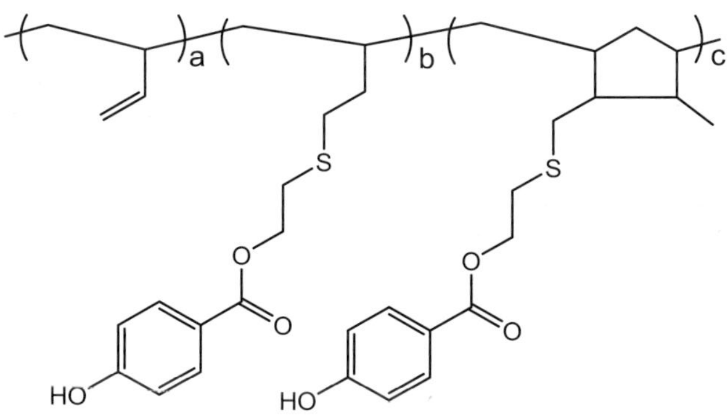

(IV)

3. Mao[3] prepared telechelic silane multiarm polyethylene glycol polymers (V) as surface coatings to inhibit adsorption of nonspecific proteins of cells to the surface of biomedical devices.

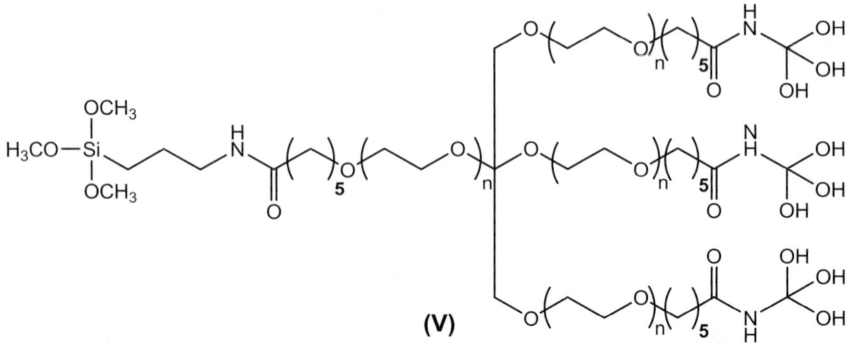

(V)

4. Telechelic polymers having phenolic hydroxyl groups at both ends (VI) or a benzoxazine ring terminus (VII) were prepared by Ishida[4] and used in compositions to improve polymer elongation properties.

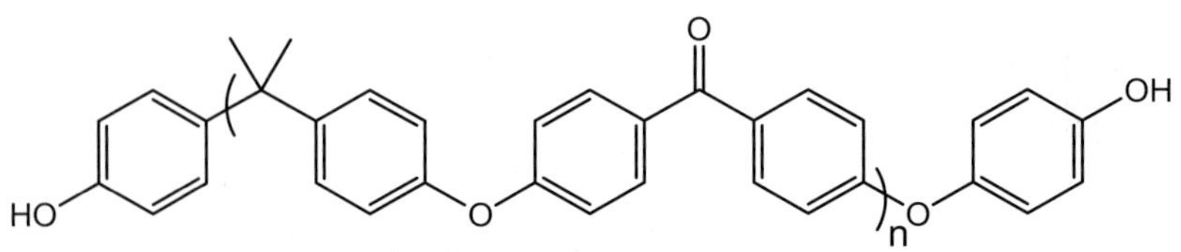

(VI)

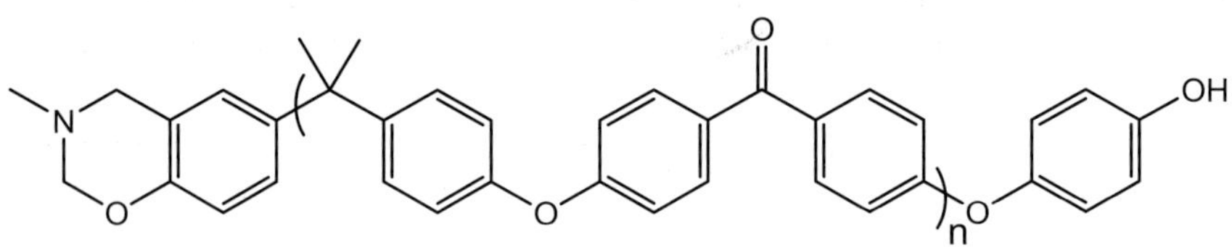

(VII)

REFERENCES

1. Duane D. Fansler et al, US Patent 7,557,177 (July 7, 2009)
3M Innovative Properties Company (St. Paul, MN)
2. Ralph L. David et al, US Patent 7,847,019 (December 7, 2010)
California Institute of Technology (Pasadena, CA)
3. Guoqiang Mao et al, US Patent 7,501,157 (March 10, 2009)
Accelro Technology Corporation (Denver, CO)
4. Hatsuo Ishida et al, US Patent 7,709,591 (November 4, 2010)
Sekisui Chemical Co., Ltd. (Osaka, JP)
Case Western Reserve University (Cleveland, OH)

I. CURING AGENTS

Imidazolium Ionic Liquids

Title:	Catalyst for curing epoxides
US Patent:	8,053,546 (November 8, 2011)
Author:	Georg Degen et al.
Assignee:	BASF SE (Ludwigshafen, DE)

Significance: *N,N′*-Dialkylimidazolium ionic salts containing acetate, formate, and methylcarbonate counterions were found to be effective in catalyzing epoxide polymerization reactions. When 5 wt% of a selected ionic salt was reacted with a 95 wt% mixture of a 1:0.9 composite of glycidyl bisphenol A and methylhexahydrophthalic anhydride, storage stabilities at 25 and 80 °C were dramatically improved. In addition, the heat of reaction of additized composites was approximately 40% greater than unadditized mixture reaction analogs.

Additive Names

1-Ethyl-3-methylimidazolium acetate (I)
1-Butyl-3-ethylimidazolium acetate (II)
1,3-Dimethylmethylimidazolium acetate (III)
1-Butyl-3-methylimidazolium formate (IV)
1-Ethyl-3-methylimidazolium methylcarbonate (V)

Safety

Although imidazolium-based ionic salts pose modest adverse health effects, they may potentially behave as eye and skin irritants.
Inhalation or vapors may potentially cause respiratory tract irritation.

Additive Structures

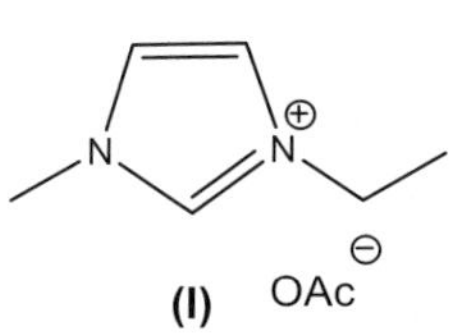

(I)

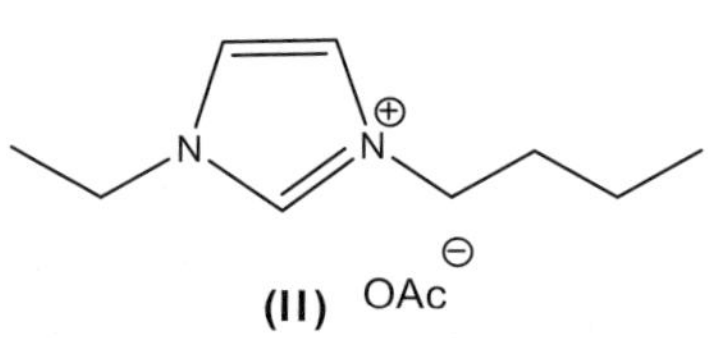

(II)

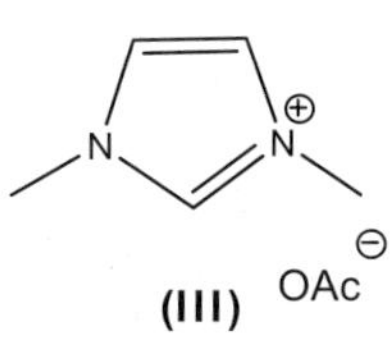

(III)

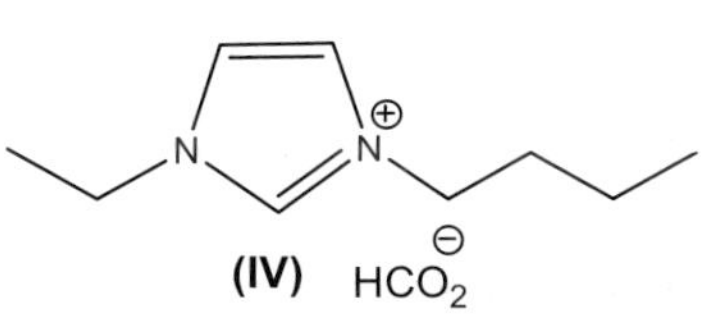

(IV)

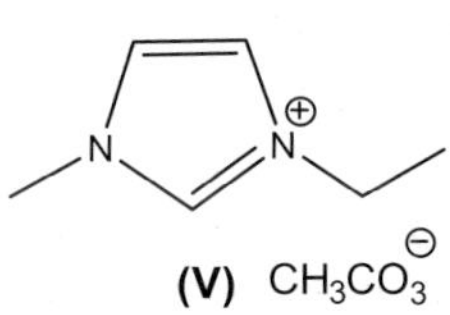

(V)

Additive Preparation

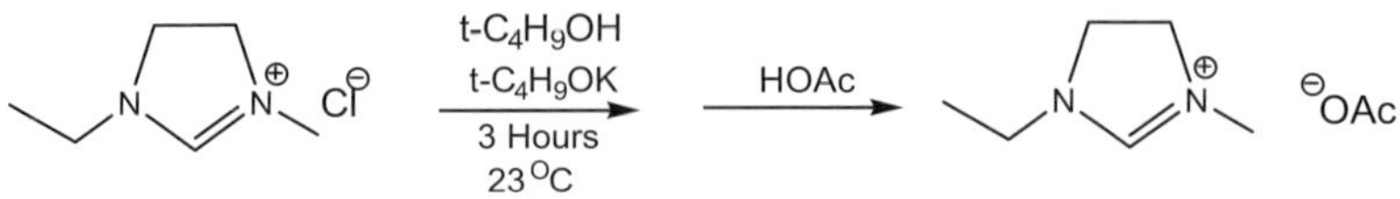

Additive Synthesis

1. *Preparation of 1-ethyl-3-methylimidazolium acetate according to the method of Maase*[1]

A reactor was charged with 92.6 g of potassium *t*-butoxide and 1000 ml of *n*-butanol and then heated to 60 °C to dissolve. The solution was then cooled to an ambient temperature and treated with 120.9 g of molten 1-ethyl-3-methylimidazolium chloride whereupon potassium chloride immediately precipitated from the solution. The mixture was then stirred for 3 h at an ambient temperature and then filtered. The precipitate was washed with *n*-butanol and 74.7 g of filtrate was isolated. A portion of the filtrate was titrated against 0.5 M of HCl, and the concentration of 1-ethyl-3-methylimidazolium *t*-butoxide in the filtrate was determined to be 0.666 mole. To convert 1-ethyl-3-methylimidazolium *t*-butoxide to the corresponding acetate, the filtrate was treated with a stoichiometric amount of 49.5 g of glacial acetic acid. The mixture was concentrated under reduced pressure, and an oil was isolated. The oil was then extracted with ethyl acetate, and after drying at 70 °C, the product was isolated in a 95.2% yield with a chloride content of 0.52% and a water content of 0.04%.

Additive Sources

Additives are available from the following venders:

Sigma–Aldrich, LLC
PO Box 14508
Saint Louis, MO 63178-9916
800-325-3010
http://www.sigmaaldrich.com

Additive Testing

1. *Glass Transition Temperature, Tg*

Experimental samples consisting of 5 wt% of a selected imidazolium and dicyandiamide (VI) alone or with diclycidyl ether of bisphenol A (VII) were prepared and evaluated to determine the curing properties of the experimental imidazolium salts. The Tgs of polymers were obtained by

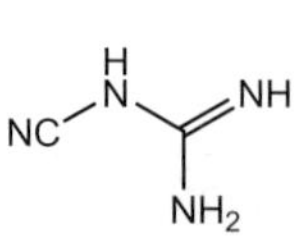

(VI)

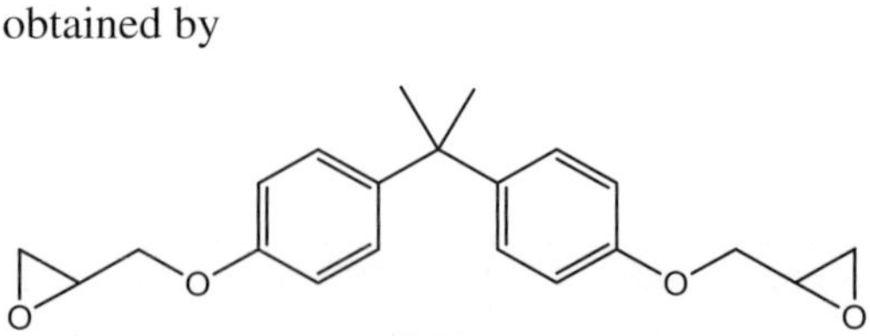

(VII)

differential scanning calorimetry using a heating rate of 30 °C/min with 20 g of samples of the uncured composition containing a selected experimental imidazolium salt. The sample consisted of a film having a thickness of 3–4 mm and was cured for 30 min at 40, 60, 80, 100, 120, and 140 °C. Testing results are provided in Tables 8 and 9.

2. *Storage Stability*

Storage stability of cured resins was obtained by measuring the relative viscosity at 25, 80, 100, and 120 °C. Properties of cured blends consisting of dicyandiamide and a selected imidazolium salt or dicyandiamide, selected imidazolium salt, and diclycidyl ether of bisphenol A are provided in Tables 8 and 9, respectively.

Test Results

TABLE 8 Physical properties of cured resins consisting of 5 wt% imidazolium salt and dicyandiamide

Additive	To (°C)	Tmax (°C)	ΔH (j/g)	Tg (°C)	Storage Stability 25 °C (days)	Storage Stability 80 °C (min)
I	73	114	495	133	2	11
II	98	117	479	141	2	14
III	96	119	475	n.d.	1	9
IV	94	124	423	138	2	20
V	107	141	538	130	4	49

TABLE 9 Physical properties of cured resins consisting of mixtures of diclycidyl ether of bisphenol A [DGEBA] and dicyandiamide [DICY] cured with 1-ethyl-3-methyl- imidazolium acetate (Additive 1)

Composition (wt%)	To (°C)	Tmax (°C)	ΔH (j/g)	Tg (°C)	Storage Stability 80 °C (s)
DGEBA + Additive 1 (100:5)	73	114	495	133	660–720
DGEBA + Additive 1 + DICY (100:3.75:1.25)	116	153	485	140	2040–2440
DGEBA + Additive 1 + DICY (100:2.5:2.5)	130	140	532	152	5140–6010
DGEBA + Additive 1 + DICY (100:1.25:3.75)	112	127	509	160	7080–9120
DGEBA + Additive 1 + DICY (100:5:5)	130	140	529	133	3380–4150

Advantages over Prior Art

Although imidazolium-based epoxide catalysts with a variety of counterions have previously been prepared, their catalytic activity typically begins at ≥190 °C. According to the current invention, 1,3-dialkylimidazolium ionic salts containing acetate, formate, or methylcarbonate counterions have a catalytic threshold of 100 °C. In addition, all ionic salts described in this invention are readily synthesized in high yields.

Notes

1. McNamara[2] prepared a curable spherical particulated sheet molding consisting of the diglycidyl ether of bisphenol A and the epoxy adduct of poly(butadiene-co-acrylonitrile) with 2-methylimidazole as the curing agent encapsulated in 4,4′-methylenebis(phenyl isocyanate).
2. Chow[3] used the toluene diisocyanate uretidone dimer (VIII) as a latent crosslinking agent in the condensation of bisphenol A diglycidyl ether with 4′-diamino-3,3′-dimethyl dicyclohexyl methane to prepare resins having a lap shear strength of at least 10.48 N/mm^2.

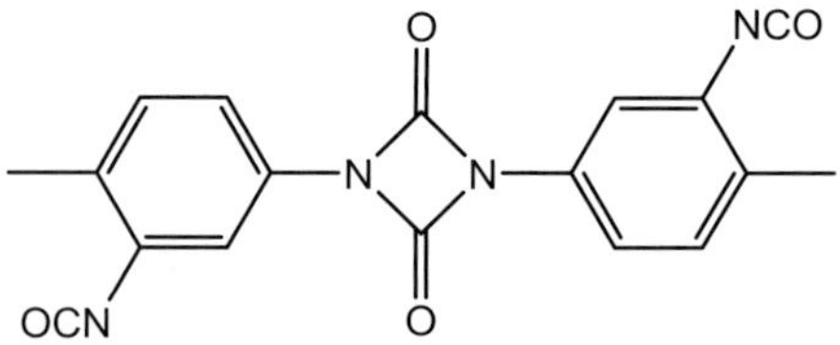

(VIII)

3. Amano[4] prepared a one-component epoxy resin composition consisting of an amineamide mixture derived from methyl 2-hydroxyisobutyrate, phenylglycidyl ether, 2-propanol, and 1,1-dimethyl hydrazine and then treating it with triethyl borate as the latent crosslinking agent.
4. Bertsch[5] prepared uncured epoxy resins consisting of epoxy-terminated poly(butadiene-co-acrylonitrile) and low molecular weight novolak resin functionalized with diglycidyl ethers and then used dicyandiamide as the latent curing agent.

REFERENCES

1. Matthias Maase et al, US. Patent 7,858,802 (December 28, 2010)
BASF SE (Ludwigshafen, DE)
2. John J. McNamara et al, US. Patent 8,044,154 (October 25, 2011)
Trillion Science, Inc. (Fremont, CA)
3. Sim Chow et al, US. Patent 7,507,778 (March 24, 2009)
Huntsman Advanced Materials Americas Inc. (The Woodlands, TX)
4. Hiroshi Amano et al, US. Patent 7,479,534 (January 20, 2009)
Ajinomoto Co., Inc. (Tokyo, JP)
5. Roert J. Bertsch et al, US. Patent 7,847,026 (December 7, 2010)
Emerald Specialty Polymers, LLC. (Akron, OH)

J. FOAM STABILIZERS

Poly[(Ethylene Oxide-co-Propylene Oxide)-b-Polysiloxane]

Title:	Process for the preparation of SiOC-linked, linear polydimethylsiloxane-polyoxyalkylene block copolymers and their use
US Patent:	7,825,205 (November 2, 2010)
Author:	Wilfried Knott et al.
Assignee:	Evonik Goldschmidt GmbH (DE)

Significance: An extremely effective polyurethane foam stabilizer additive, poly[(ethylene oxide-co-propylene oxide)-b-polydimethylsiloxane] was prepared by reacting stoichiometric amounts of poly(ethylene oxide-co-propylene oxide) diol with α,ω-hydrodimethylsiloxane. When added to polyurethane prepolymer packages, foam cell structure and porosity were fine, open, and stable.

Additive Name Poly[(ethylene oxide-co-propylene oxide)-b-polydimethylsiloxane] (I)

Safety Concerns Moderately irritating to the skin and the eyes

Additive Structures

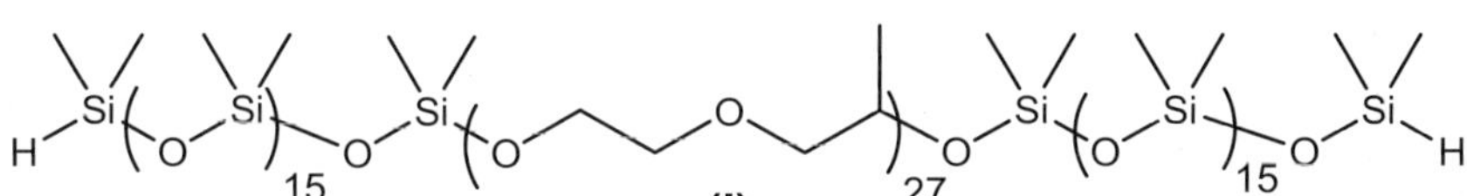

(I)

Additive Preparation

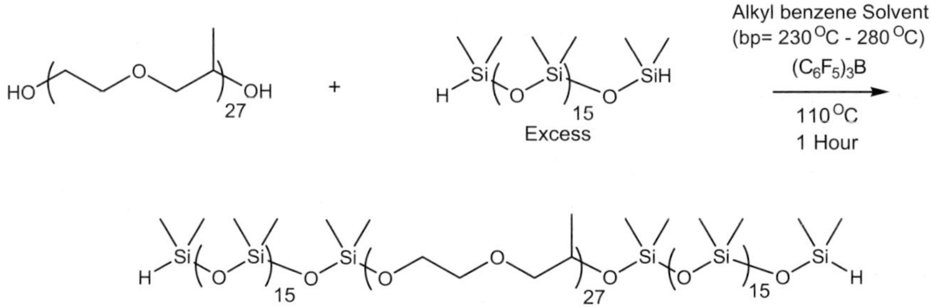

Additive Synthesis

1. *Preparation of poly[(ethylene oxide-co-propylene oxide)-b-polydimethylsiloxane]*

A 500-ml four-necked flask equipped with a stirrer, a precision glass gland, dropping funnel, internal thermometer, and reflux condenser with flexible gas-outlet tube was charged with 55.0 g of poly(ethylene oxide-co-propylene oxide)-diol having a 1:1 monomer ratio with an Mn of approximately 2800 Da, 85.3 g of linear alkylbenzenes having a boiling point range of about 230–280 °C, and 85.3 mg of tris(pentafluorophenyl)borane. The mixture was then heated to 110 °C and treated with the dropwise addition of 30.3 g of α,ω-hydrodimethylsiloxane having an average chain length of 15 in 35% molar excess. The reaction extent was monitored by the evolution of hydrogen gas, and the reaction was stopped when the evolution of hydrogen ceased Table 10.

Additive Formulation

TABLE 10 **The polyurethane prepolymer additive formulation package used to evaluate the foam stabilization properties of the experimental additive, poly[(ethylene oxide-co-propylene oxide)-b-polydimethylsiloxane]**

Component	Parts by weight
Stannous 2-ethylhexanoate	0.07
Polyol CP® 3322	30
Polyol CP® 755	70
Polyol CP® 1421	7
Water	1.95
Bis(dimethylaminoethyl)	0.20
Triethylenediamine	0.30
Dimethylethanolamine	0.20
Experimental foam stabilizer	1.20
Tolylene diisocyanate	40.30

Additive Testing

The contents of a paper cup, stannous 2-ethylhexanoate, polyol Polyol CP® 3322, Polyol CP® 755, and Polyol CP® 1421, water, amine catalysts, and the experimental foam stabilizer, were mixed for 60 s at 1000 rpm and then treated with tolylene diisocyanate and further mixed at 1500 rpm. Once foaming started, the mixture was poured into an insulated foaming box, and the cell membrane of the foam bubbles and open-pore cell structure observed. It is at this point that foam stabilization was assessed.

Foam Stabilization Testing Results

Polyurethane prepolymer packages additized with 1.2 wt% of poly[(ethylene oxide-co-propylene oxide)-b-polydimethylsiloxane] had stable foams after cooling. Measurements gave a cell number of 10 cells/cm with a porosity of 70 mm.

Advantages over Prior Art

The effectiveness of poly[(ethylene oxide-co-propylene oxide)-b-polydimethylsiloxane] as a foam stabilizer is partially the result of using the optimum stoichiometric amounts of reagents. Earlier reports of poor foam stabilization using this additive when prepared using nonstoichiometric reagent ratios may reflect the presence of unreacted reagents or a product with a broad polydispersity.

Notes

1. Venzmer[1] prepared a series of low-temperature polyurethane aqueous foam stabilizers by postreacting poly[(ethylene oxide-co-propylene oxide)-b-polydimethyl;siloxane] with butyl methacylate (II) styrene, or allyl alcohol and then polymerizing with Trigonox® D-C50.

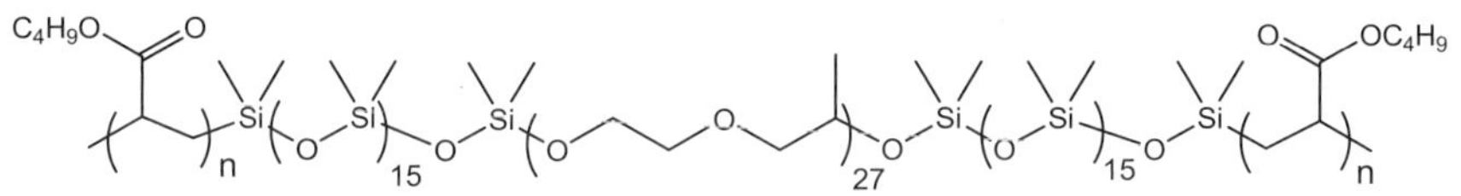

(II)

2. Stable and flexible polyurethane foams requiring the minimum use of foam stabilizers were prepared by Chan[2] by reacting poly[propylene oxide-b-(propylene oxide-co-ethylene oxide)-b-ethylene oxide] diol with 4,4′- diphenylmethane diisocyanate (III).

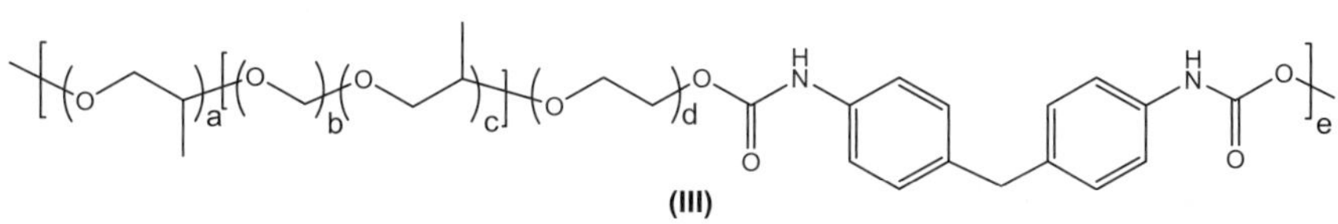

(III)

3. Boinowitz[3] prepared poly(styrene oxide-b-ethylene oxide) block copolymers (IV) initiated with 5,5-dimethylhexanol, which were effective as low-foam pigment wetting agents.

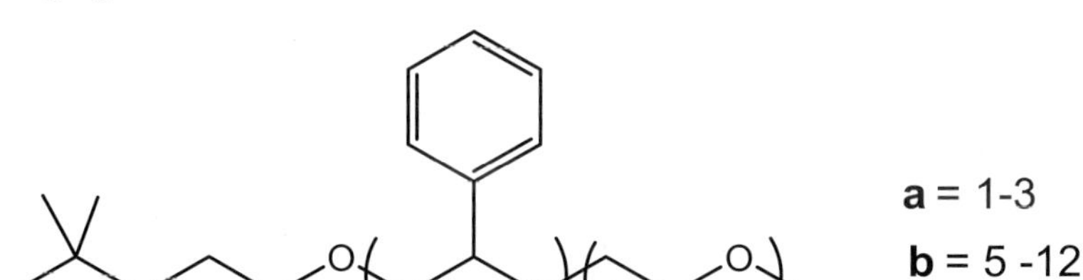

(IV)

4. Rautschek[4] prepared an aqueous antifoam composition consisting of a mixture of phenylsiloxane and a silicone resin having a terminus of 40 mol% $CH_3SiO_{1/2}$, 50 mol% SiO_2, 8 mol% $C_2H_5OSiO_{1.5}$, and 2 mol% $HOSiO_{1.5}$, which was used in detergent compositions.

REFERENCES

1. Joachim Venzmer et al, US Patent 7,759,402 (July 20, 2010)
Goldschmidt GmbH (Essen, DE)
2. Koon Yeow Chan et al, US Patent 7,790,778 (September 7, 2010)
Huntsman International LLC (The Woodlands, TX)
3. Tammo Boinowitz et al, US Patent 6,552,091 (April 22, 2003)
Goldschmidt GmbH (Essen, DE)
4. Holger Rautschek et al, US Patent 7,619,043 (November 17, 2009)
Wacker Chemie AG (Munich, DE)

K. FORMALDEHYDE SCAVENGERS FOR RESINS

Sodium metabisulfite

Title:	Reducing formaldehyde emissions
US Patent:	8,043,383 (October 25, 2011)
Author:	Kim K. Tutin et al.
Assignee:	Georgia-Pacific Chemicals LLC (Atlanta, GA)

Significance: Formaldehyde emissions from phenol–formaldehyde and urea–formaldehyde resins were eliminated by absorption into solid sodium metabisulfite dispersed on a paper towel. The formaldehyde scavenger is believed to slowly decompose into sulfur dioxide, which then condenses to form water-soluble formaldehyde sulfoxylate.

Additive Names Sodium metabisulfite (I)

Safety Sodium metabisulfite is used as an antioxidant additive in foods and cosmetics and is considered nontoxic.

Additive Structure

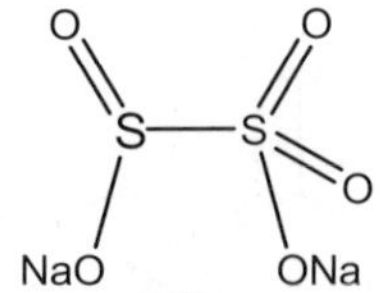

(I)

Additive Source

Sodium metabisulfite is available from

Alfa Aesar
26 Parkridge Road
Ward Hill, MA 01835
USA
Tel: 1-978-521-6300
Fax: 1-978-521-6350
e-mail: info@alfa.com

Additive Testing

1. Reduction of formaldehyde in phenol–formaldehyde resin using sodium bisulfate (Note 1)

Four pieces of B-staged laminate paper consisting of a phenol–formaldehyde resin was placed into four separate sealed containers containing a small dry paper towel previously rinsed with sodium bisulfite. A control was also prepared consisting of a phenol–formaldehyde resin containing a towel not previously rinsed with 0.1 M sodium bisulfate. All four bags were then sealed for 72 h at an ambient condition. Thereafter, the samples were analyzed by placing in a dynamic microchamber, and formaldehyde emissions were determined by the consistent steady-state technique. The control sample had an average formaldehyde emission >700 ppb while samples stored with sodium bisulfate treated paper had formaldehyde levels below the limits of detection of the instrument.

Advantages over Prior Art

There are potentially limitless consumer applications from this discovery. First, sodium metabisulfite is essentially innocuous since it is used as an additive in food and cosmetics. Second, it can be easily applied by consumers concerned about formaldehyde emissions to laminates using a damp disposable paper towel. Finally, sodium metabisulfite has an infinite shelf life so that the consumer can be assured of the product's activity after remaining unused for extended periods.

Notes

1. The route for lowering formaldehyde levels is thought to involve a two-step mechanism. First, when mixed with water, sodium metabisulfite releases sulfur dioxide, SO_2. In the second stage, when sulfur dioxide comes into contact with formaldehyde, the two reagents combine to form **formaldehyde–sulfoxylate (II).**

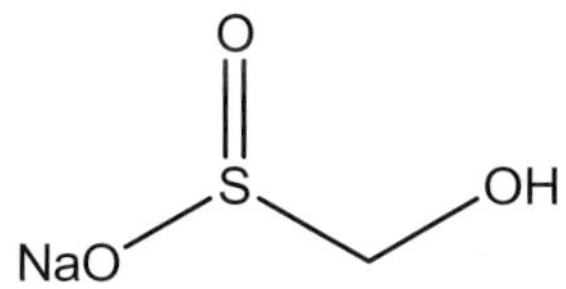

(II)

2. Olang[1] treated the surfaces of urea/formaldehyde and phenol/formaldehyde resins with urea, oil emulsion, ammonium sulfate, and silane and then baked the mixture at 400 °F to eliminate formaldehyde emissions.
3. Boyer[2] reduced formaldehyde emissions in fiberglass insulation using a mixture of tetraethylene pentamine and sodium bisulfate.
4. Stable phenolic resin polymer dispersions having low or formaldehyde-free content have been prepared by Boyle[3] using urea as the formaldehyde scavenger.

REFERENCES

1. Fatemeh Olang et al, US. Patent Application 20090004391 (January 1, 2009) **Owens Corning Fiberglas Technology, Inc.** (Summit, IL)
2. Peter Boyer et al, US. Patent Application 20080003902 (January 3, 2008) **Owens Corning Fiberglas Technology, Inc.** (Summit, IL)
3. Peter Boyer et al, US. Patent 7,671,097 (March 2, 2011) **Owens Corning Fiberglas Technology, Inc.** (Summit, IL)

L. NUCLEATING AGENTS

Bicyclo[2.2.1]Heptane-2,3-Dicarboxylate Salts

Title:	Polymer compositions comprising nucleating or clarifying agents and articles made using such compositions
US Patent:	7,786,203 (August 31, 2010)
Author:	Rob Hanssen et al.
Assignee:	Milliken & Company (Spartanburg, SC)
Significance:	An additive mixture consisting disodium bicyclo[2.2.1]heptane-2,3-dicarboxylate and 3:2,4-bis(3,4-dimethylbenzylidene) sorbitol was found to be effective as a nucleation agent for preparing plastic articles with improved clarity. This additive package was effective using low, medium, or highly crystalline polyethylene of monomodality or bimodality in extrusion, blowing, molding, compression, or extrusion blow molding applications.

Additive Name	1,3:2,4-Bis(3,4-dimethylbenzylidene) sorbitol (I) (contains up to 4% proprietary silica compound)
CAS Number	135861-56-2 135861-56-2
Safety Concerns	May cause irritation of the respiratory system. Prolonged skin contact may cause temporary skin irritation.

Additive Name	Disodium bicyclo[2.2.1]heptane-2,3-dicarboxylate (II)
CAS Number	351870-33-2
Safety Concerns	May irritate eyes although not known to cause permanent injury to eye tissue. Dust in high concentrations may irritate the respiratory system.

Additive Structures

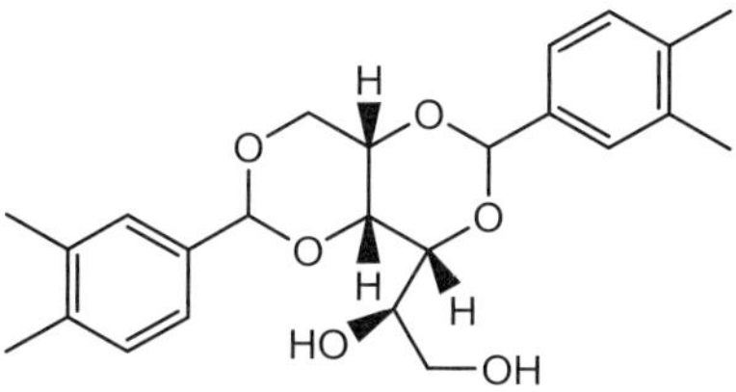

(I)

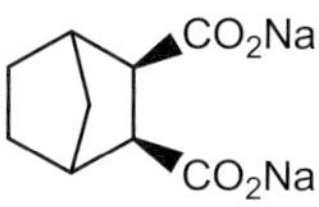

(II)

Additive Preparation: Commercial supplies of both additives are available from the following supplier.

Milliken Chemical
P.O. Box 1926
Spartanburg, SC, 29303 USA
1-864-472-9041

Additive Formulation

A blend consisting of 10 wt% of polyethylene resin was prepared by compounding 2.0 wt% of a 1:1 mixture of disodium bicyclo[2.2.1]heptane-2,3-dicarboxylate and 1,3:2,4-bis(3,4-dimethyl-benzylidene) sorbitol with 90 wt% of random poly(ethylene-co-propylene). The final compounded polymer compositions comprised 0.1 wt% by weight of both additives and 1.8 wt% of polyethylene resin.

Testing

Compounded samples of polyethylene were injection molded into 50 mil plaques using an Arburg 40-ton injection-molding machine. After storing the molded plaques for at least 12 h, the plaques were tested for haze and clarity using a BYK-Gardner Haze-Gard Plus BYK Gardner Gloss Meter, and the change in crystallization content was measured using a Perkin–Elmer DSC 7. A summary of test results for selected resin blends is provided in Table 11.

TABLE 11 Optical and physical properties of polyethylene additized with a total of 0.2 wt% nucleating agents after they were injected molded and then stored for 12 h

Resin	Clarifying Additive	Haze (%)	Clarity (%)	Tc (°C)
LDPE	Additive 1 and Additive 2	36.2	97.3	93.4
LLDPE	Additive 1 and Additive 2	38.8	95.2	100.6
HDPE	Additive 1	100	5.5	109.4

Advantages over Prior Art

The synergistic effect of this additive package particularly when used to prepare molded polyethylene thermoplastic articles with improved clarity and containing both nucleating agents is unreported, particularly when processed at elevated temperatures. This also resulted in reduced fabrication cycle times, improved stiffness, and restricted stable melt index times.

Notes

1. Li[1] improved the clarity and flow enhancement properties of polyethylene resins by coadditizing with 1,3:2,4-bis(3,4-dimethylbenzylidene) sorbitol and silica, SiO_2 particles having a mean volume diameter, D_v, of ≤20 μm with a 90th percentile, D_{90}, of ≤50 μm.
2. Grein[2] prepared high clarity and high strength propylene compositions by coadditizing with β-nucleating agents 5,12-dihydro-quino(2,3-b)acridine-7,14-dione, (III), and quino(2,3-b)acridine-6,7,13,14(5H, 12H)-tetrone (IV).

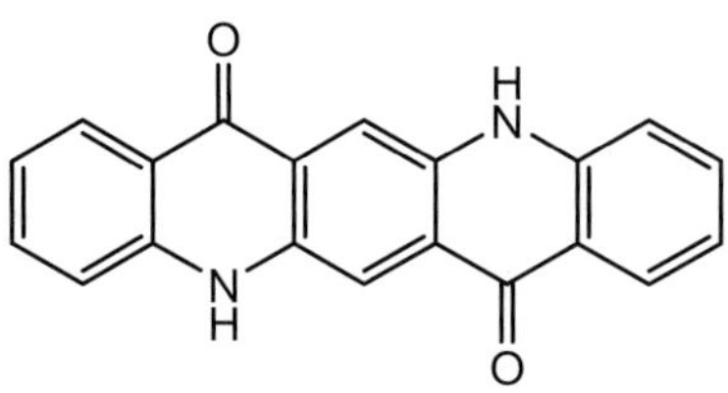

(III)

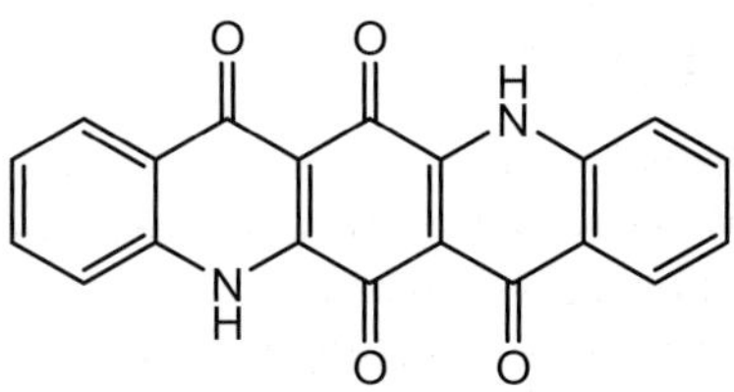

(IV)

3. Ma[3] prepared polypropylene resin compositions having excellent crystallizability, high transmittance, high clarity, low haze, and improved thermal stability by additizing with the β-nucleating agents 1,2,4,5-tetra(cyclopentylcarbonylamino) benzene or 1,2,4,5-tetra(cyclohexylcarbonylamino)benzene (V). Tetraamides were prepared by reacting the corresponding cyclic amine with 1, 2, 4, 5-benzenetetracarboxylic anhydride.

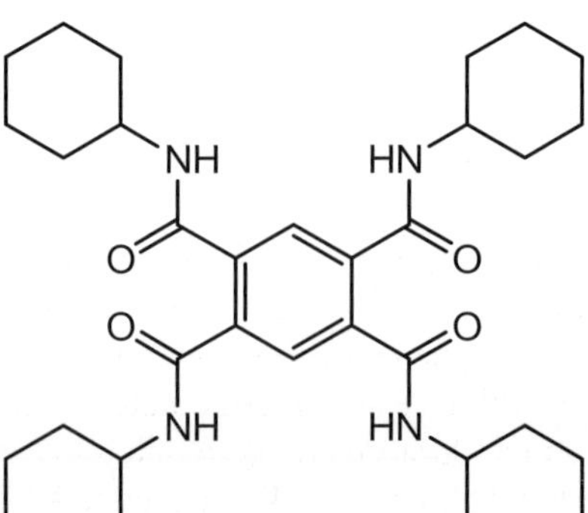

(V)

4. Schmitt[4] improved the haze and clarity in polyethylene resins using 1,3,5-tris(cyclohexylcarbonylamino)benzene (VI) and related 1,3,5-tris(cycloaryl- and arylcarbonylamino)benzene analogs.

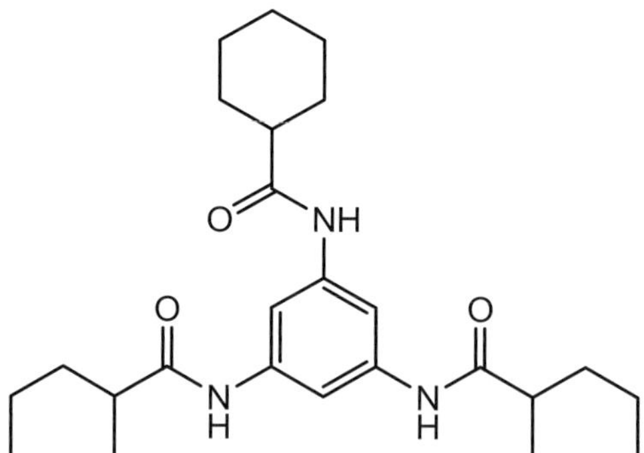

(VI)

REFERENCES

1. Jiang Li et al, US Patent 7,781,511 (August 24, 2010)
Milliken & Company (Spartanburg, SC)
2. Christelle Grein et al, US Patent 7,786,204 (August 31, 2010)
Borealis Technology Oy (Porvoo, FI)
3. Qinggao Ma et al, US Patent 7,569,630 (August 4, 2010)
Chemtura Corporation (Middlebury, CT)
4. Hans-Werner Schmidt et al, US Patent 7,790,793 (September 7, 2010)
Ciba Specialty Chem. Corp. (Tarrytown, NY)

M. PHOTOCURABLE INITIATIVES

Sulfonium Salt Initiators

Title:	Sulfonium salt initiators
US Patent:	8,067,643 (November 29, 2011)
Author:	Pascal Hayoz et al.
Assignee:	BASF SE (Ludwigshafen, DE)

Significance: Three 4-keto phenoxysulfonium hexafluorophosphates were prepared in two high-yielding synthetic steps. In the initial step, 4-thiophenylacetophenone was converted into the sulfonium chloride salt intermediate using thionyl chloride and aluminum chloride and then anion exchanged using potassium hexafluorophosphate. When used in epoxide formulations as a cationic photoinitiator, films were clear, strong, and very slightly discolored.

Additive Names 1-[4-(1-Propanonyl)thiophenyl]phenylsulfonium hexafluorophosphate (I)
1-[4-(4-Acylthiophenyl)]phenylsulfonium hexafluorophosphate (II)
1-[4-(4-Acyloxyphenyl)]phenylsulfonium hexafluorophosphate (III)

Safety Phenylsulfonium hexafluorophosphate and derivatives are considered to be extremely toxic. They are moderate to severe eye, respiratory system, and skin irritants. In addition, skin contact may result in immediate or delayed skin sensitization. These chemical agents are also considered toxic to aquatic organisms and will cause long-term adverse effects in the aquatic environment.

Additive Structures

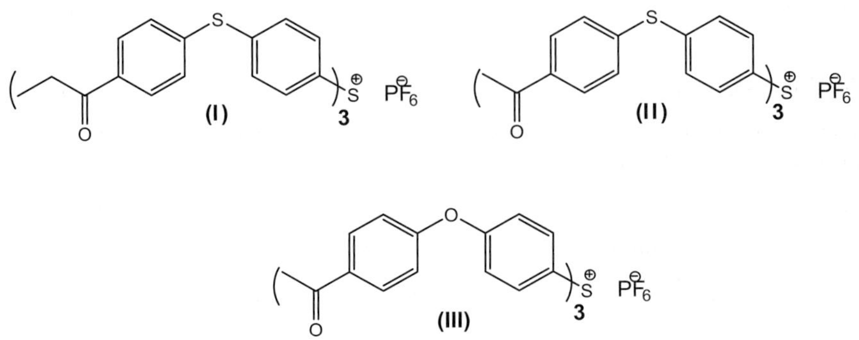

Additive Preparation

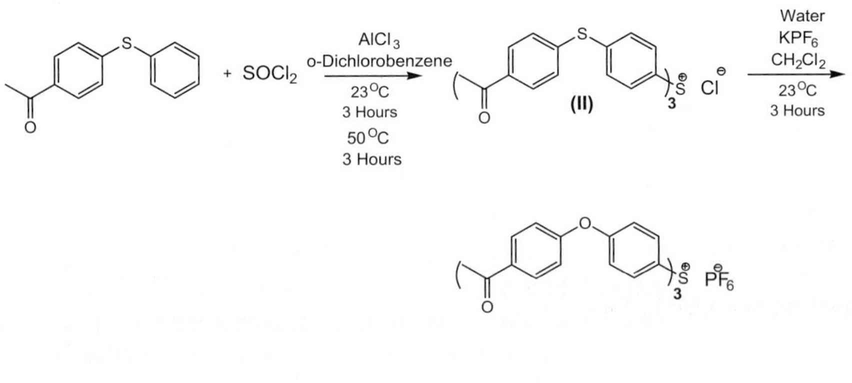

Additive Synthesis

1. Preparation of 1-[4-(4-acylthiophenyl)]phenylsulfonium chloride

A 200-ml reactor was charged with 1.54 g of thionyl chloride and 10.36 g of aluminum chloride and then treated with the dropwise addition of 9.62 g of 4-thiophenylacetophenone dissolved in 30 ml of o-dichlorobenzene at an ambient temperature. The reaction mixture was then stirred for 3 h at an ambient temperature and an additional 3 h at 50 °C. The mixture was then poured on an water–ice mixture, and the product was extracted with dichloromethane.

^{1}H-NMR (d_6-DMSO, δ) 8.05 6H d, 7.92 6H d, 7.43 6H d, 7.28 6H d, 2.59 9H s.

2. Preparation of 1-[4-(4-acylthiophenyl)]phenylsulfonium hexafluorophosphate

A container was charged with 10.8 g of the Step 1 product dissolved in 30 ml of methylene chloride and then treated with 14.17 g of potassium hexafluorophosphate dissolved in 150 ml of water. After vigorously stirring for 3 h at an ambient temperature, the organic phase was concentrated, and the product was isolated after purification by column chromatography.

^{1}H-NMR ($CDCl_3$, δ) 8.00 6H d, 7.70 6H d, 7.28 6H d, 7.13 6H d, 2.59 9H s.

Additive Testing

The composition used to assess the effectiveness of the experimental sulfonium salt additives is provided in Table 12.

TABLE 12 **Formulation used to determine the efficacy of experimental sulfonium salt additives**

Component	Treatment Level (parts)
3,4-Epoxycyclohexylmethyl carboxylate	81
3-Ethyl-3-hydroxymethyloxetane	12
ε-Caprolactane Triol	6
Byk® 307 (Silicon Surface Additive)	0.5
Felxo® Ink Varnish Formulation	100

The experimental additive was then blended into the formulation and applied to an 85-mm-thick aluminum film using a 4-mm wire bar and then placed on a fast moving conveyer belt and cured using a 120-W/cm medium pressure mercury lamp. Test results are provided in Table 13.

Test Results

TABLE 13 Effectiveness of selected sulfonium hexafluorophosphate additives as photoinitiators of polyesters using the formulation described in Table 12

Additive	Concentration (%)	Cure Speed (m/min)
I	3	170
II	4	200
III	4	200

Advantages over Prior Art

Sulfonium hexafluorophosphate cationic initiators described in the invention possess unique logistical and environmental advantages over existing materials. First, they are reactive in both thin or thick pigmented or clear coatings without the need of a sensitizer. In addition, coatings prepared using this additive are very little yellowing. A particularly favorable environmental property is that the photodecomposition products are odorless, nontoxic, and benzene-free.

Notes

1. The sulfonium salt photoinitiator adamantyl thioaromatic sulfonium tetrakis(pentafluoro)borate (IV) prepared by Hayoz[1] was particularly effective in both polymerization and free-radical crosslinking reactions and used for coating applications.

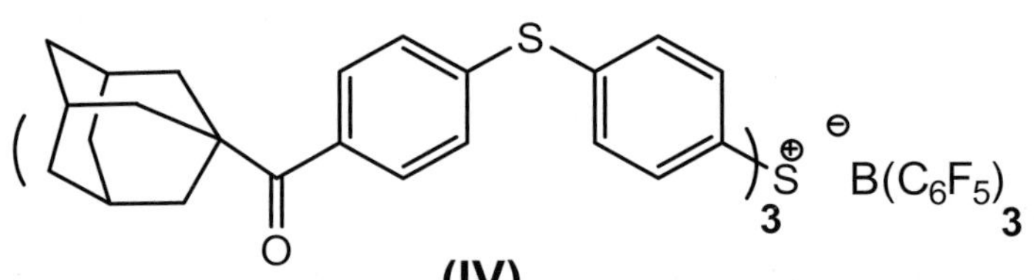

(IV)

2. Wolf[2] used diaromatic iodonium hexafluorophosphate salts (V) as the cationic photoinitiator to terpolymerize monomers (VI), (VII), and (VIII).

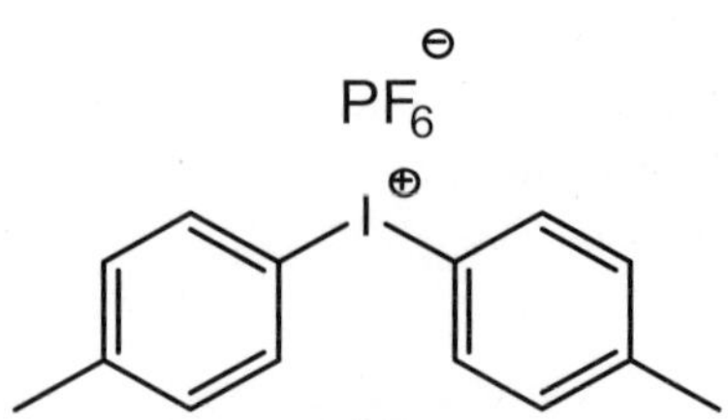

(V)

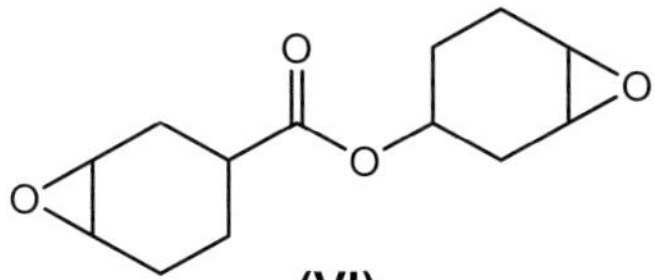
(VI)

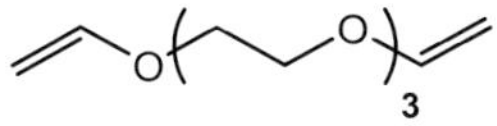
(VII)

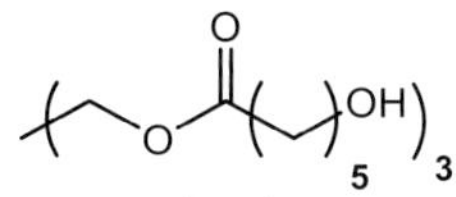
(VIII)

3. Functionalized aromatic sulfonium hexafluorophosphates (IX) and (X) prepared by Hayoz[3] were effective as photoinitiators and were used to prepare polyesters.

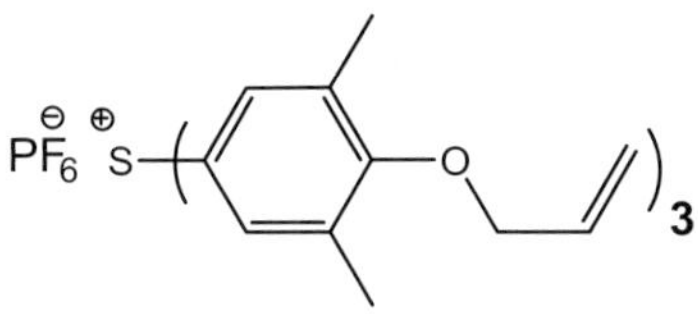
(IX)

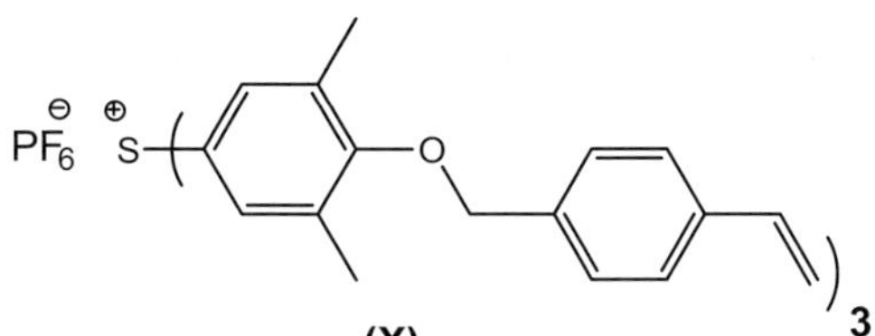
(X)

4. Nakayashiki[4] prepared dibenzothiophene sulfonium derivatives (XI), which were effective as photoacid generators and photopolymerizable agents.

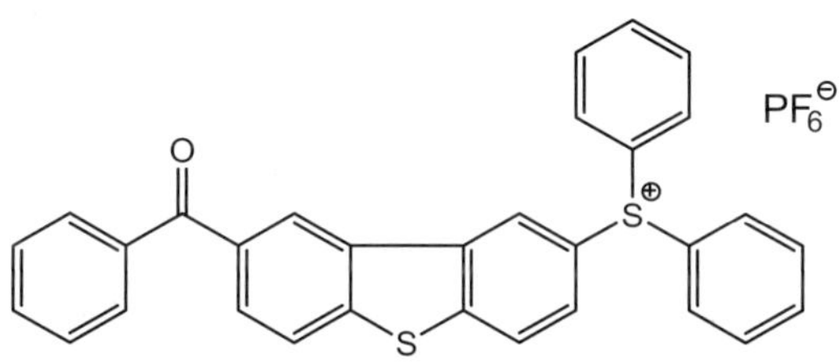
(XI)

REFERENCES

1. Pascal Hayoz et al, US Patent Application 20100297542 (November 25, 2010) **BASF SE (Ludwigshafen, DE)**
2. Jean-Pierreet Wolf et al, US. Patent Application 20070225395 (September 27, 2009) **Ciba Specialty Chemicals Corp. (Tarrytown, NY)**
3. Pascal Hayoz et al, US Patent 8,012,672 (September 6, 2011) **BASF SE (Ludwigshafen, DE)**
4. Tetsuyuki Nakayashiki et al, US. Patent 7,611,817 (November 3, 2009) **Adeka Corporation (Tokyo, JP)**

N. PLASTICIZERS

Monomeric, Dimeric, and Oligomeric Levulinic Acid Ketals

Title:	Glycerol levulinate ketals and their use
US Patent:	8,053,468 (October 25, 2011)
Author:	Sergey Selifonov
Assignee:	Segetis, Inc. (Golden Valley, MN)
Significance:	Monomeric, dimeric, and oligomeric luvulinic acid ketal polymer plasticizers were prepared by condensing levulinic acid with varying ratios of glycerol using sulfuric acid as the reaction catalyst. All ketal-containing additives were completely miscible with poly(vinylchloride), poly(3-hydroxybutyrate), cellulose acetate, and l-polylactide while lowering the glass transition temperature of each material by at least 30 °C once these are plasticized.
Additive Names	Methyl-5-(4-hydroxymethyl-1,3-dioxolanylmethylene) levulinate (I) 2-(4-Bis(3-methyl-3-propylcarboxymethoxy-1,3-1,3-dioxolanyl)) diethyl ether (II) Oligo(3-methyl-3-propylcarboxy)-4-oxymethlyene-1,3-dioxolanyl) (III)
Safety	The chief safety concern associated with this plasticizer is the presence of unreacted levulinate acid. Levulinate acid is considered extremely hazardous to the skin, inhalation, and is eye irritant. Skin contact is reported to result in scaling, reddening, or occasionally blistering.

Additive Structures

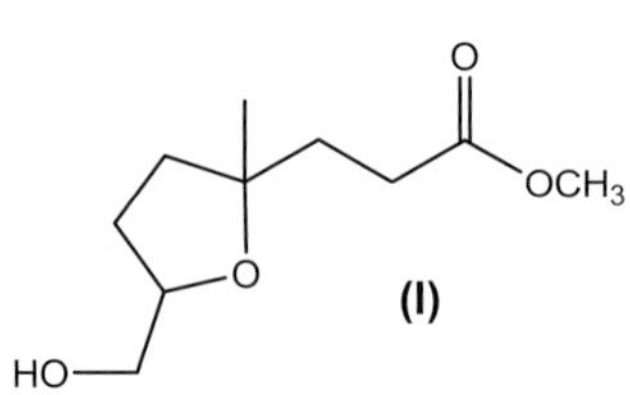

(I)

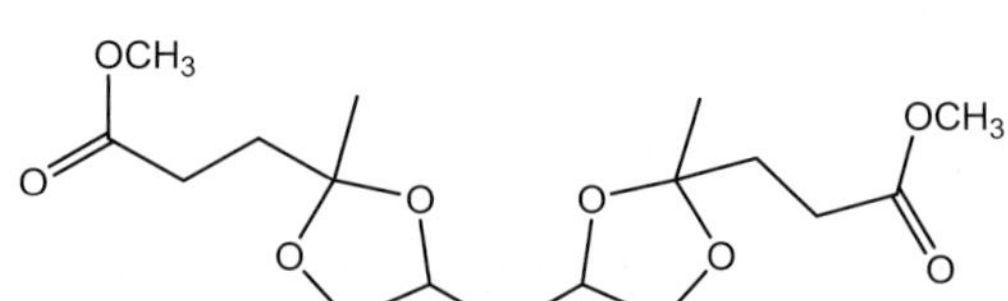

(II)

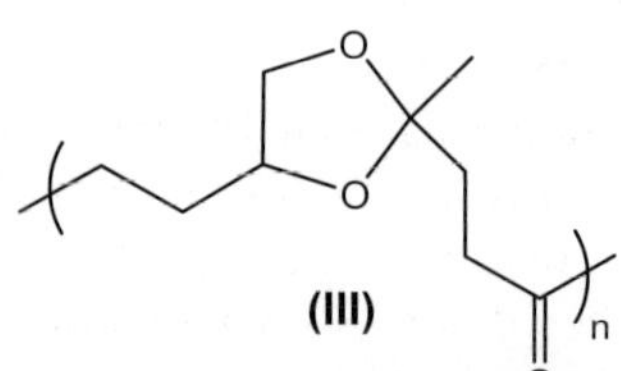

(III) n = 10 - 40

Additive Preparation

1. Monomeric Additive

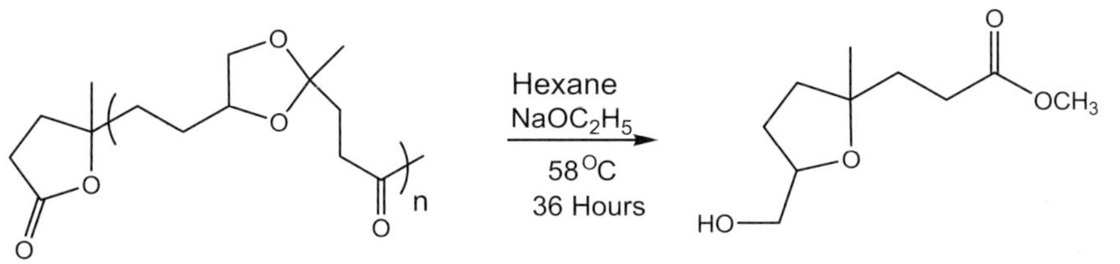

2. Oligomeric Additive

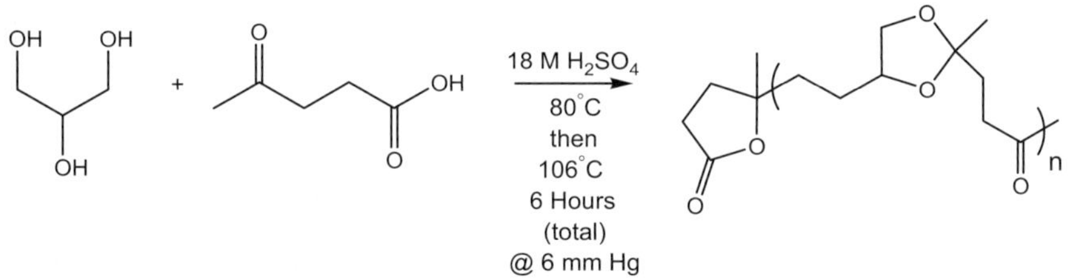

ADDITIVE SYNTHESIS

Oligomeric Plasticizer

1. Preparation of oligo(glycerol-co-levulinic acid)

A 2-liter round-bottom flask was charged with 700.1 g of 98% pure levulinic acid, 607.4 g of 99.0% pure glycerol, and 0.4 g of 18 M sulfuric acid, and it was then attached to a rotary evaporator. The rotary evaporator was adjusted to 100 rpm at 6 mm Hg, and the mixture was heated to 80 °C until 130 ml of water was collected. Thereafter, the temperature was increased to 105 °C and continued until no additional water was collected. The total heating time was 6 h. After cooling, 1097 g of a viscous, practically colorless, sticky, syrup-like liquid sparingly soluble in cold water product was isolated.

Monomeric Plasticizer

2. Preparation of methyl-5-(4-hydroxymethyl-1,3-dioxolanyl) levulinate

A reactor was charged with 20.3 g of the Step 1 product dissolved in 80 ml of methanol containing 0.4 g of sodium methoxide and then stirred at an ambient temperature while small quantities of glycerol separated to the bottom and walls of the reaction flask. The solution was then filtered through a fiberglass wool plug, neutralized by vigorously stirring for 30 min with 2 g of anhydrous potassium dihydrogen phosphate and then diluting the solution with 100 ml of methyl *t*-butyl ether. The solution was then treated with anhydrous sodium sulfate, filtered, concentrated, and 23.1 g of a clear, slightly yellowish, practically odorless liquid was isolated. Analysis of the material indicated that it consisted of 82% methyl-5-(4-hydroxymethyl-1,3-dioxolanylmethylene) levulinate.

Additive Testing of Oligo(glycerol-co-levulinic acid)

Selected Polymers

Plasticized polymer compositions of blends containing the Step 1 product were prepared by melt mixing and extrusion using (a) poly(vinylchloride) powder having an Mn of approximately 55,000 Da with and an Mw of approximately 97,000 Da having an inherent viscosity 0.92; (b) poly(3-hydroxybutyrate), having a Tm of approximately 172 °C; (c) cellulose acetate having an acetyl content of 39.8% content with an Mn of approximately 30,000 Da; and (d) l-polylactide having an inherent viscosity of between 0.90 and 1.20 Pa·s with an Mw of approximately 10,000–150,000 Da and with a Tg of 48.5 °C.

Method to Plasticize Polymers

In this process, plasticized and blended compositions were prepared in 5-g quantities by premixing cold ingredients and then extruding them through a miniature twin screw mixer-extruder chamber under nitrogen with the mixing chamber heated to between 5 and 10 °C above the melting temperature of the component with the higher melting point. The motor extruder speed was set to 100 rpm, and the samples were mixed for about 5 min and then melt extruded and isolated as flexible rods having a diameter of 3 mm. The rod was immediately cooled to an ambient temperature. Blends containing 5%, 10,%, 25%, and 50% of the experimental plasticizer were prepared using poly(vinylchloride), poly(3-hydroxybutyrate), cellulose acetate, and l-polylactide. A summary of the test results is provided below.

Test Results

1. Glass Transition Temperatures

Glass transition temperatures were determined using differential scanning calorimetry, and this indicated that all blended samples had excellent polymer compatibility, which was reflected in considerably lowered glass transition temperatures.

Polymer blends comprising poly(3-hydroxybutyrate), l-polylactide, and cellulose acetate were found to be especially compatible in a broad concentration range with the Step 1 product. These blends had significantly lowered glass transition points in comparison with their nonplasticized analogs. The Step 1 product, however, was determined to have up to approximately 10 wt% compatibility and reduced Tg by 15–30 °C.

Advantages over Prior Art

The use of monomeric, dimeric, and oligomeric ketal plasticizers derived from levulinic acid is unreported. These agents were effective as plasticizers in a wide variety of polymers including poly(vinylchloride), poly(3-hydroxybutyrate), cellulose acetate, and l-polylactide as indicated by the Tg being lowered in each material by 10–30 °C.

Notes

1. Uyama[1] prepared a polymer plasticizer consisting of a polyester polyol generated by reacting castor oil with l-lactide.
2. Mezzalira[2] plasticized polyvinyl chloride with monomeric and polymeric adipates to prevent cracking when used in garden hoses and to enhance cold flexibility.
3. Maschmeyer[3] plasticized polyvinyl chloride, homopolymers and copolymers of ethylene, propylene, butadiene, and polyvinyl acetate with the hydrogenation product of diisononyl phthalate, Vestinol 9® (IV), and di-*n*-butyl phthalate, Vestinol C® (V).

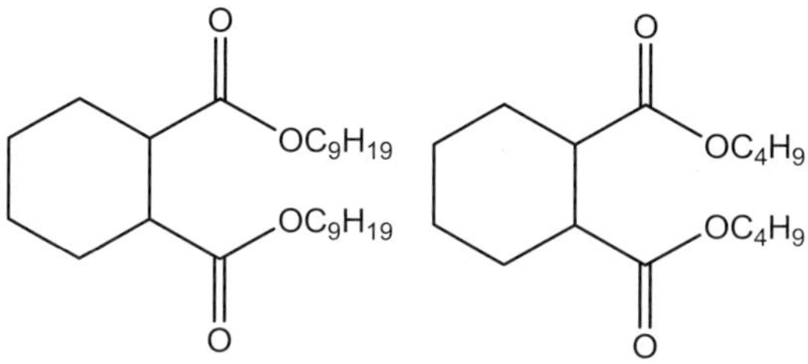

(IV) (V)

4. Arendt[4] used a plastisol consisting of butyl benzyl phthalate or a mixture containing 65 wt% of diethylene glycol dibenzoate and 35 wt% of dipropylene glycol dibenzoate containing 2,2,4-trimethyl-1,3-pentanedioldiisobutyrate as a auxiliary diluent for processing a dispersion of polyvinyl chloride having an average particle size of 0.5–5μm.

REFERENCES

1. Hiroshi Uyama et al, US. Patent 8,058,470 (November 15, 2011) **Bio-Energy Corporation (JP)**
2. Rinaldo Mezzalira et al, US. Patent 8,057,877 (November 15, 2011) **Fitt S.p.A. (Sandrigo, IT)**
3. Dietrich Maschmeyer et al, US. Patent 8,044,229 (October 25, 2011) **Evonik Oxeno GmbH (Marl, DE)**
4. William D. Arendt et al, US. Patent 8,034,860 (October 11, 2011) **Eastman Specialties Holdings Corporation (Kingsport, TN)**

O. PLASTICIZERS

Functionalized Single-Walled Nanotubes

Title:	Method for functionalizing carbon naontubes utilizing peroxides
US Patent:	7,976,816 (July 12, 2011)
Author:	Valery N. Khabashesku et al.
Assignee:	William Marsh Rice University (Houston, TX)

Significance: Single-walled carbon nanotubes (SWNT) were functionalized with ethanoic and β-propanoic acids by heating with either bis-succinic or bis-glutaric peroxides at 70 °C for 7 days. Post-reacting SWNT-g-ethanoic acid and g-β-propanoic acids with aliphatic or cycloaliphatic diamines generated the corresponding amides.

Additives

SWNT-g-diethyleneaminosuccinamide (I)
SWNT-g-diethyleneaminoglutaramide (II)
SWNT-g-dicyclohexylmethyleneaminosuccamide (III)
SWNT-g-dicyclohexylmethyleneaminoglutaramide (IV)

Safety

Both nanotubes and modified nanotubes are eye and respiratory irritants.
In the case of eye contact, immediately flush the eyes with copious amounts of water for at least 15 min. If inhaled, remove to fresh air. If unable to breathe, give artificial respiration. If breathing is difficult, provide oxygen.

Additive Structures

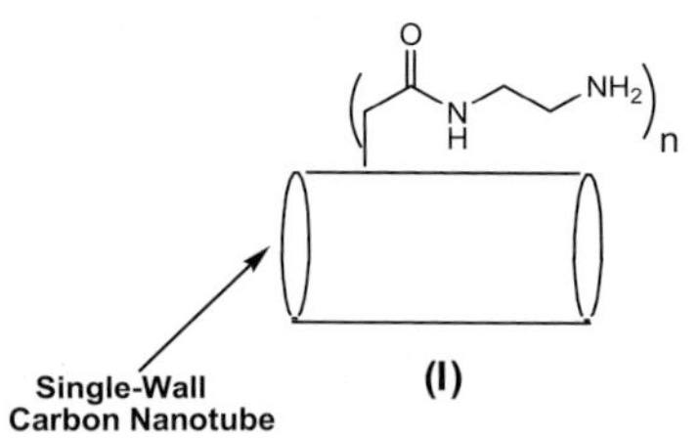

(I)

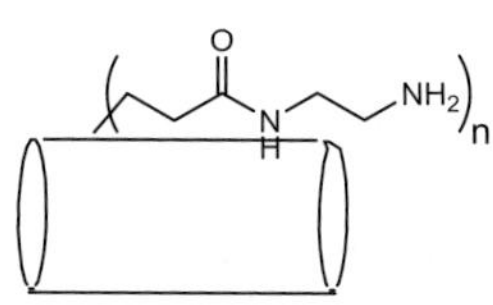

(II)

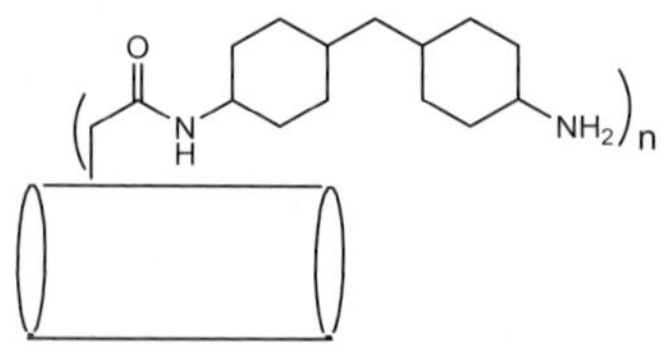

(III)

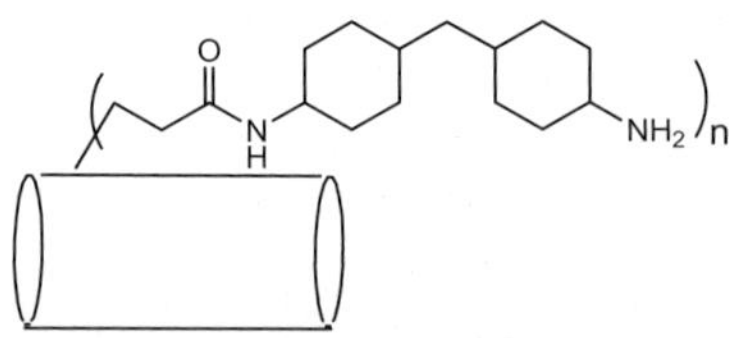

(IV)

Additive Preparation

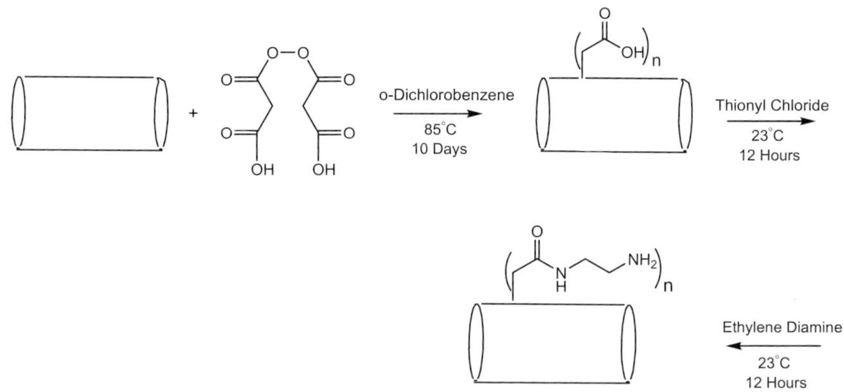

Additive Synthesis

1. Preparation of SWNT-g-succinic acid

A 250-ml flask containing 50 ml of dry o-dichlorobenzene was treated with 50 mg of purified single-wall carbon nanotubes and then sonicated for 30 min to form a suspension. The mixture was then heated to 85 °C for 10 days while adding 0.5 g of succinic peroxide every day. Thereafter, the suspension was cooled to an ambient temperature and poured into a 500 ml Erlenmeyer flask containing tetrahydrofuran and then sonicated for 15 min to form a solution. The solution was then filtered through a 0.2-μm pore PTFE membrane, and the product was collected on the membrane and resonicated for 20 min in 100 ml of ethyl alcohol. The suspension was refiltered, the product washed with ethyl alcohol, and the product was isolated after vacuum drying at 70 °C.

2. Preparation of SWNT-g-triethylenediaminoglutaramide

A dry 100-ml flask containing 20 ml of thionyl chloride was treated with 20 mg of the Step 1 product, and the mixture was stirred for 12 h at an ambient temperature. The mixture was then vacuum filtered through a 0.2-μm pore membrane, and the solid precipitate was collected on the membrane washed with dry acetone and then air dried. The precipitate was then placed into a 100-ml flask containing 20 ml of ethylenediamine and stirred for 12 h at an ambient temperature and then poured into a large amount of ethyl alcohol and sonicated for 10 min. The sonicated mixture was filtered through a 0.2-μm pore membrane, and the precipitate was flushed with ethyl alcohol and then dried overnight in a vacuum oven at 70 °C, and the product was isolated.

Testing

Testing information or results are not supplied by the author.

Advantages over Prior Art

The current method addresses an ongoing concern for nondestructively functionalizing the sidewall of inert single-walled carbon nanotubes. It is particularly propitious and useful that the sidewall was functionalized with aliphatic carboxylic acids since they are easily derivatized.

Notes

1. Wu[1] prepared the nanophase carbon black-g-3,9-bis{1,12-dimethyl-2-[β-3-*t*-butyl-4-hydroxy-5-cresol]propionyl-oxy-}ethane)}-2,4,8,10-tetraoxaspiro[5,5] undecane, (V), as an antioxidant additive in rubber and plastic compositions.

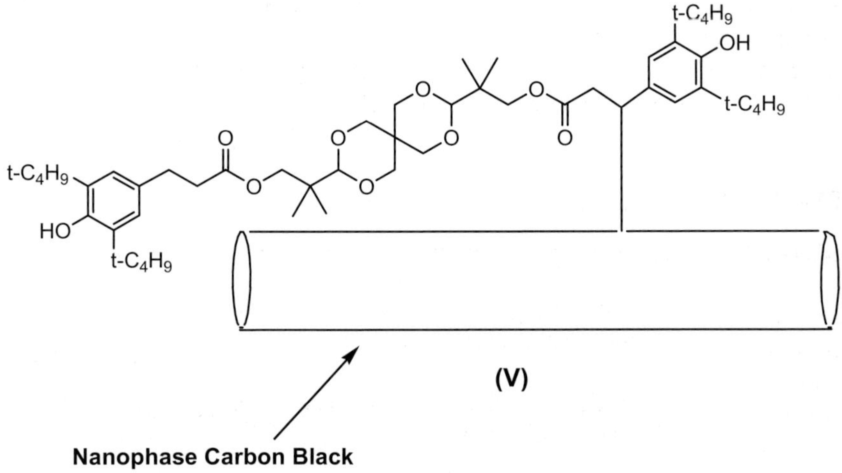

(V)

2. Inagaki[2] prepared single-walled nanotubes crosslinked with bisferrocene-based derivatives (VI) as a method for improving the electrical conductivity of nanotube composites.

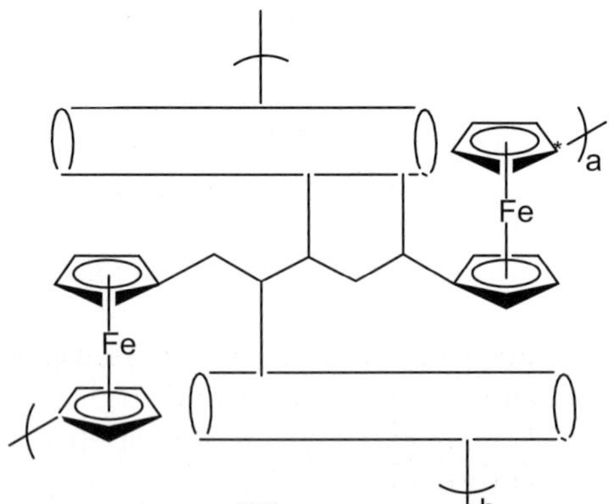

(VI)

3. Mizu[3] prepared nanotubes containing β-1,3-glucan polysaccharide-modified schizophyllan, which were effective as cell and protein recognizing agents for cholerae toxin and influenza virus.

4. Tour[4] prepared polystyrene having enhanced strength by anionically initiating the polymerization using lithiated single-walled carbon nanotubes.

REFERENCES

1. Chifei Wu et al, US Patent 7,951,310 (May 31, 2011)
   East China University of Science and Technology (Shanghai, CN)
2. Yoshio Inagaki et al, US Patent 7,943,110 (May 17, 2011)
   Fuji Film Corporation (Tokyo, JP)
3. Masami Mizu et al, US Patent 7,935,683 (May 3, 2011)
   Japan Science and Technology Agency (Saitama, JP)
   Mitsui Sugar Co., Ltd. (Tokyo, JP)
4. James M. Tour et al, US Patent 7,879,940 (February 1, 2011)
   William Marsh Rice University (Houston, TX)

P. PHOTOINITIATORS

Grafted Benzophenone or Thioxanthone derivatives

Title:	Polymeric photoinitiators
US Patent:	7,799,884 (September 21, 2010)
Author:	Donald E. Herr et al.
Assignee:	Henkel AG & Co. KGaA (Duesseldorf, DE)
Significance:	Polymeric Type II class photoinitiators were prepared by grafting benzophenone or thioxanthone derivatives onto poly(1,3-butadiene). When these materials were used in ultraviolet curing compositions to prepare hot melt films, conversions of up to 79% were obtained in 6 s.

Additive Name	Poly(butadiene)-g-4-[(1,1,3,3-tetramethylethyldisoloxane)]-1-propyletherbenzophenone (I)
CAS Number	Unavailable
Safety Concerns	Unavailable

Additive Name	Poly(butadiene)-g-2-[(2-(3-mercaptoproponate)]-1-ethylether-3,4-dimethylthioxanthone (II)
CAS Number	Unavailable
Safety Concerns	Unavailable

Additive Name	Poly[(butadiene)-g-(*N*-(4-benzophenone-succinimide)-succinic acid], (III)
CAS Number	Unavailable
Safety Concerns	Unavailable

Photoactive Intermediates

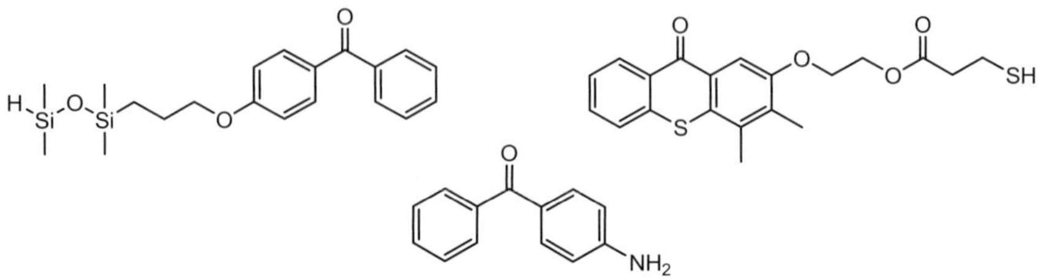

Additive Structures

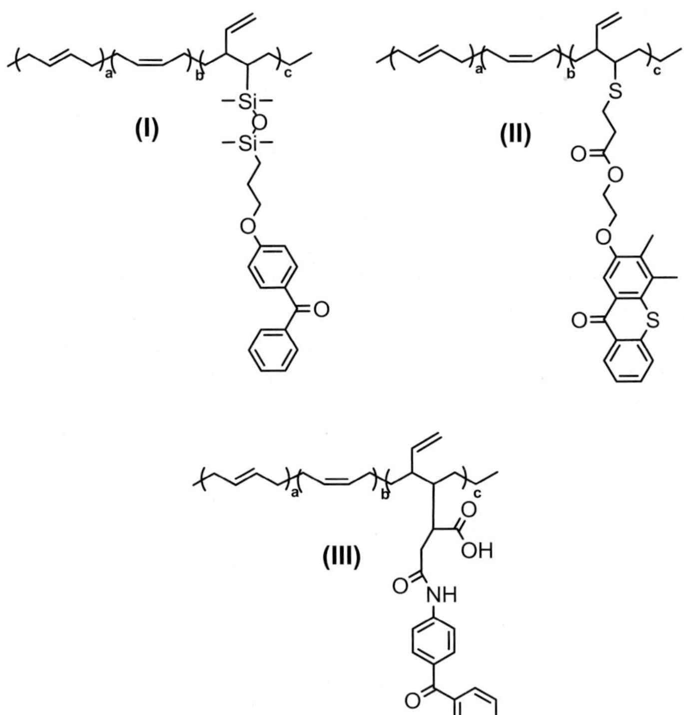

Additive Preparation

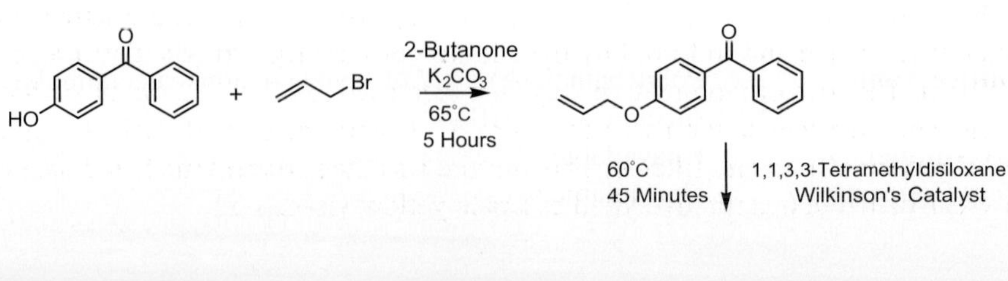

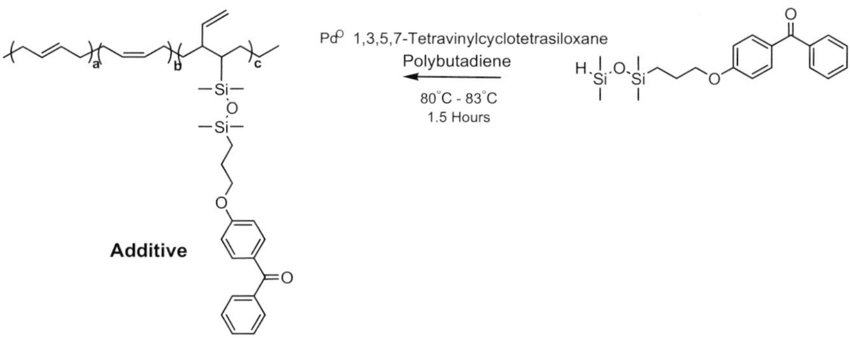

Additive Synthesis

Step 1. Preparation of 4-allyloxybenzophenone

A reactor was charged with 186.7 g of 4-hydroxybenzophenone dissolved in 700 ml of 2-butanone containing 195 g of potassium carbonate and then treated with 123 ml allyl bromide over 30 min using an addition funnel. The reaction temperature was then increased to 65 °C for 6.5 h and the contents were then filtered and extracted using 500 ml of hydrochloric acid. The organic layer was isolated, dried with magnesium sulfate, and the product was isolated in a 92% yield as a pale yellow solid.

Step 2. Preparation of SiH-functional benzophenone derivative

Under a protective atmosphere of dry nitrogen, a reactor was charged with 200 g of the Step 1 product dissolved in 150 ml of THF and then treated with 740 ml of 1,1,3,3-tetramethyldisiloxane dissolved in 100 ml THF. The reaction temperature was then raised to 50 °C, and 22 mg of chlorotris(triphenylphosphine) rhodium dissolved in 5 ml of 1:1 v/v allyloxybenzophenone/THF solution was slowly added over 45 min. The temperature was then increased to 60–65 °C. When GC evidence indicated that the reaction was completed, the mixture was cooled to 35 °C and treated with activated carbon and stirred for 30 min. The slurry was filtered, and a pale yellow solution was isolated. The mixture was concentrated in vacuo, and a yellow oil was isolated in 104% because of the presence of the impurity 1,1,3,3,5,5-hexamethyltrisiloxane in commercial grade of 1,1,3,3-tetramethyldisiloxne starting material.

Step 3. Preparation of poly(butadiene)-g-4-[(1,1,3,3-tetramethylethyldisoloxane)]-1-propyletherbenzophenone

A four-necked 5-l flask equipped with a mechanical stirrer and reflux condenser was charged with 734 g of poly(butadiene) (Ricon® 130; Sartomer) dissolved in 1,100 ml of toluene and then warmed to 50 °C. An addition funnel was then charged with 328.8 g of the Step 2 also at 50 °C. The solution was then treated with the addition of a solution of 5 ml containing 1.8 g of platinum1,3,5,7-tetravinyl-1,3,5,7-tetramethylcyclo-tetrasiloxane and followed by the addition of the Step 2 product over a period of 90 min. The reactor temperature was then increased to 80–83 °C for an additional 90 min, and then the solution was cooled to 35 °C and treated with activated carbon and stirred for 60 min and filtered. The mixture was then concentrated, and the product was isolated in quantitative yield as a pale yellow viscous oil.

Additive Formulations

The initial hot melt composition consisted of 100 g of poly(styrene-b-1,3-butadiene-b-styrene) rubber, poly(styrene-b-1,3-butadiene) rubber, tackifier resins, oil plasticizer, and a stabilizer package, which was melt blended with 2 wt% of the Step 3 product at 300 °F. The blend was then cured using a 300-W medium pressure mercury arc bulb.

A second hot melt composition consisting of 100 g of poly(styrene-b-1,3-butadiene-b-styrene) rubber, hydrogenated tackifiers, saturated oil, and an antioxidant package, which was melt blended with 4.5 g of 1 wt% of either the Step 1 or Step 2 product.

Test Results

Hot melt compositions containing experimental polymeric photoinitiators cured approximately 33% faster and at 75% lower treatment levels than when compositions were cured with Irgacure® 651 (IV).

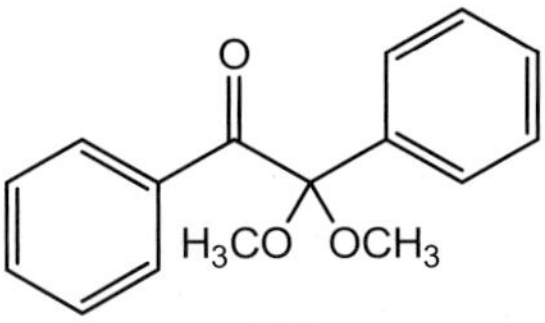

(IV)

Advantages over Prior Art

The use of high efficiency poly(1,3-butadiene) bound type II photoinitiators in curable composites eliminates the presence of extractable components or odors in the cured resin. In addition, polymeric photoinitiators have high compatibility with the resin systems and are more easily intimately blended. Their use in hot melt adhesives and coatings is particularly desirable because polymeric photoinitiators are thermally stable in the processing temperatures. Finally, type II photoinitiators based on poly(vinyl benzophenone) and acrylated benzophenone derivatives are reported to have low photoefficiencies.

Notes

1. Addition curable hot melt formulations of photoinitiators grafted to polybutadiene (V) were described by Herr.[1]

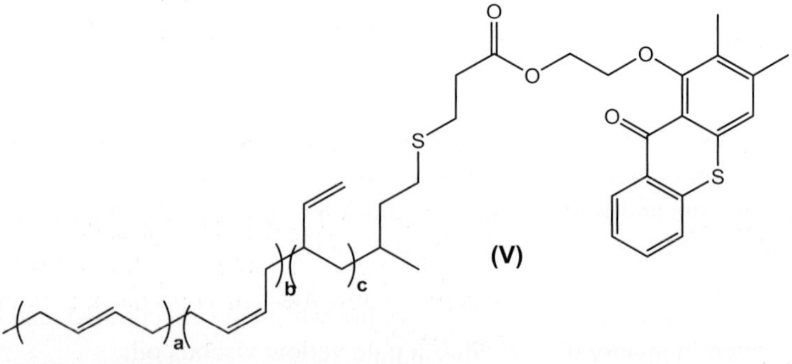

(V)

2. Polymerizable photoinitiator methacrylate monomers containing benzophenone (VI) were originally prepared by Skoulchi[2] and used in ultraviolet curable compositions.

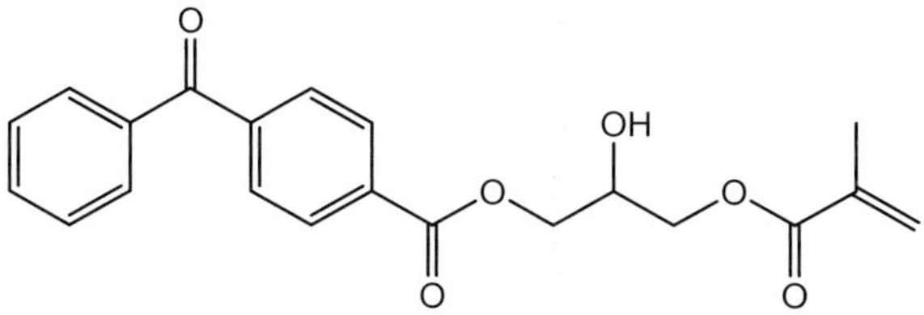

(VI)

REFERENCES

1. Donald E. Herr et al, US Patent 7,795,364 (September 14, 2010) **Henkel AG & Co. KGaA (Duesseldorf, DE**
2. Martin Skoulchi, US Patent 3,429,852 (February 25, 3,429,852) **National Starch and Chemical Corporation (NY, NY)**

Q. POLYMER COMPOSITES

Nanoparticles With Halloysite Surface

Title:	Polymeric composite including nanoparticle filler
US:	7,888,419 (February 15, 2011)
Author:	Sarah M. Cooper et al.
Assignee:	NaturalNano, Inc. (Pittsford, NY)
Significance:	A polymer nonoparticle composite filler was prepared by the surface modification of Halloysite by dry blending with up to 1 wt% benzyltrimethylammonium chloride. When melt blended with Nylon-6 or with a mixture of polypropylene and poly(propylene-g-maleic anhydride), the mechanical and thermal properties, respectively, of the blend improved.
Additive Name	Halloysite-benzyltrimethylammonium chloride
Safety Concerns	Halloysite is considered a nontoxic clay. Benzyltrimethyl-ammonium chloride may be harmful if absorbed through the skin and is an eye irritant.

Additive Structure

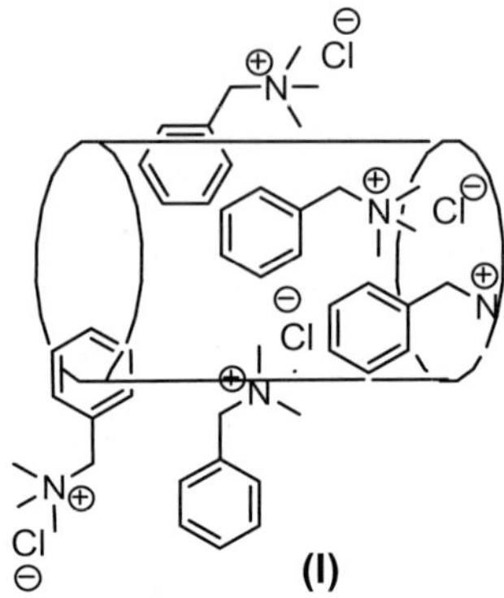

(I)

Additive Preparation

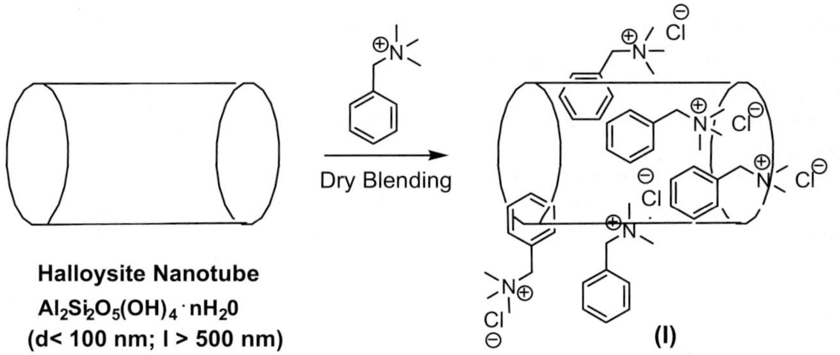

Additive Synthesis

1. Preparation of Halloysite-benzyltrimethylammonium chloride

Halloysite was dry blended with about 0.5–1% benzyltrimethylammonium chloride, and then the surfaced air was milled to break up the halloysite particles and assure dispersion in the composite. The material was then dried under vacuum for 16 h, and the product was isolated.

Testing

Testing for Nylon-6

Nylon-6 was ground in a Retsch model ZM100 grinder and combined with 5% of the Step 1 product in a Hauschild model DAC 150 FV speed mixer. The blended powder was then extruded at 250 °C through a DSM Twin-Screw microcompounder under flowing nitrogen using a screw speed of 100 rpm. Composites were then injection molded at 250 °C to obtain samples for mechanical testing. Instron mechanical testing was conducted using the ASTM standard D638 protocol using Type IV specimen dimensions and an Instron crosshead speed of 5 mm/min. Mechanical test results are provided in Table 14.

1. Mechanical and Thermal Testing for a Polypropylene and Poly(propylene-g-Maleic anhydride) Mixture

The Step 1 product was mixed with polypropylene having an Mn > 400,000 Da and poly(propylene-g-maleic anhydride) containing 4.2 wt% of anhydride having an Mn > 10,000 Da using a DSM twin screw microcompounder at 200 °C for 5 min and then microinjected molded at about 200 °C. The samples were evaluated both mechanically and for thermal decomposition properties. Decomposition test results are provided in Table 15.

Test Results

1. Mechanical Testing for Nylon-6 are Provided in Table 14

TABLE 14 Test results for Halloysite-benzyltrimethylammonium chloride polymer nanoparticle composite filler of blends indicating that the modulus of elasticity and yield strength of blends are greater than that of the polymer matrix without filler or using unmodified Halloysite without a significant sacrifice in elongation

Component	Modulus (GPa)	Yield Strength (MPa)	Elongation (%)
Nylon-6	1.5	81	1.7
Nylon-6 + 5wt% Untreated Halloysite	1.9	83	0.23
Nylon-6 + 5wt% Halloysite-benzyl-trimethylammonium chloride	2.3	87	1.51
Nylon-6 + 5wt% Platy Clay	2.5	87	1.51

2. Mechanical and Thermal Testing for polypropylene/poly (propylene-g-maleic anhydride) mixture

Advantages over Prior Art

Halloysite surface modified with benzyltrimethylammonium chloride was designed to form intimate blends with polar polymers. The results indicate that considerably less nanoparticle composite was needed to improve the selected properties of a polymer. This approach further suggests that Halloysite composite filler can

TABLE 15 Thermal properties of selected polymers additized with 5 wt% of the Step 1 product, Halloysite-benzyltrimethylammonium chloride, evaluated by DCS at a heating rate of 10 °C/min. Thermal analysis was terminated after a 1.5% sample weight loss had occurred

Component	Td (°C)
Polypropylene	390
Polypropylene + 5wt% Untreated Halloysite	412
Polypropylene + 5wt% Halloysite-benzyltrimethylammonium chloride	420
Polypropylene/Poly(propylene-g-maleic anhydride) + 5wt% Untreated Halloysite	370
Polypropylene/Poly(propylene-g-maleic anhydride) + 5wt% Halloysite-benzyltrimethyl-ammonium chloride	420

also be used in improving selected properties of nonpolar polymers by surface modifying with a quarternary amine containing one or more extended hydrophobic components.

Notes

1. Satoh[1] prepared a composite filler having heat resistance, moldability, and good impact properties by using the functional filler for a polymeric alcohol having an Mn of approximately 1000 Da prepared by treating the surface with poly-d-lactic and then melting and kneading to obtain a resin composition in which the functional filler was homogeneously dispersed.
2. A composite filler designed to both decrease the melting point and retard polymer crystallization as a result of the presence of single-walled nanotubes was prepared by Krishnamoorti[2] who dispersed nanotubes in high or low molecular weight polyethylene glycol in the presence of the surfactant lithium dodecyl sulfate.
3. Unmodified nanoclay composites useful as fire retardants were prepared by Lee[3] by treating the polymer or the surface of nanoparticulates with a mixture of di-*n*-octyl-dimethylammonium chloride, dimethyl bis(hydrogenated tallow), ammonium chloride, methyl benzyl bis(hydrogenated tallow), and methyl tris(hydrogenated tallow alkyl) chloride.
4. Rubinsztajn[4] prepared useful resin compositions for use as acoustic lens materials by means of nanoparticulate metal oxide powders in an amount required to obtain a final density in the range of about 1.55–1.65 g/cm^3. When these particles were blended into the proprietary curable silicone resin RTV®615 (Momentive Performance Materials) containing approximately 4 vol% of treated fumed silica (Degussa Corporation), the acoustic properties of the resin improved.

REFERENCES

1. Masahiro Satoh et al, US Patent 7,902,304 (March 8, 2011)
   Dai-Ichi Kogyo Seiyaku Co., Ltd. (Kyoto, JP)
2. Ramanan Krishnamoorti et al, US Patent 7,893,148 (February 22, 2011)
   University of Houston (Houston, TX)
3. Biing-Lin Lee et al, US Patent 7,884,150 (February 8, 2001)
   Teknor Apex Company (Pawtucket, RI)
4. Slawomir Rubinsztajn et al, US Patent 7,902,294 (March 8, 2011)
   General Electric Company (Niskayuna, NY)

R. POLYMER COMPOSITES

Polycarbovinylsilazane

Title:	Preceramic silazane polymer for ceramic materials
US Patent:	7,842,774 (November 3, 2010)
Author:	Michael A. Kmetz
Assignee:	United Technologies Corporation (Hartford, CT)
Significance:	The preceramic agent, polycarbovinylsilazane, was prepared in a single step and in quantitative yield by condensing trichlorovinylsilane with ammonia. Upon pyrolysis at 1000 or 1500 °C, it was converted into a glassy or crystalline nitride, respectively, in up to 80% yield.
Additive Name	Polycarbovinylsilazane (I)
Safety Concerns	Information on the human health effects from exposure to this substance is limited. However, it is closely related to materials that can potentially cause eye and skin irritation. Excessive inhalation of vapors can irritate the respiratory tract and possibly lead to fluid accumulation in the lungs.

Additive Structure

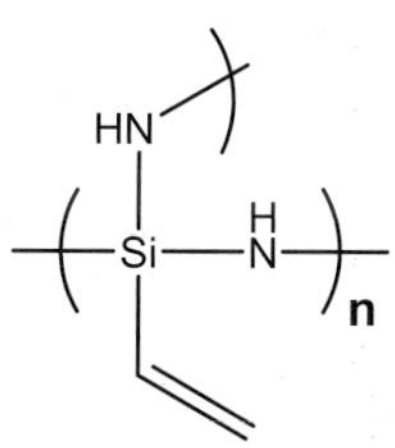

(I)

Additive Preparation and Pyrolysis

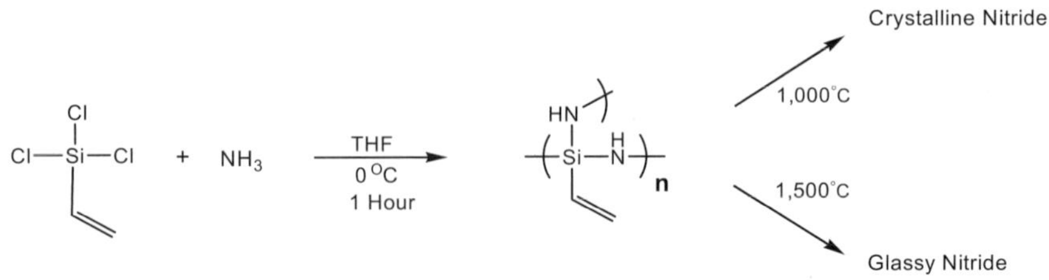

Additive Synthesis

1. Preparation of polycarbovinylsilazane

A flask was charged with 10 g of 97% trichlorovinylsilane and 100 g of tetrahydrofuran and then cooled to between about –10 and 10 °C and treated with a steady stream of dry ammonia for 1 h. During the entire amination process, the mixture was thoroughly stirred while maintaining a reaction temperature of approximately 0 °C. When the reaction was completed, the mixture was filtered to remove ammonium chloride, concentrated, and the product was isolated as a pale yellow liquid having a density of approximately 2.0 g/cm^3 with an Mn of approximately 884 Da.

2. Pyrolysis of polycarbovinylsilazane

The Step 1 product was heated to about 1000 °C at a heating rate of 1 °C/min under a nitrogen atmosphere, and a glassy ceramic material was isolated in an approximately 80% yield.

Advantages over Prior Art

The method for preparing the preceramic polymer of the current invention is versatile and can be easily modified to prepare analogs. In addition, the process uses inexpensive reagents, does not require purification of the intermediate, and the intermediate is generated in a quantitative yield. Finally the preceramic polymer could be used to prepare either a glassy or crystalline ceramic material depending on the pyrolysis temperature.

Notes

1. **Krosche**[1] prepared an SiBCN ceramic by pyrolyzing the crosslinked polymer formed by reacting dichloroborylmethyldichlorosilylamine (II) with methylamine.

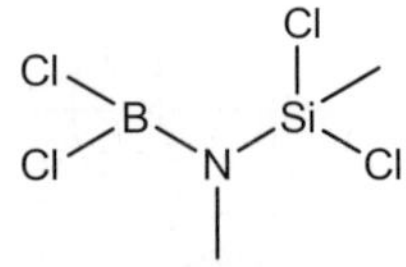

(II)

2. Wan[2] prepared inorganic block copolymers (III), which upon pyrolysis at 1000 °C initially formed nanoscale ceramics and secondary ceramic structures at 1400 °C.

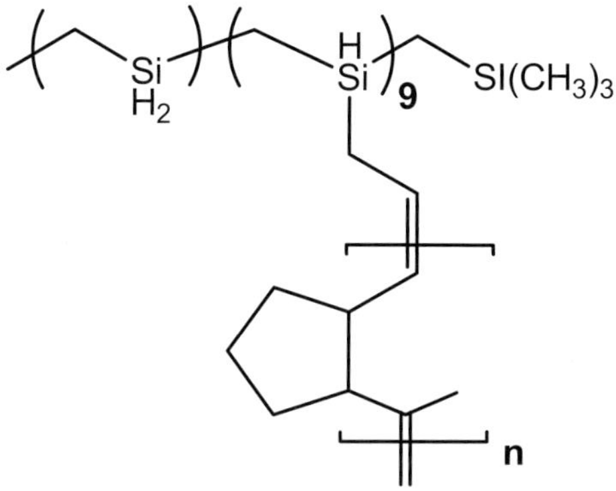

(III)

3. Sato[3] prepared a high purity, dense, and strongly heat resistant synthetic quartz glass by pyrolyzing hexamethyldisilazane (IV) from 1300 to 1500 °C.

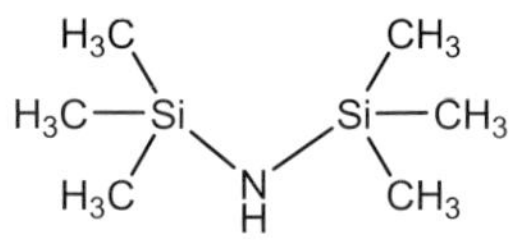

(IV)

4. Jeong [4] developed a high-yielding method for preparing polysilazane (V), which entailed reacting dichlorosilane with ammonia in pyridine. Wafers were the coated with polysilazane dissolved in dibutylether and then annealed at 800 °C.

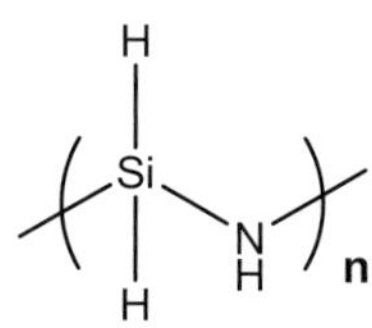

(V)

5. Pope[5] devised a method to prepare silicon carbide ceramic films by the ultraviolet photo crosslinking of the preceramic polymer intermediate, allylhydridopolycarbosilane (VI) followed by pyrolysis.

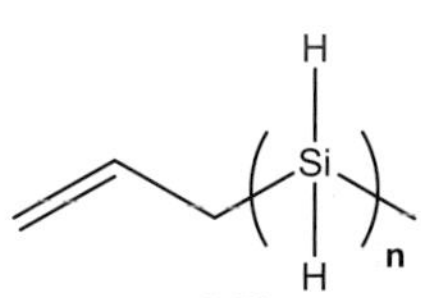

(VI)

REFERENCES

1. Matthias L. Krosche et al, US Patent 7,795,461 (September 14, 2010)
   Max-Planck-Gesellschaft zur Forderung der Wissenschaften e.V. (Munich, DE)
2. Julin Wan et al, US Patent 7,709,574 (May 4, 2010)
   General Electric Company (Niskayuna, NY)
3. Tatsuhiro Sato et al, US Patent 7,841,211 (November 30, 2010)
   Shin-Etsu Quartz Products Co., Ltd. (Tokyo, JP)
4. Eun Chang Jeong, US Patent 7,776,990 (August 17, 2010)
   Stahnek Co., Ltd. (KR)
5. Edward J. A Pope et al, US Patent 7,326,381 (February 5, 2008)
   Global Strategic Materials, LLC (Westlake Village, CA)

S. RUBBER LATEX PRESERVATION

Chitosan Amides or Inulin Esters

Title:	Natural rubber latex preservation
US Patent:	7,989,546 (August 2, 2011)
Author:	Karl Booten et al.
Assignee:	Beneo-Orafti S.A. (BE)
Significance:	Hydrophobic chitosan saccharamides and inulin saccharinates were found to be effective in improving the mechanical and colloidal stability of natural rubber latexes at 0.5–0.75 wt% treatment levels. Synergistic effects of these additives with the industrial standard, Inutec® SP1, also resulted in an improvement in colloidal stability.
Additives	Chitosan nonylamide (I) Chitosan decylamide (II) Chitosan undecylamide (III) Chitosan dodecylamide (IV) Inulin undeconate (V) Inulin dodeconate, (VI) Inulin trideconate (VII)
Safety	Chitosan and inulin derivatives are considered nontoxic and are used in cosmetic and pharmaceutical formulations.

Additive Structures

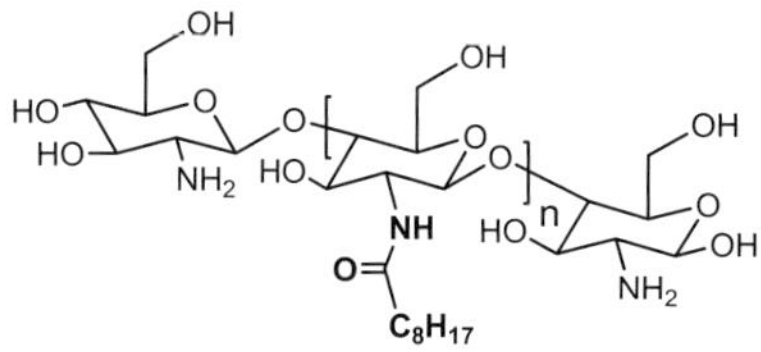

(I)

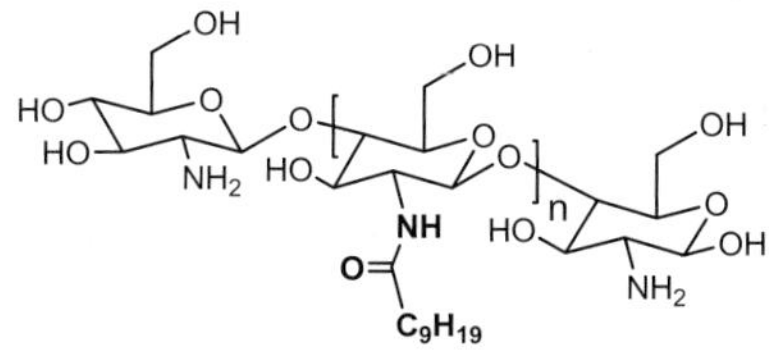

(II)

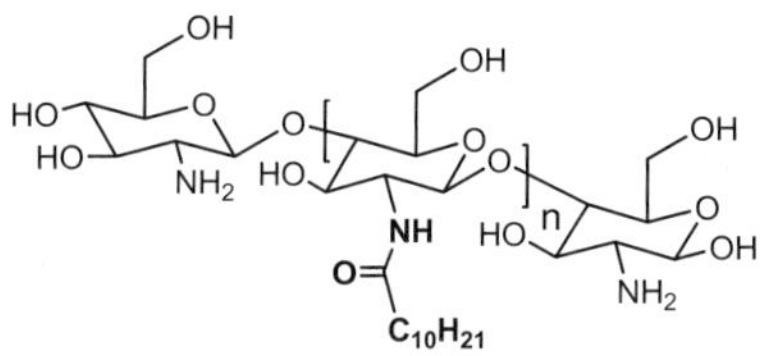

(III)

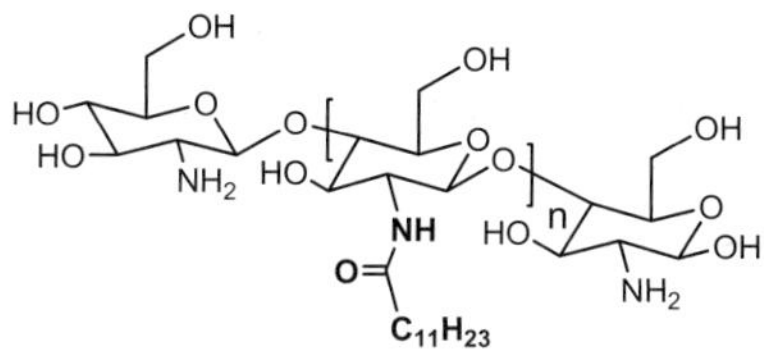

(IV)

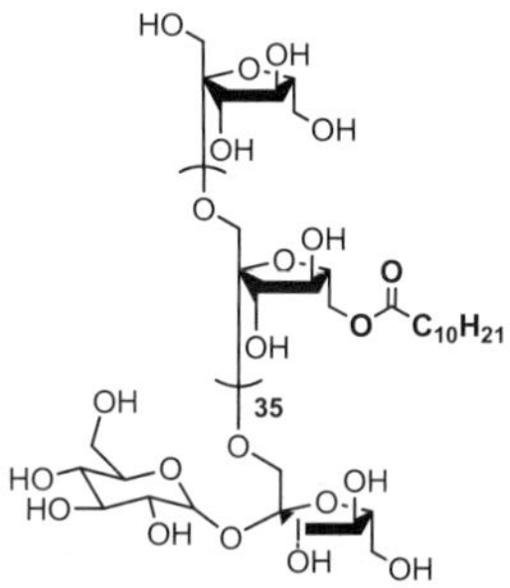

(V)

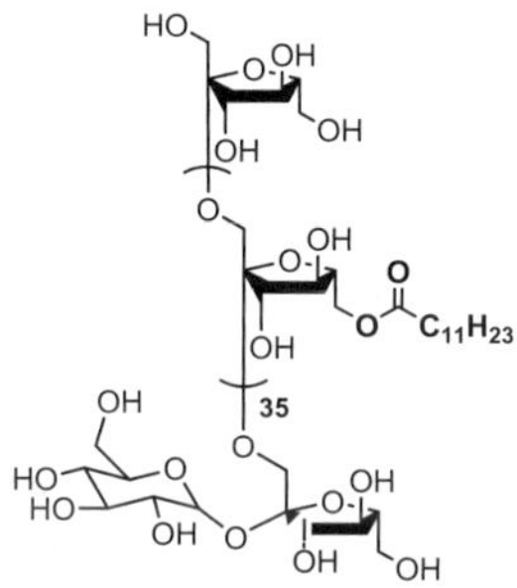

(VI)

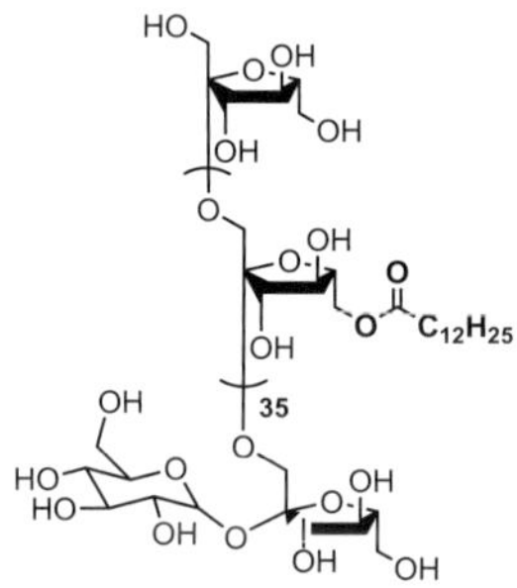

(VII)

Additive Preparation

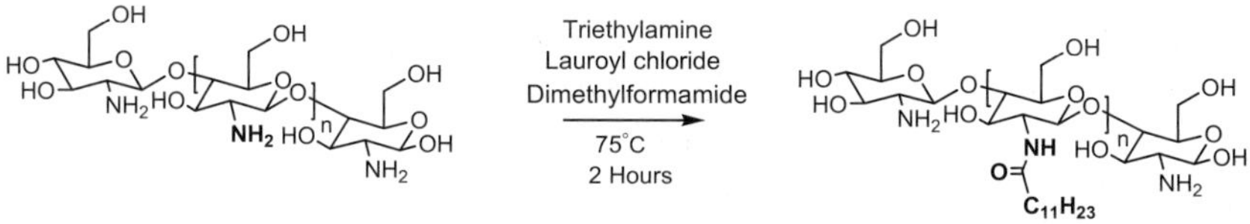

Additive Synthesis

1. Preparation of chitosan dodecylmide according to the method of Matsuda[1]

A dry 100-ml round-bottom flask containing 30 ml of anhydrous pyridine and 150 ml of dimethylformamide containing 35 mg of chitosan and triethylamine was treated with 26.64 mg of lauroyl chloride at an ambient temperature. The mixture was then heated to 75 °C for 2 h and then treated with ethyl alcohol saturated with sodium acetate. The precipitate which formed was isolated and thoroughly washed with ethyl alcohol and then dried, and the product was isolated.

Testing

1. Mechanical Stability

The mechanical stability of natural rubber latex treated with 0.5 wt% of a selected additive was determined according to the ISO 35 protocol using a latex mixture stabilized with 0.2% w/w on total latex ammonia and from 0.25% to 1.0% w/w on dry rubber with inulin laurylcarbamate (Inutec® SP1). In this test, centrifugal forces were applied, and the time the dispersion remains stable was measured, and the end point of the stability was determined by the first appearance of flocculum. Test results are provided in Table 16.

2. Colloidal and Mechanical Stabilities

The colloidal and mechanical stabilities of natural rubber latex were evaluated using the ISO 35 testing protocol. In this test method, the latex was treated with 0.2% ammonia and either 0.5 wt% or 0.75 wt% of a selected experimental additive. Colloidal and mechanical test results are provided in Tables 17 and 18.

Test Results

1. Mechanical Stability

TABLE 16 Mechanical stability of rubber latex determined by centrifuging rubber latex treated with 0.2% wt% ammonia, Inutec® SP1 (VIII) shown below, and 0.50 wt% of a selected additive

	Mechanical Stability Time (s)				
Latex Composition	Additive 2	Additive 3	Additive 4	Additive 6	Additive 7
Comparative (Unadditized)	1350	1500	1591	1604	1453
Latex + 0.25 wt% Inutec® SP1	2020	2018	1862	1941	1670
Latex + 0.25 wt% Inutec® SP1	2133	2317	2189	2276	1723
Latex + 0.25 wt% Inutec® SP1	2850	2849	2850	2865	1998
Latex + 0.25 wt% Inutec® SP1	3600	3600	3600	3600	2480

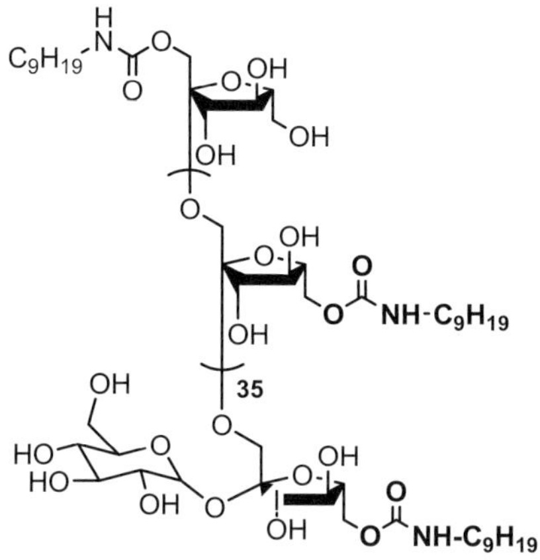

(VIII)

2. *Colloidal and Mechanical Stabilities*

TABLE 17 Colloidal stability of preserved rubber latex treated with 0.2% ammonia and additized only with a selected experimental hydrophobic saccharide. Lower values are preferred

	Colloidal Stability (months)				
Latex Additive	**Initial**	**Month 1**	**Month 2**	**Month 3**	**Month 4**
Comparative (Unadditized)	0.037	0.041	0.053	0.054	0.057
Latex + 0.50 wt% Additive 3	0.037	0.037	0.040	0.052	0.049
Latex + 0.50 wt% Additive 4	0.037	0.037	0.037	0.043	0.055
Latex + 0.50 wt% Additive 7	0.037	0.033	0.041	0.049	0.052
Latex + 0.75 wt% Additive 1	0.037	0.041	0.053	0.054	0.057
Latex + 0.75 wt% Additive 2	0.037	0.030	0.044	0.048	0.047
Latex + 0.75 wt% Additive 5	0.037	0.033	0.040	0.049	0.054
Latex + 0.75 wt% Additive 7	0.037	0.039	0.040	0.049	0.059

TABLE 18 Mechanical stabilities of preserved rubber latex treated with 0.2% ammonia and additized only with a selected experimental hydrophobic saccharide. Higher values are preferred

	Mechanical Stability (months)				
Latex Additive	Initial	Month 1	Month 2	Month 3	Month 4
Comparative (Unadditized)	1469	1026	1098	1326	1566
Latex + 0.50 wt% Additive 3	1469	1494	1644	1734	1872
Latex + 0.50 wt% Additive 4	1469	1644	1812	1968	1818
Latex + 0.50 wt% Additive 7	1469	1638	1836	2100	2418
Latex + 0.75 wt% Additive 1	1469	1026	1098	1326	1566
Latex + 0.75 wt% Additive 2	1469	1709	1940	2100	1992
Latex + 0.75 wt% Additive 5	1469	1668	2299	2522	2555
Latex + 0.75 wt% Additive 7	1469	1878	2303	2284	2406

Advantages over Prior Art

The key advantage for using chitosan saccharamides and inulin saccharinates in latex formulations is that vulcanized blends generate the desired shape with a minimum amount of latex material being wasted. In addition, experimental additives can be readily prepared using inexpensive precursors, and their colloidal effectiveness surpasses the industry standard, Inutec® SP1.

Notes

1. Stevens[1] prepared *N*-octylcarbamate derivatives of Glucidex® D.E 0.2, that is, maltodextrins and dried glucose syrups, which was obtained from starch by hydrolysis, followed by purification and spray drying with Glucidex® D.E. 47 grades. This derivative was effective in enhancing the thermal and colloidal properties and used in detergent, emulsifier, emulsion stabilizer, foaming agent, foam stabilizer, dispersant, and wetting agent formulations.
2. In an earlier investigation by Stevens[2] GLUCIDEX® D.E. 47 *N*-*n*-dodecyl carbamate was prepared and determined to be effective in improving the mechanical and colloidal properties of paint and related colloidal mixtures. The experimental protocol used in preparing this reagent is provided below:

A sample of 10g of GLUCIDEX®. D.E. 47 was dissolved under stirring at about 70 °C in 14 ml of dry DMF and was then treated with the dropwise addition of 1.49 ml of *n*-dodecyl isocyanate and stirred at 70 °C for 24 h. The mixture was then cooled to an ambient temperature and then added to 100 ml of stirred dry diethyl ether and stirred for 1 h. A white precipitate was formed, which was isolated by filtration, treated with methylene chloride and then dried, and the product was isolated. When the agent was used in paint formulations, the surface tension, interface tension, and emulsifying properties of carbamates improved the mechanical and colloidal stability of the paint formulation.

3. De Leenheer[3] prepared polydispersed saccharide compositions using fructose oligosaccharides. The method for preparing this colloidal stabilizing additive is provided below:

A fructan-containing raw material containing inulin was extracted from chicory having an average degree of polymerization of 27 fructose, glucose, and fructose-glucose, which is marketed under the name SIGMA® From this inulin, a solution of 10% dry matter was isolated by adjusting the pH to 8, and the solution was heated to 90 °C for 15 min to obtain a clear solution. The clear solution was then cooled to 65 °C before buffering it to a pH of 5.4. Next, 0.6 unit of A. Ficuum endo-inulinase enzyme (NOVO) was added to the dry matter-free inulin during a 24-hour treatment period while the temperature was maintained at 60 °C. The enzymatic hydrolysis was then stopped by bringing the hydrolyzed solution to boiling temperature and raising the pH to 8. The polydispersed saccharide solution was clarified using activated charcoal and desalted. The solvent for the polydispersed solution was then evaporated, and a syrup of 75% of dry matter-free inulin product was isolated and used as a colloidal stabilizer.

4. In an earlier investigation of Stevens,[4] a series of inulin *N*-*n*-alkylcarbamate derivatives were prepared that were effective as colloidal stabilizers when used in detergent, emulsifier, emulsion stabilizer, foaming agent, foam stabilizer, dispersant, and wetting agent formulations.

REFERENCES

1. Christian Victor Stevens et al, US 7,829,599 (November 7, 2010) **Tiense Suikerraffinaderij N.V.** (Tienen, **BE**)
2. Christian Victor Stevens et al, US 7 7,485,689 (February 3, 2009) **Tiense Suikerraffinaderij N.V. (BE)**
3. Leen De Leenheer et al, 7,084,131 (August 1, 2006) **Raffinerie Tirlemontoise S.A. (BE)**
4. Christian Victor Stevens et al, 6,534,647 (March 8, 2003) **Daenekindt, Lucien** (Gijzegem-Aalst, **BE**)

T. THERMOPLASTIC RESINS MONOMERS

2,3-Dichloroquinoxaline Derivatives

Title:	Thermoplastic polyphenoxyquinoxaline and a method for preparing the same
US Patent:	8,071,710 (December 6, 2011)
Author:	Haci Bayram Erdem et al.
Assignee:	The University of Akron (Akron, OH)
Significance:	Thermoplastic polyether copolymers were prepared by condensing 2,3-dichloroquinoxaline with bisphenol A, hexa-fluorobis phenol A, bisphenol S, or 1,4-hydroquinone in the presence of sodium carbonate. Polyethers derived from this process were clear and transparent and had an inherent viscosity of at least 0.57 Pa·s with glass transition temperatures of ≥279 °C.

Additive Name 2,3-Dichloroquinoxaline (I)

Safety 2,3-Dichloroquinoxaline is considered a toxic reagent. It is considered a moderate skin irritant but a severe eye irritant. Since it is a moderate respiratory irritant, breathing dust, fumes, gas, mist, vapors, and spray should be avoided.

Additive Structure

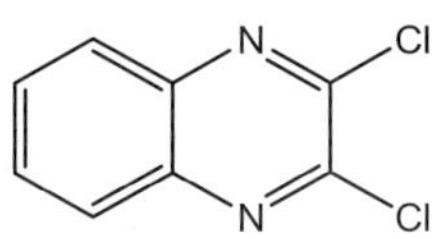

(I)

Additive Preparation

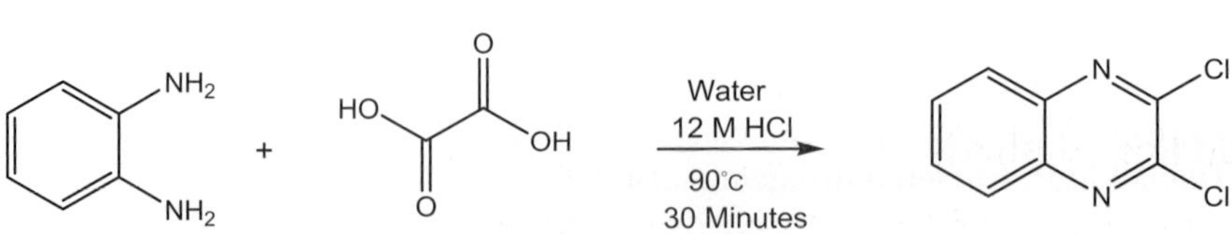

Polymer Preparation

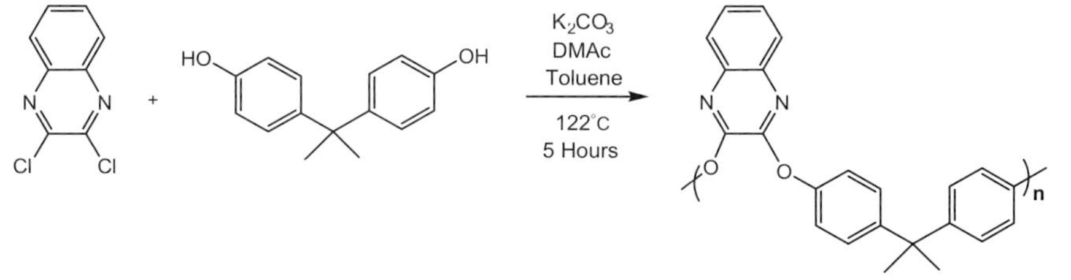

Thermoplastic Copolymer Names

Poly(2,3-quinoxalinoxy-co-fluorene) (II)
Poly(2,3-quinoxalinoxy-co-bisphenol B), (III)
Poly(2,3-quinoxalinoxy-co-hexafluorobisphenol A), (IV)
Poly(2,3-quinoxalinoxy-co-hydroquinone), (V)
Poly[(2,3-quinoxalinoxy-co-hexafluorobisphenol A)-co-(2,3-quinoxalinoxy-co-hydroquinone)] (VI)
Poly(2,3-quinoxalinoxy-co-bisphenol A) (VII)

Thermoplastic Resin Structures

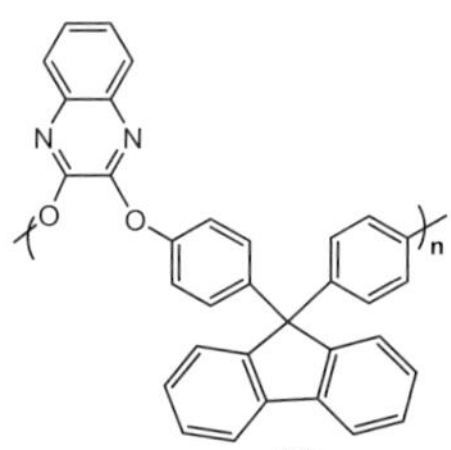

(II)

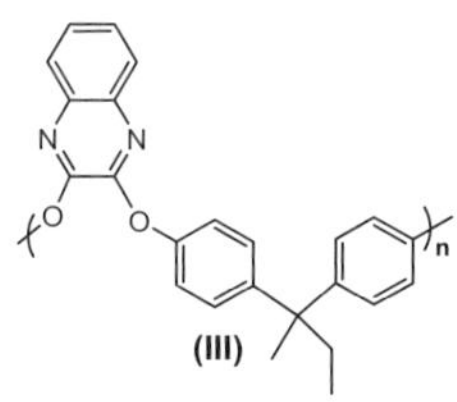

(III)

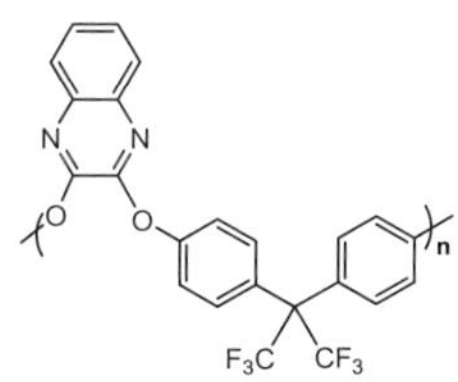

(IV)

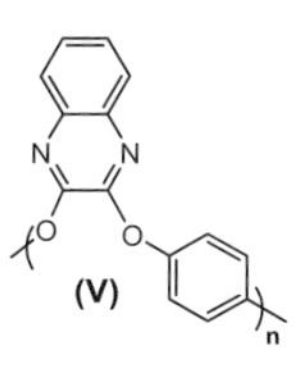

(V)

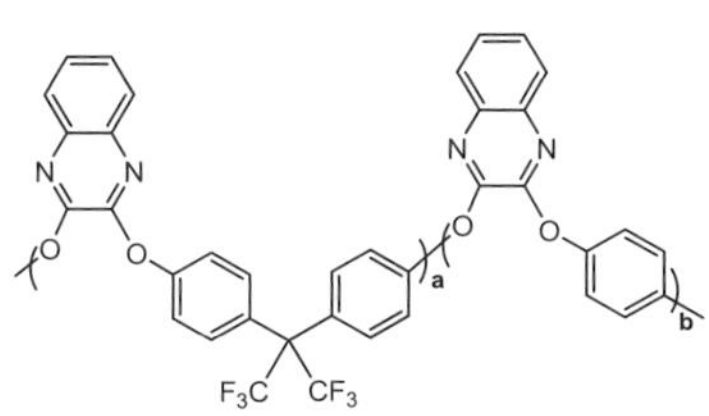

(VI)

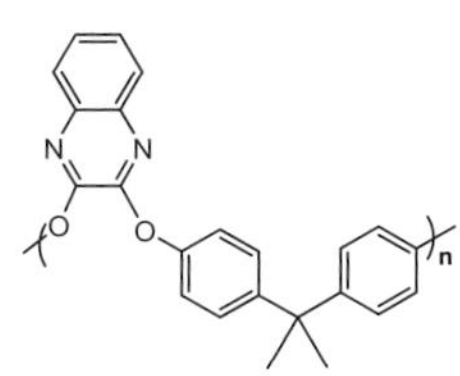

(VII)

Additive Synthesis

1. Preparation of 2,3-dichloroquinoxaline

A 2-liter flask was charged with 280.0 g of oxalic acid dihydrate and 1 liter of distilled water and then heated to 90 °C to dissolve the contents. The mixture was then treated with 400 ml of 12 M hydrochloric acid followed by the addition of 220.0 g of o-phenylenediamine and heated for 30 min at 90 °C. The mixture was then cooled to an ambient temperature and off-white crystals were collected by filtration. After washing first with water and then with methanol followed by drying the material under reduced pressure, the product was isolated in a 96% yield as off-white needles having an mp>350 °C.

Polymer Synthesis

2. Preparation of poly(2,3-quinoxalinoxy-co-bisphenol A)

A 100-ml three-necked flask equipped with a mechanical stirrer, Claisen arm fitted with a nitrogen inlet tube, Dean–Stark trap, condenser, and an exit gas bubbler was charged with 1.9904 g of the Step 1 product, 2.2829 g of bisphenol A, 2.7674 g of anhydrous potassium carbonate, 16 ml of dimethylacetamide, and 8 ml of toluene. The flask was then placed into an oil bath preheated to 122 °C and heated for 5 h, and a viscous solution was isolated. During the polymerization, toluene was incrementally added to maintain the azeotropic removal of water. The product was then diluted with 20 ml of dimethylacetamide and precipitated in 600 ml of a 5:1 water/acetic acid mixture. After filtration, the precipitate was washed with water and then methanol and then redissolved in 50 ml of chloroform acidified with 2–3 ml of acetic acid. The polymer was then reprecipitated in 500 ml of methanol, washed with methanol, and then stirred in boiling water for 2 h. After filtering, the product was washed with water and methanol and then dried in a vacuum oven at 120 °C to a constant weight, and 3.27 g of the product was isolated. The physical properties of the product are given below.

$[\eta]=0.66$ g/dL (*N*-Methylpyrrolidinone @ 30 °C)

Tg = 193 °C

Testing and Testing Results

1. Comprehension Molding

Compression molding was conducted at 300 °C under 1000 psi for 5 min using 2 g of the Step 2 product and a slightly yellow, transparent, and tough film was isolated.

2. Stress–Strain

Stress–strain measurements was performed on the Step 2 product according to ASTM D882 using a film prepared from chloroform. The tensile strength of the film was 104 MPa, and its tensile modulus was 3.3 GPa.

Advantages over Prior Art

Quinoxaline-containing polymers have previously been prepared using bis-α-carbonyl derivatives or aromatic tetraamines. In both cases, however, inherent viscosities and glass transition temperatures were too low to be used in molding applications. In another instance, polyquinoxalines were prepared by condensing 1,4-diglyoxalyl benzene with 3,3′-diaminenenzidine but nonquinoxaline segments were also present in the polymer. The current invention is distinguished from prior art in that it is able to prepare purified moldable thermoplastic quinoxaline copolymers, and the method is both simple and quantitative.

Notes

1. Banerjec[1] prepared polysiloxane/polyimide thermoplastic block copolymers (VIII) having a siloxane content of up to 30 wt%, which had low flammability, high temperature stability, high glass transition temperature, low flexural modulus, high tensile elongation, and high impact strength.

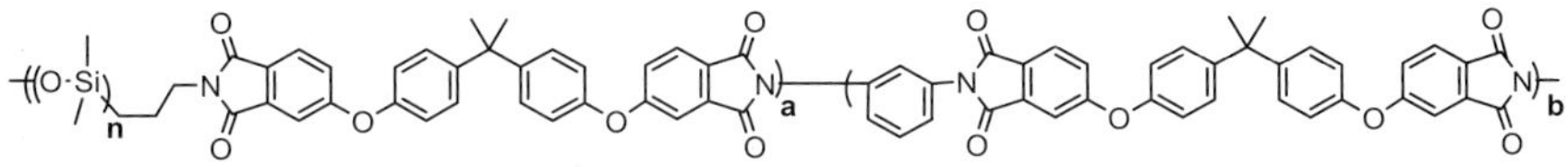

n = 10 - 50

(VIII)

2. Bala[2] prepared poly(dimethylsiloxane) thermoplastic elastomers (IX) containing pyrimidin-2-one termini, which formed supramolecular polymers through extensive intermolecular hydrogen bonding on the polymer termini (X).

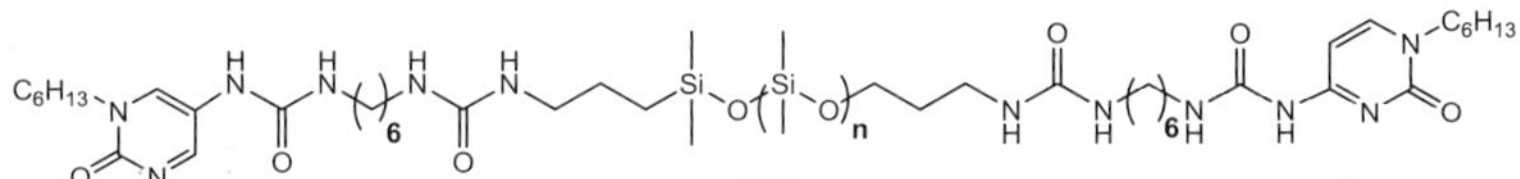

(IX)

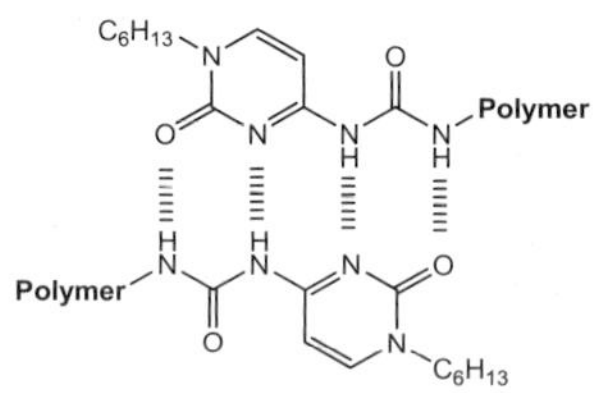

(X)

3. Dershem[3] prepared a series of malimides (XI)–(XIV), which were used to prepare thermally stable thermoset resins for use in the electronic packaging industry.

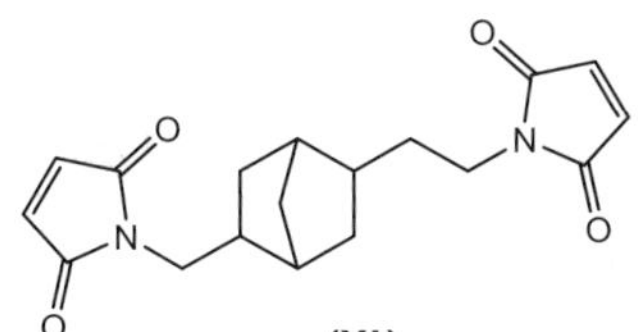

(XI)

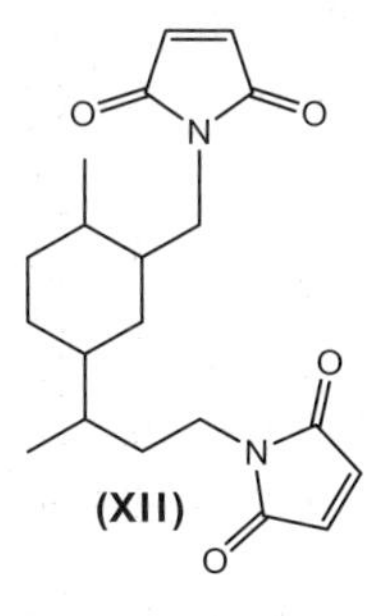

(XII)

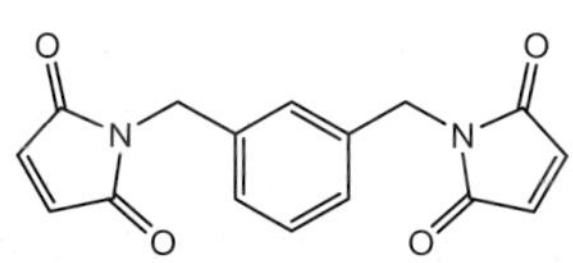

(XIII)

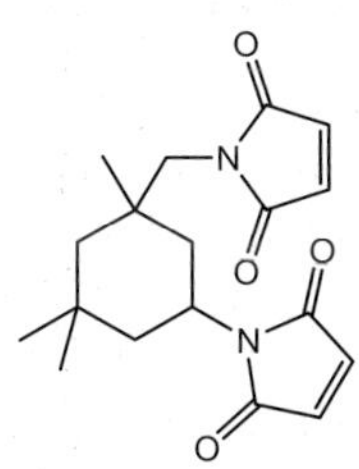

(XIV)

4. Kodama[4] prepared 2,3-naphthyldiacrylate derivatives (XV) as curing agents to prepare thermally stable resins for nanoimprints having improved dry etching resistance properties.

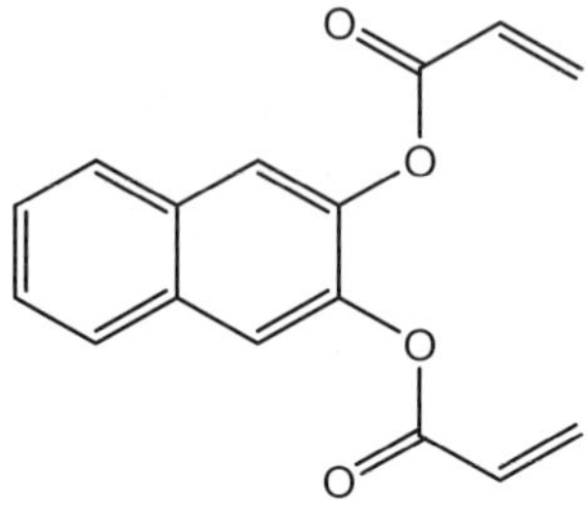

(XV)

REFERENCES

1. Susanta Banerjee et al, US Patent 8,071,693 (December 6, 2011) **Sabic Innovative Plastics IP B.V. (NL)**
2. Kason Bala et al, US. Patent 8,076,426 (December 13, 2011) **The Secretary of State for Defence (GB)**
3. Stephen Dershem, US Patent 8,043,534 (October 25, 2011) **Designer Molecules, Inc. (San Diego, CA)**
4. Kunihiko Kodama et al, US. Patent 8,025,833 (September 27, 2011) **FUJIFILM Corporation (Tokyo, JP)**

U. ULTRAHIGH MOLECULAR WEIGHT POLYMERS

Bridged Metallocene Catalysts

Title:	Process for manufacturing high to ultrahigh molecular weight polymers using novel bridged metallocene catalysts
US Patent:	8,034,886 (October 11, 2011)
Author:	Tim Dickner et al.
Assignee:	Ticona GmbH (DE)

Significance: Five monodimethylsilandiyl and bis(2-alkyl-4-alkylaryl-indenyl)-zirconium dichloride ethylene polymerization additives were prepared containing tri-isobutylaluminum as cocatalyst anchored on to a silica substrate. Polyethylene prepared using these agents had viscosity average molecular weights between 0.7×10^6 to 2.7×10^6 g/mol. Processability of polyethylene was also improved because of the narrow molecular distribution.

Additive Names Dimethylsilandiyl-bis(2-methyl-indenyl) zirconium dichloride (I)
Dimethylsilandiyl-bis(2-isopropyl-indenyl) zirconium dichloride (II)
Dimethylsilandiyl-bis(2-ethyl-4-(4-*t*-butylphenyl)-indenyl) zirconium dichloride (III)
Dimethylsilandiyl-bis(2-isopropyl-4-(4-*t*-butylphenyl)-indenyl) zirconium dichloride (IV)
Dimethylsilandiyl-(2-ethyl-4-(4-*t*-butylphenyl)-indenyl)-(2-isopropyl-4-(4-*t*-butylphenyl)-indenyl) zirconium dichloride (V)

Safety Transition metal-based polymerization catalysts are considered toxic, strong eye and skin irritants, and are corrosive. When fully activated with an alkyl aluminate cocatalyst, they are severe skin, respiratory, and eye irritants.

Additive Structures

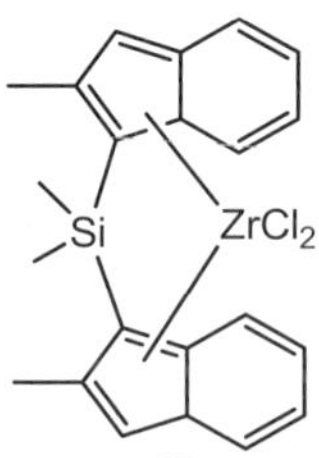

(I)

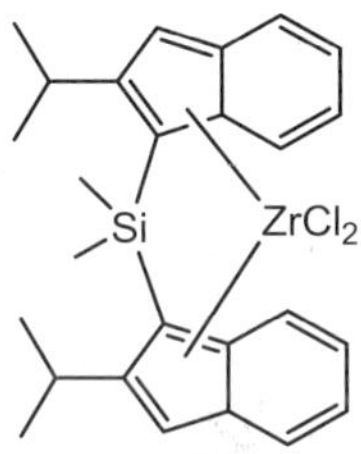

(II)

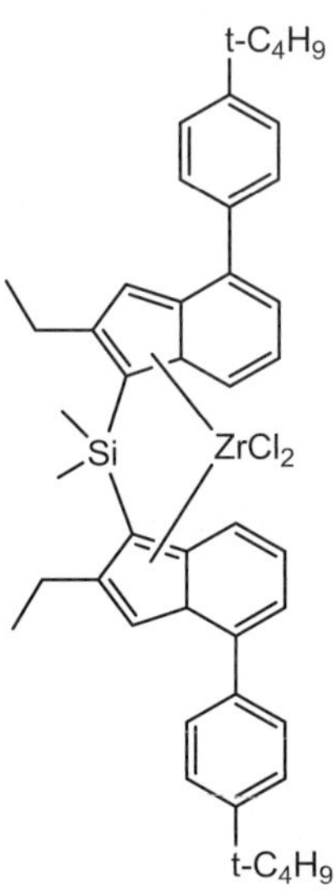

(III)

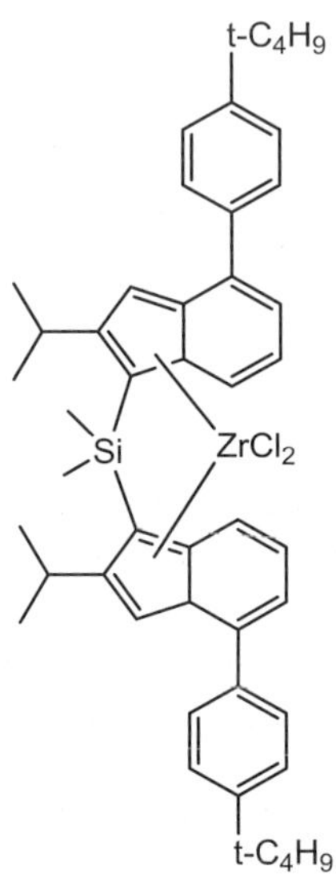

(IV)

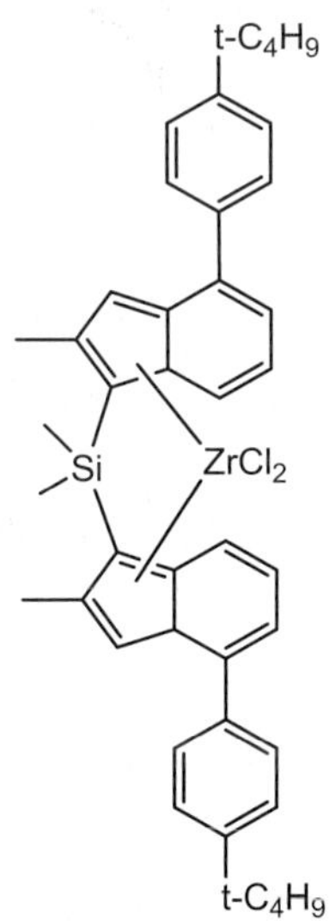

(V)

POLYMERIZATION PROCESS

Preparation of Supported Catalysts for Polymerization Additives

1. Activation with cocatalyst

In an annealed flask under an inert gas, 0.128 mmol of a selected additive was dissolved in 20 ml of toluene and then mixed with 30% of methyl aluminoxane dissolved in toluene, and the mixture was stirred for 1 h at an ambient temperature.

2. Support

In an annealed flask under an inert gas, 6 g of dried silica suspended in 30 ml of purified toluene was slowly stirred and treated with the Step 1 activated polymerization additive. The mixture was then stirred for 10 min and then concentrated, and the supported activated polymerization additive was isolated containing <5% residual moisture.

Preparation of High Molecular Weight Polyethylene

A 2-liter steel autoclave was charged with 15 mmol of tri-isobutylaluminum and then heated to 70 °C and treated with ethylene pressure until a 10-bar pressure was obtained. Nine micromoles of a selected polymerization additive suspended in EXXSOL® was injected, and the mixture was polymerized for 1 h. The solution was cooled to an ambient temperature, filtered, dried at 80 °C under a vacuum, and the polymer was isolated. The effectiveness of each experimental additive prepared using the Step 2 procedure is provided in Table 19.

TABLE 19 Viscosity average molecular weights resulting from the polymerization of ethylene at 70 °C and 10 bar of ethylene pressure catalyzed by experimental additives with triisobutylaluminum. Viscosity average molecular weights were measured in decalin

Polymerization Reaction	Polymerization Additive	Polyethylene Yield (g)	Viscosity Average Molecular Weight (1×10^6 g/mol)
1	I	70	0.7
2	II	61	2.2
3	III	64	1.1
4	IV	98	1.8
5	V	114	2.7

Advantages over Prior Art

Although bridged metallocene polymerization additives are extensively reported as a method for preparing ultrahigh homo and copolymers, the current invention introduces two improvements. First, heterogeneous metallocene polymerization additives of the current invention can be quantitatively removed after the polymerization has been completed. In addition, polyethylene prepared using polymerization additives had a molecular weight distribution of between 3 and 6 as opposed to similar heterogeneous analogs, which are reported to have molecular weight distributions between 3 and 30. The narrower polydispersity results in better and more streamlined polymer processing.

Notes

1. The preparation of the metallocene compositions of the current invention is described by Burkhardt[1] and is illustrated in Eq. 1.

Eq. 1.

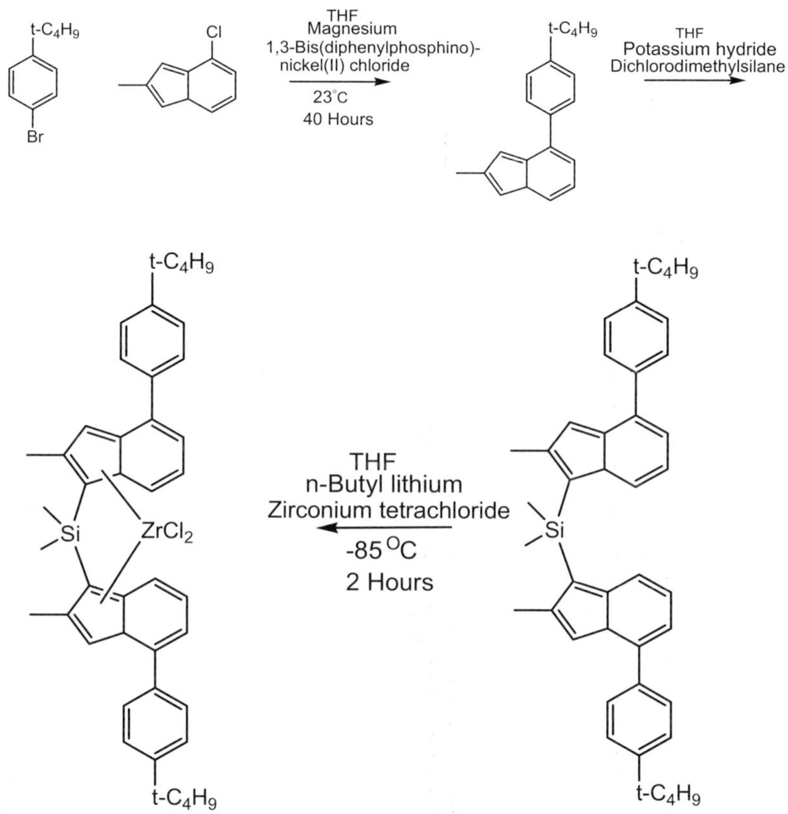

2. Paczkowski[2] prepared highly activated metallocene catalysts by coupling dimethylsilanediyl-(2-methyl-4-(4′-*t*-butylphenyl)-1-indene)(2-isopropyl-4-(4′-tert-butylphenyl)-1-indene) of the current invention with methyl aluminoxane as cocatalyst.
3. Fuchs[3] prepared polypropylene containing a high isotactic pentad content with a polydispersity of four using dimethylsilandiyl-bis(2-ethyl-4-(4-*t*-butylphenyl)-indenyl) zirconium dichloride of the current invention with cocatalysts consisting of either trimethylaluminum, tris(2,4,4-trimethyl-pentyl)aluminum, tris(2,3-dimethylbutyl)aluminum, or tris(2,3,3-trimethylbutyl) aluminum.

4. Resconi[4] polymerized the homopolymer of 1-butene having a very narrow molecular distribution using dimethylsilyl-bis(2-methyl-cyclopento[2,3-b]thiophen-6-yl)zirconium dichloride (VI) with triisobutylaluminum as cocatalyst.

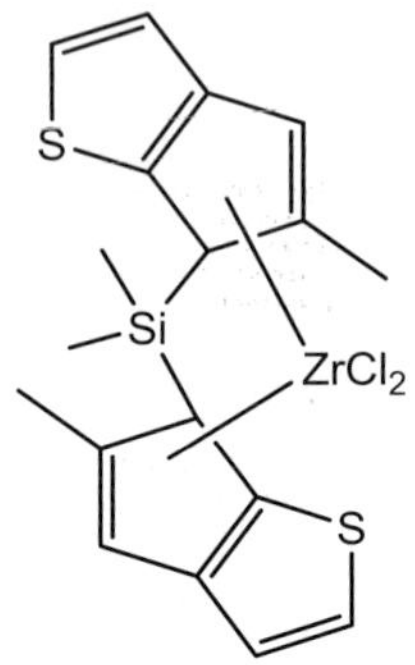

(VI)

REFERENCES

1. Terry J. Burkhardt et al, US. Patent 6,903,229 (June 7, 2005) **ExxonMobil Chemical Patents Inc. (Houston, TX)**
2. Nicola S. Paczkowski et al, US. Patent 7,169,864 (January 30, 2007) **Novolen Technology Holdings, C.V. (The Hague, NL)**
3. Alexander Fuchs et al, US. Patent 8,030,426 (October 4, 2011) **Basell Polyolefine GmbH (Wesseling, DE)**
4. Luigi Resconi, US. Patent 8,030,421 (October 4, 2011) **Basell Polyolefine GmbH (Wesseling, DE)**

Chapter 28

Powder Coating Additives

A. IMPACT RESISTANCE AND LOW OFF-GASSING ADDITIVES

Polyester Block Copolymers

Title:	Polymer additives for powder coatings
US Patent:	7,816,451 (October 19, 2010)
Author:	William H. Retsch, Jr. et al.
Assignee:	PPG Industries Ohio, Inc. (Cleveland, OH)
Significance:	Highly branched polyesters consisting of poly[(trimethylolpropane-co-adipic acid)-b-ε-caprolactone] and poly[((trishydroxylethyl)isocyanurate-co-isophthalic acid)-b-ε-caprolactone] were prepared in a high yield by step growth polymerization. When used as powder coating additives, solvent outgassing was reduced by up to 35%, and curing dwell time and temperature requirements were lowered.

Additive Name	Poly[(trimethylolpropane-co-adipic acid)-b-ε-caprolactone] (I) Poly[((trishydroxylethyl)isocyanurate-co-isophthalic acid)-b-ε-caprolactone] (II)
Safety Concerns	All materials are potentially harmful because of the presence of residual adipic and isophthalic acid both of which are severe respiratory tract irritants. Residual trimellitic anhydride and diisocyanates are both irritants and may act as sensitizers.

Next Generation of International Chemical Additives. DOI: http://dx.doi.org/10.1016/B978-0-444-53788-1.00028-0

Additive Structures

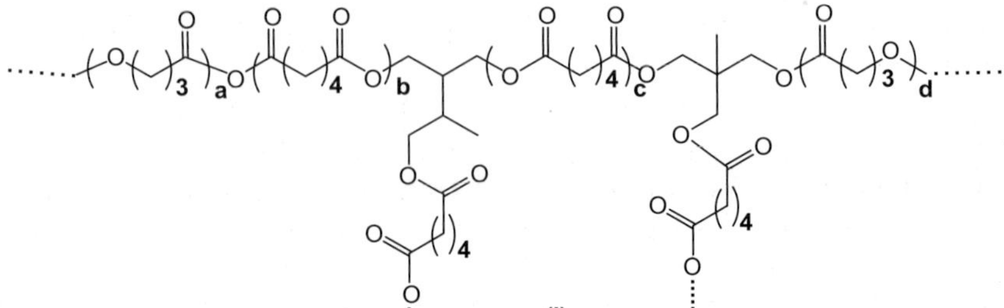

(I)

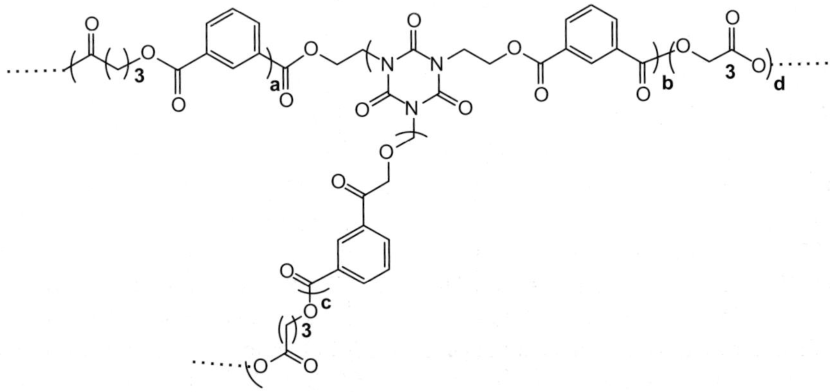

(II)

Additive Preparation

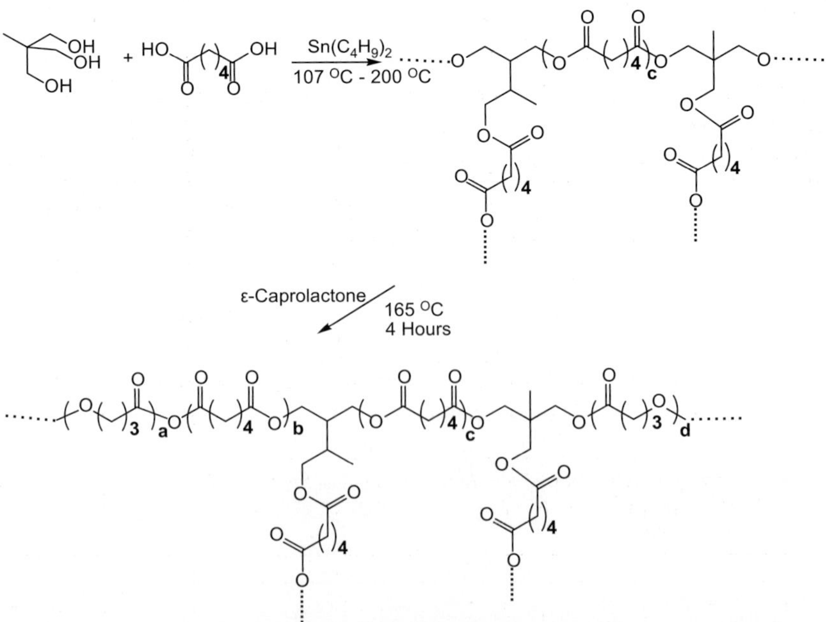

Additive Synthesis

1. Preparation of highly branched polyester consisting of trimethylol propane, adipic acid, and ε-caprolactone

A reactor was charged with 100 g of trimethylol propane, 81.1 g of adipic acid, and 5.6 g of dibutyl tin and then heated to about 107 °C. Heating was increased to 200 °C and continued until water removal was stopped and the mixture was then cooled to an ambient temperature. The intermediate was then treated with 2606.9 g of ε-caprolactone and then heated to 165 °C for 4 °h and then recooled. The polymer had an Mw of 28,237 Da, Mn of 5348 Da with a polydispersity of 5.3, a Tm of 53.7 °C while no Tg was observed.

Additive Testing

1. Impact resistance and off-gassing testing

Experimental star polymers were processed for about 20 s in a Prism blender at 3500 rpm and then extruded through a B&P Process Equipment and Systems 19 mm corotating twin screw extruder at 450 rpm at temperatures between 100 °C and 115 °C. The ensuing chip that formed was milled and had a median particle size of about 30–50 μm on using a Hosokawa Micron Powder Systems Air Classifying Mill I. The samples were then electrostatically sprayed using a Nordson corona onto cold unpolished rolled steel panels supplied by ACT Laboratories as B-1000 P60 DIW. Thereafter, panels were baked in electric Despatch LAD series ovens for a dwell time of 25 min at 325 °F and had a dry film thickness of 2.5–3.0 °mils. Test results are provided in Table 1.

Additive Test Results

TABLE 1 Effects on the impact resistance and off-gassing of panel surfaces coated with selected experimental star powder coatings of varying compositions. High impact resistance and off-gassing thicknesses are preferred

Sample	Monomer-1 (g)	Monomer-2 (g)	Monomer-3 (g)	Impact Resistance (in-lbs)	Off-Gassing Thickness (mil)
1	Trimethylol (100)	Adipic Acid (81.1)	ε-Caprolactone (2606.9)	60/50	6.6
2	Trimethylol (100)	Hexahydrophthalic Anhydride (132.1)	ε-Caprolactone (2606.9)	160/160	8.5
3	Tris(hydroxylethyl)isocyanate (100)	Isophthalic Acid (47.6)	ε-Caprolactone (2053)	160/160	9.4

(continued)

TABLE 1 Effects on the impact resistance and off-gassing of panel surfaces coated with selected experimental star powder coatings of varying compositions. High impact resistance and off-gassing thicknesses are preferred—*Cont'd*

Sample	Monomer-1 (g)	Monomer-2 (g)	Monomer-3 (g)	Impact Resistance (in-lbs)	Off-Gassing Thickness (mil)
4	Cyclohexyl-dimethylol (126) and Trimethylol (12.6)	Isophorane Diisocyanate (113.3) and 1,6-Hexane Diisocyanate (48.6)	ε-Caprolactone (2606.9) and Hexahydro-phthalic Anhydride (2.7)	160/160	7.8
5	Ethylenedi-amine (71)	Trimellitic Anhydride (75.7)	ε-Caprolactone (1366.7)	160/160	11.6

Advantages over Prior Art

Hyperbranched or star polymers mimic dentritic-like structures but with a less controlled architecture. The effectiveness of star polyesters as coating additives is a result of the presence of a high density of surface active groups from the polymers adhering to the surface and internal cavities in the powder. This type and extent of additive–surface interaction is limited or absent in linear or moderately branched polyester coating agents.

Notes

1. Caiozzo[1] prepared a powder surface consisting of the reaction product of hexamethylene diisocyanate uretdione (III) with hydroxy propyl carbamate (IV) and crosslinked with melamine (V).

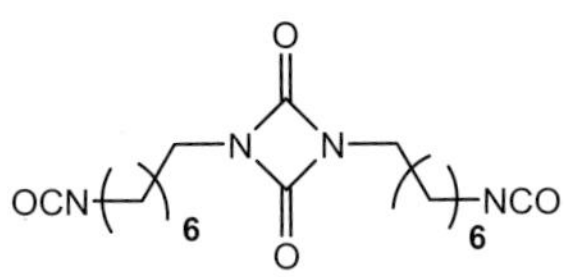

(III)

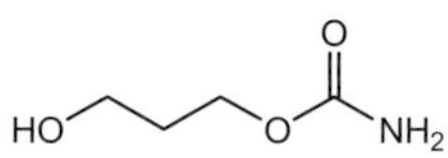

(IV)

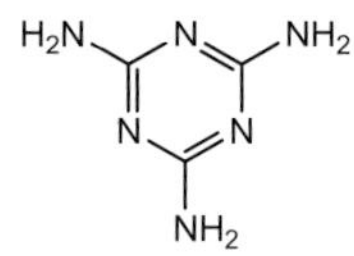

(V)

2. A composition additized with the high degassing agent, 3-[(2-oxo-1,2-diphenylethoxy)-carbonyl]-propanoic acid (VI) was prepared by Gaglani[2] and used as a degassing agent in film-forming curable coatings compositions in powder coatings.

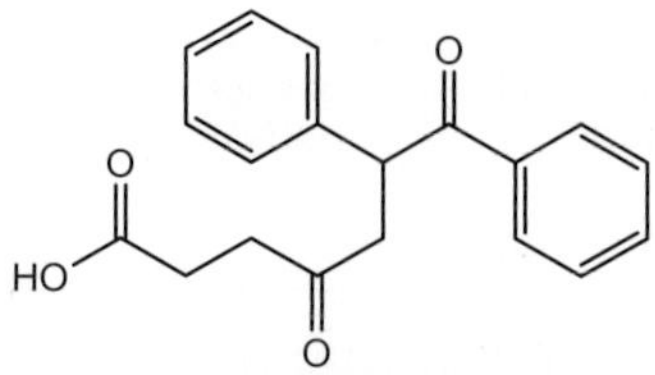

(VI)

3. Chasser[3] prepared the curable powder coating (VII) consisting of an epoxy resin condensed with *N*-methyl ethanolamine, which was effective as a curable powder coating.

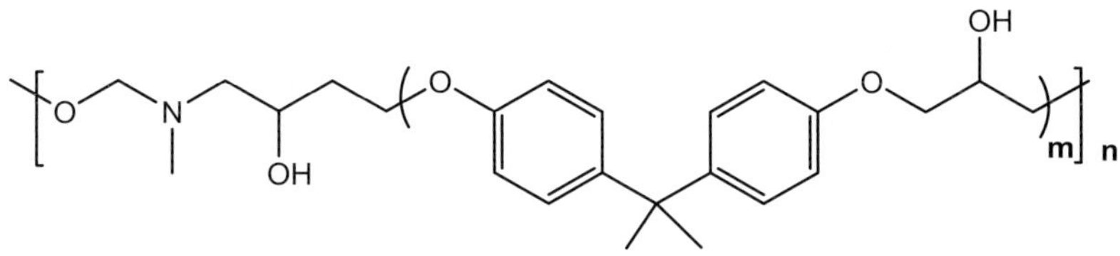

(VII)

4. A multicomponent resin composition consisting of diglycidyl ether of bisphenol A, bisphenol A, propylene glycol butyl ether, triphenylphosphine reacted with polyethyleneoxide diol and diethanolamine and then postreacted with polymethylene polyphenylisocyanate was prepared by Tazzia[4] and used in low outgassing powder coating applications.

REFERENCES

1. Nicholas Caiozzo et al, US Patent Application 20100298490 (November 25, 2010)
BASF Corporation (Wyandotte, MI)
2. Kamlesh Gaglani et al, US Patent 7,842,745 (November 30, 2010)
Troy Corporation (Florham Park, NJ)
3. Anthony M Chasser et al, US Patent Application 20060292374 (December 28, 2006)
PPG Industries (Pittsburgh, PA)
4. Charles L. Tazzia et al, US Patent 7,842,177 (November 30, 2009)
BASF Coatings GmbH (Muenster, DE)

Chapter 29

Surfactants

A. ANTIMICROBIAL SURFACE MODIFICATION

Perfluorosorbitol Esters

Title:	Fluorosurfactants
US Patent:	8,067,625 (November 29, 2011)
Author:	Wolfgang Hierse et al.
Assignee:	Merck Patent Gesellschaft Mit Beschrankter Haftung (Darmstadt, DE)

Significance: Fatty acids functionalized with fluorinated amines, ethers, or sulfanyl groups were prepared and then used to form monoesters with sorbitol and sucrose. Fluorinated sugar surfactants could be oxidatively or reductively degraded into nontoxic and nonpersistent residues. Despite having a fluorine content of ≤3.8 wt%, fluorinated sorbitol and sucrose esters displayed high surface activities.

Additive Names

Sorbitol-(E)-10-pentafluorosulfanyldec-9-enecarboxylic acid ester (I)
Sucrose-(E)-10-pentafluorosulfanyldec-9-enecarboxylic acid ester (II)
Sorbitol-7-(3,3,3-trifluoropropoxy)heptanoic acid ester (III)
Sucrose-7-(3,3,3-trifluoropropoxy)heptanoic acid ester (IV)
Sorbitol-10-*N*-di-trifluromethyldecanoic acid ester (V)
Sucrose-10-*N*-di-trifluromethyldecanoic acid ester (VI)

Safety There are no toxicity data available on the current fluorine-containing fatty acid ester surfactants.

Next Generation of International Chemical Additives. DOI: http://dx.doi.org/10.1016/B978-0-444-53788-1.00029-2

Additives

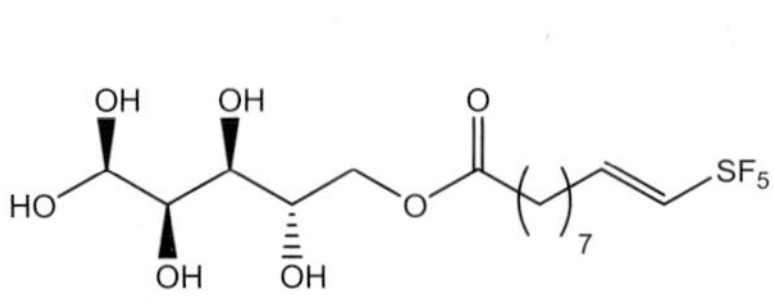

(I)

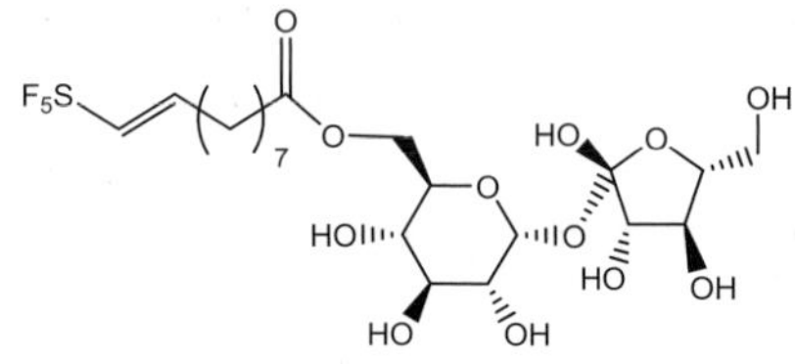

(II)

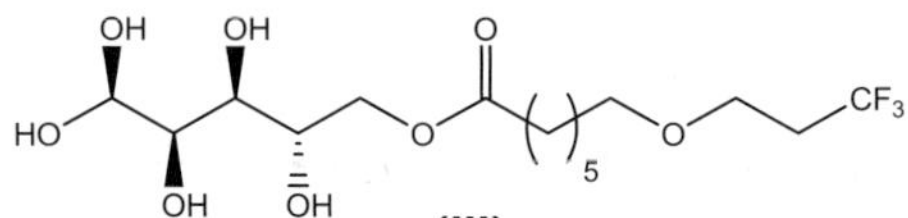

(III)

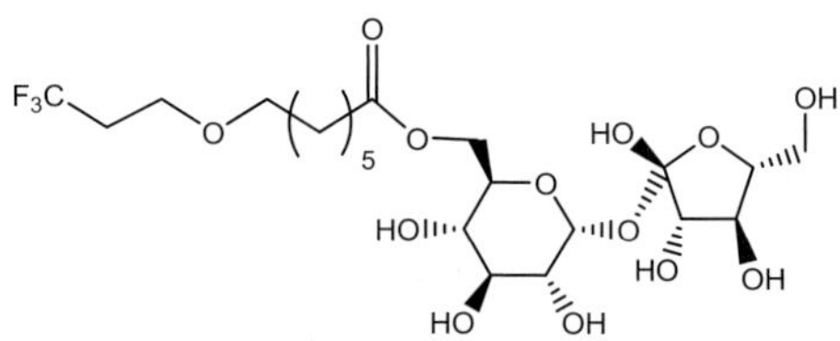

(IV)

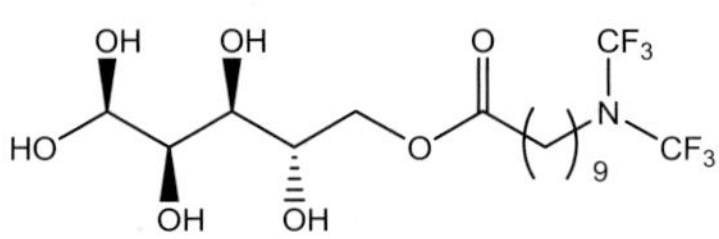

(V)

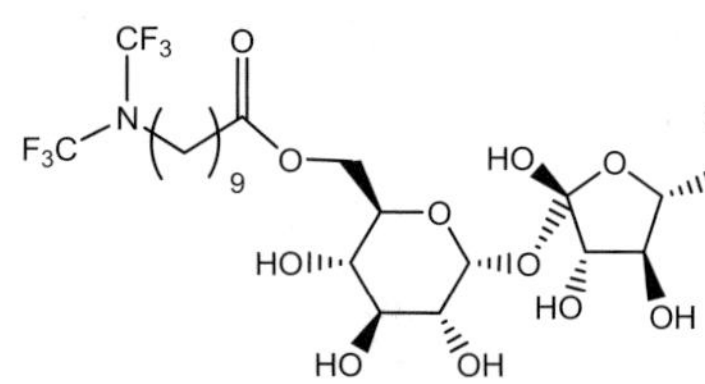

(VI)

Additive Preparation

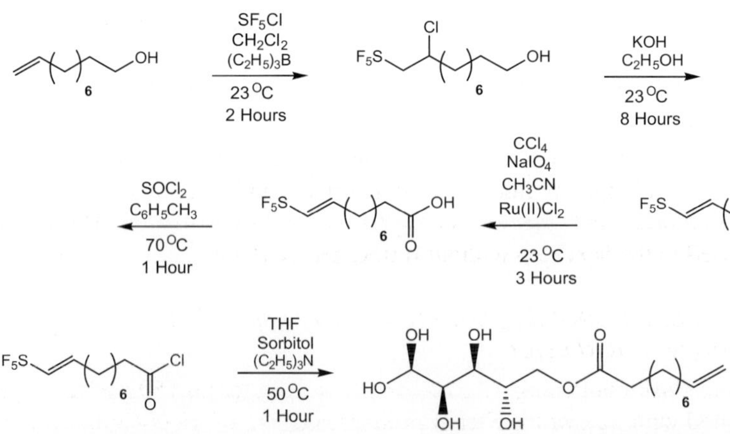

Additive Synthesis

1. *Preparation of 10-pentafluorosulfanyl-9-chloro-decanol*

A 250-ml single-neck round-bottom container was charged with 15 g of 9-decenol dissolved in 250 ml of methylene chloride and then cooled to −40 °C and treated with 27 g of sulfur chloride pentafluoride and 2 ml of 1 M triethylborane. Additional triethylborane was added until the evolutional of gas stopped and the reaction mixture was no longer warm. The mixture was then stirred at an ambient temperature for an additional 2 h and then reacted with a mixture of saturated sodium bicarbonate containing ice. The solution pH was then adjusted to 10 using sodium hydroxide, and the aqueous phase was separated and washed twice with methyl *t*-butyl ether. The collected organic phases were then extracted once with brine, dried with sodium sulfate, filtered, and the product was isolated after the solvent was removed using a rotary evaporator.

2. *Preparation of (E)-10-pentafluorosulfanyldecenol*

A 250-ml single-necked flask containing a reflux condenser was charged with 9 g of the Step 1 product dissolved in 120 ml of ethanol and 4.75 g of potassium hydroxide and then stirred overnight and concentrated. The concentrate was then treated with water and methyl *t*-butyl ether, and the phases were separated. The aqueous phase was then extracted thrice with methyl *t*-butyl ether and then washed with brine. After drying with sodium sulfate and filtering, 8.3 g of a yellowish liquid was isolated after distillation.

3. *Preparation of (E)-10-pentafluorosulfanyldec-9-enecarboxylic acid*

A 250-ml round-=bottom flask was charged with 11.3 mmol of the Step 2 product dissolved in a solvent mixture comprising 40 ml of carbon tetrachloride, 40 ml of acetonitrile, and 50 ml of water. The mixture was then treated with 5.44 g of sodium metaperiodate and 234 mg of ruthenium(III) chloride and then stirred at 23 °C for 3 h. The mixture was then treated with 50 ml of methylene chloride, and the phases were separated. The aqueous phase was further extracted twice with 50 ml of methylene chloride, and extracts were combined. After drying with sodium sulfate, the solution was filtered and then distilled, and the product was isolated as an oily residue.

4. *Preparation of (E)-10-pentafluorosulfanyldec-9-enecarboxylic acid chloride*

A reactor was charged with 30 g of the Step 3 product and 100 g of toluene and then treated with 24 g of thionyl chloride. The mixture was then heated to 70 °C for 1 h and excess thionyl chloride, and the solvent was removed by distillation. The resultant product was used in the next step without further purification.

5. *Preparation of sorbitol (E)-10-pentafluorosulfanyldec-9-enecarboxylic acid ester*

A round-bottom flask containing 18 g of sorbitol suspended in 150 g of tetrahydrofuran was treated with 32 g of the Step 4 product and 10 g of triethylamine and then heated for 1 h at 50 °C. After filtration and purification, the product was isolated.

Testing

Testing data are not provided by author.

Advantages over Prior Art

Fluorinated surfactants described in this invention have three advantages absent in existing surface active fluorinated agents. First, all materials produced using this method are surface active despite having very low fluorine contents. Second, sorbitol and sucrose fluorinated monoesters are soluble in polar media, particularly water and alcohol. Finally, oxidative or reductive degradation of these agents does not generate persistent organofluorine degradation products, particularly fluorinated carboxylic acids.

Notes

1. Phenyl (VII) and naphthyl (VIII) analogs of the current invention were prepared by Hierse[1] and used as antimicrobial surface agents.

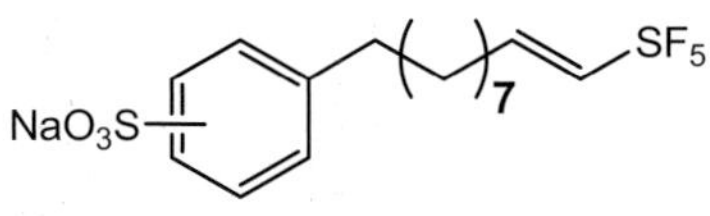

(VII)

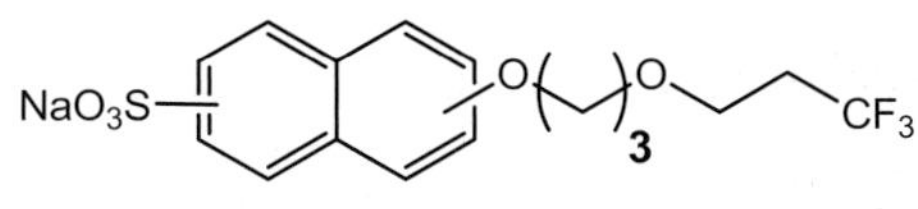

(VIII)

2. Hierse[2] prepared fluorinated aminoalcohols (IX) and sulfanyl alcohols (X), which had surface activities similar to that of classical fluorosurfactants but upon oxidative or reductive degradation did not leave behind any persistent organofluorine degradation products.

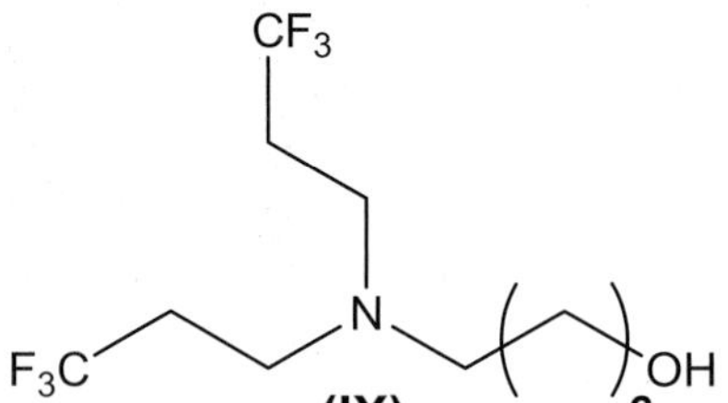

(IX)

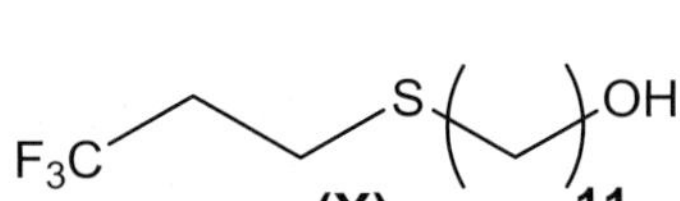

(X)

3. Murphy[3] prepared low fluorine content polyoxyalkylene glycol diamide surfactants (XI), which were effective in lowering both surface and interfacial surface tension in liquids.

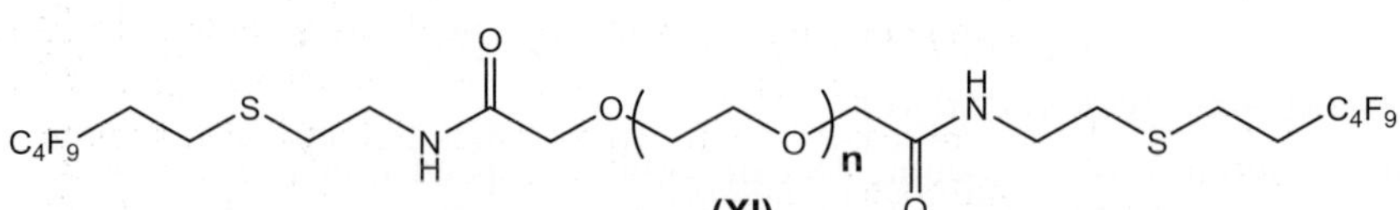

(XI)

n = 30 - 400

4. Peng[4] prepared perfluoroalkoxyphosphate polyethylene oxide surfactants (XII), which at a treatment level of 0.1% by weight in water had a surface tension of 22 mN/m and which were used in urethane paint formulations.

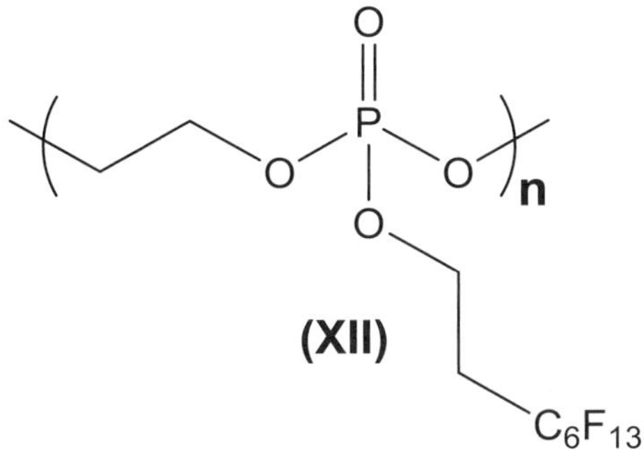

(XII)

REFERENCES

1. Wolfgang Hierse et al, US Patent Application 20100152081 (June 17, 2010) **Merck Patent Gesellschaft Mit Beschrankter Haftung (Darmstadt, DE)**
2. Wolfgang Hierse et al, US. Patent Application 20090320718 (December 31, 2009) **Merck Patent Gesellschaft Mit Beschrankter Haftung (Darmstadt, DE)**
3. Peter Murphy et al, US Patent 8,026,289 (September 27, 2011) **E.I. du Pont de Nemours and Company (Wilmington, DE)**
4. Sheng Peng et al, US. Patent 8,022,238 (September 20, 2011) **E.I. du Pont de Nemours and Company (Wilmington, DE)**

B. BIODEGRADABLE SURFACTANTS

Perfluorosurfactant Salts

Title:	Fluorosurfactants
US Patent:	8,008,358 (August 30, 2011)
Author:	Peer Kirsch et al.
Assignee:	Merck Patent GmbH (Darmstadt, DE)
Significance:	Biodegradable low molecular weight sodium alkyl sulfonates containing either a single trifluoromethoxy or pentafluorosulfuranyl terminus were prepared in four steps in a good yield. Despite the low fluorine content, the water surface tension of solutions containing these agents was significantly improved over the nonfluorinated sodium sulfonated analogs.

Additive Name	Sodium 10-trifluoromethoxydecane-1-sulfonate (I) Sodium 6-chloro-7-(pentafluorosulfuranyl)heptane-1-sulfonate (II) Sodium 7-(pentafluorosulfuranyl)hept-6-ene-1-sulfonate (III)
Safety	Perfluoromethoxy- and pentafluorosulfuranyl sodium alkyl sulfonates are considered moderate skin irritants and cause severe serious eye irritation. Repeated exposure to dust may cause respiratory irritation.

Additive Structures

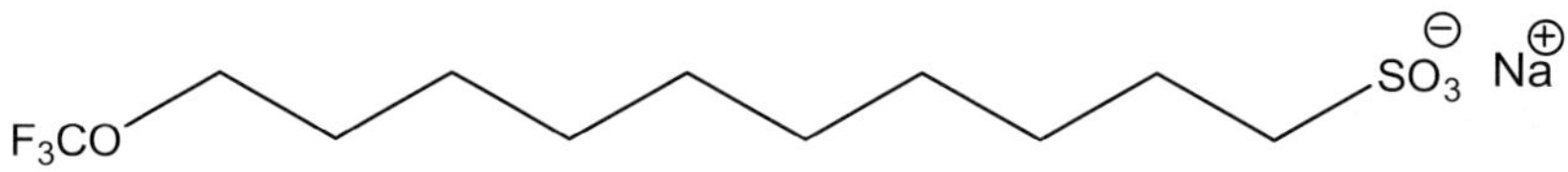

(I)

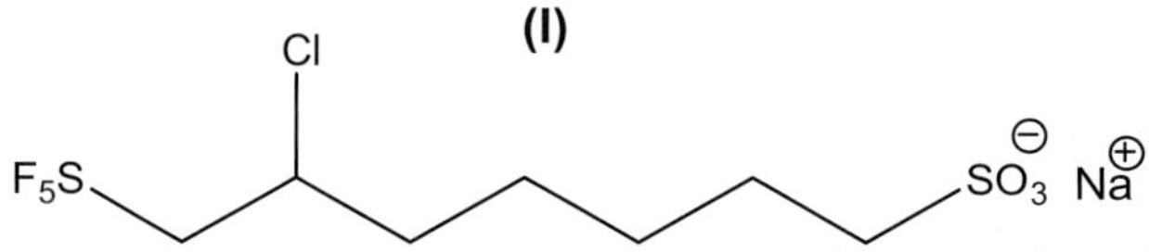

(II)

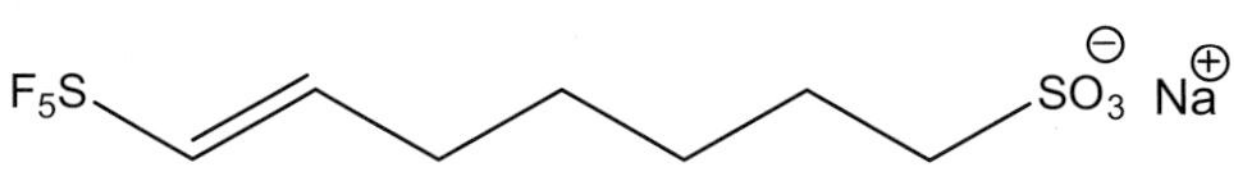

(III)

Additive Preparation

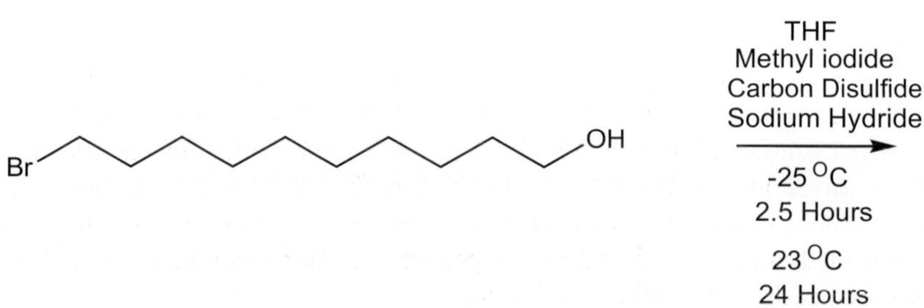

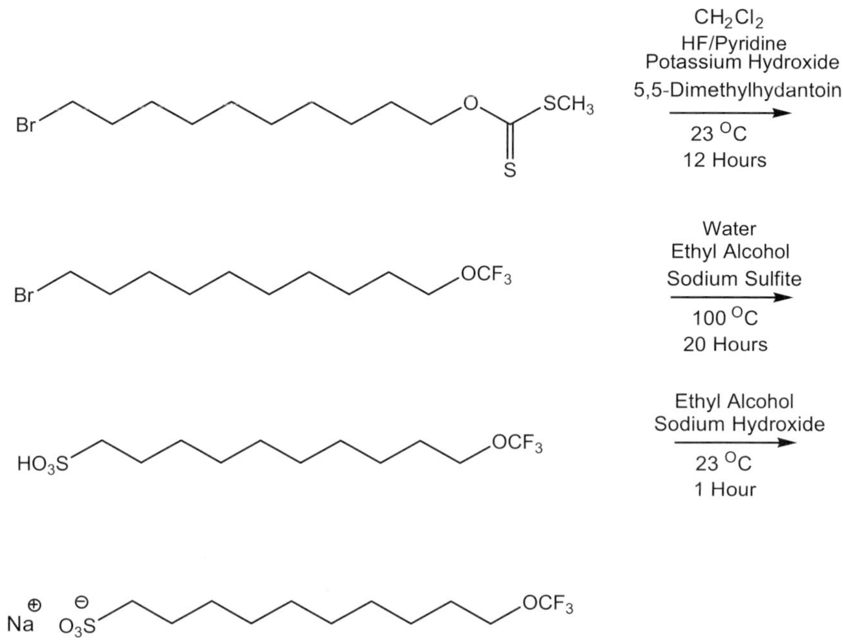

Additive Synthesis

1. *Preparation of 10-bromodecyl methyl dithiocarboxylate*

A four-necked glass apparatus was charged with 200 ml of tetrahydrofuran and 10.15 g of sodium hydride and then cooled to −25 °C and treated with the dropwise addition of 50 g of 6-bromo-1-decanol dissolved in tetrahydrofuran. The mixture was then stirred at an ambient temperature for 2 h and then recooled to −25 °C and treated with the dropwise addition of 32.1 g of carbon disulfide and stirred for 2 h at 0 °C. The mixture was then recooled to −20 °C and treated with the dropwise addition of 35.9 g of methyl iodide and then slowly warmed to an ambient temperature and stirred for 24 h. After quenching the mixture with 200 ml of 10% ammonium chloride, the phases are separated, and the organic phase was washed and evaporated to dryness, and the product was isolated.

2. *Preparation of 1-bromo-10-trifluoromethoxydecane*

A round bottom container with a suspension of 102 g of 1,3-dibromo-5,5-dimethylhydantoin dissolved in 420 ml of methylene chloride was treated with the dropwise addition of 200 ml of 65–70% HF/pyridine at −76 °C followed by the addition of 37 g of the Step 1 product dissolved in 50 ml of methylene chloride. The reaction mixture was then stirred at an ambient temperature for 12 h, and the pH was increased to 10 using aqueous potassium hydroxide. The reaction mixture was then diluted with water and methyl *t*-butyl ether, filtered, then dried, and the product was isolated by column chromatography using heptane.

3. Preparation of 10-trifluoromethoxydecane-1-sulfonic acid

A 250-ml one-necked flask was charged with 6.8 g of the Step 2 product and 3.7 g of sodium sulfite dissolved in 40 ml of water and 40 ml of ethanol and was then heated to 100 °C for 20 h. The mixture was then cooled and extracted with a 1:1 mixture of methyl *t*-butyl ether/heptane. The aqueous phase was acidified to a pH of 1 and then further extracted with methyl *t*-butyl ether. The combined organic phases were dried over sodium sulfate, filtered, and the product was isolated after evaporating the solvent.

4. Preparation of sodium 10-trifluoromethoxydecane-1-sulfonate

A mixture of 1.15 g of sodium hydroxide dissolved in 60 ml of ethanol and 6.8 g of the Step 3 product was refluxed for 1 h, and the product was isolated as a colorless solid.

Testing

1. Biochemical Degradability

The biochemical degradability of the experimental agents was determined by the Zahn–Wellens test in accordance with the publication by the European Commission: Classification, Packaging and Labeling of Dangerous Substances in the European Union, Part II – Test Methods, Annex V – Methods for the Determination of Physicochemical Properties, Toxicity and Ecotoxicity. Test results are provided in Table 1.

2. Surface Tension

The surface tension of experimental surfactants was determined using a Model K12 Kruss tensiometer at 20 °C with between 0.5 and 3.0 g/l surfactant dissolved in deionized water. Test results are provided in Table 2.

Test Results

1. Biochemical Degradability

TABLE 1 Biodegradability of sodium 10- trifluoromethoxydecane-1-sulfonate after approximately 1 month

Testing Period (days)	Free Fluoride (mg/l)	Degradation (%)
0	1.1	0
2	2.6	0.7
5	1.1	4.5
8	9.6	55
9	11.5	59
22	30.9	72
28	43.1	76

2. Surface Tension

TABLE 2 Surface tension of water additized with sodium 10-trifluoromethoxydecane-1-sulfonate, in comparison with its nonfluorinated analog, sodium decane sulfonic acid

Sample	Concentration of the Additive in Water (g/l)	Concentration of the Additive in Water (mol/l)	Surface Tension (mN/m)
Sodium 10-trifluoro-methoxydecane-1-sulfonate (Additive 1)	0.5015	1.527×10^{-3}	59.95
Sodium 10-trifluoro-methoxydecane-1-sulfonate (Additive 1)	1.0000	4.60×10^{-3}	52.55
Sodium 10-trifluoro-methoxydecane-1-sulfonate (Additive 1)	2.0005	6.093×10^{-3}	42.61
Sodium decane sulfonic acid (**Analog**)	1.0020	4.101×10^{-3}	66.13
Sodium decane sulfonic acid (**Analog**)	2.0015	8.192×10^{-3}	57.70
Sodium decane sulfonic acid (**Analog**)	3.0010	1.228×10^{-2}	52.61

Advantage over Prior Art

Existing perfluoroalkyl and perfluoroalkyloxy surfactants readily degrade into carboxylic acid derivatives, which ultimately generate serious ecological problems. The current invention addresses this pressing concern by identifying and preparing very low fluorine content additives that have physical properties comparable with perfluoroalkyl surfactants but which are oxidatively degraded into low fluorine-containing carboxylic acids.

Notes

1. In a multistep process, Peng[1] prepared the perfluoro surface active agent, $C_4F_7CH_2CF_2CH_2CH_2SO_3H$, which at 0.20 wt% in water, had a surface tension of 21.7 dyne/cm at 23 °C.
2. Yamane[2] prepared coatings consisting of polyperfluoroethers and organosiloxanes (IV), which had low surface energies, excellent stain wiping and good adhesion which were used in mobile phones, portable music players, car navigation equipments, ATMs, and eyeglass lenses.

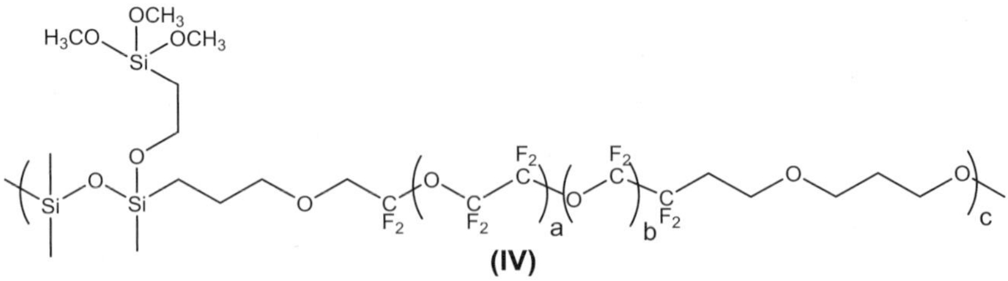

(IV)

3. Hinter[3] developed a method for removing perfluoroalkyl carboxylic acids dissolved in waste water by treating the solution with adsorbent packages containing activated carbon particles.
4. In an earlier investigation by Hinter,[4] fluorinated carboxylic acid surfactants contained in exhaust gas streams were removed by passing the gas through a scrubber liquid containing an aqueous solution of potassium carbonate.

REFERENCES

1. Sheng Peng et al, US. Patent 7,989,568 (August 2, 2011)
   E.I. du Pont de Nemours and Company (Wilmington, DE)
2. Yuji Yamane et al, US. Patent 8,013,096 (September 6, 2011)
   Shin-Etsu Chemical Co., Ltd. (Tokyo, JP)
3. Klaus Hintzer et al, US. Patent 7,795,332 (September 14, 2010)
   3 M Innovative Properties Company (St. Paul, MN)
4. Klaus Hintzer et al, US. Patent Application 20070015937 (January 18, 2007)
   3 M Innovative Properties Company (St. Paul, MN)

C. HYDROLYTICALLY STABLE SURFACTANTS

Trisiloxane Polyethylene Oxide Derivatives

Title:	Hydrolysis-resistant organomodified trisiloxane surfactants
US Patent:	7,935,842 (May 3, 2011)
Author:	George A. Policello et al.
Assignee:	Momentive Performance Materials Inc. (Albany, NY)

Significance:	Three sterically hindered trisiloxane polyethylene oxide derivatives were prepared which were hydrolytically stable in aqueous solutions having pH ranges from 2 to 11. Spreading diameters and surface tensions of these additives after 28 weeks of hydrolysis testing outperformed all existing commercially available trisiloxanes and phenolic-based surfactants.

Additives 1,5-Di-*t*-butyl-1,1,3,5,5-pentamethyl-3-oligoethyleneoxidetrisiloxane (I)
1,5-Di-isopropyl-1,1,3,5,5-pentamethyl-3-oligoethyleneoxidetrisiloxane (II)
1,1,1,3,5,5-Heptamethyl-3-oligoethyleneoxidetrisiloxane (III)

Safety Trisiloxane surfactants are generally considered nontoxic. Direct contact may cause very mild irritation. No significant skin or respiratory irritation is expected from a single short-term exposure. Repeated or prolonged exposure may cause very mild irritation. These agents are also considered very low ingestion hazards.

Additive Structures

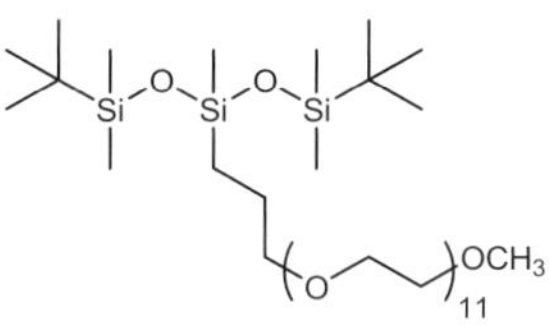

(I)

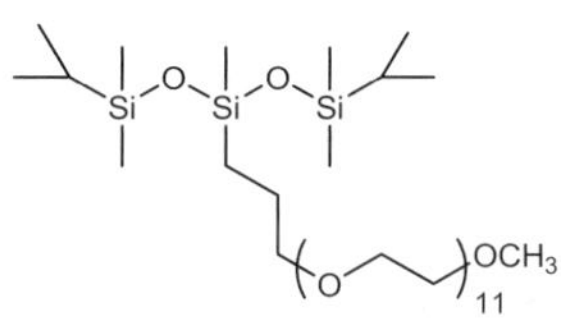

(II)

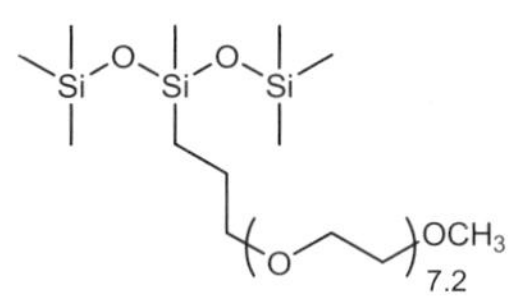

(III)

Additive Preparation

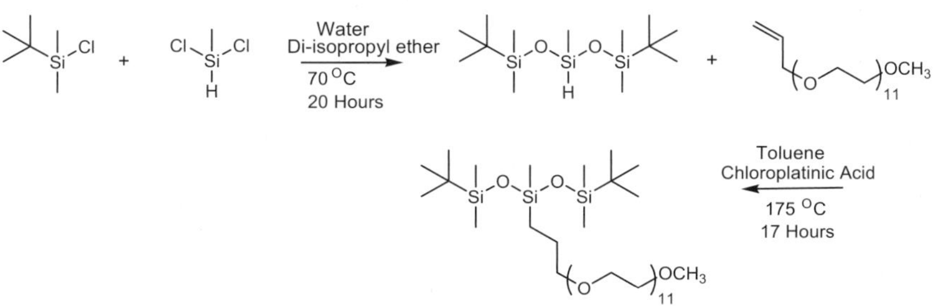

Additive Synthesis

1. *Preparation of 1,5-di-t-butyl-1,1,3,5,5-pentamethyltrisiloxane*

A mixture consisting of 100 g of *t*-butyldimethylsilylchloride and 46 g of methyldichlorosilicone hydride dissolved in 150 ml diisopropylether was put into an addition funnel and then added dropwise at an ambient temperature over 1 h to a flask containing 150 g of water and 250 ml of diisopropylether. The mixture was then refluxed at 70 °C for 20 h, and the organic phase was isolated. The solution was washed thrice with 100 ml of water containing 25 g of sodium bicarbonate and then with water alone. The reaction product was dried using sodium sulfate, filtered, and concentrated using rotor evaporator. The product was then distilled under reduced pressure, and 63 g of the product was isolated having a GC purity of 97%.

2. *Preparation of 1,5-di-t-butyl-1,1,3,5,5-pentamethyltrisiloxane polyethylene oxide according to the method of Bailey*[1]

A round-bottom flask was charged with 60 g of toluene and 39 g of $CH_2{=}CH{-}CH_2{-}(OCH_2CH_2)_{7.2}OCH_3$ and then azeotropically distilled. The dried material was then treated with 24.4 g of the Step 1 product and two drops of chloroplatinic acid dissolved in ethyl alcohol and heated to 175 °C for 17 h. After cooling to an ambient temperature, the reaction mixture was treated with activated carbon and then filtered. The mixture was concentrated and heated to 150 °C at 1 mmHg to remove volatiles and the product was isolated.

Testing

1. *Surface Tension*

Surface tension was measured using a Kruss surface tensiometer containing a sand blasted platinum blade as the sensor. Solutions of the experimental agents were prepared at 0.1 wt % in deionized water containing 0.005 M NaCl. Test results are provided in Table 3.

2. *Spreading Properties*

Spreading properties were determined by applying a 10-μl droplet of a selected experimental agent dissolved in water to a polyacetate film and measuring the spreading diameter after 30 s at 22–25 °C with a relative humidity of between 50 and 70%. Test results are provided in Table 3.

3. *Hydrolytic stability*

Hydrolytic stability was determined using a 0.5 wt% aqueous solution of a selected experimental agent over a pH range from pH 2 to pH 11. Test results are summarized below.

Test Results

1. *Surface tension*

2. *Spreading Properties*

TABLE 3 Surface tension and spreading diameter testing results for experimental additives and four commercially available surfactants. Surface tension properties of experimental agents consistently outperformed commercial additives at similar treatment levels

		Spread Diameter (mm)			
Additive	Surface Tension (mN/m)	0.05 wt% Surfactant	0.1 wt% Experimental Surfactant	0.2 wt% Experimental Surfactant	0.4 wt% Experimental Surfactant
Additive 1	23.1	9	11	12	15
Additive 2	2.23	10	13	16	25
Additive 3	20.7	18	31	48	56
Comparative Trisiloxane-1	20.9	34	53	51	25
Comparative Trisiloxane-2	20.6	37	53	50	35
Comparative Trisiloxane-3	23.6	nd	nd	nd	6
Octylphenol ethoxylate	31.8	nd	9	nd	10

3. *Hydrolytic stability*

After 28 weeks of exposure to a pH range of 2–11, all experimental additives remained unchanged. By contrast, comparative trisiloxane surfactants showed a 54% hydrolysis after 48 h at pH 10. After 1 week these samples and references were 100% hydrolyzed at pH 4 and pH 10.

Advantages over Prior Art

A strategic limitation associated with existing trisiloxane compounds and their derivatives as surfactants is that they are resistant to hydrolysis only in the narrow pH range from 6 to 7.5. By introducing bulky alkyl groups into the trisiloxane, the current invention has broadened the hydrolytic stability range from pH 2 to 11 for what appears to be an indefinite time period.

Notes

1. Sherman[2] improved the thermal and hydrolytic stability of polydimethylsiloxanes by converting them into polydimethylsiloxane oxalamides (IV), which were then used as silicate tackifying resins.

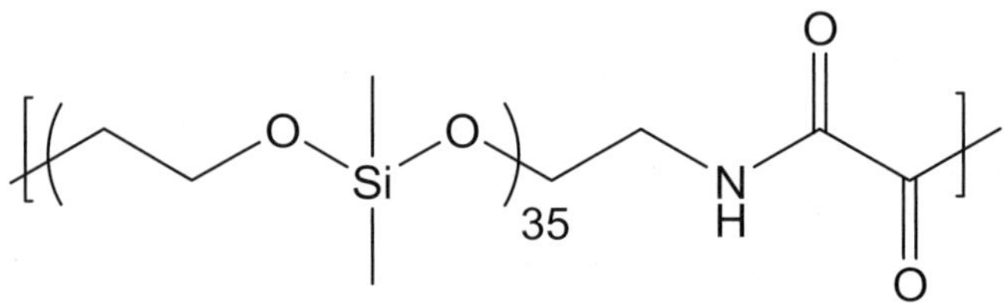

(IV)

2. Dhara[3] prepared hydrolytically stable polydimethylsiloxane carbonate (V) intermediates for use in preparing hemocompatible and protein resistant applications by condensing phosgene with bisphenol A and then postreacting the intermediate with selected polydimethylsiloxanes.

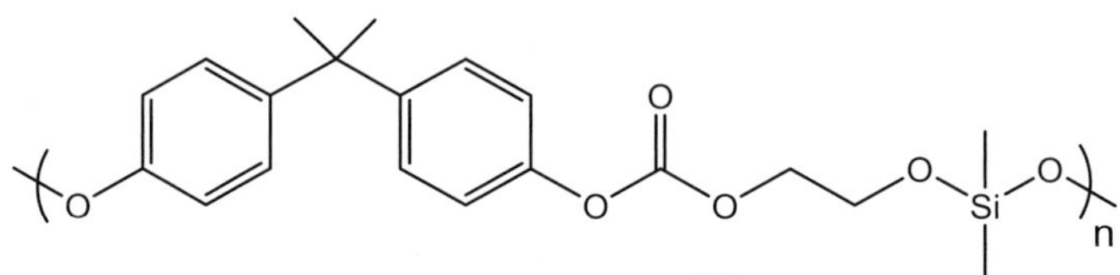

(V)

3. Knott[4] prepared thermally and hydrolytically stable poly(dimethylsiloxane-b-(ethylene oxide-co-propylene oxide) block copolymers (VI), which were used in polyurethane foam stabilizers.

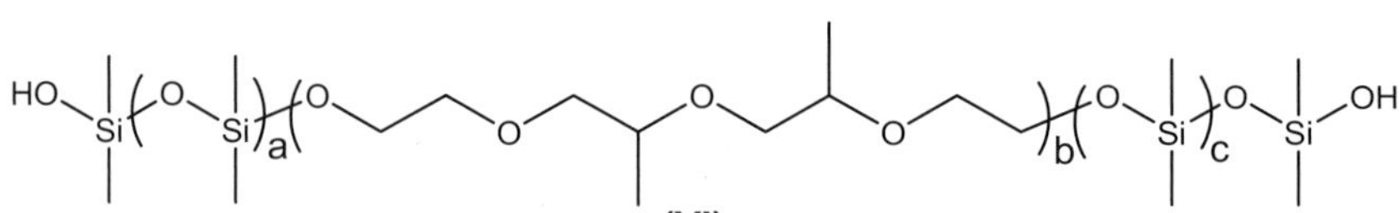

(VI)

4. Awasthi[5] incorporated polycarbosiloxane methacrylate monomers (VII) in hydrogel contact lens polymers to enhance the hydrolytic stability and restrict the surface tension value to between 70 and 72 dynes/cm.

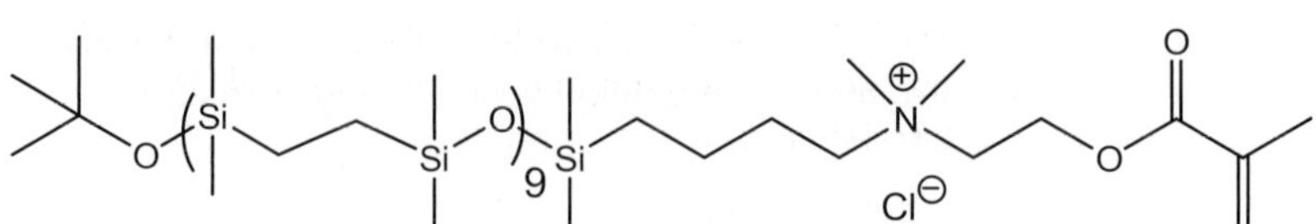

(VII)

REFERENCES

1. Donald L Bailey, US Patent 3,299,112 (January 17, 1967)
Union Carbide Corporation (New York, New York
2. Audrey A, Sherman et al, US Patent 7,915,370 (March 29, 2011)
3M Innovative Properties Company (St. Paul, MN)
3. Dibakar Dhara et al, US Patent 7,888,447 (February 15, 2011)
Sabic Innovative Plastics IP B.V. (NL)
4. Wilfried Knott et al, US Patent 7,825,209 (November 2, 2010)
Evonik Goldschmidt GmbH (DE)
5. Alok K. Awasthi et al, US Patent 7,915,323 (March 29, 2011)
Bausch & Lamb Incorporated (Rochester, NY)

D. HYDROLYSIS RESISTANT AQUEOUS SURFACTANTS

Carbodisilane surfactants

Title:	Hydrolysis resistant organomodified silylated ionic surfactants
US Patent:	7,879,916 (February 1, 2011)
Author:	Mark D. Leatherman et al.
Assignee:	Momentive Performance Materials Inc. (Albany, NY)
Significance:	Carbodisilane surfactant compositions resistant to hydrolysis in aqueous solutions having a pH range from 2 to 12 were prepared that lowered the surface tension by approximately 45% over existing materials. Spreading properties of carbodisilane surfactants in water and solutions containing ammonium chloride or sodium chloride were typically 200% greater than that of conventional surfactant products.
Additive Names	3-({3-[Dimethyl-(2-trimethylsilanyl-ethyl)-silanyl]-propyl}-dimethyl-amino-)-propane-1-sulfonate (I) [**Additive-1**] 1-Diethylamino-3-{3-[dimethyl-(2-trimethylsilanyl-ethyl)-silanyl]-propoxy}-propan-2-ol (II) [**Additive-2**] 1-{3-[Dimethyl-(2-trimethylsilanyl-ethyl)-silanyl]-propoxy}-3-[2-(2-hydroxy-ethoxy-ethoxy)-ethylamino]-propan-2-ol (III) [**Additive-3**] 1-{3-[Dimethyl-(2-trimethylsilanyl-ethyl)-silanyl]-propoxy}-3-isopropylamino-propan-2-ol (IV) [**Additive-4**]

Safety Concerns

Silyl amine derivatives are considered to be very mild eye and skin irritants. When handling these agents wear eye/face protection and suitable gloves. In case of contact with the eyes or the skin, rinse the contaminated areas with cool water.

Additive Structures

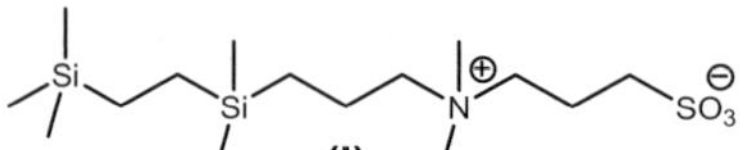

(I)

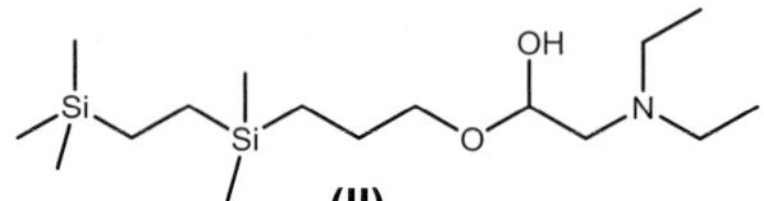

(II)

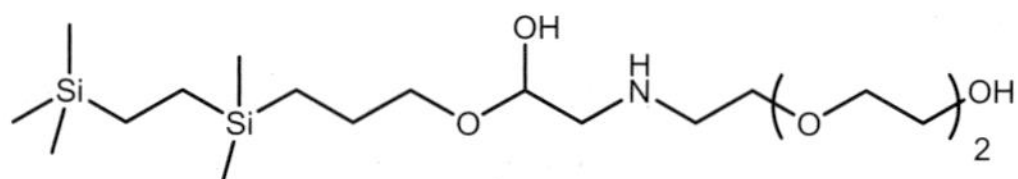

(III)

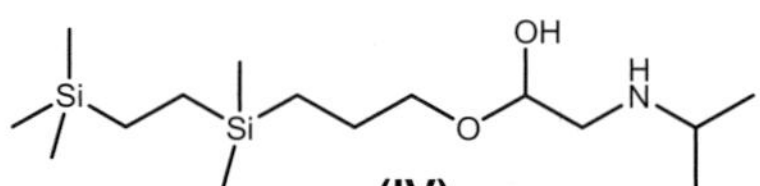

(IV)

Additive Preparation

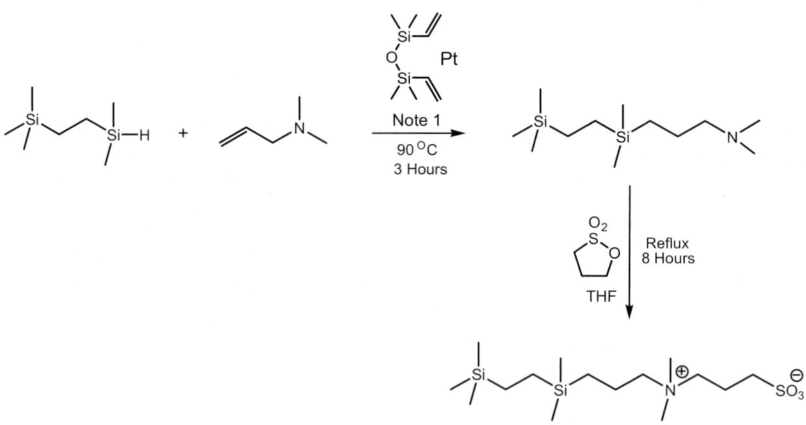

Additive Synthesis

1. Preparation of N,N-dimethyl aminopropyl pentamethyl carbodisilane

A 100-ml Schlenk flask was charged with 16.0 g of pentamethyl carbodisilane and 20 μl of platinum 1,3-divinyl-1,1,3,3-tetramethyldiloxane complex using a 0.3 wt %

solution in xylene and then heated to 90 °C and treated with the dropwise addition of 9.35 g of *N,N*-dimethyl allyl amine over 20 min. The mixture was then maintained at 90 °C for 3 h while monitoring the reaction extent by ^{1}HNMR. The mixture was then concentrated, and 17 g of the product was collected as a colorless oil having a BP = 109–111 °C at 15 mmHg.

2. *Preparation of 3-({3-[dimethyl-(2-trimethylsilanyl-ethyl)-silanyl]-propyl}-dimethyl-amino)-propane-1-sulfonate*

A reaction kettle was charged with 2.0 g of the Step 1 product and 1.10 g of 1,3-propanesultone dissolved in 15 ml dry THF. The mixture was heated to reflux overnight and then concentrated, and 3.03 g of the product was isolated as a white solid.

Testing

1. *Surface Tension*

Surface tension studies were performed by applying a 0.1 wt% 10-μl droplet of a selected experimental surfactant solution to a polystyrene surface and then measuring the spread diameter after 30 s at 50–70% relative humidity between at 22 and 25 °C. Test results are provided in Table 4.

2. *Spread Diameters*

Spread diameters were determined in deionized water containing 2 M NH_4Cl solution and 10 wt% NaCl solution additized with 0.1 wt% of a selected experimental surfactant with an automatic pipette to provide droplets of reproducible volume. Test results are provided in Table 4.

Test Results

TABLE 4 **Summary of surface tension and spreading diameters of 0.1 wt% solutions of selected hydrolysis-resistant experimental ionic surfactants**

		Spread Diameters		
Sample	Surface Tension (mN/m)	Water (mm)	2 M NH_4Cl (mm)	10 wt% NaCl (mm)
Additive-1	31.4	5	nd	17
Additive-2	22.9	4	40	nd
Additive-3	22.3	34	8	nd
Additive-4	20.7	5	37	nd
Silwet® L-77* (**Reference**)	31.8	9	nd	nd

**Trisiloxane ethoxylated surfactant containing 8.5 polyoxyethylene repeat units is available from GE Advanced Materials, Wilton, Conn.*

Advantages over Prior Art

Aqueous liquid compositions containing trisiloxane or higher homologs are extremely effective as both wetting and spreading agents. Their effective use, however, is limited to solutions having a pH range of 6–7.5, below or above which surfactant hydrolytic decomposition occurs. In addition to substantially increasing the hydrolytic stability to a pH range of 2–12, carbodisilane of the current invention is superior as a wetting and spreading agent.

Notes

1. Pentamethyl carbodisilane (V) was previously prepared by Leatherman[1] as illustrated in Eq. 1:

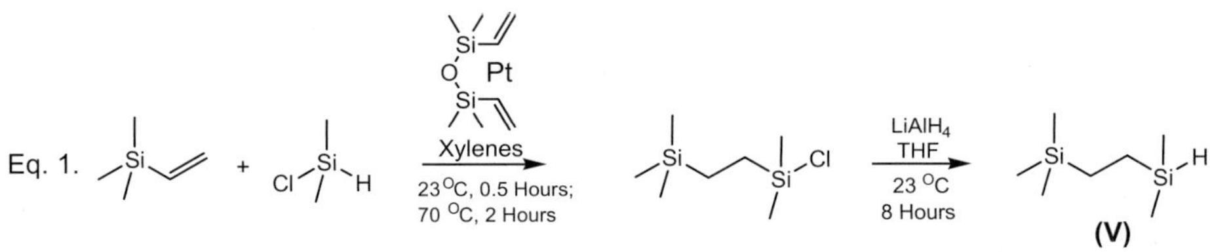

2. Hydrolysis resistant organodisilanes (V) were prepared by Wagner,[2] which were also effective as surface spreading agents.

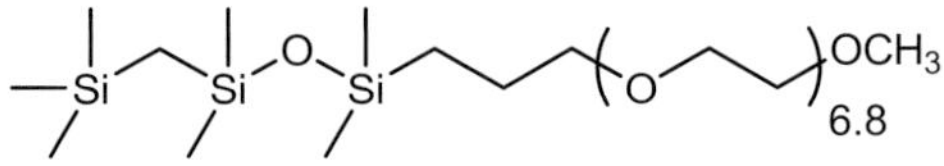

(VI)

3. Hydrolysis-resistant disiloxanes useful for extreme environment conditions were previously prepared by Leatherman[3] and are illustrated below.

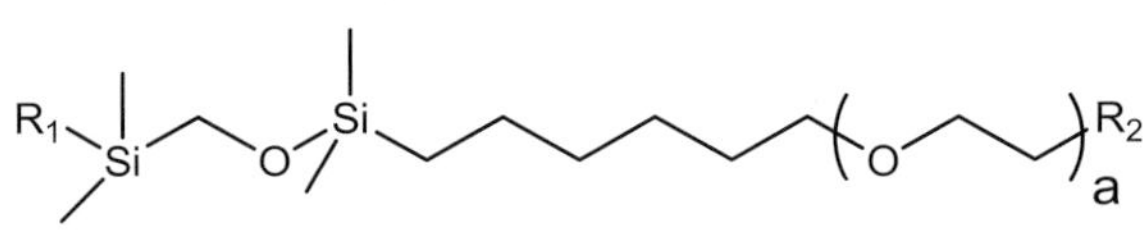

R_1	R_2	a
$(CH_3)_2CHCH_2$–	CH_3	7.5
$(CH_3)_2CHCH_2$–	CH_3	11
$(CH_3)_3C$–	H	11
$(CH_3)_2SiCH_2CH_2$–	H	7.5
$(CH_3)_2SiCH_2CH_2$–	CH_3	11
$(CH_3)_2SiCH_2CH_2$–	H	6.3

4. Super-spreading gemini silicone surfactant compositions (VII) were prepared by Wang[4] and used in agricultural and pesticide applications.

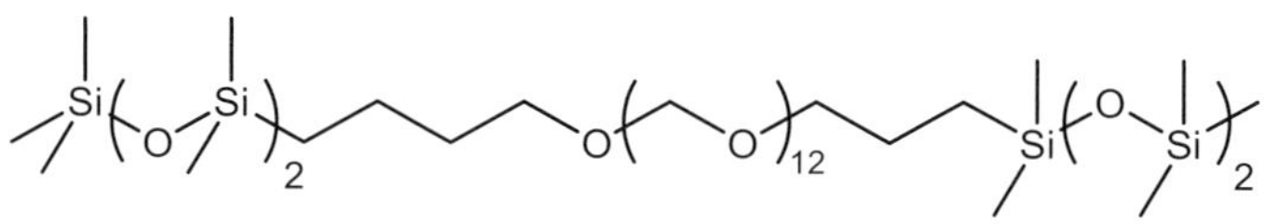

(VII)

REFERENCES

1. Mark D. Leatherman et al, US Patent 7,700,797 (April 20, 2010)
Momentive Performance Materials Inc. (Albany, NY)

2. Roland Wagner et al, US Patent 7,872,053 (January 18, 2011)
Momentive Performance Materials GmbH & Co., KG (Leverkusen, **DE**)

3. Mark D. Leatherman et al, US Patent 7,652,072 (January 26, 2010) and Mark D. Leatherman et al, US Patent 7,645,720 (January 12, 2010)
Momentive Performance Materials Inc. (Albany, NY)

4. Dong Wang et al, US Patent 7,601,680 (October 13, 2009)
Momentive Performance Materials Inc. (Albany, NY)

Chapter 30

Textile Additives

A. ANTIMICROBIAL FIBERS

Silver Polyacrylate Terpolymers

Title:	Method for fabric treatment at low pH
US Patent:	7,927,379 (April 19, 2011)
Author:	Stephanie Nussbaum Cottrell et al.
Assignee:	Rohm and Haas Company (Philadelphia, PA)
Significance:	The antimicrobial agent, poly(vinyl imidazole-ter-ammonium acrylate-ter-silver acrylate), was prepared and then converted into a polyester fabric by condensing with trimethylolpropane triglycidyl ether. Once fashioned into a fabric, the material was effective against *Staphylococcus aureus* and *Klebsiella pneumonia*.
Additive	Poly(vinyl imidazole-ter-ammonium acrylate-ter-silver acrylate) (I)
Safety	Silver-containing textiles are considered safe and are used in under garments in the treatment of atopic dermatitis.

Additive Structure

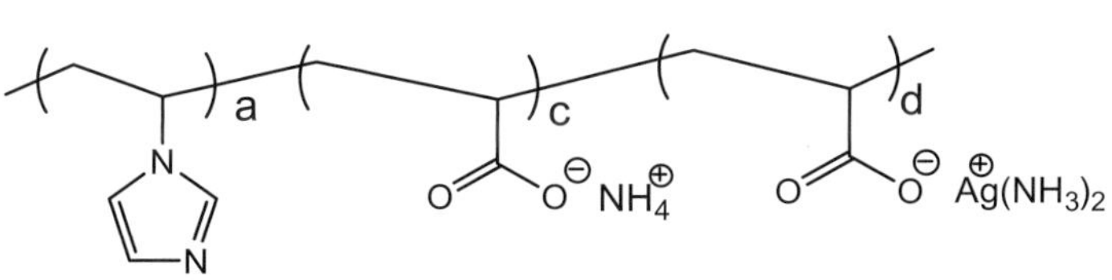

(I)

Next Generation of International Chemical Additives. DOI: http://dx.doi.org/10.1016/B978-0-444-53788-1.00030-9

Additive Preparation

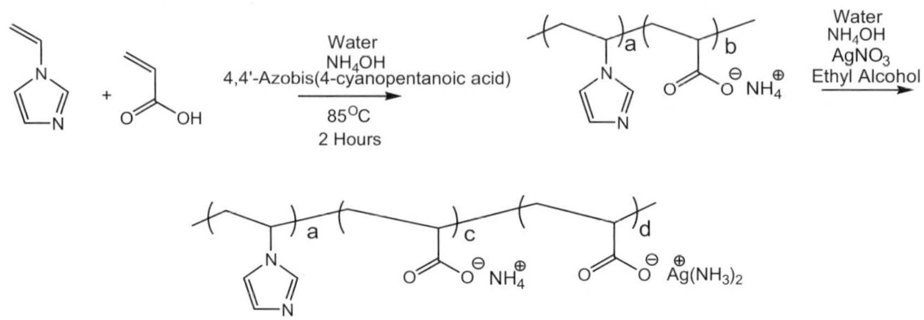

Additive Synthesis

1. Preparation of poly(vinyl imidazole-co-ammonium acrylate)

After heating a 1-liter reaction vessel containing 300 ml of deionized water under nitrogen to 85 °C, 2.1 g of 29% ammonium hydroxide and 4.62 g of 4,4′-azobis(4-cyanopentanoic acid) were treated followed by an additional 207.7 g of deionized water, 138.5 g of vinyl imidazole, and 15.4 g of acrylic acid. The addition container was then rinsed with 144 g of deionized water and the contents were added to the reaction vessel over 30 min. The reaction mixture was heated to 85 °C for 15 min and then further treated with a second addition of 34.5 g of deionized water, 0.48 g of 29% ammonium hydroxide, and 0.77 g of 4,4′-azobis(4-cyanopentanoic acid) at 0.34 g/min over 90 min while continuing to maintain the reaction temperature at 85 °C for an additional 30 min. Thereafter, the mixture was cooled to an ambient temperature, and a clear pale yellow solution containing a 20.5% solid solution having a pH of 7.6 with a viscosity of 124 cPs was isolated.

2. Preparation of poly(vinyl imidazole-ter-ammonium acrylate-ter-silver acrylate) according to the method of Ghosh[1]

A beaker was charged with 3 g of the Step 1 product dispersed in 17 g of deionized water and then treated 17 g of 95 wt% of ethanol and stirred. The mixture was then further treated with an aqueous solution of 0.44 g of silver nitrate dissolved in 5 g of deionized water and then stirred, and a white precipitate was formed. Finally, the reaction mixture was treated with 4.4 g of a 5 wt % solution of ammonium hydroxide, and a clear light yellow colored solution containing 0.53 wt % silver was isolated.

Testing

1. Fabric formulation

A solution was prepared consisting of 2.5 g of the Step 2 product dissolved in 346 g of water containing trimethylolpropane triglycidyl ether, and the pH was adjusted to 5.3 using 0.30 g of 2-amino-2-methyl-1-propanol. The coating material was then passed through the bath solution and then passed through two nip rolls to remove excess solution. The dried substrate was then converted into 100% polyester knit fabric with a typical basis weight of 3.5 oz/yd^2 and used for antimicrobial testing.

2. *Antimicrobial efficacy of silver-containing fabric*

The bacteriostatic activity of the treated polyester fabric was measured using AATCC Method 100. In this method, the bacteriostatic activity is evaluated by placing 1.0 ml of 1×10^5 organisms of a selected culture in direct contact with the experimental fabric for a 24-hour incubation period at 35 °C diluted with Dey Enley broth. Thereafter, the number of surviving organisms was determined by a plate count, and the percent reduction was calculated. Test results are provided in Table 1.

Test Results

TABLE 1 Antimicrobial efficacy of silver containing terpolyester fabrics derived from the Step 2 product against *S. aureus* and *K. pneumoniae*

Experimental Agent	Trimethylol-propane Triglycidyl Ether (g/g Additive)	Silver Content (ppm)	Log Reduction *S. aureus* (ATCC 6538)	Log Reduction *K. pneumoniae* (ATCC 4352)
Reference	0.17	0	−3.3	−3.8
5	0.17	309	2.3	3.9
6	0.35	309	>3.8	>3.9
7	1.10	309	>3.8	>3.9

Advantages over Prior Art

The antimicrobial efficacy of the silver-containing fabric of the current invention is a result of the silver biocide remaining after repeated laundering. While silver polymer complexes in combination with a variety of epoxides are extensively reported, after one or more laundering, the silver precipitates away from the polymer, and the fabric becomes biologically inactive. The current investigation has addressed this serious flaw.

Notes

1. Nussbaum Cottrell[2] prepared antimicrobial surface finishes consisting of poly(butyl acrylate-ter-1-vinylpyrrolidone-ter-ammonium acrylate-ter-silver acrylate) (II) crosslinked with trimethylolpropane and triacylate.

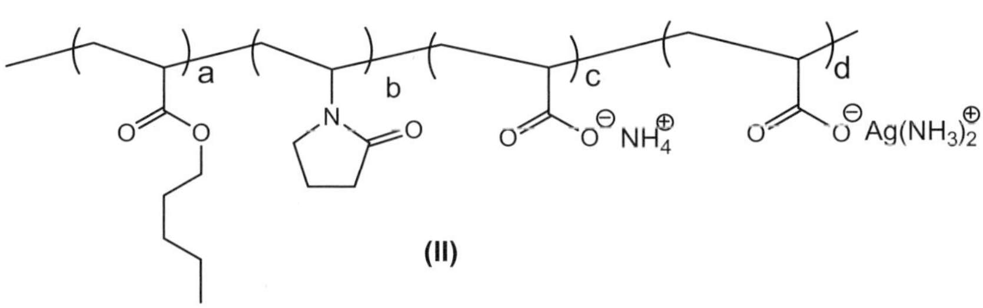

(II)

2. Patel[3] prepared a wound dressing composite consisting of the antimicrobial fabric polyhexamethylene biguanide (III) containing a coating of polymannuronic acid (IV) zinc.

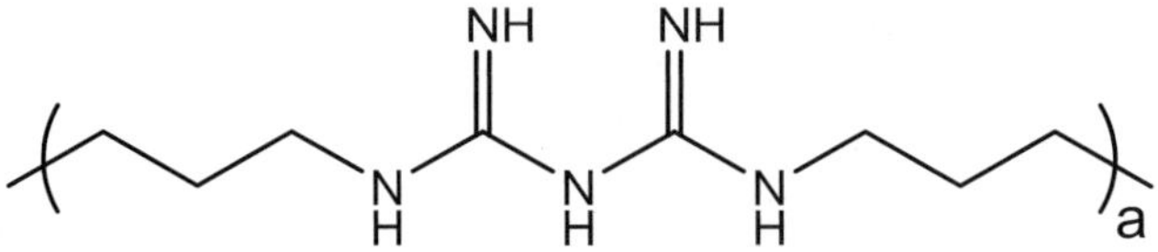

(III)

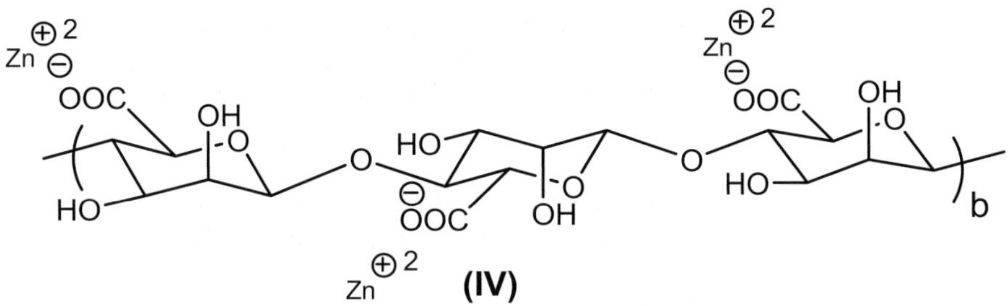

(IV)

3. Antimicrobial textiles were prepared by Rega[4] consisting of the reaction product of dimethylsiloxane diol having an Mn of approximately 1500 Da terminated with hexamethylene diisocyanate and then postreacted with polyl(trimethylene carbonate) diol and blended with either a zinc-containing silica glass powder or a silver/zirconium phosphate.
4. Antimicrobial films were prepared by Robinson[5] by coextruding polyethylene terephthalate with silver, copper, or zinc phosphate.

REFERENCES

1. Tirthankar Ghosh et al, US Patent Application 20050227895 (October 13, 2005) **Rohm and Haas Company (Philadelphia, PA)**
2. Stephanie Nussbaum Cottrell et al, US Patent 7,335,613 (February 26, 2008) **Rohm and Haas Company (Philadelphia, PA)**
3. Harish A. Patel et al, US Patent 7,799,965 (September 21, 2010) **Tyco Healthcare Group LP (Mansfield, MA)**
4. Joseph Rega, US Patent Application 20110086940 (April 14, 2011) **Joseph Rega (Wilmington, MA)**
5. Julian Neal Robinson et al, US Patent Application 20110081530 (April 7, 2011) **DuPont Teijin Films US Limited Partnership**

B. TEXTILE SHADING COMPOSITIONS

Azo Dye Conjugates

Title:	Shading compositions
US Patent:	7,909,890 (March 22, 2011)
Author:	Hans Kramer et al.
Assignee:	The Procter & Gamble Company (Cincinnati, OH)

Significance: A nonbleaching method designed to improve fabric whitening is described. In this method, in a molecular compound, the effective photocatalyst, phthalocyanine, and a monoazo dye conjugate were used to additize a laundering detergent. After repeated laundering, clothing appeared whitened with little evidence of violet–blue fabric discoloration.

Additives

Phthalocyanine mono-azo naphthyl dyestuff-1 (I)
Phthalocyanine mono-azo naphthyl triazine dyestuff-2 (II)
Phthalocyanine di-(azo naphthyl) dyestuff-1 (III)

Safety

The products are not classified as dangerous according to EC directives and national regulations. In the event of acute exposure, wash off the area with plenty of soap and water.

Additive Structures

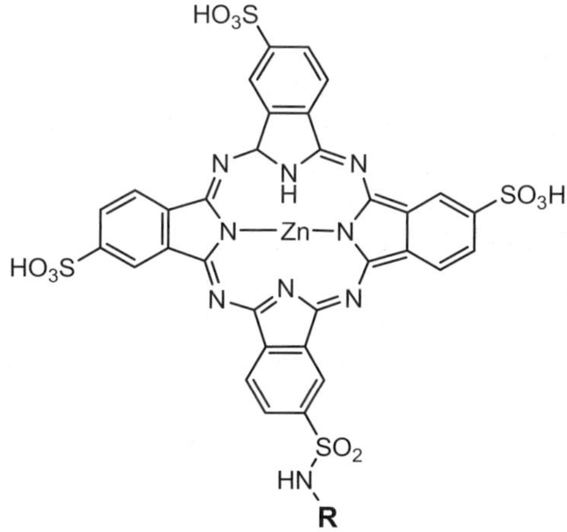

where R is

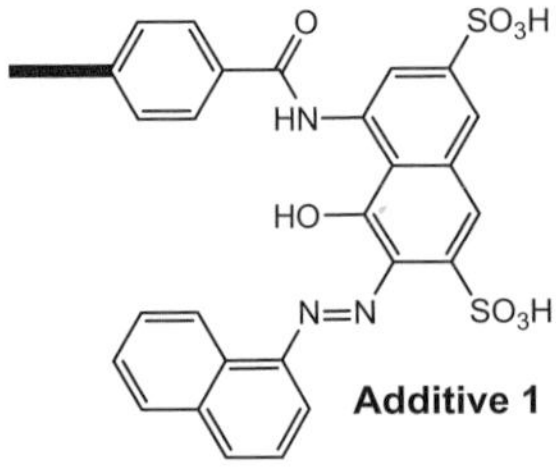

Additive 1

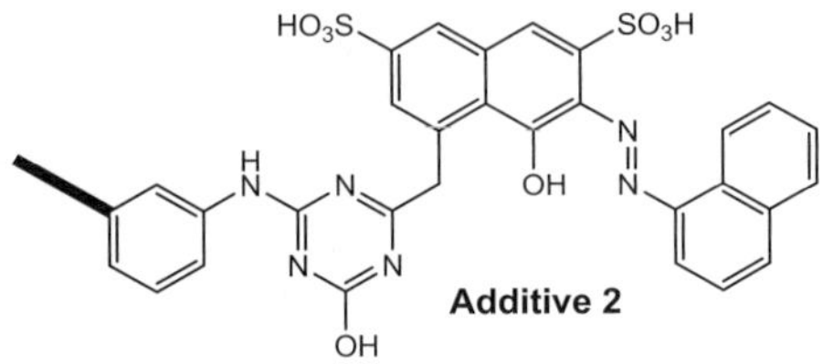

Additive 2

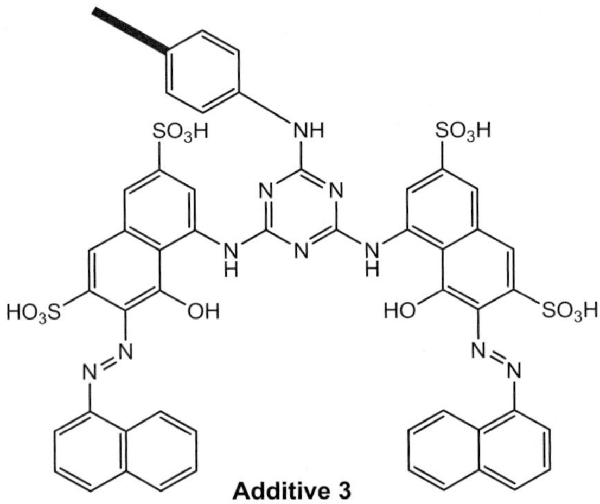

Additive 3

Additive Preparation

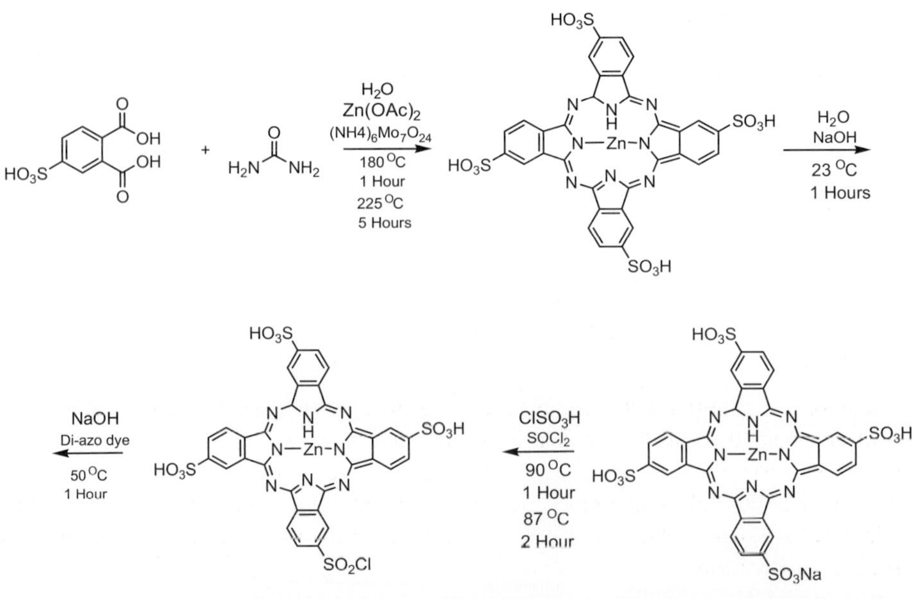

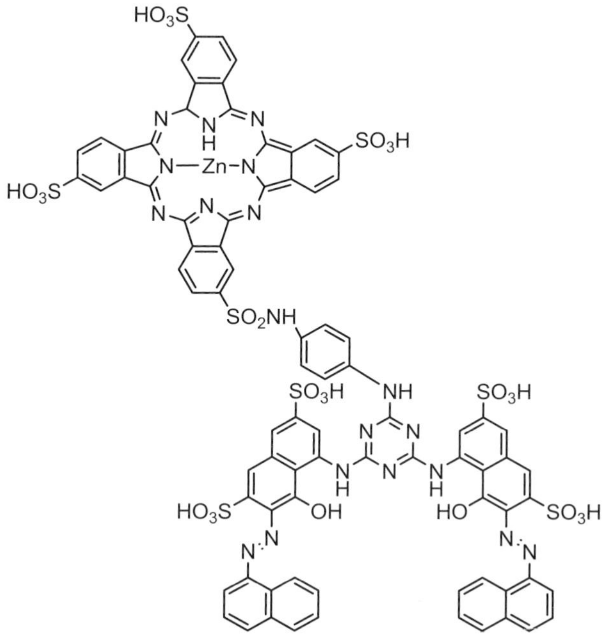

Additive Synthesis

1. Preparation of zinc phthalocyanine 4-tetrasulfonic acid

A reactor was charged with 15 g of a 50 wt% solution of 4-sulfophthalic acid, 2 g of zinc acetate, 0.2 g of ammonium molybdate, and 5 ml of distilled water and then stirred together until all the solids were dissolved. Thereafter, 20 g of urea was added to this solution, and the mixture was stirred until most of the urea had dissolved, and then the temperature was raised to 180 °C for 1 h. The reaction temperature was then increased to 225 °C for 5 h, cooled to an ambient temperature, and 11–12 g of crude product was isolated. The crude sample was purified by methanol treatment and dialysis, and 2 g of the product was isolated having λ_{max} = 669 nm and ε = 120,000–139,000.

2. Preparation of zinc phthalocyanine 4-tetrasulfonate sodium

The Step 1 product was suspended in water and treated with aqueous sodium hydroxide and then concentrated, and the salt was isolated in quantitative yield.

3. Preparation of zinc phthalocyanine monosulfonyl chloride derivative

A reactor was charged with 2.0 g of the dried Step 2 product and 15 ml of chlorosulfuric acid and heated to 90 °C for 90 min while adding 2.6 ml of thionyl chloride dropwise. The mixture was then heated to 87 °C and stirred for 2 h. After 2 h, the reaction mixture was cooled to an ambient temperature and poured onto ice and a precipitate which formed was isolated by filtration and then washed with water and used without further purification.

4. Preparation of zinc phthalocyanine triazine dye conjugate

A wet cake of the crude moist Step 3 product was dissolved in ice-cold water and an aqueous solution of 0.0195 mol of the triazine dye was added within 5 min. The reaction mixture pH was adjusted to 7 using 32% sodium hydroxide and the mixture was heated to 50 °C for 2 h while stirring. The solution was then cooled to an ambient temperature and stirred for an additional 8 h while maintaining the reaction mixture pH of 7 using 32% sodium hydroxide. During this period, the reaction extent was monitored by TLC using a mixture consisting of 25 ml of *N*,*N*-diethylamine, 33 ml of 3-methylbutanol, and 17 ml of 25% ammonium hydroxide. The reaction was determined to be completed when no unbound dye was detectable. The reaction mixture was then purified by dialysis to remove sodium chloride, and the product was isolated as a blue/violet solid.

Testing

Laundering Experiments

A bleached cotton fabric was washed with a detergent mixture containing 0.04 wt% of a selected zinc phthalocyanine dye conjugate for 15 min at 30 °C. The fabric was then exposed to a tungsten lamp for up to 120 min, and the absorption at 670 nm was measured. Test results are provided in Table 2.

Test Results

TABLE 2 Shading effects on white fabrics laundered with detergents additized with experimental phthalocyanine dyestuffs

Conjugate Dye	Amount of Dye Remaining Adsorbed after Washing (%)	Textile Color
Zinc phthalocyanine 4-tetrasulfonic acid (**Reference**)	44	White
Azo Dye Conjugate-(I)	69	Enhanced whiteness
Azo Dye Conjugate-(II)	49	Enhanced whiteness
Di-azo Dye Conjugate-(III)	54	Very Light blue/violet

Advantages over Prior Art

The current invention addresses the problem of conjugate dye accumulation, which causes fabric discoloration by using the highly activated phthalocyanine photocatalyst in detergents containing shading composition additives. In addition, despite repeated fabric laundering, the degree of whiteness remained improved and enhanced.

Notes

1. Nakahama[1] dispersed 15- to 45-nm-sized pigments in water for use in ink jet printing applications using copper (II) phthalocyanine 4-tetrasulfonic acid.
2. Hohener[2] prepared bisazo phthalocyanine dyestuff conjugate derivatives (IV), which were used as additives in detergent compositions to give fabrics a fade-resistant violet color.

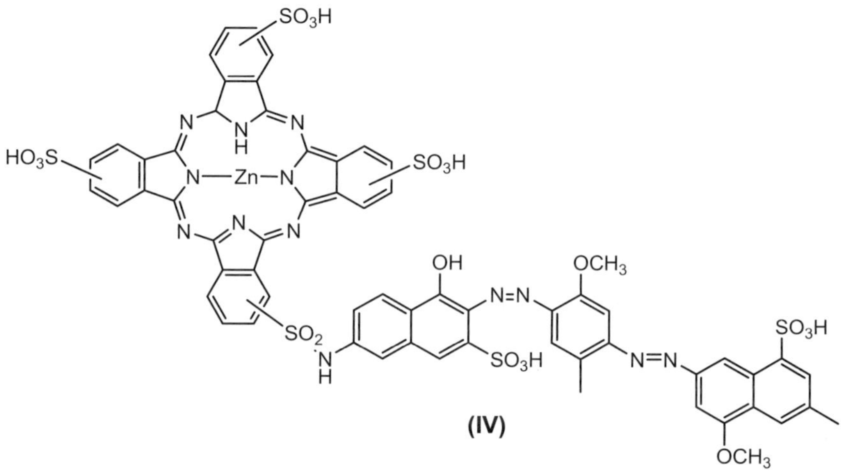

(IV)

3. Han[3] prepared benzodifuraone-based fabric dyes (V), which had good fixing rates and excellent fastness properties by condensing α-hydroxy benzeneacetic acid derivative with selected benzaldehyde derivatives.

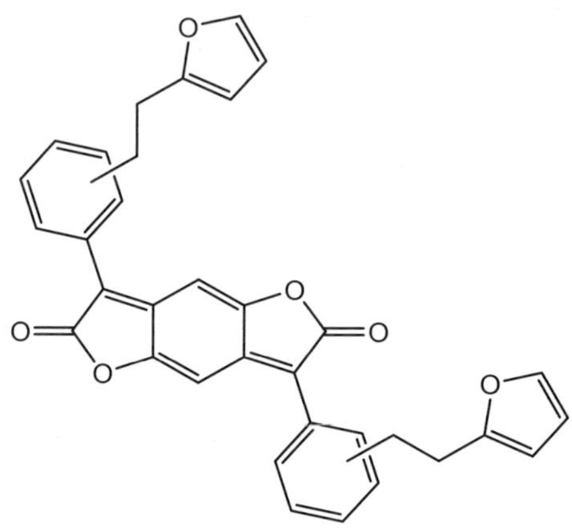

(V)

4. Savio[4] additized detergent compositions with negatively charged pigments having a diameter of 700–1500 μm for use in fabrics containing Monastral Violet (Violet 19), Pigment Scarlet (Red 60), Blue cobaltous aluminate, or Ultra Marine Blue pigment as a method for lightly dying fabrics.

5. Patel[5] prepared phthalocyanine derivatives with selected triazine diamines (VI), and (VII) for use in ink-jet printer compositions, which resulted in a significantly improved image fastness.

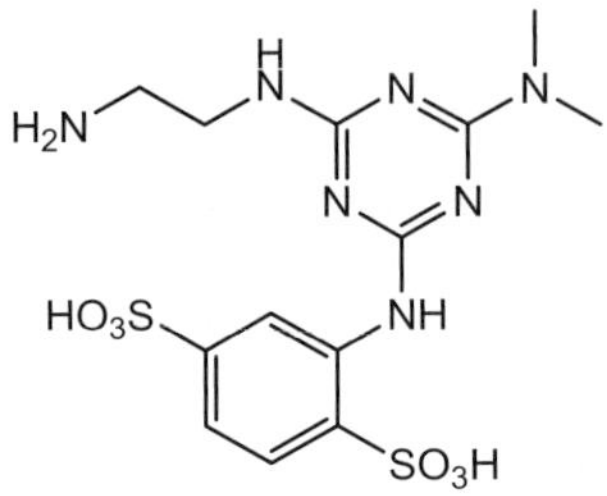

(VI)

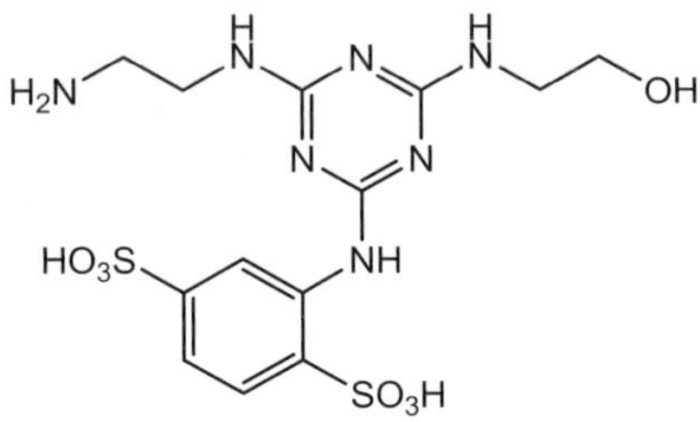

(VII)

REFERENCES

1. Kazumichi Nakahama et al, US Patent 7,914,616 (March 29, 2011)
   Canon Kabushiki Kaisha (Tokyo, JP)
2. Alfred Hohener et al, US Patent 7,799,098 (September 21, 2010)
   Ciba Specialty Chemicals Corp. (Tarrytown, NY)
3. Man-joon Han et al US Patent 7,737,286 (June 15, 2010)
   Kyung-In Synthetic Corporation (KR)
4. Larry Savio Cardozo et al, US Patent 7,645,729 (January 12, 2010)
   The Procter & Gamble Company (Cincinnati, OH)
5. Prakash Patel, US Patent 7,922,799 (April 12, 2011)
   Fujifilm Imaging Colorants Limited (Manchester, GB)

C. THERMALLY STABLE FIBER COATINGS

Elastomeric Poly(carborane-siloxane-acetelyene)s

Title:	Synthesis of elastomeric poly(carborane-siloxane-acetelyene)s
US Patent:	7,923,523 (April 12, 2011)
Author:	Teddy M Keller et al.
Assignee:	The United States of America as represented by the Secretary of the Navy (Washington, DC)
Significance:	Liquid oligo(dimethylsiloxane-ter-acetylene-ter-*m*-carborane) was used to prepare a flexible and oxidative protective ceramic coating on Zylon® [poly(*p*-phenylene-2,6-benzobisoxazole)] fibers. The coating was initially heated to 400 °C to cure and then cermatized by pyrolyzing at 1000 °C.

Additive Poly(*m*-carborane-co-1,4-dimethyldisiloxane-1,4-butadiyne) (I)
Poly(*m*-carborane-co-1,4-dimethyltrisiloxane-1,4-butadiyne) (II)

Safety The toxicological properties of this substance have not been fully investigated and are largely unknown. Contact with residual *m*-carborane or dimethyldisiloxane may result in gastrointestinal distress including nausea, vomiting, and diarrhea if swallowed. Oral, inhalation, and dermal exposure to residual hexachlorobutadiene may result in fatty liver degeneration, epithelial necrotizing nephritis, central nervous system depression, and cyanosis.

Additive Structures

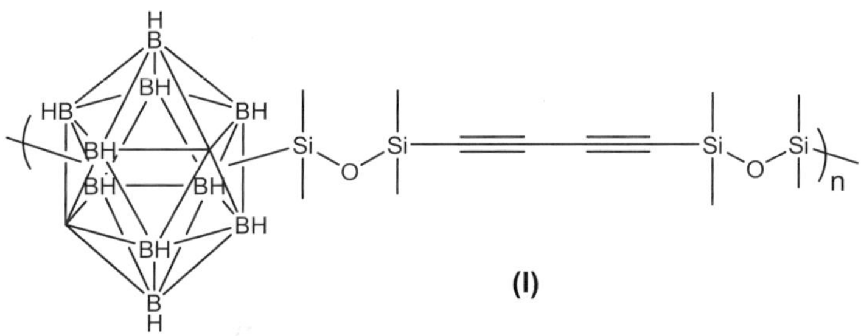

(I)

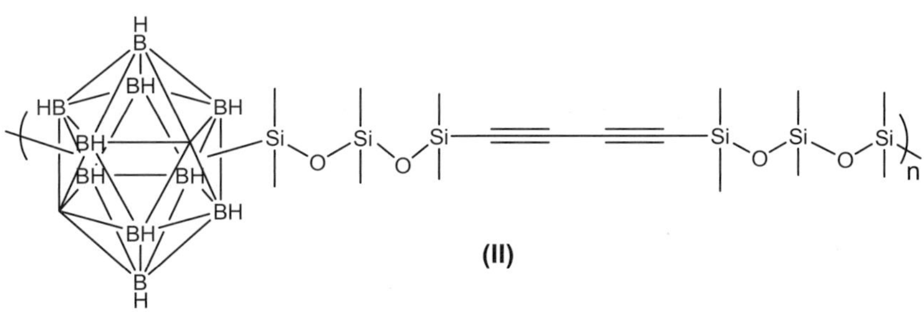

(II)

Additive Preparation

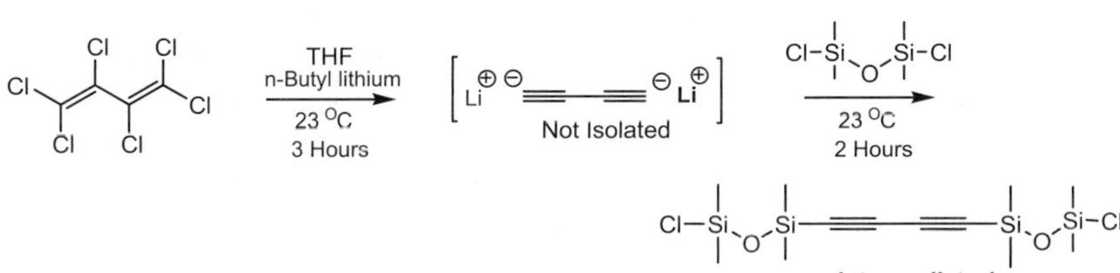

Intermediate-1

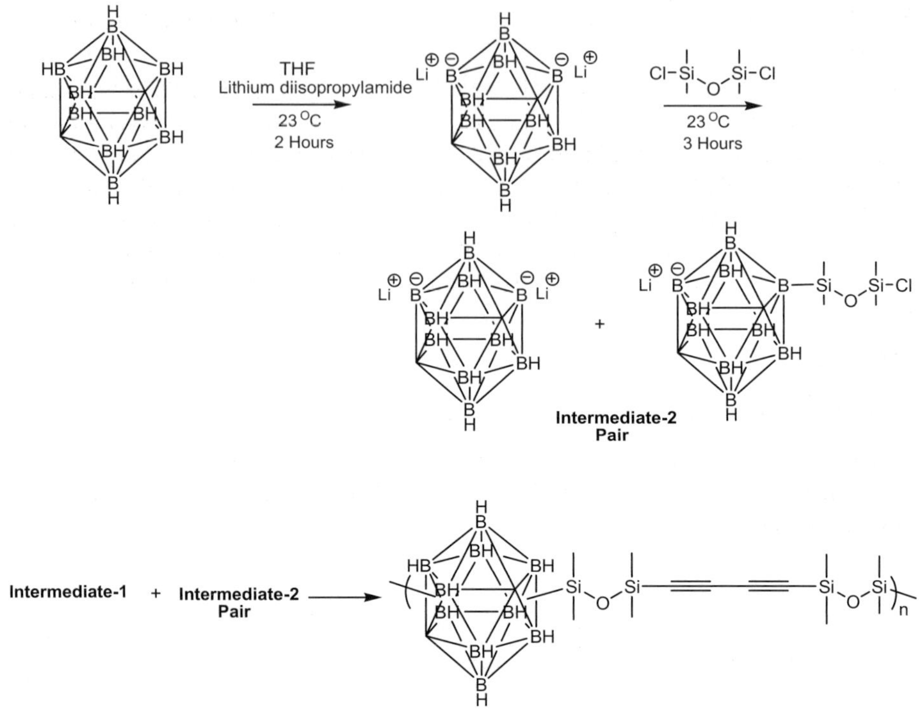

Additive Synthesis

1. *Preparation of 1,4-(chlorodimethyldisiloxane)-1,4-butadiyne*

A flame-dried 250-ml flask was charged with 10 ml of THF and then cooled to −78 °C and treated with 2.77 ml of a 2.5 M solution of *n*-butyl lithium in hexanes. The solution was then treated with the dropwise addition of 27 ml of hexachlorobutadiene over 10 min and then slowly warmed to an ambient temperature and stirred for 3 h. Thereafter, the solution was recooled to −78 °C and treated with the dropwise addition of 0.68 ml of dichlorotetramethyldisiloxane. The solution was then slowly warmed to an ambient temperature and stirred for 2 h and then filtered to remove precipitated lithium chloride and the product was isolated.

2. *Preparation of lithiated* m*-carborane-1,4-dimethyldisiloxane*

A 250-ml flame-dried flask was treated with 0.5 g of *m*-carborane dissolved in 1 ml of THF and then cooled to −78 °C and treated with the dropwise addition of 3.5 ml of lithium diisopropylamide resulting in the formation of a reddish orange solid. Once the mixture was warmed to an ambient temperature, the solid redissolved in THF and the solution concentrated and the residue was redissolved in 2 ml of THF and then cooled to 0 °C. The mixture was then treated with 0.34 ml of dichlorotetramethyldisiloxane using a gas-tight syringe and then stirred for 3 h, and the product mixture was isolated and used without purification.

3. *Preparation of poly(m-carborane-co-1,4-dimethyldisiloxane-1,4-butadiyne)*

The Step 1 product was cannulated into the flask containing the Step 2 product, and the mixture was stirred for 2 h and then treated with a few drops of dichlorotetramethyldisiloxane and the reaction mixture stirred for an additional hour. The solution was then poured into 50 ml of cold saturated ammonium chloride and extracted thrice with 75 ml of diethylether. The combined extracts were washed with 100 ml of brine and then dried over anhydrous sodium sulfate and filtered through a 1-inch pad of Celite. After the solvent was removed, the product was isolated under vacuum as a dark brown viscous oil.

Testing

Coating Procedure

A coating solution was prepared by dissolving 0.5 g of the Step 3 product in 5 ml of methylene chloride. A tow of Zylon®, poly(*p*-phenylene-2,6-benzobisoxazole) (III) fibers was then cut into

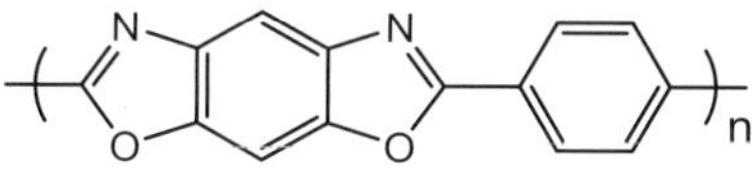

(III)

Zylon®

0.5" pieces each weighing approximately 10 mg a piece and then dried at approximately 400 °C for 4 h in a TGA apparatus under nitrogen and stored. The tow was then immersed into the coating solution for several minutes, and excess solution was drained off. The coated tow was then placed over a hot air blow dryer for 40 s to remove residual solvent and then reweighed to obtain the weight of the coating. The coated fiber was then cured at 250 °C for 30 min and at 400 °C for 2 h under nitrogen and then reweighed to determine the coating weight. After three coatings had been applied to the tow, the fiber was pyrolyzed at 1000 °C in air. The properties of coated Zylon® are discussed below.

Test Results

Weight loss during curing the initial reaction product at 250 °C and at 400 °C for was found to be negligible. After pyrolysis, the coating material remained intact on Zylon®.

Advantages over Prior Art

High acetylene content analogs of oligio(dimethylsiloxane-ter-acetylene-ter-*m*-carborane) have previously been prepared, which formed thermosets on curing. As a result, their utility as protective coating agents was limited to materials where a fixed or rigid object was needed. This investigation has demonstrated that by curing and pyrolyzing oligio(dimethylsiloxane-ter-acetylene-ter-*m*-carborane) containing ≤15 mol% of acetylene content, a flexible and protective surface coating can be obtained. The modified coating agent can now be used to coat surfaces where flexibility is required, including fibers and wires.

Notes

1. Laskoski[1] prepared ceramic–carbonaceous solids for coating fibers by pyrolyzing copolymers consisting of oligomeric divinyl aromatic ethers (IV) and divinyl carborane siloxane (V) crosslinked with tetrakis(trimethylsilyl)-silane.

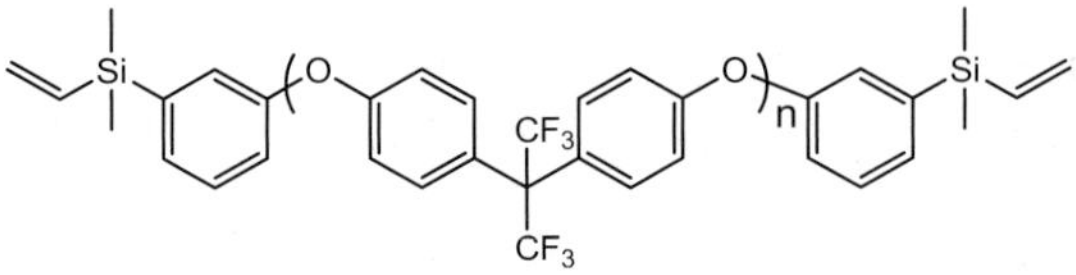

(IV)

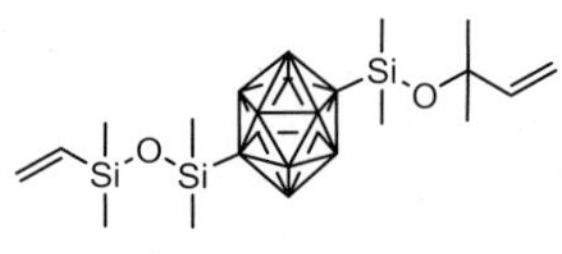

(V)

2. Keller[2] produced high-temperature coating materials by the pyrolyzing poly(dimethyl-siloxane-carborane-acetylene-ferrocene) (VI) at 1500 °C.

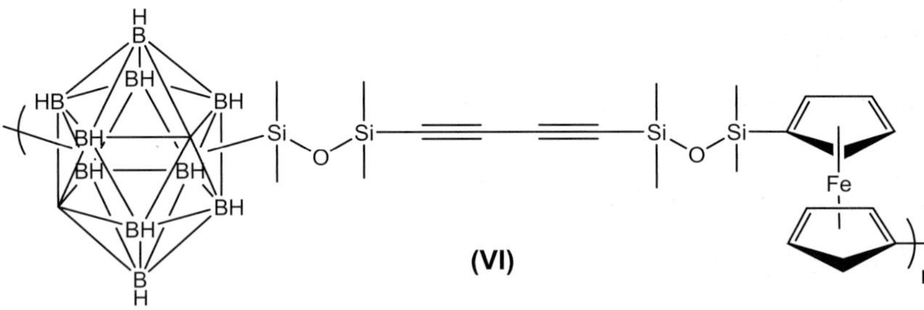

(VI)

3. Keller[3] prepared flexible thermosets and ceramics by pyrolyzing crosslinked poly(dimethyl-siloxane-co-carborane) (VII), which were used to coat fibers.

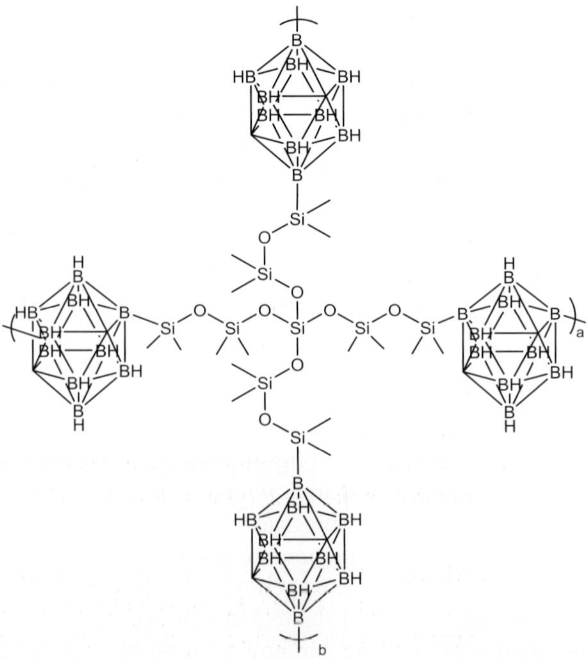

(VII)

4. Nilsson[4] increasing the strength of porous ceramic fibers by coating them with boron anhydride and then pyrolyzing the dried coating in air at 1400 °C for 2 h.

REFERENCES

1. Matthew Laskoski et al, US Patent 7,897,715 (March 1, 2011)
The United States of America as represented by the Secretary of the Navy (Washington, DC)
2. Teddy M. Keller et al, US Patent 7,579,430 (April 22, 2009)
The United States of America as represented by the Secretary of the Navy (Washington, DC)
3. Teddy M. Keller et al US Patent 7,723,454 (March 25, 2010)
The United States of America as represented by the Secretary of the Navy (Washington, DC)
4. Robert T. Nilsson et al, US Patent 7,381,681 (June 3, 2008)
Dow Global Technologies Inc. (Midland, MI)

D. WEATHERABLE FABRICS

Tulipalin Copolymers

Title:	Tulipalin copolymers
US Patent:	7,932,336 (April 26, 2011)
Author:	James Pickett et al.
Assignee:	SABIC Innovative Plastics IP, B.V. (Bergen op Zoom, NL)
Significance:	Coating additives poly(styrene-ter-methylmethacrylate-ter-α-methylene-γ-butyrolactone and poly(styrene-co-methylmethacrylate-co-α-methylene-γ-butyrolactone-co-acrylonitrile) were prepared that improved the weatherability and solvent resistance when applied to fabrics and surfaces. Both additives were quantitatively prepared by free-radical polymerization methods and contained from 30 to 47 wt% α-methylene-γ-butyrolactone (tulipalin).
Additive	Poly(styrene-ter-methylmethacrylate-ter-α-methylene-γ-butyrolactone) (I) Poly(styrene-co-methylmethacrylate-co-α-methylene-γ-butyrolactone-co-acrylonitrile) (II)
Safety	Although these materials are not hazardous under the criteria of the OSHA Hazard Communication, gases and fumes produced during the thermal processing of this material may irritate the eyes, skin, or the respiratory tract.

Additive Structures

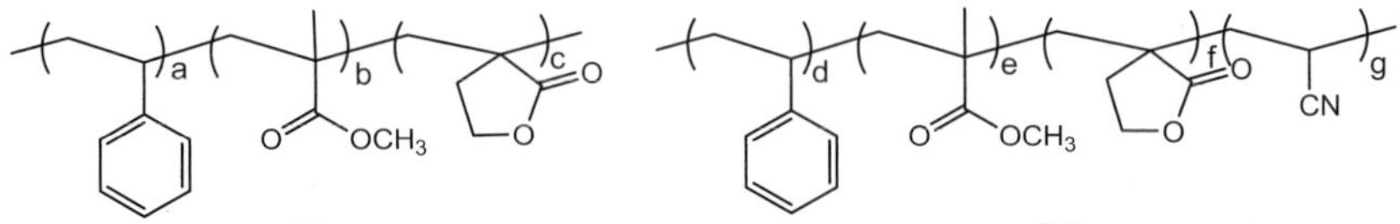

Additive Preparation

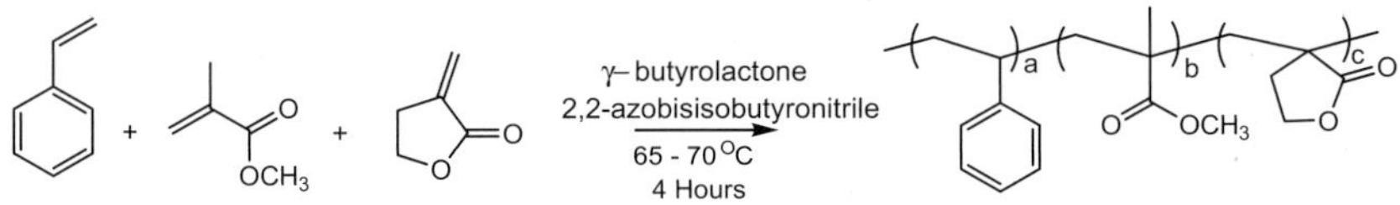

Additive Synthesis

1. *Preparation of poly(styrene-ter-methylmethacrylate-ter-α-methylene-γ-butyrolactone)*

A vacuum flask was charged with the required amount of styrene, α-methylene-γ-butyrolactone, methyl methacrylate, and 2,2-azobisisobutyronitrile as the free-radical initiator. The total concentration of reagents was kept at approximately 30 wt% solids to avoid a reaction exotherm. The flask was then degassed twice through a freeze–thaw cycle with a nitrogen purge when thawing and then heated in an oil bath to between 65 and 70 °C for 4 h. The reaction mixture was then cooled to an ambient temperature, and the mixture was precipitated by pouring into methyl alcohol, and the polymer was isolated by filtration. The polymer was dried by heating in a low-temperature vacuum oven overnight, and the molecular weight was determined using a Series 200 GPC. A summary of reaction stoichiometries and corresponding physical properties of selected terpolymers is provided in Table 3.

TABLE 3 Physical properties of selected polymers containing α-methylene-γ-butyrolactone (tulipalin), styrene, methylmethacrylate, and acrylonitrile

Reaction	Styrene (wt%)	Acrylonitrile (wt%)	MMA (wt%)	Tulipalin (wt%)	Mw (K)	PDI	Tg (°C)
22	49.64	9.77	9.76	30.82	98.70	2.88	131.20
29	37.51	0	31.69	30.80	176.65	1.9	128
33	34.40	7.92	29.45	28.22	227.42	1.96	126
34	48.35	4.38	0.72	46.55	219.66	2.22	134
35	34.48	9.96	18.92	36.63	207.77	1.65	133
36	25.94	4.96	43.65	25.45	170.52	1.87	126

Testing

1. *Weatherability*

Polymer powder samples were compression molded into films of approximately 100 μm thick at a temperature 160 °C with a pressure of 4000 psi in a Carver press. The films were then mounted on an aluminum frame and exposed to an Atlas Ci4000 xenon arc Weatherometer. The xenon arc lamp had a CIRA inner filter and a soda lime glass outer filter to simulate sunlight. The samples were continuously irradiated at an irradiance of 0.75 W/m^2/nm at 340 nm having an air temperature of 35 °C with a relative humidity of 30%. The samples were also subjected to 30 min of water spray once a week. Color measurements were made using a GretagMacbeth ColorEye 7000A spectrometer in the transmission mode according to ASTM D-1925. Test results are provided in Table 4.

2. *Solvent Resistance*

The experimental films were tested for resistance to strong bases and organic solvents using polystyrene-co-acrylonitrile (SAN) and poly(methylmethacrylate-styrene-co-acrylonitrile) (MMASAN) as references. Experimental polymer films were prepared according to the Weatherability procedure and then placed in a crystallization dish and 25 μl of a selected chemical solvent pipetted onto the film surface. The sample was then placed in an oven at 65 °C for 1 h, and the sample was evaluated. Test results are provided in Table 5.

Test Results

1. *Weatherability*

TABLE 4 Weatherability testing of selected experimental agents after continuous irradiation by a Atlas Ci4000 xenon arc. Lower numbers are preferred

Reaction Numbers of Experimental Sample	Delta YI after 1500 h Exposure to Xenon Arc	Delta YI after 2250 h Exposure to Xenon Arc
22	0	0
29	0	0
33	0	0
34	0.1	0.1
35	0.1	0.1
36	0.1	0.1
Poly(methylmethacrylate-styrene-co-acrylonitrile) (MMASAN)	2.4	3.5
Polystyrene-co-acrylonitrile (SAN)	0.5	0.6

2. Solvent Resistance

TABLE 5 Chemical resistance of experimental α-methylene-γ-butyrolactone terpolymer films. SAN and MMASAN are provided as references

Reaction Numbers of Experimental Sample	10% Sodium Hydroxide	Toluene
22	Pass	Pass
29	Pass	Pass
33	Pass	Pass
34	Pass	Pass
35	Pass	Pass
36	Pass	Pass
Poly(methylmethacrylate-styrene-co-acrylonitrile (MMASAN)	Pass	Fail
Polystyrene-co-acrylonitrile (SAN)	Pass	Fail

Advantages over Prior Art

The use of polymers containing α-methylene-γ-butyrolactone as fabric and coating additives to improve the weatherability properties and solvent resistance is unreported. Coating additives containing α-methylene-γ-butyrolactone copolymerized with styrene methylmethacrylate where their applications is limited to forming heat resistant resins or to enhance mar scratch resistance, respectively.

Notes

1. Terpolymers consisting of poly(methylmethacrylate-ter-methylacrylate-ter-γ,γ-dimethyl-α-methylene-γ-butyrolactone) (III) were prepared by Feiring[1] and then spin coated onto photoresist substrates and used as protecting films.

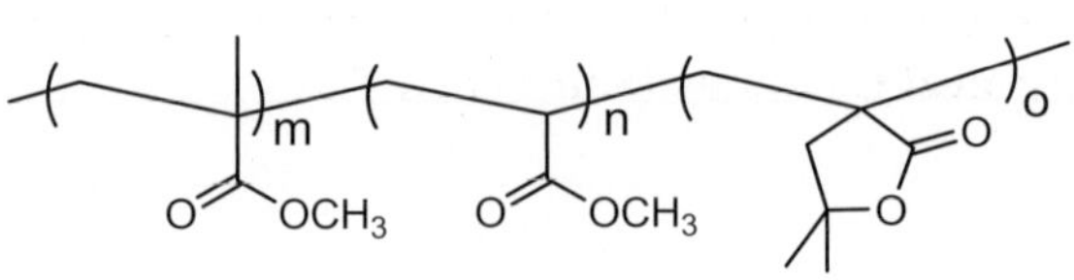

(III)

2. Davis[2] prepared weatherable thermostable polymers having improved flow by incorporating polydimethylsiloxane into a resorcinol ester polycarbonate (IV).

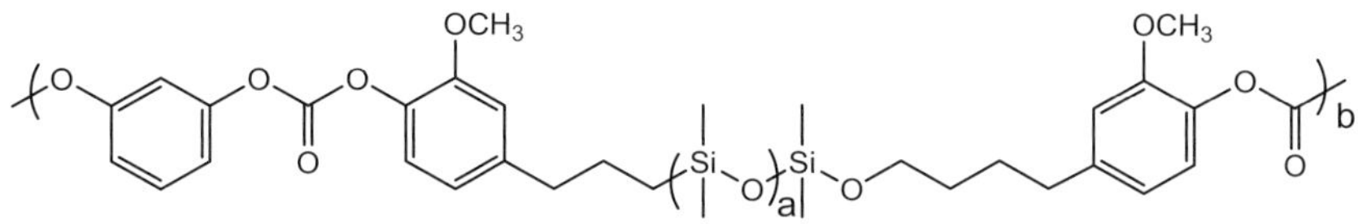

(IV)

3. Polymers (V) obtained by condensing adipic acid, bisphenol-A, and phosgene terminated by 4-cumylphenol were prepared by Mahood[3] and used to fabricate weather and ozone resistant fibers.

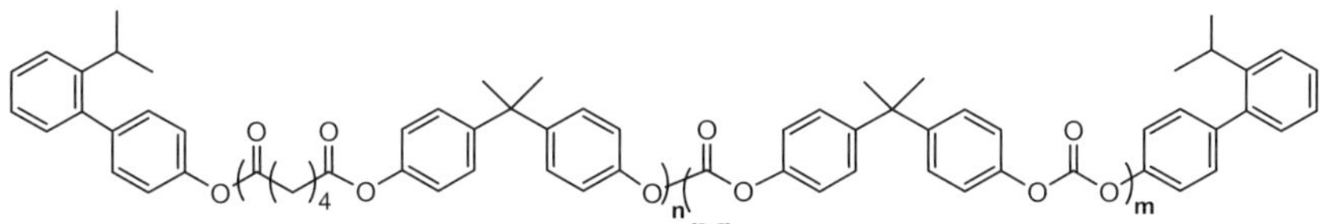

(V)

4. Weather resistance polyester compositions (VI) were prepared by Crawford[4] by the transesterification of 2,2,4,4-tetramethyl-1,3-cyclobutanediol, 1,4-cyclohexanedimethanol, and ethylene glycol with dimethyl terephthalate.

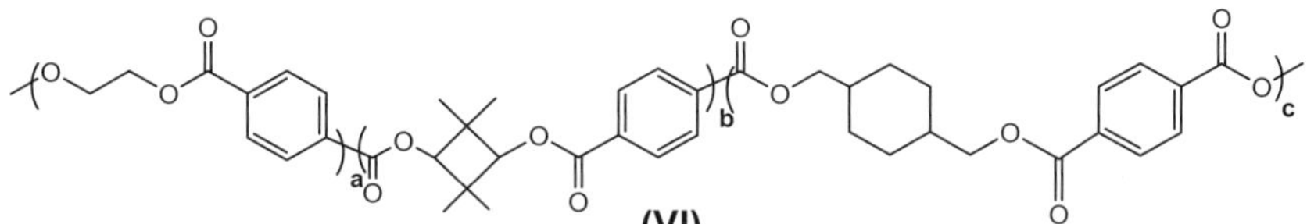

(VI)

REFERENCES

1. Andrew E Feiring et al, US Patent 7,166,416 (January 23, 2007) **E. I. du Pont de Nemours and Company (Wilmington, DE)**
2. Gary C Davis et al, US Patent 7,838,602 (November 23, 2010) **Sabic Innovative Plastics IP B.V. (NL)**
3. James Alan Mahood et al, US Patent 7,709,592 (May 4, 2010) **Sabic Innovative Plastics IP B.V. (NL)**
4. Emmett Dudley Crawford, US Patent 7,737,246 (June 15, 2010) **Eastman Chemical Company (Kingsport, TN)**

Chapter 31

Water Treatment

A. CATIONIC FLOCCULANTS

Bis(dimethylamino)-2-Propylmethacrylate

Title:	Process for the preparation of (meth)acrylate diammonium salts and their use as monomers for the synthesis of polymers
US Patent:	7,799,944 (September 21, 2010)
Author:	Piet Vanneste et al.
Assignee:	Taminco N.V. (Ghent, BE)
Significance:	High purity 1,3-bis(dimethylamino)-2-propyl acrylate quaternized with either benzylchloride or methyl chloride was prepared by inverse emulsion copolymerization with acrylamide. These high molecular weight cationic copolymers were effective as flocculants especially in the treatment of industrial and municipal waste water.

Additive Name	1,3-Bis(dimethylamino)-2-propyl methacrylate (I)
CAS Number	21476-57-3
Safety Concerns	Moderately irritating to the skin and the eyes. If the material comes in contact with the skin, wash thoroughly with soap and water.

Additive Name	Methacrylic acid methylester
CAS Number	80-62-6
Safety Concerns	Respiratory and skin irritant. Lachrymator. May act as a sensitizer by skin contact. Typical exposure limits in the range 25–50 ppm.

Next Generation of International Chemical Additives. DOI: http://dx.doi.org/10.1016/B978-0-444-53788-1.00031-0

Additive Name	1,3-Bis(dimethylamino)-2-propanol
CAS Number	5966-51-8
Safety Concerns	The toxicological properties of this material have not been investigated. Use appropriate procedures to prevent opportunities for direct contact with the skin or the eyes and to prevent inhalation.

Additive Name	Methyl chloride
CAS Number	74-87-3
Safety Concerns	Inhalation of high concentrations may cause dizziness and interfere with the normal heart rhythm. Repeated exposure to this material can result in toxicity to the liver and kidneys. Flammable.

Additive Name	Benzylchloride
CAS Number	100-44-7
Safety Concerns	Very hazardous in the case of skin or eye contact. Inhalation of the spray mist may produce severe irritation of respiratory tract characterized by coughing, choking, or shortness of breath.

Additive Name	Acrylamide
CAS Number	79-06-1
Safety Concerns	Hazardous in the case of skin or eye contact, ingestion, or inhalation. Severe overexposure can result in death.

Additive Names	1,3-bis(trimethylammonium)-2-propylmethacrylate dichloride (I) 1,3-bis(dimethylbenzylammonium)-2-propylmethacrylate dichloride (II)

Additive Structures

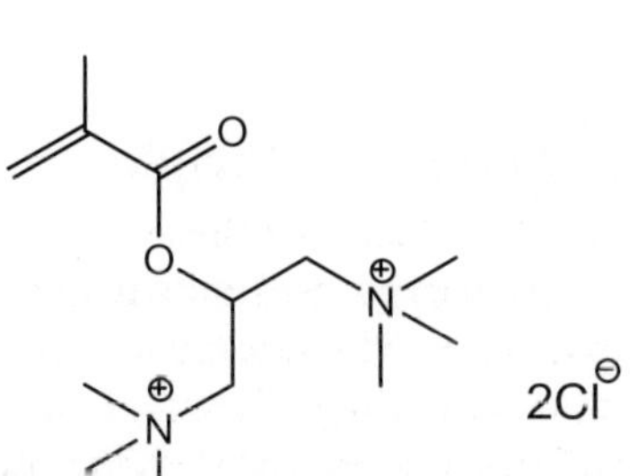

(I)

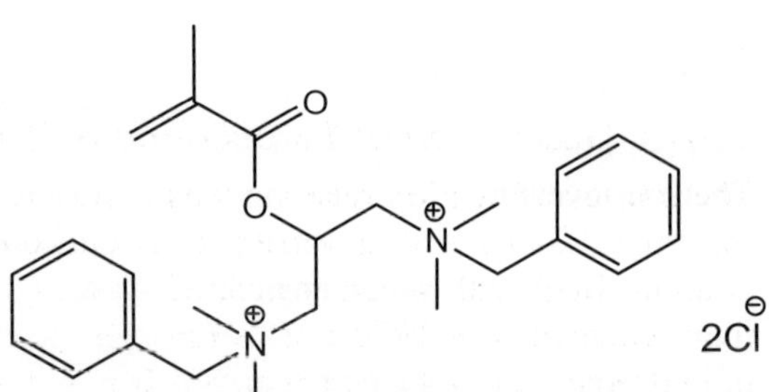

(II)

Additive Preparation

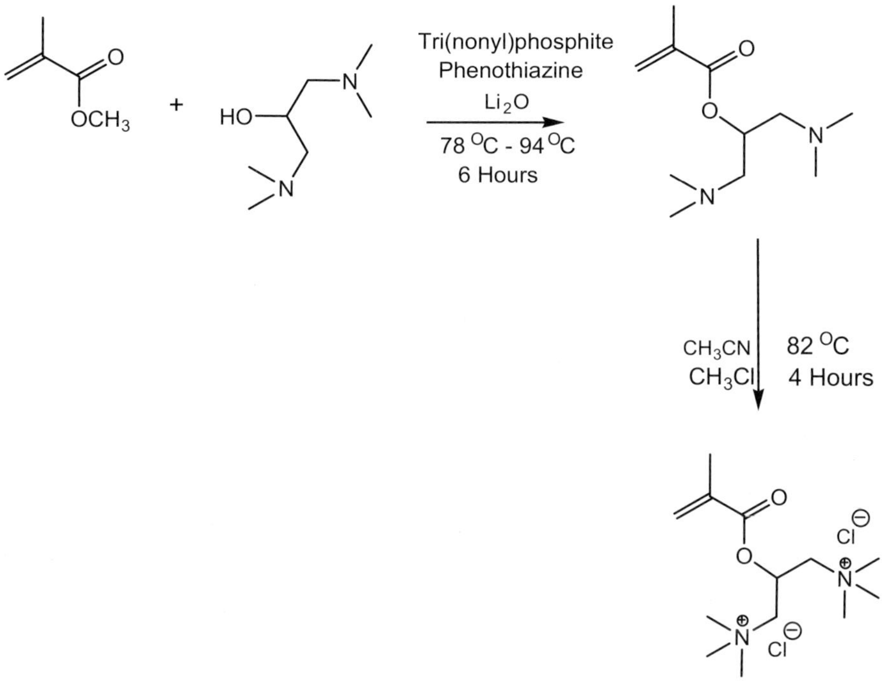

Copolymerization of Flocculant Additives

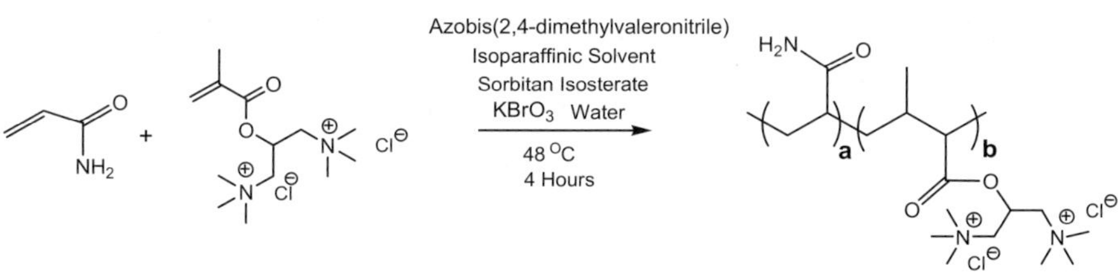

Additive Synthesis

Step 1: Preparation of 1,3-bis(dimethylamino)-2-propylmethacrylate

The transesterification reaction was performed in a 3-liter jacketed Sovirel glass reactor equipped with a stirrer, temperature probe, air sparge, and an adiabatic column filled with seven structured metal gauze packings. The reactor was initially charged with 1950 g of methacrylic acid methylester, 1.4 g of methyl ether of hydroquinone, 1.4 g of phenothiazine, and 1.4 g of tris(nonylphenyl) phosphite and then heated to 96 °C. This mixture was then treated with 95 g of methacrylic acid methylester containing traces of water and 4.5 g lithium oxide with an air flow rate of 3 l/h sparged through a sintered metal diffuser. Over a period of 100 min, 447 g of 1,3-bis(dimethylamino)-2-propanol was introduced, and methanol was distilled off at 78–94 °C over 6 h. The reaction mixture was then cooled to an ambient temperature and then filtered through a Buchner filter covered with a thin layer of silica to remove the catalyst. Unreacted methacrylic acid methylester was removed by distillation under reduced pressure, and 663 g of crude product was isolated. The material was purified using a fractional distillation column containing 16 structured Sulzer DX SS packings at 0.5 kPa. A top fraction of 31 g was obtained between 60 and 91.5 °C consisting of 1,3-bis(dimethylamino)-2-propyl acrylate and 1,3-bis(dimethylamino)-2-propyl methacrylate, which was recycled. The heart cut distilled at 91.5 °C, which provided the product in 61% and >99.5% purity. The collected product was then stabilized with 500 ppm of MEHQ.

Step 2: Preparation of 1,3-bis(trimethylammonium)-2-propylmethacrylate dichloride

Quaternization reactions were performed in a 2-liter jacketed Buchi AG glass autoclave equipped with a stirrer, plunger, temperature, and pressure probe equipped with a bottom valve. The reactor was initially charged with 193 g of the Step 1 product dissolved in 800 g of acetonitrile and then treated with 115 g of methyl chloride over a period of 1 h and heated to 82 °C. Thereafter, white solids begin to precipitate, and the mixture was heated for an additional 23 h and then cooled to an ambient temperature and excess methyl chloride removed using a nitrogen purge. The precipitate was isolated by filtration, air dried, and the product isolated in a 50% yield containing <100 ppm of acetonitrile.

Step 3: Preparation of poly(1,3-bis(trimethylammonium)-2-propylmethacrylate dichloride)-co-acrylamide by inverse emulsion copolymerization

The aqueous phase consisted of 300 g of acrylamide dissolved in 182 g of demineralized water, 200 g of a 50% aqueous solution of the Step 2 product, 0.2 g of ethylenediaminetetraacetic acid, and 0.2 g of potassium bromate. After stirring for 30 min, the mixture was treated with 0.7 g of lactic acid and 5 g of adipic acid and a final pH of 3.5 was obtained. The oil phase was prepared in a 2-liter stainless steel beaker equipped with a mechanical stirrer and contained 240 g of Isopar® M (isoparaffinic solvent), 18 g of Montane 70®, and 4.6 g of Atlas G-1086®. While the oil phase was vigorously mixing, the aqueous phase was quickly added to the oil phase. The mixture was then emulsified for 30 s at 8000 rpm using a homogenizer and then transferred to a 1-liter explosion proof reactor equipped with a 3 blade axial flow impeller. The emulsion was continuously purged with a nitrogen flow for 45 min and then heated to 40 °C and treated with a 0.5 g of solution of 0.2 g of azobis(2,4-dimethylvaleronitrile) dissolved in xylene. Thereafter, the reaction was maintained at 40 °C for 2 h, and then the temperature was increased to 48 °C over 3 h. An additional 0.12 g of free-radical catalyst was then added, and the polymerization was continued for 4 h. After an additional hour, 1.2 g of sodium metabisulfite dissolved in 3 g water was added, and the reaction temperature was further increased to 55 °C for 1 h. The resulting emulsion contained <100 ppm of unreacted acrylamide, which showed no coagulum content and where the isolated product had an intrinsic viscosity of 11.0 dl/g.

Additive Formulation

Optimum cationic flocculant properties for the Step 3 copolymer, poly(1,3-bis(trimethylammonium)-2-propylmethacrylate dichloride)-co-acrylamide was obtained when 40% of 1,3-bis(trimethylammonium)-2-propylmethacrylate dichloride had been incorporated into the copolymer.

Test Results

The flocculation properties of the Step 3 product was compared with that of Alpinefloc®, a commercial cationic polymer prepared by polymerizing impure 1,3-bis(trimethylammonium)-2-propylmethacrylate dichloride with acrylamide using a standard test consisting of a suspension of 6.6 g of kaolin in 1 l of water. Testing indicated that the flocculation properties of the Step 3 product were consistently 30% better than those of Alpinefloc®.

Advantages over Prior Art

Synthetic efforts to prepare high-purity 1,3-bis(dimethylammonium)-2-propyl methacrylate dichloride have been unsuccessful. When impure 1,3-bis(dimethylammonium)-2-propyl methacrylate was copolymerized with a variety of copolymers, cationic antiflocculants with inconsistent activities were obtained. By contrast, copolymers containing high-purity 1,3-bis(dimethylammonium)-2-propyl methacrylate dichloride resulted in the formation of well-defined materials with consistently enhanced antiflocculant properties.

Notes

1. Pressley[1] prepared antiflocculant polymer compositions by transesterifying methylmethacrylate with poly(ethylene oxide) (III) and then polymerizing the product and blending with titanium dioxide.

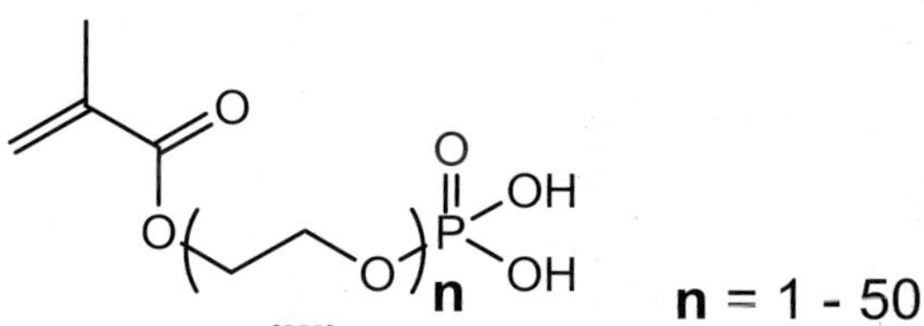

(III)

2. A one-pot method for preparing polymeric antiflocculant agents was used by Wester[2] by condensing methacrylic acid, acrylic acid, and monomethylpolyethylene glycol (IV).

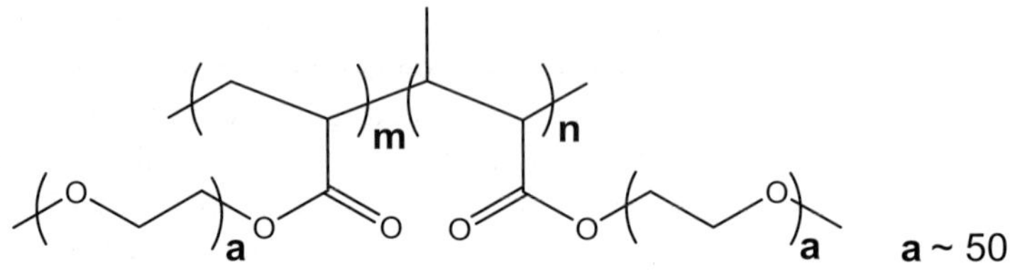

(IV)

3. Strominger[3] prepared antiflocculants polymers (V) consisting of the reaction product of acrylic or methacrylic acid with fatty acid esters where the polymeric agent was used to clarify industrial water systems.

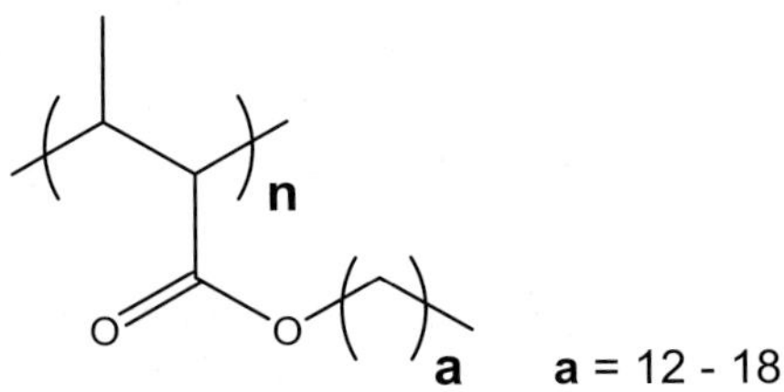

(V)

4. Palonen[4] devised a method for removing lignin from pulp water by using a solution of Kempac® 18 (Bepharm, Ltd.) consisting of polyaluminum hydroxide chloride, $(Al(OH)_m(Cl)_{3n-m})_n$.

REFERENCES

1. Ozzie Moore Pressley et al, US Patent 7,803,858 (September 28, 2010) **Rohm and Haas Company (Philadelphia, PA)**
2. Bettina Wester et al, US Patent 7,737,224 (September 14, 2010) **BASF Aktiengesellschaft (Ludwigshafen, DE)**
3. Michael G. Strominger et al, US Patent 7,776,958 (August 17, 2010) **Nalco Company (Naperville, IL)**
4. Heikki Palonen et al, US Patent 7,794,604 (September 14, 2010) **Kemira Oyj (FI)**

B. WATER PURIFICATION ADDITIVE

Chelating Agents

Title:	Chelating compounds and immobilized tethered chelators
US Patent:	7,932,326 (April 26, 2011)
Author:	Robert A. Yokel et al.
Assignee:	The University of Kentucky Research Foundation (Lexington, KY)

Significance: Di and tripodal chelating agents were prepared containing hydroxylamine termini which quantitatively removed metal ions in mixed solvents containing water. Optimum metal-binding properties of these ligands were observed when chelating agents were anchored to an insoluble polymer matrix.

Additives

Tris(2-*N*-hydroxyaminocarboxyethylmethyl ether) methyl phenylsulfonamide (I)
Di(2-*N*-hydroxyaminocarboxyethylmethyl ether methyl)methyl-phenylsulfonamide (II)
Polystyrene-g-(tris(2-*N*-hydroxyaminocarboxyethyl-methyl ether)methyl sulfonamide) (III)
Polystyrene-g-(di(2-*N*-hydroxyaminocarboxyethyl-methyl ether methyl)methyl sulfonamide (IV)

Safety Ligands and macroporous crosslinked polymers are very weakly acidic and are very mildly irritating to the eyes. In the event of eye contact, immediately wash out with plenty of water for at least 15 min. In the event of ingestion, do not induce vomiting.

Additive Structures

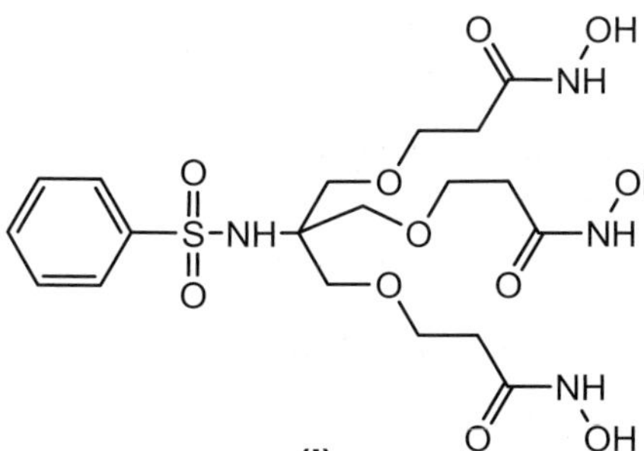

(I)

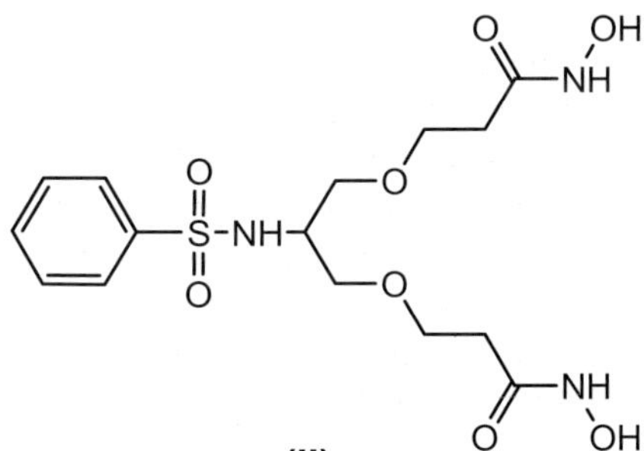

(II)

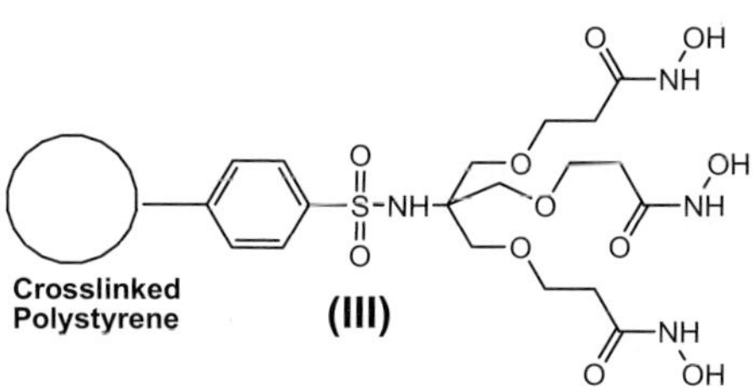

(III)

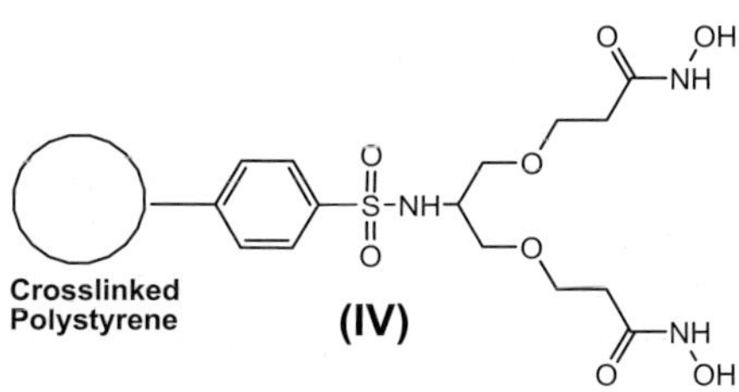

(IV)

Additive Preparation

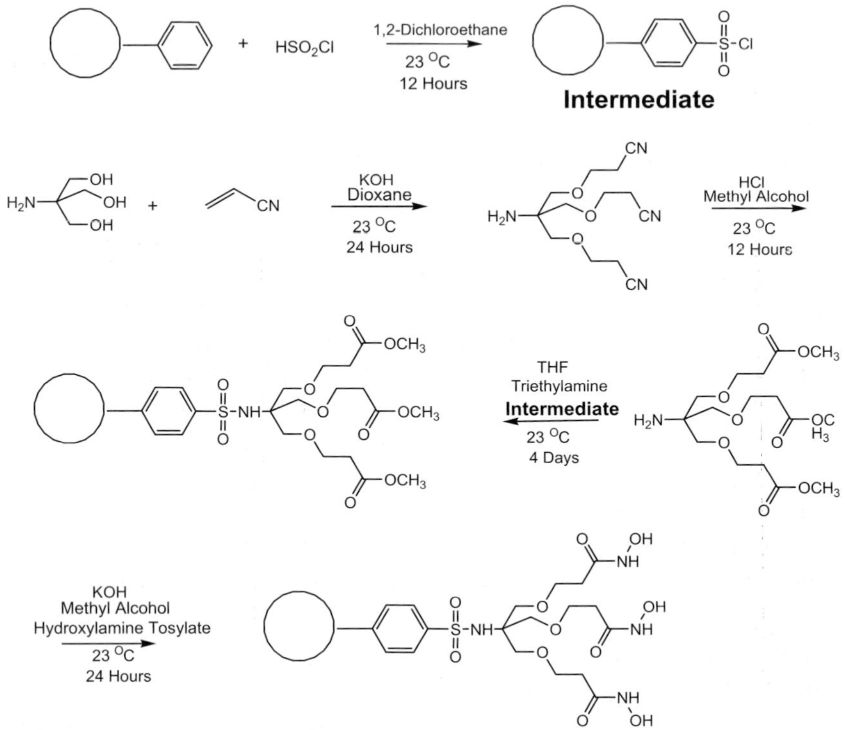

Additive Synthesis

1. *Preparation of polystyrene-g-sulfonyl chloride*

A reactor was charged with 35 g of macroporous styrene–divinylbenzene copolymer having an average pore diameter of 40 Å, 100 ml of 1,2-dichloroethane, and 160 g of technical grade chlorosulfonic acid and then stirred at an ambient temperature for 12 h. The product was then filtered using a glass frit and rinsed twice successively with methylene chloride and THF and a final wash with methylene chloride. The polymer was vacuum dried and stored at low temperature under argon.

IR (neat) (cm^{-1}) 3521, 1369, 1171
Anal. Found: C, 57.17; H, 5.50; S, 10.23; Cl, 8.49;
Calculated loading S, 3.22 mmol/g, Cl, 2.39 mmol/g.

2. *Preparation of tris(2-cyanoethylmethyl ether)aminomethane*

A solution of 50.0 g of tris(hydroxymethyl)aminomethane, 2.3 g of KOH, and 150 ml of 1,4-dioxane was treated with the dropwise addition of 71.17 g acrylonitrile over a 1-hour periodide and then stirred at an ambient temperature for 24 h. The solution was then made acidic to a pH of 2 by the addition of dilute hydrochloric acid and then extracted thrice with 100 ml of methylene chloride. The extract was dried over sodium sulfate, concentrated, and the product was isolated in a 33.5% yield.

IR (neat) (cm^{-1}) 3588, 3368, 2251
^{1}H-NMR ($CDCl_3$) δ 3.68 (t, J=6.0 Hz, 6H), 3.44 (s, 6H), 2.61 (t, J=6.0 Hz), 1.68 (br s, 2H)
^{13}C-NMR ($CDCl_3$) δ 118.2, 72.7, 65.9, 56.3, 19.0
HMS (El, MH+) calcd for $C_{13}H_{21}N_4O_3$: 281.16147; Found: 281.16138

3. *Preparation of tris(2-methoxycarboxyethylmethyl ether)aminomethane*

Dry hydrogen chloride gas was passed through a solution of 52.6 g of the Step 2 product dissolved in 150 ml of anhydrous methyl alcohol until the solution was saturated with HCl and then refluxed overnight. After the solution was cooled, the solution was filtered to remove ammonium chloride, and the filtrate was concentrated to give a gum. The gum was then dissolved in THF, filtered, the solution was concentrated, and the product was isolated in a 55.0% yield.

IR (neat) (cm^{-1}) 3394, 1735
^{1}H-NMR ($CDCl_3$) δ 3.69 (t, 6.31 Hz, 6H)
^{13}C-NMR ($CDCl_3$) δ 172.4, 69.0, 67.1, 59.6, 52.0, 34.7
HMS (El, MH+) calcd for $C_{16}H_{30}NO_9$: 380.19217, found: 380.19205.

4. *Synthesis of polystyrene-g-(tris(2-methoxycarboxyethylmethyl ether)methyl sulfonamide)*

A reactor was charged with 2.0 g of the Step 1 product dissolved in 50 ml of THF, 7.58 g of the Step 3 product dissolved in 30 ml of THF and 2.0 g of triethylamine and then swirled for four days at an ambient temperature. The mixture was then filtered, and the polymer was washed successively with THF, water, THF, methylene chloride, and dried under vacuum, and the product was isolated.

IR (neat) (cm^{-1}) 3494, 1732, 1169
Anal. Found: C, 58.96; H, 6.81; S, 8.22; N, 2.66; calculated loading: S, 2.57 mmol/g, N, 1.90 mmol/g

5. *Synthesis of polystyrene-g-(tris(2-N-hydroxyaminocarboxyethylmethyl ether)methyl sulfonamide)*

A slurry of 1.7 g of the Step 4 product dissolved in 40 ml methyl alcohol was treated at an ambient temperature with 4.02 g of hydroxylamine tosylate and 2.14 g potassium hydroxide and then swirled for 12 h. The mixture was filtered, and the polymer was washed successively with methyl alcohol, water, and methyl alcohol . The resin was then further swirled with dilute acetic acid for an hour and washed successively with methanol, water, THF, then dried under vacuum, and the product was isolated.

IR (neat) (cm^{-1}) 3479, 1644, 1173
Anal. Found: C, 54.65 H, 5.57; S, 9.51; N, 1.59; Loading: S, 2.97 mmol/g, N, 1.14 mmol/g.

Testing

Metal Binding Constants

Acid dissociation constants for the ligands were determined by potentiometric titration in 0.1 M KNO_3 at 25 °C. Binding constants for di and trivalent metal ions were also determined by potentiometric titration in 0.1 M KNO_3 at 25 °C. Binding constants for the monomeric Additive 3 and polymeric Additive 5 are provided in Tables 1 and 2, respectively.

Test Results

TABLE 1 Summary of the aluminum binding constants in a variety of polar solvents using Additive 3

Solution	Volume (ml)	pH	Resin (mg)	Aluminum (μmolar)	Al Removed (%)	$Log\beta_{110}$
0.1M Methane Sulfonic Acid	100	5.0	50	9.26	97.4	18.9
0.1M Methane Sulfonic acid	5	5.0	50	185	99.2	18.2
0.15 mM Ferron	3	5.0	25	150	78	19.2
10 mM EGTA*	1	5.7	50	14.9	58	18.8
0.23M $Ca(glu)_2$**	1	5.3	50	115	10	18.5
0.108M $Ca(glu)_2$	1	5.2	50	53.8	18	18.5
0.05M $Ca(glu)_2$	1	5.2	50	25.0	44	18.7
0.023M $Ca(glu)_2$	1	5.3	50	11.5	72	18.9
0.23M $Ca(glu)_2$	0.5	4.3	250	338	40	19.0

*1,10-Diaza-4,7-dioxadecane-1,1,10,10-tetraacetic acid.
**Calcium gluconate.

TABLE 2 Binding constants for metal complexes and the experimental Additive 5

Variable	Fe^{+3}	Al^{+3}	Cu^{+2}	Ni^{+2}	Zn^{+2}	Mn^{+2}	Ca^{+2}
$Log\beta_{111}$* **(Additive 4)**	27.60	26.27	23.61	19.10	19.13	17.06	13.34
$Log\beta_{110}$** **(Reference)**	23.78	21.44	--	10.73	10.13	8.95	3.71

*Metal–ligand binding constant for an analogous ligand
**Metal–ligand binding constant for the Step 4 product.

Advantages over Prior Art

Most chelating ligands are constructed on an amine tripodal platform. These include tris(2-aminoethyl)amine, tris(3-aminopropyl)amine, and nitrilo-triacetic acid. When chelating ligands are prepared based on these amines and then anchored onto an insoluble polymer matrix, however, a tertiary nitrogen is formed which diminishes the chelating potential of that ligand. The current invention addresses this concern by introducing a method of incorporating a di and tripodal ligand onto a polymer without generating tertiary nitrogens.

Notes

1. Atamna[1] used Uroporphyrinogen I (V) as a chelating agent to lower or eliminate toxic levels of Al^{+3}, Hg^{+2}, Cd^{+3}, Mn^{+2}, and Zn^{+2} present in blood.

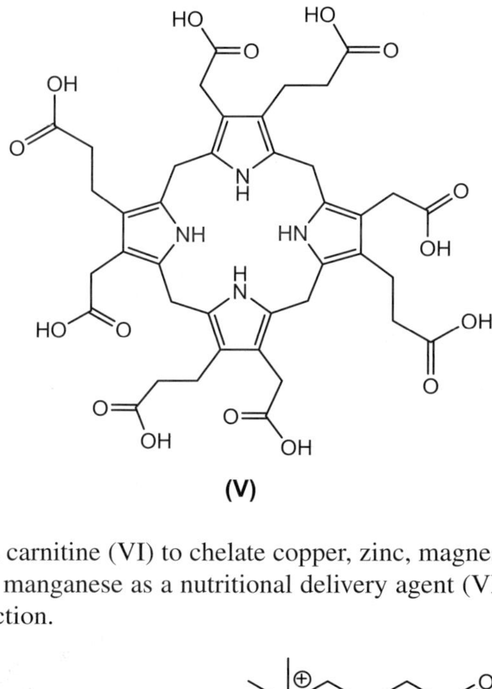

(V)

2. Ashmead[2] used carnitine (VI) to chelate copper, zinc, magnesium, calcium, iron, chromium, and manganese as a nutritional delivery agent (VII) for the treatment of alcohol addiction.

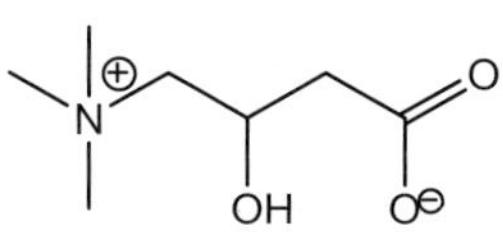

(VI)

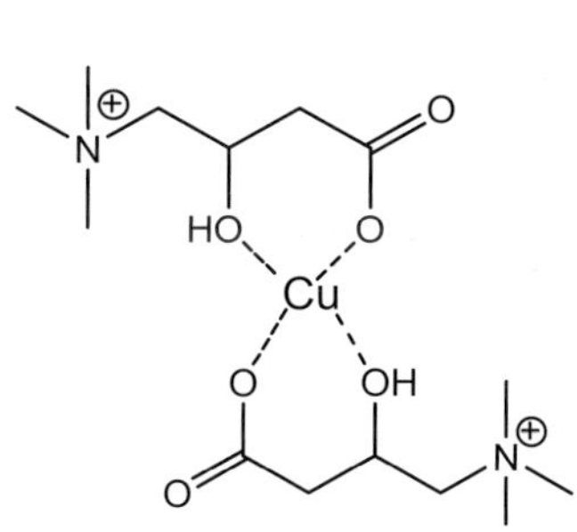

(VII)

3. Water-soluble polyethylene glycol zinc chelators (VIII) were prepared by Magda[3] and used in pharmaceutical applications.

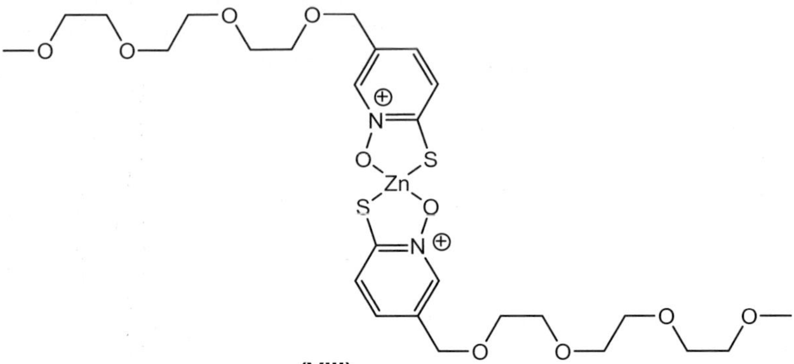

(VIII)

4. Diallo[4] used G4 dendritic poly(amidoamine) macromolecules terminated with ethylene diamine to remove Co^{+2}, Ag^{+1}, Fe^{+3}, Ni+2, ClO_4^{-1}, and NO_3^{-3} from contaminated water.

REFERENCES

1. Hani Atamna, US Patent 7,906,499 (March 15, 2011)
Children's Hospital & Research Center at Oakland (Oakland, CA)
2. H. DeWayne Ashmead et al, US Patent 7,910,137 (March 22, 2011)
Albion International, Inc. (Clearfield, UT)
3. Darren Magda et al, US Patent 7,838,536 (November 23, 2010)
Jonathan L. Sessler (Austin, TX)
4. Mamadou S. Diallo, US Patent 7,758,755 (July 20, 2010)
California Institute of Technology (Pasadena, CA)

C. MUNICIPAL WASTE WATER FLOCCULANTS

Quaternary Copolymers of Polymethylmethacrylate

Title:	Process for the preparation of (meth)acrylate diammonium salts and their use as monomers for the synthesis of polymers
US Patent:	7,799,944 (September 21, 2010)
Author:	Piet Vanneste et al.
Assignee:	Taminco N.V. (Ghent, BE)
Significance:	High purity 1,3-bis(dimethylamino)-2-propyl acrylate quaternized with either benzylchloride or methylchloride was prepared by inverse emulsion copolymerization with acrylamide. These high molecular weight cationic copolymers were effective as flocculants especially in the treatment of industrial and municipal waste water.

Additive Name	1,3-Bis(dimethylamino)-2-propyl methacrylate (I)
CAS Number	21476-57-3
Safety Concerns	Moderately irritating to the skin and the eyes. If the material comes in contact with the skin, wash thoroughly with soap and water.

Additive Name	Methacrylic acid methylester
CAS Number	80-62-6
Safety Concerns	Respiratory and skin irritant. Lachrymator. May act as a sensitizer by skin contact. Typical exposure limits are in the range 25–50 ppm.

Additive Name	1,3-Bis(dimethylamino)-2-propanol
CAS Number	5966-51-8
Safety Concerns	The toxicological properties of this material have not been investigated. Use appropriate procedures to prevent opportunities for direct contact with the skin or the eyes and to prevent inhalation.

Additive Name	Methylchloride
CAS Number	74-87-3
Safety Concerns	Inhalation of high concentrations may cause dizziness and interfere with normal heart rhythm. Repeated exposure to this material can result in toxicity to the liver and kidneys. Flammable.

Additive Name	Benzylchloride
CAS Number	100-44-7
Safety Concerns	Very hazardous in the case of skin or eye contact. Inhalation of the spray mist may produce severe irritation of the respiratory tract characterized by coughing, choking, or shortness of breath.

Additive Name	Acrylamide
CAS Number	79-06-1
Safety Concerns	Hazardous in the case of skin or eye contact, ingestion or inhalation. Severe overexposure can result in death.

Additive Names	1,3-bis(trimethylammonium)-2-propylmethacrylate dichloride (I) 1,3-bis(dimethylbenzylammonium)-2-propylmethacrylate dichloride (II)

Additive Structures

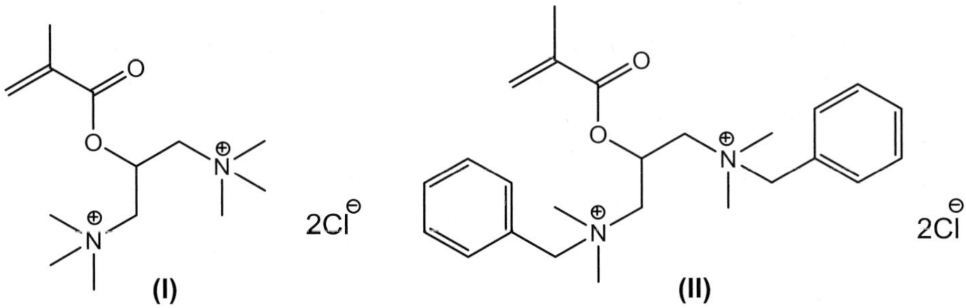

Additive Preparation

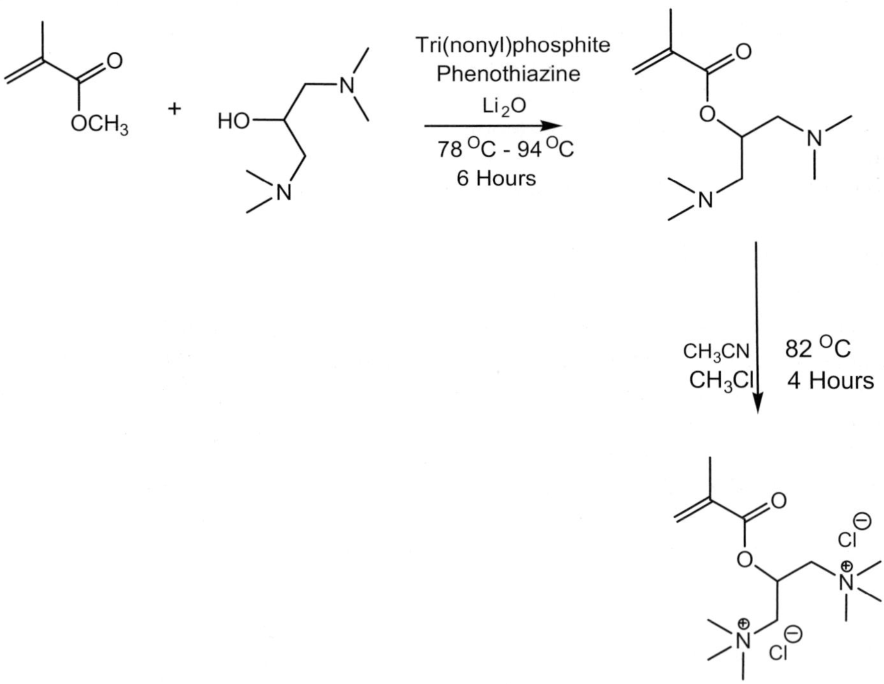

Copolymerization of Flocculant Additives

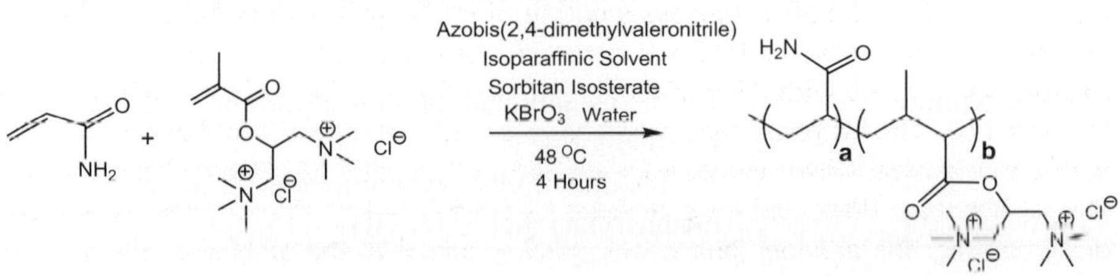

Additive Synthesis

Step 1: Preparation of 1,3-bis(dimethylamino)-2-propylmethacrylate

The transesterification reaction was performed in a 3-liter jacketed Sovirel glass reactor equipped with a stirrer, temperature probe, air sparge, and an adiabatic column filled with seven structured metal gauze packings. The reactor was initially charged with 1950 g of methacrylic acid methylester, 1.4 g of methyl ether of hydroquinone, 1.4 g of phenothiazine, and 1.4 g of tris(nonylphenyl) phosphite and then heated to 96 °C. This mixture was then treated with 95 g of methacrylic acid methylester containing traces of water and 4.5 g of lithium oxide with an air flow rate of 3 l/h sparged through a sintered metal diffuser. Over a period of 100 min, 447 g of 1,3-bis(dimethylamino)-2-propanol was introduced, and methanol was distilled off at 78–94 °C over 6 h. The reaction mixture was then cooled to an ambient temperature and then filtered through a Buchner filter covered with a thin layer of silica to remove the catalyst. Unreacted methacrylic acid methylester was removed by distillation under reduced pressure, and 663 g of the crude product isolated. The material was purified using a fractional distillation column containing 16 structured Sulzer DX SS packings at 0.5 kPa. A top fraction of 31 g was obtained between 60 and 91.5 °C consisting of 1,3-bis(dimethylamino)-2-propyl acrylate and 1,3-bis(dimethylamino)-2-propyl methacrylate, which was recycled. The heart cut distilled at 91.5 °C, which provided the product in 61% and >99.5% purity. The collected product was then stabilized with 500 ppm MEHQ.

Step 2: Preparation of 1,3-bis(trimethylammonium)-2-propylmethacrylate dichloride

Quaternization reactions were performed in a 2-liter jacketed Buchi AG glass autoclave equipped with a stirrer, plunger, temperature, and pressure probe equipped with a bottom valve. The reactor was initially charged with 193 g of the Step 1 product dissolved in 800 g of acetonitrile and then treated with 115 g of methyl chloride over a period of 1 h and heated to 82 °C. Thereafter, white solids begin to precipitate, and the mixture was heated for an additional 23 h and then cooled to an ambient temperature, and excess methyl chloride was removed using a nitrogen purge. The precipitate was isolated by filtration, air dried, and the product isolated in a 50% yield containing <100 ppm of acetonitrile.

Step 3: Preparation of poly(1,3-bis(trimethylammonium)-2-propylmethacrylate dichloride)-co-acrylamide by inverse emulsion copolymerization

The aqueous phase consisted of 300 g of acrylamide dissolved in 182 g of demineralized water, 200 g of a 50% aqueous solution of the Step 2 product, 0.2 g of ethylenediaminetetraacetic acid, and 0.2 g of potassium bromate. After stirring for 30 min, the mixture was treated with 0.7 g of lactic acid and 5 g of adipic acid and a final pH of 3.5 obtained. The oil phase was prepared in a 2-liter stainless steel beaker equipped with a mechanical stirrer and containing 240 g of Isopar® M (isoparaffinic solvent), 18 g of Montane 70®, and 4.6 g of Atlas G-1086®. While the oil phase was vigorously mixing, the aqueous phase was quickly added to the oil phase. The mixture was then emulsified for 30 s at 8000 rpm using a homogenizer and then transferred to a 1-liter explosion proof reactor equipped with a 3 blade axial flow impeller. The emulsion was continuously purged with a nitrogen flow for 45 min and then heated to 40 °C and treated with a 0.5 g of solution of 0.2 g of azobis(2,4-dimethylvaleronitrile) dissolved in xylene. Thereafter, the reaction was maintained at 40 °C for 2 h, and then the temperature was increased to 48 °C over 3 h. An additional 0.12 g of free-radical catalyst was then added, and the polymerization was continued for 4 h. After an additional hour, 1.2 g of sodium metabisulfite dissolved in 3 g water was added, and the reaction temperature was further increased to 55 °C for 1 h. The resulting emulsion contained <100 ppm of unreacted acrylamide, which showed no coagulum content and where the isolated product had an intrinsic viscosity of 11.0 dl/g.

Additive Formulation

Optimum cationic flocculant properties for the Step 3 copolymer, poly(1,3-bis(trimethyl-ammonium)-2-propylmethacrylate dichloride)-co-acrylamide, was obtained when 40% of 1,3-bis(trimethyl-ammonium)-2-propylmethacrylate dichloride had been incorporated into the copolymer.

Test Results

The flocculation properties of the Step 3 product was compared with that of Alpinefloc®, a commercial cationic polymer prepared by polymerizing impure 1,3-bis(trimethylammonium)-2-propylmethacrylate dichloride, with acrylamide using a standard test consisting of a suspension of 6.6 g of kaolin in 1 l of water. Testing indicated that the flocculation properties of the Step 3 product were consistently 30% better than those of Alpinefloc®.

Advantages over Prior Art

Synthetic efforts to prepare high purity 1,3-bis(dimethylammonium)-2-propyl methacrylate dichloride have been unsuccessful. When impure 1,3-bis(dimethylammonium)-2-propyl methacrylate was copolymerized with a variety of copolymers, cationic antiflocculants with inconsistent activities were obtained. By contrast, copolymers containing high purity 1,3-bis(dimethylammonium)-2-propyl methacrylate dichloride resulted in the formation of well-defined materials with consistently enhanced antiflocculant properties.

Notes

1. Pressley[1] prepared antiflocculant polymer compositions by transesterifying methylmethacrylate with poly(ethylene oxide) (III) and then polymerizing the product and blending with titanium dioxide.

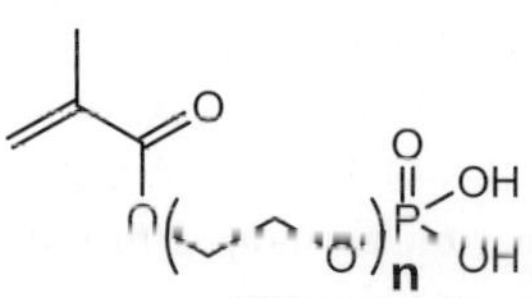

n = 1 - 50

(III)

2. A one-pot method for preparing polymeric antiflocculant agents was used by Wester[2] by condensing methacrylic acid, acrylic acid, and monomethylpolyethylene glycol (IV).

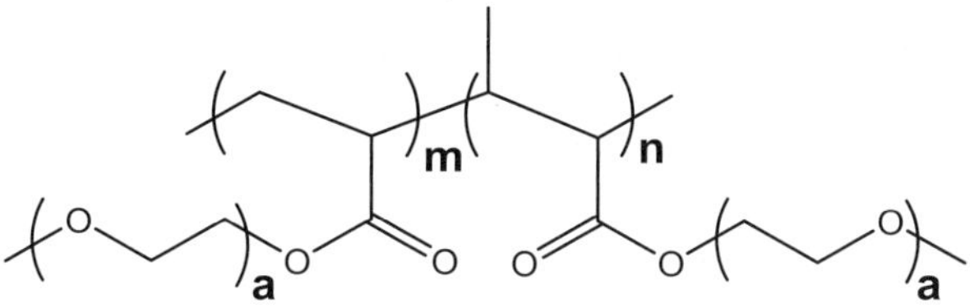

a ~ 50

(IV)

3. Strominger[3] prepared antiflocculants polymers (V) consisting of the reaction product of acrylic or methacrylic acid with fatty acid esters where the polymeric agent was used to clarify industrial water systems.

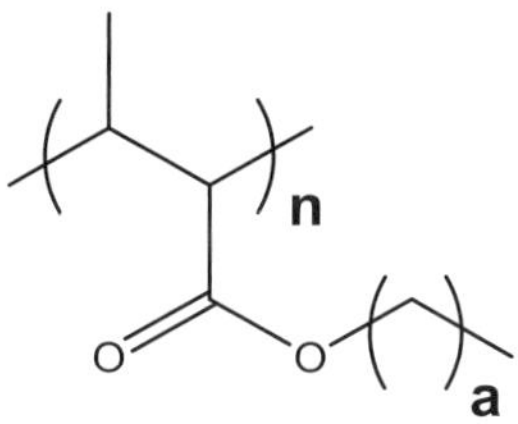

a = 12 - 18

(V)

4. Palonen[4] devised a method for removing lignin from pulp water by using a solution of Kempac® 18 (Bepharm, Ltd.) consisting of polyaluminum hydroxide chloride, $(Al(OH)_m(Cl)_{3n-m})_n$.

REFERENCES

1. Ozzie Moore Pressley et al, US Patent 7,803,858 (September 28, 2010) **Rohm and Haas Company (Philadelphia, PA)**
2. Bettina Wester et al, US Patent 7,737,224 (September 14, 2010) **BASF Aktiengesellschaft (Ludwigshafen, DE)**
3. Michael G. Strominger et al, US Patent 7,776,958 (August 17, 2010) **Nalco Company (Naperville, IL)**
4. Heikki Palonen et al, US Patent 7,794,604 (September 14, 2010) **Kemira Oyj (FI)**

Index

Note: Page numbers followed by *f,* indicate figure and *t,* indicate table.